T0146297

THE PAPERS OF THOMAS A. EDISON

THOMAS A. EDISON PAPERS

Paul Israel
Director and Editor

Thomas E. Jeffrey
Senior Editor

Rachel M. Weissenburger
Business Manager

Associate Editors
Louis Carlat
Theresa M. Collins

Assistant Editors
David Hochfelder
Brian C. Shipley

Indexing Editor
David Ranzan

Visiting/Consulting Editors
Amy Flanders
Blaine McCormick

Editorial Assistant
Alexandra R. Rimer

Linda Endersby

Graduate Assistants
Lindsay Frederick Braun
Kelly Enright

Undergraduate Assistant
Michelle Tang

BOARD OF SPONSORS (2006)

Rutgers, The State University of New Jersey
    Richard L. McCormick
    Ziva Galili
    Ann Fabian
    Paul Clemens
National Park Service
    Maryanne Gerbauckas
    Michelle Ortwein
New Jersey Historical Commission
    Marc Mappen
Smithsonian Institution
    Harold Wallace

EDITORIAL ADVISORY BOARD (2006)

Robert Friedel, University of Maryland
Louis Galambos, Johns Hopkins University
Susan Hockey, Oxford University
Thomas P. Hughes, University of Pennsylvania
Ronald Kline, Cornell University
Robert Rosenberg, John Wiley & Sons
Marc Rothenberg, Joseph Henry Papers, Smithsonian
    Institution
Philip Scranton, Rutgers University / Hagley Museum
Merritt Roe Smith, Massachusetts Institute of Technology

FINANCIAL CONTRIBUTORS

We thankfully acknowledge the vision and support of Rutgers University and the Thomas A. Edison Papers Board of Sponsors.

This edition was made possible by grant funds provided from the New Jersey Historical Commission, National Historical Publications and Records Commission and The National Endowment for the Humanities. Major underwriting of the present volume has been provided by the Barkley Fund, through the National Trust for the Humanities, and by The Charles Edison Foundation.

We are grateful for the generous support of the IEEE Foundation, the Hyde & Watson Foundation, the Martinson Family Foundation, and the GE Foundation. We acknowledge gifts from many other individuals as well as an anonymous donor, the Association of Edison Illuminating Companies, and the Edison Electric Institute.

THE PAPERS OF THOMAS A. EDISON

Volume 6

*Thomas A. Edison, in
an illustration published
in 1882.*

Volume 6

The Papers of
Thomas A. Edison

ELECTRIFYING NEW YORK
AND ABROAD

April 1881–March 1883

VOLUME EDITORS

Paul B. Israel

Louis Carlat

David Hochfelder

Theresa M. Collins

Brian C. Shipley

EDITORIAL STAFF

Lindsay Frederick Braun

Kelly Enright

Amy Flanders

Elva Kathleen Lyon

Alexandra R. Rimer

SPONSORS

Rutgers, The State University of New Jersey

National Park Service, Edison National Historic Site

New Jersey Historical Commission

Smithsonian Institution

THE JOHNS HOPKINS UNIVERSITY PRESS
BALTIMORE

© 2007 Rutgers, The State University
All rights reserved. Published 2007
Printed in the United States of America
9 8 7 6 5 4 3 2 1

The Johns Hopkins University Press
2715 North Charles Street
Baltimore, Maryland 21218-4363
www.press.jhu.edu

No part of this publication may be reproduced, stored in a retrieval system, or transmitted in any form by any means—graphic, electronic, mechanical, or chemical, including photocopying, recording or taping, or information storage and retrieval systems—without permission of Rutgers, The State University, New Brunswick, New Jersey.

The paper used in this book meets the minimum requirements of the American National Standard for Information Sciences—Permanence of Paper for Printed Library Materials, ANSI Z 39.48-1984.

Library of Congress Cataloging-in-Publication Data
(Revised for volume 3)

Edison, Thomas A. (Thomas Alva), 1847–1931
    The papers of Thomas A. Edison

    Includes bibliographical references and index.
    Contents: v. 1. The making of an inventor, February 1847–June 1873—v. 2. From workshop to laboratory, June 1873–March 1876—v. 3. Menlo Park. The early years, April 1876–December 1877.
    1. Edison, Thomas A. (Thomas Alva), 1847–1931. 2. Edison, Thomas A. (Thomas Alva), 1847–1931—Archives. 3. Inventors—United States—Biography. I. Jenkins, Reese.
TK140.E3A2   1989   600   88-9017
ISBN 0-8018-3100-8 (v. 1. : alk. paper)
ISBN 0-8018-3101-6 (v. 2. : alk. paper)
ISBN 0-8018-3102-4 (v. 3. : alk. paper)
ISBN 0-8018-5819-4 (v. 4. : alk. paper)
ISBN 0-8018-3104-0 (v. 5. : alk. paper)
ISBN-10: 0-8018-8640-6 (v. 6. : alk. paper)
ISBN-13: 978-0-8018-8640-9 (v. 6. : alk. paper)

A catalog record for this book is available from the British Library.

Edison signature on case used with permission of the McGraw-Edison Company.

TO THE MEMORY OF RICHARD P. McCORMICK

Contents

Calendar of Documents		xiii
List of Editorial Headnotes		xxii
Preface		xxiii
Chronology of Thomas A. Edison, April 1881–March 1883		xxx
Editorial Policy and User's Guide		xxxv
Editorial Symbols		xl
List of Abbreviations		xli
–1–	**April–June 1881** (Docs. 2074–2117)	3
–2–	**July–September 1881** (Docs. 2118–2160)	88
–3–	**October–December 1881** (Docs. 2161–2202)	196
–4–	**January–March 1882** (Docs. 2203–2250)	311
–5–	**April–June 1882** (Docs. 2251–2307)	446
–6–	**July–September 1882** (Docs. 2308–2343)	579
–7–	**October–December 1882** (Docs. 2344–2386)	680

–8– **January–March 1883**
(Docs. 2387–2417)   752

Appendix 1. Edison's Autobiographical Notes   811

Appendix 2. Isolated Lighting Plant Installations,
May 1883   833

Appendix 3. Specifications of Dynamos Produced at
the Edison Machine Works, April 1881–March 1883   838

Appendix 4. Cable Name Codes, 1881–1883   841

Appendix 5. Edison's Patents, April 1881–March 1883 843

Bibliography   855

Credits   869

Index   871

Contents   xii

Calendar of Documents

Doc.	*Date*	*Title*	*Page*
Docs. 1–340		Volume 1	
Docs. 341–737		Volume 2	
Docs. 738–1163		Volume 3	
Docs. 1164–1651		Volume 4	
Docs. 1652–2073		Volume 5	
2074.	1 April 1881	From Thomas Logan	8
2075.	3 April 1881	Notebook Entry: Consumption Meter	9
2076.	5 April 1881	Philip Dyer to Samuel Insull	10
2077.	5 April 1881	Technical Note: Telephony	11
2078.	7 April 1881	To Armington & Sims	12
2079.	7 April 1881	From George Gouraud	13
2080.	13 April 1881	To George Barker	15
2081.	14 April 1881	From John Lawson	16
2082.	14 April 1881	Notebook Entry: Dynamo	17
2083.	15 April 1881	To Pietro Giovine	21
2084.	17 April 1881	To Francis Jehl	21
2085.	18 April 1881	From Francis Upton	22
2086.	19 April 1881	To George Gouraud	25
2087.	19 April 1881?	Memorandum: Incandescent Lamp Patent	26
2088.	20 April 1881	To Norvin Green	27
2089.	20 April 1881	From Francis Upton	28
2090.	24 April 1881	To George Gouraud	30
2091.	April 1881	Draft Agreement with Sigmund Bergmann and Edward Johnson	31
2092.	1 May 1881	Samuel Insull to John Kingsbury	33
2093.	c. 1 May 1881	Memorandum: Ore Milling	37
2094.	2 May 1881	To Edward Acheson	38

2095.	4 May 1881	From Francis Upton	*38*
2096.	5 May 1881	From Charles Hughes	*40*
2097.	11 May 1881	R. G. Dun & Co. Credit Report	*42*
2098.	13 May 1881	To Edward Acheson	*45*
2099.	14 May 1881	To Longworth Powers	*47*
2100.	17 May 1881	Technical Note: Electric Light and Power	*48*
2101.	20 May 1881	To George Barker	*51*
2102.	20 May 1881	Technical Note: Electric Light and Power	*53*
2103.	23 May 1881?	To Theodore Puskas	*58*
2104.	27 May 1881	To Drexel, Morgan & Co.	*59*
2105.	31 May 1881	Charles Batchelor to Theodore Puskas	*60*
2106.	3 June 1881	From Calvin Goddard	*62*
2107.	3 June 1881	Samuel Insull to George Gouraud	*63*
2108.	6 June 1881	Samuel Insull to Hamilton Twombly	*67*
2109.	8 June 1881	Edward Nichols Memorandum: Incandescent Lamp	*67*
2110.	10 June 1881	From George Barker	*68*
2111.	10 June 1881	Charles Batchelor to Theodore Puskas	*70*
2112.	11 June 1881	Samuel Insull to George Gouraud	*75*
2113.	13 June 1881	From Francis Upton	*80*
2114.	14 June 1881	From Theodore Puskas and Joshua Bailey	*81*
2115.	18 June 1881	To Naomi Chipman	*83*
2116.	Spring 1881?	Prose Poem	*84*
2117.	22 June 1881	From Edison Lamp Co.	*86*
2118.	1 July 1881	Samuel Insull to George Gouraud	*92*
2119.	5 July 1881	To George Gouraud	*94*
2120.	12 July 1881	From Otto Moses	*96*
2121.	15 July 1881	To Thomas Logan	*99*
2122.	15 July 1881	Notebook Entry: Dynamo	*103*
2123.	20 July 1881	Notebook Entry: Dynamo	*105*
2124.	25 July 1881	Notebook Entry: Dynamo	*108*
2125.	29 July 1881	Memorandum: Pearl Street Central Station	*111*
2126.	July 1881	Production Model: Isolated Plant Z Dynamo	*117*
2127.	2 August 1881	Draft to George Gouraud	*117*
2128.	2 August 1881	From Otto Moses	*120*
2129.	5 August 1881	From Francis Upton	*123*
2130.	7 August 1881	To John Randolph	*125*
2131.	8 August 1881	Notebook Entry: Dynamo	*126*
2132.	10–12 August 1881	From Theodore Puskas and Joshua Bailey	*128*
2133.	11 August 1881	From Otto Moses	*131*
2134.	11 August 1881	Notebook Entry: Dynamo	*133*
2135.	12 August 1881	From Charles Batchelor	*134*
2136.	19 August 1881?	Draft to Edward Johnson	*136*
2137.	19 August 1881	From Eduard Reményi	*138*

2138.	19 August 1881	From Francis Upton	*139*
2139.	20 August 1881	From Francis Upton	*141*
2140.	22 August 1881	Charles Batchelor to Sherburne Eaton	*142*
2141.	24 August 1881	To Charles Batchelor	*143*
2142.	26 August 1881	From Otto Moses	*144*
2143.	28 August– 10 September 1881	Draft Caveat: Consumption Meter	*148*
2144.	29 August 1881	To Egisto Fabbri	*156*
2145.	29 August 1881	To Egisto Fabbri	*157*
2146.	29 August 1881	To William Hazen	*159*
2147.	30 August 1881	From Charles Batchelor	*159*
2148.	6 September 1881	From Charles Batchelor	*162*
2149.	8 September 1881	To Charles Batchelor	*168*
2150.	13 September 1881	To Charles Batchelor	*178*
2151.	13 September 1881	To Sherburne Eaton	*184*
2152.	14 September 1881	Agreement with Henry Villard	*185*
2153.	17 September 1881	To Joshua Bailey	*186*
2154.	17 September 1881	From Francis Upton	*187*
2155.	18 September 1881	From Charles Batchelor	*189*
2156.	20 September 1881	To George Gouraud	*191*
2157.	23 September 1881	To Joshua Bailey	*191*
2158.	25 September 1881	To George Gouraud	*192*
2159.	25 September 1881	From Francis Upton	*193*
2160.	29 September 1881	To Francis Upton	*194*
2161.	2 October 1881	To Edward Johnson	*200*
2162.	4 October 1881	To Edward Johnson	*201*
2163.	c. 5 October 1881	Standard Electric Consumption Meter	*204*
2164.	6 October 1881	From Calvin Goddard	*205*
2165.	9 October 1881	To Edward Johnson	*206*
2166.	12 October 1881	From Joshua Bailey and Theodore Puskas	*209*
2167.	12 October 1881	Notebook Entry: Electrical Distribution System	*212*
2168.	15 October 1881	To Addison Burk	*213*
2169.	17 October 1881	From Charles Porter	*214*
2170.	18 October 1881	To Drexel, Morgan & Co.	*215*
2171.	20 October 1881	Samuel Insull to Sherburne Eaton	*217*
2172.	23 October 1881	To George Gouraud	*219*
2173.	23 October 1881	From Grosvenor Lowrey	*220*
2174.	25 October 1881	To John Michels	*229*
2175.	26 October 1881	From Sherburne Eaton	*230*
2176.	26 October 1881	Samuel Insull to Louis Glass	*231*
2177.	27 October 1881	From Francis Upton	*238*
2178.	30 October 1881	To Edward Johnson	*238*
2179.	31 October 1881	From Charles Batchelor	*239*

2180.	3 November 1881	From Edward Johnson	*241*
2181.	4 November 1881	Samuel Insull to George Gouraud	*248*
2182.	5 November 1881	From Grosvenor Lowrey and Charles Batchelor	*250*
2183.	6 November 1881	To Francis Upton	*252*
2184.	7 November 1881	To Hinds, Ketcham & Co.	*253*
2185.	8 November 1881	To Thomas Whiteside Rae	*255*
2186.	10 November 1881	To George Gouraud	*255*
2187.	23 November 1881	To Edward Johnson	*257*
2188.	23 November 1881	From George Barker	*272*
2189.	23 November 1881	From Sherburne Eaton	*273*
2190.	27 November 1881	To Edward Johnson	*275*
2191.	28 November 1881	From Francis Upton	*282*
2192.	28 November 1881	Memorandum: Electric Light and Power Patents	*284*
2193.	28 November 1881	Technical Note: Incandescent Lamp	*289*
2194.	November 1881	George Van Ness Journal Orders	*290*
2195.	2 December 1881	Samuel Insull to Henry Villard	*292*
2196.	14 December 1881	From Charles Batchelor	*293*
2197.	20 December 1881	To Edward Johnson	*296*
2198.	21 December 1881	Samuel Insull to Naomi Chipman	*298*
2199.	29 December 1881	To W. H. Patton	*298*
2200.	29 December 1881	From Edward Johnson	*299*
2201.	31 December 1881	To Charles Batchelor	*302*
2202.	31 December 1881	To Edward Johnson	*304*
2203.	2 and 4 January 1882	To Edward Johnson	*317*
2204.	5 January 1882	From John Ott	*330*
2205.	7 January 1882	To French Minister of Posts and Telegraphs	*331*
2206.	9 January 1882	To Christian Herter	*332*
2207.	9 January 1882	To Edward Johnson	*333*
2208.	9 January 1882	From Charles Batchelor	*333*
2209.	9 January 1882	From Francis Upton	*335*
2210.	11 January 1882	From Lemuel Serrell	*336*
2211.	15 January 1882	To Edward Johnson	*337*
2212.	17 January 1882	From Francis Upton	*346*
2213.	18 January 1882	From Leslie Ward	*348*
2214.	20 January 1882	From Charles Batchelor	*349*
2215.	20 January 1882	Samuel Insull to Charles Batchelor	*351*
2216.	23 January 1882	To Edward Johnson	*355*
2217.	23 January 1882	To William Preece	*361*
2218.	23 January 1882	From Otto Moses	*362*
2219.	27 January 1882	Memorandum: Incandescent Lamp Patent	*364*
2220.	30 January 1882	Samuel Insull to George Gouraud	*366*
2221.	3 February 1882	Samuel Insull to Edward Johnson	*369*

2222.	c. 3 February 1882	Memorandum to Richard Dyer	*373*
2223.	6 February 1882	To Charles Clarke	*376*
2224.	6 February 1882	Memorandum to Richard Dyer	*377*
2225.	10 February 1882	From William Preece	*379*
2226.	13 February 1882	From Edward Johnson	*379*
2227.	13 February 1882	From John Ott	*384*
2228.	15 February 1882	To Edward Johnson	*386*
2229.	18 February 1882	To George Gouraud	*394*
2230.	20 February 1882	Notebook Entry: Pyromagnetic Generator	*396*
2231.	21 February 1882	To Theodore Waterhouse	*398*
2232.	21 February 1882	From José Navarro	*399*
2233.	23 February 1882	Mary Edison to Samuel Insull	*400*
2234.	25 February 1882	To Edward Johnson	*401*
2235.	26 February 1882	From Edward Acheson	*401*
2236.	26 February 1882	From Edward Johnson	*404*
2237.	February 1882	Memorandum of Conversation: Electric Railway	*407*
2238.	February 1882	Production Model: Central Station Dynamo	*414*
2239.	7 March 1882	To Edward Johnson	*414*
2240.	8 March 1882	From Charles Porter	*416*
2241.	13 March 1882	Telegrams: Samuel Insull to/from Mary Edison	*417*
2242.	13 March 1882	Technical Note: Voltage Regulation	*420*
2243.	15 March 1882	Drawings: Pearl Street Central Station	*429*
2244.	21 March 1882	Samuel Insull to Edward Johnson	*430*
2245.	23 March 1882	To M. Russell	*433*
2246.	27 March 1882	William Meadowcroft Memorandum: Ore Milling	*433*
2247.	28 March 1882	From Charles Batchelor	*435*
2248.	28 March 1882	From Francis Jehl	*436*
2249.	29 March 1882	From Joshua Bailey and Theodore Puskas	*439*
2250.	29 March 1882	From Sherburne Eaton	*444*
2251.	c. 1 April 1882	Memorandum: Electric Railway Patents	*449*
2252.	2 April 1882	Notebook Entry: Facsimile Telegraphy	*451*
2253.	4 April 1882	To Alfred Reade	*453*
2254.	6 April 1882	To Charles Batchelor	*454*
2255.	6 April 1882	To Edward Johnson	*454*
2256.	7 April 1882	From Joshua Bailey	*455*
2257.	8 April 1882	Samuel Insull to Pitt Edison	*456*
2258.	9 April 1882	From Edward Johnson	*457*
2259.	10 April 1882	Samuel Insull to Charles Batchelor	*473*
2260.	11 April 1882	To Francis Upton	*478*
2261.	12 April 1882	From Sherburne Eaton	*478*
2262.	12 April 1882	Draft Patent Application: Electrochemistry	*479*

2263.	19 April 1882	Samuel Insull to Edward Johnson	*481*
2264.	26 April 1882	To Edward Johnson	*483*
2265.	2 May 1882	To Charles Batchelor	*485*
2266.	3 May 1882	To Charles Batchelor	*486*
2267.	8 May 1882	To Charles Clarke	*487*
2268.	8 May 1882	To Sherburne Eaton	*488*
2269.	9 May 1882	From Charles Clarke	*489*
2270.	9 May 1882	From Edward Johnson	*491*
2271.	10 May 1882	Notebook Entry: Miscellaneous	*502*
2272.	11 May 1882	To Charles Clarke	*503*
2273.	11 May 1882	To Thomas B. A. David	*506*
2274.	11 May 1882	To Sherburne Eaton	*507*
2275.	11 May 1882	To Sherburne Eaton	*510*
2276.	12 May 1882	Notebook Entry: Electric Light and Power	*512*
2277.	13 May 1882	To Edison Lamp Co.	*517*
2278.	18 May 1882	To Sherburne Eaton	*518*
2279.	18 May 1882	From Charles Clarke	*520*
2280.	18 May 1882	From Drexel, Morgan & Co.	*523*
2281.	20 May 1882	Technical Note: Dynamo	*524*
2282.	22 May 1882	To Edward Johnson	*526*
2283.	24 May 1882	To Francis Upton	*527*
2284.	24 May 1882	Notebook Entry: Electric Light and Power	*528*
2285.	27 May 1882	From Edward Johnson	*529*
2286.	29 May 1882	To Sherburne Eaton	*531*
2287.	29 May 1882	To Uriah Painter	*537*
2288.	29 May 1882	Samuel Insull to Calvin Goddard	*538*
2289.	31 May 1882	Edison Electric Illuminating Co. Memorandum: Central Station System	*539*
2290.	1 June 1882	To Sherburne Eaton	*542*
2291.	2 June 1882	Draft Patent Application: Incandescent Lamp	*543*
2292.	4 June 1882	From Edward Johnson	*544*
2293.	7 June 1882	Agreement with Charles Dean and Charles Rocap	*548*
2294.	8 June 1882	Notebook Entry: Incandescent Lamp	*549*
2295.	8 June 1882	Notebook Entry: Electric Light and Power	*550*
2296.	9 June 1882	To Lyman Abbott	*552*
2297.	11 June 1882	From Francis Upton	*553*
2298.	12 June 1882	To Sherburne Eaton	*554*
2299.	14 June 1882	From George Bliss	*559*
2300.	14 June 1882	From Charles Clarke	*561*
2301.	16 June 1882	From Charles Rocap	*562*
2302.	19 June 1882	To Charles Batchelor	*563*
2303.	21 June 1882	From Sherburne Eaton	*564*

2304.	24 June 1882	From Charles Dean	*569*
2305.	26 June 1882	From Charles Batchelor	*570*
2306.	27 June 1882	Draft Patent Application: Incandescent Lamp	*571*
2307.	June 1882	Draft Caveat: Electric Light and Power	*572*
2308.	1 July 1882	Memorandum: Electric Lighting Patents	*582*
2309.	5 July 1882	To Francis Upton	*583*
2310.	5 July 1882	To Arnold White	*584*
2311.	6 July 1882	Telegrams: From/To Samuel Insull	*585*
2312.	6 July 1882	From Francis Upton	*587*
2313.	6 July 1882	Notebook Entry: Incandescent Lamp	*589*
2314.	17 July 1882	From Charles Batchelor	*590*
2315.	19 July 1882	To Joshua Bailey and Theodore Puskas	*592*
2316.	19 July 1882	From Charles Clarke	*593*
2317.	20 July 1882	From Arnold White	*594*
2318.	21 July 1882	From Sherburne Eaton	*597*
2319.	21 July 1882	From Francis Upton	*600*
2320.	21 July 1882	Notebook Entry: Incandescent Lamp	*603*
2321.	23 July 1882	To Sherburne Eaton	*604*
2322.	28 July 1882	To William Hammer	*607*
2323.	31 July 1882	From Sherburne Eaton	*609*
2324.	1–c. 16 August 1882	Notebook Entry: Incandescent Lamp	*612*
2325.	2 August 1882	To Charles Clarke	*621*
2326.	7 August 1882	From George Hamilton	*622*
2327.	10 August 1882	To Charles Batchelor	*623*
2328.	13 August 1882	Technical Note: Consumption Meter	*625*
2329.	21 August 1882	From Francis Upton	*627*
2330.	22 August 1882	To Emma McHenry Pond	*628*
2331.	24 August 1882	To Frederick Lawrence	*628*
2332.	25 August 1882	From Charles Clarke	*629*
2333.	28 August 1882	Harry Olrick to William Mather	*630*
2334.	29 August 1882	From George Bliss	*636*
2335.	31 August 1882	To Charles Batchelor	*639*
2336.	1 September 1882	To Calvin Goddard	*642*
2337.	1 September 1882	Samuel Insull to Miller Moore	*643*
2338.	5 September 1882	Anonymous Article in the *New York Herald*	*644*
2339.	12 September 1882	To Arnold White	*649*
2340.	13 September 1882	To the Société Électrique Edison	*654*
2341.	21 September 1882	Alden & Sterne to Samuel Insull	*654*
2342.	21 September 1882	Frank Sprague to Edward Johnson	*655*
2343.	28 September 1882	Samuel Insull to Charles Batchelor	*668*
2344.	2 October 1882	From Camille de Janon	*683*
2345.	4 October 1882	To Spencer Borden	*684*
2346.	c. 9 October 1882	Memorandum: Incandescent Lamp Patents	*685*

2347.	10 October 1882	Samuel Insull to Harry Olrick	695
2348.	12 October 1882	To Thomas Logan	697
2349.	c. 13 October 1882	Memorandum: Incandescent Lamp Patent	697
2350.	14 October 1882	From Archibald Stuart	698
2351.	16 October 1882	Charles Batchelor to Samuel Insull	700
2352.	17 October 1882	From George Bliss	702
2353.	20 October 1882	To Archibald Stuart	704
2354.	25 October 1882	To Charles Batchelor	704
2355.	c. 27 October 1882	Memorandum to Richard Dyer	706
2356.	30 October 1882	Samuel Insull to Charles Batchelor	706
2357.	October 1882?	Draft Memorandum to Edison Electric Light Co., Ltd.	711
2358.	1 November 1882	From George Bliss	715
2359.	c. 1 November 1882	Memorandum to Henry Seely	716
2360.	2 November 1882	Samuel Insull to Seth Low	718
2361.	3 November 1882	Samuel Insull to David Burrell	719
2362.	3 November 1882	Samuel Insull to Henry Draper	720
2363.	c. 6 November 1882	Memorandum: Incandescent Lamp Patent	721
2364.	8 November 1882	To George Bliss	722
2365.	8 November 1882	To George Bliss	722
2366.	10 November 1882	From Charles Batchelor	724
2367.	c. 13 November 1882	To Othniel Marsh	725
2368.	15 November 1882	Samuel Insull to Sherburne Eaton	725
2369.	17 November 1882	To Giuseppe Colombo	726
2370.	c. 20 November 1882	Technical Note: Incandescent Lamp	729
2371.	21 November 1882	To Louis Rau	730
2372.	27 November 1882	From Sherburne Eaton	730
2373.	29 November 1882	Samuel Insull to William McCrory	732
2374.	November 1882?	Draft Memorandum to Edison Electric Light Co., Ltd.	734
2375.	November 1882?	Draft Memorandum to Edison Electric Light Co., Ltd.	738
2376.	1 December 1882	From Charles Batchelor	738
2377.	c. 4 December 1882	Draft Patent Application: Primary Battery	741
2378.	c. 4 December 1882	Draft Patent Application: Storage Battery	742
2379.	11 December 1882	To Joshua Bailey	743
2380.	12 December 1882	Draft to Charles Speirs	745
2381.	13 December 1882	To Charles Batchelor	746
2382.	18 December 1882	From Josiah Reiff	747
2383.	21 December 1882	From Sherburne Eaton	747
2384.	21 December 1882	Notebook Entry: Incandescent Lamp	748
2385.	27 December 1882	From Pitt Edison	749
2386.	27 December 1882	J. Pierpont Morgan to Sherburne Eaton	750
2387.	4 January 1883	From Francis Upton	755

2388.	5 January 1883	To Francis Upton	*757*
2389.	c. 5 January 1883	Essay: Storage Battery	*757*
2390.	8 January 1883	From New York Department of Taxes and Assessments	*761*
2391.	10 January 1883	To Henry Rowland	*761*
2392.	16 January 1883	Charles Batchelor to Sherburne Eaton	*763*
2393.	16 January 1883	Sherburne Eaton Report to Edison Ore Milling Co.	*765*
2394.	19 January 1883	R. G. Dun & Co. Credit Report, Edison Lamp Co.	*771*
2395.	1 February 1883	And Edward Johnson and Francis Upton Draft to Sherburne Eaton	*771*
2396.	1 February 1883	To Arnold White	*774*
2397.	1 February 1883	Francis Upton to Edward Johnson	*776*
2398.	3 February 1883	Notebook Entry: Incandescent Lamp	*778*
2399.	8 February 1883	To Joshua Bailey	*780*
2400.	10 February 1883	To Charles Rocap	*784*
2401.	13 February 1883	From Cyrus Brackett	*784*
2402.	14 February 1883	From Richard Dyer	*785*
2403.	17 February 1883	Samuel Insull to Edward Johnson	*788*
2404.	3 March 1883	To Charles Batchelor	*790*
2405.	4 March 1883	Draft Patent Application: Electrical Distribution System	*791*
2406.	5 March 1883	To Charles Batchelor	*792*
2407.	5 March 1883	To Edward Johnson	*793*
2408.	5 March 1883	Samuel Insull to Seth Low	*795*
2409.	6 March 1883	Samuel Insull to Edward Johnson	*796*
2410.	8 March 1883	From Francis Upton	*799*
2411.	8 March 1883	Notebook Entry: Electric Light and Power	*802*
2412.	8 March 1883	Notebook Entry: Consumption Meter	*804*
2413.	13 March 1883	To George Ballou	*804*
2414.	13 March 1883	From Charles Batchelor	*805*
2415.	27 March 1883	Notebook Entry: Electric Light and Power	*806*
2416.	29 March 1883	Samuel Insull to Charles Batchelor	*808*
2417.	29 March 1883	Memorandum: Village Plants	*809*

List of Editorial Headnotes

Title	Page
Menlo Park Laboratory Reports	7
Lamp Fabrication	42
Edison and Public Relations at the Paris Electrical Exposition	95
Paris Exposition "C" Dynamo	100
Isolated Plant Dynamos	112
Standard Electric Consumption Meter	201
Innovation and Quality Control at the Edison Lamp Works	234
Examination of Edison's British Electric Lighting Patents	315
Central Station "Jumbo" Dynamo	410
Dynamo Voltage Regulation	418
Pearl Street Central Station	423
Edison Electric Light Company Bulletins	487
Sherburne Eaton Reports	506
Isolated Lighting Company Defect Reports	526
Edison's Manufacturing Operations	659

Preface

With his move from Menlo Park, New Jersey, to New York City at the end of March 1881, Edison shifted his focus from research and development to the commercialization of his electric lighting system. This volume chronicles Edison's central role in the enormous effort required to manufacture, market, and install electric lighting systems in the United States and abroad. Standard studies of this period emphasize the inauguration of the commercial electric utility industry at the Pearl Street central station, but Edison and his associates audaciously operated on a global scale, not just focusing on the major cities of North America and Europe, but reaching simultaneously from Appleton, Wisconsin, to Australia, through the Indian subcontinent and East Asia, to Central and South America.

Edison initially devoted his energies to the four manufacturing plants that produced electrical equipment, lamps, and underground conductors. He controlled two of these concerns, the Edison Lamp Company and the Edison Machine Works, and had large stakes in the other two, the Electric Tube Company and Bergmann & Company. Employing more than a thousand workers, these shops operated as partnerships with close associates from his Menlo Park laboratory, who also managed the day-to-day operations. The manufacturing shops were independent of the Edison Electric Light Company and their expanding operations put heavy demands on Edison's personal finances.

Producing durable, standardized incandescent lamps in large quantities and at low cost presented Edison with his greatest manufacturing challenge. It was for this reason that he had established the lamp factory in Menlo Park in the second half of 1880 as part of the research and development effort. The experience of those who had learned the art of making lamps in the laboratory was vital to the early manufacturing effort, as was the work of the laboratory's machine shop, which provided instruments, other equipment, and repairs for the factory. Although Edison spent little time at the lamp factory, he received frequent reports from superintendent Francis Upton regarding the extensive experiments on lamp design and manufacture, which served as the basis for Edison's numerous patents on these subjects. When he discovered a notable decline in lamp durability in the fall of 1881, he virtually lived at the lamp factory for ten days until he had improved the manufacturing process so that the vacuum and the lamps would last longer than ever. At this time the lamp factory's output was intended for isolated lighting plants. This was also true of the other factories except for the Tube Works, which manufactured the underground conductors that were laid starting in September for the Pearl Street central station district in lower Manhattan.

Although Edison believed that the future of electric lighting lay in central station service, he did not expect the Pearl Street station to be ready for almost a year and acknowledged the short-term importance of lighting individual buildings to promote his incandescent system. Isolated lighting also provided a market for the large manufacturing capabilities he would need to construct central stations. He and the chief engineer of the Edison Electric Light Company, Charles Clarke, therefore developed a series of standardized dynamos to advance this business, which expanded rapidly. Isolated incandescent lighting plants soon began to provide light in textile mills, where open-flame lighting was dangerous, and to replace gas and kerosene lights in a wide range of industrial and commercial establishments. Although unwilling to participate in the manufacturing operations, the Edison Electric Light Company did develop and run the isolated business.

As at the lamp factory, Edison's other manufacturing enterprises were the site of most of the ongoing experimental work necessary to improve the commercial incandescent lighting system. Because these efforts were dispersed in the shops, many of the experimental records have not survived. Often the only records that remain are Edison's memoranda and drafts for the nearly 200 patent applications he filed during this period. The improvements embodied in these patents not only led to improved devices and processes, but the patents themselves were intended to block potential rivals.

The relatively small number of technical records from this period is also a product of Edison's increasing involvement in business operations. Because of this and the fact that some correspondence was carried on in Edison's name by Samuel Insull, who served not only as his secretary but also his business manager, it is more difficult than ever in this period to gain a clear view of Edison himself.

As work proceeded on the Pearl Street station and the isolated business continued to grow in the United States, promotional efforts in Paris and London prepared the way for expansion into Europe. Edison entrusted Charles Batchelor with the responsibility for his exhibit at the International Electrical Exposition, which opened in Paris at the end of August 1881. This exhibit generated widespread favorable publicity as the European scientific and technical community witnessed for the first time the ambitious scope of the Edison system. Edison soon sent Edward Johnson to London to promote his system there and received accolades for his exhibit at the Crystal Palace Electrical Exhibition the following winter.

Negotiations begun in the last months of 1881 soon resulted in the formation of companies in Britain and France to exploit Edison's electric light system. In London, Johnson worked to organize the Edison Electric Light Company Ltd., but his primary mission was to set up a demonstration central station to convince prospective British investors and consumers of the reliability and economy of the Edison system. He and Egisto Fabbri, a partner in Drexel, Morgan & Company (which controlled Edison's British patents), selected a site on London's famous Holborn Viaduct, where the station opened in April 1882. Edison also urged George Gouraud, his longtime business representative there, to market isolated lighting plants in British colonies and other countries outside Europe and Britain. This led to the formation of Edison's Indian & Colonial Electric Light Company.

In France, Joshua Bailey, previously involved in negotiating the combined telephone company that included Edison's patents, took the lead in forming manufacturing and operating companies for the commercial development of Edison's electric light system in Europe: the Société Industrielle et Commerciale Edison to manufacture lamps and other components of the Edison system, the Compagnie Continentale to organize and license electric light companies throughout continental Europe, and the Société Électrique to build central stations in France. Charles Batchelor acted as Edison's representative in these companies and also had charge of the lamp factory and machine shop at Ivry-sur-Seine just outside the Paris city walls. Elsewhere in Europe, Bailey and Theodore Puskas negotiated for isolated plants and utility companies in smaller cities not covered by the Compagnie Continentale, including Strasbourg and Milan. In order to coordinate the exploitation of his patents in various European countries, Edison organized a patent holding company, the Edison Electric Light Company of Europe. Egisto Fabbri was among the investors in this company, purchasing a large block of shares on behalf of Drexel, Morgan. Fabbri was actively involved in the negotiations in Paris and London, and also organized companies in South America and Mexico. Edison's foreign operations involved an expanding network of financiers and banks in Europe, often facilitated through the agency of Fabbri and Drexel, Morgan.

During negotiations for the formation of the British and European companies, both Johnson and Batchelor expressed concern over the state of Edison's foreign patents. Edison agreed to a systematic review by outside legal experts of his British lighting patents, which led to his submitting five specifications for disclaimer in order to protect his carbon-filament lamp. He also hired patent attorney Richard Dyer to take charge of all his patent affairs. Dyer worked closely with Edison to draft his U. S. and foreign specifications. He sent foreign specifications to agents abroad, who were familiar with local patent laws, for revision and filing.

The stress of overseeing four major factories, solving a myriad of major and minor technical problems, supervising the installation of the Pearl Street station and miles of underground conductors, and negotiating the organization of companies to operate throughout the world wore out Edison. After vacationing with his family in Florida during winter 1882, he fell ill and remained in bed most of the second half of April. During this time he delegated his business affairs to Samuel Insull, who had the task of integrating the production of the shops with the demand for their goods throughout the United States and abroad.

The rapid growth of the lighting business led Edison to expand his production facilities. The lamp factory moved from Menlo Park to Harrison, New Jersey (across the river from Newark), in spring 1882 and began to employ female workers to reduce the costs of its large labor force. During that summer Edison expanded the Machine Works and soon had about 800 men employed in the production of dynamos. In September he formalized his partnership in Bergmann & Company, which soon moved into a large six-story shop, with Edison taking the top floor for his laboratory.

After spending most of the summer of 1882 at Menlo Park, Edison moved back to New York in order to be close to the nearly completed Pearl Street station. Edison and Clarke had designed steam dynamos for central stations; in July, six of these massive machines were installed and tested and the underground network was completed. Over the next month the system was tested discreetly and found to work. Then on 4 September, after giving the station a final inspection, Edison went to the office of J. P. Morgan to inaugurate it—and the commercial electric utility industry. It was almost four years to the day since he had begun intensive research into electric lighting.

The Pearl Street station seems more significant in retrospect than it appeared at the time. Although central stations were Edison's vision and the future of electric light and power, the large number of isolated plants that preceded Pearl Street meant that commercial electric lighting had already lost its novelty. Nonetheless, there was a demand for lighting in smaller towns and cities for which the expensive underground system in New York was infeasible. This led Edison to develop what he termed the Village Plant System, which used overhead conductors. The first Village System station began operating in Roselle, New Jersey, in January 1883. By March, Edison had decided to set up a separate company to promote and install small central stations throughout the United States.

The progress of the Thomas A. Edison Papers depends on the support of many individuals and organizations, including the Sponsors, other financial contributors, academic scholars, Edison specialists, librarians, archivists, curators, and students. Representatives of the four Sponsors have assisted with this volume and the editors thank them for their continuing concern and attention. The strong support of public and private foundations and of their program officers has sustained the project and helped it remain editorially productive.

Preparation of this volume was made possible in part by grants from the Division of Research Programs (Scholarly Editions) of the National Endowment for the Humanities, an independent federal agency; the National Historical Publications and Records Commission; the New Jersey Historical Commission; the Charles Edison Fund; the National Trust for the Humanities; as well as through the support of Rutgers, The State University of New Jersey, and the National Park Service (Edison National Historic Site). The editors appreciate the interest and support of the many program officers and trustees, especially Elizabeth Arndt, Michael Hall, Timothy Connolly, Marc Mappen, Mary Murrin, Howard Green, John P. Keegan, Malcolm Richardson, Ann Orr, and Francis X. O'Brien. Any opinions, findings, conclusions, or recommendations expressed in this publication are solely those of the editors and do not necessarily reflect the views of any of the above federal foundations or agencies, the United States Government, or any other financial contributor.

The Edison Papers project is indebted to the National Park Service for its multifaceted support. The editors express particular thanks to Marie Rust, John Maounis, and Nancy Waters of the Northeast Region, and to Maryanne Gerbauckas, Theresa Jung, Roger Durham, Leonard DeGraaf, Douglas Tarr, Edward Wirth, Karen Sloat-Olsen, Linda Deveau, Gerald Fabris, Joseph De Monte, Charles Magale, Marilyn Kyles, and Clarence Askew of the Edison National Historic Site in West Orange, New Jersey.

Many associates at Rutgers University have contributed significantly to the Edison Papers. The editors are grateful to President Richard L. McCormick; Philip Furmanski, Executive Vice President for Academic Affairs; and Holly M. Smith, Dean of the Faculty of Arts and Sciences, along with her dedicated staff, especially Barry V. Qualls, Dean for Humanities; Robert Wilson, Executive Vice Dean; and Barbara Lemanski, Associate Dean for Policy and Personnel. We would especially like to thank Paul Kuznekoff, Director of Development for the Faculty of Arts and Sciences. In addition we appreciate the efforts of business manager Rebecca Steika and the support provided by John Amodeo, Thomas Vosseler, and the staff of the Faculty of Arts and Sciences Computer & Network Operations Group, especially unit computing manager David Motovidlak. The editors value the support of colleagues and staff in the History Department, especially Michael Adas, Rudy Bell, Alastair Bellany, Luigi Andrea Berto, John Chambers, Paul Clemens, Ziva Galili, Ann Gordon, Reese V. Jenkins, Matt Matsuda, Philip J. Pauly, James W. Reed, Susan Schrepfer, Virginia Yans, Dawn Ruskai, Candace Walcott-Shepherd, and Mary DeMeo. We also want to thank Michael Geselowitz and his staff at the IEEE History Center as well as members of the Rutgers University Libraries, notably Marianne Gaunt, Thomas Frusciano, Ron Becker, Tom Glynn, Jim Niessen, Rhonda Marker, and the Interlibrary Loan Office. A special thanks is due Michael Siegel, staff cartographer in the Department of Geography, who prepared the maps in this volume. At the Rutgers University Foundation we have had the assistance of Carol P. Herring, Linda Corcoran, John Pearson, and Jacqueline Perkel-Joseph. We also want to thank Ronald Thompson and Steve Miller of the Division of Grant and Contract Accounting, Donna Foster of the Office of Research and Sponsored Programs, Stephen J. DiPaolo and Joseph M. Harrigan of the Controller's Office, and Allison Thomas of University Relations.

Many scholars have shared their insights and assisted the editors in a variety of ways. For this volume, notable help came from Brian Bowers (who provided the list of Edison's British patents in Appendix 5), Andrew Butrica, Jane Mork Gibson, James H. Graebner, Cathy Moran Hajo, Gregory Jankunis, Esther Katz, Claudio Pavese, Alvin J. Salkind, Hal Wallace, and Mira Wilkins.

Institutions and their staff have provided documents, photographs, photocopies, and research assistance. The staff of the Henry Ford Museum & Greenfield Village has supplied many valuable services; Marc Greuther, in particular, fielded innumerable questions and brought to the editors' attention important documentation about Edison dynamos. The editors also gratefully acknowledge Christopher Baer at the Hagley Museum and Library; Ronald Brashear at the Smithsonian Institution's Dibner Library of the History of Science and Technology; Charles Greifenstein and Valerie-Anne Lutz, both of the American Philosophical Society; Janet Linde, archivist for the New York Stock Exchange; and the reference staff of the New York Public Library's Science Industry and Business Library.

Staff members, interns, students, and visiting editors not mentioned on the title page but who have contributed to this volume include Thomas E. Jeffrey, Grace Kurkowski, Helen Endick, David Ranzan, Margaret O'Connell, Blaine Mc-Cormick, Rick Mizelle, and the late Tony Cautillo, who is missed.

As always, the project has had the benefit of the superb staff of the Johns Hopkins University Press. For this volume, the editors are indebted to Robert J. Brugger, Julie McCarthy, and Ken Sabol.

Chronology of
Thomas A. Edison

April 1881–March 1883

1881	
3 April	Begins designing rotary electric meter.
7 April	Contracts for Armington & Sims steam engine for large dynamo.
14 April	Begins designing disk dynamo.
20 April	Assigns Edward Acheson to learn details of lamp manufacture before going to Paris.
April	Enters into informal partnership with Sigmund Bergmann and Edward Johnson for manufacture of lamp sockets and accessories.
	Begins making specific arrangements for exhibit at Exposition Internationale de l'Électricité in Paris.
	Directs Alfred Haid to conduct experiments on coating filaments with various forms of carbon.
	Directs Charles Hughes to resume experiments on preservation of perishable foods in a vacuum.
9 May	Purchases buildings in Harrison, New Jersey, for enlarged lamp factory.
12 May	Charles Batchelor sails from England to New York to begin work on exhibits for the Paris Exposition.
May	Constructs magnetic ore separator for recovering iron ore.
June	Travels to Quogue, Long Island, to inspect operation of ore separator.
	Begins testing large dynamo for Paris Exposition.
	Samuel Insull begins dictating some outgoing correspondence to a typist.
15 July	Discharges most remaining employees at Menlo Park laboratory and machine shop.
	Completes reconstruction of large dynamo armature.

18 July	Receives nearly $55,300 in partial liquidation of his telephone interests in Great Britain.
19 July	With Patrick Kenny, prepares British patent specification for improvements in facsimile telegraphs.
c. 21 July	Sends Charles Batchelor to Paris to oversee exhibit at Paris Exposition.
July	Begins designing standard dynamo models for isolated lighting plants.
	Devises alternative armature winding patterns to reduce internal arcing.
8 August	Conducts experiments on reducing resistance between dynamo commutator bars and brushes.
11 August	Representatives in Paris begin legal action against display of Hiram Maxim's incandescent lamps at International Electrical Exposition.
23 August	Approves terms for consolidating control of his telephone patents in Europe with those of other inventors.
25 August	Edison lighting exhibit at International Electrical Exposition commences.
c. 28 August	Travels to Michigan to meet Mary Edison.
August	Encourages his representatives to cultivate relationships with respected technical writers in France and Great Britain.
	Begins construction at 255–257 Pearl Street for New York central station plant.
	Begins renovation of factory in Harrison, New Jersey.
	Directs Charles Hughes to conduct preliminary survey for new electric railroad line at Menlo Park.
7 September	Large dynamo shipped to Paris.
c. 7 September	Returns from trip to Michigan.
14 September	Agrees with Henry Villard to build new experimental electric railroad.
21 September	Sends Edward Johnson to London to oversee commercial development of lighting system in Great Britain.
22 September	Begins laying underground conductors for New York central station district.
23 September	Proposes terms for financing construction of European factories for electric lighting equipment.
September	Edison Ore Milling Co. begins using iron ore separator at Quonocontaug, Rhode Island.
2 October	Instructs Edward Johnson to begin disclaimer process to strengthen British electric lighting patents.
4 October	Successfully tests large dynamo for London demonstration central station at Holborn Viaduct.
18 October	Foyer of Paris Opera illuminated by Edison installation.

21 October	Direct-connected "C" dynamo begins operation at International Exposition.
25 October	Gives notice of intention to end financial support for *Science*.
6 November	Agrees in principle to terms for establishing Edison electric light and manufacturing companies in Europe.
12 November	Begins period of intensive experimentation at lamp factory.
1 December	Recalls John Branner from South American vegetable fiber search.
20 December	Mary Edison hosts party at Menlo Park.
Fall	Establishes Testing Room at the Edison Machine Works.
27 December	Receives 25 U.S. patents.
c. 28 December	Vacates New York City apartment, returning to Menlo Park home.
December	Devises audible voltage indicator for isolated lighting installations.
	Appointed officer of the French National Order of the Legion of Honor.
1882	
17 January	Edward Johnson unofficially inaugurates Edison's exhibition at Crystal Palace.
January	Construction of underground conductor network in New York halted by winter weather.
1 February	Hires patent attorney Richard Dyer.
	Fundamental British telephone patent upheld in court.
2 February	Compagnie Continentale Edison (holding company) and companies for manufacturing and isolated lighting in France organized.
14 February	Sends Francis Jehl to London with standard electric meter.
	Ships second "C" dynamo to Holborn Viaduct in London.
20 February	Arrives in Washington, D.C.
22 February	Returns from Washington.
	Mary Edison arrives in Aiken, South Carolina.
25 February	Electrical Exhibition opens at Crystal Palace in London.
1 March	Travels to Florida to join family on vacation.
15 March	Edison Electric Light Co., Ltd., organized in London.
28 March	Returns to Menlo Park from Florida.
March	Negotiates royalty payment on dynamos to Siemens Brothers, London.
Winter	Writes detailed suggestions for revisions of British electric light and power patents.
1 April	Production stops at Edison Lamp Co. factory in Menlo Park.
9 April	Orders rail car from J. G. Brill & Co. for electric railway experiments.

13 April	Central station demonstration plant on Holborn Viaduct formally opens.
21 April	Receives settlement of British telephone accounts.
April	Incapacitated by illness for several weeks.
	Experimental electric freight locomotive completed.
c. 8 May	Moves back to Menlo Park.
12 May	Resumes designing voltage-reduction devices for high voltage distribution system.
29 May	Declares opposition to proposed changes in U.S. patent law.
31 May	Edison Lamp Co. resumes production in Harrison, N.J.
May	Purchases Menlo Park factory building from Lamp Co.
	Directs tests of new K dynamo.
	Agrees in principle to terms establishing Edison's Indian & Colonial Electric Light Co. for British colonies.
2 June	Drafts patent application for cellulose lamp filament.
15 June	Approves filing disclaimers of key British electric lighting patents.
c. 21 June	Edward Johnson returns from England.
Spring	Designs high voltage direct current distribution system.
1–6 July	Borrows over $37,000 from Drexel, Morgan & Co.
8 July	Leaves on sailing trip.
c. 14 July	Reaches Montreal.
18 July	Returns to Menlo Park.
July	Experiments with chemical aids to lamp exhaustion.
	Designs three-wire electrical distribution system.
	Investigates malfeasance of patent attorney Zenas Wilber.
	Construction of New York central station largely completed.
8 August	Enlarges responsibilities of patent attorney Richard Dyer.
22 August	Receives 20 U.S. patents.
August	Selects Roselle, New Jersey, for demonstration "Village Plant" electrical system.
2 September	Executes formal partnership agreement with Sigmund Bergmann and Edward Johnson.
4 September	Begins commercial operation of New York central station.
19 September	Receives 31 U.S. patents.
Summer	Charles Batchelor opens Edison factories near Paris.
c. 26 September	Successfully operates two Jumbo dynamos in tandem for first time.
1 October	Leases house at 25 Gramercy Park South, New York City, for family.
October	Establishes laboratory on top floor of Bergmann & Co. building in New York.
10–11 November	Visits Boston.
15 November	Attends National Academy of Sciences dinner in New York.

16 November	Opens Pearl Street station to National Academy of Sciences.
c. 13 December	Visits Boston.
18 December	Loses lawsuit by Lucy Seyfert; $5,065 judgment entered.
December	Suspends experiments on storage batteries.

1883

19 January	First village plant opens at Roselle, New Jersey.
c. 1 February	Sends Edward Johnson to London for consultation with Edison Electric Light Co., Ltd.
15 February	Travels to Boston to begin establishment of an illuminating company.
19 February	Begins a trip to the South.
27 February	Returns from the South ahead of schedule.
29 March	Agrees to form Edison Construction Department with Samuel Insull and Edward Johnson.

Editorial Policy and User's Guide

The editorial policy for the book edition of Thomas Edison's papers remains essentially as stated in Volume 1, with the modifications described in Volumes 2–5. The additions that follow stem from new editorial situations presented by documents in Volume 6. A comprehensive statement of the editorial policy will be published later on the Edison Papers website (http://edison.rutgers.edu).

Selection

The fifteen-volume book edition of Thomas Edison's papers will include nearly 6,500 documents selected from an estimated 5 million pages of extant Edison-related materials. For the period covered in Volume 6 (April 1881–March 1883), the editors have selected 344 documents from approximately 10,700 extant Edison-related documents. Most extant documents detail Edison's inventive work and his business relationships; very few are concerned with his family or other personal relationships.

The editors have sought to select documents that illuminate the full range of Edison's thought and activities in this period. Those published here are primarily by or to Edison, his surrogate Samuel Insull, or others working in concert with him or on his behalf. Also included is some third-party correspondence which documents key events or illustrates the context in which he worked. Much of Edison's correspondence from this time details his business relationships, and here the editors have desired to present letters that either provide considerable summary detail or contain information about a wide range of issues and which can be annotated by reference to other related materials. There are fewer available technical records from this time period than in periods covered by previous volumes. This is both because of the nature of Edison's work and because technical records created at his factories, some of which would have documented research done directly under his orders, did not become part of his personal papers and were subsequently destroyed.

As in the other volumes, the editors have selected a few key artifacts. This volume also includes an engineering drawing of one artifact that no longer exists: the central station generating plant in New York City.

Transcription

The transcription policies used in preceding volumes have been followed in the present case, with the following additions and emendations. Edison often punctuated his sentences in very idiosyncratic ways, including the use of wholly nonstandard marks and identical marks for obviously different purposes. For the sake of intelligibility, the editors have transcribed Edison's punctuation with conventional typographic characters according to their understanding of his intent. Samuel Insull often made shorthand remarks on Edison's correspondence, but like other docket notes, these have not been transcribed or noted. Because of the profusion of documents in this volume created by secretaries, those written by an unidentified scribe are no longer described as "in an unknown hand"; however, where the editors can determine this identity, it is so indicated. For clarity, expressions of time are routinely transcribed in standard form.

Annotation

In the endnotes following each document, citations are generally given in the order in which the material is discussed. However, when there are several pieces of correspondence from the same person or a run of notebook entries, these are often listed together rather than in the order they are discussed to simplify the reference.

References to the Digital and Microfilm Editions
The editors have not provided a comprehensive calendar of Edison documents because the vastness of the archive makes preparation of such an aid impractical. Their annotations include, however, references to relevant documents in the Edison Papers digital and microfilm editions; the volume may therefore serve as an entree into these publications.

The Edison Papers website (http://edison.rutgers.edu) contains approximately 175,000 images from the first three parts of the microfilm edition of documents at the Edison National Historic Site. There are also nearly 5,000 additional images not found on the microfilm that come from outside repositories. Citations to images in the digital edition are indicated by the acronym *TAED*. The citations themselves are in two forms. Those to specific documents, such as a letter, are in an alpha-numeric code (e.g., D8104ZBU). Those to documents found in volumes such as notebooks are indicated by both an alpha-numeric code and a set of image numbers (e.g., N249:6, 140). In a few instances one or more specific images in a lengthy document is referred to by an image number or numbers in parentheses (e.g., W100DEC002 [image 7]). All of these images can be seen by going to the Edison Papers homepage and clicking on the link for "Single Document or Folder" under "Search Methods." This will take the user to http://edison.rutgers.edu/singldoc.htm, where the images can be seen by putting the appropriate alpha-numeric code in one of the two boxes to retrieve either a document or a folder/volume. If retrieving a folder/volume, the user should click on "List Documents" and then "Show Documents" in the introductory "target" for that folder/volume. Then click on any of the links to specific documents in the folder/volume and put the appropriate image number in the box under the "Go to Image" link. Putting image numbers in that box when viewing any document will take you to the specific image number.

The digital edition contains a number of other features not available in the book or microfilm editions, including lists of all of Edison's U. S. patents by execution date, issue date, and subject; links to pdf files; and a comprehensive chronology and bibliography. Other materials, such as chapter introductions from the book edition and biographical sketches, will eventually be added. Material from outside repositories, including items cited in this volume, will continue to be added. Under arrangements being made at press time, this volume and its predecessors will eventually be published in a searchable, interactive format on the web.

References to the microfilm edition are indicated by the acronym *TAEM* and refer to the appropriate reel and frame numbers (e.g., 60:167).

This volume refers to letterbooks used not only for Edison's general correspondence but also for specific purposes or by particular individuals or companies. These are grouped in a subseries of Miscellaneous Letterbooks within the Letterbook Series. As with other letterbooks, citations to the Miscellaneous Letterbooks are to the book and page, with one exception. Each page of the Cable Book (LM 1) typically contains transcriptions of several telegraphic messages. For this book, a letter designation after the page number indicates the relative position of each message (e.g., LM 1:108A).

Headnotes

Because of the limited scholarly literature and the technical complexities discussed in many of the documents, there are more introductory headnotes in these volumes than is common in historical documentary editions. This is particularly the case in the present volume. Each chapter begins with a brief introduction that highlights Edison's personal, technical, and business activities during that period. Within chapters there are occasional headnotes that appear before documents (see List of Editorial Headnotes); these are used for several purposes. Artifacts and drawings without accompanying text are always preceded by headnotes (e.g., "Standard Electric Consumption Meter"). In addition, the editors also use this apparatus to discuss particular technical issues (e.g., "Dynamo Voltage Regulation"); or to describe the characteristics of a set of documents (e.g., "Sherburne Eaton Reports"). A headnote may also provide an overview of activities that are otherwise referenced only in scattered documents and endnote references (e.g., "Edison's Manufacturing Operations").

Just as the chapter introductions and headnotes serve as guides for the general reader, discursive endnotes often contain annotation of interest to the general reader. These notes also include business or technical details likely to be of more concern to the specialized reader. In general, the editors have provided more detailed information for technical issues that have received little scholarly attention than for topics that are already well treated in the secondary literature.

Document Titles

Because most of the technical documents in this volume pertain to electric lighting, the editors have used more specific subject headings to distinguish among, for example, electric lamps, dynamos, and batteries.

Citations

Citations to endnotes of documents published in previous volumes are made somewhat differently than in the previous volumes. References are now given to the appropriate document and note numbers, rather than to volume, page, and note. This is in order to facilitate eventual publication in an interactive online format. To further simplify citations, references to headnotes and chapter introductions in previous volumes are given by volume and page numbers.

For simplicity, citations to extensive correspondence over a long period are made generally to the author name(s) in the digital and microfilm editions as *TAED* or *TAEMG[uide]*, s.v. "name." Citations of exceptionally long letters now include page numbers. Citations to contemporary periodicals are made to the original print editions, although our sources have included electronic versions such as those provided by JSTOR and ProQuest.

Appendixes

As in Volumes 1–5 we include relevant selections from the autobiographical notes that Edison prepared in 1908 and 1909 for Frank Dyer and Thomas Martin's biography of Edison (see App. 1). Unlike previous volumes, there is no longer a comprehensive list of Edison's U.S. patents, which are available on the Edison Papers website. This period was one of extraordinary patenting activity on Edison's part, however, and Appendix 5 represents that activity in several ways.

There are three new appendixes in this volume. Appendix 2 consists of tabulations of Edison isolated lighting plants in use by May 1883. It is intended to indicate the extent and variety of this type of lighting, which is often difficult to discern in the historical record. Appendix 3 is a table of specifications of dynamos produced by the Edison Machine Works. It is provided in order to lift some of the burden of technical detail from the document annotation. Appendix 4 is a list of cable name codes used by Edison and his associates in transatlantic telegraphic correspondence.

Errata

Errata for previous volumes can be found on the Edison Papers website at http://edison.rutgers.edu/berrata.htm.

Editorial Symbols

~~Newark~~   Overstruck letters
   Legible manuscript cancellations; crossed-out or overwritten letters are placed before corrections
[Newark]   Text in brackets
   Material supplied by editors
[Newark?]   Text with a question mark in brackets
   Conjecture
[Newark?][a]   Text with a question mark in brackets followed by a superscript letter to reference a textnote
   Conjecture of illegible text
⟨Newark⟩   Text in angle brackets
   Marginalia; in Edison's hand unless otherwise noted
[   ]   Empty brackets
   Text missing from damaged manuscript
[---]   One or more hyphens in brackets
   Conjecture of number of characters in illegible material

Superscript numbers in editors' headnotes and in the documents refer to endnotes, which are grouped at the end of each headnote and after the textnote of each document.

Superscript lowercase letters in the documents refer to textnotes, which appear collectively at the end of each document.

List of Abbreviations

ABBREVIATIONS USED TO DESCRIBE DOCUMENTS

The following abbreviations describe the basic nature of the documents included in the sixth volume of *The Papers of Thomas A. Edison:*

AD	Autograph Document
ADf	Autograph Draft
ADfS	Autograph Draft Signed
ADS	Autograph Document Signed
AL	Autograph Letter
ALS	Autograph Letter Signed
D	Document
Df	Draft
DS	Document Signed
L	Letter
LS	Letter Signed
M	Model
PD	Printed Document
PL	Printed Letter
TD	Typed Document
TL	Typed Letter
TLS	Typed Letter Signed
X	Experimental Note

In these descriptions the following meanings are assumed:

Document Accounts, agreements and contracts, bills and receipts, legal documents, memoranda, patent applications, and published material, but excluding letters, models, and experimental notes

Draft   A preliminary or unfinished version of a document or letter

Experimental Note   Technical notes or drawings not included in letters, legal documents, and the like

Letter   Correspondence, including telegrams

Model   An artifact, whether a patent model, production model, structure, or other

The symbols may be followed in parentheses by one of these descriptive terms:

abstract   A condensation of a document

copy   A version of a document made by the author or other associated party at the time of the creation of the document

fragment   Incomplete document, the missing part of which has not been not found by the editors

historic drawing   A drawing of an artifact no longer extant or no longer in its original form

letterpress copy   A transfer copy made by pressing the original under a sheet of damp tissue paper

photographic transcript   A transcript of a document made photographically

telegram   A telegraph message

transcript   A version of a document made at a substantially later date than that of the original, by someone not directly associated with the creation of the document

STANDARD REFERENCES AND JOURNALS

Standard References

ABF	*Archives Bibliographiques Française*
ACAB	*Appleton's Cyclopaedia of American Biography*
ANB	*American National Biography*
BDUSC	*Biographical Directory of the United States Congress*
Col. Ency.	*Columbia Encyclopedia*
Congr. Dir.	*Congressional Directory* (cited as Congress.Session [edition date])
DAB	*Dictionary of American Biography*
DBB	*Dictionary of Business Biography*
DBE	*Deutsche Biographische Enzyklopädie*
DBF	*Dictionnaire de Biographie Française*
DBI	*Dizionario Biografico degli Italiani*

DSB	*Dictionary of Scientific Biography*
Ency. Brit. 1911	*Encyclopedia Britannica,* 11th edition
Ency. Chgo.	*Encyclopedia of Chicago* (www.encyclopedia.chicagohistory.org)
Ency. NJ	*Encyclopedia of New Jersey*
Ency. NYC	*Encyclopedia of New York City*
Gde. Ency.	*Grande Encyclopédie Inventaire Raisonné des Sciences, des Lettres et des Arts*
London Ency.	*London Encyclopaedia*
NCAB	*National Cyclopedia of American Biography*
NDB	*Neue Deutsche Biographie*
NGD	*New Grove Dictionary of Music and Musicians*
OED	*Oxford English Dictionary*
Oxford DNB	*Oxford Dictionary of National Biography*
WGD	*Webster's Geographical Dictionary*
WI	*World of Invention*
WWWS	*World Who's Who in Science*

Journals

ASME Transactions	*American Society of Mechanical Engineers Transactions*
Sci. Am.	*Scientific American*
Sci. Am. Supp.	*Scientific American Supplement*
Teleg. J. and Elec. Rev.	*Telegraphic Journal and Electrical Review;* formerly *Telegraphic Journal*

ARCHIVES AND REPOSITORIES

In general, repositories are identified according to the Library of Congress MARC code list for organizations (http://www .loc.gov/marc/organizations). Parenthetical letters added to Library of Congress abbreviations were supplied by the editors. Abbreviations contained entirely within parentheses were created by the editors and appear without parentheses in citations.

CtY	Yale University, Sterling Memorial Library, New Haven, Conn.
DeGH	Hagley Museum and Library, Greenville, Del.

DSI–MAH	Archives, National Museum of American History, Smithsonian Institution, Washington, D.C.
(HuBPo)	Foundation of the Postal and Telecommunication Museum, Budapest, Hungary
Hummel	Charles Hummel, Wayne, N.J.
MdBJ	Special Collections, Milton S. Eisenhower Library, Johns Hopkins University, Baltimore, Md.
MdCpNA	National Archives and Records Administration, College Park, Md.
MH–BA	Baker Library, Harvard Business School, Boston, Mass.
MiDbEI	Library and Archives (Edison Institute), Henry Ford Museum & Greenfield Village, Dearborn, Mich.
MiDbEI(H)	Henry Ford Museum & Greenfield Village, Dearborn, Mich.
NhD	Dartmouth College Library, Hanover, N.H.
NjWOE	Edison National Historic Site, West Orange, N.J.
NN	Manuscripts and Archives Division, New York Public Library, New York, N.Y.
NNNCC–Ar	Division of Old Records, New York County Clerk, New York City Archives, N.Y.
(NNNYSE)	New York Stock Exchange, Office of the Secretary, Archives, New York, N.Y.
NNPM	Pierpont Morgan Library, New York, N.Y.
PHi	Historical Society of Pennsylvania, Philadelphia, Pa.
(UkLIEE)	Institution of Electrical Engineers Archives, London, UK
(UkLMA)	London Metropolitan Archives, London, UK

MANUSCRIPT COLLECTIONS AND COURT CASES

Accts.	Accounts, NjWOE
Batchelor	Charles Batchelor Collection, NjWOE

Böhm v. Edison	*Böhm v. Edison*, Patent Interference File 7943, RG-241, MdCpNA
Comm. Arr.	Committee on Arrangements, General Files, NNNYSE
CR	Company Records, NjWOE
DF	Document File, NjWOE
Edison & Gilliland v. Phelps	*Edison & Gilliland v. Phelps*, Patent Interference File 10369, RG-241, MdCpNA
Edison v. Lane v. Gray, et. al.	*Edison v. Lane v. Gray v. Rose v. Gilliland*, Patent Interference File 8028, RG-241, MdCpNA
Edison v. Sprague	*Edison v. Sprague, Lit.*, NjWOE
EP&RI	Edison Papers & Related Items, MiDbEI
Force	Martin Force Papers, NjWOE
Fredericks	Joseph D. Fredericks Papers, NjWOE
HAR	Henry Augustus Rowland Collection, MdBJ
Hendershott	Gary Hendershott, Little Rock, Ark.
Hodgdon	Ernest Hodgdon, Derry, N.H.
Jehl Diary	Typescript copy of Francis Jehl's diary, EP&RI
Kellow	Richard W. Kellow File, Legal Series, NjWOE
Lab.	Laboratory notebooks and scrapbooks, NjWOE
Lbk.	Letterbooks, NjWOE
LDS	FamilySearch website of the Church of Jesus Christ of Latter-day Saints (www.familysearch.org/)
Lit.	Litigation Series, NjWOE
LM	Miscellaneous Letterbook, NjWOE
	1 Cable Book (1881–1883)
	3 Insull Letterbook (1882–1883)
	5 Ore Milling Co. Letterbook (1881–1887)
Marsh	Othniel Charles Marsh Papers, CtY
Meadowcroft	William H. Meadowcroft Collection, Special Collections Series, NjWOE
Miller	Harry F. Miller File, Legal Series, NjWOE
Morgan	Private Letters, J. P. Morgan, Dec. 31, 1880–Dec. 7, 1887, NNPM
Pat. App.	Patent Application Files, RG-241, MdCpNA

Pioneers Bio.	Edison Pioneers Biographical File, NjWOE
PPC	Primary Printed Collection, CR, NjWOE
PP&L	Pennsylvania Power and Light Co. Papers, DeGH
PS	Patent Series, NjWOE
Scraps.	Scrapbooks, NjWOE
Sprague	Frank J. Sprague Papers, NN
TI	Telephone Interferences (Vols. 1–5), NjWOE; a printed, bound subset of the full Telephone Interferences
TP	Theodore Puskas Collection, HuBPo
UHP	Uriah Hunt Painter Papers, PHi
WHP	William H. Preece Papers, UkLIEE
WJH	William J. Hammer Collection, DSI-MAH
Young	Charles A. Young Papers, NhD

ELECTRIFYING NEW YORK AND ABROAD
APRIL 1881–MARCH 1883

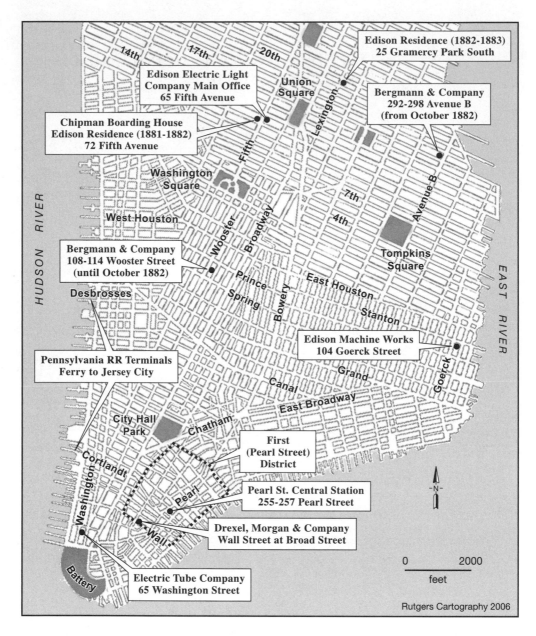

Edison Residence (1882-1883)
25 Gramercy Park South

Edison Electric Light
Company Main Office
65 Fifth Avenue

Bergmann & Company
292-298 Avenue B
(from October 1882)

Chipman Boarding House
Edison Residence (1881-1882)
72 Fifth Avenue

14th   17th   20th   Union
Square   Lexington

Fifth

Washington
Square

Broadway   7th   4th

Avenue B

West Houston

Wooster

Bergmann & Company
108-114 Wooster Street
(until October 1882)

Prince   East Houston   Tompkins
Square

Spring   Bowery   Stanton

Desbrosses

Edison Machine Works
104 Goerck Street

Pennsylvania RR Terminals
Ferry to Jersey City

Canal   Grand   Goerck

East Broadway

City Hall
Park   Chatham

First
(Pearl Street)
District

Cortlandt

Washington

Pearl   Pearl St. Central Station
255-257 Pearl Street

Wall   Drexel, Morgan & Company
Wall Street at Broad Street

Battery

Electric Tube Company
65 Washington Street

HUDSON RIVER

EAST RIVER

-N-

0   2000
feet

Rutgers Cartography 2006

*Lower Manhattan with
points of significance to
Edison's life and work,
1881–1883.*

1

April–June 1881

After having moved to New York in late winter, Edison concentrated in the spring and summer on commercializing his electric light system in the United States and abroad. Though he had experience as an independent manufacturer from his days in Newark, Edison had never faced organizational and financial challenges of this magnitude.

Geographic dispersion compounded the intrinsic organizational difficulties. The Edison Machine Works and the Electric Tube Company, both established in recent months, operated separate shops in lower Manhattan several dozen blocks from each other and from Edison's office on Fifth Avenue just below 14th Street. His lamp works, the most complex and difficult to manage of the manufacturing operations, remained miles away at Menlo Park. The factory depended on the expertise of a cadre who had learned the art of making lamps in his laboratory, and on the machine shop there which provided instruments, other equipment, and repairs. Edison spent little—if any—time there. He instead relied on superintendent Francis Upton for information about extensive ongoing experiments in lamp design and manufacture, particularly on standardizing the resistance of lamps. Foreman Thomas Logan provided regular reports from the machine shop. As his rapidly expanding shops hired new workers, Edison relied on his manufacturing partners and a small number of other close associates to oversee their operations. In addition to Upton and Logan, Edison's circle included John Kruesi at the Edison Tube Co. and Charles Dean at the Machine Works. Absent for part of this time was his most intimate colleague, Charles Batchelor, who was abroad on a rare vacation and did not return until late May. Edison's voluminous postal and cable correspondence passed through the hands of Samuel Insull, who had emigrated from England in February to become his personal secretary.

The manufacturing plants were financially independent of the Edison Electric Light Company and their expansion put heavy demands on Edison's personal finances. As the principal partner, he not only faced assessments on his shares, particularly from the lamp factory, but also occasionally covered assessments of his other partners and loaned money to meet payroll. In early May he contracted to buy a large plant in Harrison, New Jersey, where he intended to move the lamp factory. He became involved in a new enterprise when he formed a partnership with Sigmund Bergmann and Edward Johnson for the manufacture of lamp fixtures and other accessories. Edison's efforts to liquidate his substantial British telephone interests had been delayed by misunderstandings in the cable correspondence with George Gouraud, his longtime London agent. This prolonged the transactions and reduced the amount he could expect to realize. At the end of May he was "entirely out of funds" and arranged for an advance from Drexel, Morgan & Company to help cover the costs of a planned London demonstration central station.[1] To Edison's relief a long dispute over his telephone patents in South America was resolved in June, bringing him $14,000.[2]

Amid the press of business Edison continued to work on improvements to his lighting system. Between 17 May and 25 June he executed twenty-six U.S. patent applications related to electric lighting or power, of which all but four eventually issued.[3] His experimental large dynamo at Menlo Park had proved the practicality of the bar armature (constructed with heavy metal bars instead of wire windings) and of connecting the machine directly to a high speed Porter-Allen steam engine. The engine itself was problematic, however, and in April Edison contracted for one from another maker, Armington and Sims, with an option for twenty-four more for central station service. In late spring or early summer he began to plan an even larger dynamo for display at the autumn 1881 Exposition Internationale de l'Électricité in Paris, where he hoped to introduce his system to potential investors and a wide European public. He also began designing an entirely different form of dynamo with an induction disk divided in radial sections. This may have been a response to concern about infringing the broad dynamo patents of Zénobe Gramme or Werner Siemens.

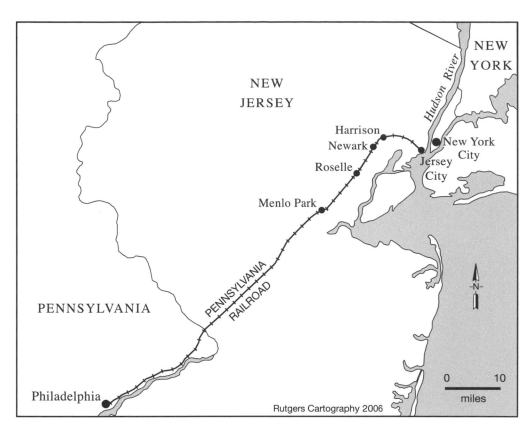

New York City and New Jersey towns and cities associated with Edison's career, 1881–1883.

Edison's incandescent lamp, a wholly new technology which was still fairly difficult and expensive to manufacture, presented many opportunities for improvement. He hoped that an experimental process, begun in February and designed to cut filaments from pressed plumbago sheets, would reduce costs, but in April it became evident that these filaments were not durable. Edison was able to salvage something from the experience, however, by having the factory's carbonizing molds made more cheaply from pressed plumbago. In April and much of May, chemist Alfred Haid carried out experiments at Menlo Park on coating lamp filaments with different forms of carbon, including those deposited by volatile hydrocarbons. In June Edward Nichols, who headed the factory's testing department, apparently found a simpler and cheaper way to seal lamps from the pumps after evacuation. Edison even investigated the feasibility of a reusable lamp globe that could be opened to replace the filament, then evacuated and sealed again. At the same time he and Francis Upton worked to resolve the numerous equipment failures and mishaps that interrupted the factory's production. He also began making plans for detailed displays of his lamps and manufacturing processes at the Paris exhibition.

Edison continued trying to improve the design of a most important part of the commercial electric light and power system—the meter. In early April he began working on a continuously rotating meter that would register current by the deposition of copper without requiring the removal and weighing of metal plates. Apparently dissatisfied with this meter's accuracy with weak currents, he directed Francis Jehl to undertake extensive experiments with different deposition solutions.

Edison largely delegated preparations for his Paris exhibition to Charles Batchelor and Edward Acheson. One significant step he took on his own, however, was to offer to help pay the expenses of George Barker. Barker, a prominent physicist with whom Edison had a long association, was a commissioner to the exposition and a delegate to the Paris Electrical Congress. He declined the offer but did ask Edison to install an isolated lighting plant at his Philadelphia home before he left for Europe. (An Edison isolated plant was also being set up in the new home of financier William H. Vanderbilt during the summer.) Edison more openly courted physicist Charles Young by offering a $300 retainer for his assistance in future patent litigation.[4]

In other activities, Edison arranged to use the magnetic ore separator (devised to separate gold and platinum from iron-bearing rock and sand) for the first time to concentrate iron ore itself. A machine was installed on the south shore of Long Island by the end of May, and Edison inspected its operation in June. In April he directed his assistant Charles Hughes to resume experiments at Menlo Park on preserving perishable foods in a vacuum. When George Gouraud received some of the experimental tubes in London, he pronounced the process "a complete failure." Edison, however, was reportedly "highly delighted" that the problem was too complex to be solved readily and promised to persist.[5] Gouraud was also active in trying to establish telephone companies throughout the world.

Edison lived in a boarding house across the street from his office at this time. His young family joined him there for part—but not all—of the period. The relatively small number of documents that Edison created and Samuel Insull's active role in his affairs make the inventor's precise whereabouts and activities hard to specify.[6] His uncharacteristic invisibility in the daily press contributes to this difficulty.

1. Doc. 2104.

2. José Husbands to TAE, 3 June 1881, DF (*TAED* D8147R; *TAEM* 59:712); Doc. 2031 n. 1.

3. See App. 5. Claims and drawings from the unsuccessful applications are in Patent Application Casebook E-2536:332, 370, 386, 392, PS (*TAED* PT020332, PT020370, PT020386, PT020392; *TAEM* 45:728–31). One other application from this interval, for a magnetic ore separator, issued as U.S. Pat. 263,131.

4. TAE to Young, 2 May 1881, Young (*TAED* X009BA). Young and his physicist colleague at Princeton, Cyrus Brackett, were placed on retainer by the Edison Electric Light Co. by June 1882, as was Barker before the end of that year. Sherburne Eaton to TAE, 23 June 1882 and 2 Jan. 1883; both DF (*TAED* D8226ZAD, D8327A; *TAEM* 61:300, 66:703).

5. Doc. 2118.

6. The editors attribute to Edison the authorship of fewer than two hundred documents in this period.

MENLO PARK LABORATORY REPORTS
Doc. 2074

Edison now spent most of his time in New York and could not easily remain as familiar with work in his Menlo Park laboratory and machine shop as he had previously. Thomas Logan, who took charge of the machine shop when Edison left Menlo Park in February 1881, began submitting frequent (often daily) summaries of activities in the shop.[1] It is not clear when Logan began doing this but he had evidently started by early March,[2] and Doc. 2074 is his first extant report. He continued to supply this information, much of which concerns preparations for the Paris International Electrical Exposition and the fabrication and repair of equipment for the lamp factory, until mid-July, when Edison largely closed the laboratory and shop (see Doc. 2121). Logan's terse reports differ strikingly from the discursive journal entries about laboratory and shop activities that Charles Mott wrote for Edison from March 1880 until early 1881.[3] Edison did not ordinarily make written replies to Logan, but a handful of specific directives, several of them undated, from about this time are in Box 37, EP&RI.

1. These reports are in Menlo Park Laboratory—Reports (D-81-37) and Menlo Park Laboratory (D-82-44), both DF (*TAED* D8137, D8244; *TAEM* 59:59, 63:184). The lamp factory at Menlo Park also made frequent reports of work there (see Doc. 2117).

2. Samuel Insull referred to one such communication in a letter to Logan on 8 March. In the meantime, Charles Mott prepared several

"Daily Reports" for Edison. Insull to Logan, 8 Mar. 1881, Lbk. 9:60 (*TAED* LB009060A; *TAEM* 81:33); Mott to TAE, 21, 22, and 23 Feb. 1881, all DF (*TAED* D8137A, D8137B, D8137C; *TAEM* 59:60, 62, 64).

3. See headnote, Doc. 1914.

–2074–

From Thomas Logan

Menlo Park, N.J., Apr 1th 1881.[a]

Dear Sir

the work for to[b] day consists of porter engine shaft   it is now finished and runing all the afternoon   porter is here has been[b] all day went away to night[1]   engine run very well so far porter says to let her run 10 hours at 300 and then speed her up to 400 and 500

another 55 armature[2] came from L[amp]F[actory] that is two that is under repair

makeing pipes for pumps of LF

working on large armature for LF.

"   " small clamp machine "

"   " annealing machines LF.

winding magnets for ore miller.[3]

working on dies for hydroulic press[4]

work for five H.P. moter[5]

the copper discs have come that was ordered[6]   yours Respectfully

Thos Logan[7]

ALS, NjWOE, DF (*TAED* D8137D; *TAEM* 59:66). Letterhead of T. A. Edison. [a]"Menlo Park, N.J.," and "1881." preprinted. [b]Obscured over-written text.

1. Charles T. Porter introduced the first successful high-speed stationary steam engine in 1862; at this time he was vice president of the Southwark Foundry in Philadelphia, the exclusive manufacturer of the engine. Docs. 1936 n. 2 and 2006 n. 2.

Edison adopted the Porter-Allen engine for his direct-connected dynamo. The engine and dynamo assembly was completed in February 1881, but the bearings consistently ran hot (see Docs. 1936, 1956, 2015, and 2057). Thomas Logan's laboratory reports (see headnote above) document his efforts to address the problem by improving lubrication and ventilation, but in May he found that the engine was misaligned with the armature shaft bearings. The armature short-circuited on 21 May and was not repaired (Logan to TAE, 13 May 1881, DF [*TAED* D8137ZAG; *TAEM* 59:81]; Jehl Diary, 23 May 1881).

2. That is, a dynamo or motor armature having a resistance of .55 ohm.

3. In 1880, Edison devised a machine employing large electromagnets to separate ferrous sand or gravel from surrounding material (see Docs. 1921, 1938, and 1950). The machine under construction was Edison's first for commercial use.

4. This equipment may have been related to efforts to manufacture lamp filaments of pressed graphite (see Doc. 2057 n. 3). Logan spent so much time on the dies that Edison asked him in mid–April to explain the difficulties he was encountering. TAE to Logan, 13 Apr. 1881, Lbk. 8:167 (*TAED* LB008167; *TAEM* 80:891).

5. Although the purpose of this motor is uncertain, another report from Logan on 11 April giving details of the winding of the motor armature appears in a scrapbook devoted to work on Edison's electric railway. Logan to TAE, 11 Apr. 1881, Cat. 2174, Scraps. (*TAED* SB012AAV; *TAEM* 89:296).

6. These were for Edison's new disk dynamo armature (see Doc. 2082).

7. Thomas Logan became foreman of the machine shop at Edison's Menlo Park laboratory complex in February 1881, having worked there for three or four years. Doc. 2062 n. 1.

–2075–

Notebook Entry:
Consumption Meter

[New York,] April 3 1881

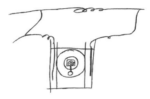

Copper cylinder revolving in a Sulphate Copper Solution with a weighted counter within it so that when cylinder revolves counter will stand still, deposit of Cu on one side & taken off other[1]

TAE   J.F.O.[2]

X, NjWOE, Lab., N-81-03-09:41 (*TAED* N206:19; *TAEM* 40:670).

1. This type of meter is similar to one in a caveat that Edison filed in October 1881 based on a draft he wrote in late August and early September (Doc. 2143). In September 1882 he applied for a patent on a rotating meter with flat paddle blades radiating from the axis. That application was placed in interference with an earlier one by John Sprague; it ultimately issued to Edison in July 1889 as U.S. Pat. 406,825. During the interference proceeding, Edison testified about the construction and operation of the device shown in the October caveat:

> It consists of a copper cylinder partially immersed in a cell containing sulphate of copper. The cylinder is on pivots so as to permit of rotation. On opposite sides of the cylinder are copper electrodes in close proximity to the surfaces of the cylinder. The current passing to one electrode takes copper from it and deposits it on one side of the cylinder; it then passes through the cylinder leaving it on the other side across the sulphate of copper to the electrode. In leaving it takes copper from the cylinder; thus one side of the cylinder be-

comes heavy while the other side becomes lighter, thus throwing the cylinder continuously out of balance and producing rotation of the cylinder, a counting apparatus being connected to the same records the number of revolutions.

Edison acknowledged that copper deposition meters in general worked well with strong currents but were "very unreliable" when used "with very weak currents, which are necessarily used in electric lighting meters." This may have been the rationale for ongoing experiments made by Francis Jehl in April and May with different copper solutions, apparently with the goal of standardizing resistance in the cells under varying conditions. Edison's testimony, pp. 6–8, *Edison v. Sprague*, Lit. (*TAED* QD008:6–7; *TAEM* 46:295–96); N-81-03-18:73–85, N-81-03-11:70–98, Lab. (*TAED* N230:35–41, N236:36–50; *TAEM* 41:425–31, 573–87).

2. John F. Ott (1850–1931) was an expert machinist employed by Edison from 1870 to 1920. During much of that time Ott was Edison's principal experimental instrument maker and trusted laboratory assistant. See Docs. 623 n. 1 and 1321; "Ott, John," Pioneers Bio.

−2076−

Philip Dyer to Samuel Insull[1]

Menlo Park, N.J., April 5 1881[a]

Dear Sir,

If possible will you please send check for Mr Chas. Batchelor's[2] assessment of $1200.[3] We will need it this week or part of it for Pay Roll.

We certainly will have to make another assessment soon or borrow some money.[4]

I think it very poor policy to run this on borrowed capital for it certainly will be some time before we can make any money on our lamps at .35¢ each.[5]

If you can't send check please inform us, so we can make other arrangements. Yours very truly

Edison Lamp Co[6]   Dyer[7]

Will you please let us have H. V. Adams letter, relating to Platina wire.[8] We will return it to you. D.

ALS, NjWOE, DF (*TAED* D8124G; *TAEM* 57:1030). Letterhead of Edison Electric Lamp Co. [a]"Menlo Park, N.J." and "188" preprinted.

1. Samuel Insull became Edison's personal secretary in February 1881. See Doc. 1947 esp. n. 2.

2. Charles Batchelor had been Edison's chief experimental assistant since 1873. He was a partner in the lamp company. See Docs. 264 n. 9 and 2039 n. 1.

3. At the suggestion of Francis Upton, the partners were assessed a total of $12,000 (of which $1,200 fell to Batchelor) on 22 February to help meet the factory's obligations. Edison had a 75% stake in the partnership, Batchelor 10%, Upton 10%, and Edward Johnson 5%. Upton to William Carman, 20 Jan. 1881; Philip Dyer to Insull, 31 Mar. 1881; Dyer

to TAE, 6 Apr. 1882; all DF (*TAED* D8124A, D8124E, D8124H; *TAEM* 57:1025, 1028, 1032).

4. Another assessment totaling $8,500 was made a week later. Dyer to TAE, 12 Apr. 1881; Dyer to Batchelor, 12 Apr. 1881; both DF (*TAED* D8124K, D8124J; *TAEM* 57:1036).

5. This was the price at which the factory sold its lamps to the Edison Electric Light Co. under terms of its patent licensing agreement. See Doc. 2039 n. 1.

6. The Edison Electric Lamp Co. was established in November 1880 to manufacture lamps under license from the Edison Electric Light Co.; its name changed about this time to the Edison Lamp Co. See Docs. 2018 and 2050.

7. Philip S. Dyer (1857–1919) had been bookkeeper for the Edison Electric Lamp Co. since January 1881 and later became its secretary. Dyer's testimony, p. 7, *Sawyer & Man v. Edison*, Lit. (*TAED* W100DDA:1).

8. H. V. Adams was a New York agent for Kolbe & Lindfors, a firm in St. Petersburg, Russia, affiliated with Johnson, Matthey & Co. Adams wrote on 7 March to offer prices for various types of platinum wire. Edison wrote on his letter: "Upton is there anything in this I believe we better stick to Johnson & Mathay for wire for sealing in but for other purposes this might do if cheaper." Upton made his own marginal notation in response that "We are not using Pt. wire for anything else than sealing through the glass. Experiments are dangerous and I should not like to change." Adams wrote again in October that he was "very sorry to learn from your favor of Aug 4th . . . that you had made new arrangements with a London firm for the supply of Platinum, and would respectfully ask you to inform me when you are again in the markets." The London firm of Johnson, Matthey & Co., one of the world's major platinum refiners, had supplied the metal to Edison since 1878. Adams to TAE, 13 July (attachment), 7 Mar., and 9 Oct. 1881; TAE to Upton, 7 Mar. 1881; Upton to TAE, 7 Mar. 1881; all DF (*TAED* D8123ZDA, D8123T, D8123ZED, D8123T1, D8123T2; *TAEM* 57:928, 796, 972, 797); McDonald 1960, 219, chap. 15; Doc. 1604 n. 5.

–2077–

Technical Note: Telephony

[New York,] April 5, 1881

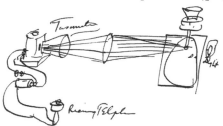

Telephony by Light[1] by use Tasimeter[2]

TAE

X, NjWOE, Lab., Cat. 1147 (*TAED* NM016:21; *TAEM* 44:245).

1. In 1879 and 1880 Alexander Graham Bell explored the possibility of telephony by light. His photophone design exploited the fact that selenium's electrical resistance varies with the intensity of incident light. Bell presented his research at scientific meetings and in journals. How much Edison knew of this work is not clear, but Menlo Park scrapbooks contain several articles by Bell and others on the influence of light on selenium. Bruce 1973, 254, 335–43; Bell 1880; Bell 1881a; Bell 1881b; Bell 1881c; Bell 1881d; Siemens 1876; Fletcher 1877; "Action of Light on Selenium," *Teleg. J. & Elec. Rev.*, 15 Jan. 1878, Cat. 1028:8, Scraps. (*TAED* SM028008c; *TAEM* 25:122).

2. Figure labels are "Tasimeter," "Receiving Telephone," and "Light." Edison invented what he termed the tasimeter, a device to measure heat emitted by radiant objects, while working on carbon telephone transmitters in 1878. It used a hard rubber rod which, when heated, expanded against a carbon button. This pressure changed the resistance of the button, which was placed in a galvanometer circuit, thereby providing a means to measure the heat absorbed by the rod. Edison unsuccessfully tried using it to measure the heat from the solar corona during an 1878 eclipse and later made the instrument available without royalty to scientists, although it was seldom used. See Docs. 1316, 1329, 1364, 1401, and 1444; and Eddy 1972.

–2078–

To Armington & Sims[1]

[New York,] 7th April [188]1[2]

Gentlemen

You may proceed to build the standard steam engine to develope one hundred and twenty five horse power with one hundred and twenty pounds boiler pressure cutting off at the most economical point with a capacity to give fifty per cent more than this power cutting off later in the stroke,[3] the engine to make three hundred and fifty revolutions per minute   Engine to be delivered at the Edison machine works 104 Goerck Street New York[4] within seven weeks from today. The engine not to exceed in cost inclusive of everything ~~of everything~~ two thousand dollars.[5]

If this engine after full test at Goerck Street in connection with a Dynamo is perfectly satisfactory to me I will give you an order [for?][a] twenty four more like it if the price and time of delivery workmanship and steam economy is satisfactory and I will advance the sum of six thousand dollars as it may be wanted on the said order for twenty four engines.[6]

In case I want any more of these engines I am to have the privilege of building as many of these engines as I desire provided orders are given to you for them sufficient to keep your works running to their full capacity whether such works are increased in the future or not, you to accept such orders subject to the above mentioned conditions as to price steam economy &c. Yours truly

Thos A Edison

LS (letterpress copy), NjWOE, Lbk. 8:153 (*TAED* LBoo8153; *TAEM* 80:888). Written by Samuel Insull. ᵃIllegible.

1. Pardon Armington and Gardiner Sims formed a partnership in the late 1870s in Lawrence, Mass., for the manufacture of steam engines. In 1880, they brought out a stationary engine that was not designed to operate at high speeds but, with its sensitive shaft governor, rigid frame, and overhanging cylinder, was easily adapted to do so. Edison's initial contact with Armington & Sims is not known, but presumably came after the Southwark Foundary declined his order in December. To accommodate Edison, Armington & Sims redesigned their machine for high speeds, incorporating some features of the Porter-Allen engine. The firm relocated to Providence, R.I., in January 1882. See Doc. 2006 n. 2; Bowditch 1989, 84–85; Armington & Sims to TAE, 1 Oct. 1881 and 20 Jan. 1882, both DF (*TAED* D8129ZBO, D8233B; *TAEM* 58:283, 61:1001).

2. Edison's draft of this letter on the same day is identified by a docket note as "Edison Machine Works Draft Contract   Armington Sims." It contains essentially the same provisions as this document except that it specified six weeks and an advance payment of $500. Armington & Sims accepted these terms on 9 April. Draft agreement with Armington & Sims, 7 Apr. 1881; Armington & Sims to TAE, 9 Apr. 1881; both DF (*TAED* D8129ZAP, D8129ZAQ; *TAEM* 58:257, 259).

3. The cut-off refers to the point in the piston's stroke at which the intake valve closes, cutting off the supply of fresh steam to the cylinder. Cutting off steam early in the stroke allows it to work expansively, maximizing economy; doing so later increases power. Knight 1881, s.v. "Cut-off."

4. Edison established this manufactory for dynamos and other heavy electrical equipment in late February or early March 1881. See Docs. 2055 and 2060.

5. Edison wanted the engine for his dynamo at the Paris Exposition (see headnote, Doc. 2122).

6. Edison advanced $1,000 to the builders in May. According to Charles Clarke's later recollection, Edison purchased about a dozen of the large engines. Armington & Sims to TAE, 27 May 1881, DF (*TAED* D8129ZAW; *TAEM* 58:264); Samuel Insull to Armington & Sims, 28 May 1881, Lbk. 8:278 (*TAED* LBoo8278; *TAEM* 80:943); Clarke 1904, 47–55.

–2079–

From George Gouraud

London April 7th 1881ᵃ

Dear Sir,

I beg to advise you that in pursuance of your letter of the 7th ult. I sold on your account 3000 United Telephone Shares @ 6⅝—[1] Owing to my not knowing exactly how many shares would come to you, in consequence of the arrangements with the Edison Coy of London not having been completed, I was obliged to sell for a late date; but as the negociations with the Gower Bell Telephone Coy were about to fall through I thought you would profit by my selling at once; indeed you did so, as upon Gower's[2] withdrawal from the negociations the shares went down to £6. Since then, however, the negociations have again been resumed on a satisfactory basis, and the Shares have risen slightly above the price at which you sold—[3] On the whole I think you cannot but be well satisfied, having regard to all the circumstances of the case. Yours very truly

G. E. Gouraud[4]   F.G[rigg].[5]

LS, NjWOE, DF (*TAED* D8149J; *TAEM* 59:926). Letterhead of George Gouraud. [a]"London" and "18" preprinted.

1. Edison's letter is Doc. 2060. The United Telephone Co., Ltd., was formed in 1880 by the merger of the Edison Telephone Co. of London and the Telephone Co., Ltd., which controlled Alexander Graham Bell's British interests. Edison had been told to expect about 3,000 stock shares in the United firm to settle his interest in the old Edison company when arrangements with investors in the antecedent companies were completed (see Docs. 1942 and 2046 n. 2). Because the shares were not yet issued, however, Gouraud had contracted to sell them at a later date. He explained that "you will have understood [that] . . . I do not get the money until I can deliver the shares. Insull will explain to you fully how this is done" (Gouraud to TAE, 9 Apr. 1881, DF [*TAED* D8149L; *TAEM* 59:930]).

2. Frederick Allen Gower (d. 1884) was a Rhode Island newspaper editor who became involved in promoting Alexander Graham Bell's telephones in France. He was party to protracted negotiations to combine his and other French telephone interests with Edison's in 1879 and 1880. Bruce 1973, 227, 235, 246; Docs. 1888 and 1983; *NCAB* 9:216.

3. Gouraud explained that the Gower-Bell Co., which had been manufacturing telephones that United Telephone claimed infringed its patents, had agreed in principle to combine with that firm. Frederick Gower, who reportedly had recently taken control of the former company, abruptly withdrew from negotiations and attempted to reorganize it as the British Gower-Bell Telephone Co. Gouraud took credit for thwarting this effort and forcing Gower to resume negotiations. The talks quickly resulted in an agreement to create the Consolidated Telephone Construction and Maintenance Co., of which Gouraud was to be a director, and which would be the sole supplier of instruments to United Telephone for domestic or export use. Consolidated Telephone also made separate arrangements to manufacture telephones for the Oriental Telephone Co. and the Edison Gower-Bell Telephone Co. of Europe. Gouraud to TAE, 9 Apr. 1881; F. R. Grigg to TAE, 9 Apr. 1881 with enclosed advertisements and prospectus; Consolidated Telephone Construction and Maintenance Co. agreement with Edison Gower-Bell Telephone Co. of Europe, 23 Aug. 1881; all DF (*TAED* D8149K,

D8149M, D8149M5, D8149M6, D8149M, D8148ZBY; *TAEM* 59:928, 932, 934–35, 850).

4. George Gouraud was an American business agent in London with whom Edison had dealt since 1873. He had interests in Edison's electric light and power patents in Britain and elsewhere. Docs. 159 n. 7, 1344, 1365, 1532, 1612, 1698, and 1978.

5. F. R. Grigg had replaced Samuel Insull as Gouraud's secretary. Gouraud to TAE, 28 Apr. 1881, DF (*TAED* D8149P; *TAEM* 59:938).

–2080–

To George Barker

[New York,] 13th April [188]1

My Dear Barker[1]

I have your letter of yesterdays date.[2]

I suppose you know that Gouraud my English agent has been appointed a Commissioner and my friend Walker[3] of the Gold & Stock Tel Co & now Consul at Paris is to be Executive Commissioner. Blaine offered to appoint me a Commissioner but I was reluctantly obliged to decline.[4]

If by any possible means I can further your wishes I shall be most happy to do so.

As for myself all that I [-----][a] want is "[full blooded?][b] truth." The trouble is I cannot always get this   Yours very truly

Thos A Edison

LS (letterpress copy), NjWOE, DF (*TAED* LB008168; *TAEM* 80:892). Written by Samuel Insull. [a]Interlined above and illegible. [b]Illegible.

1. George Barker, professor of physics at the University of Pennsylvania, had a long association with Edison. See Doc. 500 n. 8; also Hounshell 1980.

2. Barker's letter has not been found, but presumably it dealt with his upcoming duties in Paris. He was a commissioner to the Electrical Exposition and member of the committee for testing incandescent lights, and also a delegate to the related Electrical Congress. *DAB*, s.v. "Barker, George Frederick"; see also Doc. 2148 n. 6.

3. George Walker (1824–1888) was a vice president of the Gold and Stock Telegraph Co. and a director and vice president of Western Union. He served as the U.S. consul general in France from 1880 to 1887. Doc. 1173 n. 7; *ACAB*, s.v. "Walker, George"; Reid 1879, 533, 539, 626, 632.

4. James G. Blaine (1830–1893) was Secretary of State for nine and a half months in 1881, resigning soon after President Garfield's death. In March 1881, Edison declined Blaine's "very flattering offer" because his work to commercialize the electric light required his "immediate and undivided attention." *ANB*, s.v. "Blaine, James Gillespie"; TAE to Blaine, 11 Mar. 1881, DF (*TAED* LB008049; *TAEM* 80:860).

From John Lawson

Menlo Park, N.J., April 14 1881[a]

Dear Sir:

Mr. Upton[1] writes me to commence work on the plating of the carbons by contract

He agrees to furnish me with 500 carbons per day and I return them to him plated at the rate of $1.^5/10$¢ per carbon, and when the number reaches 600 per day I shall plate at the rate of $1.^4/10$¢ per carbon with 700 per day $1.^3/10$¢ per carbon and so on until the number reaches 1000 per day and then the price becomes 1¢ per carbon as[2]

Now if I agree to this I wish to have it understood that it[b] has no connection with the original contract; that I do not commence on the original contract until I can be furnished with 1000 carbons per day as agreed upon—[3]

If this is satisfactory to you please let me know by return post. Yours very truly

John W. Lawson[4]

ALS, NjWOE, DF (*TAED* D8123ZAU; *TAEM* 57:847). Letterhead of Edison Electric Lamp Co. [a]"Menlo Park, N.J.," and "188" preprinted. [b]Interlined above.

1. Francis Upton, a mathematician and physicist, joined Edison's laboratory staff in 1878 and became one of his principal assistants in electric light research. Upton was a partner in the Menlo Park lamp factory, which Edison placed in his charge at the beginning of 1881. See Docs. 1568 n. 1 and 2039 n. 1, and *TAEB* 5 chap. 9 introduction.

2. John Lawson's task was to electroplate carbon lamp filaments to the lead-in wires. At Edison's direction, he and others began researching this technique at the end of 1880 and by February had developed a satisfactory commercial process which eliminated the need for expensive screw clamps. See Docs. 2050 and 2061.

Lawson's contract has not been found. It is not clear whether he was to perform the work himself or if he also had authority to hire and manage others for this job. That type of arrangement, known as inside contracting, was common at this time in U.S. factories, particularly in metalworking. Semi-skilled and skilled workers about this time typically earned about two or three dollars per day in a six-day workweek (Hounshell 1984, 49–50; Derks 1994, 13–14). Entries made sometime after mid-May in a notebook kept at the lamp factory indicate that it took one man five hours to place 600 carbons in the plating bath and another five hours to remove, wash and dry them; the deposition process itself required about six hours (Menlo Park Notebook #144:11, 21, Hendershott [*TAED* B015:8, 11]).

3. Nothing is known of the prior contract.

4. John Lawson was a self-taught chemist whom Edison hired as a laboratory assistant in January 1879. He later took charge of the carbonizing and electroplating departments at the lamp factory. See *TAEB* 5:5–6 and App. 2; "Lawson, John W.," Pioneers Bio.

Notebook Entry:
Dynamo

Dynamo[1]

[A]

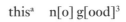

[B]

[C][2]

[D]

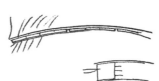

[E]

this[a]   n[o] g[ood][3]

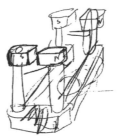

[F]

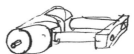

[G][4]
[H]
[I]

[J]

[K]

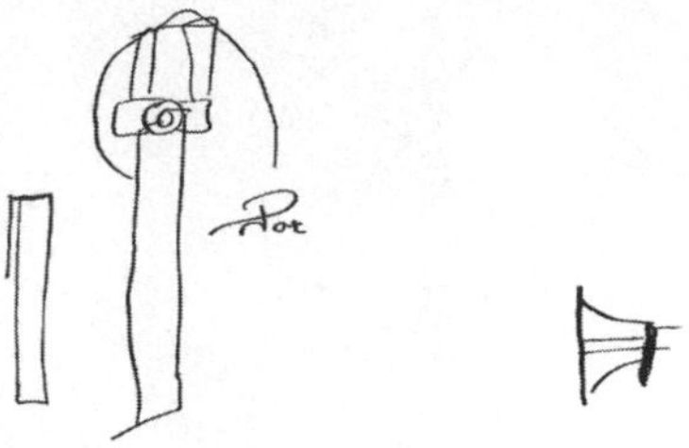

[L][5]

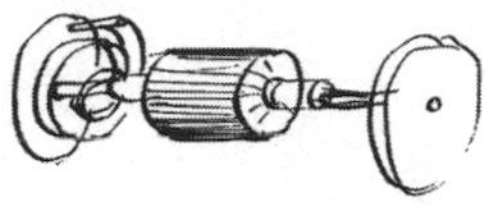

[M]

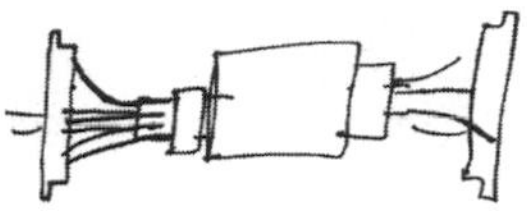

[N]

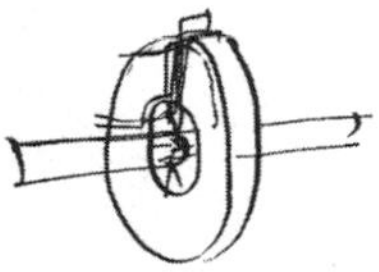

[O]

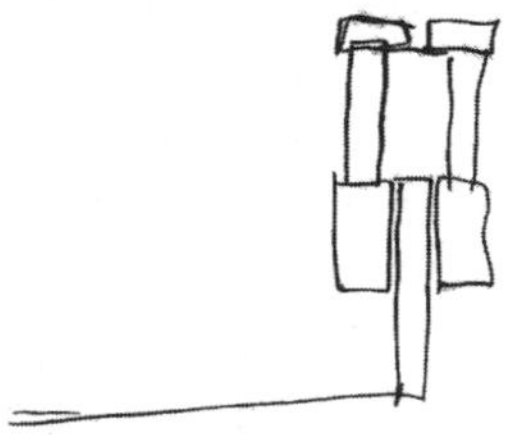

[P]

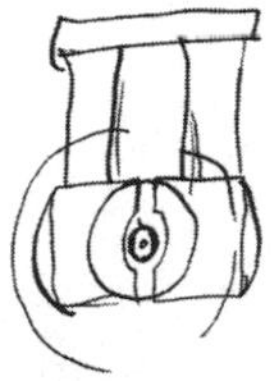

[Q][6]

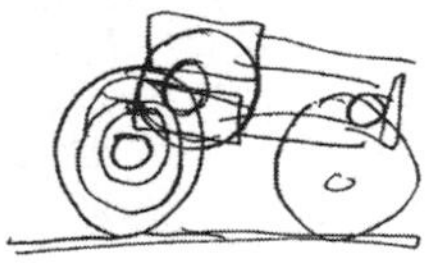

[R]

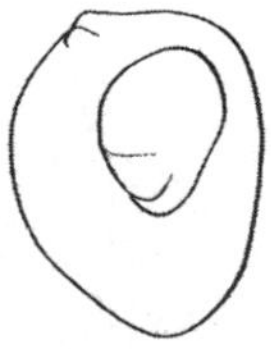

TAE

X, NjWOE, N-81-03-09:55, Lab. (*TAED* N206:26; *TAEM* 40:677). Miscellaneous doodles and calculations not reproduced. ªFigure overwritten on faint or erased sketch.

1. This notebook entry is the first extant evidence of Edison's sustained effort to conceptualize a disk armature dynamo. In a patent application executed on 3 June (U.S. Pat. 263,150), he stated that this design was for a generator

in which the iron core of the armature will not be necessary and the loss of power caused by the heating of the same will be avoided, only the inductive portion of the armature being passed between the poles of the exciting magnet or magnets, which poles can consequently be brought close together, so as to produce an intense magnetic field . . . [It] will generate a continuous current of high electro-motive force in the same direction without the use of pole-changers. All the inductive portions of the armature will be constantly in circuit, and the internal resistance of the machine will be exceedingly small.

I accomplish this object by constructing the armature in the shape of a disk or plate like that used by Arago in his experiments, but divided into radial sections. These radial sections, which form the inductive portion of the armature, are preferably naked copper bars connected together by insulating material and attached rigidly to the driving-shaft by an insulating-hub.

In the last paragraph of the specification Edison noted that "By dividing the disk into radial sections or bars and connecting them, so as to generate a continuous current, a much higher electro-motive force can be obtained than by the use of the simple undivided disk employed by Arago in his experiments." The general construction of this machine was shown in another patent application that Edison completed at the end of June (U.S. Pat. 263,143).

The principle of induction in a metallic disk rotated in a magnetic field was the basis of the earliest generators. Edison may have had this idea in mind by late March when he inquired about the validity of Zénobe Gramme's ring armature patent, which was drawn broadly enough to apply to a solid disk (see Doc. 2071). On 13 April he made sev-

Patent drawing showing the radial bar disk armature (F) rotated on the shaft (A) and insulated hub (B) between the poles (N, S). Concentric rings (H) on the periphery connect each bar with another on the opposite side.

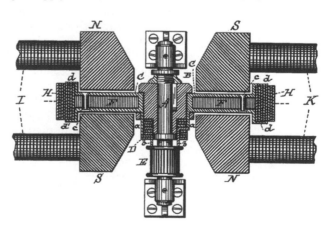

eral dynamo sketches, one of which may be of a disk arrangement (Cat. 1147, Lab. [*TAED* NM016:23–24; *TAEM* 44:247–28]). By mid-April the Menlo Park shop had started to build a full-size disk machine. The armature consisted of twenty-eight radial conducting bars; it measured about eleven inches in diameter and weighed only about thirty-three pounds. Thomas Logan made frequent references to its construction in laboratory reports until 12 May, when he indicated that the armature was being run at 1200 revolutions. Francis Jehl conducted the first recorded test on 19 May, when he measured 54 volts while running the machine at 1,200 revolutions (Jehl 1937–41, 889; Logan to TAE, 12 May 1881, DF [*TAED* D8137ZAF; *TAEM* 59:81]); Jehl Diary, 19–20 May 1881.

2. This figure represents a different orientation of the disk. On 25 October Edison executed two related patent applications for a direct-current disk dynamo having field magnet pole pieces "chambered or hollowed out" to accommodate the armature. U.S. Pats. 263,148 and 264,646; see also Doc. 2228 n. 10.

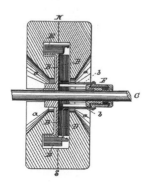

Edison's patent drawing of a disk armature within the cavities of the field pole pieces.

3. The orientation of the field poles shown here would induce currents of opposite polarity in each half of the rotating disk. In drawings for his first two disk dynamo patents, Edison reversed the orientation of one field magnet from its representation here. This would induce currents of the same polarity in each half, thereby providing a continuous current.

4. Drawings G, H, and I show alternative armature designs. G appears to represent an arrangement like that Edison sketched in more detail on 18 May showing a hub around which were attached flat "Radial bars like a printers chase [with] iron bet[ween]." Figure H is unclear. Figure I is related to a patent application that Edison executed on 24 June. In that design, copper bars comprising the induction surfaces were attached to the lateral faces of a core disk "constructed by winding strips of iron and paper together, in spiral form . . . such core preventing the circulation of magnetic currents therein and the loss of energy caused by the generation of heat in such core. The effect is assisted by the fact that the iron portion of the core does not cut the lines of force at right angles." Cat. 1147, Lab. (*TAED* NM016:64, 78, 90; *TAEM* 44:288, 302, 314); U.S. Pat. 263,143.

5. Drawings L and M appear to show armature end plates and their connections to drum-type induction windings. One or both may consist of a metal strip wound into a spiral; cf. Doc. 2102.

6. This rough sketch may show the adaptation of the dynamo as a railroad locomotive motor. In his first disk dynamo patent application, Edison noted that "the novel features of this machine are equally well applicable to electric engines and motors" and would have "the advantage of great lateral compactness, enabling me to use the engine on a narrow railway-car without projecting over the sides." Edison made several sketches on 27 April illustrating this application. U.S. Pat. 263,150; Cat. 1147, Lab. (*TAED* NM016:36–37; *TAEM* 44:260–61).

Edison's 18 May sketch of a disk armature composed of radial bars.

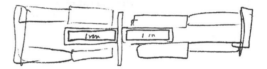

–2083–

To Pietro Giovine[1]

[New York,] April 15 [188]1

Dear Sir

Your exquisite Canto reached me promptly.[2] You say you have given your Canto immortality by using my Phonograph to record it; but your modesty forbids you from saying that I will live in the Canto like a fly in amber   Yours

Thomas A Edison

ALS (letterpress copy), NjWOE, Lbk. 8:183 (*TAED* LB008183; *TAEM* 80:897). A typed transcript is in DF (*TAED* D8104ZAW2; *TAEM* 57:93).

1. Nothing is known of Pietro Giovine, who wrote to Edison from the city of Bra, near Turin, in northwestern Italy. Giovine to TAE, 1 Apr. 1881, DF (*TAED* D8104ZAW1; *TAEM* 57:90).

2. Giovine wrote on 1 April (see note 1) that he had composed an homage to Edison and recorded it on the phonograph, which (according to an English translation of his letter) he called "the noblest prize of our century." He enclosed a copy with a favorable newspaper clipping (neither of which has been found).

–2084–

To Francis Jehl

New York, April 18[7] 1881[a]

Francis:[1]

You must at once test the meter[2] so I can go ahead and get some made but at same time continue the other experiment of the drop on the straight & taper wire;[3] It is of the utmost importance that I have the experimental[b] determinations at once; please hurry— Yrs

T.A.E.

ALS (facsimile), Jehl 1937–41, 666. Letterhead of Edison Electric Light Co. [a]"New York," and "188" preprinted. [b]Multiply underlined.

1. Francis Jehl became an experimental assistant in 1879, working primarily on vacuum pumps and lamps. Doc. 1685 n. 2, *TAEB* 5 App. 2.

2. Edison's standard electric consumption meter consisted of an electrolytic cell with two metal plates in an electrolytic solution. It was placed in a shunt circuit so that a small portion of the current flowing to the consumer passed through the cell. This current caused ions to move from one plate (the anode) and be deposited on the other (the cathode). That plate could be weighed to calculate how much current the customer had received. This principle was relatively simple, but designing a meter that would register accurately in general service required extensive trials (see e.g., Docs. 1852, 1893, 1912, and 2065). Francis Jehl had been experimenting with meters on 14 April; he returned to them on 18 April to quantify the change in the deposition cell's resistance under the action of a current (N-81-03-11:70–98, Lab. [*TAED* N236:36–50; *TAEM* 41:573–87]).

3. Edison likely wished Jehl to experiment in regard to using tapered conductors to help prevent a voltage drop in the lines of the New York distribution system, a plan he had conceived in 1880 (see Doc. 1789 esp. n. 1); Jehl's subsequent recollection of this document supports this inference (Jehl 1937–41, 665, 821–23). Francis Upton had made extensive calculations in March on the savings in copper by using tapered rather than straight conductors (see Doc. 2068). On 12 April, Jehl had been measuring the voltage drop in simulated lamp circuits having 100 ohm resistance coils in parallel. His notes are not clear and the editors have no evidence of further trials at this time (N-81-03-11:61–69, Lab. [*TAED* N236:31–35; *TAEM* 41:568–72]). Jehl had recently tested the conductivity of an alloy that John Kruesi intended for connecting the underground lines; about this time he was also checking the conductivity of wire and the resistance of insulation for the Ansonia Brass & Copper Co. (N-81-04-06:1–2, 12–16; N-81-03-11:53–60; Lab. [*TAED* N223:1–2, 7–9; 236:27–31; *TAEM* 41:37–38, 43–45, 564–68]; Jehl to TAE, 7, 11, and 25 Apr. 1881, all DF [*TAED* D8136H, D8136I, D8136L; *TAEM* 59:16–17, 23]; TAE to Ansonia Brass & Copper Co., 28 Apr. 1882, Lbk. 8:221A [*TAED* LB008221A; *TAEM* 80:911]).

–2085–

From Francis Upton

Menlo Park, N.J., April 18 1881[a]

Dear Mr. Edison:

Dr. Haid[1] has not yet furnished us with the treated plumbago loops.[2] He promises them daily.

~~Atcha~~ Achison has made five thousand loops and perfected a system by which they can be made. He is willing to throw up the contract and anxious to come into the factory to learn the process of making lamps so that he can go to France.[3] I think it would be advisable to stop making plumbago loops for the present and let him come down here.

I wish you would not ask us to make but few of these lamps at present, since it breaks in on the system of making A and B regular carbons.[4] I want to get one thousand a day of these so as to see a profit before trying to manufacture new styles.

I have written to Japan that we will wait until Moore returns before deciding.[5] We have enough fibres on hand to keep us running four months at least, and fifty thousand more are on the way. They have cut 710.000 Matake[6b] and will cut 100.000 R[7] so that we can feel sure of a supply.

Last week we met with an accident. Dr. Nichols[8] had run the wires in the photometer room under a covering on the floor. The valve of the steam pipe leaked and let water over them. The magnet or exciting lines were eaten off lowering the E.M.F. In a very few minutes a new line was run, but those minutes had let the pump stop so that the mercury jumped up in a number of pumps. Fifteen pumps were cracked mostly in the fall tube where the mercury pounds and where they would have gone in a few days. The mercury jumped up into 150 lamps making them resistance lamps.[9] When I went in the next day to see you I did not think to mention the accident in the few moments I had of conversation.

I did not write you as I have taken all precautions to prevent it happening again. When it happened Kerite wire had been ordered and it was Dr. Nichols intention to run all the lines overhead with it. This has now been done.[10] Dyer said you had a report of something very bad. I do not think it was as it [could?][c] was a defect that we had seen and were going to remedy, though a ~~lek~~ leaky valve and water were not thought of. I can write you if you wish all the trials I have, for scarcely a day passes without a new "bug" showing itself. I work at them and intend to show you as a result 1000 lamps or more coming from here all with good vacuums well plated clamps and low resistance carbonization and no trouble.[11]

Hammer[12] will send you in some curve sheets in a day or two that show we are making fair lamps.[13]

There are no great extremes few go at first and we have had none last so exceedingly long as some of those tested at the laboratory. I am going to work up some curves of the various processes on the pumps.

I am going to try to go in to New York this P.M. to see the Illumination[14]   Yours Truly

F. R. Upton

ALS, NjWOE, DF (*TAED* D8123ZAZ; *TAEM* 57:853). Letterhead of Edison Electric Lamp Co. [a]"Menlo Park, N.J.," and "188" preprinted. [b]Interlined above. [c]Canceled.

1. Alfred Haid (b. 1843?) was a Ph.D. analytical chemist who worked for Edison for about four years starting in June 1879. After leaving Edison's employ he was a freelance consulting chemist in Rahway, N.J. Doc. 1754 n. 1 and *TAEB* 5 App. 2; Haid to Insull, 12 Sept. 1884; Haid to TAE, 17 Dec. 1885 and 3 Nov. 1888; all DF (*TAED* D8403ZGK, D8513ZAE, D8805AIQ; *TAEM* 71:262, 77:428, 121:595); Haid's testimony, p. 21, *Böhm v. Edison* (*TAED* W100DEC:22); Jehl, 1937–41, 258, 263, 269, 334.

2. Edison had inquired on 16 April whether Upton had "got any plumbago carbons in the lamps ready   I would like to have some experiments made as to how high they will go & as to their life   If you have two of these lamps to spare please send them on to me." A lamp factory notebook indicates that about this time Haid was coating fiber and paper filaments with various substances, including plumbago. John Howell

tested several batches of these filaments on or about 22 April. Loops cut from pressed plumbago were also being prepared for the factory (as noted below) and may have been treated, at least experimentally, with hydrocarbons as described in Doc. 2087. On 28 April Upton sent Edison "six plumbago lamps," whose filaments evidently were coated with gas carbon. In early May Haid reported results of extensive experiments on the ideal temperature of gasoline for depositing carbon. TAE to Upton, 16 Apr. 1881, Lbk. 8:180 (*TAED* LBoo8180; *TAEM* 80:895); Cat. 1301 (order nos. 291, 299–301, 303), Batchelor (*TAED* MBNoo7:25, 26; *TAEM* 91:318–19); Upton to TAE, 28 Apr. 1881; Haid to TAE, 3 May 1881; both DF (*TAED* D8123ZBI, D8123ZBJ; *TAEM* 57:868–69).

3. Edward Goodrich Acheson began working for Edison as a draftsman in 1880 and soon became involved with lamp research. After devising a way to cut usable filaments from sheets of pressed plumbago, he contracted with Edison to produce 30,000 of them at two cents apiece. He recalled years later that he volunteered to break this agreement because the filaments burned out too quickly to be useful commercially. On 20 April, Edison instructed him to "go into the Lamp Factory and learn the lamp business in all its details." Acheson went to Paris in July to help first with Edison's exhibit at the International Exposition and later with setting up the lamp factory. Docs. 2057 n. 3 and 2069 n. 2; Philip Dyer to TAE, 4 Apr. 1881, DF (*TAED* D8124F; *TAEM* 57:1029); TAE to Acheson, 20 Apr. 1881, Lbk. 8:193 (*TAED* LBoo8193A; *TAEM* 80:900); Acheson 1965, 22; see Docs. 2128 n. 1 and 2235.

4. The "A" carbons were designed to operate at approximately 110 volts; the "B" carbons at about 55 volts (either in isolated plants or in a series of two lamps in a 110 volt central station circuit). Under the classification system adopted in December 1880, the former were six inches long and produced sixteen candlepower. The latter were three inches long and gave eight candlepower. The standard filaments were cut (before carbonizing) to .008 × .017 inch and operated at eight "A" (16 candlepower) lamps to the horsepower. Thinner carbons of 0.008 × 0.0135 inch did not last as long but ran at 10 to the horsepower and became standard for "A" lamps in 1882. TAE to Arnold White, 20 July 1882, Lbk. 7:729 (*TAED* LBoo7729; *TAEM* 80:737); Doc. 2027 n. 2; Jehl 1937–41, 811; Doc. 2312 n. 5; Howell and Schroeder 1927, 65; see also headnote, Doc. 2126.

5. Edison had dispatched William Moore to Japan in October 1880 to procure bamboo splints for the lamp factory. By mid–March Edison had decided that the splints being received were unsatisfactory, and Moore was recalled soon thereafter. The decision to which Upton refers may have been about the type of bamboo to be supplied, about which there was evidently some confusion on the part of the firm with which Moore had contracted (Doc. 2002 nn. 6 and 5; TAE to Upton, 15 Apr. 1881, Lbk. 8:175 [*TAED* LBoo8175; *TAEM* 80:894]). Inventories of scores of samples sent by Moore and separate comments on them by laboratory or lamp factory staff are in Fredericks.

6. Presumably a variant transliteration of madake, the species of bamboo which Edison had selected for his lamps. See Doc. 1993 n. 2.

7. This notation is unclear.

8. Edward Leamington Nichols was a chemist and physicist whom

Edison hired in October 1880. Nichols organized the lamp testing department at the Menlo Park factory. Doc. 2065 n. 7 and *TAEB* 5 App. 2.

9. Electric power drove an Archimedes screw pump which lifted mercury above the vacuum pumps. It is not certain how this damage would have occurred as a result of the short circuit on the line to the field magnets of the driving motor. It is likely, however, that the stoppage allowed the mercury to fall back, creating a vacuum above the mercury pumps and reversing the flow through them. (For a similar accident, see Doc. 1950 and also Docs. 1816, 1926, and *TAEB* 5:767–72.) In June, Edison sketched a vacuum pump in which the contraction for regulating the flow of mercury was separate and detachable from the fall tube. In a patent application completed a few weeks later, he stated that one purpose of this arrangement was to prevent breakage of the delicate glass contraction should the mercury jump up the tube (Cat. 1147, Lab. [*TAED* NM016:83; *TAEM* 44:307]; U.S. Pat. 263,147). A resistance lamp, according to Upton, was a defective lamp unsuited for lighting but which could be used to provide a standard electrical resistance (Upton to TAE, 20 Apr. 1881, DF [*TAED* D8123ZBA; *TAEM* 57:857]).

10. A similar accident happened about a week later when faulty insulation caused a short circuit and a small fire. Eighteen pumps were broken and 150 lamps spoiled. Upton promised to have all the electrical cables removed and thoroughly inspected. Upton to TAE, 25 Apr. 1881, DF (*TAED* D8123ZBE; *TAEM* 57:863).

11. A few days later, Upton cautioned Edison about "a number of leaky lamps owing to breaking in of new hands." He tried to accelerate production by evacuating two lamps on each pump but this arrangement proved prone to leaks. In order to meet his goal, Upton made plans to run the factory at night. Upton to TAE, 21 Apr. 1881, DF (*TAED* D8123ZBC; *TAEM* 57:859).

12. William Hammer joined Edison's Menlo Park staff in December 1879 and remained connected with Edison lighting interests until 1890. At this time he was chief electrician at the lamp factory. See Doc. 1972 n. 7 and *TAEB* 5 App. 2.

13. Hammer seems to have been involved with two types of lamp curves used at the factory. One type was a graphical determination of the candles per horsepower given by each lamp at a specific voltage (see Doc. 2061 n. 1). The other was a representation of the failure rate of a batch of lamps run at a particular voltage (see headnote, Doc. 2177).

14. The residential building at 65 Fifth Ave. occupied by the Edison Electric Light Co. had recently been lighted by electricity. A few evenings before, Edison had shown the installation to invited guests. "Edison's Light for Houses," *New York Times*, 16 Apr. 1881, 8.

–2086–

To George Gouraud

[New York,] Apr 19. 81

Noside London[1]

Only object selling Uniteds necessity immediate cash[2] notify Batchelor return not later than May fifth   Very important[3]

L (telegram, copy), NjWOE, LM 1:10A (*TAED* LM001010A; *TAEM* 83:877). Written by Charles Mott.

1. Cable code for George Gouraud; see App. 4.

2. Gouraud wrote on 9 April stating that he did not yet have the proceeds from selling Edison's shares in the United Telephone Co. (see Doc. 2079 n. 1); this document is Edison's reply. Gouraud acknowledged his telegram on 20 April and explained that he had interpreted Edison's original instructions to mean he should make immediate arrangements to sell in a declining market. He also cabled the same day asking if he should repurchase the shares, now valued above eight pounds. Edison answered immediately, "No. Want money." Gouraud explained in May that he expected Edison would have known the shares had not been issued because "Insull wrote the letter in which you instructed me to sell them   I could not but suppose that you were writing with the full knowledge of what you were doing, as he is so fully conversant with all the circumstances of the case." He then promised to try to obtain a cash advance and, in late May, arranged a $10,000 advance on the settlement of royalties owed by the Edison Telephone Co. of Glasgow, which was subsumed into the United Telephone Co. Gouraud to TAE, 20 and 28 Apr., 21 and 24 May 1881; statements of Glasgow royalties, 24 May 1881; and statements of Gouraud account; all DF (*TAED* D8149N, D8149P, D8149R, D8149U, D8105ZZA [images 4, 7–9]; *TAEM* 59:935, 938, 955, 962; 57:345); Gouraud to TAE, 20 Apr. 1881; TAE to Gouraud, 20 Apr. 1881; LM 1:10B, 10C (*TAED* LM001010B, LM001010C; *TAEM* 83:877); TAE to Gouraud, 24 Apr. 1881, Lbk. 8:208 (*TAED* LB008208; *TAEM* 80:906).

3. Edison presumably wanted to begin making arrangements with Charles Batchelor for the Paris exposition. Batchelor cabled that it would be "difficult" to change his plans; he sailed from Liverpool with his family on 12 May. Batchelor to TAE, 26 Apr. 1881, DF (*TAED* D8135B; *TAEM* 58:872); "Saloon Passengers," Cat. 1241, item 1583, Batchelor (*TAED* MBSB21582X; *TAEM* 94:630).

–2087–

Memorandum:
Incandescent Lamp
Patent

New York, [April 19, 1881?[1]][a]

⟨303⟩[2b]

~~In my application~~

Draw up patent on ~~comp~~ sheets of ~~C~~plumbago Graphatodial[c] Silicon Boron & Zirconium, moulded by pressure[c] preferably in a hydraulic press to the thickness of $8/1000$ ~~to~~ or less, Then placing a great number of these sheets in a closed[d] flask and brought up to incandescence in a furnace[c] & then the vapor of a hydrocarbon passed over the sheets so as to deposit ~~a~~ [----][e] Carbon upon them. afterwards the sheets are placed under a ~~die~~ punch & ~~one~~ a die & a flexible Carbon [pr--][e] of the proper shape punched out of the sheet. The ends being broader for[c] clamping—[3] The incandescent might be punched out of the sheets[f] & used without depositing [-][e] Carbon upon[c] them but

I prefer to so deposit because there is less breakage in punching and placing in the lamp, ~~and~~.

R[esistance] increased by charcoal.[g]

Can use any kind of carbon but more difficult than graphitorial carbon.—[g]

Charcoal & petroleum residues or solution of solid hydrocarbon[h]

AD, NjWOE, DF (*TAED* D8142ZAG; *TAEM* 59:313). Letterhead of Edison Electric Light Co. [a]"New York," preprinted; date multiply inscribed by handstamp. [b]Written in a different hand. [c]Obscured overwritten text. [d]Interlined above. [e]Canceled. [f]"of the sheets" interlined above. [g]Followed by dividing mark. [h]Paragraph written in left margin.

1. Date taken from handstamp when document was received.

2. This is the case number assigned by Edison's patent attorney to the resulting application. This document summarizes the essential points of a patent application that Edison executed the same day. The application was twice rejected and Edison substantially modified the claims (though not the text) before it issued in August 1882 as U.S. Pat. 263,145 (see Pat. App. 263,145). Also on 19 April Edison made a drawing marked "plumbago" which apparently shows a loop cut from a square blank (Cat. 1147, Lab. [*TAED* NM016:27; *TAEM* 44:251]). In February 1881 Edison had filed an application (Case 290) for lamp filaments cut from pressed plumbago. It was rejected and subsequently abandoned; the claims and a brief description are extant (Serial No. 27,191, "Carbons" in Abstracts of Edison's Abandoned Applications [1876–1885], p. 2; Patent Application Casebook E-2536:278; both PS [*TAED* PT004:3, PT020278; *TAEM* 8:528, 45:726]).

3. Edison's standard filament design since late 1879 incorporated relatively broad ends to facilitate attachment to the lead-in wires. See Doc. 1850 n. 5.

–2088–

To Norvin Green

[New York,] 20th April [188]1

My Dear Dr Green[1]

I want to make the exhibition at the Exposition Electrique at Paris.[2] Would you oblige me with the loan of one set of Quadruplex Apparatus,[3] one Universal Stock Printer,[4] one private line printer.[5]

I will return them in good condition and will be responsible for the same.[6]

If you can oblige me please give the bearer (Mr. Insull) an order for the same   Yours very truly

Thos A Edison

LS (letterpress copy), NjWOE, Lbk. 8:196 (*TAED* LB008196; *TAEM* 80:903). Written by Samuel Insull.

1. Norvin Green (1818–1893) was president of the Edison Electric Light Co. and of the Western Union Telegraph Co. Docs. 1168 n. 4, 1494 n. 4, and 1576; *ANB,* s.v. "Green, Norvin."

2. Edison had been planning since January to participate in the Exposition Internationale de l'Électricité (see Doc. 2045). The Exposition took place in Paris from 10 August to 15 November 1881 and was attended by nearly 900,000 visitors. For general descriptions of the Exposition see Fox 1996; Beauchamp 1997, 160–65; Heap 1884; and Prescott 1884, 282–303; also clippings in Cats. 1068 and 1069, Scraps. (*TAED* SM068, SM069; *TAEM* 89:34, 143).

3. Edison's 1874 quadruplex designs provided the first practical way to transmit four independent messages (two each way) on a single telegraph wire. See Docs. 348 nn. 9 and 16, 515 n. 2; Israel 1998, 78–80, 97–104.

4. Edison developed his Universal Stock Printer in 1871 for the Gold and Stock Telegraph Co., a subsidiary of Western Union, and it served as the standard ticker design for several years. See headnotes, Docs. 195 and 211; Israel 1998, 63, 71–73.

5. The Universal Private-Line Printer was an outgrowth of Edison's close relationship with the Gold and Stock Telegraph Co. Perfected in 1872, the instrument was a combination of a keyboard transmitter, by which a sender could press keys for each letter and number, and a printer of the same design as the Universal Stock Printer. See headnote, Doc. 262; Israel 1998, 62.

6. Although Edison asked to borrow the instruments, Western Union billed him for two quadruplex sets in August and October 1881. In January 1883, Western Union again submitted a bill for $412.50 for the quadruplex instruments. Edison replied that the instruments were sequestered in the Customs House along with the rest of his exhibition materials, and he promised to return them when he could secure their duty-free release. William Hunter to TAE, 19 and 30 Aug., 5 Oct. 1881; Western Union to TAE, 17 Jan. 1883; all DF (*TAED* D8135ZAT1, D8135ZBA1, D8135ZCG, D8373D; *TAEM* 58:998, 1019, 1089; 70:1123); Edison to John Van Horne, 18 Jan. 1883, Lbk. 15:178A (*TAED* LB015178A; *TAEM* 82:97).

–2089–

From Francis Upton

Menlo Park, N.J., April 20 1881[a]

Dear Mr. Edison:

I saw Mr. Barnett[1] yesterday regarding the Peters works.[2] There is a good chance to buy. He says that they will give very liberal terms, and that we can probably secure the place for next fall for $5,000.

Then partial payments can be made. He wants an offer from you of something.

We can use the buildings and the location for labor is very good. There is no question but that the buildings cost much more than we shall have to give. They claim that they can show $136,000 in bills. I wish you would see it.[3]

I think the Passaic[b] works[4] are not very promising the labor market cannot approach Newark. There nearly everything is for sale which looks bad. Yours Truly

Francis R. Upton

~~Why not offer~~[c]

ALS, NjWOE, DF (*TAED* D8123ZBB; *TAEM* 57:858). Letterhead of Edison Electric Lamp Co. [a]"Menlo Park, N.J.," and "188" preprinted. [b]Obscured overwritten text. [c]Written in left margin.

1. John Burnett was associated with Arthur Devine, a real estate agent and the Commissioner of Deeds in Newark. Letterhead, Burnett to TAE, 28 Apr. 1881, DF (*TAED* D8123ZBH; *TAEM* 57:867).

2. The Peters Manufacturing Co. produced oilcloth. Constructed in 1877, the plant occupied an entire block and consisted of three four-story buildings, the burned ruins of a fourth one, and various outbuildings; it had been subject to frequent fires. The main structures reportedly enclosed ten times as much floor space as the Menlo Park factory. Although the bill of sale and related correspondence situated the property in East Newark, it apparently lay within the contemporary boundaries of adjacent Harrison, N.J. Jehl 1937–41, 814–15; Shaw 1884, 1253; *Ency. NJ*, s.vv. "East Newark," "Harrison"; see, e.g., John Burnett to TAE, 4 May 1882; TAE agreement with Peters Mfg. Co., 9 May 1881; both DF (*TAED* D8123ZBL, D8123ZBP; *TAEM* 57:874, 879). For a photograph of the factory some years later see headnote, Doc. 2343.

3. Upton had contemplated moving the lamp factory since at least early March, principally to obtain cheaper labor. Edison evidently visited the Peters property before 28 April. In early May Burnett reported that he believed the Peters Co. would reluctantly accept Edison's offer of $50,000 for buildings that had cost $136,000. Edison contracted to buy the property for $52,250 on 9 May, and the Lamp Co. raised funds from its partners for a down payment. He agreed to pay $5,000 in cash and give notes due at intervals, with the final payment of $30,000 to be made in two years. See Doc. 2061; Burnett to TAE, 28 Apr. and 4 May; TAE memorandum, 4 May 1881; agreement with Peters Manufacturing Co., 9 May 1881; Philip Dyer to TAE, 17 May 1881; Dyer to Samuel Insull, 21 May 1881; statement of account, n.d.; Peters to TAE, 5 Nov. 1882; all DF (*TAED* D8123ZBH, D8123ZBL, D8123ZBM, D8123ZBP, D8124N, D8124O, D8105ZZA [image 10], D8123ZFB; *TAEM* 57:867, 874, 875, 879, 1044–45, 345, 995).

4. Nothing is known of this prospective site.

[New York,] 24th Apl [188]1

Dear Sir,

I cabled you on the 17th inst as follows:—

"Cable credit three thousand dollars fruit experiments" not receiving any reply I cabled you again on 22nd,

"Answer cable fruit experiments immediately" and at once you replied

"Drexels pay"[1]

I have received from Drexel Morgan & Co[2] $3000 which sum I credit you on this account

I propose to [try?][a] the experiment this season of sending over the fruit to England so as to test the market as to the practicability of going into the business on a large scale in 1882.

The basis on which I will do the thing with you is that you shall furnish the money to conduct experiments, pay patent charges, in fact all the expense incidental to getting the business started you also undertake[b] to obtain whatever capital may be necessary to work the business   In consideration of which I will give you one half of the proceeds I derive from the undertaking

Please obtain immediately full information as to what fruits it would desirable to send to England from here what quantities the market would [take?][a] and what prices we could probably get   also the same information as to American game.[3]

I think that for this year we had better confine our efforts to game & peaches but we probably send incidentally some other fruits   Yours truly

Thos A Edison

LS (letterpress copy), NjWOE, Lbk. 8:209 (*TAED* LB008209; *TAEM* 80:907). Written by Samuel Insull. [a]Illegible. [b]Obscured overwritten text.

1. These transcriptions, with minor discrepancies, are the full text of cable messages exchanged with George Gouraud, who was in Paris when Edison's first cable arrived. Perishable foods had been packed in sterilized and hermetically sealed container since the beginning of the nineteenth century but there was no commercial process for preserving food in a vacuum; this correspondence marks the renewal of Edison's interest in such a method (Muller 1991, 123–29; see Doc. 1986). Gouraud's interest in the subject was likely encouraged by what proved to be a long agricultural crisis in Britain caused by a series of poor harvests, epidemic livestock disease, and a growing reliance on imports including, from about 1880, refrigerated meat from Australia. Gouraud promised to send a power of attorney and formal agreement governing his financial participation in this project but these have not been found (TAE to Gouraud, 17 and 22 Apr. 1881, LM 1:9F, 10D [*TAED* LM001009F, LM001010D; *TAEM* 83:876–77]; Gouraud to TAE, 22

and 23 Apr. 1881, both DF [*TAED* D8104ZBA, D8104ZBC; *TAEM* 57:112, 114]; Perry 1974, chap. 2; Ó Gráda 1994, 169–70, figs. 6.2–3).

2. The New York firm of Drexel, Morgan & Co., established in 1871, had acted as Edison's bankers since late 1878. The firm was heavily involved in financing Edison's electric light and controlled his lighting patents in Great Britain. See Docs. 1239 n. 2, 1494 n. 4, 1570, 1612, 1648, and 1649.

3. On efforts to preserve meat see Doc. 2118.

–2091–

Draft Agreement with Sigmund Bergmann and Edward Johnson

[New York, April 1881]

This Agreement entered into this   day of April 1881 by and between S. Bergmann[1] party of the first part Thomas A Edison party of the second part and Edward H. Johnson[2] party of the third part all of the City, County, and State of New York

Witnesseth the said Bergmann party of the first part is possessed of a Manufactory fully equipped for the manufacture of Electrical machinery at 108 Wooster Street in the City of New York and is desirous of entering into the manufacture of special[a] appliances connected with Electric Lighting (see sheet!)[3b] and whereas the said Bergmann also desires more capital to enable him to manufacture and carry a stock of such appliances[4] and whereas the said Johnson of the third part is entitled to a share in the profits arising from the said business of the said Bergmann by a previous contract[5] and by this Contract of twelve parts of each and every dollar ~~which out~~ of the profits made by the said Bergmann in the said business and whereas the said Edison is will to contribute the sum of seven thousand five hundred dollars for the further developement of the [said?][c] ~~business of the said Bergmann~~ aforesaid for which the said Edison is to receive forty four parts of each dollar of profit earned in consideration of furnishing the said sum of seven thousand five hundred dollars the[d] said Bergmann is to receive forty four parts of each and every dollar and the said Johnson is to receive the remaining twelve parts of each and every dollar as aforesaid   It is hereby agreed between the parties of the 1st 2nd & 3rd parts[e] that the said Edison shall contribute the said sum of seven thousand five hundred dollars which he shall deposit with Drexel Morgan & Coy Bankers[a] of this city for use in the said business within six months of the date of the execution of this agreement[f] [-- --][g] Subject to Bergmanns ~~Draft~~ order[a] $2500 on signing this contract & $1000 the 1st day of Each succeeding month until the Amount[d] herein named is fully paid in[h] in the character of a special partner and that his liability shall not exceed the above named seven thousand five hundred dollars and the said Edison agrees to use his best endeavours to "Promote"[i] ~~sell the said appliances~~ business[a] ~~manufactured by~~ of[a] the said Bergmann as far as lays in his power and the said Bergmann on his part agrees to use due diligence and to give ~~his whole time and undivided attention~~ all requisite time & attention[j] to the said business and he shall receive a salary of twelve ~~thousand~~ hundred dollars a year as salary for the superintendence of the said business which shall in all cases be deducted before the profits are declared and the said Johnson on his part agrees to use his best endeavours and inventive talent to devise articles of utility in Electric Lighting and capable of being manufactured by the said Bergmann and the said Johnson further[d] agrees to use his best[d] endeavours to ~~sell or cause to be sold the appliances in connection with Electric Lighting manufactured by the~~ Promote the business of the[k] said Bergmann and it is further agreed that should the business be closed up at any time from any cause whatsoever the said Edison shall receive a sum equal to the amount of the money he has put in the business prior to division of ~~the~~ any[a] surplus there may be[6]

~~And that the said Bergmann shall likewise receive~~

Df, NjWOE, DF (*TAED* D8101C; *TAEM* 57:7). Written by Samuel Insull; marginal notations probably written by Edison. [a]Interlined above. [b]"(see sheet!)" written in left margin. [c]Canceled and interlined above. [d]Obscured overwritten text. [e]"between the parties . . . parts" interlined above. [f]"six months . . . agreement" interlined above. [g]Canceled. [h]"[-- --] Subject to . . . fully paid in" written in left margin and followed by shorthand notation. [i]Interlined below. [j]"all . . . attention" written in left margin. [k]"Promote . . . of the" written in left margin.

1. Sigmund Bergmann had worked as a machinist in Edison's Ward St. shop in the early 1870s and in 1876 opened his own shop on Wooster St. in New York City. He did business as S. Bergmann & Co. and later as S. Bergmann; after about this time, the concern was known as Bergmann & Co. In 1878 and 1879 he manufactured phonographs and large numbers of telephones and related equipment for Edison. See Docs. 313 n. 1, 1177, 1790 n. 8, 1813 esp. nn. 3 and 5; headnote, Doc. 1195.

2. Edward Johnson was a former telegraph operator who had promoted various Edison inventions for many years. He was a partner in the Edison Lamp Co. and had been working on isolated lighting plants and the design of sockets and fixtures. See Doc. 272 n. 13, *TAEB* 1–5, passim.

3. Not found.

4. There is a fragmentary draft contract which appears to lay out terms by which Bergmann & Co. would manufacture and stock small items such as lamp sockets, fixtures, and switches in advance of actual orders. Draft agreement with Bergmann & Co., 1881, DF (*TAED* D8101I; *TAEM* 57:17).

5. Johnson had entered into a silent partnership with Bergmann by the middle of 1879, the terms of which are unknown. See Doc. 1790.

6. A formal partnership agreement was not adopted until September 1882 (see Doc. 2343). That was a revision of another draft agreement between Bergmann and Johnson, dated 13 April 1881, defining the partnership in more specific terms than this document (13 Apr. 1881 draft of 2 Sept. 1882 agreement, DF [*TAED* D8201U; *TAEM* 60:20]). Edison, Johnson, and Bergmann appear to have abided by the spirit of this document in the meantime (Bazerman 1999, 278).

–2092–

*Samuel Insull to
John Kingsbury*

MENLO PARK, N.J. Sunday, 1st May, 1881.

My Dear Kingsbury:[1]

I was immensely glad to get your letter[2] of some day I know not, as I am writing this at Menlo Park, and the letter from you is in my desk at 65 5th Ave., N.Y.

Mr. Edison and myself came out here last night to spend the Sunday. We mistook the time the train started and as a consequence we only got within six miles of this [place] and came on in a conveyance the exact character and title of which I cannot tell you, as it was so dark that I could not see the concern with that clearness necessary to an exact description. My description of the country must for the same reason go by default.

I am stopping at Edison's house today and shall go back to N.Y. in the morning. Edison's people are A No. 1 and make it very pleasant for me. This morning Mrs. Edison placed a fine pair of grey ponies at my disposal, and I flew along the rough Jersey road with a comfort only to be attained with the assistance of American ponies attached to the light vehicles which abound here.

Your letter was most acceptable. I was wondering whether you had forgotten me altogether, and I am glad to see that you have not. Your assumption that I get all the news is quite misplaced, and your letter gave me information for which I was thirsting. Just go into a little more detail the next time you write me.

A few days after I came here I called on the people controlling the electric pen here (The Western Electric Mfg. Co.) and found out the state of affairs in Australia.[3] They are friends of friends of mine, and as I have met most of their principal people in London I was on good terms with them right away. They told me they had written your brother offering him the sole agency and after my explanations said they would work with him the more cordially.

You ask me about Electric Light. Well I have seen 700 lights burning, the current generated from the same dynamo-electric machine for the whole lot, all of them getting their current from the same mains (i.e., street cables) of no less than eight miles in length.[4] Edison gets eight lights or thereabouts of 16 candles each per indicated horsepower, which allows of his competing with gas. Into the *details* of the cost I cannot go, as it is not told to anybody. Suffice it to say that here in New York he can produce light and get a handsome profit on it at a charge to the consumer which would ruin the gas companies. There is not, however, that vast difference between the cost of the two lights which will allow him to be utterly oblivious of his friends, the gas producers; but his estimates show that he can compete with them and do it at a handsome profit. Besides he can furnish power by means of electric motors, which will give him an enormous pull over the gas companies as he will not have the greater part of his plant lying idle during 365 working days of the year, as the gas companies with but very slight exception must, as the business is at night; but he can sell electricity for power purposes by day, which means that his plant is never idle, his capital is never running to waste, but is always earning money by night and by day alike. Edison will work just as the gas companies do. He will have central stations where the current will be generated (probably one station of about 15,000 lights to each square mile). This current will be conveyed along the streets underground by means of copper wire embedded in two-inch iron pipes insulated with a special form of insulation of his own invention. Branch pipes will be led into each house, and the electricity, whether for light or power (to us it is all the same), will be sold by means of a registration on an electric meter, which is the most ingenious and yet the simplest thing imaginable. The district which he will light up first in New York has about 15,650 lights in the various buildings in the district and a great deal of power varying in amounts.[5] He is getting contracts just as fast as his canvassers apply for them, and we have large gangs of men wiring the houses in anticipation of the time when we can lay our mains, erect our dynamo machinery and light up. I suppose this district will be all lighted up in from three to four months, and then you [will] see what you will see. You will witness the amazing sight of those English scientists eating that unpalatable crow of which Johnson used to speak in his letters to me when I was in the old country.[6]

Menlo Park is practically abandoned. All experiments are

finished; all speculation on the probable results are dismissed; and Edison thinks and so does everyone else who has looked into the matter, that success is assured. Of course time alone can prove this. As for myself, I am not competent to judge but I can use my eyes, can see the success with which the houses, fields, roads and Depot have been illuminated here, and I can see nothing to disprove the assertions. His lamps last about 400 hours; at all events that is the estimate by a time test, i.e., by running them at about four times their ordinary candle power until the carbons break; but this estimate is every day falsified, and experience points to the conclusion that the life of his lamps will be *much longer* than the estimate. As for rivals, Edison has but little fear, *in fact,* none from them. I have seen how Maxim's lamps[7] go, and his utter want of a system by means of which alone can success be attained, and Swan[8] we put in about the same category, but as he is a fellow countryman of mine, I will spare you the plain language used towards him.

To carry out the gigantic undertaking of fighting the gas companies we have much to do. A great difficulty is to get our machinery manufactured. This Mr. Edison will attend to himself.[9] He personally has taken very large works for this purpose, where he will probably within the next six months have 1,500 men at work. The various parts of the machines will be contracted out, one firm making one part in large quantities, another firm another part and so on. At Mr. Edison's works ("Edison Machine Works"), all these parts will be assembled and put together. Then there is the lamp factory, in which Mr. Edison owns almost all the interest, for manufacturing lamps and which is now turning out one thousand lamps a day, the Electric Tube Company (of which I am secretary and Mr. E. president) for manufacturing our street mains. So you can imagine what Mr. Edison has to do, as he is the mainspring and ruling spirit of everything. And you can imagine also what I have to do as his private secretary. We work every night till the small hours, and today (Sunday) is the first Sunday I have not been at the office; and even here we are at work, as between the intervals of writing this letter I am taking notes of a lot of data he wants before I go to bed to tonight. I have got right in with Edison, sit in the same room with him, assist him in everything, and am his private secretary in every sense of the word. People say that he likes me very much; but time must be left to prove this. Johnson says my success is assured, and last, but not by any means the least, I am also absolutely satisfied that I did the right thing in coming here.

Please find out for me and let me know at the earliest possible moment the exact price per 1000 ft. at which gas is *sold by all* the various companies in London and also the price per ton at which the various kinds of *steam* and household coal can be purchased there in larger quantities. Do me the very great favour of getting this out to me *at once* as I have promised to get it, as I dispute some figures furnished here.

[Two short paragraphs, relating principally to personal matters, are omitted here.][10]

With kind regards to your cousin and uncle and hoping to hear from you soon on above points, believe me   Very sincerely yours,

SAMUEL INSULL.

Address me as follows   65 Fifth Avenue, New York, U.S.A.

PL (transcript), Insull 1915, xxxvi. First and last numbered pages (of 12) reproduced in facsimile on facing page, indicating it was written on letterhead of the laboratory of Thomas Edison. Minor typographical disparities from the facsimile have not been reproduced or noted.

1. John Kingsbury had worked for George Gouraud from early 1879 to 1881. He handled publicity and advertising affairs of the office, and at this time was associated with Kingsbury & Co., a London advertising agency. He later became connected with the Western Electric Co. and was the head of the office there. In 1915 he published a major history of the telephone (Kingsbury 1915). Kingsbury to Insull, 19 July 1882, DF (*TAED* D8240ZAD; TAEM 63:43); Insull 1915, xxxv; see also Doc. 1587 n. 7.

2. Kingsbury's letter has not been found.

3. Edison sold his rights to the electric pen in 1877 to Western Electric Manufacturing Co. He continued to collect royalties but had agreed in 1880 to accept a reduced rate for a year (see Docs. 817 and 1882). Western Electric's arrangement in Australia is unknown, but Francis Welles, who was connected with the company's pen business, was acting as Edison's agent there in connection with the telephone (TAE power of attorney to Welles, 5 Apr. 1880 [*TAED* X012L1H]; TAE power of attorney to Welles, 2 Mar. 1880; Welles to TAE, 15 June 1881; both DF [*TAED* D8046K, D8150ZAD; *TAEM* 55:686, 59:1036]).

4. Edison had constructed a small central station demonstration system at Menlo Park in the latter part of 1880. At this time he would have used the Porter-Allen direct-connected dynamo. See *TAEB* 5:875, Docs. 1972, 2038, 2057, and 2074 n. 1.

5. The *Electrician* published a map in early 1882, likely based on canvass data available by this time, indicating a potential demand for about 18,000 lamps and 3,200 horsepower in the First District. "The Installation of the Edison Light in New York," *Electrician* 8 (1882): 124; see headnote, Doc. 2243.

6. For attitudes of English scientists toward Edison's electric light experiments see, for example, Docs. 1602 n. 1 and 1751 n. 5 and Dyer, n.d., 99–111.

7. Electrical inventor Hiram Maxim (1840–1916) began experimenting with electric lighting about the same time as Edison. He devised and sold (through the United States Electric Lighting Co.) an incandescent carbon filament lamp that Edison regarded as a copy of his own. See Docs. 1617, n. 4, 2021, and 2033; Friedel and Israel 1986, 193–94.

8. Joseph Swan (later Sir Joseph), a British chemist and inventor, claimed to have devised an incandescent electric light consisting of a slender carbon filament in an evacuated glass globe prior to Edison; however, he did not publish his results until the latter part of 1880. See Doc. 2022 n. 6.

9. On the organization and scope of Edison's manufacturing operations see headnote, Doc. 2343.

10. The facsimile reproduction of the last page of this letter in Insull 1915 includes part of the latter paragraph. Having evidently made a personal request, Insull concluded that he would "leave it to you to do as best you can knowing that you will do anything for me which will help me here to secure the good will of those I am associated with."

–2093–   [New York, c. May 1, 1881[1]]

Memorandum: Ore   Ore Seperator=
Milling   Wants Machine Dynamo etc & man sent to Quogue[2]—soon
                           as possible; Ship via Long Island RR=
                           Distance 78 miles fm NY=
                           send man there first=
                           agrees to supply 100 tons to machine every day dont rain=[a]
                           pays my man 12 per week. Keeps machine & Dynamo good order—& pays 5¢ for every ton run through machine=[3]

AD, NjWOE, DF (*TAED* D8138ZAM; *TAEM* 59:153). [a]"over" follows as page turn.

1. Samuel Insull indicated on this document that it was related to correspondence with Henry Haines of the Magnetite Mining Co. in New York. No prior correspondence with Haines has been found, but on 4 May he telegraphed Edison to send an ore separator. Haines to TAE, 4 May 1881, DF (*TAED* D8138G; *TAEM* 59:117).

2. A town on the southern coast of Long Island.

3. This document is the first evidence of Edison following up a request to use his magnetic separator for recovering commercially useful iron ore rather than for separating iron from gold or other precious metals, the machine's original purpose. Iron smelters had used magnetic means, at least experimentally, to enrich ore since the early 1830s. Haines subsequently made several inquiries about the arrival date of the equipment and of George Hickman, a metalworker employed by Edison. Edison promised to ship the machinery on 5 May; the separator was operating by 30 May, when Sherburne Eaton inspected it. Eaton's favorable report prompted the company to dispatch prospectors to search for deposits of iron sands that could be worked profitably. Gordon 1996, 70, 240; Haines to TAE, 4, 8, and 10 May 1881, all DF (*TAED* D8138G,

D8138H, D8138I; *TAEM* 59:117–18); Eaton to Edison Ore Milling Co., 30 Oct. 1881, CR (*TAED* CG001AAI1; *TAEM* 97:411).

–2094–

To Edward Acheson

Menlo Park, N.J., 2nd May 1881.[a]

Mr Atchison

Please come up to the Laboratory & bring one of those nickel [-][b] moulds in which they bend the fibre to carbonize it and press a piece of plumbago the thickness of the mould. ~~It~~ It is I believe ⅛ of an inch and then hollow it out for the nickel piece to allow the carbon to draw up. After you have got it have Dr Haid pass the gas over it. I want to see if we cannot make these little plated moulds out of plumbago using the nickel piece to put straight on the fibre. If we could use these it would save a deal of money[1]   also try some experiments on getting the best mixture of litherage & glicerine also the right proportions of plaster of Paris for the sockets of the lamps

We are lame on these points   Yours

Edison   I[nsull]

L, MiDbEI, EP&RI (*TAED* X001J1AD). Written by Samuel Insull; letterhead of T. A. Edison. [a]"Menlo Park, N.J.," and "1881." preprinted. [b]Canceled.

1. The nickel carbonizing molds contained a sliding weight, also of nickel, to keep the filament legs straight as they shrank (see Docs. 1961, 1966, and 1973). John Howell recorded an order to test the durability of "Carbons out of graphite forms" about this time; on 16 May Edison sketched several "plumbago moulds for Carbonizing" (Cat. 1301 [order no. 314], Batchelor [*TAED* MBN007:27; *TAEM* 91:320]; Cat. 1147, Scraps. [*TAED* NM016:58; *TAEM* 44:282]). Carbonizing molds of pressed plumbago were made at the Menlo Park machine shop and tried at the lamp factory by 20 May, when Upton reported encouraging results. By late June the new forms were being used extensively, but Upton found that the nickel weights used to stretch the carbons deteriorated after only a few uses. Thomas Logan to TAE, 17 May 1881; Upton to TAE, 20 May and 27 June 1881; all DF (*TAED* D8137ZAI, D8123ZBX, D8123ZCQ; TAEM 59:82, 57:889, 917); see also Docs. 2308 and 2319.

–2095–

From Francis Upton

Menlo Park, N.J., May 4 1881[a]

Dear Mr. Edison:

We have tried ~~ana~~ number of lamps where the clamp was heated and found the results so good that each day we shall make 300 lamps in that manner.[1] If after a few days trial it prooves all right we shall make all lamps that way.

Dr. Haid hias just sent down 75 of the finest bamboos I have ever seen treated. They are now being put in ~~carbon~~ clamps. He says he can make 100 in two hours in the gas furnace he has above. I put the argument about paper thus to him. A lamp costs 20 cts. The difference between a paper loop and a bamboo loop is about 2 cts. in cost. As the life of a lamp depends entirely on the carbon if ~~that~~ the bamboo[b] is 10% better we have saved the 2 cts. Bamboo is far more than ~~20%~~ 10% better than paper.[2]

None of the clamps are now bending since I have had them inspected. Monday we made 880 lamps, mostly Bs[c] Tuesday 560 stopping at 7–30 P.M. Today I expect 700 lamps 400 of them As stopping at 10 P.M.

I do not want to run nights until I have about 2000 lamps ahead of the pumps so that I can change for several nights in succession.

As we have only 200 pumps running and light only 80 lamps at a time I feel quite sure that I can get 1000 lamps from the three lines with one gang, when we have more pumps and current.

Bradley[3] has his fixture for making socket under trial.

We have made some few changes but feel sure that we can make a fine socket.

Achison is trying various mixtures and Bradley is going to see how those with promise work in practice.

Dr. Nichols is making a full set of tests to see if the resistance of the lamps is lower.

I will send you his results in a day or two. We know that it has come down at least three ohms in two weeks, due to better carbonizing and methods on the pumps.[4]

I think it would be advisable to make our bamboo loops thinner so as to get 100 Ohms in place of 95[d] that the last lot have tested. We find great difficulty in being certain as to resistance as we have no standards here. Yours Truly

Francis R. Upton.

ALS, NjWOE, DF (*TAED* D8123ZBK; *TAEM* 57:871). Letterhead of Edison Electric Lamp Co. [a]"Menlo Park, N.J.," and "188" preprinted. [b]"the bamboo" interlined above. [c]"mostly Bs" interlined above. [d]Obscured overwritten text.

1. About the last week of April or beginning of May, John Howell recorded in a lamp factory notebook an order (no. 304) for "special heated clamps" and another (no. 305) for fifty "clamps to be hardened by being held in a flame." Cat. 1301, Batchelor (*TAED* MBN007:26; *TAEM* 91:319).

2. Paper, from which Edison had fashioned lamp filaments in the fall of 1879, was the subject, at least briefly, of renewed experimentation. John Howell noted on 31 May the preparation of "paper carbons from Dr Haid new process of treatment." Nothing is known of the process. Cat. 1301 (order no. 366), Batchelor (*TAED* MBN007:33; *TAEM* 91:326).

3. James Bradley, a machinist, began working for Edison in Newark in 1872. He was in charge of the fiber department at this time. Doc. 1080 n. 4, *TAEB* 5 App. 2; Jehl 1937–41, 815; "Bradley, James J.," Pioneers Bio.

4. Edward Nichols reported directly to Edison a week later that the average resistance of 200 A lamps manufactured about 28 April was 94.6 ohms. This indicated "a falling off of the average resistance of the lamp since March, off about 5 ohms." Nichols to TAE, 11 May 1881, DF (*TAED* D8123ZBT; *TAEM* 57:885).

–2096–

From Charles Hughes

Menlo Park, N.J., May 5th 1881.[a]

Mr. T. A. Edison.

Have exhausted 14 bulbs containing strawberries.[1] The first two taken off at 3 A.M. Sunday are evidently spoiled although retaining some of their natural color the water in them has come out into the bulb and left the berries soft. Vacuum 5 hours.

Nos 3 & 4   Taken off pumps at 12:20 P.M. May 2d 3½ hour vacuum are fairly kept and some of them which do not touch the glass and are not crowded are fresh as ever. These were exposed to the air 30 hours longer than 1 & 2 before being operated upon.

No 5   Containing five Florida strawberries taken off Pump at 11:45 P.M. May 2d 2½ hours vacuum look very fair some of them as fresh as when put in

No 6   same as No 5 but wrapped in Tissue paper which dont seem to make any difference

No 7   strawberries 5 Days picked sealed with out exhausting at 3 A.M. same time and the same berries as Nos 1 & 2 look about as fresh as they do.

No 8   same berries as Nos 5 & 6 exposed to the air while 5 & 6 were being exhausted 2½ hours. Taken off Pump at ~~10:50~~ 1:05 AM. May 3d 2¼ hours vacuum are quite fresh where they are not pressed against the glass.

No 9   same as No 8   Vacuum not so good   Berries just about as good. Moral high vacuum not absolutely necessary.

No 10   strawberries 4 Days picked put on pump without Phosphoric Chamber[2] with straight tube instead of bent   Got good vacuum on 2½ hours   Took off at 9:45 P.M. May 3d are quite fresh

No 11　same as No 10 lost vacuum

No 11　Charleston berries 4 days off vines very nice taken off at 2 P.M. May 4th 4 hours vacuum are quite fresh

No 12　same as no 11 equally fresh.

No 13　same strawberries as Nos 11 & 12 sealed without exhausting at 4 P.M. May 4th and are apparently just as fresh. Moral! Dont exhaust at all.

No 14　same berries as 11, 12, & 13　2 hour vacuum vacuum very fair. Berries are quite fresh. Tried six bulbs with Tomatoes. They all burst in about an hour each. Sealed some without exhausting　Dont use Phosphoric Chamber as it takes up too much moisture and makes mush of it. Think the hole spoken of by Mr Gouraud[3] was made by too high vacuum and that he will find that the water in the bulb contains all that is good of the Peach. Am open for suggestions　Yours Truly

Chas. T. Hughes.[4]

ALS, NjWOE, DF (*TAED* D8104ZBG; *TAEM* 57:122). Letterhead of T. A. Edison. Expressions of time have been standardized for clarity. a"Menlo Park, N.J.," and "1881." preprinted.

1. Hughes recorded the results of ninety-one fruit preservation experiments he conducted between 30 April and 19 August 1881 in N-81-04-30:1–75, Lab. (*TAED* N306:1–38; *TAEM* 41:1156–93). These experiments were a continuation of work done the previous August by Edison and Ludwig Böhm (see Doc. 1986). Edison executed a patent application for this process on 11 December 1880; it issued in October 1881 as U.S. Patent 248,431. By the end of 1881 Edison had spent a total of about $1,880 on the fruit preservation experiments (Ledger #5:488, Accts. [*TAED* AB003:241; *TAEM* 87:643]).

2. The phosphoric chamber contained anhydrous phosphoric acid to remove water vapor from the evacuated container, a technique used in lamp manufacturing. See Docs. 1714 n. 2 and 1816 n. 5 and Hawley 1987, s.v. "phosphoric anhydride."

3. Gouraud took samples to England in August 1880 (see Doc. 1986). He reported in April that the one remaining peach had "retained its form perfectly. The skin only is somewhat broken and shrivelled. There is about half a teaspoonful of water in the tube　There is a round hole about the size of a green pea in one part of the peach. The glass is very cold." Gouraud to TAE, 23 Apr. 1881, DF (*TAED* D8104ZBC; *TAEM* 57:114).

4. Charles T. Hughes (c. 1847–1909) entered Edison's employ about October 1879. Recently he had been working on the electric railway, probably because of his prior experience in the railroad business. His year of death is stated incorrectly in Doc. 1965 n. 5. The editors also stated erroneously there and in *TAEB* 5 App. 2 that Hughes later claimed to have started working at Menlo Park on 21 October 1879; he only recalled beginning there about the time of Edison's first successful incandescent carbon lamps.

–2097–

R. G. Dun & Co.
Credit Report[1]

[Newark?][2] May 11/81

123.[3] Menlo Park NJ & 5th Ave NY   (Thos A Edison?
[pr?][a]   He states that he is entirely out of debt & has 50mf[4] his
own money, has a lamp factory & some other private enterprises, but he has paid cash for every thing & does not intend
asking any credit

D (abstract), New Jersey, Vol. 52, p. 399, R. G. Dun & Co. Collection,
Baker Library, Harvard Business School. [a]Illegible.

1. R. G. Dun & Co., established in 1841 as the Mercantile Agency,
was by 1871 one of two major credit-reporting firms in the United
States. Norris 1978; *ANB*, s.v. "Dun, Robert Graham"; Doc. 252 n. 1.

2. R. G. Dun clerks transcribed individual reports filed by agents into
large ledgers organized by state and county. In 1871 the firm opened an
office in Newark; it is presumably there that the ledger in which this report appears was maintained. Doc. 252 n. 1.

3. This number refers to the agent making the report. The book containing the matched codes and names has been lost.

4. That is, $50,000.

LAMP FABRICATION   Doc. 2098

In this document Edison outlined a display for the Paris
Exposition that would present the most up-to-date lamp components and demonstrate their relationship to the finished
article. The display would also demonstrate the improved fabrication of lamps, now reduced to a series of manual factory
routines. The exhibit became the source of published descriptions of the lamp fabrication process.[1]

Preparation of the bamboo carbons required about a dozen
discrete operations, starting with sawing the cane to specified
lengths. Each segment was split several times into narrow
strips from which the hard outer covering was removed, then
the remaining fibrous portion was shaved to the requisite thickness. The resulting thin strip was placed between carefully
mated metal blocks that served as guides for trimming it to the
proper width, including protrusions at each end for clamping.
After being carbonized in air-tight molds, the filaments were
electroplated to short copper supports at each end to which
the lead-in wires were later attached.[2]

Assembly of the lamp globe was also labor-intensive. Glass
blowers first heated each open-ended bulb or flask received
from the Corning Glass Works, drawing out a short cone or
"tit" at the closed end. To this they fused a thin tube by which

the bulb would later be attached to the vacuum pump. In separate operations two short platinum conducting wires were inserted through a tapered glass stem and then a flange was formed on the stem. The end of the stem was heated and the softened glass pressed around the platinum to form an airtight seal. The platinum was soldered at one end to copper wires electroplated to the filament, and at the other end, to the lead-in wires.[3] The completed tube assembly was then inserted into the bulb and sealed at the flange.[4] While the assembled bulb was evacuated on the mercury pump, the filament was heated electrically, eventually reaching a high incandescence, to drive off occluded gases. After being sealed off the pump, a plaster base was molded to the bottom.[5]

Edison's eagerness to present the stages of lamp manufacture in such detail indicated his confidence in the strength and breadth of his patents. Absent from published accounts, however, were references to any but the most general techniques for working with the bamboo, glass, and wire. Specific methods and apparatus for shaping and carbonizing filaments or evacuating bulbs, for example, were not included in the exhibit, nor were the final stages of manufacturing; however, descriptions of manufacturing routines and standards in several departments are given in a notebook compiled around this time principally by Edward Acheson and Edward Nichols.[6]

1. Heap 1884 (see pp. 175–78) was the official U.S. report of the Exposition; Prescott 1884 (167–76) incorporated similar text and clearer reproductions of the drawings. An anonymous typescript at NjWOE from the period 1910–1912 describes and illustrates these processes in rich detail. "A History of the Development of the Incandescent Lamp," 15–59, NjWOE.

2. The shaping processes were illustrated in a series of seventeen figures in the official report. The blanks for full-size carbons (**no. 10**) and those from which two half-size filaments could be cut (**no. 13**) are shown (Heap 1884, 178, fig. 161; reprinted in Jehl 1937–41, 629 and with descriptive text in Prescott 1884, 173, fig. 87). The carbonization process, consisting of an initial heating for six to eight hours and a second, briefer heating, is described in Jehl 1937–41 (811).

3. By this time Edison had adopted (and applied for a patent on) the practice of using short lengths of platinum wire to pass through the glass stem between the lead-in wires and copper filament supports, because the coefficient of expansion of platinum was the same as that of glass. On 20 May he executed an application covering manufacturing processes for soldering the wire segments by means of a blow pipe. U.S. Pats. 248,419 and 251,544; see also Cat. 1147, Lab. (*TAED* NM016:51, 59; *TAEM* 44:275, 283).

4. These processes for preparing the globe are largely embodied in a

Steps in processing bamboo cane (1) into finished filaments (11, 12, 14, 15, 16, and 17).

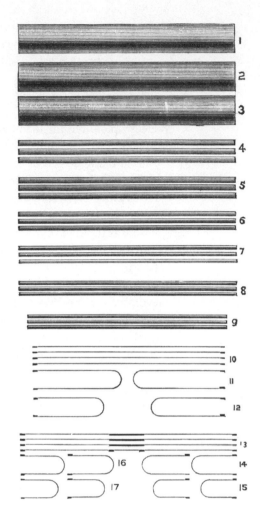

U.S. patent application that Edison executed on 21 April 1881 (U.S. Pat. 266,447). The manipulations of the glass and wire were illustrated in fourteen figures in the official report (Heap 1884, 178; reprinted in Prescott 1884, 170–71 and Jehl 1937–41, 630).

The most recent form of lamp shown was roughly cylindrical. This shape put the walls of the globe at a uniform distance from the carbon, which reportedly reduced both the likelihood of the glass cracking and the "loss of light through refraction and internal reflection." Heap 1884, 176–78.

5. According to undated notes kept by William Hammer, probably from about this time, the socket formula consisted of 1 part of glue, 300 parts water, 250 parts of a solution of glue and water, and 400 parts plaster of Paris. The mixture set in 1¼ hours. Ser. 1, Box 14, Folder 9, WJH.

6. Notebook #144, Hendershott (*TAED* B015); see also headnote, Doc. 2177. Comparable information about the preparation of carbons was included by William Hammer in the undated notebook apparently kept in the first half of 1881. Hammer also noted problems observed in

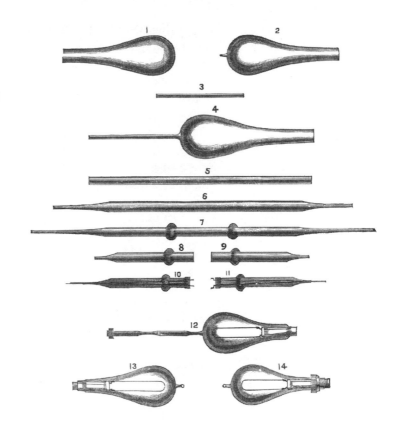

Stages in the fabrication of the glass globe and assembly of the lamp, showing the bulb as received from the Corning Glass Co. (1) and the inside tube (5) inserted with the filament into the bulb (12).

Edison exhibited this roughly cylindrical lamp as his most up-to-date form of globe.

lamps, including discoloration of the glass, deformation of the filament, and blistering of the plated clamps. Ser. 1, Box 14, Folder 9, WJH.

–2098–

To Edward Acheson

New York, 13th May *1881*[a]

Dear Sir,

Please prepare a case of sample fibres similar to the ones prepared by Hammer & which he sent here.[1] One case should show all the different fibres as from South America bamboo & all[b] in their natural state and as they appear after shaving. Another case should show all the different kinds of bamboo cut in lengths between the joints and not split. Another case should show a single cylinder of bamboo, then the first splitting and the first shaving, the second shaving, and third shaving then some of them bent round then the different sizes of carbon carbonized. In this same case I desire to show some of the bamboos wound in the form of a spiral not carbonized and also carbonized[2] Another case should show the globe as it comes from the Glass House and then with the tit on it for exhausting and then with the tit on and also cut off ready for sealing, then the tube taper at each end forming the inside part, then the little flange blown on another. Another one with a open mouth then with it squeezed together with the wires in it, then the inside parts solid in the globe—in fact I want to show the whole process of manufacture[3]

These various parts should be securely fastened by proper cement with glass covers in deep cases   You can procure at my expense what necessary sundries &c you may desire in New York. I think duplicates of each should be made in case of breakage—in fact I think you had better make three—one for this office and two to go to Paris[4]   Yours truly

T A Edison

⟨This should be done at once as I have other Paris work for you. E⟩

LS, MiDbEI, EP&RI (*TAED* X001J1AE); letterpress copy in Lbk. 8:249 (*TAED* LB008249; *TAEM* 80:931). Written by Samuel Insull; letterhead of T. A. Edison. ᵃ"*New York*," and "*188*" preprinted. ᵇ"as from South America bamboo & all" interlined above by Edison.

1. One of these cases may have been among the Edison-related materials that William Hammer exhibited at the 1904 World's Fair in St. Louis. This item, reportedly assembled by Hammer at Menlo Park in 1880, was a collection of about three dozen experimental filaments, including paper, cardboard, pressed plumbago, bamboo, and other fibers, some of them treated with hydrocarbon vapors. *Edisonia* 1904, 114–15.

2. Edison's first patent application for an incandescent lamp with a carbon filament, filed in early November 1879, specified a filament wound in a tight spiral. This was the shape Edison had been using in experiments with platinum wire "burners" until October of that year (U.S. Pat. 223,898; see Doc. 1818). Edison filed a patent application (Case 323) on 24 June 1881 for a spiral carbon. This was rejected and later abandoned, but in December he filed another application for a lamp with "a great length of carbon coiled and arranged in such a way that a small radiating surface only will be exposed" so as to minimize energy lost as heat for a given light-producing length. (Another, unstated, advantage would be a uniformly spherical distribution of light.) Six spiral bamboo lamps were tested at the factory on 11 May. Edison also included "3 or 4" lamps with spiral bamboo filaments in a set of instructions given to Francis Upton on 16 May (Patent Application Casebook E-2536:370, Patent Application Drawings [Case Nos. 179–699], PS [*TAED* PT020370, PT023:33; *TAEM* 45:729, 850]; U.S. Pat. 379,770; Cat. 1301, Batchelor [*TAED* MBN007:30; *TAEM* 91:323]; Cat. 1147, Lab. [*TAED* NM016:43; *TAEM* 44:267]).

3. See headnote above.

4. Acheson may have received additional instructions later. He wrote to Samuel Insull on 14 June that he would "be pleased to prepare the case you wish." He asked that James Bradley be instructed to send him "the material to make it of. Tell him to send 6 samples of each operation in

making both the 6 ins and 3 in carbons." Acheson to Insull, 14 June 1881, DF (*TAED* D8123ZCM; *TAEM* 57:912).

<table>
<tr><td>

–2099–

To Longworth Powers

</td><td>

[New York, May 14, 1881[1]]

Hiram Powers Junior Sculptor Florence[2]

Have some instrument maker exactly[a] duplicate ~~with great exactness~~ model of Antonio Paccinottis electro magnetic machine[3] [----][b] in University of Pisa. Ship Express. Draw expenses through Drexel Morgans. Is Paccinotti living. Answer twenty words paid[4]

Edison

</td></tr>
</table>

L (telegram, copy), Lbk. 9:108 (*TAED* LB009108; *TAEM* 81:39). Written by Samuel Insull. [a]Interlined above. [b]Canceled.

1. This cable was written above a related message from Edison on the same date (see note 3). Lbk. 9:108 (*TAED* LB009108A; *TAEM* 81:39).

2. Edison was referring to Longworth Powers (1835–1904), a portrait sculptor and photographer and son of the famous American sculptor Hiram Powers. Edison's first known contact with him was in August 1880, when Powers asked for an unspecified business appointment. Powers produced about forty portrait busts, including one of Edison, the plaster model of which is in the National Museum of American Art (acc. no. 1968.155.69). Powers to TAE, 5 Aug. 1880 and 16 Jan. 1887; TAE to Powers, 30 Dec. 1887; TAE to Gouraud, 19 Mar. 1888; all DF (*TAED* D8004ZEI, D8704AAD, D8717ACQ, D8818AGK; *TAEM* 53:193; 119:109, 644; 122:194); Wunder 1991.

3. In 1860 Italian physicist Antonio Pacinotti constructed a generator with an armature consisting of a series of coils connected together to form a continuous coil or circuit, a commutator consisting of several insulated metal strips or rods forming a cylinder, and two or more brushes to collect current from the commutator. His machine was the first to produce a steady and continuous current; this form of commutator became known generally as the Pacinotti type. In 1870 the French engineer Zénobe Gramme constructed a dynamo on this principle (as did Pacinotti three years later, apparently independently). Gramme's European and U.S. patents broadly claimed or specified these three features of Pacinotti's machine, and the Gramme company in France won an infringement suit against the German engineer Werner Siemens for his use of commutator brushes. See Docs. 1489 n. 6 and 2071 n. 1; *DSB*, s.v. "Pacinotti, Antonio"; Dredge 1882–85, 1:cxxix; "The International Exhibition and Congress of Electricity at Paris," *Sci. Am.* 45 (1881): 377; U.S. Pat. 120,057; Thompson 1902, 312; Glaser 1881, 4894–95.

In March Edison had asked Theodore Puskas to investigate whether Pacinotti had published a description of his generator prior to Gramme's patent, which would have invalidated it. Pacinotti had in fact published such an account in 1864 and on 14 May Edison cabled Puskas to obtain a copy. On 16 May Edison inquired of patent attorney Lemuel Serrell whether Gramme's patent had expired in Canada and several European countries because of nonpayment of taxes, which also would have nullified the U.S. patent. He stressed that this was "very important & should be attended to at once." At the end of the month Puskas mailed the published description of Pacinotti's machine. The validity of Gramme's patent with respect to prior publication by Pacinotti was also the central legal issue in an infringement action brought about this time by a British dynamo manufacturer against rivals. Doc. 2071; TAE to Puskas, 14 May 1881; TAE to Serrell, 16 May 1881; Lbk. 9:108A, 110 (*TAED* LB009108A, LB009110; *TAEM* 81:39, 40); Puskas to TAE, 31 May 1881, DF (*TAED* D8120ZAP; *TAEM* 57:619); Editorial note, *Teleg. J. and Elec. Rev.* 9 (1881): 215.

4. Powers replied by cable on 15 May, "Paccinonti is living   am writing him   hope send model quickly   happy serve you." On 26 May Powers wrote more fully that he had contacted Pacinotti, who had moved from Pisa to Cagliari, and had "ordered the exact facsimile of his Electro Magnetic Machine—begging all speed to be used." On 28 June he apologized for the delay, explaining that Pacinotti himself was constructing the model and was planning to bring it personally to Powers in Florence in August. Powers finally received and shipped the model in late September. Powers to TAE, 15 and 26 May, 28 June, 18 Oct. 1881, all DF (*TAED* D8120ZAM, D8120ZAN, D8120ZAQ, D8132ZAY; *TAEM* 57:615–16, 620; 58:509).

–2100–

Technical Note: Electric Light and Power

[New York,] May 17 1881—

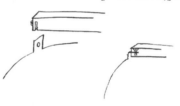

~~pa~~Make detailed drawing of present steam Dynamo— Claim coupling of Dynamo with Engine= Swing pillow block= Spiral spg on brushes—[1]

this method making Commutator.

patent plate here

brass=

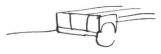

Zinc bases, nonmagnetic bolts= taper rods screw in End—[2]

this too

Covering surface iron with mica= mica between german silver=

patent= Station meters   places[a] to put it[3]

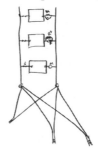

this too[4]

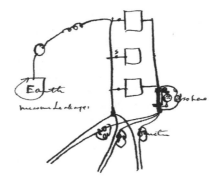

$^1/_{1000}$ curr shunted[5]

⟨316⟩[b]   Moulding socket on Lamp   Fig 1[6]

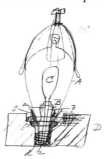

Make[a] figure 2 a section of lower end of lamp and ~~socket~~ neck socket complete, as basis for claim on plaster of Paris socket.

Witness   S. D. Mott[7]   TAE

X, NjWOE, Lab., Cat. 1147 (*TAED* NM016:61; *TAEM* 44:285). Document multiply signed and dated. [a]Obscured overwritten text. [b]Written by Samuel Insull and multiply underlined.

1. These drawings show the attachment of dynamo induction bars to the armature end plates. Edison completed a patent application for a slightly different form of attachment on 24 August, intended to provide great physical strength and low resistance (U.S. Pat. 264,647). Reducing sparks between the commutator bars and brushes was a long-standing concern for Edison and other dynamo designers. Edison assigned Charles Mott in mid-May to search the published technical literature for "the exact words all writers use in describing how to set the commutator brushes," particularly with reference to the magnetic lines of force (N-81-05-14, Lab. [*TAED* N213; *TAEM* 40:883]). Nothing is known of the proposed spiral spring arrangement; however, Edison filed applications for other means to reduce sparking during 1881 (U.S. Pats. 263,149 and 425,763).

2. These drawings also apparently represent means of attaching induction bars to the armature end plates.

3. At the end of May Edison executed a patent application for an arrangement of meters in an electrical distribution system for several purposes: to verify the total amount of current generated, to ascertain "the condition of the system and to determine the amount of leakage." Leaks to ground could be measured by a meter placed in "a line from ground to the return or negative main feeding-conductor, the leakage of the system that goes to earth producing the flow of an equal current from the ground to the generators." In a paragraph added during the examination process, Edison explained that meters placed at the central station and in the feeders prior to the consumption circuits could be used to measure "the current that leaks from one side of circuit to the other through the insulation without going to earth." This application issued in 1890 as U.S. Pat. 425,760.

4. Text is "Earth"; "measure leakage"; "meters" (below); and "also here" (right).

5. Text is "German silver" and "$^1/_{1000}$ curr. shunted."

6. Edison executed a patent application (Case 316) on 20 May for a socket molded to the lamp but which could be readily removed. The first of the two patent drawings is essentially the same as Edison's sketch; the second was made according to the instructions below. The application contained four claims, the first three of which pertained to a lamp with a base molded on the neck. These were deleted during the examination process and replaced by one for "the method of making the bases for and attaching the same" to lamps. The last of the original claims, also relating to the mold itself, was retained. Pat. App. 251,549.

7. Samuel Dimmock Mott began working for Edison in mid-1879, principally as a draftsman, and became an accomplished inventor in his own right. See Doc. 1985 n. 7.

To George Barker

My Dear Barker

I will find out when that was published about the chalk Battery & let you know as soon as possible.[1] I would be glad to see your friend Stanley when he calls[2]

What machine do you propose to use for lighting your house   If you have one please get the electro motive force of [it?][a] with the nearly saturated constant field with the machine open also give me the internal resistance of the bobbin and the resistance of the field. I may have to make your lamps to suit your machine.[3]

We make 16 candle lamps of 100 to 110 volts also eight candle lamps 100 to 110 volts also eight candle lamps from 50 to 55 volts also 16 candle lamps with from 50 to 55 volts

We have not given any of the manuscript of the book you [refer?][a] to to the printer yet but we will give you proof sheets as soon as we get them.[4] I would thank you for a copy of your Smithsonian Scientific[b] notes when for the Smithsonian Report when published[5]

If there is any other Scientist that you want me to send lamps to I will do so with pleasure.[6] I am very delighted to hear of Professor Shorts success at Denver.[7]

I tested a Dynamo last week of a new form but the same size as the one you saw at Menlo Park and it gave remarkable results. Got [74?][a] volts with an internal resistance of [$^3/_{1000}$?][a] of an ohm.[8] This is just 27 times better than has ever been done before by Siemens[9] or anyone else and the local action[10] action has been brought down to a point so low that it can almost be left out of any calculations. Hence the efficiency of this machine is up somewhere near the conductivity of the copper The other advantage is that the whole apparatus is insulated with mica; no covered wire is used

By a new phenomenon which I have lately discovered I am able to prolong the life of the lamps and shall change their economy from 3600 foot pounds for sixteen candles to 3000 foot pound for 16 candles.[11] The Lord only knows where the economy is going to stop! Telegraph me when you will be here[12]   Yours very truly

Thos A Edison   I[nsull]

LS (letterpress copy), NjWOE, Lbk. 8:269 (*TAED* LB008269; *TAEM* 80:939). Written by Samuel Insull. [a]Illegible. [b]Interlined above.

1. In 1879 Edison discovered that friction between a piece of platinum and the chalk cylinder in his electromotograph telephone receiver

produced an electric current. He constructed an instrument with four cylinders to demonstrate this principle; it was described and illustrated in the July 1879 issue of *Scientific American*. See Doc. 1738 n. 5; also Jehl 1937–41, 285–86.

2. Edison may have meant William Stanley (1858–1916), who recently had been an electrical assistant to inventor Hiram Maxim at the United States Electric Lighting Co. in New York and about this time took a similar position with Edward Weston. There is no record of a visit by Stanley. *DAB*, s.v. "Stanley, William."

3. Edison had promised in March to have Barker's house wired for electric lighting. TAE to Barker, 17 Mar. 1881, Lbk. 8:88 (*TAED* LB008088; *TAEM* 80:870).

4. It is not known to what publication Barker had referred.

5. This was likely Barker 1881 (286–88), the first of several surveys of "Recent Scientific Progress" in physics and chemistry that Barker prepared for the annual report of the Smithsonian Institution regents. Barker described Edison's carbon lamp filament and summarized results of several tests in 1880 by noted scientists (including himself) of the efficiency of Edison's lamp and generator (see Docs. 1910, 1914, 1916, and 1927).

6. Edison had sent Barker a half dozen lamps in March for experiments on the durability of filaments treated with hydrocarbons. He gave Barker another half dozen lamps in April to exhibit to the National Academy of Science. TAE to Francis Upton, 17 Mar. 1881; TAE to Edison Lamp Co., 13 Apr. 1881; Lbk. 8:89, 167A (*TAED* LB008089, LB008167A; *TAEM* 80:871, 891).

7. Electrical inventor Sidney Short (1858–1902) became professor of physics and chemistry and vice president of the University of Denver soon after graduating from Ohio State University in 1880. Edison was involved in telephone patent interference proceedings against Short. *DAB*, s.v. "Short, Sidney Howe"; see Doc. 1804.

8. On 12 May, when Thomas Logan reported that the new disk dynamo was completed (see Doc. 2082 n. 1), Edison instructed him to "have dynamo and engine ready to test lights Sunday," 15 May. The only extant record of tests with the machine about this time is Jehl's 20 May report that it had a resistance of .02165 ohm and, at 1200 rpm, produced up to 66 volts (with 29 volts on the field magnets). TAE to Logan, 12 May 1881; Jehl to TAE, 20 May 1881; Lbk. 8:246, 7:160 (*TAED* LB008246, LB007160; *TAEM* 80:928, 551).

9. This was most likely Werner Siemens (1816–1892) or the German firm he headed, Siemens and Halske. Of the five Siemens brothers who distinguished themselves in electrical technology and the complex of manufacturing firms they operated, he was the most closely identified by Edison with dynamo design and construction. See Docs. 811 n. 2, 1489 n. 5, and 1851.

10. That is, eddy currents induced in the armature core and lost as heat.

11. The "new phenomenon" is unknown.

12. Edison had recently expected to see Barker in New York but did not realize until days later that he had, in fact, been there. He later explained that he had briefly left the office, instructing Samuel Insull "to send for me should you call. I shall endeavour to find out who is respon-

sible for the gross negligence which caused you so much inconvenience and which has caused no slight annoyance to myself." Barker came to New York again at the end of May, but Edison was unable to see him. TAE to Barker, 16 May and 2 June 1881, Lbk. 9:111, 8:294 (*TAED* LB009111, LB008294; *TAEM* 81:41, 80:949).

–2102–

Technical Note:
Electric Light and
Power[1]

[A]

[New York,] May 20 1881

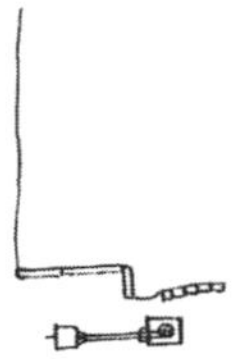

[B][2]

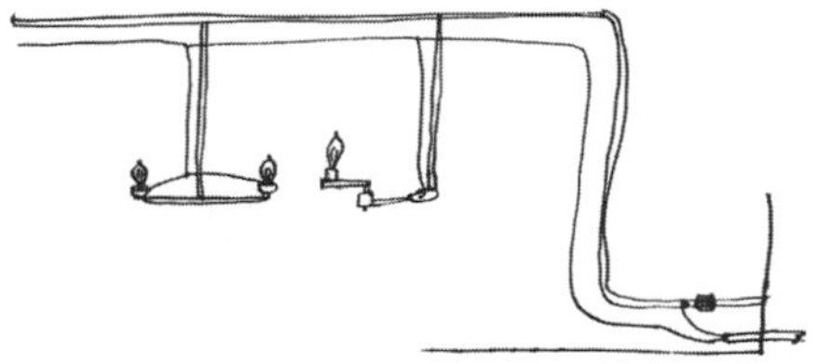

[C][3]

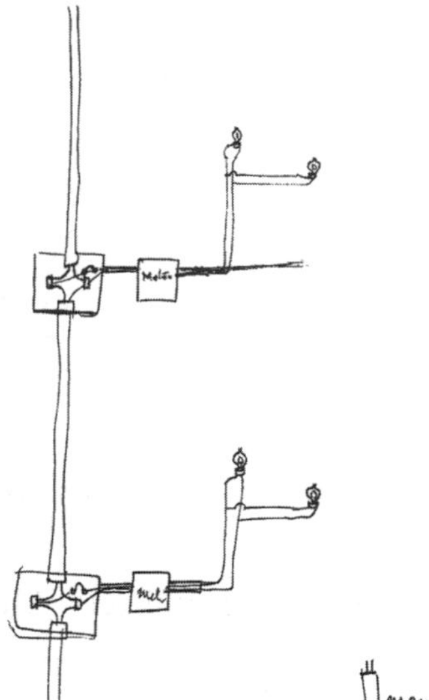

Case No. 324[a]
Using Gas pipes in House only[4]

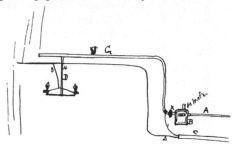

Insulating joint put ~~at~~on Gas pipe at Gas meter   C is the Electric ma[i]n G the gas pipe—

Long bulb sort can be cut off & used many times[5]
325[b]

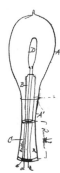

X is Rubber   Moulded in polished mould [hence?][b] slight taper   Heat radiated   Temp not enough to affect [mo--][6b]
329~~2~~[a]   322[a]
Fig 1

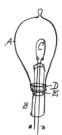

Fig 2

platina seal   Soldering two platinum rings together=[7]

~~Put in~~   Taken apart by by heat or acid (nitric acid)
[D][8]

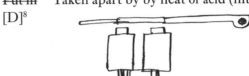

[E][9]

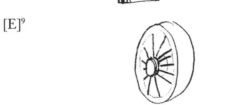

Disc Dynamo— this A   Field—[10]

[F][11]

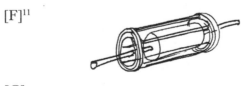

[G]

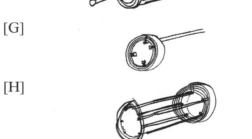

[H]

[I]

[J]

this
[K]

[L]

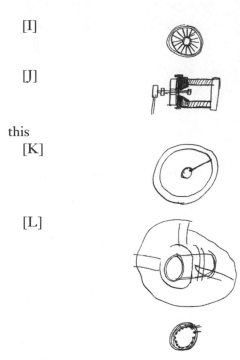

Thisᶜ way in reg machine for connecting bars by rings instead of plates

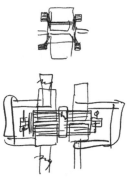

disc principle applied to Reg mach   band in middle but only 1 Commutator[12]

Witness   S. D Mott   TAE

X, NjWOE, Lab., Cat. 1147 (*TAED* NM016:66, 68, 72, 69, 73–74, 67, 70–71; *TAEM* 44:290, 292, 296, 293, 297–98, 291, 294–95). Document multiply signed and dated; miscellaneous numerals and doodles not transcribed. ᵃFollowed by dividing mark. ᵇIllegible. ᶜObscured overwritten text.

1. This document consists of loose pages pasted into a laboratory scrapbook, not necessarily in the order in which they were created. In assembling this document, the editors have somewhat altered the sequence of pages to provide greater internal coherence. For convenience

and the sake of clarity, they have also assigned letter designations to figures without substantial accompanying text.

2. This figure illustrates the principle of using functional gas pipes within a house as electrical conductors for one side of a circuit. Edison executed a patent application (Case 324) for this arrangement on 21 May, in which he specified the need for an insulated joint at the meter so that the current would not flow through the meter to ground. U.S. Pat. 251,551.

3. Figure labels are "Meter," "meter," and (lower right) "main."

4. This drawing closely resembles the one accompanying Edison's U.S. Pat. 251,551. B at the lower right is labeled "gas meter"; the insulated joint is x.

5. Text is "seal here." On this day Edison executed a patent application for an electric lamp that could be taken apart, repaired with a new filament, and "again exhausted and sealed at a less cost than the first expense of manufacture." This was accomplished by making the filament support, or inner glass tube, "with a sufficiently large ground surface with which fits closely a ground surface of similar size on the globe or bulb, one or both of the surfaces being, if necessary, first covered with a viscous substance, which fills the interstices and makes an air-tight joint. The lamp is exhausted, as usual, and the pressure of the atmosphere holds the surfaces together, the viscid substance requiring no packing to hold it in place." The patent illustrated three arrangements to accomplish this, one of which is substantially that shown in the first drawing here. One of these was probably like the "ground glass Lamp" that Edison received from the lamp factory on 16 May. U.S. Pat. 341,644; Cat. 1147, Lab. (*TAED* NM016:42; *TAED* 44:266).

6. Text at bottom is "c," "a," "b," at lower left "rubber"; at lower right "Rubber" overwritten by "this." c is the knob or enlargement at the bottom of the elongated globe; C is a tapered rubber stopper. On 24 May Edison executed a patent application (Case 325) for a lamp "adapted to be taken apart when the carbon is destroyed, and all the parts except the carbon filament again utilized." The lamp had an elongated globe with a slight taper. The lead-in wires were passed through a rubber stopper; as the lamp was evacuated, atmospheric pressure would "force the rubber stopper into the globe, and this pressure . . . serves to make the stopper close tight around the wire-support, and also to hug close to the surface of the globe-neck." The drawing in the resultant patent is essentially the same as in this document. U.S. Pat. 251,543.

7. On 21 May Edison executed a patent application (Case 322) for a lamp that could be taken apart and reassembled with a new carbon. The lamp was sealed by soldering together platinum rings around the circumference of the globe and inner support tube. The drawings accompanying the resultant specification are essentially the same as those in this document. U.S. Pat. 264,657.

8. The purpose of this magnet and armature is unclear.

9. Edison had recently resumed work on the disk dynamo. This sketch is probably related to the connections between each pair of radial induction segments and the concentric bands around the armature's perimeter through which the current was collected. Edison's first patent application for a disk machine, which largely pertained to the armature construction, specified that these connections were to be made by metal

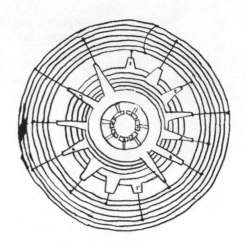

Edison's 21 May sketch for draftsman Samuel Mott's use in preparing the first disk dynamo patent application. The published patent drawing shows each radial induction bar with its own discrete connector to the peripheral concentric bands, although the specification states that a construction with forked bars, like that shown here, would be stronger.

straps for greater strength. Cat. 1147; N-81-05-21:2–10, 13–25, 112; both Lab. (*TAED* NM016:39–40, 55; N201:2–11, 17; *TAEM* 44:263–64, 279; 40:456–65, 471); U.S. Pat. 263,150.

10. The arrangement of the field magnets and their hollowed-out pole pieces in this drawing of the entire machine resembles that shown in Edison's first disk dynamo patent application. U.S. Pat. 263,150.

11. Drawings F, G, and H show the connections between induction bar pairs made through a series of concentric copper rings at each end. On 21 May and 4 June Edison instructed Samuel Mott to make drawings for a patent application, which he executed on 25 June. The previous arrangement was prone to heating (see Doc. 2122). Cat. 1147, Lab. (*TAED* NM016:77, 89; *TAEM* 44:301, 313); U.S. Pat. 251,537.

12. Text is "n[o]g[ood]," "band," and "ng." It is unclear how this machine would work.

–2103–

To Theodore Puskas

New York [May 23, 1881?][1]

Puskas[2] Paris=

Obtain from Herz[3] in your own name unqualified but not exclusive license[a] to use Gramme machines this country for three years fifteen per cent royalty by giving one thousand dollars advance royalty[4]   assign to me personally   important answer=[5]

Edison=

TD (telegram), HuBPo, TP (*TAED* Z400CA). Typed in upper case.
[a]Mistyped.

1. This conjecture is based on a docket notation, presumably by an archivist.

2. Hungarian promoter and inventor Theodore Puskas had been involved with promoting Edison's inventions in Europe since 1877. Docs. 1153 nn. 4–5, 1498, 1731, 1736, and *TAEB* 5 chaps. 3–6, passim.

3. Cornelius Herz was a promoter and apparent confidence man, reportedly a doctor and former member of the San Francisco Board of Health. He held the rights to Edison's quadruplex telegraph in several European countries, and later became editor of *La Lumière Électrique*. Herz was living in Paris at this time, from where Charles Batchelor reported he was "pretty well thought of here but you can judge very little as I dont think he has any money himself but simply pulls the ropes for some syndicate." See Docs. 1172 n. 16 and 1725; George Prescott to TAE, 29 June 1881; Batchelor to TAE, 9 Aug. 1881; both DF (*TAED* D8135ZAJ, D8135R; *TAEM* 58:904, 961).

4. Edison asked Puskas at the end of March to contact Herz, who had purchased a license to Zénobe Gramme's U.S. Patent 120,057 for generators on the basis of a 10 percent annual royalty. One of Edison's sources indicated that Herz had forfeited the license but others contradicted this. Edison wanted Puskas "to obtain through some outside party whom you may select, a control over this license, or, at least, obtain a sub-license, for the life of the patent. It will not do for either you or myself to be known in connection with it but perhaps you can select some reliable party outside . . . to ascertain upon what terms he can secure from Herz either the whole license, or a sub-license to make and sell the Gramme Magneto Electric Machine in the United States." On Edison's advice, the Edison Electric Light Co. declined to purchase the Gramme patent in 1879. In April 1881 the Gramme Electrical Co. was formed by the Brush, Fuller, and United States (Maxim) companies, and this firm purchased the Gramme patent rights. The Edison Electric Light Co. declined to join the Gramme Electrical Co. at that time but did so in March 1882. An 1883 court ruling in an unrelated infringement suit held that Gramme's U.S. patent had expired at the end of 1880 with the lapse of a related Austrian patent. TAE to Puskas, 30 Mar. 1881, TP (*TAED* Z400BZ); Gramme Electrical Co. circular, 24 Apr. 1882; "Electric Light Monopoly," *New York Times*, 27 Apr. 1882; both Cat. 1327, items 182, 184; Batchelor (*TAED* MBSB52182, MBSB52184; *TAEM* 95:185, 187); Edison Electric Light Co. Bulletin 9, 15 May 1882, CR (*TAED* CB009; *TAEM* 96:706); Passer 1953, 41.

5. Edison inquired at the end of May what royalty Herz might want but Puskas replied, "no result with Herz yet." TAE to Puskas, 31 May 1881; Puskas to TAE, 31 May 1881; both DF (*TAED* D8120ZAO, D8120ZAP; *TAEM* 57:617, 619).

–2104–

To Drexel, Morgan & Co.

[New York,] 27th May [188]1

Dear Sirs,

Referring to the conversation you had with Mr Insull this morning I beg to state that since the order for the Electric Light Plant for England (to cost $5000) was given me[1] I have increased the size of my Engines and Dynamos so that instead of giving from 700 to 800 Lights as given by the one at Menlo Park they will probably give about 1200 Lights and hence the cost of same has been materially increased   In addition to this as I understand that the necessity for the machine being sent immediately is urgent I am sparing no expense to produce it by the required time: have men now working night and day and am obliged to pay double wages for night work.[2] I have also ordered a much better class of chandelier in order to make it in every respect a first class exhibition—much better than was allowed for in my estimate to you. Under these circumstances I do not think the outfit will cost less than $8000. However whatever may be [there?][a] shall be no profit whatever to me. I have been building this plant with my own money and would not on any account have called on you for an advance were it not that I am entirely out of funds. I do not think that I shall need to call on you for the balance until everything is completed as I have made arrangements to borrow money which will probably carry me over until machinery is done[3]   Yours truly

L (letterpress copy), NjWOE, Lbk. 8:276 (*TAED* LB008276; *TAEM* 80:942). Written by Samuel Insull. [a]Illegible.

  1. The order for this demonstration plant is Doc. 2049.
  2. The immediacy was conveyed at least in part by George Gouraud, who wanted to begin planning for manufacturing plants in England like those Edison had established in the United States. He advocated this as the only way to "occupy the field in advance of others," especially Joseph Swan, noting that "Already people have begun to speak here of Swan's light and Edison's light, instead of Edison's light and Swan's light if there be such a thing as the latter." A few days later Gouraud cabled that Swan was "pushing every where" and that Hiram Maxim had just arrived with his incandescent lamp. He advised that Charles Batchelor be directed to set up a few Edison lamps in public view. Edison did not permit Gouraud to make a small exhibition of lamps, but did allow him to advertise in his name for potential investors. Gouraud to TAE, 26 and 30 Apr. 1881; Drexel, Morgan & Co. to TAE, 28 May 1881; all DF (*TAED* D8133K, D8133M, D8133R; *TAEM* 58:608, 611, 619).
  3. Probably the advance arranged by Gouraud on the settlement of Edison's British telephone interest. See Doc. 2086 n. 2.

–2105–

Charles Batchelor to Theodore Puskas

[New York,] May 31st, [188]1.

Dear Sir:
   On my return I informed Mr. Edison of the conversations I had with the chief of installation, Mr. Breguet,[1] about obtaining the best locations in the Exposition building, for his exhibits. He decided to select the two rooms on the gallery to the right of the staircase, which will give us a space of 72 feet square, to be filled with a most elaborate collection of his inventions.[2]

Mr. Edison addressed a letter to the Assistant Secretary of State,[3] making a requisition for this, and also asking the privilege to light the space in the gallery[a] occupied by the central room 72 × 72 feet, immediately at head of the staircase; the six rooms to the right, each 72 × 36 feet, and the six to the left of same size; also the staircase down to the doorway, and a few lamp-posts immediately outside. He will ~~ahev~~ have 1,000 lights to distribute in this area. He also enclosed a diagram of the Exposition Building and asked for a space 12 × 18 feet, with passage-way round on the ground floor, in which to operate his steam dynamo.

The Ass't. Sec'y of State replied he will make every effort to secure it.

Will you confer with Mr. Breguet about this as soon as possible, and inform us by cable of the consummation of the matter?

At the same time, please let your Secretary accurately design the space allotted to us, and mail a plan as soon as practicable.[4]

In the interim, will you be kind enough to exercise your taste in selecting about six elegant lamp-posts, to be placed in front of the Palais de l'Industrie, if we are permitted to do so?

Each will contain a group of six lights. As it will produce, perhaps, the first impression of our capacity for lighting—and as first impressions are most lasting—you see how much will devolve upon you.

I think every thing has wonderfully improved during my absence.[5] The electric light is absolutely perfect. The lamps in our offices have been burning since they started, five and one half hours a night, for two months and not one lamp has broken.

Edison is well, and just as full of great ideas as ever. Truly, yours,

TL (carbon copy), NjWOE, DF (*TAED* D8135H; *TAEM* 58:882). [a]Mistyped.

1. Antoine Breguet (1851–1882) was an electrician and instrument maker who served as director of installation for the Exposition Internationale de l'Électricité in Paris. Doc. 2045 n. 1.

2. A floor plan and list of items in Edison's exhibit, taken from the Exposition catalog, is in Edison brochure, "Exposition Internationale D'Électricité," 1881, DF (*TAED* D8135ZDC; *TAEM* 58:1154); see also Doc. 2111.

3. John Hay (1838–1905), diplomat and author, was Assistant Secretary of State from 1878 until shortly after President Garfield's death in

September 1881. Neither Edison's letter to Hay nor the reply has been located. *ANB*, s.v. "Hay, John Milton."

4. See Doc. 2114.

5. Batchelor sailed to England on 5 February 1881 and was in his native Manchester in April. This was evidently a rare working vacation for Batchelor, whose health had "given out" from the exertion of starting the lamp factory, resulting in Francis Upton becoming superintendent at the beginning of 1881. Batchelor began his return voyage to New York on 12 May. Thomas Carnelley to Batchelor, 25 Apr. 1881, Batchelor (*TAED* MB059; *TAEM* 92:400); "Saloon Passenger List," Cat. 1241, item 1575, Batchelor (*TAED* MBSB21573; *TAEM* 94:628); Jehl 1937–41, 813; TAE to Frank McLaughlin, 3 May 1881; Lbk. 8:226 (*TAED* LB008226; *TAEM* 80:914).

–2106–

From Calvin Goddard

New York, June 3 1881[a]

Dear Sir

The Lamp Company has called upon us to approve numerous orders for lamps heretofore drawn upon it by yourself, Mr Johnson and Dr Moses[1] and has requested that hereafter all orders shall be approved by this Company, under the terms of the contract between the Lamp Co and this Company.[2] This is quite right, and I shall at once prepare the requisite blanks, and if you will in future send your orders to me, I will see that they are sent to the Lamp Co in proper form and will thus be able to keep a record in this office, which is absolutely necessary to the proper transaction of the business

Be good enough in each case to name the purpose for which the lamps are required, that we may know how to dispose of the bills   Yours truly

C. Goddard   Secy[3]

LS, NjWOE, DF (*TAED* D8126Z; *TAEM* 58:34). Letterhead of Edison Electric Light Co. [a]"New York," and "188" preprinted.

1. Otto Moses (1846–1905) was an analytical chemist whom Edison hired in 1879. Edison had recently appointed him to assist Charles Batchelor at the International Exposition in Paris. See Doc. 1754 n. 6 and *TAEB* 5 App. 2; TAE to Georges Berger, 3 May 1881, DF (*TAED* D8135E; *TAEM* 58:877).

2. This contract, ratified on 8 March, governed relations between the Edison Electric Light Co. ("this Company") and the Edison Lamp Co. It stipulated that "In all business connected with the supplying of lamps the company will act as the agent both of the licensee and of the said Edison. No order shall be received by the said Edison directly from any licensee; all orders shall pass through the company, and all drafts for payment shall be made upon the licensee through the company as such agent; all shipments and deliveries shall be made only under the direc-

tion and control of the company." TAE agreement with Edison Electric Light Co., 8 Mar. 1881, Defendant's depositions and exhibits, 5:2352–57, *Edison Electric Light Co. v. U.S. Electric Lighting Co.*, Lit. (*TAED* QD012E:139–41; *TAEM* 47:999–1001); see also Doc. 2039.

3. Calvin Goddard (d. 1892) was secretary of the Edison Electric Light Co. from its inception and held the same position with the Edison Electric Illuminating Co. of New York. See Doc. 1535; Obituary, *New York Times*, 11 Apr. 1892, 5.

–2107–

*Samuel Insull to
George Gouraud*

New York, June 3rd. 1881.

My Dear Colonel:—

I duly received your kind favor of May 21st. which reminded me that I had abstained from writing to you for almost a month. My apology for this must be the great press of business that we have to deal with, but for the future I hope to be able to send you a letter weekly as I promised.[1]

English Telephone.

I have read your letter addressed to Mr. Edison dated May 21st. In it you seem somewhat to have mixed up my letter of the 29th. of April with his letter sent somewhere about that date. This rather confused him and he required some explanation from me as to what I had been writing to you. I would prefer my letters to you not being quoted in any letters addressed to him.[2] I am very glad that the Construction Company met with so great a success and trust that the dividends it will earn will, to some extent, justify the enormous premium at which the shares now stand.[3] If you can hurry the settlement of the Edison reversion in the London Company any you will confer a favor on Mr. Edison as he would no doubt be very glad to receive the funds due from that quarter.[4] I suppose, however, like all other matters that get into the English law courts, will require some considerable time as our legal authorities in the old country seem to aim at the point of "how not to do a thing." Your letter with enclosures relating to the Glasgow Company has just arrived. You will no doubt hear from Mr. Edison on this matter in the course of a mail or so.[5]

Oriental Company.

If I have time I shall write you an official letter to night explaining the status of the Le Gendre matter, but I may here state that Mr. Edison considers the Le Gendre contract null and void as he has done nothing whatever to push Telephone matters although he has had a contract for about a year. Mr. Edison is by no means afraid of Le Gendre troubling us any, but you may rely on Edison doing his very best in the matter. You may probably have thought that we quite overlooked this matter as you heard so little from us for so long a time, but all that time Mr. Edison was doing his very best to straighten the matter out, and I myself was also occupied considerably in attending to this matter. You must not think that simply because you do not hear from me that he and myself have entirely overlooked the matter.[6]

Electric Light, England.

You wrote Mr. Edison some time back stating what your ideas were as to how the Light Company should be completed in England, namely, that the whole of the stock should be issued; that money for actual work done should be obtained by the issue of debentures.[7] Mr. Edison did not look upon this proposition with favor at all. He seems very adverse to overloading an[a] Electric Light Company by issuing stock in this manner. The fight with the Gas companies will be quite hard enough even if we obtained money on the stock itself, but how it would be if the thing was loaded down by issuing stock working on debentures it is hard to say. The plant for England will be of the very best. The Steam Dynamo which is to be sent will be capable of running no less than twelve hundred lights, and the outfit will in every part be of the most finished character. The workmen at Edison's machine works are working night and day to get this out, so that not a moment is being lost. Mr. Edison is highly impressed with the great importance of meeting his opponents in England by an exhibition of his light at the earliest possible moment. It is practically decided as to Johnson's going to England. He will in all probability give his undivided attention to the preparation of the plant from next week forward, and you well know that when Johnson[a] takes up a thing not a stone will be left unturned in order to get it out by the required time. The exhibit in London will no doubt be a credit to the name of Edison.

The Electric Congress, Paris.[8]

As I wrote you some time back Mr. Edison is thoroughly aroused to the necessity of making the best possible exhibit at Paris. He will have no less than ninety two groups of Exhibits besides his Electric Light. The Automatic, quadruplex, stock reporter, telephone, phonograph, electric pen, in fact every thing so far as it is possible that he has brought out will be exhibited at Paris.[9a] I have seen a good many of the exhibits which are already finished. They are got up in extremely nice style, and I think that the English and other inventors will have a hard job to show anything to equal it. I know that you have for a very long time past wished to see something of this character on the other side of the Atlantic so that justice might be done to Mr. Edison and his inventions, and, unless the electrical public are very greatly prejudiced against him, they cannot but be convinced of the practicability and universality of his inventions. The Electric Light outfit[a] for Paris will be equally complete as the one just going to London. In fact it will be an exact counter part of it so that at one and the same time London and Paris will be able to see Mr. Edison's light in its perfection.[a]

I will write you fully as to other matters by next week's mail.[10] Very faithfully yours,

TL (carbon copy), NjWOE, DF (*TAED* D8104ZBK; *TAEM* 57:133). [a]Mistyped.

1. Gouraud had apologized for failing to acknowledge Insull's recent letters, adding: "Don't think I value your letters the less because of this omission. The correspondence must necessarily be one-sided, as I have so little of interest to write as compared with what you have." This document is the first extant typed letter from Insull, a form used for many subsequent letters to Gouraud. He later explained that he did not use the typewriter himself but instead dictated to an unidentified assistant. Gouraud to Insull, 21 May 1881; Insull to Gouraud, 8 July 1881; both DF (*TAED* D8149T, D8104ZBW; *TAEM* 59:960, 57:162).

2. Insull's 29 April letter has not been found. Gouraud's 21 May letter to Edison concerned the misunderstanding regarding his disposal of Edison's stake in the United Telephone Co. (see Docs. 2079 and 2086). Gouraud quoted a passage from Edison's 24 April letter concerning reports of changes in the price of the company's stock. He then incorrectly attributed to Edison's letter an explanation, to which he objected at length, of the causes of this fluctuation. Gouraud to TAE, 21 May 1881, DF (*TAED* D8149R; *TAEM* 59:955); TAE to Gouraud, 24 Apr. 1881, Lbk. 8:208 (*TAED* LB008208; *TAEM* 80:906).

3. Gouraud cabled Edison "Construction brilliant success" on 12 April, shortly after the capitalization of the Consolidated Telephone Construction and Maintenance Co. According to a news clipping, its shares stood at 50 percent over par value at the end of April. Gouraud had enrolled Edison for 250 shares, some of which he promptly contracted to sell. Gouraud to TAE, 12 Apr. 1881, LM 1:9B (*TAED* LM001009B; *TAEM* 83:876); *Money Market Review*, 30 Apr. 1881, enclosure with Gouraud and Edward Lane to TAE, 3 May 1881; Gouraud to TAE, 21 May 1881; both DF (*TAED* D8149Q, D8149S; *TAEM* 59:953, 959).

4. Edison's reversionary interest was his contractual stake in the future value of the Edison Telephone Co. of London. Disagreement over how to settle this interest delayed merger of the company into the United Telephone Co. in 1880, and disputes over its ultimate value persisted into 1881; Edison had instructed Gouraud to liquidate the account in March. See Docs. 1933 esp. n. 2, 1942, 1954, 2046, and 2060.

5. Gouraud wrote on 24 May enclosing a statement of terms for settling Edison's royalties from the Edison Telephone Co. of Glasgow. He also enclosed a check for $1,676.50 for this purpose ($10,000 having already been advanced). Edison replied on 5 June, objecting to deductions for commissions to Michael Moore and a business partner that he thought were Gouraud's responsibility. Gouraud to TAE, 24 May 1881, enclosing statement of Glasgow royalties, 24 May 1881; both DF (*TAED* D8149U, D8105ZZA [image 4]; *TAEM* 59:962, 57:345); TAE to Gouraud, 5 June 1881, Lbk. 8:306 (*TAED* LB008306; *TAEM* 80:953).

6. William Le Gendre was connected with a New York bank that advertised extensive foreign business. In May 1880 Edison assigned him the right to sell telephones in Japan, China, Korea, and Hong Kong, but by March 1881 he considered the contract nullified by Le Gendre's failure to transact any business. The Oriental Telephone Co. was organized in early 1881 to exploit the telephone inventions of Edison and Alexander Graham Bell in portions of Asia, Australia, and southeastern Europe (see Docs. 1794 and 2056). Although an exception had been made in the new company's Japanese rights, in April Gouraud urged Edison to repudiate the Le Gendre agreement entirely. His correspondence on this matter became more urgent in May, after he learned that one Joseph Morris had approached the Oriental Co. in Hong Kong claiming to be Edison's agent under authority of that contract. Nothing is known of Morris except that he reportedly enjoyed the favor of the governor of Hong Kong (Gouraud to TAE, 5 Apr. and 25 May 1881; W. G. Hall to Gouraud, 26 July 1881; all DF [*TAED* D8150M, D8150Y, D8150ZAN; *TAEM* 59:1013, 1028, 1050]). The dispute over Le Gendre's rights dragged on for months and was one of several that delayed implementation of the Oriental Co.'s contracts with Edison and Bell; it is not clear how the issue was resolved. See correspondence from Le Gendre and Gouraud, and between Gouraud and W. G. Hall, a company agent in Hong Kong, in Telephone—Foreign—U.K.—Oriental Telephone Co. (D-81-50), DF (*TAED* D8150; *TAEM* 59:988); and from Gouraud in LM 1 (*TAED* LM001; *TAEM* 83:872).

7. Gouraud wrote on 26 April that he hoped to counter the formation of Joseph Swan's company by organizing (with Drexel, Morgan & Co.'s approval) "an Edison Company for each of the principal towns in order to enable me to interest leading local people, still keeping control ourselves. I propose to issue the entire capital fully paid and to issue Debentures to represent the working capital giving as much of the fully paid stock with the Debentures as may be necessary to raise the capital." Gouraud to TAE, 26 Apr. 1881, DF (*TAED* D8133K; *TAEM* 58:608).

8. Although Insull conflated it here with the International Exposition, the International Congress of Electricians was a separate, concurrent event. The Congress considered for the first time the international adoption of electrical standards. The body accepted, at least on an interim basis, the British Association's recommended system of units based on the ampere, volt, and ohm. Beauchamp 1997, 160−61; Tunbridge 1992, 34−40.

9. A complete list is in Doc. 2111.

10. See Doc. 2112.

–2108–

Samuel Insull to Hamilton Twombly[1]

[New York,] 6th June [188]1

Dear Sir,

There was a slight mishap with the machine, which was to be sent to Mr Vanderbilt's house, on Saturday.[2] It will probably be put right by tomorrow or next day when I will advise you as to when Mr Edison will make the proposed exhibit.[3]

Mr Edison has ordered the men to run wires into the Dome.

The lamps have already arrived at the house   Yours truly
Saml Insull   Sec

ALS, NjWOE, Lbk. 8:313 (*TAED* LBoo8313; *TAEM* 80:959).

1. Hamilton McKown Twombly (1849–1910) was an investor and manager in many companies, including Western Union; he was also an incorporator of the Edison Electric Light Co. Docs. 1308 n. 8, 1467, 1490.

2. William H. Vanderbilt (1821–1885), railroad tycoon and financier, was among the original investors in the Edison Electric Light Co. (*ANB*, s.v. "Vanderbilt, William Henry"; Doc. 1497). At this time he was building an ornate residence at 640 Fifth Ave., at the corner of Fifty-first St. in a somewhat disreputable block. Known as the "Triple Palace," it was to provide separate homes for himself and his wife and for two married daughters. The interior was designed by Christian Herter as a showcase for Vanderbilt's extensive art collection (Foreman and Stimson 1991, 309–23). The "Dome" referred to below has not been identified but may have been the high vaulted ceiling in the art gallery. Edison recalled years later that Vanderbilt entrusted the electric light arrangements to Twombly; planning for the installation began at Menlo Park in late 1880 (App. 1.B.14; *TAEB* 5, chap. 8 introduction).

3. Edison planned to make an exhibition of the lights in mid-June, but nothing is known of the event. The following spring, Vanderbilt ordered the plant disconnected before the installation was complete because his wife objected to the machinery in the basement; there was also a small fire, which Edison attributed to temporary wiring. Samuel Insull to Twombly, 12 June 1881, Lbk. 8:300 (*TAED* LBoo8300A; *TAEM* 80:951); "The Doom of Gas," *Boston Herald*, [May 1882], Cat. 1016, Scraps. (*TAED* SMo16B; *TAEM* 24:86); App. 1.B.14.

–2109–

Edward Nichols Memorandum: Incandescent Lamp

Menlo Park, N.J., June 8 1881[a]

Memorandum—[1] On the value of the single double sealed tips—for lamps—

One hundred of each kind, under Order No's 297 and 298 respectively: were tested for Volts &c. at 16 candles, April 28.; laid aside for more than a month and retested June 6.—[2]

Most of the lamps remained unchanged. One lamp of each

kind lost vacuum completely. Three of each kind showed a decided loss of vacuum—(causing a change of E.M.F of over 10 Volts.)

There were no other lamps which had shown any decided change.—

From this test no advantage of the double sealed tips is apparant. Before deciding that the single seal is perfectly satisfactory, another careful test would be necessary.[3]

E. L. Nichols

ADS, NjWOE, DF (*TAED* D8123ZCI; *TAEM* 57:907). Letterhead of Edison Electric Lamp Co. [a]"Menlo Park, N.J.," and "188" preprinted.

1. Francis Upton sent this document to Edison the next day without comment. Upton to TAE, 9 June 1881, DF (*TAED* D8123ZCH; *TAEM* 57:905).

2. John Howell recorded these orders in a lamp factory notebook on 22 April. The first was for "One Hundred (100) Lamps reg. A. size carbons finished without additional glass tip to be carefully tested and laid aside for (6) six months to test vacuums." The second order was the same except the lamps were to be "finished with additional glass tips." An entry in another notebook at the factory, possibly made by Edward Acheson, shows nearly identical mean results of electrical measurements on these lots (Cat. 1301 [orders 297–98], Batchelor [*TAED* MBN007:25; *TAEM* 91:318]; Menlo Park Notebook #144:30, Hendershott [*TAED* B015:15]). The additional glass tips were presumably the small blobs of molten glass applied where the lamps were sealed off the pump. Edison adopted this practice in January 1880 after experiencing trouble maintaining the vacuum in laboratory lamps (see Doc. 1887 n. 3).

3. There is no record of such a test, and it is not known if the factory altered its practice.

–2110–

From George Barker

Philadelphia June 10. 1881.

My dear Edison:—

I am very much obliged for your letter of the 2d and am very glad to have the report I heard, positively contradicted. I have waited before replying to see how I was coming out financially. I have made my collections pretty closely and I believe I shall be able now to get through without putting any of my friends to inconvenience. I feel very much indebted to you of course for your generous offer. At one time it looked as if I should have to ask some assistance from my friends in the way of a loan; and then it was that I wrote to you.[1] But I think you will appreciate the advantage of being independent in such a matter as this. I believe now I have money enough for Mrs. Barker and myself to go on and spend the summer pleasantly. You do not need to be told that I shall be glad to do for you in every way all that I can in Paris. Bachelor I know will give me every facility for showing your exhibits and proving to the scientific men there the high character of your inventions. If there are any special points you wish me to make, please send me a memorandum of them.

I was in hopes that I could have lighted my house with the Edison light before I went to the Electrical Congress because I desired to say that my house here had been so lighted. But no wire has yet come for the conductors. And no switches, lamps, chandeliers, sockets, or fixtures of any kind have appeared.[2] Since I saw you I have been over the route more carefully and I think that by the way I shall have to go, it will take not less than 2000 feet of wire. I know however that you are especially busy just at this time getting your splendid collection of exhibits ready for Paris. So I am anxious not to have you put yourself to inconvenience on my account. If however you can send the wire, the men are here ready to put it up.

It may be that I shall be in New York on Tuesday next and if so, I shall call to see you. I will let you know at what time to expect me.

I hear our friend Robinson is talking of casting his lot with you.[3] I should be glad if he could make himself so useful to you as to secure for himself a permanent position.

Many thanks for the Latimer Clark cells you ordered made for me.[4] I shall be glad to have them to standardize my condenser with.

If I do not see you again you have my best wishes for a pleasant summer. Cordially yours

George F. Barker.

ALS, NjWOE, DF (*TAED* D8135J; *TAEM* 58:888).

1. The editors have not identified the "report" to which Barker referred. Edison replied on 2 June to Barker's 20 May letter (not found) that he was "absolutely certain that no officer of the Edison Electric Light Co made any such statement as you heard ascribed to him. I presume you must have heard the remark at the United States [Electric Lighting] Companys offices & if so it was probably made with the sole object of seeing what you would say   Anyway the statement that any of our officers made such a remark is without foundation   If you do not desire to accept the proposition I made you of course I would not press it upon you but as I have already stated to you I do not wish you to do anything for me that is not strictly in accordance with the truth." The proposition referred to was presumably Edison's prior offer to "fix your Parisian trip to your satisfaction." TAE to Barker, 2 June and 29 Apr. 1881, Lbk. 8:294, 225 (*TAED* LB008294, LB008225; *TAEM* 80:949, 912).

2. Edison replied that the wire should already have reached Barker's house but it had been ordered again. He also indicated that the installation was unlikely to be completed before Barker left, unless the lamps were hung temporarily by the wires without fixtures. TAE to Barker, 17 June 1881, Lbk. 8:325 (*TAED* LBoo8325; *TAEM* 80:961).

3. Heber Robinson managed the Western Union office in Philadelphia. He evidently applied to the Edison Electric Light Co. for the right to form an illuminating company in Camden, N.J., where he lived. Doc. 1238 n. 9; Robinson to TAE, 31 July 1883, DF (*TAED* D8340ZBW; *TAEM* 68:410).

4. The British electrician Latimer Clark devised a mercury-zinc battery which produced current at a relatively constant 1.457 volts. The Clark cell was one of several forms of so-called standard batteries used in situations requiring steady voltage over a period of time. Edison was having these cells made for Barker at Menlo Park. Dredge 1882–85, 2:9; TAE to Barker, 17 June 1881, Lbk. 8:325 (*TAED* LBoo8325; *TAEM* 80:961).

–2111–

Charles Batchelor to
Theodore Puskas

[New York,] June 10, [1881]

Dear Sir:

Our exhibit will consist of:—[1]

1. 150 Horse-power Boiler.
2. Dynamo and Engine for 1000 lights.[2]
1. Small dynamo.[3]
1. 5 Horse-power motor with countershaft.
1. Sewing machine motor, and sewing machine.[4]
1. Motor driving pump.
1. Motor and fan.[5]
1. New principle dynamo.[6]
1. Main current regulator and appliances.[7]
1. Set of apparatus for measuring resistance of lamps.
1. Set apparatus for measuring[a] economy of lamps.
1. Photometer[a] and appliances.[8]

Lamps.[9]

200. 3 light chandeliers.
100. 2 light chandeliers.
50. 1 light chandeliers.
50. Swing brackets for lamps.
100. Rigid brackets for lamps.
12. Regulator meters for the light.[10]
1. Meter by deposit.[11]
Safety catches etc.[12]

1. Private line Telegraphic printing system.
1. Stock quotation printing telegraph system.
1. Automatic fac-simile telegraph system.[13]
2. Systems of Domestic Telegraphy.[14]
1. Quadruplex system of telegraphy.
2. Duplex systems of telegraphy.[15]
1. Automatic chemical Morse system of telegraphy.
1. Automatic   "   Roman letter system of telegraphy.[16]
1. Motograph Telephone system.[17]
1. Combination Telephone system.[18]
1. Phonographic-Telephonic system.[19]
1. Musical Telephone system.[20]

Samples of all pipes and service boxes used on the Edison system.

1. Mirror Webermeter.[21]
1. Tasimeter and appliances.
1. Odorscope and appliances.[22]
1. Resistance box for strong currents.
1. Thermo Galvanometer.[23]
1. Current regulator, (fluid-bridge.)[24]
1. Set Leyden jars with high vacuum.
1. Carbon resistance box.
1. Telegraphic carbon strip Relay.[25]
1.   "   Motograph Relay.[26]
1.   "   pressure Relay.[27]
       "   expansion Relay.[28]
1. Telephone repeater[29]
1. Telephone switch-board.
1. Set Electric Pens and duplicating presses.[30]
1. Chalk battery.
1. Condenser in vacua.
1. Segar lighter.[31]
1. Motograph gong.[32]
1. Set apparatus for illustrating "Etheric Force."[33]
1. Magnetic motograph.[34]

A number of pieces of apparatus illustrating the different methods of increasing and decreasing the resistance of a closed circuit by carbon contact and illustrating the progress of the Edison telephone transmitter as universally used.

Cases of samples of Bamboos, Carbonized articles, processes of manufacture, etc etc etc.[35]

Photographs,—Books, etc etc etc.[36] Very respectfully yours,

L (typed transcript), NjWOE, DF (*TAED* D8135I; *TAEM* 58:884). Third page typed on letterhead of T. A. Edison. ᵃMistyped.

1. Edison sketched a number of items from this list in a notebook (N-81-05-23, Lab. [*TAED* N187:57–66; *TAEM* 40:201–10]). During May, Charles Batchelor compiled a separate, more extensive illustrated list of possible items for Edison's exhibit; some of his entries are noted below (N-81-01-00:121–39, Lab. [*TAED* N212; *TAEM* 40:774]). Batchelor's list of "what I still want   July 18th 1881," just before his departure, is in DF (*TAED* D8135Z; *TAEM* 58:928).

2. This was the C dynamo; see headnote, Doc. 2122.

3. Contemporary illustrations indicate this was a Z dynamo. Authoritative accounts of the Exposition mentioned this machine while giving detailed descriptions of the much larger Jumbo, creating a tendency for subsequent writers to conflate the two. Batchelor to TAE, 30 July 1881, DF (*TAED* D8135ZAE; *TAEM* 58:945); du Moncel 1881, 4920; Heap 1884, 59–61.

4. Edison developed a sewing machine motor run by generator current in 1879 and exhibited it at Menlo Park that December. See Docs. 1800 and 1866.

5. Batchelor's 23 May sketch of an "Electric fan to work on the system" shows an artificial human hand which a motor would wave back and forth. N-81-05-23:23, Lab. (*TAED* N212:12; *TAEM* 40:785).

6. This was a small model of the disk dynamo. It was not completed until early 1882 (see Doc. 2228), although a large version was constructed in May 1881 (see Docs. 2082 esp. n. 1, and 2150).

7. For Edison's work on dynamo regulation see headnote, Doc. 2036.

8. Edison sent several pieces of apparatus for measuring lamp resistance, economy, and intensity, including two calorimeters for testing lamp economy and a photometer for measuring luminous intensity (N-81-05-23:9, Lab. [*TAED* N212:5; *TAEM* 40:778]). An illustration and description of Edison's testing bench at the exhibition appeared in du Moncel 1881, 4921.

9. A list of lamp fixtures and chandeliers sent to Paris is in N-81-05-23:220–22, Lab. (*TAED* N212:98–99; *TAEM* 40:871–72). For contemporary illustrated descriptions of Edison chandeliers and brackets see Catalogue and Price List of Edison Light Fixtures, 1883, PPC (*TAED* CA002C; *TAEM* 96:185).

10. Apparently a transcription error; Batchelor specified in his note-

Edison's testing bench for electric lights.

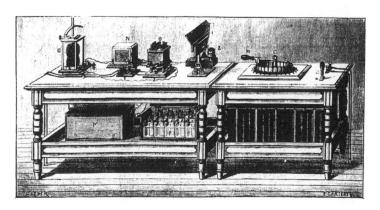

book "12 Regular Meters." N-81-05-23:155, Lab. (*TAED* N212:79; *TAEM* 40:852).

11. Edison had devised several forms of meter embodying the electro-deposition of copper or other metal on a plate; it is not known which he sent. See headnote, Doc. 2163.

12. The safety catch was a fuse consisting of a short piece of lead wire which melted to open the circuit if the current exceeded a certain value.

13. This is the facsimile telegraph that Edison had been developing with Patrick Kenny since 1878 (see Doc. 2252). Kenny apparently accompanied the instrument to Paris (Batchelor to Edison, receipt, 31 Mar. 1882, DF [*TAED* MBLB4188; *TAEM* 93:605]). For published accounts of the facsimile at the Paris Exhibition see "Télégraphe Autographique De M. Edison," *La Lumière Électrique* 5 (1881): 418–21; and "The Edison Exhibit at the Palais de Industrie," *Engineering* 32 (1881): 343.

14. In 1874 Edison formed the Domestic Telegraph Company to exploit his district telegraph system. See *TAEB* 2:121–22 and Docs. 615 (and headnote), 653 and 654.

15. Duplex telegraphs were capable of transmitting two simultaneous telegraph signals, one in each direction, on a single telegraph wire. For Edison's work on duplex telegraphy see headnote, Doc. 275 and *TAEB* 1, chap. 11, and 2, chaps. 1–5, passim.

16. For Edison's work on automatic chemical systems of telegraphy, which were capable of printing both Morse code characters and Roman letters, see *TAEB* 1 and 2, passim, and Fox 1879.

17. Edison's telephone receiver was based on his 1874 discovery of the electromotograph principle—a change in friction caused by electro-chemical decomposition. He devised experimental receivers beginning in 1877 and commercial instruments in 1879. See Docs. 873 and 888–889.

18. In the summer of 1877 Edison devised a combination telephone transmitter and electromotograph receiver. He developed a commercial version for the British market in 1879. See Docs. 962, 1681 (and headnote), 1784 (and headnote).

19. Edison's search for a means to create permanent records of telephone transmissions led to the invention of the phonograph in 1877; a few months later, he experimented with using the device to send a recorded message between two telephone stations. See Docs. 969, 1159 and 1227.

20. The musical telephone was an electromotograph receiver with a large resonating box; it was used for public demonstrations. See Docs. 889 n. 1 and 1211.

21. The mirror webermeter, a device used to measure current, was based on a galvanometer developed by Sir William Thomson. It consisted of a small mirror with a magnetized needle on its back, suspended vertically in a thin circular coil of insulated wires. Passage of a current through the wires deflected the needle, causing a beam of light reflected by the mirror to move along a horizontal graduated scale. *DSB*, s.v. "Thomson, William"; Knight 1881, s.v. "Galvanometer."

22. The odoroscope detected the presence of minute amounts of water and hydrocarbon vapor. It was similar to the tasimeter, except that it used a strip of gelatin instead of hard rubber. In an 1889 lecture to the Franklin Institute of Philadelphia, William Hammer described the odoroscope as "exceedingly sensitive to moisture, a few drops of water

thrown upon the floor in a room containing the instrument, sufficing to throw a spot of light off the scale when projected by the Thompson Galvanometer, in circuit with it. The carbon button is employed in a manner similar to the Tasimeter arrangement. Barometers, Hygrometers, and similar instruments of great delicacy can be constructed upon this principle, and it may be employed in determining the character or pressure of certain gasses and vapor" (Hammer 1889). An undated sketch appears in N-81-01-00:125, Lab. (*TAED* N187:59; *TAEM* 40:203). For other descriptions see Ministère des Postes et des Télégraphes 1881, 162; Jehl 1937–41, 188; and Dyer and Martin 1910, 590.

23. The thermogalvanometer measured heat by registering the change in resistance of a long platinum wire. Batchelor sketched designs for this in N-81-01-00:123, N-81-05-23:3, Lab. (*TAED* N187:58, N212:2; *TAEM* 40:202, 775).

24. It is unclear how this regulator worked. Edison made an undated sketch of it, specifying water and mercury, and Batchelor sketched the same device in May. N-81-01-00:121, N-81-05-23:45, Lab. (*TAED* N187:57, N212:22; *TAEM* 40:201, 795).

25. An undated sketch by Edison appears to illustrate this device. An armature actuated by an electromagnet would press on a strip of carbon, presumably affecting the local circuit. N-81-01-00:135, Lab. (*TAED* N187:64; *TAEM* 40:206).

26. Edison devised this instrument in 1874 to re-transmit signals using his electromotograph principle. It allowed greater sensitivity than that of conventional magnetic relays. The instrument sent to Paris used electromotograph chalk cylinders like those in the commercial electromotograph telephone receiver. It is illustrated in Jehl 1937–41, 65; see also Docs. 881, 913, and 1171 n. 7.

27. A telegraphic relay based on the variability of plumbago with pressure. See Docs. 885, 887, and 926.

28. The operation of this device is not clear, but it appeared to use a length of platinum wire in a circuit to open or close a contact as it expanded with heat. N-81-05-23:129, N-81-01-00:135, Lab. (*TAED* N212:64, N187:64; *TAEM* 40:837, 208).

29. Repeaters were used to overcome the attenuation of electrical signals over long distances. The instrument devised by Edison in 1877 employed a magnetic receiver on the incoming line to vibrate the diaphragm of a carbon transmitter. The transmitter varied the resistance in the primary winding of an induction coil, inducing a current in the secondary winding which passed to the outgoing line (see Doc. 1277 n. 4). Edison's undated sketch of a somewhat different instrument for Paris is in N-81-01-00:133, Lab. (*TAED* N187:63; *TAEM* 40:207).

30. The electric pen was a document duplicating system developed by Edison in 1875. The operator used a rapidly vibrating needle to prick small holes in paper, making a stencil that could produce hundreds of copies. See Doc. 721 (and hcadnote).

31. Batchelor included in his list a "Cigar lighter to be worked on the system." Edison had thought about and roughly sketched such a device in 1876 but evidently did nothing with it at that time. He promised Batchelor in September that one was almost ready, but nothing more is known of it. In October 1883, Batchelor reported experimenting in Paris on a battery-powered lighter. That fall, Edison included a wall-mounted

lighter in his exhibit at the electrical exhibition in Vienna. It had a thin asbestos disk sewn across with platinum wire; lifting the attached handle closed a circuit between the platinum wire and an unspecified power source. N-81-05-23:17, NS-76-01, Lab. (*TAED* N212:9, NS7601A3; *TAEM* 40:782, 7:347); Doc. 2150; Batchelor to Société Électrique Edison, 10 Oct. 1883, Batchelor (*TAED* MBLB3177; *TAEM* 93:458); "Electrical Cigar Lighter," *Sci. Am. Supp.* 16 (1883): 6557; see also Edison Electric Light Co. Bulletin 20 (p. 7), 31 Oct. 1883, CR (*TAED* CB020; *TAEM* 96:867).

32. Nothing is known of this device.

33. During experiments with rapidly cycling relays in the fall of 1875, Edison noticed sparking between the magnet core and armature, which he interpreted as a new "etheric" force. On 23 May 1881, Batchelor sketched an "Etherscope and Etheric apparatus." See Docs. 665–666, 668–670, 678–680; N-81-05-23:21, Lab. (*TAED* N212:11; *TAEM* 40:784).

34. Batchelor's sketch of this device shows an iron motograph cylinder on one pole of a bar magnet. A coil around the same end of the bar would vary the magnetic field according to the strength of the line current, thereby varying the attraction (and friction) between the cylinder and iron "rubber." Edison's nephew Charles Edison made one of these instruments at the Menlo Park laboratory in early 1879. N-81-05-23:43, Lab. (*TAED* N212:21; *TAEM* 40:794); Edison's testimony, TI 1:81–82; Batchelor's testimony, TI 1:247; Edison's exhibit, TI 2:502; (*TAED* TI1:58–59, 97; TI2:450; *TAEM* 11:61–62, 100, 629).

35. See headnote, Doc. 2098.

36. Photographs sent to Paris included those of the Edison Electric Light Co. headquarters at 65 Fifth Ave., the Machine Works on Goerck St., the lamp factory and laboratory at Menlo Park, the new lamp factory at Harrison, and an illustration of the electric railroad from the *New York Daily Graphic*. Edison also sent six bound sets of his U.S. patents, two bound sets of his British patents, and two bound sets of all British and French patents pertaining to electric lighting, as well as a scrapbook of periodical articles on electric lighting. This last book included the 21 December 1879 *New York Herald* that made public his electric light system (see Doc. 1868 n. 3) and an article from the *Philadelphia Ledger* about the lighting plant of the steamship *Columbia*. N-81-05-23:109–13, Lab. (*TAED* N212:54–56; *TAEM* 40:827–29).

–2112–

Samuel Insull to George Gouraud

New York, June 11th. 1881.

My Dear Colonel

I beg to confirm dispatch to you of letter of June 3rd.[1] which I hope you will receive in due course.

Fruit Experiments.

From your letters to Mr. Edison you seem to have assumed that the question of preservation of Fruit was absolutely settled last year and that there would be no further difficulty about the matter.[2] As a matter of fact it requires considerable experimenting and although Mr. Edison proposes[a] to ship some fruit and other[a] things to you during this season it has not yet proved a perfect success. Of course he intends working on it until it is,[a] but I thought it as well to make these remarks in case you should be under the impression that the question is settled beyond a doubt. One of Mr. Edison's assistants has been down to Delaware and made arrangements there with a peach farmer by which he will be able to put the fruit up in carboys, exhaust them and ship them right from the farm itself. As soon as the season commences you will no doubt receive some considerable trial shipments and in the mean time you will get from us various experimental[a] packages.[3]

Ore Milling.

I have been down to Quogue this week with Mr. Edison to see his magnetite ore separator in operation. You may remember receiving from me about 6 months ago specifications for patents to be taken out in several foreign countries your consideration for doing so being that you were to receive a certain interest in the invention in those countries. What they are doing here is to pass the black sand, which is found in very large quantities on the coast of Long Island and which contains about twenty per cent of the purest ore, through the magnetic separator and the results they get are of the best character. You may remember that Mr. Edison originally experimented with this invention with the idea of abstracting from gold tailings all that is magnetic and so more easily work gold, but he finds that instead of throwing the magnetic portion away it can be worked up into the very best iron equal to Swedish bars[4] at a cost which makes it far more valuable than working the gold tailings. To the best of my recollection when you received the specifications for the patents very little was done in the matter. I think that we filed the the preliminary specifications in several countries and I feel certain that nothing more[b] was ever done in the matter. I think it would be well for you to look into this matter and get these patents straightened out as wherever black sand is to be found this invention will be very valuable indeed.[5]

Telephone General.

I sent you a number of letters the early part of this[a] week on the subject of Telephone matters and should be very glad if you could settle them up as quickly as possible.[6] With reference to that question of the Glasgow royalty Mr. Edison seems very decided in his opinion, and I cannot say that I disagree with him. You well know Johnson's views on this subject and mine are pretty nearly identical with his. We have heard[a] nothing from you for some time with reference to the European Telephone Company and are consequently considerable in the dark as to how this matter is going on.[7] Several times Mr. Edison has asked me if I have heard anything[a] from England about it, but as I was to hear only[b] from either Mr. Bailey or yourself, and, as Mr. Bailey has been such a short time in Europe,[8] I have received no information at all upon the subject.

Pond Indicator.[9]

When I first got here I wrote you that I would get you some information as to the present state of the Pond Indicator. My time has been so very fully occupied ever since my arrival that it was not until a very few days back that I was able to go and make an inspection of that very useful and complete invention[a] The ~~connection~~ Indicator[b] is very complete in every respect whether it be simply for messenger service or for use in connection with the Telephone Exchange system. We certainly have nothing in England equal to it, in fact they seem very far before us here in all such matters. I am told that when you saw the instrument it was much[c] more complicated whereas now any child could operate it in the easiest possible manner. Mc.Kenzie has invented what he terms The Universal Signal Transmitter which will enable a subscriber to send any[a] signal to the Telephone Exchange from one up to a thousand, or it could be arranged to go beyond that.[10] This is done by such a simple movement that it is impossible for a mistake to be made and if it were adopted throughout the Telephone[a] system in England would undoubtedly greatly increase the reliability of the service. I think you would do well to look this matter up as it must play an important part in the future of the Telephone.[11] I understand that it will be exhibited at the Paris Exposition and it is probable that our friend Mc.Kenzie will be in charge.[12]

There is very little to say to you with reference to the preparations for the exhibit at Paris. Mr. Batchelor is hurrying forward as quickly as possible and hopes to get every thing there before the opening! Electric Light matters here are about the same.[a] Our men are still wiring houses with astonishing rapidity with the canvassers at the same time getting just as many orders as they like to apply for. Your exhibition plant for London is well under way and will be shipped a very little after the time that the one starts for Paris.[13] Very Faithfully yours,

(Samuel Insull)[d]

TL (carbon copy), NjWOE, DF (*TAED* D8104ZBN; *TAEM* 57:139).
ªMistyped. ᵇInterlined above by hand. ᶜHandwritten in right margin.
ᵈHandwritten.

1. Doc. 2107.

2. Gouraud had written on 23 April that he expected the $3,000 he had provided Edison would "be expended in great part in actual putting up of stuff? As so large an amount could scarcely be required for experiments, having regard to the successful experiments made last summer and the great simplicity of the process." Gouraud to TAE, 23 Apr. 1881, DF (*TAED* D8104ZBC; *TAEM* 57:114).

3. Edison and Charles Hughes expanded the preservation experiments in June and July. Nothing is known of the plan for Delaware but Hughes wrote Edison on 3 June that he had arranged with a Maryland farmer "for the privilege of erecting my apparatus on his place. He has the finest orchard with the nicest varieties" and easy access to a rail line. On 17 June Insull informed Gouraud that Edison planned to use "a considerable portion" of the $3,000 for "the appliances requisite to start this business," including "a four horse power engine, and Archimedean screw in connection with mercury pumps, [and] a moveable frame building." By 8 July Hughes had built a frame building at Menlo Park which could be readily reconstructed at the orchard where peaches were to be packed into evacuated carboys for shipment to New York. By the end of July, Gouraud was pleased that "preparations are being made on a very business-like scale" but he still wanted to see "tangible results" and asked to have "a complete statement" of expenditures to date. He also doubted the wisdom of shipping large quantities of peaches to England until Edison had perfected the process. Hughes to TAE, 3 June 1881; Insull to Gouraud, 17 June and 8 July 1881; Gouraud to Insull, 30 July 1881; all DF (*TAED* D8104ZBJ, D8104ZBQ, D8104ZBW, D8104ZCE; *TAEM* 57:132, 148, 162, 172).

4. Iron produced in Sweden was valued for the high quality resulting both from the native ore's low phosphorous content and the use of charcoal instead of contaminant-laden coal or coke in smelting. Gordon 1996, 176, 179.

5. Gouraud promised Edison at the end of July that he would look into the foreign patents. There was evidently confusion as to who would pay the fees, and Insull requested John Randolph to send from Menlo Park copies of all correspondence with Gouraud about ore milling. It is not known what, if anything, happened regarding these foreign patents. Gouraud to TAE, 30 July 1881; Insull to Randolph, 16 Aug. 1881; both DF (*TAED* D8138Q, D8136ZAH; *TAEM* 59:128, 44).

6. TAE to Gouraud, all 5 June 1881, Lbk. 8:306, 308, 311A (*TAED* LB008306 [Glasgow royalty], LB008308 [Oriental Telephone], LB008311 [London settlement], LB008311A [acknowledgment]; *TAEM* 80:953, 955, 958).

7. Gouraud had been working for months to consolidate the interests of the Edison Telephone Co. of Europe, which controlled the inventions of Edison, Elisha Gray, and several others in continental Europe outside of France (see Doc. 1731) with the parties controlling Alexander Graham Bell's patents. Edison was to receive £20,000 cash and £40,000 in stock under a tentative agreement he approved in February, consummation of which depended in part on his transmitting through Drexel,

Morgan & Co. relevant patent assignments and contracts (Gouraud and Theodore Puskas to TAE, 5 and 9 Feb. 1881; TAE to Gouraud, 11 Feb. 1881; all DF [*TAED* D8148X, D8148ZAD, D8148ZAF; *TAEM* 59:770, 781, 783]). Gouraud made several inquires about these documents before they were sent in April, after which Insull admonished him to "remember that Mr Edison is not always able to make other people move as quickly as he does himself." On 9 April Gouraud reported that he hoped to bring the combined company before the public in London and Paris simultaneously. The most recent information from him was on 25 April, when he wrote that legal counsel would likely approve the arrangements (TAE to Gouraud, 2 Apr. 1881, CR [*TAED* CH001AAG; *TAEM* 97:550]; TAE to Gouraud, 4 Apr. 1881; Insull to Gouraud, 20 Apr. 1881; Lbk. 8:137, 195 [*TAED* LB008137, LB008195; *TAEM* 80:883, 901]; Gouraud to TAE, 9 and 25 Apr. 1881, both DF [*TAED* D8149L, D8148ZBH; *TAEM* 59:930, 815]; see also correspondence from Gouraud in Telephone—Foreign—Europe (D-81-48), DF [*TAED* D8148; *TAEM* 59:743]).

8. Nothing is known of Joshua Bailey prior to his involvement in early 1878 with marketing Elisha Gray's telephone in Europe; he and Theodore Puskas subsequently agreed to combine the Gray and Edison telephone concerns in France. In April 1879 Bailey acquired an interest in the Edison Telephone Co. of Europe and later that year began trying to consolidate telephone interests throughout the rest of continental Europe. Bailey resided in Paris but had been in the United States from at least late March until about 6 May. During that visit he made arrangements regarding his and Puskas's interests in Edison's telephone and electric lighting in Europe. See Docs. 1213, 1449, 1731, and 1826; Bailey memorandum, 24 Mar. 1881; Bailey to Insull, 7 May 1881; both DF (*TAED* D8148ZAP, D8148ZBK; *TAEM* 59:798, 820).

9. Chester Pond obtained patents on devices to give visual signals of incoming calls at a telephone switchboard. In 1880 Edison acquired a two-thirds interest in Pond's patents with the understanding that he would help develop the instruments. He assigned Gouraud half this interest in certain foreign countries. See Doc. 1975.

10. Edison learned telegraphy from James MacKenzie as a young boy (see Doc. 1975 nn. 1 and 4). MacKenzie had an unspecified association with Chester Pond at this time. Edison assigned him a portion of his own interest in Pond's indicator, and MacKenzie did at least some development work at Menlo Park in 1880. The device described here was likely related to the "automatic transmitter for electrical indicators" patented by Pond in 1880. In January 1881 Edison ceded back to MacKenzie and Pond responsibility for improving the indicator in exchange for one-third of any profits they might realize. He sold his entire interest in the system in several countries to the nascent Pond Indicator Co. of Europe for $2,500 in June. TAE agreement with MacKenzie and Chester Pond, 12 Jan. 1881; TAE to Calvin Goddard, 17 June 1881; both DF (*TAED* D8104H, D8104ZBP1; *TAEM* 57:40, 145); U.S. Pat. 235,569.

11. Insull suggested this course again later, but Gouraud declined to pursue the Pond instrument for England because "it seems to drag behind so much that I am afraid it will be superseded before we get the benefit of it." Insull to Gouraud, 17 June 1881; Gouraud to Insull, 20 July 1881; both DF (*TAED* D8104ZBQ, D8104ZCC; *TAEM* 57:148, 170).

12. The Pond Indicator Co. exhibited both the indicator and the MacKenzie transmitter. France. Ministère des Postes et des Télégraphes 1881, 75.

13. Insull reported a week later that he had "just now come from the works and am sure that if you could only see the system upon which things are managed there, the great rush with which our superintendent is trying to turn out the necessary plant for England, you would have confidence that it is utterly impossible for any shop either on this or the other side of the Atlantic to do the work quicker." Insull to Gouraud, 17 June 1881, DF (*TAED* D8104ZBQ; *TAEM* 57:148).

–2113–

From Francis Upton

Menlo Park, N.J., June 13 1881[a]

Dear Mr. Edison:

I see you have marked 45 cts. as the price which we shall charge Col. Gouraud's lamps to Drexel, Morgan & Co. I think 50 cts. is as low as we can furnish them.[1] Though we are making lamps now at a cost of less than 35 cts counting the whole number made yet as we can only sell certain Volts it leaves us with a large number on hand.[2] If these are not paid for from those we sell we shall run at a loss for some time to come.

Besides this the English Company[3] have now the full benifit of all the money we have expended to perfect the manufacture, and has not as yet paid anything towards it. If we give them the lamps at a low figure we cannot ever make them pay anything towards these expenses. More than this, if without their asking we reduce our price, we have less inducement to offer for them to give us a larger order and more[b] liberal terms.

Am I to understand your order duplicating the Paris order to cover all the experimental lamps.[4] Yours Truly

Francis R. Upton.

ALS, NjWOE, DF (*TAED* D8123ZCL; *TAEM* 57:910). Letterhead of Edison Lamp Co. [a]"Menlo Park, N.J.," and "188" preprinted. [b]Added later in different ink.

1. The Edison Lamp Co. had contracted to sell lamps for domestic use at 35 cents apiece; some allowance was also made to compensate the company for research and development that subsequently lowered manufacturing costs (see Doc. 2039 n. 1). In April, when Edison had the factory send forty lamps to William Thomson at the University of Glasgow for his own experiments, Philip Dyer suggested that because "we have been to considerable trouble about picking these out, we ought to charge more than 35¢ each." Edison replied that the factory should bill them to Drexel, Morgan & Co. at 60 cents and explain that the extra work of making a special lot justified a surcharge. In the meantime, Francis Upton billed them at $1.50 each. William Thomson to TAE, 12 Mar. 1881; Philip Dyer to TAE, 14 and 16 Apr. 1881; all DF (*TAED* D8133E,

D8123ZAV, D8123ZAY; *TAEM* 58:600, 57:849, 852); TAE to Edison Lamp Co., 15 Apr. 1881, Lbk. 8:173 (*TAED* LB008173; *TAEM* 80:893).

2. Only a portion of lamps manufactured, though otherwise satisfactory, had the proper resistance to give their rated candlepower at a particular voltage. Those that did not were withheld for use as resistance lamps or for special installations. On manufacturing costs cf. App. 1.B.54.

3. Drexel, Morgan & Co. controlled Edison's lighting patents in Britain until negotiations with organizers of the Edison Electric Light Company, Ltd., concluded in early 1882.

4. Edison's directive to Upton has not been found. Charles Batchelor placed Edison's original undated order in a notebook amid his own detailed list of items for the Exposition, following a notation to "Get Uptons order for Paris given him by Edison." Edison requested Upton to make (and charge to the Edison Electric Light Co. of Europe) 9,500 standard lamps of several types, a handful of specific experimental or demonstration lamps, and "any other nice thing you think would be interesting at Paris." N–81-05-23:133, 143–47, Lab. (*TAED* N212:66–67, 73–75; *TAEM* 40:840–41, 847–49).

–2114–

*From Theodore Puskas
and Joshua Bailey*

London June 14th 1881[a]

Dear Sir,

Tuesday last, Mr Walker advised us of having received from the U.S. Com[missioner] a letter enclosing one from you regarding exhibiting and lighting space in the Exposition,[1] and offered to go with us, to call on Mr Berger [2] and express the strong official interest felt in your exhibition. M. Berger showed the most obliging disposition, but was embarassed by the fact that the space referred to in your letter had been given to other exhibitors. He explained that at the time of Mr Batchelor's visit the applications for space were so few that his main concern was how to fill it up, but that since that time the situation had entirely changed, and it had been found necessary to cut down exhibitors and to refuse many applications altogether. Also your attention at the time is called to the fact that M. Puskas took more space than your cable authorised him to take, and you cabled him that you would try to fill what he had taken.[3] Under these circumstances Mr Berger explained that it would not be possible to consider the question of 12 salons out of a total of 24 even to you. But he proposed to apply to the Syndicate[4] to give up to you for lighting, the grand staircase, and the space at the head of the staircase, and for exhibiting purposes as well,[b] a part of the latter space.

Today M. Berger informs us that the Syndicate has given up to you the space desired as above, both for lighting and exhibition purposes. It is possible that the space now given for lighting will take up the whole of the lamps at your disposition but if not M Berger suggests that you might take one or two rooms of the Museum of Decorative Art, which adjoins your salon, and the lighting of which would form a most attractive exhibition

Enclosed herewith you will find ground plan of the floor allotted to the Exhibition which we explain perhaps unnecessarily to guard against error.[5] The staircase comes up on each side the landing with an intermediate landing, the whole width occupied being 12 metres and the stairs themselves being 4 metres wide. On each side there are at the foot of the staircase at the intermediate landing and at the head of the staircase two marble pillars at both sides making twelve places in all for candelabras lighting the staircase. aAlso in face of the landing on each side there are two pillars for candelabra, making in all 16. The Commission offers to furnish you with lamp-posts for the Candelabra (if these will serve) from which the lamp fixtures will be removed and you can put in such and so many branches as you may see fit.

The landing itself as you will remark is 16 metres by 12 metres wide and it is 12 metres from the floor to the ceiling

The measurement of the salons accorded you is (the partition being removed) 24 metres by 16 metres and it is 7½ metres high.

The above is the space available for lighting and you will remark that you are in one case along side of the "Maxim," and in the other along side of the "Swan"

For exhibition you have the salon 24m × 16m and a space say three to five[c] feet wide by 16 metres long on the landing of the staircase. The vestibule at the foot of the staircase will be lighted by the Syndicate but this will be quite separated by draperies at the foot of the staircase so as not to interfere with your light, and at the great space on the landing your neighbor will be the "Maxim" from whom you are entirely cut off except for the communicating doorway.

Mr Berger suggested your putting up quadruplex on the space at head of staircase. You will select, bearing in mind that no machinery or heavy weight can be placed there, and that it is the most prominent place in the Exposition whether for lighting or Exhibition purposes.

In the plan of the rez de chaus[s]ée[6] is shown place for Dynamo & Boiler   Very respectfully & truly yours

Puskas and Bailey. per E Lane.[7]

LS, NjWOE, DF (*TAED* D8135L; *TAEM* 58:892). Letterhead of 6 Lombard St. ᵃ"London" and "188" preprinted. ᵇ"as well" interlined above. ᶜ"to five" interlined above.

1. Robert Hitt (1834–1906) was a journalist, congressman, and diplomat. As U.S. Commissioner General for the Paris International Electrical Exhibition, he served as an intermediary between American exhibitors and French exhibition officials for issues such as the allocation of exhibit space. Neither Edison's letter to Hitt nor Hitt's to George Walker has been found. Hitt to TAE, 26 and 27 Apr. 1881, both DF (*TAED* D8135C, D8135D; *TAEM* 58:874–76); *ANB*, s.v. "Hitt, Robert Roberts."

2. Georges Berger (1834–1910), the Commissioner General of the Paris International Exposition, was responsible for allocating exhibit space. Berger had directed the Paris Universal Exposition of 1878. Trained as an engineer, Berger was also a scholar of French painting. *DBF*, s.v. "Berger (Paul-Louis-*Georges*)"; see also Exhibitions—Paris Electrical Exhibition (D-81-35), DF (*TAED* D8135; *TAEM* 58:870).

3. On 3 May Edison told Berger that he had authorized Puskas and Bailey to "make all necessary arrangements in regard to space and position of apparatus" at the Exhibition. Edison's cable to Puskas regarding exhibit space has not been found. TAE to Berger, 3 May 1881, DF (*TAED* D8135E; *TAEM* 58:877).

4. During the summer of 1881 Puskas and Bailey secured the support of a group of French investors to exploit Edison's lighting system in France. This syndicate, led by Paris financiers Charles Porges and Elie Léon, placed 200,000 francs at Edison's disposal to fund his exhibit at the Electrical Exposition. The syndicate was the nucleus of Edison's lighting businesses in France and elsewhere on the Continent. Puskas and Bailey to TAE, 14 July 1881, DF (*TAED* D8135X; *TAEM* 58:924); Edison Electric Light Co. of Europe report, 7 Mar. 1884 (p. 4), CR (*TAED* CE001003; *TAEM* 97:209); Israel 1998, 215.

5. The enclosed plans (not reproduced here) showed the exhibit areas (some numbered) on the ground and first floors; spaces assigned to Swan and Maxim were marked in pen. (*TAED* D8135L, images 5 and 6; *TAEM* 58:896–97).

6. That is, the ground floor.

7. Edward Lane was evidently a clerk in Gouraud's office at 6 Lombard St. Lane to Insull, 25 Aug. 1882, DF (*TAED* D8213S; *TAEM* 60:673).

–2115–

To Naomi Chipman[1]

[New York,] 18th June [188]1

Madam

Referring to the enclosed Bill I think there must be some mistake in the allowance for "absence"[2]

For ten days the only people at home were myself and the Governess.[3] Then there has been only four of us at the house since 28th April & you were notified this would be so two or three days before. When we took the rooms it was upon the understanding that allowance would be made when the number boarding was less[4]

Kindly let me have a corrected statement[5] & I will send you check   Yours truly

Thos A Edison

LS (letterpress copy), NjWOE, Lbk. 8:320 (*TAED* LB008320; *TAEM* 80:960). Written by Samuel Insull.

1. Naomi Chipman operated a boardinghouse at 72 Fifth Ave., across the street from Edison's offices at 65 Fifth Ave., just south of 14th St. Edison moved to New York in February 1881, but it is not clear when he contracted with Chipman or brought his family to New York. *Trow's* 1881, 256, 457; Jehl 1937–41, 924; Israel 1998, 231.

2. The bill has not been found, but Chipman charged Edison $125 a week with a $2.50 rebate per person per day for absences. Chipman to TAE, 30 Sept. 1882, DF (*TAED* D8204ZHC; *TAEM* 60:309).

3. The unidentified governess presumably was hired to care for the three Edison children: Marion (eight years), Thomas, Jr. (five years), and William Leslie (thirty-three months).

4. Neither Edison's notification of his family's absence in April nor a prior lease agreement has been located.

5. A corrected statement has not been found. Edison and Chipman continued to disagree over the terms of his lease until November 1882; in one instance Edison attempted to stop payment on a rent check. Chipman to TAE, 30 Sept. and 8 Nov. 1882, both DF (*TAED* D8204ZHC, D8204ZIO; *TAEM* 60:309, 377); TAE to Bank of the Metropolis, 22 and 23 Nov. 1882, Lbk. 14:466, 472A (*TAED* LB014466, LB014472A; *TAEM* 81:1023, 1025).

–2116–

Prose Poem[1]

[New York, Spring 1881?[2]]

[--------------------------------------]ᵃ

Boulevard St Antoine[3] that damnable merchants of inhumanity Citenian[4] wharfrats. Why Centenus[5] dost run a line already greased from Sirus[6] to Capella with angularity, whereon[b] ten million[c] devils slide down to the farthermost sag and piss into pendemonum   Tell me winged soldier of Hell[c] ⊖if in the farthermost ends of infinity warted demons with cavernous mouths spit saliva on the balls of [-----]ᵈ the firmament to produce deluges, hast seen the Juif errant[7]   he mocks ~~destru~~ the angel of destruction, amuses himself by letting off fire works in powder mills. Citronella this damnd perfume on the Vine Clad Hills of Andulusia[8] wafted as from a garden filled with Red hat giraffes a Rain of Boullion Soup descended on the thirsty soil and the wells and [--]ᵈ snake holes puked ~~from the~~ miasmatic water from the richness of the rapast, prithee

tells us[b] Centenus, we of the finite minds, if Vesuvius consti-
pated from derangement of the terranian[c] kidneys will ever
vomit up again to engulf ~~the~~ more[b] Earthlice   They tell me
the milky way is formed of stars & planets so innumerable & its
diameter so great that [iff?][d] if Adam was one and all the sands
of the sea were cyphers they would not express a preceptible
segment of the grand ring—

Doth god have a pegasus & use the milky way as a hitching
ring—

Whizzozririzing an asteroid has run plump into a favorite
angel. This Crank lump of matter will be used for [---- ----
-----][d] water closets in Hades[e]   two angels at a candy pull,
they streatch the viscous molasses from Vega to Acturus at
every pull, flies as big as Japan buzzed around it.

A Bowery angel smoking a palm tree stubbed his toe on a
comet,[c] and pimples came out on his toe nail as big as ~~a~~ moun-
tains, he swore so much that God made eight new planets out
of the conversation & peopled and fauna'd[c] & flora'd[c] them
eccentrically   The almighty has a vein of humor   he made
these planets & peopled them to give amusements to beings on
the rest of the celestial plantation. The men were 800 miles
long & 1/4 inch thick   they slept on telegraph poles, and animals with bodies as big as a pea with 900[c] eyes each as big as
a saucer lived on these long men by catching them by the feet
& sucking them in like maccaroni[f]

In the most charming nook in Paradise amid surroundings
more beautiful than are[b] imaginatively conceivable, in a palace
dwell Gabreal & Evangeline,[9] each endowed with a capacity
capable of enjoying that which DeQuincy,[10c] had he a brain 300
miles in diameter full of opium would never imagine—

AD, NjWOE, Lab., N-81-04-30:103 (*TAED* N306:45; *TAEM* 41:1200).
[a]First line canceled and followed by dividing mark. [b]Interlined above.
[c]Obscured overwritten text. [d]Canceled. [e]"water closets in Hades" interlined above. [f]Followed by dividing mark.

1. The circumstances surrounding the creation of this document are
entirely unknown. Although highly unusual, the document is not un-
precedented in form or tone; see Doc. 523.

2. Charles Hughes made most of the entries in this notebook in the
late spring and summer. Edison wrote this document amid a number of
unused pages following a single entry dated 20 October. However, it lies
between undated pages of ambiguous notes and drawings that may pertain to the disk dynamo armature that he designed in April, May, and
June. These are followed first by Edison's genealogical notes and then a
Hughes entry dated 12 June.

3. This Paris street is a thoroughfare to the site of the Bastille. In

A Tale of Two Cities, Charles Dickens portrayed the St. Antoine section of Paris near the Bastille as an overcrowded slum, which it was during the French Revolution and in Dickens's own time. Sanders 1988, 43–44.

4. Edison's meaning is unclear. He may have been using an obscure form of cithern or cittern, a stringed instrument similar to the zither popular in the sixteenth and seventeenth centuries. The instrument was frequently decorated with a grotesquely carved head, from which arose "cittern head" and "cittern-headed" as terms of contempt. *OED*, s.vv. "citern," "cithern."

5. Centaurus.

6. Sirius.

7. Literally, "wandering Jew."

8. Andalusia, a fertile region of southern Spain. *WGD*, "Andalusia."

9. Henry Wadsworth Longfellow's 1847 epic poem *Evangeline* told the story of the heroine and Gabriel, separated on their wedding day by the British expulsion of the Acadians from Nova Scotia. Calhoun 2004, chap. 9.

10. Thomas De Quincey (1785–1859), English critic, essayist, and exponent of German philosophy, was most noted for his 1821 work, *Confessions of an English Opium-Eater. Oxford DNB*, s.v. "De Quincy, Thomas."

–2117–

From Edison Lamp Co.

Menlo Park, N.J., June 22 1881[a]

Dear Sir.

The following is the report of the life of lamps now burning in the 16 candle power test.[1]

At 5.30 A.M. June 22.

Set of 10 Reg. A. $(8 \times 17) = 831$ h. 54 min   (7 still burning) Still give $15\frac{1}{2}$ to 16 <u>available</u> candle power[2]

Set of 10 Reg A $(8 \times 13\frac{1}{2}) = 324$ h 24 min. (8 still burning) Give 16 candles full.—

Set of 10 Reg B $(8 \times 13\frac{1}{2}) = 129$ h 25 min   (8 still burning)

We have been giving this report almost daily to Mr Acheson under the supposition that it was to go to you; and so failed to make direct report. It seems however that his report was for Mr Batchelor and we shall therefore report hereafter daily by mail.[3] Yours truly

Edison Lamp Co   per N[ichols].

L, NjWOE, DF (*TAED* D8125B; *TAEM* 57:1098). Written by Edward Nichols on letterhead of Edison Electric Lamp Co. [a]"Menlo Park, N.J.," and "188" preprinted.

1. A test of lamps with the standard "A" .008 × .017 inch filaments reached 551 hours on 3 June, when Charles Batchelor noted that two had failed; this group ultimately averaged 1,425 hours. A separate regimen

for the high efficiency .008 × .013½ inch filament "A" lamps reached seventy-eight hours that day with two failures; this batch averaged 844 hours. Cat. 1237:61–63, Batchelor (*TAED* MBN006:11–12; *TAEM* 91:214–15).

2. The actual candlepower would have been measured by a photometer. Lamps gradually accumulated a deposit of carbon throughout the inside of the globe, a phenomenon known as electrical or carbon carrying. In measuring the effective illumination from his lamp, Edison was implicitly referring to the difficulty of obtaining the rated candlepower from gaslights in practical operation, which he had investigated in September 1880. See headnote, Doc. 1898; and Doc. 1990.

3. These reports have not been found. Francis Upton intermittently included information about lamp life and electrical characteristics in letters to Edison, particularly around this time (Edison Electric Lamp Co.—General [D-81-23], DF [*TAED* D8123; *TAEM* 57:756]). Late in the year John Marshall began making more frequent reports of 48 candlepower tests; at least some of these were for experimental lamps. Marshall and others continued to supply this information in early 1882 (Edison Electric Lamp Co.—Lamp Test Reports (D-81-25); Edison Lamp Co.—Lamp Test Reports (D-82-32); both DF [*TAED* D8125, D8232; *TAEM* 57:1090, 61:955]). John Howell made numerous records of lamp lifetime tests in factory notebooks in 1881 and 1882 (Cat. 1301, 1302, 1303, all Batchelor [*TAED* MBN007, MBN008, MBN009; *TAEM* 91:293, 365, 438]).

July–September 1881

Edison had firmly established himself in New York by the summer of 1881. From there he oversaw development of a large exhibit for the Exposition Internationale de l'Électricité in Paris; arrangements for electric lighting in Europe and Great Britain; and preparations for his most prized goal, an electric light and power system in Manhattan.

Edison entrusted Charles Batchelor with the day-to-day responsibilities of putting together his Paris exhibit. Many of the planned items represented technologies in which he was no longer actively involved (like the telegraph, telephone, and phonograph) or those such as incandescent lamps and small generators, which he no longer considered experimental.[1] The large dynamo, a prototype of what Edison planned for the New York central station, was a different matter and required his personal attention.

This dynamo, built by the Edison Machine Works, was the largest in the world. It was constructed according to the design of an experimental machine made at Menlo Park early in the year, and Edison believed it would operate about 800 lamps. When he tested it in June and July, however, he found several serious faults. It ran too hot, was prone to arcing in the armature, produced low voltage, and sparked badly at the commutator. The last problem, he later admitted, made him fear he could not take off the heavy current the machine could generate. However, solutions to the sparking and low voltage were relatively straightforward. After a series of experiments on the Paris dynamo and two smaller ones to find the causes of heating and arcing, he ordered the massive armature to be rebuilt entirely. This required more than a week of day-and-night labor that even Edison called "a terrible job."[2] When finished, the machine performed to his expectations. He also gained from the experience a more thorough understanding of the electrical characteristics of his dynamos, which enabled him to improve the design of his smaller ones as well.

Edison took a number of significant steps in August and September toward developing the electric lighting business in foreign countries. Contrary to his usual reluctance to take legal action in defense of his patents, he instructed his representatives to seek an injunction in a French court against Hiram Maxim's incandescent electric lights at the Exposition. This was done successfully, and Edison next wanted to pursue Joseph Swan, but his advisor Grosvenor Lowrey counseled against it. Edison also urged his agents in Paris and London to cultivate friendly relations with prominent technical writers, with the expectation that they would be paid retainer fees. His exhibit attracted considerable favorable attention, and in late September he declined an invitation to visit Paris.[3] Because Batchelor and others warned that the display made widespread patent infringement likely, Edison was anxious to lay the administrative and financial foundations to manufacture equipment and sell lighting plants on the Continent. Financier Egisto Fabbri and Joshua Bailey presented rival proposals from prospective investors. Fabbri withdrew and Edison negotiated with Bailey, pushing for a large investment in factories. Edison suggested making, the financing contingent on successful operation of the large dynamo and the New York central station. He and Bailey reached an accord in October that led in early 1882 to the formation of manufacturing and operating companies for the commercial development of his electric light system in Europe.[4] In the meantime, he began planning for a demonstration central station in England. Charles Clarke started designing a 1,000-lamp dynamo for this plant in mid-August. At the end of September, Edison dispatched Edward Johnson to make arrangements in London. He also urged George Gouraud, his business representative there, to market isolated lighting plants in British colonies and other countries outside Europe and Britain.

Having selected the financial district of lower Manhattan for his first commercial electric system, Edison threw himself into the task of building the network to light its brokerages, banks, offices, and newspaper publishers. He selected two adjoining buildings on Pearl Street, below Fulton Street near the unfinished Brooklyn Bridge, for the generating plant. The

Edison Electric Illuminating Company took title to the properties on 3 August and Edison promptly began outfitting them. On the afternoon of 22 September, a crew a few blocks away at the corner of Peck Slip and South Street began excavating for the underground conductors.[5] Exactly what form those cables would take was still the subject of experiment by John Kruesi. Edison also continued to refine the accuracy and reliability of his consumption meter.

Because Edison did not expect the central station to be ready for almost a year, isolated lighting for individual buildings was an immediate concern. He needed a small standardized dynamo to advance this business. Drawing on his July experiments with the large Paris machine, he and Clarke modified the distinctively tall dynamo developed at Menlo Park, designating the new 60-lamp machine the Z model. The first production unit was installed in September, and dozens followed in the next year.

At the lamp factory, output reached nearly 1,000 lamps per day. This was greater than demand, and Francis Upton contemplated a two week layoff. Planning began on rehabilitating the buildings for the new lamp factory in Harrison,[6] where Edison and Upton hoped to reduce labor expenses and eventually expand production. John Branner, a botanist whom Edison sent to South America in late 1880, continued his search for plant materials suitable for lamp filaments. Upton and his staff continually experimented with ways to decrease the cost of lamps and increase their durability and efficiency. They reported results to Edison, who did not participate directly in experiments at this time but clearly had an active role in the factory's management. The price of lamps (and later those of dynamos and other necessary equipment) started to cause discord among Edison-related interests, especially the new foreign firms.

The expansion of Edison's manufacturing capabilities continued to strain his personal finances. Some relief came in mid-July, when he received about $55,000 in partial liquidation of his British telephone interests. The amount disappointed him, and he carried on a running dispute with George Gouraud over the balance he believed was owed. He looked forward to remuneration from the merger of his European telephone interests with those of Alexander Graham Bell and Frederic Gower, although progress from a preliminary to final agreement was slow. Formation of the Oriental Telephone Company to control the patents of Edison and Bell in British colonies proceeded even more fitfully because of Edison's prior disposition of his Hong Kong patent rights.

In other activities, Edison discharged all but a handful of workers at his Menlo Park laboratory in July. Charles Hughes remained and continued to experiment on vacuum preservation of perishable foods. Having finished that work, Hughes made a preliminary survey for a new electric railroad in late August. A few weeks later, financier Henry Villard agreed to underwrite construction of an experimental railroad at Menlo Park.

Samuel Insull continued to play a vital role in Edison's business, particularly his extensive transatlantic cable and postal correspondence. Insull's ubiquity in these affairs prompted George Gouraud to ask him to "discriminate between your personal letters and letters written specially at Edison's request."[7]

Edison and his family remained at a Fifth Avenue boarding house near his offices. In August he was introduced to famed violinist Eduard Reményi, who proclaimed his company "intellectual heaven" and invited him to several concerts.[8] Mary Edison spent an unknown length of time in Michigan, and in late August, Edison went there to bring her home, returning about 7 September. Nothing more is known about this trip except that it may have coincided with a calamitous wildfire on 4–6 September in the region around Port Huron, where Edison's father, brother, and other family members lived. The fire burned a million acres in five counties and killed almost 300 people. Edison gave one hundred dollars to a relief fund for survivors.[9]

1. Batchelor's list of equipment for this exhibit is Doc. 2111.

2. Doc. 2122.

3. TAE to George Gouraud, 25 Sept. 1881, Lbk. 9:141 (*TAED* LB009141; *TAEM* 81:50).

4. Joshua Bailey and Theodore Puskas to Edison Electric Light Co. of Europe, 1882, DF (*TAED* D8228ZAX; *TAEM* 61:714).

5. William Hammer memorandum, 22 Sept. 1881, Ser. 2, Box 23, Folder 1, WJH.

6. Upton to TAE, 6 July 1881, DF (*TAED* D8123ZCV; *TAEM* 57:924).

7. Gouraud to Insull, 30 July 1881, DF (*TAED* D8104ZCE; *TAEM* 57:172).

8. Doc. 2137.

9. On Edison's trip see Doc. 2149. "Major Post-Logging Fires in Michigan: the 1880's," Michigan State University, Department of Geography course 333 (http://www.geo.msu.edu/geo333/fires.html);

Dunbar 1980, 408; Pyne 1982, 199–206; Port Huron Executive Committee for Relief to TAE, 21 Sept. 1881, DF (*TAED* D8104ZDA1; *TAEM* 57:211).

–2118–

*Samuel Insull to
George Gouraud*

New York July 1st. 1881.[a]

My Dear Colonel:—

I have written you weekly for some time past, but have not received any note from you as to whether you have got my letters.[1] I presume, however, that they have all arrived and that you will see whether you can find time to write me. When you have a little time to spare I should like a line from you giving me some general information as to how matters progress in England.

Fruit Experiments.

Your letter relating your experience with the chops, and steaks and strawberries came duly to hand.[2] Mr. Edison was really amused by it, and he wished me to say to you that he is highly delighted that the first attempt should not prove successful as it would be really too bad to find that he was working on a thing[b] that any man could make go all right, and that one [----][c] guarantee against infringers when he does get the thing to work in proper order will be that they will have the same difficulty in getting an efficient process, and will in fact have to go through precisely the same ~~process~~ experience as he always expects to have to undergo when experimenting on some new invention. But, of course, the fact that because those chops emitted such an abominable smell will not deter him from continuing his experiments which will no doubt turn out all right eventually. From my previous letters you will have understood that your report as to the condition of what you receive does not surprise him in the slightest degree.

Electric Light.

The plant for England is being pushed ahead as quickly as possible. We have the Porter-Allen engine which ~~yo~~ is to operate the machinery already in our shop and the dynamo is far on the way to completion, and as Johnson is giving the whole of his time to seeing that this equipment is rushed through you may depend that not a moment will be lost. We had the Paris machine running a few days ago for a short time, and a splendid sight it was to see the thing running.[3] The dynamo is, of course, the largest in the world and will put into the shade every thing that opponents can bring both in the way of economy and efficiency. Yours will be if anything, better than the Paris machine, in asmuch as Edison will[d] profit by his experience in the manufacture of the Paris machine, and if there should happen to be any slight defects in that they will be rectified so far as the London outfit[a] is concerned. Our lamps are showing a wonderful record. The way in which Mr. Edison arrived at the conclusion that the life of the lamp was from three to four hundred hours is as follows. He would set up a number of sixteen candle lamps at about fifty candle power and run them at this state of incandescence until they played out. Now, of course, this was no fair test as the lamps are only made to run at sixteen[a] candle and to run them at fifty would entirely alter the conditions. The first real test as to how long the lamps would last at sixteen candles is now being made.[4] Up to the present moment they average about seven hundred and thirty hours each. One went out after burning seventy hours. Another went out at one hundred and sixty. Another went out somewhere in the neighborhood of three hundred. The fourth[a] one has just died. The remaining six of the original ten lighted have up to the present been burning nine hundred and ninety hours. How long they will last, of course time alone can prove. The longer they do last, of course, the better is the average life of the lamp, and anyway it would appear that the average life will be considerably above one thousand hours. These ten lamps were just taken out from our ordinary stock and therefore the average will be an absolutely fair one. We are watching this test with very great interest indeed. You will appreciate the very great importance of getting the life of our lamps as long as possible as it must considerably affect the economy of the system. The lamps in this building still hold out. As I have already told you, we lit up on the 24th. of March and not a single lamp has yet given way.

Telephone.

They seem a long time settling up that Edison Telephone Company of London liquidation. If you could send Mr. Edison the proceeds of the sale of his shares it would be of very great advantage to him at this time. What with his lamp factory, machine works, tube company and laboratory he requires a very great deal of money, and if he could collect what is due him on English Telephone rightaway it would save him some considerable anxiety. I will send forward the release which Renshaw sent us to be signed by Edison and Johnson by next mail.[5] I propose sending it to you as there are a number of blanks[a] that require to be filled in and we prefer that the release should be in your hands until this is done. Very truly yours,

(Insull)[e]

TL (carbon copy), NjWOE, DF (*TAED* D8104ZBU; *TAEM* 57:156). [a]Mistyped. [b]"n" interlined above by hand. [c]Canceled. [d]Second "l" added by hand. [e]Handwritten.

1. Gouraud had responded to Doc. 2107 but had not acknowledged Doc. 2112 or Insull's 17 June letter. Gouraud to Insull, 17 June 1881; Insull to Gouraud, 17 June 1881; both DF (*TAED* D8104ZBP, D8104ZBQ; *TAEM* 57:143, 148).

2. Gouraud reported that a lamb chop, a porterhouse steak, and

a few strawberries, have come safely to hand— I opened the tubes a few days after their arrival, but the result was anything but satisfactory— The stink emitted from the "Chop" and "Steak" was to such a degree noxious, that I was unable to remain in the room & it was some time before I recovered from its bad effects— The strawberries appeared to be on the point of decomposition & the tube contained a good deal of water. I regret to say that the process, as applied to these instances at least, has proved a complete failure. [Gouraud to TAE, 15 June 1881, DF (*TAED* D8104ZBO; *TAEM* 57:142)]

3. See headnote, Doc. 2122.
4. See Doc. 2117.
5. Alfred George Renshaw of the law firm Renshaw & Renshaw was acting as trustee for Edison's interest in the defunct Edison Telephone Co. of London (see Doc. 1954). In sending the form to Insull, he explained that the number of shares Edison was to receive in the successor company was still subject to the liquidator's payment of a few outstanding claims. Insull returned the executed release on 8 July. Renshaw to Insull, 16 June 1881; Insull to Gouraud, 8 July 1881; both DF (*TAED* D8149W, D8104ZBW; *TAEM* 59:965, 57:162).

–2119–

To George Gouraud

[New York,] July 5th. [1881]

My Dear Gouraud:

Referring to your favor of the 15th ult., I have no feeling particularly against Mr. Preece[1] excepted that I think[a] he acted "pretty rough."[2b] I think the whole matter about that microphone controversy arose from my utter inability to conceive that a man of science could see any difference between the principle of the microphone and the principle of my carbon telephone, which are exactly similar as [I][c] think Mr. Preece will now admit. I therefore think that at the time I had considerable justification in going for him, but I see that a great many men quite as scientific as he, have made the same mistake and could not see that the carbon telephone and the microphone were one and the same thing   Give Mr. Preece any lamps that he may want   [I?]ᶜ ordered some time [ago?]ᶜ sufficient to fill your requirements for such purposes. Very truly, yours,

Thos A Edison   I[nsull]

TL (letterpress copy), NjWOE, Lbk. 8:356 (*TAED* LBoo8356; *TAEM* 80:966). Typed in upper case. ᵃ"I think" interlined above by Samuel Insull. ᵇQuotation marks added in pen. ᶜIllegible.

1. William Preece was the electrician and assistant chief engineer for the British postal telegraph system. He and Edison enjoyed a cordial relationship until a bitter and public falling out in 1878 over Preece's role in advancing David Hughes's claims to the invention of the microphone. See Docs. 1331, 1338, 1346–48, 1366–67, 1370, 1375, 1378, and 1398; Baker 1976, 176–77; Israel 1998, 157–60.

2. Gouraud wrote that "Preece manifests great interest in your light, and I understand gives it precedence over all others, and is likely to prove a strong ally. It is highly desirable to cultivate this and so bury an old hatchet. Sir Wm. Thomson advocates this feeling and I shall act upon it unless you advise me to the contrary which I sincerely trust you will not." Preece had also told Gouraud "that he has in the Post Office all the different kinds of lamps including Swan's and Maxim's, and that he would like a couple of yours—and for which I have promised to ask you, not feeling myself at liberty to do so without your approval." Samuel Insull wrote extensive shorthand notes on the back of Gouraud's letter which are presumably the basis for Edison's reply. When Gouraud received the reply, he wrote back that "it would do a great deal of good" to show it to Preece. In the meantime, Gouraud had given two lamps to Preece, whom he reported "seems to be very highly pleased with them. It looks as though he were going to be a useful card to us." Gouraud to TAE, 15 June and 19 July 1881, DF (*TAED* D8133W, D8133Y; *TAEM* 58:621, 626).

EDISON AND PUBLIC RELATIONS AT THE PARIS ELECTRICAL EXPOSITION   Doc. 2120

Edison's exhibit at the Paris Electrical Exposition generated widespread favorable publicity.[1] Otto Moses and Charles Batchelor were his main representatives at the Exposition and both strove to create favorable impressions among scientists, the press, and prominent visitors. Moses, fluent in French and German, made connections with European scientists and journalists. The relationship he established with the celebrated French electrician Théodose du Moncel proved crucial. Du Moncel, a senior editor of the influential French journal *La Lumière Électrique,* was initially lukewarm to Edison's incandescent light. With help from Theodore Puskas and Joshua

Bailey, Moses succeeded in obtaining his services as press agent and advocate for the Edison interests for the sum of 1000 francs a month for the three months of the Exposition and 10,000 francs to be paid by the syndicate of Edison investors in France. Du Moncel wrote a flattering account of Edison's lighting exhibit which appeared in *La Lumière Électrique* and in translation in the *Scientific American Supplement*.[2]

Charles Batchelor, with his detailed knowledge of Edison's lighting system, managed the exhibit. He maintained and repaired equipment, supervised Edison's technical team, and gave Edison frequent reports about competitors' displays. He also answered visitors' questions about particular elements of the exhibit and solicited applications for lighting plants from impressed viewers. Noteworthy visitors to the Edison rooms included King Kalakaua of Hawaii, French Prime Minister Léon Gambetta, and French economist Léon Say.[3]

1. For a summary of the publicity activities of the Edison interests at the Exposition see Fox 1996 (164–65). Fox attributes a substantial portion of the success of Edison's exhibit to the publicity efforts on his behalf. Scores of articles about the exhibit and the Exposition in general were kept in two scrapbooks for Edison. Cats. 1068 and 1069, Scraps. (*TAED* SM068, SM069; *TAEM* 89:34, 143).

2. Du Moncel 1881.

3. See Docs. 2142 and 2147; Moses to TAE, 7 Sept. 1881, DF (*TAED* D8135ZBI; *TAEM* 58:1038).

–2120–

From Otto Moses

Paris, July 12. 81

My dear Sir:

The following is an[a] abstract of Armengaud[1] and Du Moncel[2] conversations referred to in my last.—[3] Mr. Bailey introduced me to Armengaud for the purpose of giving him an insight into your system, and also to remove certain prepossessions of mind by which it could easily be seen he was affected. I spent two or three hours with him profitably. He knew nothing whatever of the merits of the case, and appeared to be more ignorant of the subject than one would have supposed possible after the many conversations he must have had with Messrs. Bailey and Puskas. I enlightened him without in the slightest wounding his 'amour propre.'[4] He was delighted with the systematic appearance of all your electric light inventions, no doubt because his mind, freshly steeped in telephone affairs, was impressed with the necessity for[b] a thoroughness in all such extensive undertakings. With much self-satisfaction he mentioned that Bell lost his French patents through him; and he also said that Mr. Edison had taken precautions which saved him; to this fact I ascribe the apparent ease with which he accepted my statements about your extreme prudence in patent affairs. He had elaborated a comparative table of the patents granted you and Maxim, and he showed that 'over-claims' or rather a mixture of claims on different subjects, had caused the Government to demand a separation of the claims ~~M~~made by Maxim. Before this was done, however, publication had[b] taken place, and so the matter stood. Armengaud said this might give rise to grave questions. I told him that he might rest assured his estimate of your caution was correct; and I cited the case of your not allowing two copper plated carbons to go into Italian hands while any doubt existed about your patents being protected there; also your getting Serrell to cable to Europe to know if the disc dynamo had been protected, before allowing me to describe it.[5]

I am to meet Armengaud at intervals in order to compare the english and french patents, and to post him on any points upon which he may want general information. A. is very busy, has a half dozen clerks, and affects the hurried air of a man with business too large for his clerical force. He listens attentively and at the same time gives orders about business, bounces around, telephones his clients, and makes memoranda. He is Attorney for several electric light companies, and you know his official connection with the telephones. Altogether he is agreeable, polite, and he makes an honorable impression; still Mr. Bailey speaks of employing counsel with him.

Du Moncel. When I arrived I presented the letter I showed you, to Dr Herz who seems to have a preponderating influence with the "Lumière Electrique." He has really been of service to me. He sent me 'La Lumierè Electrique' from the beginning, introduced me to every one of the writers on that paper, and finally took me up to Du Moncel's house to introduce me. He left me alone immediately with that gentleman, and I talked Edison &c to him for two hours. I will not of course tell you what I said it would simply be disgusting to you. One fact I mentioned however, which I must tell. He was showing me a coil of several hundred feet of copper wire closely wound on a bobbin without insulation, with compressor plates at the ends worked by screws, for purpose of demonstration that pressure longitudinally increased conductivity; when I said "Permit me, Monsieur le Comte, (he is a count, you know) to mention a little incident. I was one day talking to Mr. Edison about the wonderful properties of his carbon button, when he said to me he had just been told of your having also spoken in your great work, of the ~~latter~~[c] varying conductivity of carbon.[6] He (you) then expressed most naively his inability to verify the statement of his informant as he could neither speak nor read french, and asked me to translate passages from your work" This seemed to touch the man; for he thawed immediately and before I left asked the privilege of describing your lamp and offered to carry on the experimental verifications in the laboratory of the Institute. He can be easily had to read a paper on the subject before the Institute which would be subject to revision and correction; however I do not think there is any more sting in him after what I said during a very long visit. He said he would call on me soon.

By the by, I never got the letters of introduction I was to have brought. In France you know it is necessary to be identified and endorsed. Even Mr. Bailey who is well known here, takes with him your the[a] letter ~~of~~ authorizing him to act for you at the Exposition, whenever he wishes to transact business for your account. By the way I am received however, on the mention of my business I am always sure of a cordial hearing.

I am just in receipt of Maj. Eaton's[7] letter informing me of the unavoidable delay in shiping the great dynamo.[8] I will reply to him on the subject and to do so must close in order to mail by today's steamer via "Queenstown."[9] Faithfully yours,

Otto A Moses

ALS, NjWOE, DF (*TAED* D8135W; *TAEM* 58:920). [a]Interlined above. [b]Obscured overwritten text. [c]Interlined below.

1. The brothers Jacques-Eugène (1810–1891) and Charles (1813–1893) Armengaud were patent agents and consulting engineers. It is unclear to which Moses referred; in Doc. 2071 n. 2 the editors conjectured that Edison was dealing with the elder brother, but Charles was his agent for at least nine French patents obtained in 1882 and 1883. *DBF*, s.vv. "Armengaud, Jacques-Eugène" and "Armengaud, Charles."

2. Count Théodose-Achille-Louis du Moncel (1821–1884) was a scientific instrument maker, engineer, administrator for the state-owned French telegraph system, and a senior editor of the influential French electrical journal *La Lumière Électrique*. Doc. 1248 n. 13; Fox 1996, 164.

3. Moses wrote on 10 July that he had "seen Armengaud and duMoncel It is sufficient to say that in the case of both 'I came, I saw, and will conquer.'" He gave few details of his conversations but reported that Armengaud "talks Maxim to me and I talk system and piracy to him. He wanted me to telegraph for the papers in M[axim]. vs. Edison." Moses to TAE, 10 July 1881, DF (*TAED* D8135U; *TAEM* 58:908).

4. Pride or self-esteem.

5. Lemuel Serrell became Edison's patent attorney in May 1870. He

continued to act in that capacity until January 1880, when George W. Dyer and Zenas Wilber took charge of his new patents. Serrell continued to act for Edison in regard to prior U.S. applications and interferences and foreign patents. On 29 April, Edison directed him to find out whether disk dynamo patents had been properly filed in Europe before allowing a published description of it to appear. See *TAEB* 1:196, n. 2 and Doc. 1270 n. 16; TAE to Serrell, 29 Apr. 1881, DF (*TAED* D8142ZAL; *TAEM* 59:318).

6. Du Moncel 1878, of which Edison had tried to obtain an English translation in 1879. See Doc. 1738.

7. Sherburne Eaton was an attorney and important Edison business associate. He became vice president and general manager of Edison Electric Light Co. in early 1881 and president in October 1882. He was also a director (1880–1884) and vice president (1881–1884) of the Edison Electric Illuminating Co. of New York and president of the Edison Electric Light Co. of Europe. In the late 1880s he became Edison's personal attorney. Israel 1998, 210–28; *NCAB* 7:130.

8. The "C" or "Jumbo" dynamo; see headnote, Doc. 2122.

9. Neither Eaton's letter to Moses nor Moses's reply has been found.

–2121–

To Thomas Logan

[New York,] 15th July [188]1

Dear Sir,

When you have finished the Lamp Co work that you are actually engaged on & Batchelors work (with the exception of the new non magnetic Dynamo)[1] please discharge every man in the Laboratory and the Machine Shop except Alfred Swanston[2] who will act as night watchman   Of course Dr Haid Randolph[3] & the men working for Mr Claudius[4] will remain.

I will procure a four horsepower Baxter Engine[5] and have it sent to Menlo Park   You can disconnect the heavy piece of shafting from the Engine & run the shop with the Baxter Engine. You will remain, as you are, doing whatever work Dr Haid Hughes & Claudius may desire

Please ship to the Edison Machine Works 104 Goerck St New York immediately the large lathe[a] the large Drill Press & large plainer,[6] also the various parts of the new non magnetic Dynamo which I propose having finished at Goerck St

Yours truly

T A Edison

LS, NjWOE, DF (*TAED* D8137ZBW; *TAEM* 59:105). Written by Samuel Insull. [a]Obscured overwritten text.

1. This was the disk dynamo for the Paris Exposition, on which the shop had been working since at least 6 June. This machine was "non magnetic" in the sense that, as Edison explained in a patent, "the iron core of the armature will not be necessary" to complete the magnetic

circuit across the relatively narrow armature. Logan to TAE, 6 June 1881, DF (*TAED* D8137ZAY; *TAEM* 59:90); U.S. Pat. 263,150; see also headnote, Doc. 2074.

2. Alfrid Swanson was the night watchman and a general assistant at the Menlo Park laboratory. He had worked for Edison since 1876. Doc. 2069 n. 3, *TAEB* 5 App. 2.

3. John Randolph had worked for Edison as an office assistant and laboratory factotum since the end of 1878. Among Randolph's duties at this time was taking care of the library and answering requests for materials in it from Edison and others in New York. *TAEB* 5 App. 2.

4. Edison hired Hermann Claudius, a former engineer with the Austrian Imperial Telegraph Department, in 1880 to build scale models and made calculations for central station distribution systems. Doc. 2028 n. 2.

5. The Baxter was considered a portable steam engine because it combined the boiler and engine in a single moveable unit that did not require a permanent foundation. Portable engines were increasingly popular, as one contemporary noted, "wherever it is necessary to do work sufficiently great to pay for them, but not for permanent business." The Baxter was distinctive in being manufactured with interchangeable parts by the Colt's Fire Arms Manufacturing Co. Several thousand of them, from 2 to 10 horsepower, were in use by this time. Hunter 1985, 494–96.

6. The Edison Machine Works received this equipment by the end of August, when Charles Rocap inquired about its cost. A marginal notation indicates the lathe was $858 and the drill press $525. Rocap to John Randolph, 30 Aug. 1881, DF (*TAED* D8129ZBG; *TAEM* 58:275).

PARIS EXPOSITION "C" DYNAMO Docs. 2122, 2123, 2131, and 2134

Edison decided in the spring of 1881 to construct a dynamo for the Paris Exposition even larger than the experimental one completed at Menlo Park in February. The laboratory machine had been a partial success that demonstrated both the practicability of connecting the armature shaft directly to a steam engine and the fundamental utility of the bar armature construction. However, it proved impossible to run safely at its intended speed of 600 revolutions per minute. It also produced too much heat, owing to the relatively high resistance of armature end pieces and their connections with induction bars.[1]

The new machine was designed and built at the Edison Machine Works in New York, from which few records are extant.[2] Charles Clarke, who was largely responsible for designing the Menlo Park dynamo, did most of the planning for what was termed the "C" dynamo.[3] The machine was to have six horizontal magnet cores 57 inches long. Its armature measured 33¾ inches long and 26⁷⁄₁₆ outside diameter and consisted of 146 bars connected in pairs through half as many copper disks at each end.[4] Edison had contracted for a 125 horsepower Armington & Sims steam engine in early April, and about a month later Clarke was comparing the rotational energy of the engine's flywheel with that of the projected dynamo's components, presumably to check the dynamic compatibility of the two machines.[5] Nothing more is known of the dynamo until late June, when it was run on a circuit with resistance coils immersed in casks of cooling water.[6] Samuel Insull called it "a splendid sight" and promised that it would "put into the shade every thing that opponents can bring both in the way of economy and efficiency."[7] Docs. 2122, 2123, 2131, and 2134 are among the extant records of tests on this machine, which Edison also described and explained in Doc. 2149. Francis Jehl assisted and carried out related experiments on insulation, heat dissipation, and the heat capacity of insulated and uninsulated copper bars. He also measured the heat produced by passing bars through a magnetic field, which may have been to distinguish between magnetic and electrical causes of heating.[8] During the summer Clarke made calculations about the machine's electrical and physical characteristics, including the mechanical strain and centrifugal force on the armature.[9]

Initial tests showed the armature running too hot; it also sparked between the bars.[10] Edison decided to have it rebuilt as described in Doc. 2122, even though this meant a substantial delay getting the machine to Paris. (Until it arrived near the close of the Exposition, two much smaller dynamos powered Edison's lighting display.) He used narrower bars which could accommodate japanned paper insulation. However, the new bars increased the resistance and heating so Edison enclosed the armature and shaft in a duct through which a belt-driven blower forced cooling air. Docs. 2131 and 2134 indicate that heavy sparking at the commutator continued to be a serious difficulty.[11] Because output voltage was also lower than expected Edison added two magnets on top. A contemporary engineering account noted that this "unequal distribution of the field magnets . . . could not but tend to produce an unsymmetrical disposition of the lines of force within the field" and this apparently contributed to heavy sparking at the commutator brushes.[12] The completed machine had an armature resistance of only .0092 ohm. Driven by the Armington & Sims engine at 350 rpm, which became the standard speed for Edison's large dynamos, it could operate about 700 lamps at 16 candlepower.[13] Its construction cost Edison about $6,000 but he asked the Edison Electric Light Co. to cover the cost of related experiments, about that amount again.[14]

Edison's first "Jumbo" direct-connected dynamo, constructed for the 1881 Paris Exposition.

The dynamo required one last-minute repair, of a broken engine shaft, before it was hauled with a police escort to the docks in time to sail on 7 September aboard the *Canada*.[15] It began operating on 21 October and continued until the Exposition closed about 15 November.[16] After that, Charles Batchelor installed it at his factory near Paris; it later went to a factory in Rotterdam.[17] Even before the first "C" began its work at the Exposition, however, Edison had completed a larger dynamo and planned another, both based on the general design of the Paris machine.[18]

1. See Docs. 2057 and 2067 n. 3.

2. Charles Clarke's retrospective account of the machine's design and construction is Clarke 1904, 35–39. Francis Jehl's retrospective account is Jehl 1937–41, 970–74. It is illustrated and described generally in Dredge 1882–1885, 1:260–65.

3. Other letter designations were used for isolated plant dynamos. See headnote, Doc. 2126.

4. Clarke 1904, 37.

5. See Doc. 2078; N-81-02-20:271–77, Lab. (*TAED* N214:103–6; *TAEM* 40:1048–51).

6. N-81-04-06:20–21, Lab. (*TAED* N223:11; *TAEM* 41:47); N-81-02-20:255–67, Lab. (*TAED* N214:96–102; *TAEM* 40:1041–47). A notebook containing Clarke's records of his tests of the Paris machine in early August is in EP&RI (*TAED* X001J8).

7. Doc. 2118.

8. See Doc. 2122 n. 1; N-81-04-06:47–63, 152–63, Lab. (*TAED* N223:21–29, 73–78; *TAEM* 41:57–65, 109–14). Related tests by Jehl are scattered through this notebook.

9. At one point iron bands were wound around the armature for

strength; they quickly broke and were replaced by copper wires. In a patent application completed in August, Edison explained that because the induction bars were "raised off the [armature] core and separated by small blocks or by projections" to increase air circulation, circumferential wires were needed to hold them in place. N-81-02-20:215–23, N-81-04-06:69, both Lab. (*TAED* N214:91–95, N223:32; *TAEM* 40:1036–40, 41:68); U.S. Pat. 263,133.

10. The sparking prompted Edison to adopt an entirely different pattern of arranging the induction coils and connecting them to the commutator blocks in smaller dynamos (see headnote, Doc. 2126). This alteration was not adopted in the large machines, presumably because of mechanical considerations.

11. See Jehl 1937–41, 972–73. Edison executed a patent application for the air blower on 24 August (U.S. Pat. 263,133). In May, Charles Mott had researched for Edison the practices of other inventors in placing commutator brushes with respect to the magnetic lines of force (see Doc. 2100 n. 1).

12. According to a diagram sent to Charles Batchelor, the field magnets were arranged in two groups of four series-wired coils. Dredge 1882–85, 262; Clarke 1904, 50; William Hammer to Batchelor, 1 Sept. 1881, Ser. 1, Box 1, Folder 1, WJH; see also Doc. 2149 n. 4.

13. Clarke 1904, 37; Jehl 1937–41, 972–73.

14. The experimental costs may have included development of the machine at Menlo Park. TAE to Sherburne Eaton, 13 Sept. 1881, DF (*TAED* D8126ZAA; *TAEM* 58:35).

15. Jehl 1937–41, 973–75; TAE to Robert Hitt, 7 Sept. 1881, Lbk. 8:484 (*TAED* LB008484; *TAEM* 80:993); Otto Moses to TAE, 22 Sept. 1881, DF (*TAED* D8135ZBX; *TAEM* 58:1074); "Shipping News," *New York Herald,* 5 Sept. 1881, 10; see also App. 1.B9.

16. See Doc. 2173.

17. Clarke 1904, 36.

18. See headnote, Doc. 2238.

–2122–

Notebook Entry:
Dynamo

[New York,] July 15 1881.

After working 55 men days and 60 nights for 8 days and 8 nights we have at last made the change in bars and disk tits which the previous test showed was necessary to make the machine more practical=[1] The narrow tit between which the alternate bars could pass was found to be insufficient and heated the plate thus as shewn by the dotted line,

hence it was necessary to make the tit wide, this would then necessitate passing the bars over the tit instead of between[2]   this was done as[a] described in some other book.[3] it was a terrible job. we silver soldered the extra piece on the tit but our big

press when done will punch out the plate with tits simultaneously   The inside piece from the plate ~~kept~~ leading to the
commutator kept breaking and we connected the bars to the
bobbin thus.[4]

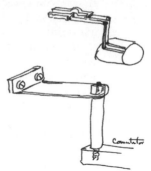

we japanned the bars—painted them with Zinc white and
Linseed oil, then wound tissue paper then[a] repainted. after assembling put mica between all the bars—put the new anti
spark device on.[5] The new foundation being done placed Engine Dynamo[b] on it. 11 AM. Engine running nicely. taking indicator diagrams, for friction, now taking with field on 108
Volts across field armature open i.e.[c] brushes off—

X, NjWOE, Lab., N-81-04-06:37 (*TAED* N223:16; *TAEM* 41:52). [a]Ob-
scured overwritten text. [b]Interlined above. [c]Circled.

1. The machine was first tried in the latter part of June (see headnote
above) but there is no other record of excessive heating. In an entry dated
14 July on the previous page, Francis Jehl recorded the armature resistance of the C dynamo as .0085 ohms; in an 11 July entry he gave it as .15
ohms. On 4 June, Jehl had conducted a series of simple tests, first measuring the resistance through an undivided armature bar, then a second
bar cut into five segments held tightly by screws, and then a third bar
with as many joints soldered together. The resistance of the first and
third was the same; that of the second slightly higher. N-81-04-06:28,
24; N-81-03-18:86–87, all Lab. (*TAED* N223:15, 13; N230:42; *TAEM*
41:51, 49; 41:432).
2. The induction bars were of different lengths so each could connect
to one in a series of copper disks at each end, where they were attached
with screws to protrusions or lugs (U.S. Pat. 431,018). In August Edison

Figure from Edison's U.S.
Patent 431,018 showing
the original method of
armature construction.
Induction bars E, of
different length, are each
connected to a disk in the
series M or M' at each end.

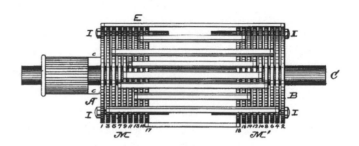

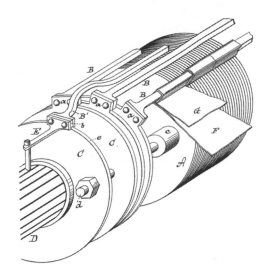

Edison's patent drawing showing the enlarged connections between armature end plates and induction bars to alleviate heating at those joints.

filed an application for the method described here of providing a larger contact surface for the joint, which entailed forming an arch at the end of alternate bars to allow them to pass over neighboring connections. To further reduce resistance and inhibit oxidation, Edison prescribed plating the surfaces with gold or silver, or coating them with mercury (U.S. Pat. 264,647).

3. Not found.

4. Text is "comutator."

5. At the end of June Edison sketched several ways to "do away with the spark on a Dynamo Machine" and a few days later drew one device in more detail (Cat. 1147, Scraps. [*TAED* NM016:94, 92, 101; *TAEM* 44:318, 316, 325]). In a patent application executed on 22 July, Edison described this as a "breaking-cylinder" that rotated with and was similar to, but was insulated from, the commutator. Current passed from the main brush, then through a number of brushes in parallel to the breaking cylinder. As the cylinder rotated, it instantaneously interrupted the circuit from the multiple brushes at the moment the main brush passed from one commutator bar to another, preventing the heavy spark there. He stated his observation that the size of spark was reduced as the square of the number of points at which the circuit was broken (U.S. Pat. 263,149). Edison had described the use of multiple commutator brushes to divide the current and thereby reduce commutator sparking in an 1880 British patent; that arrangement is also implicit in an 1880 caveat (Brit. Pat. 3,964 [1880], Cat. 1321, Batchelor [*TAED* MBP030; *TAEM* 92:212]; Doc. 2036).

–2123–

Notebook Entry: Dynamo[1]

[New York,] July 20 1881

We run armature little while & then put field on just[a] as the field was charging up.[2] Spitting commenced with loud sound across from bar to bar with $1/32$ of air space.

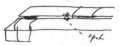

Two spits took place both starting from edge of paper or right near the paper—but across air space $1/32$ inch   The spit was small only threw up knobs about $1/32$ dia & roughened edges of bar[b]

The Commutator brushes were not on Evidently. This spitting is due to EM.F.[a] accompanied by a <u>Static</u> effect which first starts a static arc and then the other arc commences—& this is the trouble with all dynamos.

One thing is noticable & that is that the "spit" occurs at the moment the bars pass at X & never at .G.

Crossing of armatures[3]

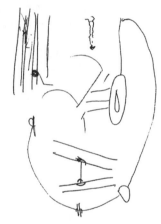

Crossing of armatures

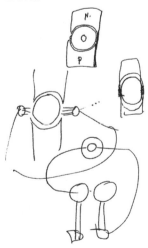

Test the kick between the insulated armature and the field magnet when the field is put on[4]

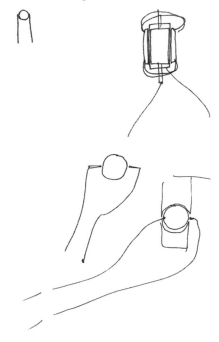

TAE

X, NjWOE, Lab., N-81-04-06:71 (*TAED* N223:33; *TAEM* 41:69). Document multiply signed and dated. [a]Obscured overwritten text. [b]Followed by "over" as page turn.

1. See headnote, Doc. 2122.

2. Edison started this test of the Paris C dynamo the previous evening but had to abort it when the iron bands around the armature broke (see headnote n. 9, Doc. 2122). Repairs were made overnight, and the machine was ready at 7 A.M. N-81-04-06:69, Lab. (*TAED* N223:32; *TAEM* 41:68).

3. These drawings are unclear. In a series of tests the next day, Edison noticed large flashes on the armature when the machine was run on a resistance circuit; he determined that the armature was crossed in three places (N-81-04-06:82–88, Lab. [*TAED* N223:39–42; *TAEM* 41:75–78]). The difficulty of preventing leakage across the high potential difference between bars led Edison to consider new ways of connecting the armature bars (see Docs. 2124 and 2150). In a patent application executed about a month later, he claimed the use of parchment paper or mica (preferably both) to insulate the induction bars. He noted that both materials were also excellent heat conductors (U.S. Pat. 264,647).

4. These sketches are also unclear but may represent attempts to measure induction effects in the stationary armature, possibly resulting from imperfectly continuous current flowing through the field coils. Francis Jehl made several measurements of the "kick" on 19 July. On 21 July, Jehl recorded an unsuccessful effort to bring a current off the front of the armature through a comb–like collector. N–81–04–06:66–68, 82, Lab. (*TAED* N223:31–32, 39; *TAEM* 41:67–68, 75).

–2124–

Notebook Entry:
Dynamo

[New York,] July 25 1881

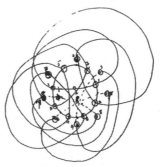

OK winding armature with the alternate Coils over the others mks dif EMF small[1]

OK[2]

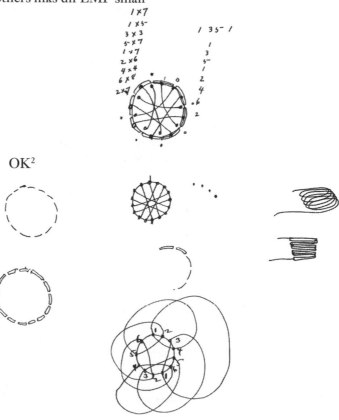

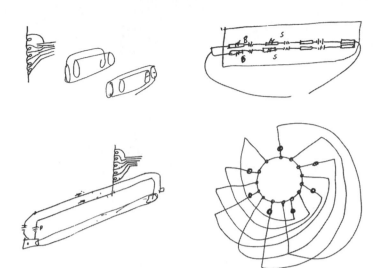

OK with odd no of Coils   Wound 1 on top other

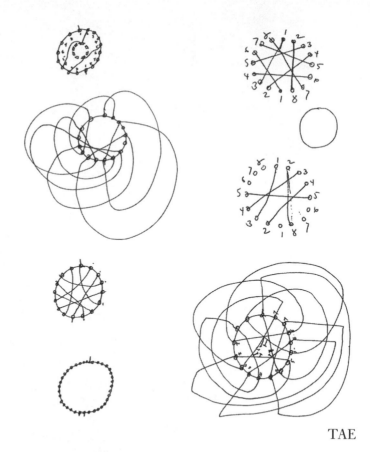

TAE

X, NjWOE, Lab., N-81-04-06:126 (*TAED* N223:60; *TAEM* 41:96). Document multiply signed and dated.

1. This notebook entry is interleaved among numerous armature winding diagrams that Edison sketched on 22 and 23 July (N-81-04-06:96–125, 140–49, Lab. [*TAED* N223:45–59, 67–71; *TAEM* 41:81–95, 103–109]). His evident satisfaction with a small difference of potential between armature bars suggests that this pattern was intended to eliminate arcing between armature bars, the difficulty noted in Doc. 2123. This drawing suggests two other innovations related to that problem. One is the placement of bars in two layers around the armature to facilitate construction of low-resistance connections at the end plates (see Doc. 2122 n. 2). Edison later explained to Batchelor that this necessitated another change, the use of an even number of commutator connections (see headnote, Doc. 2126 and Doc. 2150). This was a departure from the practice Edison adopted with his first practical dynamos in 1879 (see headnote, Doc. 1682 and Doc. 1694). It entailed an asymmetrical pattern; in addition, the coils in two pairs (**5** and **4**; **7** and **6**) appear to be connected only to their partners.

2. The meaning of the paired and single numbers accompanying this drawing has not been determined, nor is it clear that there is any relationship between the two columns.

MATTERS NECESSARY TO BE DONE, JULY 29, 1881.

Memorandum: Pearl Street Central Station[1]

Accurate determination of feeders by Claudius. CLARKE.[2]

Map complete for Kruesi[3] to lay wires. KRUESI, CLAUDIUS, CLARKE.

Arrangement with contractors and Kruesi for laying mains.[b] GODDARD.

Arrangement at Central Station for continuous testing.[b] CLARKE.

Bids[c] on iron work; on mason work; on carpenter work. EATON, HORNIG,[4] CLARKE.

Supplies of tubing, boxes, mains from tube factory. EATON.

Getting station and preparing internal structure. CLARKE, HORNIG.

Arrangement with Babcock & Wilcox[5] relative to boilers, prices etc.[b] EATON.

Meter-room[c] at Central Station. Hornig.

Appliances there. CLARKE.

Books for meter tests, records, consumption blanks for meters. GODDARD, EATON.

Regular books for doing business at Station. EATON, GODDARD.

Arrangement of staff for operating station. CLARKE, EDISON.

Water, Croton.[6] Driven wells. EATON, HORNIG, CLARKE.

Coal supply, information, data, place of storage. EATON.

Contract for meters. EATON.

Moving Greenfield's room to Central Station.[7] EATON, HORNIG, CLARKE.

Motors.— — — — — — — —[d] CLARKE, EATON, EDISON.

Appliances to utilize the present gas chandeliers and fixtures. JOHNSON, EATON.

Storage room at station for lamps, waste oil, etc.

TD, NjWOE, DF (*TAED* D8122B; *TAEM* 57:723). [a]Date taken from document. [b]Written on two lines enclosed by brace at right. [c]Mistyped. [d]Dashes handwritten.

1. The contents and style of this document suggest Edison as the most likely author, possibly in conjunction with Charles Clarke.

2. Charles Clarke (1853–1941) was a civil engineer and draftsman who joined Edison's Menlo Park laboratory staff in early 1880. He immediately became deeply involved in designing dynamos and central station systems. At this time Clarke was chief engineer of the Edison Electric Light Co. Doc. 1921 n. 3 and *TAEB* 5 passim.

3. John Kruesi worked as a machinist for Edison for many years. He

had charge of the Menlo Park machine shop before becoming manager and treasurer of the Electric Tube Co. in early 1881. Docs. 659 n. 6 and 2058 n. 2.

4. Julius Hornig was a German-educated draftsman and mechanical engineer whom Edison hired in January 1881 to help plan central stations. Doc. 1897 n. 3.

5. The New York firm of Babcock & Wilcox was a noted manufacturer of stationary boilers, especially its patented "non-explosive" water-tube design which carried a higher pressure than other types. The company had given Edison estimates for large central station steam plants in 1880. Hunter 1985, 336–39; Docs. 1897 and 2008.

6. Brought into service in 1842, the aqueduct from the Croton River in Westchester County was the principal source of potable water in Manhattan. By this time the roughly thirty-mile pipeline was operating at or above its designed capacity of 75 million gallons per day. In 1881, the region suffered a severe drought that led to fears of a "water famine" in New York. Koeppel 2000, 271–89; "Public Works in New York City," *Sci. Am.* 46 (1882): 137.

7. Edwin Greenfield was an electrician, recently with the Holmes Alarm Co. in New York. Whether this move occurred is unclear. In September he was located at 41 Ann St. in lower Manhattan. Greenfield worked with Edison for several years, during which he invented conduit and wiring devices used in isolated lighting installations. Later he became a superintendent for the Bell Telephone Co. but is noted principally for his invention of BX armored cable, which was widely used in the U.S. for decades. "Greenfield, Edwin T.," Pioneers Bio.; TAE to Charles Walton, 13 Sept. 1881, Lbk. 8:494 (*TAED* LB008494A; *TAEM* 80:997).

ISOLATED PLANT DYNAMOS   Doc. 2126

By mid-July Edison had in hand the design of a standard dynamo intended specifically for isolated lighting plants. He instructed Charles Clarke to give the Edison Machine Works "immediately the size & amount of wire to be put on the new Z machines,"[1] and Clarke began making the calculations on 11 July.[2] The first one was built between then and late September, when it entered service in a mill at Newburgh, N.Y. More than 260 were sold in the next ten months.[3]

The Z type was the first of several standard dynamos for isolated plants fabricated by the Machine Works. Each type was rated for a certain number of lamps and given an arbitrary letter designation (see App. 3). These belt-driven models were variations on the distinctively tall machines made at the Menlo Park laboratory complex in 1879 and 1880. By September 1882 the Edison Company for Isolated Lighting offered four designs. These were the L and K models, larger than the Z, and the smaller E; the Z was by far the most common at that time.[4]

It is difficult to attribute the conceptualization of any specific model to extant notes and drawings. Relatively few records, most of them by Edison or Clarke, remain from the design process. In the case of the Z machine, work cannot always be distinguished from that done simultaneously on the large central station dynamos.[5] The fact that Clarke was also designing electric motors at this time makes interpretation of the records even more problematic. However, it is possible to infer certain fundamental design features of the isolated dynamos from remaining notebooks, specification sheets, and test records.

Each machine was designed to operate either 110 volt A lamps or twice as many 55 volt B lamps by using one of two interchangeable armatures. One gave 110 volts; the other gave half the voltage with a capacity for twice the current. Because both varieties of armature had the same external dimensions, the same size and length of wire, and required the same speed and power, switching from one to the other did not require complex changes to other parts of the machine. According to an 1883 specification chart prepared for the Edison Electric Light Company (probably by John Ott), the B armatures in general were wound in half as many coils (loops) around the core as the A designs. This would reduce the induced voltage by a factor of two. Each coil or loop of the B armature had twice as many individual wires, however, presumably connected in parallel, which would double the induced current. Together, these alterations gave the B configuration a resistance one-fourth that of the A armature. William Hammer's notes indicate that to operate the B armature, the field coils were to be connected in parallel with each other instead of in series. This was evidently in order to maintain the crucial relationship between the resistance of the armature and that of the field magnets (connected in a shunt circuit with the armature). In the case of the Z machine, with two 30 ohm coils, changing connections to the B configuration reduced the field resistance from 60 to 15 ohms, or the requisite factor of four. Were this not done, it is evident that current through the magnets would be insufficient to maintain full field strength and, consequently, the desired output voltage. The terms "A" and "B" armatures were standard by October 1881, when Edison ordered a number of them made for Z machines going to England. Both forms were also available on the E, L, K, and H models by 1883.[6]

The machines were wired so that a single switch opened or

closed both the main circuit and the field circuit simultaneously. In plants with a single dynamo, the engine and machine would be brought up to speed before the switch was closed, allowing the electrical load to increase gradually as the field gained strength. The same effect could be obtained by starting with the switch closed but all the resistance plugged in the field circuit. A somewhat different arrangement was used for multiple dynamos connected in parallel. In that case, opening the armature connection to the main line did not de-energize the field magnets. Keeping the field magnets energized prevented the armature of one machine from presenting a short circuit to the others.[7]

In July 1881 Edison and Clarke departed from their established practice of using an odd number of commutator blocks. Edison observed sparking between adjacent insulated induction bars in the armature of the huge C dynamo intended for the Paris Exposition (see Doc. 2123). Evidently concerned that this could also occur in smaller machines like the Z, he experimented with two layers of induction coils in order to reduce the potential difference between successive windings. Edison stated to Charles Batchelor, without further elaboration, that this required an even number of commutator connections.[8] He sketched numerous armature winding patterns (much as he and his assistants had done in early 1879),[9] attempting to devise a pattern for an even number.[10] By 25 July he had devised acceptable arrangements that reduced the difference in potential between successive windings.[11] It is unclear how the first Z machines were wound but one with an even number was sent to Paris in September. When William Hammer recorded dynamo specifications in 1882, the Z, H, G and E had an even number of commutator blocks; his records for other models are either incomplete or damaged beyond legibility.[12]

Clarke suggested one cosmetic change to the Z in November 1881. He advised Edison to have the patterns modified to "round all the edges of the base and field . . . and introduce any features in the way of graceful curves which certainly will add much to the appearance and nothing to the cost."[13]

1. TAE to Clarke, c. 10 July 1882, N-81-07-11, Lab. (*TAED* N220:30; *TAEM* 41:32).

2. Clarke's notes and calculations indicated the new machine was to have magnets the same size as those in the demonstration dynamos built at Menlo Park in 1879–1880 (see Docs. 1727 n. 2, 1849, 1985, and 2062 n. 3), at this time retrospectively designated the "A" model. He calcu-

lated the resistance needed to produce 115 volts, and the resistance and amount of wire for 125 volts. In early August he made a dimensioned sketch of one coil, showing insulation of japanned parchment paper (N–81-07-11:1–5, Lab. [*TAED* N220:1–3; *TAEM* 41:3–5]).

3. Edison Machine Works list, 10 June 1882, DF (*TAED* D8234B1; *TAEM* 62:7); see also Edison Co. for Isolated Lighting brochure (p. 6), 1 Sept. 1882, PPC (*TAED* CA002A; *TAEM* 96:103).

4. Edison Co. for Isolated Lighting brochure, 1 Sept. 1882, PPC (*TAED* CA002A; *TAEM* 96:103); see also Dredge 1882–1885, 2:331–33.

5. See headnote, Doc. 2122.

6. Unbound Notes and Drawings (1883), Lab. (*TAED* NS83:10; *TAEM* 44:1144); Hammer notebook 8:22–48, Ser. 1, Box 13, Folder 2, WJH; TAE to Edison Machine Works, 7 Oct. 1881, DF (*TAED* D8129ZBS; *TAEM* 58:290).

7. Charles Clarke to James Bishop, 14 Mar. 1931, acc. no. 30.415.1, accession file, EP&RI. See Doc. 2203 esp. n. 30.

8. See Doc. 2150.

9. See *TAEB* 5:76–78 and Doc. 1694.

10. Clarke 1904 (50) later attributed much of the commutator sparking as well to the odd number of commutator blocks. A contemporary

Edison's K 250-light dynamo for isolated plants.

*Edison's L dynamo for
150 lamps.*

standard engineering text explained that the trouble with an odd number was that "when the brushes bear on the collector diametrically opposite to each other, the sectors do not pass simultaneously from under them. While one brush bears on the centre of a sector, the opposite brush bears on two sectors, and so short circuits the two bobbins connected therewith." Schellen 1884, 353.

11. See Doc. 2124; also Docs. 2122 n. 2, 2149, and 2150. Edison executed a patent application for the overlapping layers of induction bars in November. In the patent he made no reference to the commutator connections but referred only to improved insulation and greater simplicity of construction. U.S. Pat. 263,146.

12. See Doc. 2150 and App. 3; William Hammer notebook 8:22–48, Ser. 1, Box 13, Folder 2, WJH and also Jehl 1937–41, 978. In November, Edison instructed Edward Johnson to apply mercury to commutator blocks of small dynamos to prevent sparking. See Doc. 2190.

13. Clarke to TAE, 11 Nov. 1881, DF (*TAED* D8129ZCN; *TAEM* 58:313).

Production Model:
Isolated Plant Z
Dynamo[1]

M (base 114 x 99 cm; height 183 cm; weight 1,364 kg), MiDbEI, EP&RI, acc. no. 30.415.1.

    1. See headnote above.

–2127–

[New York,] 2nd August 1881[1]

Draft to George
Gouraud

I beg to confirm receipt of £12,000 ~~& $11,000 from you &~~ & my telegram to you asking for balance ~~U~~upon the receipt of which you sent me $11,000 which is as near as I can judge £2245[2] not being able to understand how the total amount for ~~the~~ my share of reversion could only be £14,245 I cabled you asking ~~fo~~ the total number of shares received for ~~my reversion & you cabled me~~ it & recd from you in reply[a] "Your reversion two thousand eighteen less your share trust charges[b] 186 pounds your ~~Lon~~ London and Joint shares two hundred probable further reversion two[c] hundred making oversold your account five eight two delivered from mine"[3]

My reason for cabling for above information was that from your letters to me & the explanation of your views by Insull I was under the impression that the ~~amount~~ number of shares to be received by me would be certainly 3000.[4] In your letter to me stating that you had in the face of the opposition of the Liquidator obtained the order of the Court for the division of the shares you ~~stated~~ said nothing whatever as to the reduction of the number of shares coming to me & I naturally thought that you had in sending me £12 000 held for some reason the balance which I estimated at between £7000 & £8000 whereas according to your second remittance I only had about £2245 ~~further~~ to come to me on ~~a/c of Reversion~~ this a/c[5d]

I would point out to you that I have been kept totally in the dark as to[c] this matter for some time past that is on the <u>very important</u> point to <u>me</u> of how much I was going to get out of the Reversion & I had naturally made my arrangement as to financing my business under the impression that I should get from you ~~£19~~ £19,000 or £20,000 as[c] I was led to expect from correspondence with you ~~& this has caused me very great inco~~ & the sale of shares you made and[e] not being posted as[c] to this has caused me considerable inconvenience. I may mention in this connection that although it is now almost a year since the promoters of the[f] Oriental Company first commenced working in my name I have not as yet recd any settlement as ~~per~~ arranged & I have[c] not heard at all from you as to how this stands.[6] Then as to European Telephone although I ~~understand~~ see by the English papers & from conversation with one of <u>Gowers</u> colleagues that a Company has been brought out consolidating my European interests outside France[g] I have no word from you as to the matter & your silence has placed me in a somewhat peculiar position with the Directors of the Edison Telephone Co of Europe as they repeatedly ask for information which they are certainly entitled to & which I feel I ought to be able to give them[7]

I notice in your cable above quoted that you state ~~my account is~~ I am[h] oversold 580 shares. Now in as much as you sold the shares upon[c] your own judgement I think this is somewhat incorrect. Furthermore you could not by any means construe the letter I wrote you as to the sale of my shares as authorizing you to sell ~~my shares~~ my interest in the "London & Joint shares" at the price you sold the bulk of my shares[i] & I [--][j] therefore feel that I am entitled to the full ~~value~~ [--][j] <u>present</u> market value on the 200[k] shares which you say are coming to me on this account[8]

⟨Wrote Gouraud also as to European Light shares explaining lost opportunity[9]⟩

Df, NjWOE, DF (*TAED* D8149ZAB; *TAEM* 59:971). Written by Samuel Insull. Several financial calculations on the reverse and in the margins have not been reproduced. [a]"it & recd from you in reply" interlined above. [b]"your share trust charges" interlined above. [c]Obscured overwritten text. [d]"this a/c" interlined above. [e]"& the sale . . . made and" interlined above. [f]"promoters of the" interlined above. [g]"outside France" interlined above. [h]"I am" interlined above. [i]"at the price . . . my shares" interlined above. [j]Canceled. [k]Interlined above.

1. Edison's letter has not been found; Gouraud replied to it on 12 September after returning from vacation. Gouraud to TAE, 10 and 12 Sept. 1881, both DF (*TAED* D8104ZCV, D8148ZCA; *TAEM* 57:202, 59:856).

2. Gouraud wrote on 23 July that eight days previously he had cabled £12,030 through J. S. Morgan & Co. in London. On 30 July he confirmed the text of Edison's cabled directive that day to "send balance reversionary." Gouraud responded the same day: "Drexels pay balance." On the back of one page of his draft, Edison calculated the dollar value at the rate of $4.90 per pound, the approximate exchange rate in 1881. Edison received the $11,000 on 1 August. Gouraud to TAE, 23 and 30 July 1881, both DF (*TAED* D8149Y, D8149ZAA; *TAEM* 59:967, 969); Gouraud to TAE, 30 July 1881, LM 1:17 (*TAED* LM001017A; *TAEM* 83:880); Private Ledger #2:75, Accts. (*TAED* AB006:68; *TAEM* 88:230).

3. The text of Gouraud's reply is quoted in full. Edison instructed Gouraud on 1 August to "Cable total number shares obtained in settlement" (Gouraud to TAE, 2 Aug. 1881; TAE to Gouraud, 1 Aug. 1881; LM 1:17 [*TAED* LM001017C, LM001017B; *TAEM* 83:880]). Edison had consented in July 1880 to have his London telephone interest held in a trust. The "joint shares" were presumably those which, by a separate agreement, he and Gouraud arranged to hold together in the trust, representing a substantial portion of Edison's total interest (see Doc. 1954 n. 3).

4. See Docs. 2046 n. 2 and 2079.

5. A month earlier Gouraud had "at last succeeded in getting an equitable division of the Shares of the United Company." He reported that the liquidator had tried in court "to get a division so as only to return the Shareholders' money and keep back the balance— This attempt I effectually prevented." Gouraud to TAE, 2 July 1881, DF (*TAED* D8149X; *TAEM* 59:966).

6. Gouraud had written in mid-June that the Oriental Company's settlement with Edison and the other inventors was delayed by the difficulty of getting Alexander Graham Bell's legal signature on the documents. The status of Edison's patents in Hong Kong was also unsettled, as Gouraud reminded him in a 24 June cable. Gouraud to TAE, 17 June 1881, DF (*TAED* D8104ZBP; *TAEM* 57:143); Gouraud to TAE, 24 June, LM 1:14 (*TAED* LM001014B; *TAEM* 83:879).

7. See Doc. 2112 n. 7. Edison may have been referring to Hilborne Roosevelt, a New York organ manufacturer formerly involved in the phonograph business, who was the brother of Cornelius Roosevelt, one

of Gower's partners in Paris. Samuel Insull indicated that Roosevelt was one source of Edison's knowledge of this subject, though Joshua Bailey reported from London at the end of June that plans for the new European company were complete (Insull to Gouraud, 7 Aug. 1881; Bailey to Insull, 30 June 1881; both DF [*TAED* D8104ZCF, D8148ZBV; *TAEM* 57:176, 59:846]). These arrangements were substantially altered by the end of the month; Gouraud, in his 12 September reply to Edison (see note 1), explained that "if I had written you the necessary letters to keep you advised of all the infinite variations, the hopes and fears, the ins and outs, the ups and downs, connected with the negociations I would have utterly exhausted your patience and disgusted you altogether with the whole subject, as was I, and everybody connected with it here." The Consolidated Telephone Construction and Maintenance Co. was organized in November (Gouraud and Bailey to TAE, 20 Aug. 1881; TAE to Bailey, 23 Aug. 1881; LM 1:24B, 25A [*TAED* LM001024B, LM001025A; *TAEM* 83:884]; Gouraud to TAE, 12 Sept. and 5 Nov. 1881 with unidentified clipping of 5 Nov.; TAE and Edison Telephone Co. of Europe agreement with Gouraud, Bailey, and Edison Gower-Bell Telephone Co. of Europe, Ltd., 10 Nov. 1881; Consolidated Telephone Construction and Maintenance Co. agreement with Edison Gower-Bell Telephone Co. of Europe, 10 Nov. 1881; all DF [*TAED* D8148ZCA, D8148ZCM, D8148ZCM1, D8148ZCV, D8148ZCW; *TAEM* 59:856, 878–79, 892, 897]).

8. Gouraud shortly reported two errors in the settlement of this account. In Edison's favor was the fact that shares had been sold at a higher net price (£7.1.0 after commissions) than originally stated. More than offsetting this was the fact that Edison received only 2,218 shares instead of 3,000 as estimated by the liquidator, with perhaps another 200 still to come. Gouraud proposed to share equally in the loss arising from having sold more than Edison owned but suggested waiting until the price declined to repurchase the shares. Gouraud to TAE, 4 Aug. 1881, DF (*TAED* D8149ZAC; *TAEM* 59:978).

9. Edison cabled Gouraud on 29 July: "European Light at seventy you better buy some through Drexels" and again to this effect the next day. Thinking that Edison had meant to suggest telephone stock instead, Gouraud cabled back for clarification; Edison cabled on 1 August that it was "Too late to get European light." TAE to Gouraud, 29 July 1881; Gouraud to TAE, 30 July 1881, LM 1:16C, 17A (*TAED* LM001016C, LM001017A; *TAEM* 83:880); Gouraud to TAE, 30 July and 4 Aug. 1881, DF (*TAED* D8149ZAA, D8149ZAC; *TAEM* 59:969, 978).

–2128–

From Otto Moses

Paris, Aug. 2. 81

My dear Sir:

Saturday, Mr. Batchelor arrived with his family and the rest of the staff.[1] I had an Agent to meet him and take charge of the shipments; so he came right on reaching here that night.

I am glad to say, he found us further advanced in our preparations than any other exhibitors in the Palais.

Yesterday while breakfasting with Dr Herz and Mr. De-pretz[2] I was introduced to Mr. Preece, Mr. Hughes,[3] and Sir Charles Bright.[4] A more favorable opportunity than that over the breakfast table could not have happened; and before it was over the two first gentlemen expressed themselves anxious to have all the previous unpleasantness absolutely forgotten.[5] I made an appointment to introduce Hughes to Batchelor and to have Preece present. I took occasion to tell Batchelor of it a very short while after. Prof. Barker was talking to him and said Preece had 'gone' for him on a certain occasion. I suggested hed should join us in a visit. He went, and a happy reunion was effected. So the ground is smoothed all around. Night before last 'La Lumierè Electrique' sent an Editor here with a message to the effect that as Maxim and Swan both wanted cuts printed of their lights, we should have a show. Maxim's cut was[a] a view of a picture gallery, Swan's a street in Newcastle and his store illuminated. The engravings were very good. I accepted the invitation and told the Editor I would be satisfied with a little sitting room with a three drop chandelier in it and a lady sewing or reading, and a gentleman at a desk like yours with one lamp over it.[6]

I am delighted to hear of the success of the dynamo. Berger desired something done by us[b] to illuminate the grand escalier, so we had to decide on using some form of arc lamp. He proposed Jablochkoff[7] but after deliberation I thought that would be advertising him at our expense; particularly, as the Jab. people asked 4000 francs for a months illumination with 16 of their candles, and a proviso that if we did not come to time by 1st Sept they would stay in possession;[c] so I took an arc lamp belonging to a friend of Puskas (Pilsen patent)[8] which we get for nothing during the time we use it, and can take out when we are ready. This lamp is said to be pretty good but we will only take enough of them to light the stair case, not to illuminate it on anything like the scale we are going to employ.

The Exposition is beginning to take shape; but there will not be anything like a full opening. There is so much cutting of concrete soil going on in the Palais that a cloud of dust is always filling the atmosphere. I did not pretend to open the fine instruments while this lasted so fixed everything else first. There was nothing hurt in all our shipments to date but two lamps in the carbon rheostat, two lamps in the barrel that fell from the gang plank and a bent axle in a stock printer.[9]

Du Moncel and his wife called to day at our rooms and examined all that was out. She is a blue stocking and yet very interesting—took interest in everything and said she would call every day.

Herz responded politely to the request to have the Quadruplex instruments, provided Dr Green telegraphed him to do so. He does everything he can to facilitate my personal communication with scientific people here. He proposes to help me all he can in the Lumierè Electrique. Truly yours,

Otto A Moses.

ALS, NjWOE, DF (*TAED* D8135ZAF; *TAEM* 58:953). [a]Obscured overwritten text. [b]Interlined above. [c]"they would stay in possession" interlined above.

1. Batchelor and his wife, Rosanna, had two daughters: Emma (b. 1874) and Rosa (b. 1876). In addition to Moses, Batchelor planned to have eight of Edison's assistants in Paris. These included William Hammer, Edward Acheson, Martin Force, and James Hipple. Philip Seubel had sailed earlier with Moses; Force and at least one other assistant sailed after them, landing in France on 18 July. Doc. 870 n. 1; N-81-05-23:1, Lab. (*TAED* N212:1; *TAEM* 40:774); Force to Mrs. Force, 20 July 1881, Force; see Doc. 2148 n. 10.

2. Moses was presumably referring to Marcel Deprez (1843–1918), an early innovator in electrical power generation and transmission. He did pioneering work in the transmission of direct current over long distances and in the design and characterization of direct-current motors. *DSB*, s.v. "Deprez, Marcel."

3. David Hughes was a London-born but American-educated electrician and inventor living in London. His 1878 claim to have discovered the principle of the microphone set off a bitter controversy among Edison and Hughes partisans on both sides of the Atlantic. *TAEB* 4 chaps. 3–5 passim; *Oxford DNB*, s.v. "Hughes, David Edward."

4. Sir Charles Tilston Bright (1832–1888) was a British telegraph engineer who was instrumental in the development of submarine cables, particularly the Atlantic cables linking Ireland and Newfoundland. He was a Liberal member of parliament from 1865 to 1868 and a delegate to the Paris electrical congress. *Oxford DNB*, s.v. "Sir Charles Tilston."

5. See Doc. 2119. On 9 August Batchelor told Edison that he had "a long talk" with Preece, who "seemed [to] be mighty glad to 'bury the hatchet.'" Batchelor to TAE, 9 Aug. 1881, DF (*TAED* D8135ZAJ; *TAEM* 58:961).

6. An illustration of an art gallery illuminated by electric lights, presumably Maxim's, appeared in the 20 August issue; the two Swan illustrations referred to appeared in the 1 October issue. An engraving of Edison's lamps illuminating a parlor also appeared in that issue; it was reprinted in Prescott 1884, 189. "La Lumière Électrique," *La Lumière Électrique* 4 (1881): 227; "Les Lampes Électriques à Incandescence," ibid., 5 (1881): 9–11.

7. The form of arc lamp invented by Paul Jablochkoff, a Russian emigré, was installed in a public square in Paris in 1878, to much acclaim. Doc. 1659 n. 1.

8. The Pilsen lamp was an arc light invented by the Austrian engi-

An engraving of Edison's incandescent lamp as it might be used in an upper-class parlor.

neers Ludwig Piette and Franz Krizik. It employed an iron core inside a solenoid to advance the carbon. Prescott 1884, 324–42.

9. On 10 July Moses wrote Edison that "a fortunate accident occurred in landing. The barrel containing the lamps through gross carelessness of the boatswain was allowed to fall from the gang plank mid way, and dropped six feet on the head which was staved in. I immediately examined the contents of the top layers and did not find a carbon broken!" DF (*TAED* D8135U; *TAEM* 58:908).

–2129–

From Francis Upton

Menlo Park, N.J., Aug 5 1881[a]

Dear Mr. Edison:

When your letter came asking the question how 104 Volts on 125 Ohms can give ten per H.P., I was discussing the same question.[1] Dr. Nichols tested the lamps, and said that they were 125 Ohms resistance, he also stated that they were 10 per H.P. in the calorimeter. His electrical tests showed only nine per H.P. I knew the lamps were tested by copper deposition standards and that we were keeping the marking the same.[2] That is we were marking lamps 105 Volts [–][b] week in and week out that would test 105 Volts, for I would have old lamps retested at frequent intervals.

During the past week Marshall[3] and Howell[4] have been working over all the constants of the instruments and testing the lines by means of instruments borrowed from Princeton[5] and we found that Dr. Nichols has been out in his measurements of the resistances of lamps so that they ~~run th~~ test now considerably higher than before. I am not yet prepared to give exact results yet, but can say that 140 Ohms is nearer to the resistance of the lamps than 125 Ohms.

I therefore think that you will be satisfied with 95 Volts 8 × 13½ A lamps[6] for Paris, of which we can now give you with the range you mentioned all packed of

92, 93, 94 Volts	68 lamps
95	44
96	108
97 Volts	200 lamps
98	228 "

with about 200 lamps in the photometer room.

We can give you all you may want of lamps 8 × 17.

I am sorry that we should have made any error but I knew that which ever way the truth was it would be more favorable to the lamp.

I am going to lay off Monday, Tuesday and Wednesday of next week and during that time I shall have all the connections rerun in the Photo room; we are very sure that the Volts are good and will soon be sure regarding the resistance.

We take copper as our guide and think that we have marked lamps to be relatively right.

I shall be in New York this evening. Yours Truly

Francis R. Upton.

ALS, NjWOE, DF (*TAED* D8123ZDJ; *TAEM* 57:940). Letterhead of Edison Lamp Co. [a]"Menlo Park, N.J.," and "188" preprinted. [b]Canceled.

1. Edison's letter has not been found, but see Doc. 2149. Samples from each lamp lot were routinely tested to determine the voltage required to produce the rated illumination. In addition, the factory often directly measured the resistance of lamps and, as evident in this document, calculated the number that could be operated by 1 horsepower. See headnote, Doc. 2177.

2. That is, using the rate of deposition in an electrolytic cell to measure the flow of current.

3. William Marshall began doing experimental work at the lamp factory in June 1881, three years after graduating from Rutgers College. He was in charge of the photometer room. After he died of typhoid in September, his brother John took his place. Another brother, Bryun, had worked at the Menlo Park laboratory; David Marshall later joined Edi-

son's West Orange Laboratory staff. Marshall 1930, 161–62, 151, 141–42; Upton to TAE, 26 Aug. 1881, DF (*TAED* D8123ZDP; *TAEM* 57:951).

4. John Howell (1857–1937) had done research at the lamp factory for his senior thesis at the Stevens Institute of Technology, from which he graduated in the spring of 1881. Upton hired him in early July and assigned him to make drawings of the recently acquired factory buildings in Harrison. About this time Howell took over the testing department duties of Edward Nichols, who had left in June for a teaching job. Howell had a long career in electric lighting and eventually became chief lamp engineer for General Electric. Upton to TAE, 6 July 1881, DF (*TAED* D8123ZCV; *TAEM* 57:924); "Howell, John W.," Pioneers Bio.; Jehl 1937–41, 810–11, 815; Wise 1985, 73, 121–22.

5. Upton borrowed a galvanometer from Princeton on 25 June; it is not known what other instruments he may have obtained. Upton to TAE, 27 June 1881, DF (*TAED* D8123ZCQ; *TAEM* 57:917).

6. See Doc. 2085 n. 4.

–2130–

To John Randolph

[New York,] Aug 7 1881

Johnny—

[----]ᵃ Lawson will hire 2 Laborersᵇ at the mine.[1] you can putᵇ them on pay roll; also he wants some picks & shovels= also little Carpenter work=

Edison

⟨$10. Geo Hickman⟩²ᶜ

ALS, NjWOE, DF (*TAED* D8138R; *TAEM* 59:130). ᵃCanceled. ᵇObscured overwritten text. ᶜMarginalia possibly written by Randolph.

1. On 26 July, Edison obtained the right to extract and reduce copper ore for one year, in exchange for one-sixth of any net profits, at an abandoned mine in a pasture near Menlo Park. From September through November, John Lawson undertook the problem of pumping water out of the old shaft, which had defeated efforts sixty years ago. Edison spent $700 on the work during this time. TAE agreement with Mary Ayers, C. P. Ayers Kelly, and Mary Freeman, 26 July 1881, Miller (*TAED* HM810154; *TAEM* 86:413); Lawson to TAE, 15, 17, and 20 Sept., 18, 20, and 25 Oct., 4 Nov. 1881; Samuel Insull to Lawson, 25 Oct. 1881; all DF (*TAED* D8138T, D8138U, D8138V, D8138Z, D8138ZAA, D8138ZAD, D8138ZAE, D8138ZAB; *TAEM* 59:132–35, 137, 140, 141, 139); Ledger #5:109, Accts. (*TAED* AB003:68; *TAEM* 87:471).

In response to an inquiry from geologist J. Volney Lewis in 1906, Edison recalled that he had "Found native or carbonate Copper in the shists which had sheets of Dolerite interposed in the strata— The ores were to[o] lean to pay—of the streak we worked which was about 4 feet wide the average was about ½ per cent." Lewis's historical and technical account of the site concurred with this negative assessment. Lewis to TAE, with TAE marginalia, 28 Nov. 1906, DF (*TAED* D0606ABE; *TAEM* 190:527); Lewis 1907, 153–54 and plate XXXII.

2. A tinsmith by trade, George Hickman had worked at a variety of jobs at the Menlo Park laboratory since 1880. *TAEB* 5 App. 2.

–2131–

Notebook Entry: Dynamo[1]

[New York,] August 8th 1881

Experiments on the resistance of moving contacts especially in regard to commutators and their brushes used on Dynamos—[2]

with South Amn Mac shaft run by gramme[3] at varying s[p]eeds brushes sideswise but not end on good pressure.[4] (more than ordinary

Resistance commutator still.		.055.
"	[--][a] shortckted.	.043.
"	Speed 670 Rev	.065
"	812 "	.068
	915 "	.072
	930. "	.073.

Brushes put slightly more End on— Speed.
(copy of Edison's chart)[5]
Experiments to be tried tonight[6]   Clean Commutator

1   Sims[7] to make experiments with the string on the indicator— Field off brushes off.
2   Take diagrams with boiler pressure 90   ditto 70
3rd   Run 20 minutes with brushes off—indicate[8] try heat, no blower on (Field on)
4   Cool down armature by blower then put ~~pr~~ brushes on, one brush on block the other on insulation—run 2 minutes indicate—notice heat (open circuit)
5   Put on 3 barrels[9] go to 103 volts then set commutators ahead block at a time[10] in direction of ro[t]ation notice spark and drop in volts. use the mercury brushes.
6   Try new brushes (ie) sheets without mercury
7   "   "   "   with mercury
8   Try wire on other brush flatwise
9   Set one brush on one side block ahead of the other brush on the same side   get drop.
10   Mcasure resistance wihle[b] running bet[ween] 2 brushes on the same side.

(Ex 1)   Started ~~10:15~~ 9:45[c]   B[oiler].P[ressure].   340.
We took 5 cards numbered from 1 to 5 ~~of~~ with field off Brushes off—9:57=

We now at 10:02 take cards with field off but brushes on—one brush on insulation one on middle of block=

Now we put field on. 89 volts, seperate machine— Brushes on—no current from machine= Brushes on—stopped 10:42—no blower on—worked cards off at diff boiler pressure 90 down to 60—[d]

We start the blower & take Cards to see if there is any more friction—10:46 PM=— stopped 10:53—taking 2 cards one with blower off & 1 with blower on= 60[d]

1 Barrell on 11:13 PM own field.—Hg Brushes— Theres no spark except very very slight.

with one Bbl on 103. It dropped to 94—when ~~Bb~~ another Barrell put on—2 bbls on now— shut down 27 minutes to 12

4 bbls 12:16 AM. Started

    12:42 " Shut Down armature[b] not very hot would run that way all day— Thought we saw it spark— shut down   started again 1:06 AM. to see if it would cross again. 4[b] barrells on.

                                                    TAE

⟨110 B.P..⟩

X, NjWOE, Lab., N-81-04-06:243 (*TAED* N223:117; *TAEM* 41:152). Miscellaneous calculations omitted. Expressions of time standardized for clarity; some decimal points added for clarity. [a]Canceled. [b]Obscured overwritten text. [c]"(copy of Edison's chart)" . . . ~~10:15~~ 9:45" written by Francis Jehl. [d]Followed by dividing mark.

1. See headnote, Doc. 2122. This entry is continued in Doc. 2134.

2. This document was written in several stages. After this brief introductory statement, Edison recorded results from experiments on the resistance of a dynamo commutator. Then Francis Jehl copied into the book Edison's instructions for additional experiments to be tried (the original has not been found). The last section consists of results given by Edison from related tests.

3. That is, a Gramme dynamo run as a motor.

4. This generator was built for a South American demonstration plant (see Docs. 2048 and 2144). Edison described this test in Doc. 2149.

5. Francis Jehl wrote Edison's list of experiments to be tried and began to make notes for the first trial, at which point Edison resumed his notes.

6. As evidenced by the resistance barrels and blower for cooling the armature, these experiments were to be made on the completed Jumbo dynamo.

7. Probably Gardiner C. Sims (1845–1910). Sims superintended a locomotive works when he met Pardon Armington, with whom he formed a partnership for the manufacture of high-speed stationary engines. He and Edison developed an electric torpedo, and Sims later served in various engineering capacities in the U.S. Navy and Army. Obituary, *ASME*

Transactions, 32 (1910): 1501–2; Bowditch 1989, 85; Sims-Edison Electric Torpedo Co. brochure, 1886, PPC (*TAED* CA017B; *TAEM* 96:619).

8. That is, to inscribe an indicator card giving a graphical representation of the steam engine's performance.

9. Resistance coils in barrels of cooling water.

10. That is, to shift the brushes the width of one commutator block.

−2132−

From Theodore Puskas and Joshua Bailey

Paris Aug 10[–12]/81

Dear Sir:

Cabled you yesterday to cable date of delivery of the American patents that cover the French patents delivered Paris August '79, and January February and June '80.[1] This was because that, in order to get an injunction it is necessary that the American patent, should have issued to the party asking it.[2] Our counsel in interview two days ago said that it is alleged that the Maxim patents have been issued and yours not. In the absence of copies of your patents here, it was necessary to have your formal assurance on this point.— Whether the fact that all the patents invoked by us have not yet been issued, will make a difficulty in getting injunction, remains to be seen.

Also, it is necessary that the original patents should be produced in support of the injunction. The absence of these may cause the injunction to be raised provisionally, i.e. until the production of the original patents. Should the injunction be granted and should you hear of its being raised within a few days, you need not be uneasy. The case seems a very tight one and if we get it on Maxim propose to go at once for Swan and Fox.—[3] The Leon syndicate are about to appoint two engineers to make report to them on the economy etc of the Light in your Exhibition. In interview with Mr. Leon[4] this morning the writer proposed to him that another engineer should be added to their two by us, and that the selection on the one side and the other should be mutually acceptable or a change should be made.

Mr Leon accepted this. [----][a]

12th Aug. As telegraphed last evening injunction was granted yesterday. This morning passed by the Exposition to see if the Maxim people have lamps visible to seize, and finding that they have, telephoned advice[b] to counsel who replied that seizure would be made at Exposition at two P.M. On entering office 33 Ave de l'Opera, found your cable saying "injunction against whom and on what." It was supposed that you would understand the phrase "principal infringer" used in the first cable sent you regarding the matter. The answer "Against Maxim and will be served at Exposition this afternoon unless you cable contrary" was sent you at ten this morning, and counsel was advised not to make seizure till five this P.M. to give time to receive your reply, should you desire to make one.[5]

Batchelor and Moses have been present, one or both of them, at the interviews with counsel, at which the points of the respective patents have been discussed, and are clear in the opinion that the injunction should be pushed. Enclosed herewith you will find extracts from Paris journals, which have been kept in as adv. for several days.[6]

Unless Batchelor and Moses and the Counsel who have examined the points are entirely mistaken, and your patents are valueless because you have invented nothing, the proceeding is a good one. The Exposition raises the questions between you and Maxim and others in such a way that, even if your case were a weak one and it was your interest to dodge and mystify that course is not open. If you have really invented the things claimed for you, the policy of a square attack in reply to the claims made against you is the only one to be thought of. We have no doubt you wish this, and all of us do understand.

Armengaud, who[b] is not a very positive man, and is very timid, said squarely to counsel that he considered Maxim an infringer and that he advised pursuing him. In the article "l'Exposition d'Edison" (p 161,) of the Catalogue Armengaud tells me that he put in to the article[c] the four lines which he thinks sum up your claims on the lamp question, and which he thinks you can hold against everyone, "Edison est le premièr qui ait fait [usage], et ces brevets en font foi[,] d'un filament de charbon [incandescent] <u>continu</u>, avec une resistance superieure a dix ohms, dans un vide maintenu par un globe de verre <u>continu</u> dans lequel on scelle les conducteurs metalliques."[7d]

We have paid 3000f retainer to counsel, Falertoz.[8] Yours very truly

Puskas & Bailey

ALS, NjWOE, DF (*TAED* D8135ZAK; *TAEM* 58:963). Written by Joshua Bailey. [a]Canceled. [b]Obscured overwritten text. [c]"to the article" interlined above. [d]Missing text supplied from printed copy.

1. In reply to the message from Bailey and Puskas, Edison cabled the same day a list of the relevant issued and pending American and French patents. On 10 August Puskas and Bailey asked him to "send quick official copies American patents mentioned your cable both of those issued and those not issued   also official copies those of your infringers." On

21 August Edison wired that he had sent official copies of the patents; they arrived on 7 September. Puskas and Bailey to TAE, 9 and 10 Aug. 1881; TAE to Puskas and Bailey, 9 and 21 Aug. 1881; LM 1:20A, 21A, 20B, 25C (*TAED* LM001020A, LM001021A, LM001020B, LM001025C; *TAEM* 83:882, 884); Edison Electric Light Co. of Europe memorandum, 20 Aug. 1881; Puskas and Bailey to TAE, 9 Sept. 1881; both DF (*TAED* D8135ZAW, D8135ZBN; *TAEM* 58:1003, 1056).

2. Bailey and Puskas recommended in early July that they take legal action on Edison's behalf against Hiram Maxim and Joseph Swan. After the arrival of Otto Moses, the three consulted with [Charles?] Armengaud about the feasibility of obtaining injunctions. Armengaud was confident of sustaining an injunction against Maxim on the grounds that Edison's patent claims anticipated Maxim's by about a year. Puskas and Bailey cautioned that Maxim, in particular, "should not be allowed to get possession of public opinion" and promised that an injunction would "deter capitalists" from investing in his system. Puskas and Bailey to TAE, 22 July, 2 Aug., and c. 10 Aug. 1881; Dorval memorandum, 13 Aug. 1881; Dorval and TAE memorandum, 19 Aug. 1881; all DF (*TAED* D8135ZAA, D8135ZAG, D8135ZBD, D8135ZAL, D8135ZALI; *TAEM* 58:930, 956, 1027, 967–77); Fox 1996, 179–81.

3. St. George Lane-Fox (who later adopted the surname Pitt) filed four British patent applications related to electric lighting in 1878 and 1879. He subsequently organized the Lane-Fox Electrical Co., Ltd., which licensed patents to the Edison interests in Britain. Doc. 1780 n. 15.

On 17 August Puskas and Bailey wrote Edison that they were closely examining Swan's patents and scientific publications to determine whether they could succeed in getting his exhibit enjoined. However, they were concerned that doing so might alienate prominent British participants at the Exposition, particularly members of the award jury. They also suggested that the action against Maxim made it "unnecessary so far as business reasons are concerned to be in any undue haste regarding Swann." In early September, after a review of Swan's patents and publications, they decided "to make an immediate application to the Tribunal for an order of seizure on Swan, taking just the same line of procedure as in the case of Maxim." In mid-September, however, Puskas, Bailey, and Batchelor decided to delay proceedings until Grosvenor Lowrey arrived in Paris to review the legal situation and there is no evidence of further action. Bailey and Puskas to TAE, 17 Aug., 2 and 9 Sept. 1881, all DF (*TAED* D8135ZAP, D8135ZBF, D8135ZBN; *TAEM* 58:985, 1030, 1056); Batchelor to Eaton, 18 Sept. 1881, Cat. 1331:38, Batchelor (*TAED* MBLB3038; *TAEM* 93:329).

4. Elie Léon was a French financier who furnished part of the money for Edison's exhibition at the International Exposition in Paris. In exchange, he received the right to form an Edison company for France. After the success of Edison's exhibit, Léon and Charles Porges organized a large syndicate of European investors that in turn established the Compagnie Continentale in February 1882 to control Edison's electric light business in Continental Europe. TAE and Edison Electric Light Co. of Europe agreement with Porges and Léon, 15 Nov. 1881; undated Porges report, 1881; both DF (*TAED* D8228K, D8132ZCJ; *TAEM* 61:597, 58:591); Edison Electric Light Co. of Europe report, 7 Mar. 1884 (p. 2), CR (*TAED* CE001003; *TAEM* 97:209); Fox 1996, 184–85.

5. On 11 August Bailey cabled Edison: "Injunction granted   have your official copies duly certified to produce in court." Edison replied the same day, "Against whom and on what was injunction issued." Bailey responded on 12 August: "Against Maxim and will be served at exposition this afternoon unless you cable contrary." Edison immediately instructed him to "serve injunction   now how about Swan." He also cabled separately the same day: "Important give news regarding injunction agents American Press with request to cable." Bailey to TAE, and TAE to Bailey, both 11 Aug. 1881; Bailey to TAE, and TAE to Bailey, all 12 Aug. 1881, LM 1:21C–23A (*TAED* LM001021C, LM001022A, LM001022B, LM001022C, LM001023A; *TAEM* 83:882–83).

6. These extracts have not been located.

7. This passage in English is, "Edison was the first to use and the patents show a continuous carbon filament with a resistance greater than ten ohms, in a vacuum maintained by a continuous glass globe in which are sealed the metallic conductors." Text supplied in the quotation is taken from the published version. France. Ministère des Postes et des Télégraphes 1881, 161.

8. Not otherwise identified.

–2133–

From Otto Moses

Paris, Aug. 11. 1881

My dear Sir:

The Exposition opened on the 10th in order to allow of inspection by the President[1] and his Cabinet. The exhibitors as much as possible had cleaned up the place but all their efforts did not put the machinery in motion. It was a mere formal opening. We are nearly ready however, and I believe more advanced than any of the large exhibitors. We had crowds of visitors; among them several of the Syndicate to whom the working, manufacture &c. were fully explained. They did not appear to know anything about it, and were very much interested. I have been entertained by Mr. Porges[2] at his country seat at St. Cloud and have found him, aside from his wealth and influence (which are very great) a very intelligent and enquiring man. Mr. Léon too has been very kind. He asked me yesterday to enquire about the possibility of obtaining a plant from America for his house near the Bois de Bologne. He will pay $2000 for a 40 light installation. If it could be managed it would be a good plan to let him have it; Mr. May[3] also took great interest in the matter. My relations with all these gentlemen are very pleasant and at anytime I could personally present any views you might desire to express.

Du Moncel is greatly interested in everything we have out on the tables. He and his wife pay us visits at least twice a day.

Her influence over him is very great as he seems to rely upon her observation entirely as his eyes are now very weak. He has just written an article for La Lumierè Electrique on your system which Mr. Batchelor and I revised, and which Mrs Du Moncel corrected as I called her attention to the uncertain points in the descriptive parts.[4] She has a great admiration for you and I believe she is inclined to hero worship. Mr. Bailey has no doubt written you about Du Moncel's new relations with his paper, so I will not touch on the subject.[5]

Mr. Fabbri called to see the salons the day of his arrival and expressed himself as well pleased. The tapestries on the parlor room walls[a] (as you might call the one from which the coarser instruments have been excluded) so attracted him, that he desired me to make special enquiries about their prices probably with a view to purchasing some of them. An inventory of the valuables loaned in your exhibit shows them to be worth over 350.000 francs. All this is in strong contrast with the empty look of Maxim's counter in the Salle d'honneur as he advertises the room he proposed to light up. But the blow which he did not expect came this afternoon. Count De[liliand?],[b] Berger's Secretary, came to our room and said that Mr. Maxim's agent in Paris, who represents the U.S. Electric L[ighting]. Co., had just come to his office in great consternation about the injunction which had been put upon Maxim's light. The Count said "he was sorry for him, but it was all right if it was deserved." The whole matter has caused quite a furor in the Palais. The newspapers have not yet spoken of it, simply because if you wish anything given to the public, (and there's money in it), you must pay for it. The reporters openly accost you with pen in hand (in place of pistol as brigands would do) and say they will publish such and such a notice at such and such a price. They do it as unblushingly as a bootblack would ask you for his nickle. In fact beggary is a profession here— practiced at times in silk or furs and sometimes in rags; but more generally the first. Paris is in this respect sui generis. The foreigner is the natural prey of the people by common consent. It begins as you enter Paris, or for that matter as your food comes into the city, for[a] it is taxed, in eating it you are taxed and robbed, and as you progress you do it with your hand in in[c] your pocket. Pay–pay–pay. That is the war cry here.

The newspapers must be managed, however, so I hope you will arrange some way by which we can 'go it strong' with them. Mr. Puskas arrives in N.Y. tonight and I hope things will be focussed time enough for the Congress.[6]

Mr. Batchelor is getting along all right. The instruments are all in good condition with but one or two exceptions. He hopes to have the plant running by Monday; but the official opening at night will not take place before the 27; so you may be sure everything will be working smoothly by that time. Faithfully Yours,

Otto A Moses

ALS, NjWOE, DF (*TAED* D8135ZAM; *TAEM* 58:978). [a]Interlined above. [b]Illegible. [c]Repeated at end of one page and beginning of next.

1. François Paul Jules Grévy (1813–1891) was president of the French Republic from 1879 to 1887. *Ency. Brit.* 1911, s.v. "Grévy, François Paul Jules."

2. Charles Porges was associated with the Banque Centrale du Commerce & de l'Industrie in Paris. With Elie Léon, he organized a large syndicate of investors for the control of Edison's electric light business in Continental Europe. See Doc. 2132 n. 4.

3. All that is known of E. May is that he was affiliated with the Banque Franco Egyptienne and he was involved in commercializing Edison's telephone in France. May to John Harjes, 16 Apr. 1880, DF (*TAED* HM800101; *TAEM* 86:123).

4. Du Moncel 1881.

5. On 29 July Puskas and Bailey wrote Edison that they had a "long conversation with Du Moncel yesterday morning in which he explained fully his connection with Lumiere Electrique, and agreed on certain conditions . . . to sever his connection with L.E. and go with us." They also told Edison that they were working to establish a journal and to obtain the services of du Moncel and other prominent French scientists as contributors. However, du Moncel retained his post as scientific editor of *La Lumierè Électrique,* and the Edison interests did not start a new electrical magazine. Puskas and Bailey to TAE, 29 July 1881, DF (*TAED* D8135ZAD; *TAEM* 58:939); Fox 1996, 164–67.

6. The International Electrical Congress, the first of its kind, was held in conjunction with the Exhibition.

–2134–

Notebook Entry: Dynamo[1]

[New York,] Aug 11 1881—

Tests with Indicator and weighing Coal & water on Babcock & Wilcox Boiler, Lump anthracite—Croton water=

First test simple friction Engine, no field=

Evaporation Commenced, B[oiler]P[ressure][a] 120— 10:51½ PM.— Engine started at [--][b] 11:03 PM with 3 bbls on— on test= Volts. 103=[a] Besides the 3 bbls there is 9 other lamps on.

103 Volts on armature   Average heat 129 deg in barrells—

Had[c] to increase speed of blower after 2 hours run after 4 hours had to increase it again= bobbin very hot.[d] took speed of the blower—(Large size=) 1940 Rev per minute. It will require the boreing out of $^1/_{32}$ and air equal to a speed of the big blower 2500 per minute to keep cool—

Mercury wears nearly off Commutator after 4 hours run and brushes especially X

Sparks somewhat freely as it catches stuff at .g.— 4:30 AM.[e] Mercury worn off sparks ratherly badly. 3:35

Shut down at 5:10 AM—stopped Evaporating water at 5:17 AM[d] Mercury wore off about 3 AM Sparked badly after that & spotted up Commutator pretty well= Armature when stopped not extra hot though it wouldnt do to let it get any hotter= The joints or contacts were quite hot and the bolts that goes through plates very hot= temperature of the contacts at bars 147= There was 6 ohms outside of field—

X, NjWOE, Lab., N-81-04-06:251 (*TAED* N223:121; *TAEM* 41:156). Expressions of time standardized for clarity. [a]Interlined above. [b]Canceled. [c]Obscured overwritten text. [d]Followed by "over" to indicate page turn. [e]"4:30 AM." interlined above.

1. See headnote, Doc. 2122. This entry is a continuation of Doc. 2131.

–2135–

From Charles Batchelor

Paris Aug. 12th 1881

My dear Edison,

Recd. your cables etc—[1] The exposition was opened but not a single exhibitor ready. My tressle work was not quite finished but shall be running by the 15th which will be as soon as any and before the great majority There is a terrible lot of red tape here— I also find great difficulty in getting anything done here; and not all without great expense. The exhibition has made everybody very independent and if you buy or hire anything they will not let it go without pay beforehand— I was not able to get a fast speed engine so had to hire one that only runs 81 revolutions, but of course can make that do; although it looks much larger than it would need to if I could have got a fast speed— I have to pay $200 per month[a] for the use of it and had to pay $350 to put it in place and raise stack ready for use. When I first came here our men were being boarded at a very high rate nearly $18 per week per man I have now got them a cheaper place although the best I have been able to get is $12 per week— Of course I want to keep them as near the exposition as possible— I have got a man looking round for still cheaper rates—

The Electric pen, Stock printers, District, pressure relay expansion relay, Motograph relay, Telephone duplex, Etheric force, 1 Universal private line printer (the other badly damaged but am fixing it) & all telephones; are all up and working well—

(English Light.)[b]   Fabri was here and consulted me about London— I sent Littell the boiler man[2] over there to see the places that Mr Fabri had picked out as suitable to put the boiler and Dynamo and he will report to me in a few days

I have got hold of Biggs[3] of the "Electrician" and Kempe[4] of the "Telegraphic Journal" and spent yesterday 4 solid hours talking economy of the system— They started in very skeptical but have now become very much interested— Biggs told me he was the scientific head of the Engineer (and between you and I it is a damn poor head) so I am trying to kill two birds with one stone—

DuMoncel is in every day— Dubose the Electric regulator man[5] ~~said that~~ told the Doctor[6] yesterday that ours was the only thing he had ever seen that looked like business— There is no other exhibitor here with a complete system such as ours and people are struck with the display of conductors and other plant— The two little machines look so small at one end of this immense building that I am frequently asked the question are you going to light both these large rooms with those little machines down stairs   Yours

"Batch"

ALS, NjWOE, DF (*TAED* D8135ZAO; *TAEM* 58:983). [a]Multiply underlined. [b]Written in left margin and multiply underlined.

1. On 9 August Batchelor cabled Edison that "Moncell wants cuts Street boxes and armature   can we publish." Edison gave his answer in a cable to Puskas on the same day: "Regarding injunction consult Batchelor and use your best judgment   Tell Batchelor following   second countershaft and quadruplex shipped. do you want whole or half light machines. give Moncel cuts street boxes and armatures. autographic protected   can exhibit." Du Moncel used these illustrations for his article in *La Lumière Électrique* on Edison's electric lighting system. Batchelor to TAE and TAE to Puskas, both 9 Aug. 1881, LM 1:19B–19C (*TAED* LM001019B, LM001019C; *TAEM* 83:881); du Moncel 1881.

2. G. W. Littell was an engineer for Babcock & Wilcox. Batchelor to Theodore Puskas, 14 June 1881, DF (*TAED* D8135N; *TAEM* 58:899).

3. C. H. W. Biggs edited the British journal the *Electrician*. He championed a variety of battery technologies during the 1880s and 1890s, par-

ticularly the "chloride process" which ultimately proved impractical. In July Biggs asked Edison to provide descriptions and drawings of his exhibits for the journal. Biggs to TAE, 16 July 1881, DF (*TAED* D8135Y; *TAEM* 58:927); Schallenberg 1982, 216.

4. Harry Robert Kempe (b. 1852) was a British telegraph engineer who helped to found the *Telegraphic Journal*. He wrote two handbooks on electrical testing and practice. Bright 1974, 186.

5. French electrician C. Dubos held two British patents for arc light regulators and one for a dynamo. Dredge 1882–85, 2:xlvi, lxxiv–lxxvi.

6. Otto Moses.

–2136–

Draft to Edward Johnson

[New York, August 19, 1881?[1]]

My Dear Johnson:

I have just learned that the Maxim Co[2] have by their usual misrepresentation succeeded in obtaining contract for lighting one of the Penna RR ferry boats as against us.[3] This is pretty rough when one considers that what they put on[a] is one of the most shameful infringements that ever existed that an unimportant decision in[a] the patent ofs [--------][b] upon an entirely foreign matter is represented to the Penna RR ~~as having~~ affecting our patents & giving them rights,[4] ~~and when~~ when it is a notorious fact known [far?][b] throughout the world ~~that~~ & should even have been known to a Penna Official that ~~at Menlo Park in 1879 in~~ that ~~subdivision of that~~ that "scientific impossibility" the subdivision of the Electric Light was accomplished by me, ~~and all the means & methods~~ and ~~it would~~ one would naturally suppose that [it?][b] even a novice would know that the means & methods of doing it would belong to me & not to a pirate Company who appears on the field over a year afterwards= I[b] ~~have not the slightest concern about any mon~~ do not care the slightest about any money we could make out of the Penna RR but It seems a little rough that this contract should be given [-- ------][b] ~~record for of~~ [------][b] because we would not stoop to false statements, and Especially when one considers that th~~ise~~ [-------][b] Exhibition of this very light has brought over 70 000 people over the Penna RR to see it in operation at Menlo Park[5] and[a] that long before these pirates started in business—[6] I understand you are acquainted with Mr Cassatt,[7] would you not go and place this matter before him, and Explain the lies of these infamous shysters.[8] Yours

T A Edison

ADfS, NjWOE, DF (*TAED* D8120ZAU; *TAEM* 57:625). [a]Obscured overwritten text. [b]Canceled.

1. Edward Johnson wrote or copied an amplified version of this letter, dated 19 August, based on Edison's draft, presumably to transmit to the Pennsylvania Railroad. TAE to Johnson, 19 Aug. 1881, DF (*TAED* D8120ZAV; *TAEM* 57:628).

2. The United States Electric Lighting Co., formed in 1878, controlled the lighting patents of its chief engineer, Hiram Maxim. At this time the company was promoting incandescent carbon electric lamps, which Edison considered inferior imitations of his own. Bright 1972, 47–48; Passer 1953, 147–48; see Docs. 2021 and 2022.

3. The Pennsylvania Railroad operated over leased tracks to its terminal at Jersey City, from where it ran ferries to New York. Condit 1980, 50–51.

4. Johnson referred in his version of this letter (see note 1) to "the Thermostatic Regulator—an entirely foreign matter to the present state of the Electric Lighting problem." In 1879 Edison obtained his first patent for an incandescent electric lamp (U.S. Patent 214,636), a platinum lamp with a thermostatic current regulator to prevent the metal burner from overheating. The patent was placed in interference with an application by Hiram Maxim, however, and in February 1881 the examiner awarded priority to Maxim. Edison appealed twice, but the Commissioner of Patents affirmed the ruling in late July or early August. By that time neither the Edison Electric Light Co. nor the U.S. Electric Lighting Co. foresaw any practical use for the invention. Decision in *Maxim v. Edison*, 4 Feb. 1881 (*TAED* W100DCA022); "The Platinum Lamp," *New York Tribune*, 3 Aug. 1881; "The Thermostatic Regulator," *New York World*, 4 Aug. 1881, both Cat. 1242, items 1629 and 1625, Batchelor (*TAED* MBSB31629, MBSB31625a; *TAEM* 94:640).

5. See *TAEB* 5:539–40.

6. In the longer version (see note 1), a separate paragraph was inserted here. It called the Maxim company's claim to superior operating economy "a simple fraud— It is notorious that they can get But $1/2$ the No of Lamps of a given candle power—per Horse Power that we obtain."

7. Alexander Cassatt (1839–1906), brother of noted painter Mary Cassatt, was connected with the Pennsylvania Railroad most of his working life and played crucial roles in its rapid expansion. He was general manager until 1874 and third vice president until 1880, when he was appointed first vice president. Disappointed at not having been named president, Cassatt retired in 1882. He returned to the Pennsylvania in 1899 as president, a position he held until his death. *ANB*, s.v. "Cassatt, Alexander Johnston."

8. In the longer version (see note 1) Edison referred to Johnson, who began his telegraphic career as a station agent, as "an old Penna RR man" and asked him to request a hearing: "My relations Past present & to come with the Penna Co—as a resident on their Line & a patron of no inconsiderable moment should also entitle me to this— Please see them on this subject & report to me." The editors have found no reply from Johnson or the company, but see Doc. 2216. *NCAB*, 33:475.

From Eduard Reményi

Dear Sir and Friend[1]

Since I was with Victor Hugo and Liszt I never was so much in ~~the~~ a[b] intellectual heaven as day before yesterday— I was wide awake, still I was in a dream-land, and I want to remain there, and to nourish[c] myself on that heavenly food—and in the same time I do not wish[c] to be so terribly in debt toward you—otherwise I will be soon bankrupt,—therefore prepare yourself immediately—if not sooner, to a musical assault on your doomed head—and then, only then we will be even—

My most affectionate regards to that sympathetic straightforward luminary Johnson— Your affectionate fidler

Ed Remènyi[2]

Looking at your Photo—I invent also all sorts of melodies—you bet.

ALS, NjWOE, DF (*TAED* D8104ZCN; *TAEM* 57:188). Letterhead of Westminster Hotel. [a]"New York," and "188" preprinted. [b]Interlined above. [c]Obscured overwritten text.

1. It is not known how or when Edison met violinist Eduard Reményi. Their correspondence was underway on 2 February when Reményi thanked Edison for accepting a dedication, possibly to a "Liberty Hymn" that he later sent. He also acknowledged Edison's invitation to meet him. In August and September 1881, while performing at Koster & Bial's on 23rd St. in New York, Reményi invited Edison to attend on several occasions (Reményi to TAE, 2 Feb. 1880, 19 May, 16 and 29 Aug., 13 Sept. 1881; all DF [*TAED* D8004ZAK, D8004ZDG, D8104ZCK, D8104ZCS, D8104ZCY; *TAEM* 53:67, 162; 57:185, 195, 207]). Edison later recalled that the violinist made late-night visits to his Fifth Ave. office and the Edison Machine Works to discuss philosophy and politics and, on at least one occasion, to give an impromptu performance. He offered him an option on shares of Edison Electric Light Co. stock in September 1882; in 1898, he was a pallbearer at Reményi's funeral (App. 1.B.15; TAE agreement with Reményi, 20 Sept. 1882, Miller [*TAED* HM820165A; *TAEM* 86:494]; "The Funeral of Remenyi," *New York Times,* 30 May 1898, 7).

2. Hungarian-born of Jewish ancestry, violin virtuoso Eduard Reményi (1828–1898) entered exile in the United States after participating in the unsuccessful 1848 revolution in Hungary. He toured Europe with Brahms in 1853 and became a protege of Liszt. A popular concert artist, he toured extensively through Europe in the 1860s and 1870s before again taking up residence in the U.S. about 1878. He continued to travel widely until he died while performing in San Francisco. *NGD,* "Reményi [Hoffman], Ede [Eduard]."

From Francis Upton

Dear Mr. Edison:

Wednesday I had the mercury cleaned[1]   Yesterday I started the four workers in one ~~iaisle~~ aisle with ~~oneh~~ one hundred pumps. They got off 600 A and 100 B lamps and will do better than this. Next week Friday I am going to start one hundred more pumps and put the two gangs into one and work daytimes only for all the lamps I can get.

Frank Holzer[2] has taken a contract to make 500 pumps at 40 cts. each. I shall do no repairing until this contract is finished and expect to have a number of pumps each day more than he has been giving for contract work brightens up slow men amazingly.

Holzer, Dyer and I have each given Welsh[3] a braceing up[4] and have made him work much better. His breakage has come down to the old figures about 9% once more.

Frank Holzer found it almost impossible to introduce the platinum foil into the fall tube.[5] Will. Holzer[6] will try it again.

Bradley ~~tryied~~ tried last winter[b] every way he knew of to seperate fibres from the bamboo without success. He steamed them several hours, twisted them between his fingers rolled them between plates. He will try again. This sticking of the fibres together is what makes the bamboo so good for our use. We could never cut a ribbon $^6/_{1000}$ ~~thinck~~ thick if the fibres were not held firmly together.

There is no moisture[c] in the lamps that the eye can see. I shall have certain ~~pup~~ pumps run with lamps that have been taken from Sulphuric Acid bottles and report.[7]

I have sent plumbago forms to Bergmann to plate.[8]

When you spoke about our not trying experiments I was rather taken aback. That night I counted 12 experimental pumps on the line and we have ~~tryied~~ tried over 150 forms in the last six months.

The only way with pumps is to keep at it and try the same form again and again to see if there is any change due to mercury or weather. It is the hardest of all experiments to try as a pump works well at one time then badly the[n] good and we cannot tell why.

We have now tested ~~over~~ about[d] 200 orders since starting and are putting through lamps as fast as we can not neglecting regular work.

Everything is pulling together now again and I shall show full runs in a few days.[9]

I am cutting the price on piece work constantly[10]   Bradley is bringing down the price of fibres. You know that 1000 lamps a day means a good deal and costs. 35 cents a lamp was as low a price as you could see at the start and we are ~~doing that~~ making them at a little less than that. Yours Truly

Francis R. Upton

ALS, NjWOE, DF (*TAED* D8123ZDN; *TAEM* 57:945). Letterhead of Edison Lamp Co. [a]"Menlo Park, N.J.," and "188" preprinted. [b]"last winter" interlined above. [c]Obscured overwritten text. [d]Interlined above.

1. Francis Upton had the vacuum pumps overhauled at the end of July following several days of poor operation. They worked better for a short period but on Wednesday, 17 August, he "stopped the pumps so as to clean the mercury and give them a thorough cleaning." Alfred Haid suspected the mercury was contaminated with lead and Upton thought it "best to take a decided step and clean the whole with acid." Upton to TAE, 30 July and 17 Aug. 1881, both DF (*TAED* D8123ZDH, D8123ZDM; *TAEM* 57:937, 944).

2. Frank Holzer (1859–1927) learned glass blowing from his older brother William and joined the payroll at the lamp factory in October 1880. His major duty at this time was making vacuum pumps. He spent his working life at Edison's lamp works and its successors. "Holzer, Frank," Pioneers Bio.

3. Nothing is known of Alexander Welsh before he began operating vacuum pumps at the lamp factory in the fall of 1880. He was placed in charge of carbonizing about February 1881. Upton's testimony, pp. 3255–56, *Edison Electric Light Co. v. U.S. Electric Lighting Co.*, Lit. (*TAED* QD012F3254; *TAEM* 48:134–35).

4. To summon resolution for an effort or prepare for exertion (*OED*, s.v. "Brace," 5b, c). Upton had recently expressed a desire to fire Welsh, whom he said "has proved very unsatisfactory of late. He is thoroughly lazy and given to untruths." However, Upton feared "that he will go to the opposition and that he has had his eyes quite wide open since he has been here. I can get good carbons without him and feel better satisfied" (Upton to TAE, 1 Aug. 1881, DF [*TAED* D8123ZDI; *TAEM* 57:938]).

5. The purpose of the foil is not known but may be related to arrangements Edison sketched in June for inserting short lengths of a constricted platinum tube into the supply tube in order to standardize the rate of flow among all pumps. He executed a patent application embodying this idea on 1 July. Cat. 1147, Lab. (*TAED* NM017:82–83; *TAEM* 44:306–7); U.S. Pat. 263,147.

6. William Holzer, brother of Frank, was an experienced commercial glassblower from Philadelphia whom Edison hired in 1880. He married Alice Stilwell, Mary Edison's sister. About this time he was superintendent of the factory's glassblowing department. Holzer to TAE, 6 Jan. 1880, DF (*TAED* D8014C; *TAEM* 53:488); *TAEB* 5 App. 2.

7. See Doc. 2139; this was presumably an experiment to remove moisture from the lamps using sulphuric acid instead of phosphorous anhydride, the factory's usual practice. When Upton reported results of tests on contamination in the mercury a week later, he "thought best to let the experiments with other materials than phosphorous wait. Pumps are now so important that I only want to try such experiments as will in-

crease their life. The experiments with other driers must have a long run to decide anything." Upton to TAE, 26 Aug. 1881, DF (*TAED* D8123ZDP; *TAEM* 57:951).

8. On the new plumbago carbonizing molds see Doc. 2094. Upton was still trying to determine why the nickel weights in the new forms deteriorated rapidly, but nevertheless asked Sigmund Bergmann to finish 500 more forms. Edison executed a patent application on 19 September for a plumbago carbonizing mold which he stated was durable, cheap to manufacture, and would not damage the filaments. Upton to TAE, 27 June 1881, DF (*TAED* D8123ZCQ; *TAEM* 57:917); U.S. Pat. 263,144.

9. In his 1 August letter (see note 4), Upton confessed that he had endured "the worst week nearly I have had since I have been here. Every thing was wrong and dragging and I am ashamed. This week will be much better I hope for I think I have pulled things together again."

10. See headnote, Doc. 2177.

-2139-

From Francis Upton

Menlo Park, N.J., Aug 20 1881[a]

Dear Mr. Edison:

We tried the japanned clamps yesterday. The blue lasted very long on them so that the workers complained a good deal of the delay.[1] The blue seemed to make just at the edge of the carbon and in one instance it broke the carbon itself which happens very seldom with copper clamps.

The drying by sulphuric acid does not show any advantage so far. Those that have run pumps with dried bulbs say they are no better, they can see no difference.

I find I cannot get 1000 A lamps off 200 pumps in one run of 11 or 12 hours sealing off lamps that have been worked high, even if I let tubes I know are not just right pass.

Now about reducing expenses. Considering that we can make lamps faster than the market calls for them would it not be well to cut down to one gang and run about 600 A lamps a day increasing as we can take care of them?

I know we cannot make the lamps as cheaply as we can in larger quantities per lamp but the running expenses will be less.

The great trouble I think is that we shall have to discharge a number of good and well trained men who can carry away points.

As regards the pumps I am sure that [charg?][b] the pumpers can tell what we have called good lamps and how they have been made. Yet I know they are somewhat mixed and that I can mix them more by running through experiments.

I shall try and see you Monday. Yours Truly

Francis R. Upton.

ALS, NjWOE, DF (*TAED* D8123ZDO; *TAEM* 57:949). Letterhead of Edison Lamp Co. a"Menlo Park, N.J.," and "188" preprinted. bCanceled.

1. A bluish discharge was often visible around the clamps when lamps were brought to high incandescence on the vacuum pumps. It was attributed to an emanation of gas and understood as a sign of insufficient vacuum (headnote, Doc. 1898). On 16 August, John Howell recorded an order (no. 399) for "75 Lamps with clamps japaned." Two days later he noted a batch (no. 408) of "Lamps that blue hangs in and trouble workers" (Cat. 1301, Batchelor [*TAED* MBN007:38–39; *TAEM* 91:331–32]).

–2140–

*Charles Batchelor to
Sherburne Eaton*

Paris France. August 22nd 1881

My dear Sir,

Yours of the 9th received[1] also numerous papers etc for all of which, thanks. We are progressing well and although backward somewhat we are more forward than the majority of exhibitors. The time for lighting has been set for the 27th but I am lit up every afternoon from 3 till 7. Swan has made one miserable attempt and Maxim not at all— It is the general remark of everybody that we get hold of that we seem to have a complete system whilst any other exhibitor has only a "lamp" or a "machine"— We frequently find Swan's manager in our place studying our "processes for lamps," handling sockets, safety catches etc— We shall have to lose no time now or these fellows will steal all we have and use it right under our very nose.[2] We find it is impossible to light up the garden spoken of by Messrs. Puskas and Bailey as it is impossible to have it included as part of the exhibit— Very respectfully yours

Chas Batchelor

P.S. I wish also to call your attention to the fact that of all the lamps that have been sent here not one was broken— In Havre one of our barrels dropped about 10 feet in unloading and in that one, we found 2 of the carbons broken. This speaks well for the lamp as far as packing and shipping were concerned. All the barrels were packed a little different from each other and I was asked by the Edison Lamp Co to report on their condition after opening but each seems equally as good as the other— Yours   C.B.

ALS, NjWOE, Cat. 1331:5, Batchelor (*TAED* MBLB3005; *TAEM* 93:303).

1. Eaton's letter has not been located.
2. Batchelor reiterated this to Eaton about two weeks later. After describing the various installations of Edison lights he had made in and

around the Exposition, he warned that "all we have is open to view; and in Europe the 'Light' will have to be exploited very quick or the other people will take all we have; for already Maxim puts a very small copper wire in his circuit in place of our safety catch, and Swan works his lamps in multiple arc, not in series as we suppose." Batchelor to Eaton, 4 Sept. 1881, Cat. 1331:22, Batchelor (*TAED* MBLB3022; *TAEM* 93:313).

–2141–

To Charles Batchelor

[New York,] Aug [24, 18]81[1]

Knoside[2] Paris

Adhere (U.S. [Electric Lighting] Coy) deny seizure even one of their lamps and say not one has been removed by court also that all their lamps still burn   what date did Abaft (Edison) Maxim Swan and Fox begin lighting and has each continued without break   was one Maxim lamp removed by court and where is it now   How many days since exhibition opened have above four exhibitors been lighting   How many days have Maxims arc lamps burned. Has Maxim lamps burning in Paris outside exhibition and if not would court allow it   cable full and exact facts for newspapers[3]   answer quickly[4]   Cable anything interesting daily

L (telegram, copy), NjWOE, LM 1:29A (*TAED* LM001029A; *TAEM* 83:886). Written by John Randolph.

1. This cable was entered into the cable book among other cables dated 24 August 1881.

2. Cable code for Charles Batchelor; see App. 4.

3. On 13 August several newspapers reported that the Edison interests had obtained an injunction against Maxim. By 21 August the press was offering more complete explanations of the legal situation. At least one of Maxim's lamps had been seized but a special order permitted him to continue his exhibition. Maxim's patents, like all those issued in France, were issued without any government warrant of priority; in the French system, priority would be determined in a civil lawsuit. "Electric Light Troubles: Ready to Fight for Their Patents," *New York Tribune;* untitled article, *New York World;* "Electricians Quarreling," *New York Evening Telegram,* all 13 Aug. 1881; "The Maxim Light at Paris," *New York World,* 21 Aug. 1881; "The Rival Electricians," *New York Evening Post,* 24 Aug. 1881; "Electric Squabbles," ibid., 24 Aug. 1881; "The Maxim Lights Not Seized," *New York World,* 25 Aug. 1881; all Cat. 1085:30, 32, 33; Scraps. (*TAED* SM085030a, SM085032a, SM085033b, SM085033d, SM085033, SM085033a, SM085033c; *TAEM* 89:199–200).

On 21 August Bailey cabled Edison that when Maxim's exhibitors were served with the injunction they made "so violent resistance that the Police were called in to aid the officer making seizure" and that the court had issued a summons for "Maxim to appear and answer to charge of counterfeiting Edison lamp." He and Puskas warned Edison to expect a legal battle of at least five months; proceedings continued for three years. Bailey to TAE, 21 Aug. 1881, LM 1:26B (*TAED* LM001026B; *TAEM* 83:885); Puskas and Bailey to TAE, 19 Aug. 1881; all DF (*TAED* D8135ZAR; *TAEM* 58:992); Fox 1996, 179–81.

4. Batchelor replied on the same day that he would give a fuller account on Friday, 26 August, but cautioned that Maxim and other exhibitors "don't run yet." On 26 August he cabled, "Get accurate details tomorrow early   received cables whilst showing Gambetta and party round   open tonight first time for press   grand success." The following day Batchelor reported that one of Maxim's lamps was "seized and sealed by Captain police who made Maxims agent responsible produce it in court   after making seizure court dont interfere further until decision rendered   cant find any small lamps burning outside   public not admitted after six till last night   lighting before very irregular   we have lit fully every day since twentieth." Batchelor to TAE, 24, 26, and 27 Aug. 1881, LM 1:29B, 30B, 30C (*TAED* LM001029B, LM001030B, LM001030C; *TAEM* 83:886–87).

–2142–

From Otto Moses

Paris, Aug. 26. 1881.

My dear Sir:

With the greatest pleasure I chronicle the complete success of our illumination. Last night for the first time we ran the entire capacity and I assure I never saw a more beautiful sight. The two crystal chandeliers in the center of the rooms were resplendent with $^{60}/_2$ lights each,[1] and 48 lights drooped from 12 chandeliers around them.[2] The two effects did not interfere in the least   I asked everyone whose opinion was valuable as to which would be most acceptable to Parisians and the great majority declared in favor of $^1/_2$ lights, simply however, because 8 candle lights are nearer to the bougie (candle) in appearance. Educate the eye hereafter, for the present, in salons, 8 candle power is enough. Shops will take 16 candle lamps every time.

Berger came in last night while we were illuminated. He was charmed, and paid the light profuse compliments. Capt. Eads[3] and his wife were here when he came. Capt. E. said to me when he left "Oh I would like to give ten years of my life to the study of electricity." He was enthusiastic over your Exhibit.

Mr Berger told us to-night would be considered the opening night for the government and the press. that MM. Gambetta[4] and Léon Say[5] would attend.

Edmonds,[6] Swan's partner, (whom you know) also illuminated the Salle du Congress and the adjoining lunch room. Altogether he had more light than we, but the effect was very poor, and every one[b] declared it far inferior to ours; however, it required no impartiality to discover that ours was superior, being so much better. Edmonds has a unique style of chandelier which is quite pretty, thin spider like arms curving out from a center and lamps suspend at intervals   Above each lamp there is a cap of flourescent glass like a drop shade. It is very light and pretty. The Director of the Telephone central station said to me "there is something blinding about the Swan light which you do not see in Edison's." His loops are smaller and carried to an incandescence equal at least to our 48 candle power.

Maxim's illumination was the sickliest thing you ever saw. His central light was a ring of lamps with an incandescence running at one side from about 4 up to about 12 candles on the other—and two broken lamps. Such a mess I never did see! His head man is sick and he has fallen into the hands of the Philistines.

You have received news of the accident to Lane-Fox's lamp by cable.[7] It appears that some of Swan's men were on the roof of the Palais and saw smoke issuing from one of the partitions in the reading room near our Salle. They shouted to our watchman who called the firemen always on duty in the building. The latter caught the chandelier in their hands and half dozen were sent sprawling one after the other before the cause could be made known. Berger came in great consternation to Batch and said he wanted to know what safeguards should be adopted; and when he was shown the cut off system (as Johnson proposed to call the 'safety clutch' as name we gave to Du Moncel) he was perfectly satisfied. To day the whole Palais has been speaking of the fire. Hurrah for our side! The Commission sat to day and apportioned the representation by countries on the juries. France gets 75, Belgium 11, Germany 10, England 10 U.S. 7, Italy 6, Russia 5, Sweden 5, Switzerland 4, Low Countries 3, Spain 3, Austria 5 Japan 1, &c &c. equal 75. France getting ½   Commissioners Freeman & McLean[8] protested at the smallness of the number given to US (This is literally true because besides you there is nothing much to speak of from the United States). We are organizing our campaign. Faithfully Yours

Otto A Moses.

ALS, NjWOE, DF (*TAED* D8135ZAX; *TAEM* 58:1005). [a]Obscured overwritten text. [b]Interlined above.

1. That is, sixty half lamps each of 8 candlepower luminosity.

2. Moses was mistaken about the forty-eight lamps of 16 candlepower, which were arrayed in 16 chandeliers with three lamps each. Two Z dynamos provided current for the display, one for the 120 lamps of

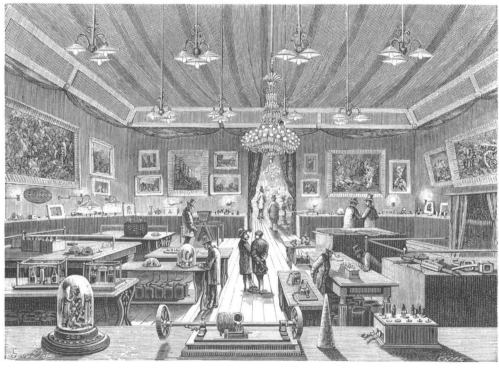

Views of Edison's two exhibit rooms at the Paris Electrical Exhibition, showing the placement of the 60-lamp and 3-lamp chandeliers.

eight candlepower and the other for the forty-eight lamps of 16 candle-power. See Doc. 2147 and Johnson to TAE and Eaton, 22 Oct. 1881 (pp.15–16), DF (*TAED* D8133ZAJ; *TAEM* 58:642).

3. James Buchanan Eads (1820–1887) was a civil engineer best known for his pioneering bridge-building techniques. He introduced the pneumatic caisson to the United States and in 1874 completed the Eads Bridge across the Mississippi River at St. Louis. More recently he opened the mouth of that river to navigation. *ANB*, s.v. "Eads, James Buchanan."

4. Léon-Michel Gambetta (1838–1882), prominent French republican statesman, served as President of the Chamber of Deputies from 1879 to 1881 and briefly as Prime Minister in late 1881 and early 1882. *Ency. Brit.* 1911 and *DBF*, s.v. "Gambetta, Léon."

5. Léon Say (1826–1896), French statesman and economist, was finance minister throughout much of the 1870s and early 1880s. *Ency. Brit.* 1911, s.v. "Say, Léon."

6. Henry Edmunds (1853–1927) was a British engineer, entrepreneur, and (since 1877) Edison acquaintance. In the early 1880s he promoted Joseph Swan's incandescent lamp. He helped secure contracts to install Swan lamps on passenger and military vessels, including the *City of Richmond*, the first passenger ship to make an Atlantic crossing outfitted with incandescent lamps, and the HMS *Inflexible*, the first Royal Navy ship so equipped. Edmunds also managed Swan's exhibit at the Paris Electrical Exposition. Swan and Edmunds made the lamp the centerpiece of their exhibition and used Brush dynamos and Faure batteries for power. While at Paris Edmunds secured for Swan an important contract to provide incandescent lighting for French naval vessels. Doc. 1205 n. 3; Tritton 1993 chaps. 4, 6–7.

7. Batchelor cabled Edison on 25 August that "one Fox lamp wires caused small fire   officials with scientists made strict search into our wiring and expressed great admiration for complete safety system" (Batchelor to TAE, 25 Aug. 1881, LM 1:30A [*TAED* LM001030A; *TAEM* 83:887]). He wrote Sherburne Eaton a fuller description on the same date:

> Lane-Fox is lighting up a reading room near ours and about ½ past eleven our watchman saw the fire and immediately called the firemen & had it attended to. Early in morning about 8 oclock I was there; and a deputation of officials and scientific men shortly after came to our rooms to examine our wiring. I went carefully with them right through our mains and branches showing and explaining our safety catch all through and after their expressing admiration at the perfect safety of the system I showed them that independent of this our wires covering would not burn and illustrated it by holding the wire in a flame until it decomposed the covering. [Batchelor to Eaton, 25 Aug. 1881, Cat. 1331:10, Batchelor (*TAED* MBLB3010; *TAEM* 93:305)]

8. Frank Freeman was an examiner in the U.S. Patent Office and later a patent attorney; T. C. MacLean was a lieutenant in the U.S. Army. A complete list of jury members and countries of origin is in "Congrès International des Électriciens," *La Lumière Électrique* 4 (1881): 417–20.

Draft Caveat:
Consumption Meter

[New York, August 28–September 10, 1881[1]]
CAVEAT ON METERS.

The object of this invention is to produce an electric meter capable of measuring in a convenient and economical manner the quantity of electricity passing in an electric circuit.

The invention consists in various devices, many of which I have tried and others which I am now engaged in experimenting upon to ascertain the best kind to meet all the conditions for practical use in my system of electric lighting.

Fig 1[a]

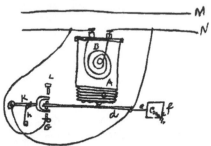

Expansion of air Meter[2]

In Fig. 1 is shown a meter which records by the expansion of the air in a closed chamber A, such expansion being due to the heating of a coil of wire carbon or other conductor B placed within such chamber. C is a flexible portion of the chamber working like that of an aneroid barometer or an accordion; the movement of this flexible portion of the chamber serves to give motion to a lever *d,* which actuating a ratchet in the counter *f* serves to count every reciprocation or vibration of the lever *d.* The wire B being in one part of the main circuit M, N, is heated upon the passage of the current, this in its turn expands the air within the chamber; this moves the lever downward when at a certain point it touches the lever K and moves it from the point L to the point G. Now the lever K being connected to one side of the wire in the chamber while the point G is connected to the other side, the contact of the two serves to shunt the current almost entirely from the wire B, thus allowing it to cool, hence the air contracts, the lever is drawn upwards and when it reaches a certain point it disconnects the lever K from G, breaking the shunt, whereupon the coil B again becomes heated and expands the air and the lever makes another vibration, the minimum current with which the lever *d* will make a complete vibration being that due to placing a single electric lamp across the circuit, the addition of more lamps will cause the air to expand more quickly, hence the lever *d* will make a greater number of vibrations per minute, the number being proportionate to the number of lamps, each reciprocation counts.

Fig 2[a]

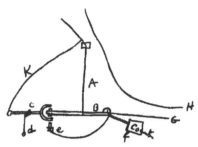

Expansion of wire Meter[3]

Fig. 2 is a modification, the expansion of the wire A, forming part of the circuit serving to replace the air chamber. Preferably this wire is enclosed in a chamber but the expansion of the air contained therein is not utilized.[4]

fig 3[a]

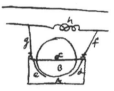

also in circuit   Continuous Meter

Fig. 3 shows a continuously counting meter upon the copper depositing principle.[5] A is a narrow trough in which rotates a disc B of copper upon the pivot C. On its opposite edges are two copper poles or electrodes. Connected to the resistance *h* in the main line K, L by the wires *g, f.* These electrodes are marked *e, d.* When a current passes through the liquid from the electrode *d*, it passes from it through the thin stratum of liquid to the edge of B nearest to it, thence through the copper disc to the other edge opposite *e*, thence through the liquid to *e*, a portion of the current, of course, passes through the liquid in the bottom of the trough but this is very small. The result of the action of the current is to take off copper from *d*, adding it to the edge of B, thus making B heavier on the side towards *d*, and at the same time copper is taken off the edge of B opposite *e* and deposited upon *e*, thus lightening the edge of the disc B opposite *e*, hence by the copper deposit one side of B is continually made heavier while the other edge is made lighter, this causes a continuous rotation of the disc which, if its shaft be connected with a counter will give the amount of current passing.[6]

fig 4ᵃ

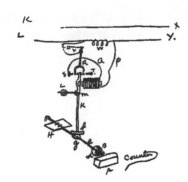

Meter

Fig. 4, shows an electro–magnet N, which vibrates a lever K pivoted at *m* and retracted by the moveable weight L. On the lower extremity of this lever is a rack *f* which engages into a pinion *g* secured to the shaft *e*. Upon the same shaft is a re-tarding fan H, and also a disc *d*, which carries a click or dog[7] B, engaging in a ratchet wheel placed on another and independent shaft, the latter shaft being a part of the counter. At every reciprocation or vibration of the lever K the shaft *e* is rotated a $\frac{1}{2}$ or $\frac{1}{4}$ turn and then brought back to its original position; but this reciprocation of the shaft *e* causes a rotation of the counter shaft in a constant direction. R is a lever which is moved by K. When a current passes through the magnet N the lever K is attracted when it reaches a certain point in its forward move-ment it separates the lever R from the point S and breaks the circuit of the magnet N, the lever K falls back and throws R against S, again closing the circuit, when the same action again takes place, the number of vibrations of K being, within cer-tain limits, proportionate to the current passing through the magnet N, it follows that the counter A will record the total current passing.[8]

fig 5ᵃ

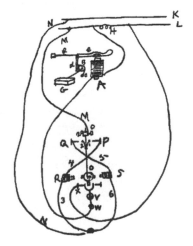

Fig. 5 shows a continuously vibrating pendulum O, secured at 20 and provided with contact springs 1 and 2, facing contact points Q, P; the point P is connected by wire 4 to the magnet R while Q is connected to the magnet S by the wire 5. The other ends of the magnet are connected together and to the line by the wire N. The pendulum itself is connected to the other portion of the line by the wire M; thus a derived or multiple arc circuit serves to work the pendulum, when the latter in its oscillation has its contact point come in contact with the point P, a current passes through the magnet R for an instant, causing it to attract the pendulum; upon the bob T of the latter there is secured a piece of soft iron on each side; hence the pendulum goes towards R; when the spring Z touches point Q the reverse action takes place and the magnet S attracts the pendulum; this continues as long as there is current on the main line K L. The pendulum itself serves to vibrate a lever V pivoted at W, and playing between contact points; the lever and points serve to open and close the circuit of a magnet A at each vibration of the pendulum; thus the lever *e* of the magnet A is vibrated regularly; upon the extremity of this lever is a pawl *d*, engaging in a ratchet wheel B. This ratchet has a click *c*, which prevents it going backward; this ratchet is on the shaft of the counter. The retractile force on the lever *e* is a stiff spring *f*. If a single lamp is put across the circuit at the ends marked L, K a current passes through the magnet A and the lever vibrates, but owing to the stiffness of the spring if it barely catches one tooth in the rachet B, thus advancing the counter shaft very slightly at each vibration. If now another lamp is put across the main circuit the current is doubled in A, and as it has more power the spring *f* bends to a greater extent and the click *d* carries the ratchet wheel forward two teeth, and so on until ten lamps are on; when this point is reached a second magnet requiring the current due to ten lamps to give its first vibration can be put in circuit, its counting being of a higher value.

fig 6[b]

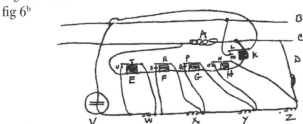

In Fig. 6 a copper depositing cell V is put across the line in multiple arc, but included in circuit with it are a number of resistances, W, X, Y, Z. These resistances are cut in and out of circuit by the movement of the levers of the electro-magnets E, F, G, H, K. The magnets K and E are so adjusted that the placing of the first lamp across the mains will allow enough current to pass to cause the magnets to attract their levers; the lever of K serves to connect the depositing cell and resistance in circuit, while the lever of E cuts out R, W, causing the current passing to be of the proper strength to deposit the amount of copper in V to represent a lamp. If now another lamp is placed across the main circuit it will cause the lever of F to be attracted, cutting out the resistance X and causing double the deposit to take place in V, and so on.

fig 7[9b]

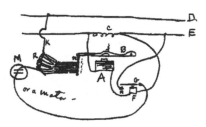

In Fig. 7 is shown an electro-magnet. A whole lever rests upon a large number of springs i, i, i, all separated from each other. When no current energizes the magnet A a resistance R is divided upon into as many coils as there are springs and a spring is connected by a wire and between each coil.

M is a copper depositing cell or electro-motor working a counter; its current is obtained by a derived or multiple arc circuit across the main and through the resistance, R. F is an electro-magnet which, when no lamps are on, open the meter circuit, thus preventing recording, but when a lamp is put in, the circuit causes F to close the meter circuit and the deposit takes place; if now two lamps are put in the lever of A comes down upon the springs with sufficient force to close the top and next spring under together, cutting out of the meter circuit a definite portion of the resistance, R, thus increasing the deposit; if three lamps are put in, then two more springs are pressed together by the action of the increased strength of current acting through A upon the lever B, and so on.

fig 8[c]

Patent Caveat

Fig. 8 shows a device which I now use in my regular meter to close the meter circuit only when a lamp is on, and to open it when no lamps are on, so that the counter electro–motive force will not cause a redissolving of the copper deposited by lamps previously on.

fig 9^{d}

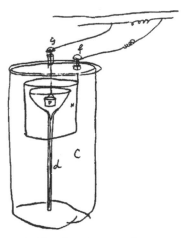

Fig. 9 shows an indicating meter where mercury is used. C is the main containing cell of glass; N a carbon electrode, p is another carbon electrode; d is a tube small at the bottom and wide at the top.

The whole of the cell is filled with a mercurial solution. When a current passes metallic mercury appears at P and drops down in the tube d as fast as formed and in proportion to the strength of the current by using an index card, the amount of mercury in the tube can be read off; by reversing the current this mercury may be made to disappear, and thus allowing of reading the total current which has passed in a given time.

fig 10^{d}

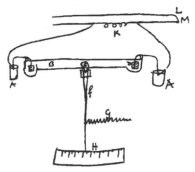

Fig. 10 shows a balanced beam cell, B, containing a mercurial solution with the electrodes at the end; the beam is balanced at F, a pointer, f, retracted by a spring, G, serves to indicate the deflection of the beam at H. A A are mercury cups, into which wires dip, which lend to the carbon electrodes in the ends of the beam cell; when the current passes mercury is taken by electrolytic action from one end of the beam and deposited at the other, thus causing it to deflect and indicate.[10] It is obvious that continuous counting could be obtained by applying the devices shown in my beam meter, for which I already have a patent.[11]

fig 11[d]

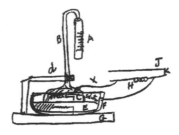

mercury meter

Fig. 11 shows two dishes; one, F, contains metallic mercury and forms one electrode, while a glass chamber, C, over the open mouth of which is stretched or placed a porous diaphragm; this chamber is also filled with metallic mercury up to the top of the tube, B.

Some mercurial solution is poured over the mercury E to allow of electrolysis; the mercury in C is connected to the main line shunt by a platina wire, X, passing through the chamber, while the mercury E is connected by another wire. When a current passes the total amount of metallic mercury in C is increased, hence it overflows into A, where its amount can be read off.

It is obvious that if instead of allowing it to fall in A, it were to fall in buckets arranged at intervals around the rim of a wheel, it would rotate the wheel and each bucket would, when it came around, deliver the mercury back into E to be again carried upwards into c, the shaft of the bucket wheel being connected to a counter a continuous counting would take place.[12] Good-night,

T. A. EDISON.

ADDENDUM[e]

[New York,] September 9, 1881

Add—In my regular deposit meter I have used plates of amalgamated zinc in a solution of sulphate of zinc, the zinc being electrically deposited and weighed.[13]

Fig. 12[14] shows an electro magnet A in the main or consumption circuit. It may instead be in a shunt therefrom. The armature lever B is retracted by spring *a* and carries a counter or a register C, operated by an exposed cog wheel *b*. Cog wheel *b* engages with the teeth of a variable gear D, which is driven at a uniform speed by clock work E, or other suitable driving mechanism. The gear D is a cylinder having rows of teeth, which vary in number, the number of teeth being regularly diminished from the bottom to the top of the cylinder. If no lamp is in circuit, the wheel *b* will be raised by spring *a* wholly above the teeth of D. If one lamp is turned on, *b* will be drawn down and will be moved by one tooth on D. If two lamps are used, *b* will be drawn down to next row which has two teeth, and so on for additional lights until the maximum number of lights for which the meter is arranged has been reached.

Witness   S. D. Mott   TAE

PD (transcript) and X (photographic transcript), NjWOE, Lit., *Edison v. Sprague,* Edison's Exhibit No. 2 (*TAED* QD008014; *TAEM* 46:301). All figures drawn by Edison on five separate dated pages; all but the fifth page are signed. [a]Figure signed and dated 28 August. [b]Figure signed and dated 7 September. [c]Figure signed and dated 6 September. [d]Figure dated 10 September. [e]Addendum is a PD (transcript).

1. Edison dated the accompanying sketches between 28 August and 10 September. This document is a transcribed version of a caveat manuscript that Edison later testified he prepared in his own hand and gave to his attorneys. It was transcribed in the course of a subsequent patent interference proceeding. The transcribed manuscript is substantially the same as the filed version of the caveat, except as noted. The caveat was executed on 23 September and filed on 4 October 1881; it was also published as part of the interference proceedings. Edison's testimony, 5–6; Edison's Exhibit Meter Caveat of October 4, 1881; *Edison v. Sprague,* Lit. (*TAED* QD008:5–6, QD008:21–27; *TAEM* 46:295–96, 311–17); Edison Caveat, 23 Sept. 1881, MdCpNA, RG-241 (*TAED* W100ABT); see also headnote, Doc. 2163.

2. Figure label at lower right is "counter."

3. Figure label at lower right is "counter."

4. In an unnumbered sketch following figure 2, Edison proposed using a "Fan to do work= gramme ring very light but large diameter to obtain leverage say 10 inches, all to weigh one pound—field in multiple arc." The Gramme motor, placed in a shunt circuit across a resistance, would operate a registering mechanism.

5. Edison executed a patent application for a modified form of this instrument in August 1882; the patent did not issue until 1889. U.S. Patent 406,825.

6. This meter is similar to that drawn by Edison in April in Doc. 2075.

7. General term for a tooth or protrusion, such as that used to engage a ratchet to prevent backward motion. Francis Jehl drew several such devices on 1 September but their purpose is unclear. Knight 1881, s.v. "dog"; N-81-09-03:9–11, Lab. (*TAED* N235:5–6; *TAEM* 41:514–15).

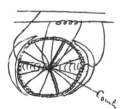

Edison's sketch of a meter, not included in the caveat, using a Gramme motor to move the recording mechanism.

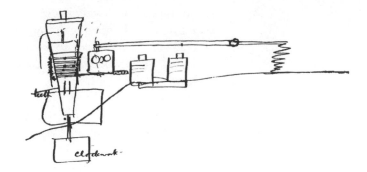

Edison's 17 May sketch for a meter caveat. The position of the relay armature determined how fast the registering mechanism would be revolved by the clockwork-driven cylinder.

8. The electromagnet, faintly marked N, is near the center of the drawing.

9. Label at lower left is "or a motor."

10. The first evidence of a meter operating on this principle is on 9 September, when Francis Jehl made a "mercury deposition cell" for measuring the rate at which mercury would move from one electrode to the other. He continued these tests on 10 and 11 September. N-81-09-03:22–23, 26–28, Lab. (*TAED* N235:12, 14–15; *TAEM* 41:521, 523–24).

11. U.S. Patent 240,678 issued to Edison on 26 April, 1881. It encompassed electro-mechanical mechanisms for registering the movement of a balance with electro-deposition plates at either end. A device for reversing the current when the beam reached its limit of motion was also provided. In August 1882 Edison applied for a patent on an alternative beam meter design having a reversing apparatus similar to that shown in figure 5 (U.S. Pat. 304,082). Edison had begun experimenting with this general form of recording meter in August 1880 (see Doc. 1974 n. 1).

12. In the version of the caveat Edison filed (see note 1), the paragraph pertaining to figure 12 immediately followed this one; the brief paragraph below about zinc plates was the last one in the body of the text.

13. See headnote, Doc. 2163.

14. The drawing referred to has not been found. The device is shown in a drawing by Edison, without lettered figure labels, dated 17 May. It was entered into the interference record. Apparently on the basis of that drawing, Edison stated in his testimony that this was the date of the caveat manuscript. Text at bottom is "clockwork"; text at lower right is "$1/200$ of ohm in mag[net] for 10 Light meter." Edison's testimony, 5–6, *Edison v. Sprague,* Lit. (*TAED* QD008:5–6; *TAEM* 46:295–96).

–2144–

To Egisto Fabbri[1]

[New York,] 29th Aug [188]1

My Dear Sir,

I have yours of 22nd August & note your brothers wishes in the matter of letters from Norway.[2]

I am sorry that there has been such a delay with the South American machines. The reason is that the machines originally ordered were made outside of my works (they were ordered before I started the works) & when I came to test them I found them so unsatisfactory that I decided to make ~~them~~ new ones which will be ready shortly[3]   Yours truly

Thos A Edison   I[nsull]

L (letterpress copy), NjWOE, Lbk. 8:474 (*TAED* LB008474; *TAEM* 80:988). Written by Samuel Insull.

1. Egisto Fabbri (d. 1894) was among the original investors in the Edison Electric Light Co. and the Edison Electric Illuminating Co. of New York. He sold out his founding interest in the firm of Fabbri & Chauncey when he became a partner in Drexel, Morgan & Co. in 1876. Docs. 1494 n. 4, 1504 n. 5, and 2037 n. 2; Obituary, *New York Times*, 27 June 1894, 8.

2. Fabbri's brother was Ernesto, with whom he helped establish Fabbri & Chauncey, in which Ernesto apparently retained an interest. Fabbri stated that his brother had received a letter from Norway that Samuel Insull had forwarded to George Gouraud, but would prefer to have similar correspondence sent to him directly. Nothing more is known about this matter. Obituary, "Egisto P. Fabbri," *New York Times*, 27 June 1894, 8; Fabbri to TAE, 22 Aug. 1881, DF (*TAED* D8131J; *TAEM* 58:394).

3. Fabbri & Chauncey, a major New York commercial and shipping firm, controlled Edison's South American electric lighting patents. The firm had ordered equipment for two 75-light demonstration plants, at least one of which was intended to sustain Edison's patents in Chile (Docs. 1504 n. 5, 1566, and 2048; Fabbri to TAE, 22 Aug. 1881, DF [*TAED* D8131J; *TAEM* 58:394]). Charles Clarke began designing the machines in February 1881. Edward Hampson & Co. built them in New York by early June. It is not clear when the new machines were completed, but the lamps were ready in October or November (N-81-02-20:1–103, Lab. [*TAED* N214:1–38; *TAEM* 40:946–83; Hampson & Co. to TAE, 4 June 1881; Philip Dyer to TAE, 1 Nov. 1881; both DF [*TAED* D8129ZAX, D8124ZAM; *TAEM* 58:265, 57:1069]).

–2145–

To Egisto Fabbri

[New York,] 29th Aug [188]1

Friend Fabbri,

You may remember the cable correspondence we had with Gouraud some time back with reference to retaining experts in case we should want them on Law suits.

I think that you should see that the Scientific writers on the Electrician, Telegraphic Journal, & especially "Engineering" are retained   Conrad Cooke[1] of the last named paper should be retained. The fees to be paid to them should be in accordance with those usually paid for such service in England. I have myself the poorest opinion of the ability of these men but they can nevertheless be ~~of~~ a source of a great deal of petty annoyance to us in the future as they most certainly have been in the past so it would be just as well to enlist them on our side as for every dollar of fees it will save us a great many dollars worth of trouble.

Matters are progressing here satisfactorily. We have taken possession of the buildings at Pearl Street where we are now putting the Boilers for the Central Station[2]   The Tube Co[3] has a large stock of street mains on hand and we shall commence [t]o[a] laying the mains in the streets very shortly.

Last night we tested the Large[b] Dynamo that is to go to Paris. The test was in every way satisfactory and we shall ship the machine by the [--------][a] leaving here for Havre on 7th September[4]

The London machine [will be sent soon I hope on the 14th September[5]   Yours truly?][a]

Thomas A. Edison

LS (letterpress copy), NjWOE, Lbk. 8:480 (*TAED* LB008480; *TAEM* 80:990). Written by Samuel Insull. [a]Faint copy. [b]Interlined above.

1. Conrad Cooke (c. 1843–1926) was an engineer who had helped the Edison Telephone Co. of London put on exhibitions in May 1879, later claiming to have been offered the position of consulting electrician. Cooke co-authored the first volume of Dredge 1882–85. Edison wrote Edward Johnson in October that Charles Batchelor reported "he has captured Conrad Cooke (Editor) & Dredge (Proprietor) of Engineering & that they will render us some kind of justice at last. Now I think you should get hold of Cooke have a good square talk with him & tell him you want him to be connected with us in an official capacity. Give him clearly to understand that he will be treated very differently to the way in which Arnold White treated him." Docs. 1741, n. 1, 1744 nn. 2–3, and 1780; TAE to Johnson, 9 Oct. 1881, Lbk. 9:174 (*TAED* LB009174; *TAEM* 81:58).

2. See Doc. 2125. The Edison Electric Illuminating Co. purchased adjoining buildings at 255–57 Pearl St., in lower Manhattan, on 3 August. It reportedly paid $30,000 for one structure and $1 for the other to the same seller. "Edison's Company Buys Property," *New York World*, 4 Aug. 1881, Cat. 1242, item 1625b, Batchelor (*TAED* MBSB31625b; *TAEM* 94:640); see also App. 1.B.54.

3. The Electric Tube Co. was incorporated in March 1881 to manufacture underground electrical conductors for Edison. See Doc. 2058.

4. The dynamo was shipped aboard the *Canada* on 7 September. See headnote, Doc. 2122.

5. This date has been inferred on the basis of Edison's 28 August cable to Charles Batchelor that the machine would sail with Edward Johnson on that day. LM 1:31A (*TAED* LM001031A; *TAEM* 83:887).

−2146−

To William Hazen[1]

[New York,] 29th Aug [188]1

Dear Sir,

I have your favour of 25th.[2] Its receipt was somewhat delayed in consequence of its being addressed to Menlo Park

I will with pleasure supply you with lamps for the purpose you name but shall have to make you special lamps. I have sent your letter on to Francis R. Upton of the Edison Lamp Co Menlo Park N.J. to whom I have given instructions as to what I want.[3] Please write him the latest moment at which the Lamps can be received at Washington

He will give you all information as to Battery &c Yours truly

Thos. A. Edison I[nsull]

L (letterpress copy), NjWOE, Lbk. 8:470 (*TAED* LB008470; *TAEM* 80:987).

1. Brig. Gen. William Babcock Hazen (1830–1887) was Chief Signal Officer of the U.S. Army from December 1880 until his death in January 1887. *ANB*, s.v. "Hazen, William Babcock"; Raines 1996, 55–62.

2. This letter has not been found.

3. On this date Edison instructed Upton to supply Hazen with some lamps of low candlepower with platinum-iridium filaments for use with battery power. Celebrated balloonist Samuel Archer King used these lamps in an ascent he made on 12 September 1881 in his balloon, the *Great Northwest.* Accompanied by five reporters and a member of the Army Signal Corps, King intended the voyage to perform atmospheric research and to demonstrate the practicality of long-distance aerial navigation. However, the balloon made a forced landing several hours later due to high winds, which destroyed the craft on the ground. King also made several ascents to conduct meteorological research for the Weather Service of the Signal Corps. TAE to Upton, 29 Aug. 1881, Lbk. 8:469 (*TAED* LB008469; *TAEM* 80:986); Hazen to TAE, 30 Sept. 1881, DF (*TAED* D8120ZAY; *TAEM* 57:636); *ANB*, s.v. "King, Samuel Archer"; Crouch 1983, 451–63.

−2147−

From Charles Batchelor

Paris— Aug. 30th 1881

My dear Edison,

Last night we experienced what the parrot said he had with the monkey "a hell of a time."[1] The half light machine stripped off the brass wires and before we could do anything it tore off the canvas and bent up the bars considerably. This occurred at 10¼ oclock. We immediately cut the belt and ran till 11 with the full light machine on the 16 3 light chandeliers. Our rooms were densely packed as they are every night but the short time we were stopped prevented any disturbance— At

11 we took out the armature and commenced the fixing of it and at 5 oclock tonight we began to run and continued till 11 without further trouble   I never let the boys leave it until it was complete having all their meals brought in to them and it was a great disappointment to some of our competitors to see us come up to the scratch 3 hours before time & run all night— These armatures run a little too close to the field and I think it would be a great deal better if you would give them a hundredth of an inch more when boring them out— Both my armatures touch the field when running full power and I am rigging up a boring bar to take a little out of the fields— In rubbing they help to increase the heat. Our exhibit is made by 2 handsome glass chandeliers with 60 lights each, driven by the $^{20}/_{1000}$ machine, and 16 of our 3 light chandeliers placed as in sketch driven by the 110 Volt machine—

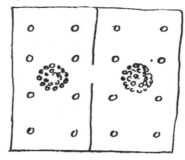

Swan lights the next room to ours with 425 lights about 300 of which are festooned round the walls the rest in pendants holding about 20 each near middle of room. His lamps are very irregular and ever since we have been here he has been weeding them out. They are very irregular indeed yet and as we have a door in our room that leads into theirs we never fail to point out this difference to visitors by first taking a good look at Swan's and then turning round to our chandeliers where we have 60 in a bunch; and aAlthough we never picked them out there is not a particle of difference in their intensity— None have busted as yet but of course it is hardly time for that— M. Leon was in for the first time yesterday and congratulated me on the success of the light as compared with others— I went round to Swan's and Maxim's with him and pointed out their defects which he appreciated highly   They can plainly see that we have got the system complete whilst others have only got a lamp or a machine, and even their lamp does not begin to compare with ours.

We have lately rubbed against the "Blarsted Royalty" in the shape of King Kalakahua[2] who expressed himself very much pleased with the light. We showed him also the singing telephone and the Motograph— We could keep these fellows easy enough to see all the apparatus but there is always such a devil of a crowd in the place that they want to get away   I have hired two men who speak French and German to do nothing else but explain the things as the boys show them— Gambetta has been here and expressed his intention of coming again in the morning to have it all explained to him   I believe he is a very large holder of gas shares— I saw a letter from Eaton to Mr Fabri in which he said you had come to the conclusion to manufacture isolated plants so as to supply those people now who will take your light when you can supply it from a station,[3] for instance "Hines and Ketchum"[4]   This I think is a very good idea   We have numerous applications here for it and if they could be supplied and terms fixed there are a great many places I am sure that it could be put in in Europe immediately— Bailey has applications from Brussels, Vienna, Lyons etc and if we do not supply them of course Swan and Maxim will. It takes so much time to make the plant and put central stations in operation that we ought to do something to secure the people beforehand   I think also that it will be very much easier to form a company in any of these cities if a few people are using it—

I hear Brush has sold his light here for Fcs 2,000,000 to Credit Lyonnaise—[5] The syndicate have appointed their three engineers to go over our plant, one of them is 'Clerac' of carbon pressure notoriety[6]   however as Bailey says he is a great friend of his and will do anything for him I suppose he is all right   We are to appoint two of whom Armengaud the patent agent is one   I have had two or three interviews with these engineers and at present we shall have to educate them up to what we have got before we make any tests but I have no doubt we shall be able to convince them that we have something worth Securing— Gas here is sold at $1.80 per 10[00][a] feet   there is an enormous lot of it used in the streets and places of amusement but the public generally do not give it the place it would have if it did not give so much heat and spoil the decorations of the house   The whole of Paris is supplied by one company who have a monopoly of the laying of mains etc which lasts 20 years more but the City reserve the right to break this contract if any better method of lighting should come up— Of all places Paris should be the best as every house is about 7 stories high with stores on the ground and private families in every floor each of whom would take the Electric light where they now use candles in preference to gas because of its high price, deleterious effects, and above all the trouble experienced with the monopolist company   Let me know your views about these isolated plants. Yours

Batchelor

ALS, NjWOE, DF (*TAED* D8135ZBB; *TAEM* 58:1020). ªPaper damaged.

1. The phrase "parrot and monkey time," popularized about this time from an unidentified "droll and salacious tale," referred to a period of quarreling. Partridge, s.v. "parrot and monkey time."

2. Kalakaua (1836–1891) was king of the Hawaiian Islands from 1874 until his death. He traveled widely and was keenly interested in public works projects and new technologies like the telephone and electrification. Moses wrote Edison that royalty and prominent political figures daily visited the Edison exhibition. He related that King Kalakaua "was at the motograph telephone with me and Batch was at the transmitter end. He (B.) was asked to sing and gave us 'Mary had a little lamb,' which delighted his swarthy Majesty very much. The thought of the narrow escape we had made, flashed upon me, and I blessed my luck—for Batch had intended, as he told me afterwards, to give 'There was an old nigger,' which is the other half of his usual répertoire, but didn't." On his return from Paris, Kalakaua visited Edison in New York at the end of September and received a full explanation of the electric lighting system. *ANB*, s.v. "Kalakaua, David Laamea"; Moses to TAE, 7 Sept. 1881, DF (*TAED* D8135ZBI; *TAEM* 58:1038); "King Kalakaua's Movements," *New York Times*, 26 Sept. 1881, 5.

3. This letter has not been found but see Doc. 2156 esp. n. 2.

4. Hinds, Ketcham & Co. was a commercial printer and one of the first businesses to adopt the Edison electric light. Its isolated lighting plant became operational in February 1881. See Doc. 2053.

5. Crédit Lyonnais was a French bank founded in 1863. It expanded dramatically; by 1880 it had become one of the largest French banks, with seventy branches within France and offices in Egypt, Turkey, Switzerland, Britain, Spain, Russia, and Tunisia. Cameron 1961, 172–73.

6. Hippolyte Clerac was an engineer with the French government telegraph service. He claimed to have anticipated Edison in discovering the principle of the microphone in 1865 by showing that a tube made of plumbago changed its resistance under pressure. Docs. 1398 and 1482.

–2148–

From Charles Batchelor

Paris Sept 6th 1881

My dear ~~Major~~ Edison

The great necessity of something being done very quick impresses itself more and more upon my mind every day. If I could make arrangements to supply I could take a couple of orders per day of about a 300 light plant each and I feel terribly galled to think we have such a good thing and we have not even our manufactories started to supply anything.[1] There is a large dry goods store here called the "Au Bon Marché."[2] They use 5000 gas lights and they want to replace them with our light, they have been to see it often and yesterday I went and inspected their building to see how it could be adapted to Electric lighting I find they have tried Jablochkoff lights and Siemens' lights at an enormous expense and gone back to gas Their gas bills average about 138,000 francs per year. I told Bailey that I was going to see this place and he communicated with Leon, after which he told me that if this could be put in writing and I would make an estimate of about what saving could be made in this store and put the same into the hands of the Syndicate they would consider this as $^{1}/_{2}$ of the test that we have got to make.[3] After visiting the store I saw how easily it could be done and proposed to the "Bon Marché" people the following:— They to write me a letter asking me whether I could light up their place and what it would cost; also stating that they would like a preliminary trial of say 300 lights (for which we should have to put a portable engine in similar to what I have at present in the exhibition) for 3 months, and to submit to them an estimate for the expense to them of such preliminary trial, that if they are satisfied with the light at the end of the 3 months they will give us an order for the whole store. To light their whole 5000 lights would take 320 horse power (the paris gas jet is not more than our half light at 9 candles). They have got excellent cellaring large enough for the boilers engines and dynamos etc and are very anxious to get it; their chandeliers are so arranged that I could very easily use them all for our light. Now in such a case as this I should advise the putting in of such a plant, as they would stop their engines and take current from the company's mains when they are laid; besides Maxim and Swan are open to all such offers and the reason they come to us is because we show the best light, and upon investigation they find we have a complete system—

Now for one more thing:— I have a large manufacturer in Rheims (the champagne district) who has been here for about a week investigating the lights; he has one part of his factory lighted by arc lights driven by 8 gGramme machines and he wanted the rest lighted by our light; altogether about 300 lights. If I could give him a price I am sure I could get the order immediately. He wanted to know whether I could use the Gramme machines he had; so I took him down stairs where I have the American Commission offices, Dolbear's telephone office,[4] a machine shop and Berger's office all lighted from one little Gramme machine (in all about 42 lights)   I showed him our own machines and explained their greater economy. If we were manufacturing here I could secure him immediately at my own price; as it is he may be gobbled by Maxim or Swan although in every case I take great care to point out every defect in the other systems— I have also a man from near Brussels wants 300 lights; one from Vienna 250, one from Strasbourg about 300 and lots more;[5] now to supply these I would make special plants driven by belt as they care[a] not for great economy— I have found that to run the small machines above 1200 makes them spark, but below 1200 with whatever load, they give no spark; so I have slowed down to 1200 and get my electro motive force by taking off a layer of wire on the magnets; you must impress this on the people using the isolated plants as it means that the life of a commutator is <u>years</u> instead of <u>months</u>— You must not think that because I write about isolated plants that I have lost sight of the primary object of this light that is, the lighting of cities in competition with gas; I merely cite cases that are outside of the gas companies (with the exception of the "Bon Marché") and which the other people will take immediately they get the chance; for your own private information.

The engineers for the Leon party are making calculations and <u>learning</u>[b] the method of Electric lighting on the multiple arc system. I think one of the best evidences of the originality of your methods is that these 3 electrical engineers have to study it out as a new problem that they have never believed possible before. We shall be ready for them to test in a day or two. We have got Berger to get from the "Conservatoire des Arts et Metiers" a dynamometer which we shall put in between the machine and the engine, we expect to get that this week; whatever tests our apparatus goes through I have got two or three influential people in the Congress who will ask that the Maxim, Swan, and others may be tested also in the same manner, and I shall lend my photometer, Calorimeters etc for such tests if they have not any, which I know they have not; and I doubt whether they would know how to use if they had.—[6]

I have always given a great deal of credit for arc-lighting to Brush[7] but now we must be on the warpath for here he runs down our light terribly and tells everybody that the Swan light is the only incandescent light that is reliable and that Edison's infringes it— When I talk on arc-lights the beauties of the Siemens and the Pillsen stand out in glaring contrast with the miserable flickering Brush light— I have had a visit from Mr Shillito[8] the manager of Lord Salisbury's[9] place and his Lordship comes to see me today— Shillito says they have 150 Swan lights burning from a large Brush machine driven by waterfall 1[½?]^c miles away   I went all through the thing with him and they will want some of your lights to replace the Swan as I should judge from what he said they were not satisfied sufficiently to put any more in   I gathered from him that they never knew what resistance the lamps were hot, that they varied considerably but that by picking them out after putting them in they could make a pretty good showing. I measured a few of our lamps for him hot and they did not vary   he was surprised at the completeness of our system but in his case economy is not the primary object as his <u>lordship</u> spends his money on science instead of horse racing— Edmunds of the Swan light had told him that they had not had one break since they came to the Exhibition, I told him that they had some go every night and called Suebel[10] and Force[11] to witness it and Suebel says come with me and we will sit down in his room a little while and see and whilst they were sitting there only a few minutes they saw three go, Shilleto could not understand this as Edmunds had told him that they were put 10 in series so I told him that that was another item he had stole from us— Maxims of course is not lighted and has not been for a week so we can make no comparisons with him—

Now what I want to know from you is=

1   What do you propose for manufacturing for France?

2   Can you supply isolated plants for other countries?

3   Do you propose to work Europe by agency or by separate Co.?

4   Will you entertain the idea of supplying these isolated plants and at what price?

5   Would you entertain the idea of putting our lamps on Gramme machines outside of France and thereby replacing arc-lights that are already running and at what price or royalty?

Also send me any information that you can that will facilitate matters.[12]

Suebel heard Mr Leutz[13] that is Siemens' manager say that he was sorry his firm did not go into the incandescent light 2 years ago but now he [intends?]^c to do so as he sees what a good thing it is and besides he intends to go for Mr Edison in Germany on the fastening the Carbon by electrolysis so we must look out for him[14]   Yours

"Batch"

ALS, NjWOE, DF (*TAED* D8132ZAK; *TAEM* 58:466). [a]Obscured overwritten text. [b]Multiply underlined. [c]Illegible.

1. In a letter written to Edison a few days earlier, Joshua Bailey and Theodore Puskas warned: "At the Exposition you have shown out entirely your hand & lots of people are taking note of everything you show, some of them making careful designs and unless you put them in practical operation at once, it will be found in full blast worked by other people and a series of Law Suits will be the result." They reported having received numerous requests for isolated installations. Charles Batchelor proposed quickly setting up isolated lighting plants or small demonstration stations in a dozen major European cities. Puskas and Bailey promised to arrange funds to do so after the Paris Jumbo dynamo had operated successfully for two weeks. Bailey and Puskas to TAE, 2 Sept. 1881 and 29 Aug. 1881, both DF (*TAED* D8135ZBG, D8132ZAG; *TAEM* 58:1033, 455).

2. The Bon Marché, one of the world's first department stores, was still among the largest and most prestigious. Its building on the Left Bank was in the midst of a long period of renovation and expansion at this time. On 5 September Bailey wrote Edison that "an application was made two days ago for the Bon Marché requiring 5000 Burners" and that Batchelor estimated the annual cost to the store for an incandescent lighting plant, including fuel and interest payments, to be 52,000 francs (approximately $10,000); the director of the Bon Marché claimed that the store consumed 100,000 francs worth of gas a year, a figure Bailey thought was lower than the actual cost. The store was wired for an Edison isolated station and 500 lamps in the middle of 1882. Miller 1981, 41–43; Bailey to TAE, 5 Sept. 1881, DF (*TAED* D8132ZAH; *TAEM* 58:459); Edison Electric Light Co. Bulletin 13, 28 Aug. 1882, CR (*TAED* CB013; *TAEM* 96:738).

3. Edison's French investment syndicate required him to achieve two benchmarks, the lighting of the Opera's foyer and the successful operation of a station powering 2,000 lamps, before they would release more funds. See Doc. 2166.

4. Amos Dolbear (1837–1910) was professor of physics and astronomy at Tufts College. He claimed to have invented a telephone that used permanent magnets, rather than electromagnets, before Bell. He exhibited several of his telephone inventions at the 1881 exhibition. Doc. 1043; France. Ministère des Postes et des Télégraphes 1881, 75–76.

5. Batchelor wrote Edison the next day that six more firms had inquired about Edison lighting plants, and he had promised them that "the policy of the company will be determined inside of a month when they will hear from us." Batchelor to TAE, 8 Sept. 1881, DF (*TAED* D8135ZBK; *TAEM* 58:1044).

6. A committee consisting of George Barker, William Crookes, A. Kundt, E. Hagenbach, and E. Mascart performed these efficiency tests. Batchelor supplied the committee with a ten-cell Daniell battery, resistance coils, Wheatstone bridge, photometer, and a Z dynamo. The committee concluded that Edison's lamps provided more illumination per horsepower than the lamps of Swan, Maxim, or Lane-Fox. Barker and Crookes published the test as Crookes 1882, a copy of which is in Ser. 3, Box 44, Folder 2, WJH.

7. Charles Brush (1849–1929) was an inventor and businessman. Originally trained as a mining engineer, he turned his attention to arc lighting in the 1870s. He developed an improved dynamo and arc light in 1877 and 1878, and he installed the first electric street lighting system in Cleveland, Ohio in 1879. He formed the Brush Electric Company in 1880, which installed street lighting systems in several cities. In 1880 he also invented a lead-acid battery independently of French inventor Camille Fauré. *ANB*, s.v. "Brush, Charles Francis"; Docs. 1489 n. 4 and 1582 n. 1.

8. Not otherwise identified.

9. Robert Arthur Talbot Cecil-Gascoyne (1830–1903), third Marquis of Salisbury, was a journalist and politician best known for his service as British prime minister for the Conservative party in 1885–86, 1886–92, and 1895–1901. He was also an early adopter of electric lighting. By December 1881 his residence, Hatfield House, had a Brush lighting plant consisting of a 16-horsepower dynamo and 117 lamps. *Oxford DNB*, s.v. "Cecil, Robert Arthur Talbot Gascoyne–"; "Electric Lighting. Fatal Accident," *Electrician*, 8 (1881): 68.

10. Philip Seubel may have been associated with Sigmund Bergmann before he sailed with Moses in June to help prepare the Edison exhibits at the Paris Electrical Exposition; he later helped install several isolated plants. Jehl 1937–41, 768, 1023; William Carman to Bergmann, 26 Jan. 1881, Lbk. 6:856 (*TAED* LB006856; *TAEM* 80:486); Batchelor to Puskas, 14 June 1881, DF (*TAED* D8135N; *TAEM* 58:899).

11. Martin Force had worked as a carpenter constructing Edison's Menlo Park laboratory and had been employed since late 1877 or early 1878 as a laboratory assistant. Doc. 1039 n. 3; *TAEB* 5 App. 2.

12. On 5 September Bailey also urged Edison to decide these issues quickly, noting that "all the arrangements relating to the establishment of the factories & for the purchase of the individual plants can be completed within 30 days from this date." On 17 September Edison cabled Batchelor to "get bids for making small dynamos   can not Turritini make our Dynamos low figure   could we use small Siemen's or Gramme machines for isolated pending establishment our works   can ship complete outfits from here including small engines for isolated to all countries where importation does not effect patents" (Bailey to TAE, 5 Sept. 1881, DF [*TAED* D8132ZAH; *TAEM* 58:459]; TAE to Batchelor, 17 Sept. 1881, LM 1:37B [*TAED* LM001037B; *TAEM* 83:890]).

13. Not otherwise identified.

14. On 19 September Edison cabled Batchelor to "make actual sales isolated countries where importation not invalidate patents especially germany. head off Siemens   complete outfits can be shipped from here our capacity getting enormous." TAE to Batchelor, 19 Sept. 1881, LM 1:41B (*TAED* LM001041B; *TAEM* 83:891).

To Charles Batchelor

[New York,] 8th Sept, 1881

My Dear Batchelor

I had to go to Michigan to get my wife five days before your big machine was ready for shipment and did not return until the machine was shipped.[1] I told Francis to write you a full description as to the ~~magnet we~~ we connect the field magnet, but it appears he wrote you a description of the whole machine.[2] For fear that he might not have put everything in I will give you a description myself.

I must tell you that when you left the first experiment with the big machine brought out the fact that it was a problem of great difficulty to take off from 600 to 800 lights from a commutator with three large brushes on each side when the brushes were put on new and at a proper angle and the ends ground so that all of them would touch the commutator and when six hundred light were put, a few minutes after, the sparking acting bad and increased so that at the end of about half an hour we were always compelled to stop the machine. Sometimes the heat due to the sparking would be so great as to melt the solder and the brushes would come to pieces. The following phenomina occured which explained why the sparking increased so: The sparks melted the ends of the copper and a film of copper got onto each end and connected to its neighbor so that at the expiration of half an hour the crust at each end was practically solid. It had no elasticity and the commutator would also get badly burnt. I say I was in a bad hole as it looked as if it were impossible to take off such a tremendous current. I then determined to try to investigate the subject thoroughly. I took one of the South American machines having only a spindle in the bearings and a complete commutator with brushes.[3] On the spindle was a pulley which was run by an electric motor so that I could vary the speed. I found that with the commutator standing still with new brushes upon which I put very hard pressure and with the commutator highly polished that there was $1/1000$ of an ohm resistance between the brushes and the commutator, or half of $1/1000$ of an ohm per brush. upon Upon rotating the shaft at different speeds I found that the resistance increased in proportion to the speed whether I started with a light pressure or a heavy pressure, viz. that the resistance was greater when the brushes were on flat than when they were end on. This is of course easy to understand because when they were on flat but a small portion of the brush is in contact. That is to say the total area of brush contact is small, while when they are end on the total area of contact is many times greater. Besides when end on each wire gets current direct from the commutator, while if on flat the current is taken off through each layer of wires one with the other. But notwithstanding all experiments the resistance with the most terrible pressure was considerable. I think the lowest resistance we could get was $^5/_{1000}$ of an ohm Now $^5/_{1000}$ of an ohm is not much resistance with a small dynamo with only sixty. That is to say it is a small factor of the total resistance but when you come to a big machine it becomes a very important factor. There seems to be a physical phenomena between moving surfaces composed of the same metal and no amount of pressure or change of form will prevent it. It is fixed and definite and the [missing text] I came to the conclusion that it would be impossible to take off such a tremendous current as we require for the big machine by means of copper brushes, as copper [missing text] without the use of an impracticably large commutator. I then thought I would try the effect of more current. I amalgamated the surface of the commutator and the brushes Presto! There came a change! The thing had no resistance noticeable whether moving or standing still. Upon the next occasion that the big machine was ready to start I amalgamted the commutators and brushes and when 600 lights were put on no sparking could be seen standing a few feet from the machine. Only by looking down between the commutator and the brushes could any sparks be detected and these sparks were blue, and were due to the effect of the machinery while the sparkes of copper would have been yellow and [missing text] I then knew that the problem was solved.[4] We ran the machine for about one hour and a half a few days afterwards we started on a a six hours test with 649 lamps on. The commutators gave no trouble whatever up to the fourth hour, when the mercury got worn off in places and a spark was apparant, but even up to the end of the test the sparks were not great. I afterwards experimented and found that the following was the best way of renewing the mercury. I took the brushes and dipped them for about six seconds in a solution consisting of about fifty parts nitrate acid and about fifty parts of water, dipping about $2^{1}/_{2}$ inches of the brushes in the solution. I then put a lot of mercury in a flat dish (perhaps it would be better to use a deep dish) dipped the brushes in the mercury for about five seconds then dipped them again in the nitrate acid solution for another five seconds, then put them under a running stream of water at such an angle that all the acid would be washed out, working the ends of the brushes so that the mercury would freely flow all over them but being careful not to get them out of position. I then took hold of the brush with both hands, held it over the mercury dish, and gave it eight or ten sharp jerks like the cracking of a whip so as to dislodge the mercury. I then lay them on blotting paper for a few minutes to get rid of the water turning them over from time to time. You will notice that the mercury becomes very black and that the copper when first dipped and then taken out looks very dirty and black much gets over it. This is nothing but oxide of mercury and when steeped in the acid bath the second time is immediately clean. After the brushes are amalgamated several hours they get dull, dry, and the mercury becomes oxidized, therefore the brushes should be amalgamted only fifteen or twenty minutes before they are put into the machine and they should be reamalgamated each time you make a fresh run with the machine. We also amalgamate the commutator. Before I left we had not burned off the commutator which had a great many bad places in it due to its use before amalgamation. I told Dean[5] to turn the commutator off before shipping machine to you. I amalgamate the commutators in the following manner. Disolve mercury in nitrate acid until the acid will soak up no more mercury. Then get a stick and rap around it a linen rag so as to form a kind of sugar tit[6] which should be quite hard   Keep this tit in the solution and then daub it several times on a second rag so as to rid it of any surplus solution as it only requires to be moist. Start the engine slowly, get your commutator clean, then take the rag arranged in the form of a tit and hold it gently on one spot on the commutator during about ten or fifteen turns of the machine. Then take it off and hold a perfectly clean rag on the same spot so as to polish it. The object of this is to get the acid off as quickly as possible as the solution is a acid solution and it is essential that no acid should get down to the mica between the bars; therefore it requires that great care should be taken that the tit should only have a minumum dampness. By performing successive operations as above you will soon get the commutator fully amalgamated and very shiney. Then you better have a superficial amalgamation [missing text] the brushes may be put on and the machine is ready to start. The brushes should be so set that on one side the ends of the brushes are exactly in the centre of the insulation between the blocks and the brushes on the other side should be exactly in the centre of the block so that there will be the same number of blocks on the one side as on the other from [missing text] to brushes and these blocks should be exactly on the neutral point which you can ascertain by following up the blocks, then the wire up to the bar over the space between the field magnets. Great care should be taken in setting these brushes so that they shall be truly end on. It is quite difficult to get them so, as sometimes the lower part of the brush will bear while the upper wires although they may look as as if they touch while they really do not so that one side of the brush will be right and other not; that is to say not at right angles to the commutator. Care should therefore be taken before starting the machine to get all the ends as far as possible to bear truly on the commutator block.

I do not know whether Dean sent you the brushes that I used. I found all brushes are liable to spread out at the edges that is to say that some of the wires at the edge would get spread out. I therefore took a copper wire and wound it around the brush about an inch or an inch and a half, if I remember right, from the edge of the holder so as to keep them together and to prevent this load of wire from gradually working down towards the commutator where it might cause trouble, I connected it to another wire the end of which I fastened round the screw which holds the brush in the holder.

I will now explain to you how to keep up the supply of mercury on the commutator. This is important and requires a little judgement. (After running a few moments the commutator will look a little dull, therefore to prevent this a clean dry rag should be held on the commutator and worked back and forth so as to keep it polished.)

To replenish mercury on the commutator take a couple of thicknesses of extremely fine woven cloth, then pour in some mercury about the size of a pea right in the centre of the cloth. Gather the cloth up like a sugar tit by twisting it. Then while the machine is running and the lamps are on you can as occasion may require, hold this sugar tit so to speak, with the globule of mercury in it on the commutator and by pressing gently the mercury will "spray out" through the cloth onto the commutator. You can see by the shine on the commutator the proper degree of pressure necessary. While holding it on the dry cloth should be held in advance of the tit and worked backwards and forwards so as to spread the mercury. In this way you can reamalgamate the whole surface of the commutator very nicely. About every five or six minutes the dry cloth should be held on the commutator so as to polish it and about every twenty or thirty minutes a slight amount of fresh mercury should be put on the commutator. You will notice when you are running with lights on a little line of sparks on different parts of the commutator especially after putting fresh mercury on. When these are seen the dry cloth should be held on and it will remove them. It is nothing more than a little film of mercury on the surface of the wires between the bars. Some times when you are putting on mercury by means of the tit you may press too hard and too much mercury get out. In this case little snappy arcs will form. Although this does no harm it is best to put it on very gently. The great thing is to keep plenty of mercury on the commutator and keep it very bright. You will find after putting on metalic mercury that a great deal of it is sprayed over on the brush holders by centrifugal forces occassioned by the movement of the armature but this does not matter. After you are through with the nights run the commutator should be wiped very dry and although I have not tried it my impression is that the brushes should be taken out and the ends put in water as if the commutator is allowed to stand for a length of time with any considerable amount of mercury on it it gets very bad as the mercury oxidises to a white solid. I found that when I had eight hundred lights on with two brushes on each side they would carry it with great facility therefore it is possible to lift a brush off if you ever have occasion to do so to fix anything without any danger and it would be even possible (you could?) accurately mark the brushes as to their position on the brush holder, their angle and other things to reamalgamate a break or take it out and fix it and put it back while the machine is running, but of course this could not be done without some mark. I should advise you to try this while the machine is standing still to see if the thing can be done with safety.

The bars are wound with four thicknesses of parchment paper each layer being japanned. We found on testing that a half inch spark from the big Ritche[7] [missing text] with a condenser on it would not penetrate this insulation, but not with standing this insulation and $\frac{1}{12}$ inch air space between each bar a spark occurred between two bars when we have 840 lights on nuy er could not get it to spark when we had 753 lights on, so you see this tendency to spark is a very extraordinary one and I am surc it is just a new (shenanagin) and therefore you should keep it to yourself. It depends entirely on the amount of current passing.

Now I advise that you do not run regularly with more than 500 or 600 lights and as you say a great many half lights can be used this will be quite sufficient. For instance you may use 250 full lights and 700 half lights. This will be equal to 600 full lights. Then you will be perfectly safe. When you make the economical test put on 700 full lights. The machine for this test will be perfectly safe with such a number. But you must look out that there are no crosses in your chandeliers or conductors as a cross would be equal to 200 or 300 extra lights and this added to the 600 you ordinarely burn would surely bring out the above referred to phenomina. This sparking which occurs between the bars does no harm so long as the arc does not stick. Every time that it has occurred so far the arc which was formed has been broken by the attraction of the lines of force at the neutral point and only a few scintillations have been thrown out at the ends of the machine but once the arc was not destroyed by the attraction of the lines of force it got large and a low resistance short circuit of the machine occurred and yanked the bars round in a lovely manner and bent them $\frac{1}{4}$ of an inch out of the true. These arcs which occurred when we had 850 lights on were immediately destroyed in the manner set forth and we continued to run, even after we had seen these arcs, for half an hour without that occurring again the [missing text] probably was covered up by the Japan running into the [missing text] the arcs are of such a nature that one would think that they are due to the wire bands from the armature striking the field magnets. We soon found that this was not so. What led to the discovery of the importance of this phenomena was this. I had concluded that the spotting was due to conduction across the mica between the bars which was the way you remember we first conducted them. Then I concluded that air being the [missing text][a] of the bar from which copper was taken and deposited on the other bar the copper being taken over an air space fully /64 of an inch. I then knew I had to deal with a static current.[8] I then went to work and had the bars Japaned finding by investigating with a [missing text] coil that Japan retards static sparks. We assembled the machine the bars being heavily Japaned but she [missing text] every time we had three barrels or 600 lights on and I found this was due to the fact that the Japan was put on unequally. We then took the bars out made them smaller and wound them as I have stated with four thickness of parchment paper. I found that parchment paper is the only kind of paper that offers great resistance to the static spark. You know that all papers are nothing more than a combination of fibers, that the spark goes right through the air spaces between the fibres and the paper offers no resistance at all, in fact it travels through the air, but if the air space is filled up with any glutinous substance so that it must pass through matter then it offers high resistance. Now in parchment paper the fibres are all agglutenous the air spaces are (clogged?), hence the spark must pass through solid matter hence the great utility of parchment paper for this purpose. These four layers as the insulation you have on the machine and you are quite safe for 600 lights providing you have no cross in the chandeliers. Perhaps you can get along with even four or five hundred full lights. It would be better to get along with a smaller number until the test for economy takes place.[9] Then you can put on 700 lights. You will notice that we have bored the field out over $\frac{1}{8}$ of an inch so there is not the slightest danger about the band touching. We were compelled to bore out the field so that we could get sufficient air through the to cool the machine. With the present velocity of the blowers the machine will not over heat itself with 750 lights on. We were compelled to pass the air through one end so that it will go out at the other end as we did not dare to arrange it otherwise. Hence the end of the armature where the air goes in after running many hours would be [missing text] perfectly cold while the end nearest the commutator would be found pretty hot especially the tits. But there is no danger even if the tits get so hot that you cannot scarcely hold your hand over. Of course the end nearest the commutator is at the disadvantage of having hot air thrown against it to cool it.

You will have to be very careful about that "cross belt." It was the [missing text] and you will have to inspect it well. I do not know whether you will need to run this machine every night or only certain nights in the week. If every night then after closing and the lights are off I should run the machine about half speed with the current altogether off so as to reduce the temperature of the armature. Otherwise it will not not be very cool when you start the next night. However do not suppose this matters very much as the speed of the blower is such that the heat never can rise to a dangerous point, without more than 700 lights are put on. I do not know whether Dean written regarding what kind of oil he uses,[10] but my impression is the best lard oil is the thing to use. I think I have heard him say so. I should only run two or three hundred lights at first until you have got your bearings nicely soaking   You will notice that the [missing text] has been changed. It pounded so that Dean put on a regular connecting rod. I want to tell you that just before we were ready to ship this machine the shaft broke off at a point where the disc on the engine connects with it. It appears Armington & Sims made this shaft of green metal.[11] It was a very lucky accident as you might have had it occur in Paris. The idea of using a green metal shaft on a high speed engine with such a terrible thrust seems [missing text] to me.[12] In twenty five hours with a big gang of men Dean had a new steel shaft in with the machine running. The dynamo then worked so nicely and satisfactorily that I told Dean that I would not be satisfied if he did not get all the thump out of the engine and I told him to overhaul the engine throughly. In this investigation he found that the disc on the engine with which the [missing text] rod is connected was cracked. He then had to turn off the wheel and shrink an iron band on. After this was done the engine started and actually the iron band broke it being made of poor iron. He then made another band of the best moor iron[13] and that is what you have on it now. You will notice that we have lagged the governor pulley so as to get the proper speed to the blower.

Now about the electro motive force. There has been some very bad miscalculations in connection with this machine all the way through. I had the cheek (!) to connect up the field magnets so as to put more foot pounds on it and the result was that with even some exterior resistance, two ohms I think, and with a speed of only 320 revolutions we got about 100 or 110 volts which is more than we needed. Your lamps are 103 volts, the loss in your conductors will probably be not more than four volts hence 107 volts is all you require. About 3 horse power is used on the field magnets; the way it was connected when you were here it had but $\frac{1}{8}$ of a horse power.

Be careful that no water or oil gets round the blower as it might work into it and throw on to the armature.

The English machine which will be shipped on the 17th of this month[14] has 106 bars instead of 146 as in your machine and it has one field magnet longer. It is to be insulated with twelve layers of Japanned parchment paper and the air will be injected in the centre a separation being made between the field magnet; it will give 1000 lights.

I forgot to say that the lamps you have are 140 ohms instead of 125 ohms resistance. This was Dr. Nichol's mistake. After making about 30,000 of these lamps every body supposing them to be 125 ohms I got to figuring one night and found that if a lamp requires 103 volts with an economy of ten per horse power it would have to have more resistance than 125 ohms or else there must be too many foot pounds on it for a ten per horse power economy. I spoke to Upton about it and upon investigation he traced it to Dr. Nichols. The Dr. has left us and is now fitting a Professor's chair in some provincial college, teaching the young idea how to shoot.[15]

Tell [missing text] to look out for water in his cylinder and always open the drain cocks and get dry steam and heat the cylinder up very slowly when first starting; otherwise he will get in trouble like we did here.

Upon receipt of Bailey's telegram about making lamps in France[16] I had Hughes ship you his Barton (?) engine[17] and archimediar pump burner and blower which he had rigged up for fruit expriments. These I believe have gone forward and I have also instructed Upton to send you fifty thousand [missing text] fibres. He could spare you some cutting moulds but I dare not send them as both the drawings machine and cutting moulds are patented in France. However with the aid of what we have sent you, you can make a very nice little start and I should get a room somewhere in the suburbs 25 ft by 30 ft or 40 ft which will be quite sufficient for the time being and start the thing as economically as possible. I suppose you ought to be able to get such a room as you require for $15. or $20. a month.

You will have to go some glass blowing place and get your globes pot blowers and your tubes drawn from the same pot. Your platina you will of course get at Johnson, Mathey and Co. of Hatton Gardens, London.[18] You should claim to get it at the same rate at which they supply it to the Lamp Co. (we have special rates with them) stating that you require it for the Lamp Co.'s use in France.

If you have trouble about getting nickel we can send you sheet nickel from here. We use plumbago covers in carbonizing that is to say the large cover which goes over the nickel forms is made of plumbago.

Upton finds that about 2% of his lamps after being put away for two or three months loose their vacuum and finds it due to the fact that the sealing is not long enough. He has therefore added $\frac{1}{8}$ of an inch to the length of this platinum so that he squeezes $\frac{1}{8}$ of an inch more glass on it and he also brings the glass in the inside part up to a more "pastey" condition that is to say he gets it very much hotter or better fused before he squeezes it onto the wires

You will probably have a very healthy time in getting cutting mould made in France. I should only get a half lamp cutting mould made at first.

I will write you further if I should have omitted anything.

Please continue to keep me well posted. Yours very sincerely

Thomas A. Edison   Written by S. Insul.

L (typed transcript), NjWOE, DF (*TAED* D8135ZBL; *TAEM* 58:1046). Typed transcript of illegible document. ᵃSeveral lines left blank.

1. The reason for Mary Edison's trip to Michigan is not known. The Jumbo dynamo left New York on 7 September but presumably was at the dock before that date (see headnote, Doc. 2122). Edison was away prior to 31 August (Philip Dyer to Samuel Insull, 31 Aug. 1881, DF [*TAED* D8124ZAB; *TAEM* 57:1058]).

2. The letter from Francis Jehl to Batchelor has not been found.

3. Edison's notes from this test are in Doc. 2131.

4. See Docs. 2131 and 2134. Edison executed a patent application on 20 August for this method of reducing sparking in large dynamos (Case 342). The application was rejected and subsequently considered abandoned. Edison had it reinstated, however, and in 1888 amended it to meet the Patent Office's objection. The new text differentiated his invention from Faraday's description "of the amalgamation of a contact spring rubbing on a continuous copper wheel. My invention however relates to commutators made up of conducting bars separated by insulation, and here the result attained by amalgamation is different. In my construction sparking occurs as the brushes make and break circuit with the different bars, and the mercury performs the function of carrying off the spark heat by its vapor thus saving the copper of the bars and brushes; for instead of copper vapor being carried off which of course would effect the destruction of the copper, mercury vapor goes off and the mercury is readily renewed from time to time." The patent issued in 1890. Pat. App. 425,763; Casebook E-2537:12, PS (*TAED* PT021012; *TAEM* 45:735); Edison's undated notes for the U.S. and foreign patents are in Undated Notes and Drawings (c. 1882–1886), Lab. (*TAED* NSUN08:136–37; *TAEM* 45:219–220).

In a retrospective analysis, Charles Clarke blamed the sparking of the early central station dynamos on the odd number of commutator blocks, an insufficient number of blocks, and an unbalanced magnetic field. He noted that mercury compensated somewhat but "the sparking was nevertheless sufficient to fill the air with mercury fumes, which so badly salivated the attendants that the method had to be abandoned, and thereafter a careful adjustment of the brushes and attention to their condition and to the surface of the commutator were relied upon to minimize its harmful effects." Clarke 1904, 50; see also headnote, Doc. 2122.

5. Charles Dean superintended the Edison Machine Works. He had worked for Edison as a machinist in Newark and Menlo Park. Doc. 1914 n. 8; *TAEB* 5 App. 2.

6. A sugar teat is "a small portion of moist sugar tied up in a rag of linen of the shape and size of a woman's nipple, given to quiet an infant when the mother is unable to attend." *OED*, s.v. "sugar teat."

7. Edison referred to a Ritchie coil, a well-known induction coil developed by Edward Ritchie. See Docs. 41 and 434.

8. See Doc. 2123.

9. See Doc. 2148 n. 6.

10. No communication from Charles Dean on this subject has been found.

11. That is, unannealed iron.

12. Francis Jehl later recalled that the shaft broke just as the engine was starting and that it was "an act of providence" that it did not break at full speed, as Charles Clarke and others were standing nearby. Jehl also recalled that "Edison shook his head and expressed his surprise (in language hardly fit to record) that cast iron had been used for a shaft of that importance." Jehl 1937–41, 973–74.

13. This reference is unclear.

14. This dynamo, intended for the demonstration central station in London, did not ship until early October. See headnote, Doc. 2122.

15. On the miscalculation by Nichols, see Doc. 2129. After leaving Edison's employ in June 1881, Nichols taught physics at Central University in Richmond, Ky., the University of Kansas, and Cornell University. Doc. 2065 n. 7.

16. On 31 August Bailey cabled Edison: "Maxim making lamps for outside   very important we do same   Batch says no difficulty about getting some out before factory started says can make few lamps and even small Dynamos if necessary   do you authorise." Bailey to TAE, 31 Aug. 1881, LM 1:32C (*TAED* LM001032C; *TAEM* 83:888).

17. A form of piston pump, used on fire engines. Knight 1881, s.v. "fire-engine."

18. Johnson, Matthey & Co. was a leading refiner of precious metals and manufacturer of platinum devices, with which Edison had dealt since 1878. See Doc. 1604 n. 5; McDonald 1960, chap. 15.

–2150–

To Charles Batchelor

New York September 13th 1881

My Dear Batch,

I shipped you six meters some days back. You had better do nothing more than show them   Fill them with blue viterol[1] and let them stand. Do not attempt to weigh the plates as it is a very delicate job and will cause you considerable trouble. I have got a Cigar Lighter which will be sent you shortly. You may have to change the size of the platinum wires. Bergmann has been about [a][a] ~~four~~[b] months making the revolving arc lamp and it is not done yet.[2] If done in any reasonable time will send them to you

The Dynamo with raidial bars is not done yet. I stopped making a large machine and am making a smaller one & will endeavour to send it to you.[3] Try and get it brought prominently out in one of the Scientific papers as a new and novel Dynamo constructed on entirely new principles and have it illustrated as[c] in the Patents where you will find how the connections are made.

I suppose if you get the Contract for lighting the whole of the Grand Opera House you had better have steam Dynamos sent you of the new type and the best plan will be not to try to use the big Dynamo you have for this purpose but to ship that to England to act as a spare machine for Johnson. I shall be able to give you all the Steam Dynamos you want by the end of December.[4]

Regarding the European Company what I am trying to do is to have a large Syndicate formed in Paris or elsewhere by Fabbri, Puskas & Bailey, or anyone else which Syndicate is to form a Parent Company for operating the Light on the Continent of Europe and to prevent this new Coy from being a purely speculative one I propose that the Company shall pay up one million of dollars for the purpose of forming a large Manufacturing Coy for making lamps Dynamos, Engines, Tubes, Chandeliers, and all appliances connected with Electric Lighting and the proposition is that the Company shall be formed Capital Twenty million dollars of which one million dollars is to be paid the present European Co in cash and nine million dollars stock. So it would amount to Two million dollars of stock being sold one of which goes as I have said to the European Company and the other million to be subscribed to the Stock of the Construction Company. This is the correct thing to do and Puskas agrees with me in this opinion. It may be that we cannot swing such a large amount but still I believe the longer we hold on the more likely we are to get it. With one million dollars in the Construction Coy run by our men and started by us we having absolute control the first year there would not be any doubt about the technical success of the enterprise and if the technical success is assured the commercial success would naturally follow and the whole thing would be a success while most inventions sent over there have been just the opposite. In my telegram of today I spoke of 5% to be given us.[5] What I mean by that is that we are to supervise and start all the factories and put them in operation and furnish all duplicate drawings and have constructed (at cost of Constn Co) here or on the other side all the special machinery and to give all improvements which we devise in our works here and of which the European Constn Co would equally get the benefit. We should have to start and supervise the manufactory in other Countries than France where under the Patent law we are compelled to manufacture within the Countries themselves. For all this it is but fair that we should receive 5% added calculated on the actual cost of the goods which should be paid us from the date when the first goods are turned out, we agreeing to turn over the factories to persons competent to do the work when our connection ceases that is except so far^c as the 5% are concerned. This sum is to reimburse us for our time our ~~expenses~~rience on this side, our drawings and experience and you will easily see that this will be a great bargain for the Constn Co as in finding out what we have learned the cost of experimenting to them would be more than five times what they^c pay us.

As to the Installation of these various works I shall have to depend on you entirely. You can have what men you want that we can spare and as to the division of the 5% I would make that perfectly satisfactory to you

The life of the lamps are very much longer than I ever expected   I have not seen the record for the last few days but the last I saw the life of the 8½ per electrical horse power was 1900 hours with a average of 1300 hours while the ten per electrical horse power were 1300 hours with an average life of some what over 700 hours: three lamps of each kind are still running but we are making very much better lamps now as we have curves at 48 candles in which the average life was 94 hours and the longest life of any one lamp was 304 hours at 48 candles. This was a 8½ per horse power set. The average life of the ten per horse power lamps is enormously less at 48 candles. But the lamps on which we are making the 16 candle record only had an average life of 12 hours at 48 candles. Since that time we have curves with the ten per horse power lamps with an average life of 22 ~~candles~~ hours^b at 48 candles.[6]

I suppose you will have some trouble over on the other side in getting people to believe the statement that the average life our ten per horse power lamps is 800 hours because there is no way to prove it. I suggest that you take the earliest opportunity to put up ten lamps of the lowest volts in a box with glass front having same sealed and start them going at 16 candles. Have a responsible person to seal the box and verify the burning times & he should be a man whom everybody has confidence in. Whether you can get such a chance I do not know but it is possible you might in the Cellar of the Grand Opera House when you make the installation there   At any rate I personally will guarantee any contract, with a penalty of twenty thousand dollars ($20,000), to put up twenty five thousand lamps (25 000) in any City in Europe, the said guarantee being that if the lamps do not average a life of seven hundred hours (700) with ten lamps per electrical horse power said lamps giving sixteen candles illumination I will forfeit the sum above named.[7]

Let me know if they are going to make a test as to the efficiency of the Dynamos. If so use the copper Rod Dynamo[8] as you will doubtless get about 95% effeciency out[c] of it. I should take out of it from nine to ten horse power. That will give you the best efficiency. Use plenty of japan around your Dynamo Bobins. It is a splendid thing   Put on a thin coat over night in between the wires or in the case of the small disc Dynamo paint around and over the bars and the discs at the end. It gets very hard and is a splendid Insulator.

I have just thought that the Dynamo I sent you is wound in a different way with a even number of commutators and I have not sent your any description how to wind them.[9] I wrote you previously that I had sent you a lot of wire so that Martin Force can wind the armatures. By next steamer I will have a model sent you showing how to wind them. You know the greatest difference of electro motive force is between any two layers of wire and as it is some what difficult to insulate each layer or section[d] by itself in the old way of winding we have divided the bobin into half the number of spaces making each space twice as wide. Now we wind each space half full going square round the machine; then we insulate the whole of the bobin and wind around again over the top. This keeps the coils which have the greatest difference of potential and tend to cross one above the other instead of side by side. Thus we are enabled to get a good insulation between these coils, but to do this we had to use an even number of coils and commutators. I will send you a wooden bobin in which the first layer is shown in white thread and the second layer with black or red thread (I do not remember which now). I have struck a new way of dealing with men now whom we are sending away say to South America. I make them wind two two complete armatures before they can go and then instead of sending an extra armature I only send wire so that if they break down they can easily rewind the mac armature.

Please let me know what are the legal results of the examination of my patents in France and also what is my legal status in England if anything has been done there by way of examination.

We cannot find that Swan ever published anything showing that he ever experimented upon a filament of carbon in high vacuo in a chamber made of glass, nor can we find any patents until after our patents were issued on such a device. Do you know anything to the contrary.

Do you believe it would be possible to get the writers of "Engineer" & "Engineering" on our side and what would be the best way to go about it. Would money in the form of a fee for an opinion on the validity of our patents as against Swan, Lane Fox, & Maxim be the right method

I hope you have got the Roman Letter Automatic working as I am sure it would be very striking. It worked very beautifully here—in fact quite astonished me.

With reference to "Abortion" I offered to pay his way over to Paris & back which he said would amount to $2000.[10] A few days before he sailed he told me he had come to the conclusion that he had better not: he would rather go untramalled and free to do anything and even if he accepted he would consider himself untramalled. My impression is that he was bought off and hired by Maxims Coy although I may be mistaken in this. You may be able to tell however by hearing what he says to others where his interests are   It is impossible to tell by talking to him. He is a very deceptive man   Very Sincerely Yours

T A Edison

Since writing above have received following

Test Lamps

Life and Average   Sept 13th 7 A.M.

Elect. H.P.	Still burning	at hours—	Average hours
8½	3	1846	1237
10	3	13389	728
20 (half lamps)	2	1154	590

⟨½ Lamps were poor set   TAE⟩

LS, NjWOE, Cat. 1244, Batchelor (*TAED* MBSB7D; *TAEM* 95:271); a letterpress copy is in DF (*TAED* D8132ZAO; *TAEM* 58:476). Written by Samuel Insull. ᵃPaper damaged; text from letterpress copy. ᵇInterlined above. ᶜObscured overwritten text. ᵈ"or section" interlined above.

1. Blue vitriol (copper sulphate, $CuSO_4$) was used as an electrolyte in many batteries. Pope 1869, 12–14.

2. This was presumably the form of lamp which Edison sketched on 24 May as a design for a "Rotating Carbon arc Lamp." Three days later Edison completed a patent application covering the use of an electric motor to rotate one of the carbons longitudinally at two or three thousand revolutions per minute so as "to make an absolutely steady arc and secure and even and smooth consumption of the carbon points." In September Edison also sketched a variation of this with a revolving magnet "so that the arc will revolve instead of the carbon." N–81-05-21:39, Lab. (*TAED* N201:16; *TAEM* 40:470); U.S. Pat. 251,538; Cat. 1147, Scraps. (*TAED* NM016:111; *TAEM* 44:335).

3. Edison referred to the small disk dynamo, which was not completed until early 1882. See Doc. 2228.

4. On the preceding day, Batchelor cabled Edison that the French government had awarded the Edison interests a contract to light the

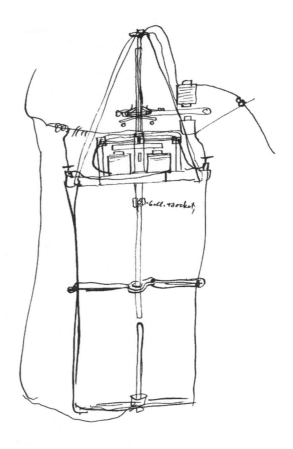

Grand Hall of the Paris Opera with 800 lamps. The government also stipulated that this would not jeopardize Edison's prospective patent rights. If the Edison interests successfully lighted this part of the Opera building by 7 October, the government promised to award them a contract to light the whole building with 8,000 lamps. Batchelor added that the "government architect would not consider any other incandescent light   Can you ship small machines to run these also thousand half lamps." On 17 September Edison cabled that he would soon send seven dynamos for this installation. Batchelor to TAE, 12 Sept. 1881; TAE to Batchelor, 17 Sept. 1881; LM 1:38B, 37B (*TAED* LM001038B, LM001037B; *TAEM* 83:891, 890).

5. Edison's cable has not been found, but the same day he wired from Menlo Park to Samuel Insull in New York to "telegraph Batchelor that he & myself should have five per cent on cash price for all goods turned out million dollar works for first ten years." TAE to Insull, 13 Sept. 1881, DF (*TAED* D8132ZAN; *TAEM* 58:475).

6. For details on lamp life tests, see Doc. 2117 and headnote, Doc. 2177.

7. Batchelor apparently marked the beginning and end of this paragraph; these marks are not found in the letterpress copy.

8. Edison referred to the bar armature, rather than one wound with wire.

9. Edison was referring to one of the Z dynamos which he shipped to Batchelor in September, and not to the large C dynamo, which used an odd number of commutator bars. During the summer Edison designed isolated plant dynamos with an even instead of odd number of commutator bars in order to reduce sparking at the brushes. See headnote, Doc. 2126 esp. n. 10.

10. Cable code for George Barker; see App. 4. On Edison's offer see Doc. 2110.

–2151–

To Sherburne Eaton

[New York,] 13th September [188]1

Dear Sir

I beg to enclose you herewith statement of expenditure[1] on account of experiments on the Central Station Dynamo amounting to $6171^{31}/100.

As you well know this experimenting has been done on the machine which is now on its way to Paris but in as much as it has been done with a view to perfecting the machines to be used in the Central Station I render the statement to the Electric Light Company and request payment under my agreement with them by which they undertake to pay the cost of my experimenting on Electric Light.[2]

If the Light Company had been obliged to pay for the machine itself as well as the experimenting the cost to them would have been about $12,000, without the machine being of any practical value to them as it would not have been possible to use it in our Central station. By the machine being taken by the European Company the Light Company is saved an expense of about $6000

This expenditure ($6171.^{31}/100) is cash paid out of my pocket and I shall be glad if you will arrange for me to be reimbursed the amount by Friday as I need the money to meet my engagements this week[3]   Very truly Yours

Thomas A Edison

LS, NjWOE, DF (*TAED* D8126ZAA; *TAEM* 58:35). Written by Samuel Insull.

1. Enclosure not found.

2. Doc. 1576. Weekly statements of expenses related to development of the electric light generally at this time, of which Doc. 1562 is an earlier example, are in Electric Light Co. Statement Book (1880–1884), Accts. (*TAED* AB032; *TAEM* 88:512).

3. That is, Friday 16 September. There is no record of a reply but see Doc. 2164.

–2152–

Agreement with
Henry Villard

New York, Sep. 14. 1881[a]

Edison will build 2½ miles of Electric Railway at Menlo Park, equipped with 3 cars, 2 locomotives, 1 for freight and one for passenger, capacity of latter 60 miles per hour.[1] Capacity freight engine 10 tons net freight, cost of handling a ton of freight per mile per horse power to be less than with ordinary locomotives. Experiments in traction and economy and practicability to be made by Edison and supervised by Villard's Engineer, if experiments successful, Villard[2] to pay actual outlay in Experiments and to treat with the Light Co. for the installation of at least 50 miles of Electric RR in the wheat regions.[3]

It is also assumed that the cost of the track will not exceed $2300. per mile, that of the locomotives and cars $800. and $250. respectively each, and the experiments not exceeding $500., unless Villard's Engineer should require further experiment.[4]

It is understood that Villard shall own the whole of the Electric Railroad, including motive power & rolling stock, if he pays for the same under this agreement.[5]

H. Villard   Thomas A Edison

DS, NjWOE, DF (*TAED* D8143C; *TAEM* 59:433). [a]Place and date written by Villard at bottom.

1. Edison had built a one-half mile electric railroad (later extended) in 1880 and continued to experiment on it through that summer. Planning for a new electric railroad at Menlo Park began in late August and September 1881, when Charles Hughes started to acquire rights of way, rails, and other equipment (Hughes to TAE, 26 Aug. and 22 Sept. 1881, both DF [*TAED* D8143B, D8143E; *TAEM* 59:431, 435]). For a general view of work on the electric railroad in 1881 and 1882, see Edwin Hammer 1904; Railroad–Electric (D-81-43 and D-82-49), both DF (*TAED* D8143, D8249; *TAEM* 59:418, 63:585); and testimony in the patent infringement case *Electric Railway Co. v. Jamaica and Brooklyn Road Co.*, Lit. (*TAED* QE001; *TAEM* 115:736). Related correspondence and technical information are in Cat. 2174, Scraps. (*TAED* SB012; *TAEM* 89:251); and N-82-03-12, Lab. (*TAED* N249; *TAEM* 41:861).

2. Financier Henry Villard (1835–1900) was a director of the Edison Electric Light Co. In September 1881 he took control of the Northern Pacific Railroad to ensure it would not threaten his coastal transportation system assembled under the Oregon Railway & Navigation Co. (for which Edison installed his first electric light plant on the steamer *Columbia* in 1880). Villard had worked as a journalist as a young man and about this time he also acquired the *New York Evening Post* and the *Nation*. Buss 1978 [1977], 189–90; Doc. 1892; *ANB*, s.v. "Villard, Henry."

3. In May 1879 Edison had sketched the rudiments of an electric railroad that would operate automatically, carrying grain to conventional trunk lines throughout the West. See Doc. 1745.

4. On 19 September Charles Hughes told John Randolph that Edison

had given instructions to "open an account with Henry Villard President of the Northern Pacific Railroad and charge everything done on or for the Electric R.R. to him." Acting through his Oregon & Trans-Continental Co., Villard made a $12,000 loan to Edison, secured by one hundred shares of Edison Electric Light Co. stock. By 1 February 1882 Edison had spent just over $10,000 on this project. Hughes to Randolph, 19 Sept. 1881, DF (*TAED* D8143D; *TAEM* 59:434); TAE agreement with Oregon & Trans-Continental Co., 2 Mar. 1882, Miller (*TAED* HM820159; *TAEM* 86:460); Electric Railway Financial Report, 1 Feb. 1882, Cat. 2174, Scraps. (*TAED* SB012ABW; *TAEM* 89:356).

5. This agreement was read into the record at the 23 September meeting of the Executive and Finance Committee of the Edison Electric Light Co. The company then granted Edison permission to experiment according these terms provided he made no arrangements for patents under its control. Edison Electric Light Co. to TAE, 28 Sept. 1881, Cat. 2174, Scraps. (*TAED* SB012ABA; *TAEM* 89:306).

–2153–

To Joshua Bailey

[New York,] Sept 17. 81

Heraclite Paris[1]

Object of factory erect square mile plants hence require two to four hundred thousand dollars to carry this out independent investment in factories[2]   Factory is the primal essential after that immense and rapid business possible otherwise long delays loss prestige competition. How can first Paris station be put up without large factory[3]   with million dollars assured hundred thousand could be used start small scale furnish isolated plants until large works started[4]   Is European Telephone definitely closed[5]

Edison.

L (telegram, copy), NjWOE, LM 1:37A (*TAED* LM001037A; *TAEM* 83:890). Written by John Randolph.

1. Cable code for Joshua Bailey; see App. 4.

2. That is, the investment required for each central station serving a square mile district would be $200,000 to $400,000, in addition to the investment in factories to equip those stations. Cf. Doc. 2148 and see Fox 1996, 184–85.

3. Bailey replied the next day: "First eight words your cable badly transmitted repeat them   think can arrange factory if rest programme satisfies you." In a cable dated 15 September (but evidently not transmitted until the next day), he and Puskas offered terms that included "Manufacturing company five million [francs] to be formed when option declared when first paris station lighted as per previous cable." They explained further in a 16 September letter to Edison that prospective backers would not make "an absolute engagement to invest so large a sum as five million of francs in a factory and to pay down five million more to the company in advance of the lighting of a station or section in

Paris." They would be willing to form a company on the strength of reports from the Paris Exposition but were uncertain "'whether on
mounting a large district complication[s] may not be developed that will
put a very different phase from that the business wears at present.' From
the point of view of the investor I could find no answer to this." On 19
September Bailey cabled: "Parties ready meet your views about factory
so far as to secure immediate and rapid manufactory." The prospective
investors also requested a "reasonable time during which no other
proposition entertained." Bailey to TAE, 18, 15, and 19 Sept. 1881, LM
1:39B, 35A, 106B (*TAED* LM001039B, LM001035A, LM001106B;
TAEM 83:891, 889, 925); Bailey and Puskas to TAE, 16 Sept. 1881, DF
(*TAED* D8132ZAS; *TAEM* 58:492).

4. Bailey wrote (and evidently also cabled) Edison on 5 September
that he and Theodore Puskas had

> been able to arrange here for 500,000 francs to put into the purchase
> and setting up the small [isolated] Plants in the principal Cities of
> Europe & for the Capital necessary for starting the lamp, Conduc-
> tor, fixtures & dynamo factories. The suggestion of the purchase of
> Plants for immediate use as above was made by Mr Batchelor— He
> says that persons are constantly engaged in handling or in making
> designs of the various parts of your exhibit relating to the light &
> that unless the Coy is ready to go into the field immediately it will
> find itself anticipated either with direct immitations or with con-
> trivances intended to evade your Patents.

He promised that these arrangements could be made in thirty days. Bailey to TAE, 5 Sept. 1881, DF (*TAED* D8132ZAH; *TAEM* 58:459).

5. In his cable the next day Bailey answered, "think European Telephone closed will confirm tomorrow." On 28 September he wired:
"Telephone finished tomorrow." Bailey to TAE, 18 and 28 Sept. 1881,
LM 1:39B, 49B (*TAED* LM001039B, LM001049B; *TAEM* 83:891, 896);
see also Doc. 2127.

–2154–

From Francis Upton

Menlo Park, N.J., Sept 17 1881[a]

Dear Mr. Edison:

I enclose a letter to Mr. Goddard regarding his request that
we select the lamps for isolated plants.[1]

I know a number of the lamps we make are not wholly
shapely but as I am trying every way to improve them I think
all should count as lamps.

I do not think 50 cts. is too large a price to charge. for we
shall not make money at that. I wish you would give Mr. Goddard the letter if you approve of it.

He said that the lamps were crooked in the sockets, this I
cannot believe to be the case, I think the trouble must have been
in the wooden socket.

We are trying small and thin tubes on the pumps and we so far think that they are going to work well   They will not be so apt to crack from the cold we hope.

Welsh has taken some high carbons and put them in nickle [-][b] covers with charcoal powdered and then covered this with plumbago and recarbonized with excellent results.

The putting a second cover over the nickle seems to be a good thing. [-][b] We will try two runs this way next week for we can save the forms and make the shrinkage smaller for the high heats make the weights stick.[2] Yours Truly

Francis R. Upton.

ALS, NjWOE, DF (*TAED* D8123ZDS; *TAEM* 57:958). Letterhead of Edison Lamp Co. [a]"Menlo Park, N.J.," and "188" preprinted. [b]Canceled.

1. Calvin Goddard's request has not been found. In reply, Upton addressed Goddard's complaints about the lamps for an unidentified isolated lighting plant. He explained that "we test every socket before sending them away to see if they are loose and throw out those that are loose. We are aware that by retrying them some can be loosened though we take every precaution that has been suggested. We think that the man who put the lamps in at Newburg [a woolen mill] must have taken hold of the glass and the plaster and tried them with a firm grip. During the past week we have made an improvement by twisting the glass where it goes into the plaster so that it can make a good contact and be irregular in the body of the plaster." He reassured Goddard that "crooked or crinkeled carbons are good so far as the lasting qualities and economy are concerned." He also explained that the factory could not afford to be more selective at the contract price of 35 cents apiece but offered that, for 50 cents, "we will give you selected lamps with straight carbons, well placed in globes and firm sockets tried with a firm twist." Upton to Goddard, 17 Sept. 1881, DF (*TAED* D8123ZDT; *TAEM* 57:960); Edison Co. for Isolated Lighting brochure, 1 Sept. 1882 (p. 6), PPC (*TAED* CA002A; *TAEM* 96:103).

2. The standard carbonizing mold devised in 1880 included sliding weights that kept the curved fiber flat and taut, but also allowed it to contract (see Doc. 1961). Alexander Welsh's experiments may have included experimental lots made two days later (nos. 450–52) of high resistance carbons "treated for resistance" by placing graphite or lamp black in the molds. Lamps made with these carbons had an unusually long average lifetime. On 27 September John Howell noted two batches (nos. 466–67) of fibers treated in charcoal, one placed in nickel forms and the other in plumbago forms. There is no record of test results from these. Cat. 1301, Batchelor (*TAED* MBN007:45, 47–48; *TAEM* 91:338, 340–41).

*From Charles
Batchelor*

My dear Edison,

As your telegram requested me to try and bring Bailey and Fabri together I may say I have been at that very thing for a week—[1] At first they would not listen to such a thing at all—but now I think I shall succeed— The proposition which Fabri had and was preparing for the company's consideration I was asked my opinion of by them. What they proposed was entirely inadequate for the working of such a thing as this is; after a couple of hours talk they abandoned it altogether and from what Mr Fabri says I should judge he considers it impossible to raise 5,000,000 francs for manufacturing purposes. I showed them that they can not do it for less. They are now at it again. Bailey is pushing ahead ~~again with the proposition~~ and as I told Fabri the other day if he only had the backing of a house like Drexel Harjes and Co it would be difficult indeed to find his equal here—[2] They know he is a hard worker but of course they feel that they would not like to be connected with him— However I expect they will eventually come together and I hope work well as this thing ought now to be settled and we started— I have met Dubois Raymond[3] and in him you have a friend of course just now almost all my time is taken with these professors as we want them to thoroughly understand our apparatus as many of them will be jurors— Proff Zetsche[4] has been here and gone to Switzerland— We have plenty of friends and also enemies— We make a lecture every morning on the apparatus and illustrate it by working it—

All the good scientific men are going to test our apparatus for publication—

Have just got the little machine and am putting it up—[5] Dont fail to send me the actual cost price at our works of the 110 & 62 volt dynamos also send me full price lists of wires (copper) as it seems dear here also costs of lamp manufacture (latest) also from Bergman and Johnson prices of everything——

I have asked the Leon people to put a man in my room at night; also one in Swan's, and one in Maxim's, unknown to us, to see what breakages there may be. I expect they have done so

Cable you today:— Getting bids for dynamos, small gramme runs 50 half[6] large Mulhouse wish put 10,000 our lamps on gramme machines their make now running arc lamps in Germany[7] fix price lamps & Royalty swan stands ready tried for week bring abatement and abdomen together think shall succeed   send 1500 half, with sockets opera   send actual cost small dynamos.[8] Yours

"Batchelor"

ALS, NjWOE, DF (*TAED* D8132ZAS1; *TAEM* 58:490). Letterpress copy in NjWOE, Batchelor, Cat. 1331:40 (*TAED* MBLB3040; *TAEM* 93:331).

1. Edison's cable has not been found. For at least several weeks Egisto Fabbri had been competing with Joshua Bailey to finance the commercial development of Edison's lighting system on the European continent. Fabbri disavowed any personal interest but wrote from Paris that "all I care for is to see the European Co. [Edison Electric Light Co. of Europe] properly represented here and its interests protected more efficiently than they can at present be with nobody with full and special authority to do so— Anybody that can do better than I think I can succeed in doing is welcome to the business." Bailey to Theodore Puskas, 4 Sept. 1881; Fabbri to TAE, 21 Sept. 1881; both DF (*TAED* D8132ZAI, D8133ZAB; *TAEM* 58:464, 629).

2. Drexel, Harjes & Co., the Paris affiliate of Drexel, Morgan & Co., was one of the leading private banks in Europe. Anthony Drexel founded it in 1868 with John Harjes, who was born in Germany of Danish parents but grew up largely in the United States before emigrating to Paris (Carosso 1987, 134–35). Batchelor expressed the same opinion of Bailey in a letter to Sherburne Eaton on this day (Batchelor to Eaton, 18 Sept. 1881, Cat. 1331:38, Batchelor [*TAED* MBLB3038; *TAEM* 93:329]).

3. Probably Emil du Bois-Reymond (1818–1896), professor of anatomy at the University of Berlin, who established the modern field of electrophysiology. He was especially noted for his work on animal electricity. His brother, Paul David (1831–1889), was a distinguished mathematician at the University of Tübingen. *DSB*, s.vv. "Du Bois-Reymond, Emil Heinrich"; "Du Bois-Reymond, Paul David Gustav."

4. Karl Zetzsche (1830–1894) was a mathematician, electrician, and postal telegraph official. Zetzsche taught telegraphy at the polytechnic school in Dresden until about this time, when he took a similar post in Berlin. He was evidently working as a patent expert for Lemuel Serrell on Edison's behalf. *DBE*, s.v. "Zetzsche, Karl Eduard"; Serrell to TAE, 27 June 1881, DF (*TAED* D8142ZBD; *TAEM* 59:338).

5. Probably the Z dynamo Edison shipped at the end of August. TAE to Batchelor, 28 Aug. 1881, LM 1:31A (*TAED* LM001031A; *TAEM* 83:887).

6. That is, 8 candlepower ("B") lamps.

7. This was probably the Mulhouse engineering and tool making firm whose workshop at the Exposition Batchelor had arranged to light with 15 half lamps. Batchelor to Eaton, 4 Sept. 1881, Cat. 1331:22, Batchelor (*TAED* MBLB3022; *TAEM* 93:313).

8. This is the full text of Batchelor's cable. Upon its receipt the cable code name "abatement" was transcribed as Fabbri, and "abdomen" as Bailey. Batchelor to TAE, 18 Sept. 1881, LM 1:40A (*TAED* LM001040A; *TAEM* 83:892); see App. 4.

To George Gouraud

Noside London[1]

At last ready furnish complete outfits and men isolated business from here without interrupting progress general distribution system   Factories completely equipped immense capacity   Take measures at once operate Isolated plants our joint countries leaving general system some regular way[2] Johnson leaves Gallia[3]   machine twenty eighth

L (telegram, copy), NjWOE, LM 1:42B (*TAED* LM001042B; *TAEM* 83:893). Written by Charles Mott.

1. Cable code for George Gouraud; see App. 4.

2. Edison authorized Gouraud in August 1880 to form electric companies for working his patents in numerous countries and territories outside of Britain and the major European nations (see Doc. 1978). In a 10 September 1881 letter that might not yet have reached New York, Gouraud admonished Edison for having decided to postpone "the introduction in any form of your lamp until you can introduce it as an entire system." He feared this policy would give advantage to Hiram Maxim and St. George Lane-Fox, who were already exhibiting incandescent carbon lamps. However, he was "more than delighted" to learn from Egisto Fabbri's correspondence with Sherburne Eaton that Edison had "at last determined to go in as extensively as possible for the isolated plants. If this policy is vigorously pursued you will soon recover whatever ground, if any, has been lost. I think it is Johnson's intention to bring one over here for my country place." Gouraud to TAE, 10 Sept. 1881, DF (*TAED* D8104ZCV; *TAEM* 57:202).

Edison had cabled Charles Batchelor the previous day to make arrangements for selling isolated plants in countries where doing so would not jeopardize his patents, especially in Germany, where he hoped to "head off Siemens." TAE to Batchelor, 19 Sept. 1881, LM 1:41B (*TAED* LM001041B; *TAEM* 83:892).

3. Edward Johnson sailed on Wednesday, 21 September, and arrived on 30 September. TAE to Charles Batchelor, 20 Sept. 1881, Lbk. 9:138 (*TAED* LB009138; *TAEM* 81:49); Johnson to TAE, 30 Sept. 1881, LM 1:49C (*TAED* LM001049C; *TAEM* 83:896).

To Joshua Bailey

Heraclite Paris[1]

Falsetto[2] has made proposition which I find is unsatisfactory. Following will beat it. Form syndicate raise three hundred thousand dollars to establish factories immediately upon success large dynamo. then upon success New York station to raise seven hundred thousand dollars more for increasing factories, of first and second sums our company shall receive fifty one hundredths of fully paid shares of manufacturing com-

pany out of which they agree pay Edison and Batchelor for unpatented improvements and cooperation factories here in
cheapening manufacturing. The ~~companies~~ syndicate to aid in
financiering companies throughout Europe The present
European company giving them one fifth proceeds Profits of
isolated business to be divided equally between manufacturing
and present European Companies The syndicate shall from
time to time and until profits come in advance such sums as
shall be mutually agreed for litigation, the press and miscellaneous expenses which shall be repaid by syndicate out of first
profits with six per cent interest.[3] Consult Batchelor.

Edison

L (telegram, copy), NjWOE, LM 1:105A (*TAED* LM001105A; *TAEM*
83:924). Written by Charles Mott.

1. Cable code for Joshua Bailey; see App. 4.
2. Cable code for Egisto Fabbri; see App. 4.
3. Charles Batchelor summarized these terms in a memorandum
after Edison instructed him in a cable on 24 September to assist Bailey
"in explaining small risk and handsome profit from isolated to syndicate
if they accept this plan." Bailey and Puskas replied on 26 September that
the offer was "received favorably" and soon promised that these terms
could be arranged in eight days. Edison cabled back on 4 October that
Fabbri would make no further offers "so you can go ahead with your
eight days option." Edison authorized Bailey to sign the necessary papers contingent upon his final ratification and also approval by Charles
Batchelor and Grosvenor Lowrey in Paris. See Doc. 2182 regarding
the final terms. Batchelor Memorandum, n.d.; TAE to Batchelor,
24 Sept. 1881; both Cat. 1244, Batchelor (*TAED* MBSB7E, MBSB7F;
TAEM 95:282–83); Bailey and Puskas to TAE, 26 and 27 Sept. 1881;
TAE to Bailey and Puskas, 4 Oct. 1881; LM 1:48B, 107A, 52B (*TAED*
LM001048B, LM001107A, LM001052B; *TAEM* 83:896, 925, 898);
TAE to Bailey, 10 Oct. 1881, Lbk. 8:191 (*TAED* LB009191; *TAEM*
81:68).

–2158–

To George Gouraud

[New York,] 25th Sept [188]1

Dear Gouraud,

Electric Accumulator

Yours of 10th inst came duly to hand.[1]

I have a patent in England on the application of the above to
Electric Lighting. It is amongst the devices which I abandoned
for something better.[2]

If you set down and figure out the matter <u>commercially</u> you
will see that in the present advanced state of the art there is
nothing in it at all that would warrant one in taking up the sub-

ject. Doubtless eventually the people investing in it will find this out to their cost.[3] Yours very truly

Thos. A. Edison   I[nsull]

L (letterpress copy), NjWOE, Lbk. 9:142 (*TAED* LB009142; *TAEM* 81:52). Written by Samuel Insull.

1. Gouraud sent two letters to Edison that day in accordance with his practice of restricting each letter to a single topic. In the one concerning electric accumulators, or secondary (storage) batteries, he reported that "a tremendous amount of stir is being made" over the Faure battery. He suspected that there were plans "to make a big thing of it" and thought there was "some degree of merit . . . in the idea per se of an accumulator, for many purposes, and the object of this letter is to ask you to just give a little thought to the subject, and let us have an 'Edison accumulator.'" Charles Batchelor recently advised Edison from Paris that "the Faure battery people here have been running Swan lights all over the exhibition." Gouraud to TAE, 10 Sept. 1881; Batchelor to TAE, 8 Sept. 1881; both DF (*TAED* D8104ZCW, D8135ZBK; *TAEM* 57:204, 58:1044).

2. One of Edison's first British patents for electric lighting included "means for storing the electric current or energy so that the same may be used as required. This is done by the use of secondary batteries, and there are devices for shifting the current from one secondary battery to another periodically so that one may be in use while the other is being charged from the main circuit." Brit. Pat. 5,306 (1878), Cat. 1321, Batchelor (*TAED* MBP015; *TAEM* 92:102).

3. See Doc. 2389. Despite this view of the subject, Edison had applied in June for a patent on an arrangement "to maintain constant the electro-motive force of secondary batteries or accumulators—that is, to reenforce their pressure as the same becomes lowered, when such batteries are used as a source of electricity for electric lights or other translating devices." The application was rejected, then substantially rewritten in 1888; it finally issued in 1890 as U.S. Patent 435,687. Edison returned periodically to storage battery experiments throughout 1882; see Docs. 2276, 2278, 2295, and 2307; also cf. Docs. 2274 and 2334.

–2159–

From Francis Upton

Menlo Park, N.J., Sept 25 1881[a]

Dear Mr. Edison:

Dyer is going into New York today. The money question is growing in importance. For example I looked round New York and found the cheapest place for straw paper. We sent them three orders which they filled immediately and then dunned us for money as the sales were for cash. We have not yet paid them as we took our bills in order. Two weeks ago we sent another order which they have not filled so that yesterday we ran out of paper.[1]

The Lehigh Co.[2] have been sending us [-][b] very bad coal, we have written them without effect for we are not prompt in payment.

We are making between 900 and 1000 lamps every working day and the running expenses are between $1700 and $1800 a week, actual money expended.

To keep running this outlay must be met and to run economically it must be met promptly.[c]

We cannot lay off without very serious[c] loss as we have been cutting down wages so that when running constantly we lose now and then hands. Shutting down for two weeks would drive away a number more.[3] Yours Truly

Francis R. Upton

ALS, NjWOE, DF (*TAED* D8123ZDY; *TAEM* 57:967). Letterhead of Edison Lamp Co. [a]"Menlo Park, N.J.," and "188" preprinted. [b]Canceled. [c]Obscured overwritten text.

1. Straw paper may have been used merely as packing. According to Jehl 1937–41 (809), finished lamps were "well-wrapped in paper" and shipped in barrels.

2. The Lehigh Valley Coal Co., a subsidiary of the Lehigh Valley Railroad, operated in the anthracite coal fields of northeastern Pennsylvania. The firm had New York offices at 21 Courtlandt St. Davies 1985, 26; *Trow's* 1881, 900.

3. Upton planned to close the factory on Monday, 26 September. This was a national day of mourning for the funeral of President Garfield, who died on 19 September. Upton to TAE, 23 Sept. 1881, DF (*TAED* D8123ZDX; *TAEM* 57:965); Ackerman 2003, 430.

–2160–

To Francis Upton

[New York,] 29th Sept [188]1

Dear Sir

I understand some men were sent out from here to the Lamp Factory without an order from me[1]

For the future please admit no one whatever to view the Factory without an order from me, and even in such a case make your explanations of working[a] the most superficial character possible[2]   Yours truly

T. A. Edison.

ADDENDUM[b]

[New York, September 29, 1881]

E[dison]. E[lectric]. L[ight Co.].

Orders from my offices of [--][c] Co to admit persons not good except countersigned by me

Edison

LS (letterpress copy), NjWOE, Lbk. 9:159 (*TAED* LB009159; *TAEM* 81:57). Written by Samuel Insull. a"of working" interlined above. bAddendum is an ALS. cCanceled.

1. Nothing more is known of this incident.
2. Cf. Docs. 2185 and 2309.

–3– October–December 1881

Having just passed the third anniversary of full-time work on an incandescent lighting system, Edison could see his first commercial central station literally taking shape in lower Manhattan. While his business affairs in the United States and abroad continued to demand attention each day, Edison had a new and pressing responsibility. As he told one appointment-seeker, he could "be found most evenings . . . in the neighborhood of Pearl St. superintending the laying of our electric light mains."[1] Edison's multiple responsibilities forced him to delegate authority more often than he was accustomed—or liked—to do. When Charles Clarke proposed a particular method of testing underground cables, Edison acceded only on the condition that Clarke "take the responsibility as to any trouble hereafter. . . . I do not at all believe it is the right course to pursue and a little more brain power on the part of someone would devise" a better method.[2] John Lieb was entrusted with supervising construction of the generating station at Pearl Street. John Kruesi ran the Electric Tube Company and had primary responsibility for laying the underground conductors. Charles Dean remained in charge of the Edison Machine Works, which completed another huge C dynamo to Edison's satisfaction in early October. By late November, the growing Machine Works was capable of turning out two dozen small dynamos per week.[3]

Edison asserted direct personal control in the vital area of lamp manufacture. He received frequent test reports from the factory in Menlo Park, and when he noticed a decline in the durability of the lamps, he called it to the attention of superintendent Francis Upton in October. Perhaps because of an indeterminate absence to attend to his gravely ill father, Upton did not correct the problem immediately.[4] Edison took things into his own hands in November and virtually lived at the factory for what he called "a grand bounce of the bugs."[5] This period of intensive experimenting lasted about ten days. Edison claimed to have slept only eighteen hours in an entire week, never once taking off his boots. He left satisfied that he had improved the manufacturing process to the point that the vacuum—and the lamps—would last longer than ever. He also approved new lamp base and socket designs, the basic elements of which have been in use ever since.

Edison periodically returned to Menlo Park to conduct experiments. In late November, for example, he began a series of experiments to develop a higher resistance lamp, especially for the British market.[6] In addition, he had facilities to tackle specific problems at the Machine Works and the lamp factory. At the former, he designed a new regulator that would protect lamps from excessive voltage, a particular concern at isolated lighting installations. The November experiments at the lamp factory led to three of the thirty-five U.S. patent applications that he executed in the last quarter of 1881. Twenty-seven patents eventually issued, including several for electric arc lighting, on which he had also been working.[7] Edison ordinarily had dozens of patent applications pending at any given time, and on 18 October he had the pleasure of cabling Charles Batchelor in Paris that the U.S. Patent Office "issued to me today twenty three patents on system Electric lighting."[8]

In Paris, Batchelor oversaw assembly of the direct-connected steam dynamo, the largest in the world, before the International Exposition closed in November. At about the same time, he made arrangements for a demonstration lighting installation at the Paris Opera. Joshua Bailey, meanwhile, successfully negotiated amended terms to finance companies for developing the electric light throughout much of Europe. When Bailey fell seriously ill just as the agreements were being formalized, Grosvenor Lowrey cabled Edison from Paris that "Batchelor understands and I advise you rely wholly on him."[9] In December, Batchelor made arrangements to acquire a factory just outside the Paris city walls to manufacture the lamps and heavy equipment that would be needed in Europe.

Edison looked especially to Great Britain for successful foreign development of the electric light. There he relied almost entirely on the entrepreneurial zeal and managerial acumen of Edward Johnson, who had reached London at the end of

September. Remembering keenly the legal problems associated with Edison's British telephone patents, Johnson urged a comprehensive review of his lighting patents by eminent authorities and a better method of filing new specifications. Edison assented to both. Edison became deeply involved in explaining his lamp patents and justifying his claims to skeptical legal minds. Johnson was particularly concerned about rival claims to the dynamo armature by the Siemens interests and to the incandescent lamp from several quarters. He moved cautiously in the first case, suspending sales of small dynamos and opening negotiations for a licensing agreement. In the second, he seized on an invitation for Edison to join a lighting exhibition at London's Crystal Palace. There he hoped to "Bust the Bubbles [Hiram] Maxim [and Joseph] Swan & others blew at Paris or else acknowledge that they are better 'showmen' than I."[10] Johnson's primary mission, however, was to set up a demonstration central station to establish the reliability and economy of the system to prospective British investors and consumers. He and Egisto Fabbri (working on behalf of Drexel, Morgan & Company) selected a site on London's famous Holborn Viaduct in October and immediately began preparations for installing the huge dynamos, the first of which was already waiting on the dock. Edison also wanted Johnson to demonstrate the system's safety because he feared that carelessness by rivals, such as that responsible for a fire on the steamer *City of Rome*, would diminish the market for electric lighting.[11]

Edison expected that he would continue to manufacture heavy equipment in the United States for the British lighting market for several years. This additional capacity could only have heightened his financial anxieties; already he had an "urgent" need of funds to cover immediate demands.[12] Some relief came with an advance of $10,000 from George Gouraud, apparently on account of his yet-unresolved British telephone interest.[13] Incorporation of the Oriental Telephone Company, to which he also looked for money, crept forward. One operation in which Edison did not have a direct financial stake was the Edison Company for Isolated Lighting, which began operating independently of the parent Edison Electric Company in November. Its manager, Miller Moore, joined Edison and officials of the parent company at a mid-November conference to discuss lowering manufacturing expenses, mainly by reducing the number of employees and their wages.[14] In a personal venture about which little is known, Edison directed his chemist John Lawson to explore an abandoned copper mine in

Menlo Park. Lawson pumped out the water using an electric motor and was able to discern ore veins, but Edison evidently did not develop the mine commercially.

With Edison away from his Fifth Avenue office frequently and irregularly, Samuel Insull continued to solidify a foundational position in his financial affairs and correspondence. Various Edison shops and companies experienced start-up problems and bickered about prices, but Insull kept Edison's personal affairs moving remarkably smoothly. Of course, even his talents could not satisfy every constituency, as when Sherburne Eaton grumbled about being uninformed of foreign developments. Edison's erratic schedule and the volume of his correspondence made it increasingly difficult for Insull to reply swiftly to letters. Most received reasonably prompt replies, but Edison's longtime acquaintance George Bliss complained that "a man would starve to death while waiting for you to answer his letters. This is not your usual way of doing business."[15]

Edison and his wife appear to have spent considerable time at Menlo Park from the end of November. Mary Edison planned a party there, apparently a dance, on 20 December.[16] The next day, Samuel Insull gave notice that the family planned to vacate their New York rooms by the end of the year. The Edisons' eight-year-old daughter Marion, who was enrolled at a boarding "School for Young Ladies" at 63 Fifth Avenue, presumably did not return to Menlo Park with her parents.[17]

1. TAE to J. C. Massa, 25 Oct. 1881, DF (*TAED* D8104ZEK; *TAEM* 57:251).

2. TAE to Clarke, 20 Oct. 1881, Lbk. 9:211 (*TAED* LB009211; *TAEM* 81:74).

3. See Doc. 2187. For a detailed accounting statement of the Edison Machine Works, including the value of the machinery and tools, monthly expenses, income by dynamo model, and profit, see Rocap to TAE, 18 Nov. 1881, DF (*TAED* D8129ZCQ1; *TAEM* 58:317).

4. Upton to Samuel Insull, 4 Oct. 1881, DF (*TAED* D8123ZEC; *TAEM* 57:971).

5. See Doc. 2187.

6. See Docs. 2192, 2197 and 2202.

7. For the unsuccessful applications, see Patent Application Casebook E-2537 and Patent Application Drawings, both PS (*TAED* PT021, PT023; *TAEM* 45:733, 818).

8. LM 1:65C (*TAED* LM001065C; *TAEM* 83:904).

9. Lowrey to TAE, 15 Nov. 1881, LM 1:99B (*TAED* LM001099B; *TAEM* 83:921).

10. Johnson to Insull, 1 Nov. 1881, DF (*TAED* D8133ZAL; *TAEM* 58:674).

11. See TAE to Johnson, 4 Nov. 1881, LM 1:88A (*TAED* LM001088A; *TAEM* 83:916).

12. See, e.g., TAE to Gouraud, 6 Nov. 1881; TAE to Batchelor, 6 Nov. and 23 Dec. 1881; LM 1:92B, 92A, 122A (*TAED* LM001092B, LM001092A, LM001122A; *TAEM* 83:918, 933).

13. Gouraud to TAE, 10 Nov. 1881, DF (*TAED* D8104ZET; *TAEM* 57:279).

14. Eaton to TAE, 9 Nov. 1881, DF (*TAED* D8126ZAJ; *TAEM* 58:64).

15. Bliss to TAE, 3 Nov. 1881, DF (*TAED* D8120ZBK; *TAEM* 57:653).

16. Insull to Pennsylvania Railroad Co., 17 Dec. 1881, Lbk. 9:451 (*TAED* LB009451; *TAEM* 81:165); Dempsey & Carroll to TAE, 17 Dec. 1881, DF (*TAED* D8114J; *TAEM* 57:545).

17. Graham School for Young Ladies to TAE, 24 Oct. 1881, DF (*TAED* D8114E1; *TAEM* 57:539).

–2161–

To Edward Johnson

[New York,] Oct. 2. 81

Johnson

Urge necessity engaging all best experts and lawiers   Also experts prepare disclamer patents where necessary[1]   Will aid this work from here   give professors lamps sent abduction[2] Urge abduction take immediate steps push isolated unpatented countries   Also abatement[3] in England Norway Sweden[4]

Edison

L (telegram, copy), NjWOE, LM 1:51A (*TAED* LM001051A; *TAEM* 83:897). Written by John Randolph.

1. Under British law, invalidation of a single claim for any reason could invalidate the entire patent; the disclaimer process permitted a patentee to narrow the scope of a final specification. Edison's electric lighting patents to date had been prepared in the U.S. and filed (without revision) by English agents (Davenport 1979, 34–46; see Docs. 1822, 1870, 1880, and 2203). In mid-November, Johnson complained that he did not believe Edison had learned from having had his telephone patents prepared in that way, and urged that

> If your future Patents are to be of any value—they should pass through a channel which would scrutinize them with reference to what has gone before— In my judgment every application for an English Patent should be made by some one having the knowledge which he could apply—of the requirements of the Patent Law of England—Just as Wilber does in re—to the U.S. = If you say so— I will at once employ a thoroughly reliable Patent Agent. [Johnson to TAE, 12 Nov. 1881, DF (*TAED* D8133ZAR; *TAEM* 58:698)]

2. Cable code for George Gouraud; see App. 4.

3. Cable code for Egisto Fabbri; see App. 4.

4. Edison agreed in March 1881 to sell to Drexel, Morgan & Co. his patents in Norway and Sweden for electric light, power, and heating. Egisto Fabbri (with Grosvenor Lowrey) was a trustee for purposes of

assigning ownership of the patents, a role he also had with respect to Drexel, Morgan & Co.'s control of Edison's electric light patents in Great Britain. TAE power of attorney to Drexel, Morgan & Co., 1 Mar. 1881; Grosvenor Lowrey to Drexel, Morgan & Co., 25 Mar. 1880; both DF (*TAED* D8132P, D8026ZBE; *TAEM* 58:431, 54:221).

–2162–

To Edward Johnson

[New York,] Oct. 4. 81

Johnson

English dynamo works to perfection thousand lights on first trial heating scarcely noticable[1]   packing now goes on Greece Thursday[2]   also three small machines

L (telegram, copy), NjWOE, LM 1:52A (*TAED* LM001052A; *TAEM* 83:898). Written by John Randolph.

    1. This was a large direct-connected C dynamo for the London central station demonstration. See headnote, Doc. 2238.

    2. Thursday fell on 6 October; the *Greece* sailed on 5 October. "Shipping News," *New York Herald*, 4 Oct. 1881, 10.

STANDARD ELECTRIC CONSUMPTION METER   Doc. 2163

A key element of Edison's design for central stations involved charging customers based on a measurement of the amount of electricity each one used. Edison's thinking on this subject was influenced by the way that gas utilities charged for gas distributed from central stations.[1] In late 1878, he conceived the basic design of his electric consumption meter, which used the action of electrolytic decomposition like that in a battery to determine how much electricity had passed through a circuit.[2] Edison experimented with two fundamental designs. His earliest forms used a single metal electrode in an electrolytic solution such as copper sulfate; current passing through the cell deposited metal ions from the solution onto the electrode. Later Edison employed two metallic plates instead. As the customer used electricity, ions from one plate (the anode) passed to the other (the cathode). Because the deposition of a metal occurs at a fixed ratio to the current strength (other conditions remaining constant), the amount of electricity used by a customer could be determined by weighing the plate on which the metal was deposited. Edison used copper electrodes in most of his meter experiments, often in a copper sulfate solution, and he filed a patent for a basic copper deposition meter in March 1880.[3] As Edison began to commercialize his system in 1881, Francis Jehl, who had carried out many of the experiments at the Menlo Park laboratory, made numerous tests at the Edison Machine Works Testing Room to improve and standardize the copper deposition meter for regular service.

Jehl began an extensive series of trials on 30 August that led Edison to substitute zinc for copper. These involved the effects of different solutions and types of plates under various conditions, and Jehl continued them throughout September and much of October.[4] Edison's addendum to Doc. 2143 indicates that he had tried "plates of amalgamated zinc in a solution of sulphate of zinc" by 9 September. Jehl followed with several days of experiments on copper plated with silver or gold, and began trying various zinc solutions within the week. The first extant record of tests of meters with zinc plates are Jehl's from 15 September.[5]

Edison's shift to zinc was prompted by the fact that polarization, a chemical change at the electrodes which produces a counter electromotive force, rendered the copper meters inaccurate. As Jehl explained in his 1882 pamphlet on the Edison meter, "An endless number of experiments on this point were made by me, under Mr. Edison's direction, employing every element known." These experiments showed that "a copper deposition cell, and some other metals, is suitable for large currents . . . but when it is required to register a very small current, such as $1/1000$ of a weber, and when the deposition cell is always on a closed circuit, it becomes necessary to use something else than copper in order to obtain accurate results." Edison determined "that by using electrodes of pure zinc, amalgamated with mercury . . . great practical accuracy is ensured when an infinitesimal quantity is desired to be measured." Jehl's experiments "terminated in the adoption of zinc plates for his [Edison's] meters."[6] On 5 October 1881, Edison executed a patent application for a meter with amalgamated zinc plates.[7] Jehl took the new instrument with him to Europe in February 1882 and the first detailed description and illustrations of it appeared in the 12 May issue of *Engineering*, about the time that his twenty-four page pamphlet on the instrument was being printed.[8] Extensive meter experiments continued throughout 1882 at the Testing Room, many carried out by George Grower.[9]

The standard meter consisted of an iron box containing two electrolytic cells, German silver short-circuiting shunts, and an Edison lamp operated by an automatic thermostat to prevent the electrolytic solutions from freezing. Each cell contained two amalgamated zinc plates, separated by ebonite blocks, suspended in a zinc sulfate solution. Current passing through a shunt circuit to each cell caused the transfer of zinc ions from one plate to the other. Plates from one cell were collected and weighed monthly. The other cell, whose shunt circuit had higher resistance and therefore carried less current, were collected quarterly as a check on the monthly measurements. Recorded in a meter book, the weights provided the basis on which to calculate the amount of light used by converting the measure of current (amperes or as they called it at the time webers) into candles.[10] Edison initially made two versions of the meter, one adapted for 25-lamp circuits and the other for 50 lamps.

1. On gas meters see Docs. 1439, 1548, 1593 n. 1, 1712, and 1900.

2. See Doc. 1622 nn. 1–2. During winter 1881, Francis Jehl experimented with a meter in which a small motor operating proportionally to the current would operate a mechanical register to indicate the amount of current. Edison applied for a patent on it in March 1881. Such meters were not sufficiently accurate in practice because of the effects of initial inertial forces when starting the motor. Jehl Diary, 26–38 passim; Mott to TAE, 21 and 22 Feb. 1881, DF (*TAED* D8137A, D8137B; *TAEM* 59:60, 62); Jehl 1937–41, 661–65; U.S. Pat. 242,901; Jenks 1889, 5–7, 36–37.

3. U.S. Pat. 251,545. Edison had included but later removed a copper deposition meter in a February 1879 patent application (Pat. App. 227,227). He also included it in his first British electric light patent (Brit. Pat. 4,226 [1878], Cat. 1321, Batchelor [*TAED* MBP013A; *TAEM* 92:107]). See also Doc. 1733 n. 1.

4. N-81-08-30, passim, Lab. (*TAED* N228; *TAEM* 41:253).

5. N-81-08-30:21–63, Lab. (*TAED* N228:11–33; *TAEM* 41:264–86).

6. Jehl 1882a, 6–7.

7. U.S. Pat. 281,352.

8. "The Edison System of Electric Illumination," *Engineering* 33 (1882): 467–69; Jehl 1882a; Jehl to TAE, 11 May 1882, DF (*TAED* D8239ZBM; *TAEM* 62:888).

9. See Docs. 2269 and 2272, n. 1; George Grower to TAE, 14 June and 7 July 1882; all DF (*TAED* D8235O, D8235Q; *TAEM* 62:44, 47). In 1936, Charles Clarke had copies made of notebooks containing records of meter trials at the Testing Room in 1882; four are at the Henry Ford Museum (Clarke Meter Notebooks nos. 1–4, EP&RI [*TAED* X001K5, X001K6, X001K7, X001K8].

10. Edison wanted customers to be charged on the same basis as gas, the charge for which was based on the amount of gas used by a 5-foot burner to produce about 12 candles. The Edison company could easily calculate the equivalent number of gas burners provided to customers on the basis that a 16-candle A lamp was equivalent to a 7-foot gas burner, which could be calculated as $^{11}/_{14}$ of an ampere. Jehl 1882a, 14–15.

*Standard Electric
Consumption Meter*[1]

[Fig. 1][2]

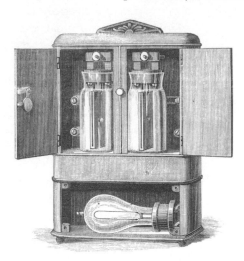

[Fig. 2][3]

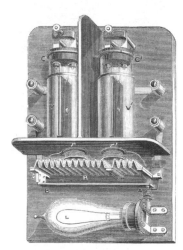

[Fig. 3][4]

M (historic drawing) (est. 23 × 34 × 10.5 cm), *Engineering* 33 (1882): 468.

1. See headnote above.

2. A meter based on this design and used in the Pearl Street central station district is at the Henry Ford Museum. MiDbEI(H), Acc. 29.1980.275.

An example of the standard consumption meter, used in Edison's New York First District.

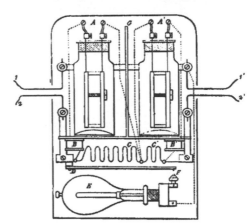

*Schematic diagram of the standard zinc meter from Francis Jehl's 1882 booklet. The bimetallic thermostatic strip **D**, located beneath the shunt short circuits **C** and **C′**, closed the circuit at **F** to keep the electrolytic solution in the cells from freezing.*

3. The short-circuiting shunts, made of German silver wire, are shown underneath the cell jars. The one on the left is connected to the monthly cell and has a resistance of 0.01 ohm while that on the right, connected to the quarterly cell, has one-quarter the resistance or 0.0025 ohm. Underneath the shunts is a bimetallic thermostat switch. At about 42° F, the switch closed a circuit to turn on the heating lamp. Jehl 1882a, 16–17; "The Edison System of Electric Illumination," *Engineering* 33 (1882): 467.

4. The two zinc plates are $3 \times 1 \times 0.25$ inches and are attached by ebony screws to two ebonite blocks that keep them .24 inches apart. "The Edison System of Electric Illumination," *Engineering* 33 (1882): 467.

–2164–

From Calvin Goddard

New York, October 6 1881[a]

Dear Sir

Strict directions have been given that hereafter no orders shall be executed for account of this Company either for labor or material except upon requisitions bearing the written approval of the Vice President[1] or Secretary[2]

You will therefore please present all requisitions to the Sec-

retary daily, as far as possible before 12 o'clock noon, accompanied by a memorandum thereof, on form supplied for that purpose,[3] to be filed in the Secretarys Office[4]   Yours Truly
                                                                C. Goddard   Secy.

ALS, NjWOE, DF (*TAED* D8126ZAD; *TAEM* 58:55). Letterhead of Edison Electric Light Co. [a]"New York" and "188" preprinted.

1. Sherburne Eaton.
2. Goddard referred to 1 October circular letters from the Edison Electric Light Co. and the Edison Electric Illuminating Co. of New York. They instructed suppliers to "execute no orders for account of this Company either for labor or material except upon requisition bearing the written approval of the Vice President or Secretary. From and after this date, no bills will be paid except for work done or material furnished in pursuance of such requisitions." Both DF (*TAED* D8126ZAC, D8122D; *TAEM* 58:54, 57:726).
3. No examples of this form have been found.
4. Goddard sent a copy of this letter the same day on behalf of the Edison Electric Illuminating Co. of New York (cf. Doc. 2106). Goddard to TAE, 6 Oct. 1881, DF (*TAED* D8122F; *TAEM* 57:728).

–2165–

To Edward Johnson

[New York,] 9th Oct [188]1

Dictated

My Dear Johnson

Look at the article in the Engineer (by Swan) on page 229 of September 23rd 81!![1] You will see the cheek of the fellow!

So far I have been able to find nothing about Swans doings in the proceedings of the New Castle Society except a publication in 78 or 79 about a Carbon Lamp, with a pencil of carbon in it.[2] I think you had better hunt up all the transactions of this Society (I forget the exact name) and also have some one go to New Castle on Tyne[a] to find a member who has attended all of the meetings of the Society during the last few years and ascertain just what he did have.[3]

You will see that the fellow states that he plated the carbon to the platinum in 1879. He is now trying to get the credit of this   The question will naturally arise why does not he do it now as it is the only practicable way to fix the carbon to the clamps. Why does he adopt his present cumbersome plan if he had the ~~platt~~ plating business two years ago. It is very evident that seeing how we do it he now wishes to claim it, having been so successful in getting his claim, to the lamp generally, recognized.[4]

It is essential that some communication should be made calling attention to the following:—

1st He did not apply for any patents until after my lamp had been announced all over the world

2nd That the records of the Society shows that he had only an old King Lamp with a pencil of carbon[5]

3rd That he now knows that we plate the carbon to the clamps and although not having even tried to patent it he announces that he had it. That he showed it to Sir William Armstrong[6] but did not mention it at the Society. Sir William Armstrong could have been easily mistaken not being an expert and also being an interested party, as I understand, makes the whole thing look very suspicious.

Of course these various reflections on reputable people are all sub rosa but I think after you have satisfied yourself as to the facts you should call attention to the matter[7]

Your orders for machines lamps and engines are all being attended to. We will take care not to mix up orders.[8]

I cannot understand why Gouraud does not give his orders.[9] He must understand that in as much as we have to supply a great many people he cannot expect to be promptly attended to unless he orders quickly   Yours very truly,

Thos A Edison   I[nsull]

L (letterpress copy), NjWOE, Lbk. 9:176 (*TAED* LB009176; *TAEM* 81:60). Written by Samuel Insull. ᵃ"on Tyne" interlined above.

1. Swan 1881.

2. Joseph Swan exhibited a carbon lamp to an informal meeting of the Newcastle Chemical Society on 19 December 1878. Bowers 1982 (113) quotes a report of the event from the 19 December 1878 *Transactions of the Newcastle Chemical Society*. The *Chemical News* noted that Swan "described an experiment he had recently performed on the production of light by passing a current of electricity through a slender rod of carbon in an enclosed globe." According to a published extract from the notes of an assistant, "Swan was endeavouring to use carbon rods, [and] for this purpose he obtained arc carbon rods from Carée of Paris. The smallest were about 2 mm in diameter and were cut into lengths of about 1 in. The central part was reduced to about 1 mm diam. for a length of about 1 in." Swan again exhibited this lamp (or one like it) before about 700 people at the Literary and Philosophical Society of Newcastle on 3 February 1879. "Proceedings of Societies," *Chemical News* 39 (1879): 168; Swan 1946, 12, 23–24; Chirnside 1979, 98; Swan and Swan 1929, 64; Bowers 1982, 113–16 surveys Swan's early public demonstrations.

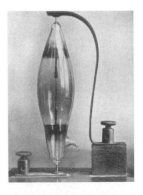

The incandescent lamp with carbon pencil that Joseph Swan exhibited in Newcastle in December 1878 and February 1879.

3. A few weeks earlier, Edison cabled Charles Batchelor to "engage man work up every detail Swans whole case for three years back to ascertain exactly what he did do." Batchelor answered that "Bailey got pretty well to bottom of Swan matter and Lowery and [Frederic Henry] Betts are now working it up still further." TAE to Batchelor, 23 Sept. 1881, LM 1:44B (*TAED* LM001044B; *TAEM* 83:894); Batchelor to TAE, 25 Sept. 1881, DF (*TAED* D8135ZCA; *TAEM* 58:1080).

4. Edison was probably referring to Swan's use of spring clamps or "forceps" to attach platinum wires to the carbonized thread in a more recent form of lamp. This was illustrated and described by the *English Mechanic* in September 1880. That article elaborated on Swan's earlier claim, made in a letter to the editor published in the 1 January 1880 issue of *Nature,* to priority on an incandescent lamp with a horseshoe-shaped filament of charred cardboard or paper ("Swan's Electric Lamp," *English Mechanic,* 10 Sept. 1880, Cat. 1015:102, Scraps. [*TAED* SM015102a; *TAEM* 24:54]; "Edison's New Lamp," *Nature* 21 (1880): 202; see also Doc. 2022 n. 6.). Swan stated in his *Engineer* article that an assistant, Charles Stearn, "undertook to mount some of my [carbonized] papers in a good vacuum, and after many failures from carbons breaking he at last succeeded in making some bulbs very highly exhausted, containing my paper carbons, attached by electrically-deposited copper to platinum strips, which carried the current in and out of the lamp." He later stated that "soon after" his February display of the carbon rod lamp in Newcastle, "and I am quite sure without knowing what I was doing Mr. Edison produced a lamp identical with mine in all essential particulars. It, too, consisted of a simple bulb from which the air had been exhausted by the Sprengel pump, and which, like mine, had no screw-closed openings nor complications of any kind, but contained simply the in-going and out-going wires sealed into the glass, with the carbon attached to them" (Swan 1881, 229).

5. E. A. King obtained a British patent (10,919) in 1845 as agent for J. W. Starr of Cincinnati for a lamp with a thin rod or plate of carbon (or platinum) heated by passage of an electric current; the conductor was enclosed in a glass globe evacuated by a mercury barometer. This was reportedly the first patent on a lamp with incandescent carbons. Dredge 1882–85, 1:xix–xx.; Swan 1881, 229; Heap 1884, 160.

6. William Armstrong (1810–1900), a distinguished inventor, particularly in firearms, was the son of a former mayor of Newcastle and a major benefactor of that city. Armstrong was president of the British Association's Section G (Mechanical Science) when Swan reportedly showed him the lamp in February 1879. Armstrong's house was the second (after the inventor's own) to be lighted by the Swan light. *Oxford DNB,* s.v. "Armstrong, Sir William George"; Swan 1881, 229; Swan 1946, 31.

7. In long letter reviewing the state of Edison's light in England, Johnson wrote on 22 October that "everywhere we are cautioned against openly assailing Swan—on two grounds— 1st   That he certainly did something with carbon in vacuo—before the date of our Patents= 2nd That he is a man of so much influence in scientific circles that to assail him will react on us severely." Johnson to Sherburne Eaton and TAE, 22 Oct. 1881 (pp. 21–22), DF (*TAED* D8133ZAJ; *TAEM* 58:642).

8. Johnson cabled a firm order on 4 October for ten dynamos to operate half lamps, ten machines for full lamps, one thousand half lamps, and one thousand full lamps. The next day he ordered "two Engines for driving two dynamos each." He later confirmed these requisitions and reminded Edison not to confuse them with George Gouraud's orders. Johnson originally intended the dynamos for isolated installations but diverted them to the Crystal Palace. Johnson to TAE, 4, 5, and 6 Oct. 1881, LM 1:52C, 54A, 54C (*TAED* LM001052C, LM001054A,

LM001054C; *TAEM* 83:898–99); Johnson to TAE, 6 Nov. 1881, DF (*TAED* D8104ZER; *TAEM* 57:260).

9. See Doc. 2172.

–2166–

From Joshua Bailey
and Theodore Puskas

[Paris,] Oct 12. 81

Edison   New York

Following are conditions under four heads   first   factory five hundred thousand francs day contract signed million additional put at disposition and company formed when foyer opera lighted[1] three & half million more when first station two thousand lamps lighted Paris at price gas   capital increased after first five millions according to requirements of business our company to have founders shares carrying half profits after five per cent interest on money shares   our company name majority board directors three years Batchelor full control half profits go to our shares and this proposition continues in all increases of capital   your plan of shares abandoned because impossible under french law   syndicate to have quarter of our founders shares   second head   exploitation syndicate furnish five hundred thousand francs for all Europe when foyer opera lighted million and half additional when first station lighted as above   the two million repaid from first profits before any division made   minimum capital any city or country two million francs every hundred thousand inhabitants   this basis is twenty percent increase on former bacles[2]   our company to have forty five per cent of capital of all companies formed   no contract to be made for sub companies or syndicates without consent of our company consent or refusal to be given within fifteen days after any contract proposed   our company has three months in which to make another contract in place of that refused   our company has our third committee of syndicate and direct representatives on boards of all companies formed   Third head   plants all towns not having more ten thousands inhabitants and all plants outside of towns having more than this number belong to our company and syndicate half profits each   fourth head   in case Paris company give license to syndicate to manufacture and make installations immediately in Paris as they desire to do our company to give six weeks delay in closing Paris contract the syndicate intends during this time to absorb Paris contract bringing Paris under terms of present contract   if you cable tomorrow acceptance three conditions with authority to sign contract am authorized to say contract will be signed within twenty four hours and first money paid in[3]

Puskabailey[4]

L (telegram, copy), NjWOE, LM 1:57D (*TAED* LM001057D; *TAEM* 83:900). Written by John Randolph.

1. The six-year-old Opera building was fitted with electric lights in conjunction with the Paris Electrical Exposition on 15 October, and the official opening occurred three nights later. The arc and incandescent lighting interests represented at the Exposition jockeyed for favorable position in lighting portions of the building. Otto Moses successfully negotiated for Edison the opportunity to light the Grand Foyer, a large and lavishly decorated room with a vaulted ceiling nearly 100 feet high. Because of the size of the room, the Edison interests installed ten chandeliers of forty-eight lamps each. However, the chandeliers were only 15 to 25 feet above the floor and could not adequately illuminate the paintings on the upper walls and ceiling, necessitating the use of Lampe Soleil arc lights. Moses regarded the installation at the Opera "as the biggest piece of advertising possible" for Edison's lighting system. At the end of November, a committee consisting of artists, the Opera's director and architect, and the French Minister of Fine Arts conducted a trial to determine which lighting system would be chosen to illuminate the interior of the whole building. The committee chose the Edison incandescent and Lampe Soleil arc lamps; a circuit diagram of the installation is in Undated Notes and Drawings (c.1879–1881), Lab. (*TAED* NSUN07:70; *TAEM* 45:191). Edison asked the Edison Electric Light Co. of Europe in September 1882 to defray his expenses for lighting the Opera and the Electrical Exposition, about $6,000 and $34,000 respectively. Moses to TAE, 22 Sept. 1881; William Meadowcroft to Bailey, 15 Sept. 1882; Bailey to TAE, 7 Oct. 1881; all DF (*TAED* D8135ZBY, D8228ZAJ, D8132ZAV; *TAEM* 58:1075, 61:679, 58:501); Edison Electric Light Co. of Europe report, 7 Mar. 1884, CR (*TAED* CE001003; *TAEM* 97:209); Fox 1996, 174–77; "L'Éclairage Électrique," *La Lumière Électrique* 6 (1882): 184.

2. Presumably, "basis."

3. Edison replied the same day that he accepted the first and third provisions unconditionally and the second "with one amendment, instead of giving syndicate twelve and half hundredths we will give them one quarter of proceeds coming to us   Exploitation funds must cover also all expenses of litigating patents but our company will pay for taking out future patents." He also consented to the fourth provision "if Paris company gives license to syndicate we will give six weeks delay in closing Paris contract   Contract now to be made with syndicate subject to existing contract with Leon" (William Meadowcroft summarized terms of the preliminary contracts with Leon in an 18 July 1881 letter to Charles Batchelor [Cat. 1244, Batchelor (*TAED* MBSB7B; *TAEM* 95:268)]). Edison authorized Bailey to sign if these terms were also approved by Charles Batchelor and Grosvenor Lowrey. TAE to Bailey, 12 Oct. 1881, LM 1:59B (*TAED* LM001059B; *TAEM* 83:901).

Bailey revised the second provision the next day. The syndicate would "offer make manufacturing company million and half francs and

The foyer of the Paris Opera House. The chandeliers held Edison lamps and, at the top of each, two Soleil arc lamps to illuminate the artwork on the vaulted ceiling.

put two million for exploitation on signing contract and to increase capital for factory as needed without limiting figure also abandon all conditions about establishment first station and about comparisons price of gas and go in squarely without hesitations or experiments." Edison consented immediately, and these were substantially the terms that Samuel Insull summarized for James Hood Wright a week later. The Compagnie Continentale Edison and two related companies were formally organized on 2 February 1882. Bailey to TAE, 13 Oct. 1881; TAE to Bailey, 13 Oct. 1881; LM 1:61A, 62A (*TAED* LM001061A, LM001062A; *TAEM* 83:902–3); Insull to Wright, 21 Oct. 1881, Lbk. 9:215 (*TAED* LB009215; *TAEM* 81:78); incorporation papers, 2 Feb. 1882, DF (*TAED* D8238R; *TAEM* 62:252).

4. Cable code for Theodore Puskas and Joshua Bailey; see App. 4.

Notebook Entry:
Electrical Distribution
System

Feeders—Energy On.[1]

Oct. 3rd ~~and~~ experiment was tried with one of the copper underground electric tubes at Edison Machine Works. The current from 800 A lamps of 112 ohms resis. with 102 volts was passed through one of these tubes, each conductor having 262,951 cir. mils. cross section. The tube was not sensibly warmed.

Resis. of lamp[a] circuit 112/800 = .14 ohms.

Current = 102 × 800/112 = 729 vebers

The resis. of one foot of tube[2] would be (two feet of copper conductor) $R = \dfrac{10.33 \times 2}{262,951}$.

Therefore the energy lost in one foot of tube is $C^2 R \times 44.3$

$729^{2b} \times \dfrac{20.66}{262,951} \times 44.3 = 1850$ ft. lbs. per foot of tube.[3]

In one lamp at 10 to the H.P. there are 3300 ft. lbs.

Therefore $^{56}/_{100}$ of a lamp was lost on one foot of conductor. The lamps in first district are to be 140 ohms, 10 per h.p. and it is intended to lose 15% on the feeders. 16,000 lamps to be supplied.

To lose 15% the total equivalent lamps will be $^{100}/_{85}$ of 16,000 = 18,824   Lost in feeders 2824 lamps. There are 20 feeders— ∴ 141 lamps on each feeder, and none should have more than ½ lamp per foot of feeder.

With all the lamps on each feeder carries the current to supply 800 lamps.[4]

Clarke

X, NjWOE, Lab., N-81-03-24:5 (*TAED* N227:3; *TAEM* 41:247). Written by Charles Clarke. Document multiply signed and dated; commas supplied in large numbers for clarity. [a]Interlined above. [b]Horizontal bar written over "729."

1. No records have been found of the 3 October experiments, probably made with the London Jumbo dynamo. On 11 October Clarke reported that this machine, with 800 lamps in circuit, "gave 102 Volts external at 320 revo. The joint resis. of 800 lamps was .14 ohms and with .0049 ohms in armature supposing no other resis. in circuit, the total E.M.F. was . . . 105.6 volts. At 350 revo. this would be 115.5 Volts." He then analyzed the effects of putting more layers in the magnet windings. Also on that date he determined physical and electrical specifications of a central station dynamo with fewer armature sections but longer field magnets than the London machine. N-81-07-11:41–55, Lab. (*TAED* N220:17–24; *TAEM* 41:19–26).

2. The term 10.33 in the following equation is the resistivity of copper expressed as the ohm-mil-foot, or ohms times cross sectional area per length.

3. The term 44.3 in this equation is a constant for converting watts to foot-pounds.

4. On 17 October Clarke calculated the number of volts needed at the generator to operate 16,000 lamps at 102 volts through feeders of 0.03126 ohm resistance. He then determined the amount of energy lost in "regulating resistance" in the system. Ten days later, in his last entry in this book, Clarke listed a decreasing series of "the resistances required to be successively in circuit with lamps so that they shall never be more than 105 volts or less than 102 volts." N-81-03-24:9–15, Lab. (*TAED* N227:5–8; *TAEM* 41:251–54).

—2168—

To Addison Burk[1]

[New York,] 15th Oct [188]1

Dear Sir,

I am very much obliged to you for your favour of the 10th calling attention to the lighting of the Baldwin Loco Works.[2]

I understand that the plant was sold to the Baldwin Loco Works in precisely the same way as we would sell to anyone else. If they like to put an Arc Light in competition with us we can hardly prevent them. More over it is a competition I am not at all afraid of as I know I can get more effective light for $1.00 by means of my Incandescent Light than it is possible to get with an Arc Light

Our own people were over at the B. L. Wks & saw the light start up & they seem perfectly satisfied with the result and report that the Partners of the Firm are very much pleased with it.

If you have not already seen the Light running there I would suggest that you call at the Baldwin Loco Wks some evening & I am sure you will notice the progress made since you were last at Menlo Park.[3]

Again thanking you for your kindness in writing me   I am Very Truly Yours

Thos A Edison

L (letterpress copy), NjWOE, Lbk. 9:201 (*TAED* LB009201; *TAEM* 81:69). Written by Samuel Insull.

1. Addison Burk (1847–1912) was assistant managing editor of the *Philadelphia Public Ledger* and had conveyed one of Edison's lamps to the Franklin Institute around the time of the December 1879 Menlo Park public lighting exhibition. *NCAB* 28:464; Obituary, *New York Times*, 28 Feb. 1912, 11; "The Edison Electric Light," *Journal of the Franklin Institute* 109 (1880): 145–56, reprinted in Complainant's Rebuttal— Exhibits (Vol. VI), 4212–26 [p. 4218], *Edison Electric Light Co. v. U.S. Electric Lighting Co.*, Lit. (*TAED* QD012G:124–31 [image 127]; *TAEM* 48:618).

2. The Baldwin Locomotive Works, in Philadelphia, was the largest

locomotive fabricator and among the largest capital equipment builders in the U.S. The firm employed about 2,900 at this time (Brown 1995, 24–27, App. A). Burk wrote that the works planned to use Edison's incandescent light "in competition with the Brush and Thompson-Houston voltaic arc lights and thinking that you might not know the conditions of the test I write to give you a friendly warning. Eighty of your lamps are hung in chandeliers in a dingy workshop in size about 250 × 100 feet. They cannot possibly light the place; eighty gas jets, thus disposed, would not do it" (Burk to TAE, 10 Oct. 1881, DF [*TAED* D8120ZBA; *TAEM* 57:641]).

3. Burk wrote again on 18 October, giving details of the installation and enclosing a clipping from the *Ledger*. While accepting Edison's claim for the economic superiority of incandescent lighting, he noted that "the trouble is to make the public understand what you are aiming at. I shall do what I can to have the Edison light given fair play in that as in other respects." He stated that the lamps were run at a very high intensity and about eleven had burned out quickly. By February 1882, they reportedly lasted an average of 600 hours. Burk to TAE, 18 Oct. 1881, DF (*TAED* D8120ZBF; *TAEM* 57:646); Edison Electric Light Co. Bulletin 3, 24 Feb. 1882, CR (*TAED* CB003; *TAEM* 96:674).

–2169–

From Charles Porter

PHILADELPHIA, Oct 17th 1881[a]

My Dear Sir:

I am in distress about the engine sent to London. My arrangement with Mr Johnson in Major Easton's presence, and to which he also agreed, was expressly that I was to be notified when it was ready to be tried, and should see personally to its being all right. How could you send it off so?[1]

Nothing will now do but for another engine to be tested by me and sent in place of it. They are duplicates, the clutches are interchangable, (or we will meet that matter any way), and this is absolutely necessary.[2] I have also several things to change, especially the lubricators. Your own interest and mine are identical in this matter. I have no confidence in the governor action. Must send an engine that I myself find perfect.

The one we were trying on Friday & Saturday will not do as it has an exceptionally long shaft. Others all alike. There is plenty of time. When can I see you?[3] Very Truly Yours

Chas. T. Porter

ALS, NjWOE, DF (*TAED* D8129ZBY; *TAEM* 58:296). Letterhead of Southwark Foundry and Machine Co. [a]"PHILADELPHIA," and "188" preprinted.

1. The 125 horsepower Porter-Allen engine and dynamo for the demonstration electric light plant were shipped on 5 or 6 October. TAE to Edward Johnson, 4 Oct. 1881, LM 1:52A (*TAED* LM001052A; *TAEM* 83:898); see also headnote, Doc. 2238.

2. The "clutches" provided a flexible coupling between the engine and dynamo, not a means of disengaging them. In their 1882 paper for the American Society of Mechanical Engineers, Edison and Porter noted the value of avoiding "a rigid connection between the engine and the armature shafts, which would require the entire series of bearings to be maintained absolutely in line." This presumably referred to their unhappy experience with the Menlo Park experimental direct dynamo (see Doc. 2074 n. 1). The first London machine instead had

> a self-adjusting coupling . . . which will permit of considerable errors of alignment without any abnormal friction being produced in the bearings.
>
> The point of difficulty was the backlash, the engine having no fly-wheel except the heavy armature itself, which was to be driven through the coupling. Provision was made for taking this up by steel keys of a somewhat peculiar form, between which the tongues of the couplings move freely, while they themselves are immovable. These keys are held between set screws threaded in wrought iron rings covering the flanges on the ends of the shaft. All the faces liable to move upon each other are oiled from a central reservoir. This coupling is a very compact affair, without a projection anywhere above its surface, and gives every promise of completely answering its purpose. [Edison and Porter 1882, 6]

3. Porter wrote Edison that he would meet one of his own employees at the Edison Machine Works on Monday, 24 October, and was "exceedingly anxious to meet you there at that time." This was changed to 25 October but nothing more is known of this meeting. Porter to TAE, 20 and 22 Oct. 1881, both DF (*TAED* D8129ZCA, D8104ZDY; *TAEM* 58:299, 57:240).

–2170–

To Drexel, Morgan & Co.

[New York,] 18th Oct [188]1

Dear Sirs

I have your favour of 17th and in reply beg to state[1]

1st   I think that ultimately we shall have to establish a Factory in England but for the first three or four stations the machinery can be supplied from here at lower cost and much better made than were we to start a Factory in England to do the work right away.

2nd   We will supply the lamps for the Isolated Business ("Z" Dynamos) at fifty cents each.[2] When a station is started I will supply them at forty cents each for the purpose of lighting up a District   I am certain it would not pay for a long time to come to start a Lamp Factory in England as freights being low and there being no duties it would be impossible to compete with our Lamp Factory here as we have established such a perfect system of manufacture and trained men who are now very skillful and experienced

3rd   I will contract to deliver twelve "Z" Dynamos per week for use in England without interfering with my supply to the Electric Light Company in this Country

4th   I will duplicate the Big London Dynamo within sixty days of the order being given.[3] We will deliver Station Dynamos at the same ~~cost~~ price as to the Light Companies here The exact cost cannot be ascertained until we have[a] made more of them.

5th   As to the Exhibition at the Crystal Palace[4] it has not yet been definitely decided to remove the Paris Exhibit to London. Mr Fabbri first communicated with me on the subject. I then wrote to Mr Johnson requesting him to consult Mr Fabbri on the subject. I stated to Mr Johnson that in as much as the exhibit ~~would~~ of my other inventions would add considerable prestige to the exhibition of the Light I thought that if the exhibition was made Mr Fabbri should provide the funds for the purpose. I have no interest personally in making such an exhibit as is asked for [~~in?~~][b] as[a] my inventions, with the exception of the Light, are all disposed of for England   My expenses in connection with preparing the instruments for the exhibit at Paris were very heavy indeed and I had to bear the whole of it myself. I shall be glad to see the exhibition made at the Crystal Palace as I believe it will greatly assist Mr Fabbri in his negotiations in London[5]   Very truly Yours

Thos A Edison

LS (letterpress copy), NjWOE, Lbk. 9:204 (*TAED* LB009204; *TAEM* 81:71); an incomplete copy is in DF (*TAED* D8133ZAH; *TAEM* 58:639). Written by Samuel Insull. [a]Interlined above. [b]Canceled.

1. Drexel, Morgan & Co. inquired about five items corresponding to the numbered paragraphs below. Samuel Insull made extensive shorthand notes on it which presumably formed the basis for this reply. Drexel, Morgan & Co. to TAE, 17 Oct. 1881, DF (*TAED* D8133ZAG; *TAEM* 58:636).

2. At this time there were no specific plans for Edison isolated lighting plants in Great Britain. By the end of October Edison was eager to start the business, but Edward Johnson thought that unwise without sufficient capital in hand. By mid-November Drexel, Morgan & Co. had agreed to commit $25,000. TAE to Johnson, 31 Oct. 1881, LM 1:82A (*TAED* LM001082A; *TAEM* 83:913); Johnson to Samuel Insull, 1 Nov. 1881; Johnson to TAE, 19 Nov. 1881; both DF (*TAED* D8133ZAL, D8133ZAU; *TAEM* 58:674, 712).

3. In response to Edward Johnson's inquiry when a second big dynamo could be sent, Edison cabled on this day "Sixty days sure." Johnson to TAE, 17 Oct. 1881; TAE to Johnson, 18 Oct. 1881; LM 1:64C, 65B (*TAED* LM001064C, LM001065B; *TAEM* 83:904).

4. This vast glass conservatory building was originally built for the

Great Exhibition of 1851 in Hyde Park. After that event it was relocated and enlarged across the Thames at Sydenham, where it became the central feature of an amusement park and was used for concerts, theater, and exhibitions. *London Ency.*, s.v. "Crystal Palace."

5. Egisto Fabbri wrote Edison in September that the manager of the Crystal Palace "had decided to open an Exhibition the same as is now going on at the Palais de l'Industrie immediately after this closes and to last from December to March— . . . Major [Samuel Flood] Page assured me that he would do everything to satisfy you as regards quantity of space and locality & hoped you would give him a favorable reply." George Gouraud forwarded a similar invitation a few weeks later, along with advice that Edison have his exhibits removed promptly from Paris after the Exposition so as not to violate French patent laws. Edison wrote Edward Johnson on 9 October that "the exhibit at Paris seems to have acted as a first class advertisement for the Light   I am quite willing that the Paris exhibit should be sent to the Crystal Palace for exhibition but in as much as it would act as a very big advertisement I think Fabbri should bear the expense. . . . I do not think he should object to this. Of course under any circumstances he would have to pay for the Electric Light part of the exhibit. See Fabbri & cable me what he will do." Edison reminded Johnson by cable on 18 October to inquire about the expenses, to which Johnson replied "Fabbri will instruct Drexels to pay." The Société Électrique Edison sent six cases of exhibit items to Johnson in early December. Fabbri to TAE, 21 Sept. 1881; Gouraud to TAE, 6 Oct. 1881; Société Électrique Edison inventory, 3 Dec. 1881; all DF (*TAED* D8133ZAB, D8133ZAD, D8135ZCT1; *TAEM* 58:629, 633, 1124); TAE to Johnson, 9 Oct. 1881, Lbk. 9:174 (*TAED* LB009174; *TAEM* 81:58); TAE to Johnson, 18 Oct. 1881; Johnson to TAE, 19 Oct. 1881; LM 1:65B, 68B (*TAED* LM001065B, LM001068B; *TAEM* 83:904, 906).

-2171-

Samuel Insull to
Sherburne Eaton

[New York,] 20th Oct [1881]

Major Eaton

Cabling. European Co[a]

"I dont think the way this cabling business is managed is right"[1]

For instance. Cable comes at 11.30. Mr Edison Mr Puskas[2] & myself discuss it. Best form of reply decided on. I go to W[estern]U[nion]. office send same by which time it is 1.30 or 2 oclock[b]   Dont you think it would be better to have a clerk & messenger boy here till the small hours of the morning waiting to reply to cables. You would then be sure of having both cables & replies thereto. And in as much as Mr Banker[3] has developed so wonderful an interest in the business I will undertake to share my bed with him so that he will be here to decide the form of reply.

Such an arrangement would be agreeable to myself & Mr Edison also I imagine as it would give us several hours extra sleep every evening.

I just mention this as a suggestion to overcome a very great difficulty.

As a matter of fact you always do see cables & this is the first time there has been any chance to complain.

The cables were answered but the copies of the[c] replies were in such a form that I was obliged to copy them out before shewing them. I ventured to attend to some very particular letters which had to go by yesterdays European mail & afterwards to an appointment which could not be delayed & I should have missed both if I had attended to the cables the contents of which being disasterously muddled were not of such great importance until explanations for which we cabled could arrive from Paris.[4]

Mr Edison too was in the building knew what the cables contained was cognizant of the answers thereto & could have helped you out of your difficulty.

I told Mr Hannington[5] I would be back at 4 o'clock or perhaps not till 5 o'clock & Mr Banker waiting therefore[d] is not chargeable to my a/c.

If you think there was any negligence on my part I would prefer you to submit it to Mr Edison

Saml Insull

ALS (letterpress copy), NjWOE, Lbk. 9:212 (*TAED* LB009212; *TAEM* 81:75). "X" written at top of page. [a]Followed by dividing mark. [b]"by which time . . . 2 oclock" interlined above. [c]"copies of the" interlined above. [d]"Mr Banker waiting therefore" interlined above.

1. Insull quoted from Eaton's 19 October complaint that cables received the preceding night had not come to his attention: "Things relating to the Europe Co. I ought to see— Mr Banker has been sitting in my room now nearly an hour, waiting for you to come in— I had to simply admit that I knew nothing abt the cables. You see it makes me appear negligent." Eaton to Insull, 19 Oct. 1881, DF (*TAED* D8127W; *TAEM* 58:155).

2. Theodore Puskas had been in the United States since at least early September. Joshua Bailey to Puskas, 4 Sept. 1881; Bailey to TAE, 5 Sept. 1881; both DF (*TAED* D8132ZAI, D8132ZAH; *TAEM* 58:464, 459).

3. New York City investor James H. Banker (1827–1885) was among the original trustees of the Edison Electric Light Co. and one of the organizers of the Edison Electric Light Co. of Europe; he was also involved in other Edison interests. See Docs. 1668 n. 2, 1731, and 1736; TAE, Theodore Puskas, Joshua Bailey, and Banker agreement with Edison Electric Light Co. of Europe, 1 Apr. 1881, DF (*TAED* D8127A1; *TAEM* 58:88).

4. Insull and his assistants did not normally note the time of receipt or transmission of trans-Atlantic telegraphic messages. The cables referred to in this exchange, however, probably included one from Grosvenor Lowrey about the relationship of the prospective European company with existing contracts for lighting Paris. Edison's reply on this subject, addressed to Joshua Bailey, stated that the incoming message was "muddled." Lowrey to TAE, 18 Oct. 1881; TAE to Bailey, 18 Oct. 1881; LM 1:66B, 67A (*TAED* LM001066B, LM001067A; *TAEM* 83:905).

5. Charles Hanington assisted Insull in the office. He remained in Edison's employ for several years and, in 1888, conducted a search for fibrous plants in Latin America. Jehl 1937–41, 948; see also Hanington's correspondence in Electric Light—General (D-88-28), DF (*TAED* D8828; *TAEM* 122:753).

-2172-

To George Gouraud

[New York,] 23rd Oct [188]1

My Dear Sir,

After several cables to Johnson & yourself I at last got yours as follows[1]

Ship immediately three sixty light complete outfits without power. Cable price and steamer   Drexels advised.
to which I today replied[2]

Three plants will cost twenty two hundred forty five dollars. Will cable shipment. Am filling your order for twenty. Drexels only instructed pay for three   Answer

Almost a month intervened between your order for twenty machines & the receipt of your above quoted cable. On the receipt of your first cablegram I ordered the stock for your machines & set my shop at work on them.[3] Your long silence caused me considerable inconvenience as your not being punctual in answering my enquiry as to how I was to get payment made no difference to my creditors who sent in their bills with that same charming regularity as is their wont.

I have very heavy accounts to meet every week which of course is natural with a very large & consequently expensive shop, my resources of manufacture are taxed to their utmost & therefore if you want machines you must back up your orders by promptly saying where I am to get my.[a] Were I a millionaire I should not be quite so sharp after such things but as I am not & have to stand personally a weekly expenses of from $20,000 to $25,000 I have to take very great care to keep myself out of what would be a very big hole.

Further more you must remember that the countries you control have got to be operated in very vigourously or else it will be the same as it has been in many instances in Telephones—others will occupy the ground before us. Yours truly

Thos A Edison

LS (letterpress copy), NjWOE, Lbk. 9:221 (*TAED* LB009221; *TAEM* 81:82). Written by Samuel Insull. "Per S.S. ~~Wyoming~~ Gallia" written to left of dateline to indicate ship carrying letter. [a]Obscured overwritten text.

1. No other copy of Gouraud's message has been found. He had cabled instructions on 24 September to "rush ten sets each [A and B plants]   cable probably earliest dates shipments"; these were for isolated lighting installations in countries outside Britain under his control (see Doc. 1978). Edison replied the next day that he could "ship three outfits end month, seven week following balance two weeks after without engines   These for parties already having power   How many engines both sizes shall I order   These made by other parties requiring time." He told Gouraud a week later to "Answer last cable isolated" and, a few days following that, requested Edward Johnson to find out why he had not responded. Johnson reported that Gouraud had waited to confer with him, but Edison cabled back in exasperation that this delay was "very prejudicial." Then in mid-October he told Gouraud that Drexel, Morgan & Co. had received no instructions to pay for this equipment. Gouraud to TAE, 24 Sept. 1881; TAE to Gouraud, 25 Sept. 1881, 5 and 18 Oct. 1881; TAE to Johnson, 2 and 7 Oct. 1881; Johnson to TAE, 6 Oct. 1881; LM 1:46A, 48A, 50D, 65A, 53A, 55B, 54C (*TAED* LM001046A, LM001048A, LM001050D, LM001065A, LM001053A, LM001055B, LM001054C; *TAEM* 83:895–97, 904, 898–99); see also Doc. 2165.

2. Edison quoted the full text of his reply on this day. LM 1:72A (*TAED* LM001072A; *TAEM* 83:908).

3. Gouraud cabled instructions a few days later for Edison to ship only three lighting plants. He explained in a subsequent letter that because "they will have to be used chiefly if not exclusively in countries where we have no patent protection, it will be necessary to proceed with caution and after we have duly felt our way." He also noted that for competitive purposes the capital outlay for these plants, which he was to supply, should be as small as possible. For that reason he also asked Edison to supply the equipment at cost (Gouraud to TAE, 26 Oct. 1881, LM 1:75C [*TAED* LM001075C; *TAEM* 83:909]; Gouraud to TAE, 29 Oct. 1881, DF [*TAED* D8104ZEM; *TAEM* 57:253]). At least some of the twenty machines were evidently completed (see Doc. 2229).

–2173–

From Grosvenor Lowrey[1]

Paris, October 23. = 1881[a]

My dear Edison

I send you by this mail the official list of awards.[2] You will know all about it before this letter reaches you, but you will never know the satisfaction which your friends here have felt in the last two days, nor be able to realise the contrast between our feelings now and four weeks ago. I keep the recountal of particulars until I see you. I am now trying hard to get away on the 29th by the Alaska but Bailey's contract is awfully in the way.[3] Fabbri does not seem to care to have me remain in England now, and I do not see that during the next month (during which I could not stay in that clime) there is anything to be done in which I could assist very much.

Two telegrams went to you yesterday one of which was as follows[4]

"You have received the highest award in the power of the jury to give and I congratulate you— (signed) Joseph Wilson Swan.

the other:—

"accept my congratulations   you have distanced all competitors and obtained a dDiploma of Honor the highest award of the Exhibition. No person in any class in which you were an exhibitor received a like award. George F. Barker—[b]

I also sendt you a despatch explaining how the five gold-medals voted you by the Subjuries were committed to into a diploma of Honor, which, though but a bit of paper is considered vastly more valuable.[5]

The first despatch was written in my presence at the breakfast table yesterday where I was by invitation of Mr. Swan, with James C Stephenson, Member of Parliament for Newcastle-on-Tyne and president of Swan's Coy[6] and Mr. Watson, solicitor, of Newcastle, a director.[7] Mr Swan said "I think I ought to have had something better than Maxim and Lane-Fox, but I admit that Edison is entitled to more than I! He added he has seen farther into this subject, vastly than I, and foreseen and provided for details that I did not comprehend until I saw his system." Swan is a man of a style as different from Maxim as you can conceive. He is a chemist and keeps a drug store I believe, but is a man of literary cultivation very modest manners, and I should think a most excellent fellow. At any rate while he said nothing that was fulsome and only in the casual way of conversation I have never heard more satisfactory expressions concerning you than I have heard from Mr Swan. I doubt if there is any man who appreciates your work better than he, and he, partly appreciates it so well, because—as another friend informs me,—Sir Wm Thomson[8] who is a friend of Mr. Swan's, constantly has warned Mr Swan and his friends that they must arrange with you or combine with you, that you are the only one who has a good system. The object of their invitation to me was, that as Mr Stephenson was on his way thro' here to Italy they wanted to discuss the question whether we were going to be open enemies, or secret friends, or what. The question was a very fair one and I tried to deal with it as prudently as I could, my belief really being that if we could, on terms satisfactory to us, and which should put the question where it belongs, secure with[c] their approbation a re-issue of our English patents and an organisation of a Company that should in some way take in both, we should then sweep the board in England. Without that we shall not get Sir Wm Thomson, although I believe we shall get Preece. Crookes, I have written to Eaton about.[9] I don't remember whether I have written about Preece with reference to his determination, that until he shall be in London and have studied the system a little more, he will be entirely independent, declaring however, meanwhile that if he does not go with us, his "mouth will be closed," and that if he does go in with us, he will put his whole soul into it. This he will do, however only[d] when he can satisfy himself that everything is sound in the technical sense, so that he can pledge his reputation to it. These English scientific men are, I find, a very close corporation, and stick together like wax, partly from affection, and partly from fear of each other. At present, Swan has got the friendship of all of them. They are proud of him, and, as you will see by Professor Forbes's letter in the "Times" of the 22d,[10] they, in the newspapers, simply assume that the Incandescent lamp enclosed in a continuous globe, etc; etc; is Mr Swan's invention the outcome of 20 years continuous work. We have no newspaper voice in that country with which to contradict this, and, as in France, you cannot buy the publication of your communications through any respectable journals.

The second cable message was written in my presence at the dinner table on the same day as the first. I shall have a good deal to say to you about it, and the writer, when I see you. Enough for me to say at present, that, as United States Commissioner he felt it his duty to get for Mr Maxim a gold medal[e] if Swan, & Lane-Fox, got medals; but that the United States Coy having two medals might have got a Diploma of Honor if he had asked for it, but he did not ask. He has in most respects satisfied me; in one or two, not quite, but the explanation is one that I cannot gainsay. He is now definitely retained by me, by a conversation I had with him the day before yesterday, after the termination of his duties as juryman, to be for Edison, day & night, at all times, in all countries, and against all persons. There will be a retaining fee to pay him when he gets home, and he will bring you a lot of figures and information concerning everything in the Exhibition, which will be very useful. I shall pay him £100 here for a few days' consultation in England with Johnson and our lawyers. Bailey will also pay him a small fee for a report here. On the whole he will not do so badly as he will expect to have considerable employment with us.[11]

His great notion is, though to form a Museum and Technical School in New York, in which you shall be the prime mover. He is however only looking to you for aid in collecting curious things suitable for such a Museum. When he first spoke of it to me, I told him of a conversation you and I had in the very early days of the Electric Light when I said to you "if we get rich out of this, let us form an Edison laboratory or School." Perhaps you will remember this. I think it was at Tarrytown.[12]

I telegraphed you last week about a point in the contract here, which telegram I will now explain, although your answer revealed to me all that I cared to know.[13] I thought, from the cables interchanged, and in which there was frequent reference to an absorption of the Paris contract, that you might suppose that when you paid to the new syndicate, the commission or interest of 28%, which they are to have in all local business in France, and had surrendered to them the manufacturing here upon the terms agreed on, that you were to have nothing more to pay to anyone. This you might naturally think because you are making both contracts with Mr Léon. But I am informed Mr Léon has no interest left; or comparatively none, in the old contract, and cannot control the owners of it. He hopes, by offering them a chance of coming into the new Syndicate to induce them to surrender the old. By the 14th section of the old they are entitled to 5% of all such business as you are now doing with the new, and will of course, claim it. You will therefore have to pay them 5% out of anything which you receive out of the new contract. I was not sure that you understood this, and was unwilling to approve of a scheme of contract which necessarily involved that, without a clear understanding. I saw by your answer to Bailey, that you had comprehended it, and as your ~~explanation~~ question would require a rather long explanation,[14] I authorised Bailey to send for me a short dispatch, substantially as follows:

"Neither syndicate demands anything and your answer gives me all the information I require"[15]

I shall hurry up the contracts and get away from here on^e Wednesday night if I can, or on Tuesday night if possible. It will be very rapid work and will greatly disappoint Bailey, for when I parted with him last night, he expected that I would remain over for another month. There are various reasons why I cannot. In the first place I have many personal and private reasons for being at home.[16] Next, I dread the coming month or two months here, fearing that I may have an attack of my old trouble and be laid up;[17] but, beyond all that it appears to me that if I stayed, it will be rather thought that I stay unnecessarily, and the inconvenience to myself, or&ᶜ value to others will not be understood. This is rather a difficult matter to speak of, because it seems like blowing one's own trumpet, and I don't know anything more disagreeable than to be thought to overvalue one's own importance. This is a risk I am unwilling to encounter under the circumstances. So I shall get through in Paris if I can. If not I will leave things in the best condition I can to have the contract executed. I will put Batchelor in the hands of a good and safe lawyer who speaks English, and the terms of the contract being all settled, there is not much risk in that, i.e., a point is arrived at, at which I can withdraw without any serious risk, although if I felt free to remain here, I can see very, very many points at which I could be of the utmost service to the European. Bailey is overworked, and excited. Batchelor is inexperiendced—first rate in his place, but of slight importance in the field to which I refer. Moses, whose mind is more fruitful, is, according to Bailey's ideas not sound, and Bailey has taken a great dislike to him, while the state of things between him and Batchelor is such as to rob Moses of most of his usefulness. I must tell you that I think the trouble between Moses & Batchelor is one of those unfortunate pieces of business in which neither party is to blame, but, which is irreconcileable.[18] Many people have remarked in my hearing how well Moses represents you in all communications with the French and German people. He is polite, assiduous, perfectly intelligent and as faithful as a man can be. I don't see how you could have got on without him or someone like him. Batchelor on the other hand suits the scientific and practical men exactly, but for all others, even those who speak English, he does not come up to the mark, does not see the point in conversation, and omits certain little things of finesse and tact which have been so necessary in a Congress like this, where one needed to make friends. This resulted only from lack of experience in such things. However, the friends have been made and you are at the top of the ladder in this Exhibition. Nobody for a moment questions now that you are the great man, and that you have contributed more to its interest and success than all others put together. Perhaps I ought to except one person. Siemens,[19] I think believes himself the great man, and Professor Forbes concurs in that opinion. However, Forbes speaks very respectfully of you in this last letter and Preece has promised to take care of him entirely.

I am feeling very well now,—better than for a long time before as is proven by the fact that I have seldom in my life worked harder and more hours than during the last month. I have scarcely been to a place of amusement or interest with my wife since we reached here.

The big dynamo started on Friday night with perfect success, with one brush on each side. There was scarcely a spark to be seen. Lord Crawford[20] looked at it for half-an-hour, and finally turned to me and said "I never saw anything run so perfectly." Mr Swan told me it was a most "wonderful" machine. Preece said it was "splendid." Your lights for the first time looked what they really are. The fact is, they have been pretty sick with the miserable little engine which Batchelor was able to provide.

Try and catch Freeman[21] as he goes thro' New York, and get his ideas about the Exhibition. I think Freeman is a very square man in every respect, and I think he is very sound on all, or nearly all the questions in which you are interested. At any rate, see him. He sails in the Britannia which sails about the time this letter goes. Bailey informs me that the Paris engineers believe your system good for 100 metres, but are doubtful whether you can distribute it at a greater distance. The editor of the Figaro told me on Saturday that he should go over to New-York to see the opening there   I advised him to wait until we had been running for a month or so, and arranged to give him notice. He got me talking for half the day in giving him points about you for an article. The correspondent of the Daily News some time ago did the same and made an article which I did not like when I saw it.[22]

Give my love to Eaton.[c] I hope you appreciate what a man you have got in him. He is of the sort not made much more frequently than a big dynamo, or than you are.

Mrs Lowrey sends her kindest love to Mrs Edison and hopes sincerely she is better than when Major Eaton wrote a few days ago.[23]

No decorations were given by the Exposition but something

may yet be done on that subject satisfactory to you! Ever Truly
Yours

Lowrey

ALS, NjWOE, DF (*TAED* D8135ZCL; *TAEM* 58:1098). ªFollowed by
dividing mark. ᵇ"George F. Barker—" interlined below. ᶜObscured
overwritten text. ᵈWritten in left margin. ᵉInterlined above.

1. Grosvenor Lowrey was general counsel for Western Union, and
his firm—Porter, Lowrey, Soren, and Stone—served as its legal department. In the fall of 1878 Lowrey became Edison's principal advisor
regarding financial support for electric light experiments and was instrumental in establishing the Edison Electric Light Co. See Taylor
1978; Docs. 1459 and 1471.

2. The list sent to Lowrey has not been found but a complete list of
awards was published in "Awards at the Paris Electrical Exhibition," *Engineering* 32 (1881): 437–38.

3. Lowrey decided to stay until the contract for the European lighting syndicate was completed. Batchelor explained to Eaton that Lowrey
was anxious to leave "but if he goes now it will probably bust this contract and I have told him so, he has therefore decided to stay and see it
out." Lowrey signed the final contract on behalf of Edison on 15 November. He cabled Edison that day summarizing affairs at his departure:
"Left Paris this morning   contract signed   Bailey sick and incapacitated for business for present   Leon party dislike Puskas and refuse to
sit in board of principal company with him   Batchelor understands and
I advise you rely wholly on him until I arrive and explain." Lowrey
arrived in New York by 3 December. Batchelor to Eaton, 27 Oct. 1881;
Edison and Edison Electric Light Co. of Europe agreement with Porges
and Léon, 15 Nov. 1881; both DF (*TAED* D8135ZCO, D8127W9F;
TAEM 58:1109, 173); Lowrey to TAE, 15 Nov. 1881, LM 1:99B (*TAED*
LM001099B; *TAEM* 83:921); Batchelor to Eaton, 3 Dec. 1881, Cat.
1239:27, Batchelor (*TAED* MBLB4027; *TAEM* 93:504).

4. The cables from Swan and Barker were copied in LM 1:69B, 69C
(*TAED* LM001069B, LM001069C; *TAEM* 83:907).

5. Lowrey telegraphed that "official list published today shows you in
the highest class of inventors   no other exhibitors of Electric Light in
that class. Swan Lane fox and Maxim receive medals in class below. the
subjuries had voted you five gold medals but general congress promoted
you to the diploma of honor class above. this is complete success the
congress having nothing higher to give." Lowrey to TAE, 22 Oct. 1881,
LM 1:71A (*TAED* LM001071A; *TAEM* 83:907).

6. James Cochran Stevenson (1825–1905) was a chemical manufac-
turer and public official. He succeeded his father in 1854 as head of the
Jarrow Chemical Company, one of the largest chemical firms in the
United Kingdom. Stevenson also played a leading role in the River Tyne
Improvement Commission and served as a Member of Parliament from
1868 to 1895. *Daily Telegraph*, 26 May 1882, Cat. 1327, item 2197,
Batchelor (*TAED* MBSB52197; *TAEM* 95:195); *Oxford DNB*, s.v.
"Stevenson, James Cochran."

7. Robert Spence Watson (1837–1911) was a politician, reformer, and
businessman. When Swan formed his first electrical lighting company

in February 1881, Watson underwrote and promoted the first issue of £100,000 of capital stock. *Oxford DNB*, s.v. "Watson, Robert Spence"; Swan agreement with Watson, 2 Feb. 1881, Edison Electric Light Co, Ltd., CR (*TAED* CF001AAA; *TAEM* 97:222).

8. Sir William Thomson (1824–1907), later Lord Kelvin, professor of natural philosophy at Glasgow University, was one of the foremost physicists and electrical engineers at this time. *DSB*, s.v. "Thomson, William"; see also Doc. 1751 n. 5.

9. William Crookes (1832–1919) was a prominent British chemist, physicist, and science publisher. His principal accomplishments lay in precise experimental technique, measurements of atomic weights, and investigations into X-ray phenomena. Edison was familiar with his writings and had subscribed to his journal *Chemical News* since 1874. *DSB*, s.v. "Crookes, William"; Israel 1998, 93; see also Docs. 1714 n. 2 and 2034.

Lowrey's letter to Eaton about Crookes has not been found. At the beginning of October Batchelor wrote Edison that Crookes was "very much opposed to Swan but . . . favorable to Maxim." Batchelor to TAE, 1 Oct. 1881, DF (*TAED* D8135ZCC; *TAEM* 58:1082).

10. George Forbes (1849–1936), a British astronomer and electrical engineer, was a member of the awards jury at the 1881 Paris Electrical Exposition (*WWWS*, s.v. "Forbes, George"; Passer 1953, 288–90; Jonnes 2003, 287–306). In an unsigned article from Paris describing the awards given to exhibitors, Forbes gave a laudatory description of Swan's incandescent lamp and implicitly credited him with priority by describing his lamp as the product of twenty years of labor following his original conception. He made no mention of other incandescent lighting exhibits. Forbes also reopened the controversy between Edison and David Hughes over of the microphone by noting that Hughes had received a personal diploma of honor as the "inventor of the microphone, which he generously presented to the public, special modifications of which have been patented by many people, so that the public has been deprived of the benefits of his generosity" ("The Electrical Exhibition," *Times* (London), 22 Oct. 1881, 10).

11. At the beginning of October Batchelor also praised Barker's work on behalf of Edison. As a Commissioner of the Exposition, Barker exercised some influence on the selection of the awards jury for incandescent lighting exhibits, ensuring that the final composition was favorable to Edison. Batchelor reported that Barker had arranged, at least initially, "so that there is not a single Englishman on the experimental jury on incandescent lamps." He concluded that Barker "is working hard for us and ought to be taken care of as the others would gladly get him if they could." Batchelor to TAE, 1 Oct. 1881, DF (*TAED* D8135ZCC; *TAEM* 58:1082).

12. Edison visited Lowrey's home in Tarrytown in September 1879. Lowrey and Barker urged Edison in 1882 to assist Columbia College in establishing an electrical engineering program, using the equipment from Edison's exhibit at the Paris Electrical Exhibition. See Doc. 1805 n. 1 and Barker to TAE, 30 May 1882, DF (*TAED* D8204ZCF; *TAEM* 60:163).

13. Edison had recently instructed Batchelor to retain Lowrey to advise on the proposed contract. On 18 October Lowrey cabled Edison:

"Seems new European syndicate cannot get Paris contract settled until after new European signed   This leaves Paris five for France outstanding future negociations   shall I consent if judge necessary to save business on basis of Baileys cables   important answer tomorrow   am acting in accord Bailey." Edison's reply has not been found. TAE to Batchelor, 11 Oct. 1881, Cat. 1244, Batchelor (*TAED* MBSB7G; *TAEM* 95:284); Lowrey to TAE, LM 1:66B (*TAED* LM001066B; *TAEM* 83:905).

14. Edison cabled Bailey on 18 October:

We did not ask or require new syndicate take up Paris contract supposing that you arranged that new syndicate have every thing Europe except what covered by Paris contract   If old syndicate comply with conditions we are bound to fulfill   Does new syndicate desire to get rid old contract and require us here to do it or do they want us to fulfill and leave them all other territory or do they want old syndicate take Paris and give up to new syndicate provincial France and the five per cent, if so do they want us to pay from our share the five per cent   In writing your cables you presume too much on our knowledge here of your negociations there. [LM 1:67A (*TAED* LM001067A; *TAEM* 83:905)]

15. This is the essence of the cable sent in Lowrey's name to Edison on 22 October. LM 1:69A (*TAED* LM001069A; *TAEM* 83:907).

16. Among Lowrey's reasons may have been the fact that his second wife, Kate Armour Lowrey, whom he married in September 1880, was pregnant with their first child, born in March 1882. Taylor 1978, 54.

17. Lowrey suffered from gout; see Doc. 1711.

18. This conflict began in late August, apparently over the extent of Batchelor's authority over Moses. In early October Batchelor wrote Edison that Moses was "a perfect failure. I have had my hands full ever since I came here dispelling ideas he has put in Baileys, Fabris and Lowery's head . . . regarding the worthlessness of our patents. . . . I have had a number of rows with him about it but he evidently thinks he is doing good for the company. . . . If he would only turn his attention to getting articles in the papers he could put in 4 or 5 everyday, but that he wont do and when I promise articles in French he says he will give them but never does." On 2 October Edison cabled Moses, "Do please work more as Batchelor wishes whom I must hold responsible." Moses replied that he was working diligently for Edison and that he had asked Batchelor "if he ever had communicated with you in a way to give rise to such a message." Batchelor denied doing so. Batchelor to TAE, 1 Oct. 1881; Moses to TAE, 9 and 31 Oct. 1881; all DF (*TAED* D8135ZCC, D8135ZCI, D8135ZCP; *TAEM* 58:1082, 1091, 1112); TAE to Moses, 2 Oct. 1881, LM 1:50C (*TAED* LM001050C; *TAEM* 83:897).

19. There were eight Siemens brothers. Lowrey probably referred to Werner von Siemens (1816–1892), who co-discovered the principle of the self-exciting dynamo in 1866. He co-founded the German firm Siemens & Halske, which moved into electric power generation and distribution and was a major force in the worldwide development of electrical industries. He may also have meant Karl Wilhelm von Siemens (1823–1883), who emigrated to Britain, where he was known as Charles William Siemens. Also instrumental in the development of the dynamo, he was a principal in laying the transatlantic cable in 1874. He headed the

London firm of Siemens Bros., part of the family enterprises, which manufactured generators, arc lamps, submarine cables, and related equipment. He was president of the British Association in 1882 and presided over several other professional societies during his career; he was knighted in 1883. King 1962, 378–79; Doc. 811 n. 2; *Oxford DNB*, s.v. "Siemens, Sir (Charles) William."

20. James Ludovic Lindsay (1847–1913), 26th Earl of Crawford and 9th Earl of Balcarres, was a prominent astronomer. He was elected fellow of the Royal Society in 1878 and president of the Royal Astronomical Society in 1878 and 1879. He served as chief British commissioner at the 1881 Paris Electrical Exposition. *Oxford DNB*, s.v. "Lindsay, James Ludovic."

21. Frank Freeman was a U.S. Patent Office examiner and member of the awards jury for the Paris Electrical Exhibition. On 27 October Batchelor wrote Edison, "I wish you would personally thank Freeman if he calls on you as he has worked indefatigably to show off our things to the best possible advantage and at all times. Also another American Commissioner Lieut Maclean if he calls on you." "Congrès International des Électriciens," *La Lumière Électrique* 4 (1881): 417–20; Batchelor to TAE, 27 Oct. 1881, DF (*TAED* D8132ZBC; *TAEM* 58:518).

22. Lowrey probably referred to a long report on the unofficial opening of the Exposition. The first paragraph stated that Edison's exhibit was "pretty well advanced, but is yet far from being so complete as the Menlo Park magicians and Mr. Batchelor, the agent, intend it be. They complained to-day of the fearful want of reliability of French workmen, who have no idea of the nature of a time engagement." "The Electrical Exhibition at Paris," *Daily News* (London), 12 Aug. 1881, Cat. 1069, item 12c, Scraps. (*TAED* SM069012c; *TAEM* 89:147).

23. Eaton's letter to Lowrey has not been found, nor has any information on the state of Mary Edison's health at this time.

–2174–

To John Michels[1]

[New York,] 25th Oct [188]1

Dear Sir,

I hereby give you notice that I will at the expiration of sixty (60) days from this date stop the further publication of "Science" if in my opinion there has not been in the meantime a great improvement in the valuation of the property[2]

Unless you receive further notice from me you will please prepare to close up your account on that date when I will dispense with the further services of yourself and your staff Yours truly

Thos A Edison

LS (letterpress copy), NjWOE, Lbk. 9:219 (*TAED* LBoo9219; *TAEM* 81:80). Written by Samuel Insull.

1. John Michels (b. 1841?) was a journalist who, with Edison's financial backing, started the journal *Science* in 1880. Doc. 1932 n. 4; Kohl-

stedt 1980, 33–35; U.S. Bureau of the Census 1965, roll M593_1019, p. 363, image 156 [New York, Ward 7, District 10].

2. Michels sent an urgent request for funds the same day. Edison had decided in February 1881 to end his financial support of *Science* but he continued to pay the journal's bills through the end of the year. It ceased publication in March 1882. Michels subsequently tried to persuade him to invest in a company to revive it but Samuel Insull responded that Edison would "not have anything further to do with the publication of 'Science'" (Doc. 2054; Michels to Samuel Insull, 25 Oct. 1881, 4 Feb. and 28 Apr. 1882; all DF [*TAED* D8144ZAU, D8251A, D8251E; *TAEM* 59:570, 63:662, 665]; Insull to Michels, 9 May 1882, Lbk. 12:255 [*TAED* LB012255; *TAEM* 81:639]). Edison promised Sherburne Eaton in July that he would settle a disputed personal account with Michels "to make him feel good." Abram Hewitt and Edward Weston considered investing in the journal but Alexander Graham Bell and Gardiner Hubbard eventually bought the title and subscription lists from Michels and resumed publication in 1883 (Sherburne Eaton to TAE, 3 July 1882, DF [*TAED* D8226ZAK1; *TAEM* 61:324]; TAE to Eaton, 3 July 1882, Lbk. 7:648 [*TAED* LB007648; *TAEM* 80:712]; Kohlstedt 1980, 35). After Edison had ended his association with *Science,* Sherburne Eaton twice discouraged proposals for the Edison Electric Light Co. to pay for planting articles there. He did so with some delicacy because about that time Michels was writing a long report on New York electric lighting enterprises for the *New York Tribune* (Eaton to TAE, 29 June and undated 1882, both DF [*TAED* D8226ZAI, D8224ZDE; *TAEM* 61:317, 139]).

–2175–

From Sherburne Eaton

New York Oct. 26. 188[1][1a]

Dear Sir:

A special meeting of the Directors of the Edison Ore Milling Company, Limited, will be held at the above address[2] on Friday October 28th, 1881 at 3.30 P.M. for the following purposes:

1. To receive report of progress at Quonocontaug Beach, R.I.[3]

2. To receive report of applications for machines.[4]

3. To fix the price for separators and conditions of sale.[5]

4. To audit bills, including bill for large amount for expenses in connection with sand separations.[6] Yours truly

S. B. Eaton

TLS, NjWOE, DF (*TAED* D8139T; *TAEM* 59:206). Letterhead of Edison Ore Milling Co., Ltd. [a]"New York" and "188" preprinted.

1. A dated and signed version of this letter was copied into a company letterbook by William Meadowcroft. LM 5:37 (*TAED* LM005037; *TAEM* 84:248).

2. The company's offices at 65 Fifth Ave.

3. The company leased access to the beach at this location in July and

had a separator working by early September. The machine did not work properly and was replaced in October by a new model which by this time was processing six to seven tons of ore daily. When Eaton summarized operations for the board in late October, he stated that the greatest problem was the location facing the open ocean because "captains of schooners will not anchor unless there is a N.W. wind blowing; thus we cannot be as regular in shipments as if there were a harbor or dock." The company had recently sent an ore sample to the Poughkeepsie Iron & Steel Co., which ordered an additional seventy-five tons of processed ore and then wanted to contract for the mill's entire output at $10 per ton. William Meadowcroft to Edison Machine Works, 1 Oct. 1881, LM 5:8 (*TAED* LM005008; *TAEM* 84:234); Sophia Pendleton agreement with Robert Cutting, Jr., 7 July 1881, DF (*TAED* D8139P2; *TAEM* 59:184); Eaton reports to Edison Ore Milling Co., 17 Jan. 1882 and 30 Oct. 1881, both CR (*TAED* CG001AAI2, CG001AAI1; *TAEM* 97:416, 411); further details of the operation can be found in Eaton's correspondence with M. R. Conley, who replaced W. H. Cheesman as superintendent in November, in Ore Milling Co. Letterbook, LM005 (*TAED* LM005; *TAEM* 84:210).

4. In his October report (see note 3) Eaton stated that "a large number of persons have made inquiries lately regarding the terms of sale for our separators," among them two companies wishing to separate pulverized iron ore. Eaton noted that experiments showed that this material "is too fine and clogs up the mouth of the hopper. Mr Edison will shortly make some improvement in the hopper to meet the exigency and we can then probably dispose of several machines." He told the board in conclusion that "we are now prepared to supply separators and to instruct parties in the use of them."

5. The directors set the price of machines, which cost $500 to manufacture, at $1000 each plus a royalty of fifty cents per ton. Purchasers were required to use the machine only for the purposes and within the territory specified in the contract. Edison Ore Milling Co. minutes, 28 Oct. 1881, CR (*TAED* CG001 [image 30]; *TAEM* 97:414).

6. Eaton submitted bills showing disbursements of $464.93 for general company expenses and $1,921.83 for operations at Quonocontaug. Edison Ore Milling Co. minutes, 28 Oct. 1881, CR (*TAED* CG001 [image 29]; *TAEM* 97:413).

-2176-

Samuel Insull to
Louis Glass

[New York,] October 26th. [1881]

My Dear Mr. Glass:—[1]

I am almost ashamed to sit down to write to you as I have given your letters the go by so long that I am thinking you have got the impression that I did not intend to answer them.[2] I hardly like to set about excusing myself so will simply say that inasmuch as I have to put in eighteen hours solid work every day, Sunday and week days, with the most charming regularity, some of my correspondence gets overlooked. However, Frank Mc.Laughlin[3] just came on here and he says if I do not write to you he will shoot me. So rather than become a martyr I will send you a few lines of explanation as to the electric light. I was just looking over the applications we have got from San Francisco for the franchise for that city and I find that there are about twenty some of which are from extremely good people so you see what you have got to start out against. But I hope that a little watchfulness when the time comes to close up a contract for San Francisco we shall be able to get it for you and your associates. I enclose you herewith a few circulars which we usually send out in answer to applications for general information and if you give them to your friends to read it may in a very few words give them some slight information of our system. I also enclose you circular asking for information as to the lighting consumption of a given city.[4] You must use your own discretion as to answering this. I would not go to much trouble about the matter. I simply send it to you to show you how systematically we set about our work. I also enclose the agreement between The Edison Electric Light Company owning the patents and The Edison Electric Illuminating Company of New York who are going to undertake the general distribution of electricity for light and power throughout Manhattan Island.[5] This will give you some idea as to the nature of the agreement which will be made as to other cities. I think that the best thing for you to do is to get just the strongest people you can and[a] put in an application for the franchise for San Francisco.[6] It would be as well to have the application sent by yourself and all your associates and address it to The Edison Electric Light Company. If you send it to mine or Mr. Edison's care I will write a note and get Mr. Edison to sign it so as to place on file the fact that he is desirous that when the franchise is granted that it should be given to you of course all things being equal as[a] to Position and solidity of your associates, but upon this point I am sure there will be no trouble. I do not think the company will enter into any very serious negotiations until we have the first station[a] started here in New York. You will appreciate how much better will be their position when[a] Mr. Edison has placed beyond doubt the practicability of electric lighting on general distribution as opposed to and in competition with gas. They will be able to get a much better price for their property. Mr. Edison is very hard at work here preparing to light up his district down town about which I told you when you were here.[7] We are making the machines down at our works at Goerck St. just as quickly as possible. We have men engaged[a] laying the street mains at the rate of from seven hundred to one thousand feet a night. (we do not work during the day at this work.) At our central station in Pearl St. we are now engaged in the installation of twenty-two hundred horse power Babcock & Wilcox Boilers and are also putting up a structure for the purpose of carrying the dynamos. We have the houses wired down in that district having obtained contracts by which we shall replace with electric lighting ninety per cent of the total gas lighting of that district which is about three quarters of a mile square. The gas companies are beginning to get somewhat scared down there now and we understand that they are returning to their customers the money deposited on the meter. One man told us that they not only returned him the money but also twenty two years interest added. Everything so far as inventing goes is absolutely completed[a] and has been for some time past. The only trouble now is in the securing of machinery and if we could get that tomorrow we could light up within a month. However as matters go at present I do not suppose we shall get started before February at the earliest. Send on your application[a] and do not think that because I have been so remiss in my attention up[a] to this time that is is chronic with me. very truly yours,

L (carbon copy), NjWOE, DF (*TAED* D8120ZBH; *TAEM* 57:649).
[a]Mistyped.

1. Louis Glass (1845–1924) was secretary of the Spring Valley Mining & Irrigation Co., a hydraulic gold mining concern in Cherokee, Calif., with which Edison made arrangements to purchase platinum-bearing tailings in 1879. Glass later became general manager of the Pacific Phonograph Co. in San Francisco and California manager of the Edison General Electric Co. Docs. 1776 n. 1, 1777, and 1844 n. 1; "Glass, Louis," Pioneers Bio.

2. In August, Glass had outlined his plans and prospective partners for a San Francisco electric lighting company. Glass to Insull, 29 Aug. 1881, DF (*TAED* D8120ZAW; *TAEM* 57:631).

3. Frank McLaughlin (d. 1907) was a telegraphic acquaintance of Edison and a former agent for the electric pen and phonograph. He became Edison's representative to California mining interests in 1879 and took up residence in San Francisco. Docs. 1776 n. 2, 1844, and 1938.

4. Neither enclosure has been found.

5. An unsigned copy of this contract, dated only 1881 and probably a printed draft, is in DF (*TAED* D8122T; *TAEM* 57:749).

6. Glass identified and described his partners in a letter to Edison in November. In a subsequent letter inquiring about the status of their application, Glass referred to obtaining a lighting franchise for the entire state of California. Glass to TAE, 26 Nov. 1881; Glass to Insull, 16 Feb. 1882; both DF (*TAED* D8121H, D8220I; *TAEM* 57:716, 60:760).

7. Frank McLaughlin wrote a letter of introduction in July for Glass to present to Edison. McLaughlin to TAE, 15 July 1881, DF (*TAED* D8139Q; *TAEM* 59:189).

INNOVATION AND QUALITY CONTROL AT THE EDISON LAMP WORKS   Doc. 2177

Edison's incandescent lamp required more attention to the manufacturing process than other components of his light and power system. Producing lamps in large quantities presented the Menlo Park factory with several closely related challenges. One was standardization to ensure that lamps approximated their rated candlepower. Another was durability. A third was cost, particularly important because the wholesale price to the Edison Electric Light Co. was fixed by contract. Responsibility for all this fell on factory manager Francis Upton, who also had to train and oversee a labor force to manufacture what until recently had been a custom-made laboratory item.[1] Many of the production problems confronting Upton were variations or extensions of those that Edison had begun to address in mid-1880 while setting up the factory.[2] Charles Batchelor presciently warned him that "there is nothing wants such continual watching as a 'finished process for manufacturing anything cheap' and . . . you must always be on the look out for some unexpected trouble."[3]

Unlike the Menlo Park laboratory notebooks, those used at the lamp factory provide little detail about the experiments conducted there and still less information about the rationales for them. Nevertheless, particulars about materials and processes can be gleaned from notebooks kept by John Howell, head of the testing department.[4] Upton's correspondence with Edison, well represented in this volume, provides an overview of the lines of investigation. An additional source is Edison's patent applications; between April and December 1881 he executed nine applications dealing principally with manufacturing processes, and filed a similar number in 1882. Several pertain to inert gas atmospheres in lamps, or to chemical means for exhausting gas or vapor. None of these methods was used in production at this time, and the extent to which the factory adopted other new techniques is not clear. A notebook probably used by Edward Acheson and Edward Nichols provides considerable detail on the testing of lamps and of routine carbonization and manufacturing processes.[5]

The factory's most important raw material—bamboo—possessed considerable natural variability. Manifestly unsatisfactory blanks were picked out by hand before carbonization, but even so, the resistance of filaments carbonized simultaneously in the same mold could differ by twenty ohms.[6] This would produce a corresponding—and unacceptable—range of intensity in finished lamps. The factory instituted a regimen for testing each lamp with a photometer and galvanometer to determine the voltage required to produce its rated candlepower. The "economy" or efficiency of special or experimental lamps was also calculated. According to detailed notes of factory operations made by Edward Acheson and Edward Nichols, this was computed either by measuring the resistance and electromotive force, or by measuring the heat given off in a calorimeter.[7] In addition to routine tests, new materials and processes were continually being evaluated. These experiments contributed to the factory's unrecoverable expenses, as Upton lamented in Doc. 2177.

The lifetime of each lamp was affected by the character of the carbon and the stability of the vacuum, among other factors. Select batches of finished lamps were routinely run at high voltage until they broke, and careful records were kept of how long this took. The "lamp curves" produced from these data graphically represented the range of lifetimes in each sample.[8] Edison interpreted these results as indicators of the quality of the manufacturing process and when the average lifetimes fell in November he thought Upton "had got away off his base & was trying to get back without informing me."[9] This set off a spate of experiments on evacuating and sealing lamps which resulted in three patent applications at the end of 1881.[10] Earlier in the year William Holzer had devised a method of making a more secure seal between the outer globe and inside support, and Edison took the unusual step of permitting him to apply for a patent in his own name in September.[11] The 16-candlepower A lamps made in 1881 reportedly lasted an average of about 3,000 hours. Blackening of the globes by carbon carrying drastically reduced their illuminating efficiency long before they broke, but both Edison and Upton considered this a problem of design rather than of manufacturing.[12]

Upton keenly appreciated the importance of lowering expenses, in part by reducing waste. He noted in September 1881 that 1,600 to 1,700 raw fibers were needed to produce 1,000 finished lamps. About 10 to 20 percent of carbonized filaments were broken or thrown out while being put in lamps. In a good week, fully 10 percent of lamps broke in the pump room, representing a significant loss of time and material. The pumps themselves routinely gave out under the pounding of mercury droplets after an average of two weeks.[13] Labor costs were an ongoing concern and helped persuade Edison to relocate the factory to Harrison.[14] Upton kept "a full account of each man's work" in the pump department and by May 1881 had established production rates and piecework pay scales for numerous discrete production tasks.[15] Glassblowing was disproportionately expensive, and when Holzer designed a process to facilitate attachment of the exhaust tube at the top of the globe, a highly skilled operation, Edison again permitted him to file for a patent in 1883.[16] Considerable effort was also given to faster means of attaining a high vacuum. The evacuation process, about five hours per lamp in June 1881, required a great investment in expensive pumps and was a production bottleneck. Upton adapted as best he could, as when he advised Edison in July 1882 to promote the half ("B") lamps because they were cheaper to produce and "help out in the pump room to fill up odd moments when there is not time to make A lamps."[17] He also conducted ongoing experiments with modified pumps and novel desiccants to speed the evacuation process.[18]

Upton had to attend to these concerns while still maintaining a high volume of production. The factory's daily output of about 1,000 pieces was high by contemporary standards in other industries and required careful attention to balance the flow of materials. Neither Upton's professional training nor his prior work for Edison had properly prepared him. He learned as he went, on at least one occasion shutting down lamp assembly until the supply of carbons could catch up. The factory was also the first industrial application of electric motive power, and inevitable malfunctions and delays further complicated his task.[19]

1. The lamp factory employed more than a hundred workers until it moved to Harrison in 1882. Edison Electric Light Co. Bulletin 11:3–4, 27 June 1882, CR (*TAED* CB011; *TAEM* 96:720); see also headnote, Doc. 2343.

2. See headnote, Doc. 1950.

3. Batchelor to Upton, 22 Aug. 1881, Batchelor (*TAED* MU054; *TAEM* 95:660).

4. John Howell recorded results of tests on both experimental and production lamps in several notebooks kept at the factory (Cat. 1301, 1302, 1303, all Batchelor [*TAED* MBN007, MBN008, MBN009; *TAEM* 91:293, 365, 438]). This information came to Edison's attention somewhat erratically (see e.g., Docs. 2117 esp. n. 3, and 2187).

5. Notebook #144, Hendershott (*TAED* B015); see also headnote, Doc. 2098.

6. Notebook #144:30, Hendershott (*TAED* B015:15).

7. Notebook #144:31–35, 61–71, Hendershott (*TAED* B015:15–17, 26–31); see also Doc. 2129. An undated notebook kept by William Hammer, evidently before Edward Nichols left the factory in June 1881, nar-

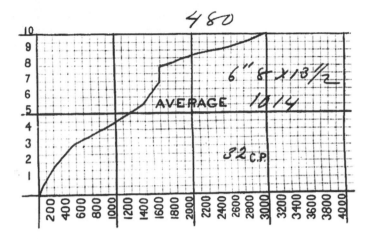

John Howell's 27 October curve sheet for a group of thirty lamps.

rates the testing procedures for experimental and production lamps. Ser. 1, Box 14, Folder 9, WJH.

8. John Howell drew curves for three groups of lamps on 27 October; the graph reproduced here was for no. 480, a batch of 30 A lamps tested at 32 candlepower. Howell manually computed the average lifetime at 1,014 minutes (cf. Doc. 2150) but did not indicate how this would correlate to average lifetime at normal operating voltage. The horizontal axis represents time (in minutes); the vertical axis is a scale of the number of inoperative lamps, so that a steeply rising curve indicates a relatively rapid succession of failures. In reference to unsatisfactory test results about this time (see Doc. 2183), Upton explained to Edison that "the first part of the curve shows what the lamp is doing the last part is generally abnormal." Cat. 1301, Batchelor (*TAED* MBN007:50; *TAEM* 91:343); Upton to TAE, 8 Nov. 1881, DF (*TAED* D8123ZFC; *TAEM* 57:996).

9. See Doc. 2187.

10. See Doc. 2183 n. 2; U.S. Pats. 266,588, 251,536 and 264,650.

11. U.S. Pat. 264,698.

12. *A History of the Development of the Incandescent Lamp*, 15–16, NjWOE; see *TAEB* 5:623–25 and Docs. 2294, 2313, 2346, 2349, and 2411.

13. Upton to TAE, 23 Sept. 1881, DF (*TAED* D8123ZDX; *TAEM* 57:965).

14. See Docs. 2061 and 2089.

15. Upton to TAE, 1 Sept. 1881, DF (*TAED* D8123ZDR; *TAEM* 57:956); Notebook #144:3, Hendershott (*TAED* B015:3).

16. U.S. Pat. 289,837.

17. Upton to TAE, 6 July 1882, DF (*TAED* D8230ZBB; *TAEM* 61:781); Howell and Schroeder 1927 (123–25) note that the rate of five hours per lamp was improved by shortening the stem between the pump body and dryer tube, so as to reduce the evacuated volume. One attendant could operate about fifty pumps.

18. See, e.g., Doc. 2139; Upton to TAE, 26 Aug. 1881 and 1 Sept. 1881, both DF (*TAED* D8123ZDP, D8123ZDR; *TAEM* 57:951, 956).

19. The output level was roughly that achieved a few years earlier by

the several factories of the Singer Sewing Machine Co., which were regarded as models of high-volume production. Although the sewing machine was in some respects more complex than the lamp, its production
depended on established metalworking practices and an existing supply
of trained labor. Hounshell 1984, 89–96; on the factory shutdown, see
Upton to TAE, 30 July 1881, DF (*TAED* D8123ZDH; *TAEM* 57:937).

–2177–

From Francis Upton

Menlo Park, N.J., ~~Sept.~~ Oct.[a] 27 1881[b]

Dear Mr. Edison:

Could not some arrangement be made to pay part of our experimental expenses?

The European Light Co. propose starting a factory and are
interested in having the life of the lamp extended and the processes cheapened.[1]

It would be very valuable to ~~hav~~ them to have full transcripts
of all our curve sheets old and new. They would do well to take
them and ask for new ones paying us so much money for each
lamp used.

England could also give something and the American Co.
would be interested in the same way.

If fifty cents a lamp, of which a complete record were taken,
could be collected from each of these companies it would be a
great help to the Lamp Co. and benifit the parties mentioned
still more for it would point out to them and us how better
lamps could be made.

There is no justice in the present arrangement in which the
Lamp Co. gives everything and takes nothing. Yours Truly

Francis R. Upton.

ALS, NjWOE, DF (*TAED* D8123ZEW; *TAEM* 57:989). Letterhead of
Edison Lamp Co. [a]Interlined above. [b]"Menlo Park, N.J.," and "188"
preprinted.

1. See headnote above.

–2178–

To Edward Johnson

[New York,] Oct 30. 81

Pho[no]s London[1]

All harping on combining Swan only brings weakness[2]   we
can swamp all infringers without litigation   give them plenty
of rope and they will kill themselves as Maxim doing here
where daily expenses seven hundred dollars no returns our
prices, economy and life lamps being impossible to infringers[3]

We want solid men plenty backbone who can wait   Then plunge ahead fearing only Gas interests[4]

Edison

L (telegram, copy), NjWOE, LM 1:79B (*TAED* LM001079B; *TAEM* 83:911). Written by John Randolph.

1. Cable code for Edward Johnson; see App. 4.

2. See e.g., Doc. 2173, which would not yet have reached Edison. In another letter still in transit, Edward Johnson reported that "[William] Preece said to me=If you could arrange with Swan you could sweep the field." Johnson to TAE, 22 Oct. 1881 (p. 22), DF (*TAED* D8133ZAJ; *TAEM* 58:642).

3. Cf. Doc. 2365.

4. Johnson wrote to Samuel Insull on 1 November that he and Egisto Fabbri were

> holding aloof from all considerations of either sale or barter with our opponents— Lowrey at Paris has to some extent been over-powered with the powerful sway of the English Incandesent Element and has lent an attentive ear to the talky talky members of the various tribes—and has even referred them to us—but Fabbri— though conservative in speech and general deportment has—and this I know—very decidedly put his foot down on anything that looks like a weakness in this direction. . . . We shall most certainly not do anything so unwise as to concede anything to people who have nothing but cheek to give us in return= Swan is the only man who merits the least consideration—and he only by virtue of the powerful influence he has command of— At present we are of the opinion that this influence is the growth of a misconception of the facts and we have resolved to test that question by an attempt to educate leading men up to a proper understanding of the main issues involved—then we feel we shall command their influential support and sympathy and Mr Swan bereft of that will be rendered bald indeed. [Johnson to Insull, 1 Nov. 1881, DF (*TAED* D8133ZAL; *TAEM* 58:674)]

-2179-

From Charles Batchelor

Paris, le Oct 31st 1881[a]

My dear Edison—

The machines at the Opera house are working splendidly but there is this trouble with all at first; after a good 5 hours run with full load[b] on they all rub on the magnet heads— there is clearance enough in the middle of the armature and all along except at the end where the commutator is and I have to file all machines[c] out on that end very considerable— You cannot tell this in the shop unless you run them 5 hours with full load—[1] Another thing; after such a run you will find lots of the brass wires a little loose and these ought to be soldered again as they are liable to damage your armature if one should accidentally get loose   I take the precaution now to solder them again in between your solderings as I think you solder too far apart—[2]

What are you doing about making other sizes   a 200 light machine ~~would be good~~ for a belt would be good?— Also for small machines?— Johnson was here yesterday and I lighted up the opera foyer for him in the afternoon—[3] The foyer is very fine but in my opinion the 250 full lights I have in the Grand Escalier at[b] the Exposition (showing off the pictures so well as they do) are a finer exhibition of lighting— There is very much better satisfaction in running a lot of lights from one big machine than bothering with a lot of belts & commutators   Yours

Batchelor

ALS, NjWOE, DF (*TAED* D8132ZBF; *TAEM* 58:523). Letterhead of Exposition Internationale d'Electricité, Lumière Électrique Edison. [a]"Paris, le" and "18" preprinted. [b]Obscured overwritten text. [c]Interlined above.

1. Batchelor worked to resolve several technical problems with the Opera dynamos from October to December. The field magnets of one of the dynamos were damaged in transit in October, although it is unclear if this caused problems during operation. On 19 October Batchelor cabled Edison that one of the armatures failed after running an hour because the "brass wire on commutator end of armature rubs . . . when full load on armature being larger there   this must be attended to." Batchelor wrote the next day to say that the spare armature for the machine did not fit because the shaft was too small. On 19 November he reported that heating in the commutators caused the copper blocks to expand and short-circuit with each other, and complained that "they dont seem to be made anything like so solid as the old machines in the commutator." On 3 December Batchelor reported that the "Opera house plant works well and all the time." Two weeks later, however, the boiler had failed, and he was "working all hours" to repair it "but it makes a very bad impression." Batchelor to TAE, 20 Oct., 19 Nov., and 19 Dec. 1881, all DF (*TAED* D8132ZBA, D8132ZBN, D8132ZBX; *TAEM* 58:512, 550, 575); Batchelor to TAE, 19 Oct. 1881, LM 1:68A (*TAED* LM001068A; *TAEM* 83:906); Batchelor to Sherburne Eaton, 3 Dec. 1881; Batchelor to TAE, 19 Dec. 1881; Cat. 1239:27, 53, Batchelor (*TAED* MBLB4027, MBLB4053; *TAEM* 93:504, 518).

2. Batchelor also directed Philip Seubel to resolder the brass binding wires in the armatures of the dynamos he was installing at the Strasbourg railroad station in January 1882. This was a persistent problem until at least November 1882. Batchelor to Seubel, 9 Jan. 1882, Cat. 1239:97, Batchelor (*TAED* MBLB4097; *TAEM* 93:545); Doc. 2366.

3. Edward Johnson wrote Samuel Insull on 1 November that "Batch has the Opera House all to himself, or will have in a day or so—all the others having received the notice I told you some time ago they would receive—to quit= Last Saturday he made a magnificent display in the

Grand Foyer—& the Director was wild over it. . . . He lit it up for me—
& It was certainly a treat even to my eyes." DF (*TAED* D8133ZAL;
TAEM 58:674).

-2180-

From Edward Johnson

London ~~Oe~~Nov 3/81.

My Dr Edison—

<u>Crystal Palace Exhibition</u>

The one thing which Paris failed most conspicuously in was
"Reliability"[1] I am therefore "going for" that "Bug" more
particularly than any other—though there are others almost as
important=

Instead of having my entire plant of 12 machines run with
<u>one</u> Engine I propose to run it with <u>three</u> (4 dynamos to each)[a]
I will draw upon my 12 machines for only about 600 Lights
maximum all machines to lead to one common circuit= I will
thus have a margin of 2 machines to go on Ere I draw upon the
"Battery" for its nominal capacity viz: 720 Lamps—

If now a machines gives trouble & I have to stop its engine I
shall stop 4 machines—thereby drawing upon the remaining
8—for my 600 Lamps or—75 Lamps Each—

I want to ask you some questions[2]

1st= Will this be too severe a strain on the 8—for a few minutes say 10 or 15—until we can start up again—or shut off
some of the Lamps—

2nd Can I sufficiently increase my Field strength in the
remaining 8 to bring my Lamps up to 16 candles—always providing my steam Power is sufficient—and will the opening of
the 4 machines one at a time—but in rapid succession be likely
to make my Belts slip?

3rd Will the stoppage of the Engine driving the 4 machines <u>without</u> opening the Dynamos—or rather—<u>before</u>
opening them result in harm to their armatures—or in over-
burdening the other 8 machines by reason of the fact that the
4 on ceasing to be active simply become low resistance arc's
across the mains? If so I will put in a switch for Each Battery
of 4—so I can in an instant open the entire lot & then[b] [~~some~~?][c]
quickly stopping my Engine—probably save myself from any
considerable wreck—such as I saw at Paris—& which was due
to lack of facility for quickly stopping the Engine—[3]

Will this sudden withdrawal of the 4 machines at one blow
of a switch—too suddenly & severely tax the capacity of the
other Eight?

~~4th~~

You see my point=

I am going to obtain it Even if I have to use 4 Engines & run 3 machines on Each— This would give me sufficient margin in all the above particulars would it not?=

I hope however you will tell me I can do it with 3—

My Idea is to demonstrate practically—but on a limited scale—your method of obtaining reliability by the subdivision of your Power= This I can do in the way I propose without the use of the Big Dynamo— In fact If I had only one Big Dynamo I would not show reliability at all=

I want the failure of a machine to show simply a momentary decrease of Light—but never a total Eclipse   I want to be able Even to demonstrate this by a practical Exhibition of the result of actually stopping any one of my 3 Engines—

Regulation[4]

You will observe that in such an arrangement a quick & ready means of increasing the field strength is important to produce a good effect— It has occurred to me that I might obtain greater Efficiency in this respect, as well as save a machine—by dispensing with the exciting machine and let Every machine make its own field— To do so I should of course have to introduce one regulator into a common lead wire from the field of Each machine thus—[5]

Questions

1st   Can the proper Resistance for such a Regulator be practically made for so many as 12 machines? Would the resistance of wire & other parts of the Regulator have to be $^1/_{12}$ of that used in the other single machine regulator—or is there a large margin?

2nd— If I stopped one Engine would the fact of the 4 machines thus ceasing to supply their own fields result in altering the conditions in respect to the other fields, and thus make it necessary to radically alter the position of my Regulator? If this is at all problematical of course I shall not attempt it= If not & you advise it—please make & ship by mail steamer the requisite Regulator & Resistances[d]

I want to illustrate "The Edison system"= and not rely solely on the Exhibition of a Lamp=

In respect to the Big machine I shall hang up Photographs (Send me a Lot)[e] of it—liberally—& accompany them with an explanation of why it is not on view at the Exhibition and append an invitation to all who may care to see it in practical operation to come to the City where it is in Daily use lighting up the Premises of Mr[b] Edison—the[b] street Lamps—shops &c of the Holborn Viaduct=[6] I shall take care that capital is made rather than lost by its absence from the Palace— I calculate by this & other means to make my London Establishment a sort of supplement to the Exhibition=

Engines   There are no High speed Engines worthy of the name in this country and I am of the opinion that I can make quite a stir with a good one of those 20 H.P. Double Belt wheel Engines—by putting it up at the Palace & driving a Couple of Half Light Dynamos with it—240 Lamps—almost Equal to Swans standards (If I bring them up a little)   This as a sort of auxilliary show to my regular outfit= I hope you have therefore sent me a good one=

Lecture   Preece is preparing his Lecture on Paris Exhibition—to deliver it before the society of arts in Decr[7]—timing it with the opening of my London shop & the Crystal Palace Exhibition=so as to create a furore— (This is his Idea— He wants some glory)[d]

Competitors

Don't give yourself any concern on this head— All that has been said about it has emanated from Mr Lowrey at Paris— but all here are of the opinion that all such talk is premature= We have fully decided to do the following things before Even so much as listening to the tempter=

1st= Make our Big demonstration & realize its Effect on the Public & on the would be amalgamators

2nd   Thoroughly explore the literature of the subject & learn whether the proportion of ownership & rights are 98 or 99 per cent Edison & 2 or 1 per cent the rest of mankind—

3rd— And in this Preece heartily concurs— Exhaust every Effort to secure to ourselves the Scientific & Popular "Influence" of which Swan so loves to Boast as now in his possession—and upon which he principally wants to trade—[8] This is rapidly coming to us Hopkinson[9] being the latest addition— (I am to see him tomorrow to fully Enlist him)—

Then if we [----][c] find that there is still Enough of Swan remaining to buy—we may possibly throw him a small sop— but that we will is exceedingly improbable—as it dawns upon those who at first advocated this Idea of amalgamation—How much we have & How little Swan has—they rapidly recede from their position & heartily concur with us in the[b] opinion

that He can be anniahilated Easier than he can be bought—
Time will resolve this view into one of whole antagonism to
anything [like?]ᶜ having the semblance of compromise—

Appropos of this:

Faure Battery—¹⁰ These people are jubilant over their Railway success—and they see a great business in this & other
fields not readily accessable to us— They have therefore been
most friendly to us—and are seeking in many ways to become
Identified with us— We notified them yesterday that we were
prepared to supply them with Lamps by the 1000 at 5 shillings
and by the 10,000 at a reduced rate= I sent them a few sockets
& Lamps   Today they called & were exultant over both=
They say they will now throw out all other Lamps—(Maxim
& Swan have thus far filled their show windows & literature)
and use only the Edison¹¹   They say that at 5/– it is impossible for Swan (or others) to compete with us—that out of 100
Swan Lamps—20 will vary within as many ohms of one resistance—20 of another & the remaining 60 will be practically
useless= Hows that?= This Explains the Savoy Theatre variation of Incandescence¹²   Mr Fabbri & I are going Sunday to
see a new Pullman train—on the Brighton Road—lit up with
Faure Battys & Edison Lamps—¹³

This sort of thing will Educate the people to our standard—
& popularize us— I am fitting up their (Faures)—shops with
our fixtures—

These people want to know if you will entertain a proposition to build a Large Dynamo say 5 or 600 H.P.—?=

Isolated Business= We have finally determined that until
we get our present work in operation—our resources are too
restricted to Engage in this business—and shall postpone its
active developement until about the 1st of January= What
with a Big show in London—another at Crystal Palace—a lecture at Kensington¹⁴ (Preeces society of arts Exhibit & Lecture)—and all the Legal & other literary work now on hand we
are of the opinion that I at least have quite as much as I can
properly attend to= of course If a customer comes along who
simply wants a machine & Lamps & will himself furnish all the
Engineering & Electrical talent to put it in practical operation—we shall not refuse to scll to him— In fact I think in this
way we shall do considerable businessᵈ

London Plant= This progresses as fast as the manual
Labor can be performed working night & Day— The Brick
shaft will require 5 weeks for its Erection= Other work goes
on meanwhile—as best it can [with?]ᶜ againstᶠ the obstacles of

official hindrances & ignorant labor— It is no small job to Plant such an outfit within the compass of the 4 walls of a City business House— Permits of all sorts have to secured—and were it not for the intervention of some of Mr Fabbris friends we would today be unable to Even so much as break ground— the permit to do so not yet having passed the various red tape routine— We are only—as I say—able to go on because our friends have by underground means got our doings ignored= More anon   au revoir

Edwd. H. Johnson   Manager

"The Edison Electric Light System" (our sign)

ALS, NjWOE, DF (*TAED* D8133ZAM; *TAEM* 58:682). [a]"(4 dynamos to each)" interlined below. [b]Obscured overwritten text. [c]Canceled. [d]Followed by dividing mark. [e]"(Send me a Lot)" written in right margin. [f]Interlined above.

1. Johnson was probably thinking not only of the Exposition but also Charles Batchelor's exhibition at the Paris Opera. He feared that Batchelor would "have a grand Bust up before many days— He is working up too close to his capacity & he hasn't a spare machine— He says himself—that he simply looks on stoically with the knowledge that if one machine should fail he is dead beat." Johnson to Samuel Insull, 1 Nov. 1881, DF (*TAED* D8133ZAL; *TAEM* 58:674).

2. Doc. 2187 includes Edison's reply.

3. Johnson probably referred to the destruction of a 52 volt (B lamp) dynamo, which Charles Batchelor blamed on a saboteur having put gravel in it. Batchelor reported that "on starting up after dinner a few days ago it immediately tore off all bars and also lugs off plates— I was right near it at the time and rushed to save it but too late— It tore every bar off clean and threw them in all directions." Johnson came to believe instead that the armature was not strong enough to withstand the strain of starting up. Batchelor to TAE, 10 Oct. 1881; Johnson to TAE, 2 Jan. 1882; both DF (*TAED* D8135ZCJ, D8239B; *TAEM* 58:1093, 62:626).

4. Doc. 2190 includes Edison's reply to this portion of the letter.

5. Text is "To one main," "To other main," and "Regulator."

6. Completed in 1869, the Holborn Viaduct was (and remains) a commercial corridor in the City of London between Holborn and Newgate St., 80 feet wide and 1,400 feet long, across the valley of the underground Fleet River and Farringdon St. It was the site of an electric streetlight exhibition in late 1878 and early 1879. In October 1880, the City of London was divided into three districts for the purpose of comparing different systems of outdoor electric illumination. *London Ency.*, s.v. "Holborn Viaduct"; "The Holborn Viaduct," *Electrician* 4 (1880): 206; Bourne 1996, 81–82.

Preparations for a central station demonstration in London had been underway for several months; by the end of July Gouraud had "received a large number of cases of the London light plant." On 22 October, the day the big dynamo arrived in port, Johnson wrote that he and Fabbri had found a suitable location at 57 Holborn Viaduct. He recounted de-

Aerial map of the Holborn Viaduct, site of Edison's demonstration London central station.

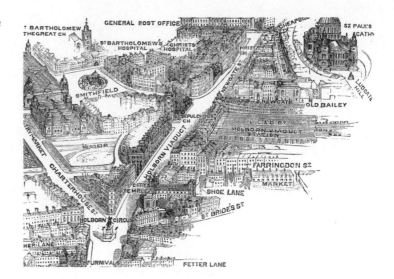

tails of their lengthy search, which had been constrained by local ordinances, noise considerations, a desire to situate the plant in the "foul Gas poluted air" of the City of London, and by laws granting gas companies exclusive license to dig in the streets for private lighting. The Edison interests agreed to light the Viaduct and establishments along it, including the General Post Office and City Temple, free for two months. At the end of that time, they promised to provide light for six months at the same price as gas. Gouraud to Insull, 30 July 1881; Johnson to TAE, Sherburne Eaton, and James Hood Wright, 22 Oct. 1881 (pp. 2–10); both DF (*TAED* D8104ZCE, D8133ZAJ; *TAEM* 57:172; 58:642); Bourne 1996, 82.

7. William Preece's lecture to the Royal Society of Arts is Preece 1881; see also Doc. 2203. Edison requested fifty copies of the lecture after its publication. TAE to Johnson, 4 Jan. 1882, LM 1:129D (*TAED* LM001129D; *TAEM* 83:936).

8. Johnson had recently outlined to Samuel Insull his plans to obtain the support of leading scientists. Two weeks later he gleefully told Edison that he had got "Sir Wm Thomson on a string (at last)." However, Thomson would not endorse Edison's lamps because, Johnson suspected, he had a prior commitment to the Swan interests. Johnson to Insull, 1 Nov. 1881; Johnson to TAE, 19 Nov. 1881; both DF (*TAED* D8133ZAL, D8133ZAU; *TAEM* 58:674, 712).

9. Electrical engineer John Hopkinson (1849–1898) set up an electrical consulting business in London in 1878, the same year he was elected a Fellow of the Royal Society. He also served frequently as an expert in patent litigation. In 1879 and 1880 Hopkinson published two fundamental papers on the theory of the dynamo, the second of which presented a graphical representation of the machine's performance that Edison and other dynamo designers subsequently adopted (see Doc. 2067). He exhibited an alternating current machine at the Paris Exposition. In 1882 Hopkinson became a consultant for the Edison company in Britain. Bowers 1990; *Oxford DNB*, "Hopkinson, John"; *DBB*, s.v. "Hopkinson, John."

London's Holborn Viaduct; arrow marks location of Edison's demonstration central station.

10. The French chemical engineer Camille Faure (1840–1898) developed the first lead-acid batteries in 1880. His major innovation was a process for covering both sides of a lead plate with a paste of lead powder and sulfuric acid. This breakthrough attracted investors as it allowed for the industrial manufacture of batteries and raised hopes they could be used in mobile lighting applications such as railroad cars and ships. Ultimately his battery system failed for lack of durability. Schallenberg 1982, 51–67.

11. Johnson had apprised Edison in October that the Faure battery company "is exceedingly anxious to do business with us   They want us to sell them Lamps of various powers—in large quantities   Mr Fabbri & I concluded that was a matter for the Lamp Factory to consider" (Johnson to TAE, Sherburne Eaton, and James Wright, 22 Oct. 1881 [pp. 25–26], DF [*TAED* D8133ZAJ; *TAEM* 58:642]). About a week later Johnson reported that the Faure company was using Edison lamps exclusively and wanted to purchase large quantities. His enthusiasm diminished, however, when he realized the batteries could operate only two hours without being recharged (Johnson to Francis Upton, 11 Nov. 1881, Upton [*TAED* MU055; *TAEM* 95:661]; Johnson to TAE, 19 Nov. 1881, DF [*TAED* D8133ZAU; *TAEM* 58:712]).

12. The Savoy Theatre was completed in 1881 in the Strand and reportedly was the first public London building to be lighted electrically. Swan's installation there eventually consisted of 1,158 lamps in six groups, operated by as many dynamos. Published accounts of it were sent to Edison and added to a scrapbook. Johnson visited the theater with William Hammer on 21 October and reported that the lamps did

October–December 1881

247

not exceed 8 candlepower; he described the installation more fully in a January letter. *London Ency.*, s.v. "Savoy Theatre"; Dredge 1882–1885, 2:642–47; "The Electric Lights on the Savoy Stage," *Daily News* (London), 29 Dec. 1881; "The Electric Light at the Savoy Theatre," *Engineering,* 6 Jan. 1882; Cat. 1243, items 1907 and 1954, Batchelor (*TAED* MBSB41907, MBSB41954; *TAEM* 95:91, 109); Johnson to TAE and Eaton, 22 Oct. 1881 (p. 30) and 6 Jan. 1882 (pp. 4–6), both DF (*TAED* D8133ZAJ, D8239B; *TAEM* 58:642, 62:626).

13. The London, Brighton, and South Coast Railway planned several new trains of Pullman cars lit by 8 candlepower Edison lamps and Faure batteries. Trains were to operate several times daily between Victoria Station and Brighton beginning 5 December. "By Pullman to Brighton," *London Telegraph,* 2 Dec. 1881; "Lighting of Railway Carriages by Electricity," *Times* (London), 2 Dec. 1881; "New Pullman Car Train," *London News,* 2 Dec. 1881; Cat. 1243, items 1790, 1792, 1796; Batchelor (*TAED* MBSB41790, MBSB41792, MBSB41796; *TAEM* 95:53–55).

14. The vast South Kensington Museum (later the Victoria and Albert Museum), dedicated to the collection of fine and useful arts from around the world, included the Patent Office Museum. *London Ency.*, s.vv. "Kensington," "Victoria and Albert Museum."

–2181–

*Samuel Insull to
George Gouraud*

New York, Nov. 4th. 1881.

My Dear Colonel:—

It is some time since I wrote you the reason being that with Mr. Johnson on the spot there was very little that I could tell you from here. I do not think matters have taken any new departure since Johnson left[1] and I merely write this so that you shall not think that I have entirely forgotten my promise to keep you posted as to progress made on this side of the water.

Edison is working very hard indeed and has been for a long time past and he has now the pleasure of seeing all his various shops running with such charming regularity that it almost becomes monotonous. Our capacity for turning out those small dynamos is almost unlimited. We have 400 now under way and when that lot is finished shall probably go ahead with another order of about the same number. Last week we turned out twenty three. This week we shall do about the same. Our Lamp Factory has a stock of 60,000 lamps so that it can keep apace with our output of machines. The main business in hand just now is the laying of street mains for the Central Station work. We do this work entirely at night and as the result of two or three night's attention paid to it by Mr. Edison he has got it systematized so that with quite a small force we can lay one thousand feet a night. I was down to the Central Station to day and saw them putting in the first boilers. These go underneath the dynamos the latter being carried on a structure built very much after the style of the Elevated Railroad here.[2]

Mr Edison is now having a new track laid for his electric railroad. It will be about three miles long on perfectly level ground and is intended for experimenting on traction, speed and such like matters. He is building a passenger locomotive which will be fitted up in splendid style and which will have a maximum capacity of one hundred miles an hour. Whether it will ever be run at this rate when finished will very much depend upon the courage of the driver. I think it would be a very good speculation to insure the lives of the passengers the first time Mr. Edison determines to run at this speed. Then he is going to build a freight locomotive which will have sufficient power to draw cars each carrying   tons of freight. I suppose that the whole thing will be finished in say three months or maybe a little longer. The road bed is all graded so that the permanent way will be completed in a very short time.

You may remember those two bronze medallion portraits of Mr. Edison working on his Phonograph which was got up by Mr. Kelly and which were sent over in a package of Telephones some time in the latter part of '79 or '80 I am not sure which.[3] Mr. Kelly was in here the other day and I promised to see if you could do anything with a number of these. He had to give up the work in order to attend to some other matters but he is now free and could let you have a supply if anything could be done. Cannot you put Mr. Kelly in communication with the London Stereoscopic Co?[4] Considering the honors which Edison has gained at Paris and the prominence his name will attain in connection with Johnson's operations in London I should think they could sell a great number of them. Mr. Kelly incidentally informed me that you did not pay for these medallions. I think the cost was to be about ten dollars each. Will you send me a check for that amount so that I can hand it to him? Talking about accounts reminds me that Wilber[5] has several times asked me to jog your memory about his fees on Ore Milling. He sent papers to you and a bill for seventy five dollars but has never received payment. The things were sent to you from Menlo Park. I think I remember them coming about August 1880.[6]

Cannot you write us a long letter, a long one I mean; one of those very long ones something after the style of those you used to be so fond of dictating to me explaining to us the whole situation in England of electric light and kindred matters and a full account of your own impressions and the impressions of

English scientists on Mr. Edison's exhibits at Paris. We have had everybody else's opinion except yours and I am sure it would please Mr. Edison very much if you were favor us with your views at very considerable length.[7]

TL (carbon copy), NjWOE, Cat. 2174, Scraps. (*TAED* SB012ABJ; *TAEM* 89:316). [a]Text missing.

1. Johnson sailed to England at the end of September. See Doc. 2156.

2. See headnote, Doc. 2243.

3. James Edward Kelly (1855–1933) was a sculptor, illustrator and painter whose primary works were Civil War monuments. He made a sketch of Edison with the phonograph which appeared in the November 1878 issue of *Scribner's Monthly* and executed a medallion of Edison's likeness in the fall of 1879. *NCAB* 25:434; "A Night with Edison," *Scribner's Monthly* 17 (1878): 88–99; Gilder to TAE, 21 June 1878; Kelly to Batchelor, 22 Sept. 1879; both DF (*TAED* D7805ZBR, D7903ZIA; *TAEM* 17:115, 49:384).

4. The London Stereoscopic & Photographic Co., a partnership of George Swan Nottage and Howard John Kennard, specialized in portrait photography and advertised itself as official photographers for several international exhibitions and for the Prince and Princess of Wales. The firm also marketed scientific novelties. In 1878 Edison arranged with this company to market his phonograph in Britain. Doc. 1237 nn. 1, 7; Israel 1998, 148.

5. Zenas Wilber (d. 1889) was a patent attorney and former patent examiner. In January 1880, he and George W. Dyer took charge of Edison's new patent applications. He was also retained by the Edison Electric Light Co. Docs. 372 n. 3, 1828 n. 1; Obituary, Aug. 1889, enclosure with Batchelor to TAE, 27 Aug. 1889, DF (*TAED* D8968AAN; *TAEM* 128:363).

6. On 30 July Gouraud promised Edison to look into the matter, but he noted that Insull's insistence "that I never paid Wilber his fee of $75, I don't quite understand, unless it refers to the understanding that I should do this based upon the understanding that I was to have my usual joint interest with you in the Foreign patents, but if my memory serves me, we found that your contract with the Milling Co was for all your rights in all parts of the world, in which case it was not expected that I should pay the costs of taking out the patents." Gouraud to TAE, 30 July 1881, DF (*TAED* D8138Q; *TAEM* 59:128).

7. A reply from Gouraud, giving his views of the Paris Electrical Exposition, has not been found.

–2182–

From Grosvenor Lowrey and Charles Batchelor

[Paris,] Nov. 5. 81

Edison N.Y.

Legal considerations compel modifications plan originally cabled abandoning syndicate and creating three companies[1] first manufacturing company capital one million five hundred thousand francs profits to be divided half to us half to capital second or exploitation company capital one million seventy two per cent profits to us. third a small plant company capital one million fifty per centum profits to us[2]   those proportions of profits secured to us in all future increases of capital by f[o]unders shares which are an approved form of security decidedly recommended by our French counsel   property of patents for light and power belong to exploitation company but remain in you for all other uses   expenses this contract as well as of all legal proceedings to be paid here   we reserve absolute right to veto without condition or penalty sites or licenses of any entire country   also right to reject any contract for a town or district with obligation to make a better within three months under penalty of five percentum on contract subsequently made for that place. Minimum capital any city or district two million francs for every hundred thousand inhabitants forty five percentum of which comes to exploitation company   contract provides organization companies within six weeks but Leon assures us capital will be ready for business on signing contract   exploitation and isolated capital to be repaid out of profits before dividends manufacturing capital by 10 percent yearly on profits   shall we approve contract on these principles if minor provisions are satisfactory to us   we advise acceptance

Lowbatch.[3]

L (telegram, copy), NjWOE, LM 1:90A (*TAED* LM001090A; *TAEM* 83:917). Written by John Randolph.

1. The complications have not been identified. Lowrey and Batchelor cabled at the end of October that a contract providing for one manufacturing company and one exploitation company was nearly ready. On 4 November they promised to telegraph a complete synopsis of the proposal the next day and urged Edison to call a special meeting of the board of the Edison Electric Light Co. of Europe for its consideration. They and Joshua Bailey and Theodore Puskas further assured him on 5 November that the contract was with "honorable parties" and that "capital of the several companies will be increased to any figure that may be shown desirable in the development of the business." Lowrey and Batchelor to TAE, 31 Oct. and 4 Nov. 1881; Lowrey, Batchelor, Bailey, and Puskas to TAE, 5 Nov. 1881; LM 1:83C, 89B, 89C (*TAED* LM001083C, LM001089B, LM001089C; *TAEM* 83:913, 916).

2. Edison cabled Joshua Bailey the next day: "we accept contract." He had already given Bailey power of attorney to sign on his behalf (TAE to Bailey, 6 Nov. 1881, LM 1:91B [*TAED* LM001091B; *TAEM* 83:917]; TAE power of attorney to Bailey, 2 Nov. 1882, DF [*TAED* D8127W1; *TAEM* 58:157]). The terms given in this document are substantially those of the 15 November agreement executed on behalf of Edison and the Edison Electric Light Co. of Europe by Bailey, Lowrey, and Batche-

lor. The other parties to the agreement, Charles Porgès and Elie Léon, were to form the "exploitation" or patent holding company within fifteen days. Its jurisdiction encompassed France and its colonies (except Paris and its environs, where a separate "working" or utility company was to be formed within six weeks), Belgium, Denmark, the German Empire, Austria and Hungary, Russia, Italy, and Spain (its colonies excluded). The contract stipulated a somewhat more complex division of profits according to a ratio of special "founders" shares and regular dividend-bearing shares. The Compagnie Continentale Edison (holding company), the Société Industrielle et Commerciale Edison (manufacturing), and the Société Électrique Edison (isolated lighting) were incorporated on 2 February 1882 (Edison and Edison Electric Light Co. of Europe agreement with Porgès and Léon, 15 Nov. 1881; incorporation papers of Compagnie Continentale Edison, Société Industrielle et Commerciale Edison, and Société Électrique Edison, all 2 Feb. 1882; all DF [*TAED* D8127W9F, D8132ZBL1, D8238R, D8238S, D8238Q; *TAEM* 58:173, 533; 62:252, 284, 227]). Fox 1996 (184–85) provides a useful narrative of the formation of these companies.

3. Cable code for Grosvenor Lowrey and Charles Batchelor; see App. 4.

–2183–

To Francis Upton

[New York,] 6th Nov [188]1

My Dear Sir,

I desire to further call your attention to the fact that the new lamps we are getting are losing their vacuums very badly indeed. The majority of the lamps we are now putting up are on the first or second night or gradually oxidise on the fourth or fifth night. I should say the average life of the lamps now furnished is about twenty five hours[1]   This is undoubtedly due to poor work on sealing and may account for the difficulty in getting vacuums. While they are cold the leakage is small but whatever it may be a change has got to be made and more care taken in sealing as we did in earlier days.[2] Yours very truly

Thos A Edison

LS (letterpress copy), NjWOE, Lbk. 9:259 (*TAED* LB009259; *TAEM* 81:92). Written by Samuel Insull; "X" at top of page.

1. Charles Batchelor complained about this time that of the 420 half lamps at the Paris Opera, five broke each night. Pointing out that many of the carbons were misshapen as well, he thought these were lamps "that the factory was glad to get rid of." Batchelor to TAE, 8 Nov. 1881, DF (*TAED* D8132ZBJ; *TAEM* 58:530).

2. Upton replied two days later that "last week some curves were made that were very poor. I have been over all our curve sheets and reexamined all the processes of manufacture to see if I could find the trouble. I admit that we have not taken curves as often as we should and as often as we shall in future. Yet in looking back I know that we have tried. In the

photo[meter] room we have had more than a share of trouble," including a succession of four different superintendents. Upton also attributed problems to having "worked so many pump orders that we felt a little sure of being all right. I feel convinced now that the only way to learn about lamps is to make life tests, and this I shall do." In closing, he promised that "Holzer is taking steps to have the lamps sealed much stronger than before and to run through some orders worked different ways. We are using platinum wire harder drawn than at one time as it makes much finer inside parts and is less apt to break. . . . I am doing all I can to improve the lamp and bring it up to the old standard and make 600 hours within the truth." Upton to TAE, 8 Nov. 1881, DF (*TAED* D8123ZFC; *TAEM* 57:996).

Although the factory did not operate on 7 or 8 November while new equipment was installed, John Howell on 8 November subjected ten lamps with "inside pts with soft platinum wires" to trials of one and two days, then kept ninety others to "notice the state of vacuum before putting to one side and also at the end of 4 weeks and 3 months by vibration test." The next day he observed the state of globes and the vacuum in 100 lamps "with the seal made with great care and in every way perfect," another batch in which "ordinary care is used in sealing" and a third group sealed "using but little care." These trials mark a period of more than a week, during which Edison largely lived at the factory (see Doc. 2187) and paid particular attention to the evacuation and sealing of lamps. The average lifetime of lamps tested at high intensity nearly doubled during this time. Cat. 1301, Batchelor (*TAED* MBN007:53–58; *TAEM* 91:346–51).

–2184–

To Hinds, Ketcham & Co.[1]

[New York,] 7th Nov [188]1

Dear Sirs,

The discrepancy between the number of lamps per horsepower which we claim and what you obtain is due entirely to the extra friction on the belting and shafting when the extra load comes on.[2] As we have indicated at least 25 different kinds of engines varying from ~~eighteen~~ one hundred & eighty[a] horsepower to ten horsepower and never got less than 12 to 14 lamps of eight candles each by the indicator card and as these engines have been indicated by many different persons who have figured the thing out both in the old manner and by planimeters[3] and have found our figures are correct we are absolutely certain the fault is at your works & not in our lamps and machines   In my mind the whole trouble is due to the failure to recognize the fact that there is an enormous loss of power when transmitted through several series of belts. Were you to put in a small Armington Engine and connect the Dynamos directly with it you would doubtless get the desired result. In fact I will guarantee twelve eight candle lights per indicated horsepower if such an engine is used and that you will have no further difficulty in the matter. If you find it inexpedient to put in a small ~~dynamo~~ engine we will take the Dynamos back and be just as good friends as ever.

Any night this week if the gentleman, who has indicated your present engine and made these calculations, has time he may come to Goerck St by informing me one day ahead and I will put up 150 eight candle lights on a small engine and if we do not get twelve lights per indicated horsepower I will agree to eat engine dynamo lamps and all[4]   Yours very truly

Thos A Edison   I[nsull]

L (letterpress copy), NjWOE, Lbk. 9:263 (*TAED* LB009263; *TAEM* 81:93). Written by Samuel Insull; "X" at top of page. [a]"one hundred & eighty" interlined above.

1. Hinds, Ketcham & Co. was a lithographer and printer of colored labels and show cards in New York. An Edison isolated lighting system, the first in a commercial establishment, was installed there in January 1881. See Doc. 2053.

2. In March 1881 Edison addressed difficulties that Hinds, Ketcham & Co. evidently had in running lamps at the proper intensity. No more recent complaint has been found, although Sherburne Eaton forwarded to Charles Batchelor a letter from the firm in August or September. Mechanical engineers were only beginning to study the measurement and reduction of frictional and slippage losses in belt transmission mechanisms at this time. Losses in even well-maintained millwork (systems of overhead shafts, pulleys, and belts) have been estimated between 10 and 50 percent. TAE to Hinds, Ketcham & Co., 16 Mar. 1881, Lbk. 9:69 (*TAED* LB009069; *TAEM* 81:34); Batchelor to Eaton, 27 Sept. 1881, Cat. 1239, Batchelor (*TAED* MBLB4001; *TAEM* 93:489); Hunter and Bryant 1991, 119–24.

3. The irregular figure drawn by the indicator represented pressure on one side of the piston with respect to time through one engine revolution. The average net pressure, given by the figure's area, was a factor in calculating horsepower. The area was customarily estimated by marking the card into a grid. A planimeter, also called a mechanical integrator, operated "by the combined sliding and rolling motion of a small measuring wheel which has a total rotation proportional to the area enclosed by the figure the periphery of which it traverses." Thurston 1890, 213–19.

4. The printers wrote a promotional letter to the Edison Electric Light Co. on 23 February 1882 attesting to the lighting system's "perfection, simplicity, and the many other good features it possesses. We have found it to be entirely free from all the faults and objectionable features of other artificial lights, and is the best substitute for daylight we have ever known and almost as cheap." That letter was subsequently published by the Edison company. Edison Electric Light Co. Bulletin 4:2–3, CR (*TAED* CB004; *TAEM* 96:676).

-2185-

To Thomas Whiteside Rae[1]

[New York,] Nov. 8 [1881]

My Dear Sir

Referring to the request, in your letter of Oct. 31, for permission, for the Members of the American Society of Mechanical Engineers[2] to visit my shops etc in the City,[3] permit me to say that during my experiments on the Electric Light &c at Menlo Park where everything was under my control visitors were always welcome to visit my Laboratory and works. Now however matters are in the hands of the Company and I very much regret in this case they object to any visitors being allowed in the works.[4] Very Truly Yours &c.

Thos A Edison   I[nsull]

L (letterpress copy), NjWOE, Lbk. 9:271A (*TAED* LB009271A; *TAEM* 81:96). Written by Charles Mott; signed for Edison by Samuel Insull.

1. Thomas Whiteside Rae was secretary of the American Society of Mechanical Engineers from 1880 to 1882. He was son-in-law to Henry R. Worthington, a leading hydraulic engineer and a Society vice president. Sinclair 1980, 23.

2. Edison was invited to help establish the American Society of Mechanical Engineers, founded in 1880 as that profession's first national organization. He apparently did not do so but became a lifetime member in May 1880. Sinclair 1980, 22–39; John Sweet to TAE, 24 Jan. 1880; American Society of Mechanical Engineers to TAE, 12 May 1880; both DF (*TAED* D8012A, D8012G; *TAEM* 53:469, 476).

3. Rae wrote concerning the Society's forthcoming meeting in New York on 3 and 4 November. The Society wished "to obtain for the members, who come from all parts of the Country, the privilege of visiting and inspecting whatever there may be of professional interest in the vicinity." He added that each member would carry a card of introduction so as "to prevent unauthorized persons from availing themselves of this opportunity." Edison's answer was based on his draft reply on the back of Rae's letter. Rae to TAE, 31 Oct. 1881, DF (*TAED* D8129ZCH; *TAEM* 58:305).

4. Cf. Doc. 2160.

-2186-

To George Gouraud

[New York,] 10th Nov [188]1

Dear Sir,

Referring to our cable correspondence as to Electric Light plants I have received from Drexel Morgan & Co notification that they are prepared to pay for 17 plants on receipt of Invoice & Bill of Lading. The Dynamos are now being packed & will be shipped this week[1]

I have duly received your favour of 29th.[2] I am rather surprised that I should have been put to the inconvenience of having machines on my hands on which I could not realize while

I awaited the receipt of your letter asking me to forego any profit on them to which as a manufacturer I am very justly entitled. You seem to forget that I have made a heavy investment of my own personal funds in order to create an establishment where I might be enabled to turn out Dynamo machines at a low price. The figures I name to you are much less than you can get them manufactured elsewhere & will enable you to sell at a price which will kill all competition. I know also[a] that we have an equal advantage in possessing a much better article to present to the public.

To provide funds to start the Edison Machine Works I have been obliged to sacrifice a great deal which course could only be justified by the extreme necessity of Building up such an establishment & were I to accede to your request I should have a very large capital quite unremunerative

As a matter of fact I do not at present make a cent on a single Dynamo sent out of my works & do not expect to for some time to come. As to the lamps I sell them to you at the same price I have to pay the Lamp Co & the sockets & other sundries are also charged at prices paid Bergmann & Co[3]

I think if you were to get estimates from other people to turn out any number of the machines you will not wonder at my surprise at receiving your letter   Yours truly

Thos A Edison

L (letterpress copy), NjWOE, Lbk. 9:283 (*TAED* LB009283; *TAEM* 81:97). Written by Samuel Insull. [a]Interlined above.

1. On 6 November Edison cabled Gouraud that he had shipped three dynamos with the remaining fourteen ready to go. He instructed him to "cable funds   Urgently want money." On the same day he cabled Johnson: "Urge ordering Z dynamos   have thirty on hand. Urgently want money." Johnson replied the following day that he and Gouraud would place more dynamo orders "soon" but that they "must sell some first." Gouraud wrote Edison two letters on 10 November. In the first he remarked that he "was not aware until after I had some little talk with Johnson that your present need for ready cash was so urgent." In the second he told Edison that he had instructed Drexel Morgan to pay for the dynamos. TAE to Gouraud, TAE to Johnson, both 6 Nov. 1881; Johnson to TAE, 7 Nov. 1881; LM 1:92B, 92C, 93B (*TAED* LM001092B, LM001092C, LM001093B; *TAEM* 83:918); Gouraud to TAE, both 10 Nov. 1881, DF (*TAED* D8104ZET, D8104ZEU; *TAEM* 57:279, 280).

In a letter specifying the bill of materials and shipping instructions for the seventeen isolated plants (consisting of Z dynamos, lamps, sockets, and other fixtures) to be sent to Gouraud, Edison informed the Machine Works that "Mr Insull will arrange for your receiving payment on these." TAE to Edison Machine Works, 9 Nov. 1881, DF (*TAED* D8129ZCL; *TAEM* 58:310).

2. On 29 October Gouraud wrote Edison that he believed that low cost was the best means to compete successfully with other lighting systems, at least until the technical superiority of Edison's system became obvious. He insisted that "the outlay for all plant must be kept as low as possible— I would suggest therefore that, having regard to the fact that there are no patents & that I furnish all necessary capital, giving you half profits, you should supply the necessary plant at cost price to yourself." Gouraud concluded that Edison would agree with him after "careful consideration." Gouraud hoped that after receiving an adjusted list of prices to reflect cost prices, he would "be able to send you a considerable order" for lighting plants. Gouraud to TAE, 29 Oct. 1881, DF (*TAED* D8104ZEM; *TAEM* 57:253).

3. In October Edison said that he would charge the English company fifty cents for lamps intended for isolated lighting plants and forty cents for those used in central station systems (see Doc. 2170). No prices specific to the British isolated lighting business have been located, but in September 1882, an Edison Co. for Isolated Lighting catalog listed a price of approximately $1,595 for a Z dynamo with 60 A lamps and $1,860 for a Z dynamo with 120 B lamps, fully equipped with wiring and fixtures. The manufacturing costs of Edison's dynamos are not known definitely but see App.3. The internal prices of Bergmann's goods are unknown; Bergmann & Co. published a catalog in 1883 with prices for sockets and fixtures. Bergmann and Co., "Catalogue and Price List of Edison Light Fixtures," 1883 (pp. 81–82); PPC (*TAED* CA002C:79–80; *TAEM* 96:263–64).

–2187–

To Edward Johnson

[New York,] 23rd Nov [188]1

My Dear Johnson,

Your lengthy letters are before me and I give you below my answers and comments thereto.

<u>Yours dated Lord Mayors Day</u>.[1] We have not determined to put B & B wires in our ditches.[2] Your information from outside sources is false   What we do want very badly is a wire but it is too late now for the First District   If Delany[3] would go ahead and get his cable going we would use it. Of course we do not care anything about interference from Induction currents What we want is a cheap wire for lighting ~~cheap~~ street lamps winding clocks &c. The other matter namely a general system of underground Telegraphs has been abandoned for the present. Let Delany show us progress. You seem to have got badly off your base on this question. I have not the slightest objection to your being interested with Delany, but what I do desire is that Delany or some one else would produce a cheap lead covered ~~wire~~ cable & then I can go ahead. I will give Delany every [invitation?][a] if he will only use diligence.[4] I am not going to use the cable made by Kruesi as it is a failure[5]

<u>Yours of Nov 6th</u>[6]   You will remember that I told Wright (of D M & Co)[7] that the extra hurry and expense was quite unnecessary as I would get money   that the Steam Dynamo would be in London and lay there absolutely useless for several weeks before it was wanted[8]   Of course reliability is the great thing now to be established and then everything will be a success and this you know is only attainable by having a spare machine.[9] The engine for the new steam Dynamo is ready and will be mounted on the base next Tuesday   They are putting the Iron armature together. The copper bars are all upset on one ~~side~~ end; the plates are all cut out and finished excepting bevelling the edges. The insulating paper is all ready. The commutator bars are all cut and they are putting the commutator together   The cores of the field magnets are all done & the finishing touches are being put on the cast iron ends. The machine will probably be shipped in twenty one days from today.[10] What did Batchelor decide about the Paris Dynamo Where is it going?[11] We are negotiating now for several Isolated places where 2500 B lamps are wanted. In these cases we intend to put in two steam Dynamos one being a spare. Insull has sent you two copies of our little book. It is the first edition and very incomplete[12]   Will send you some more in a few days. We hope soon to get out another edition which will be much more complete. Of course we do not want you to order Dynamos that you cannot dispose of but only thought we would send you a gentle insinuation that we would like to dispose of some to you.

Mr Fabbri is entirely too sensitive. He takes as slights things that are never intended as such draw wrong infrences and misunderstands motives. I think he is one of the nicest men to get along with I ever knew but is so fearfully sensitive

About the sale of lamps to the Faure Company you know if we put up lamps in our District and invest money to light them we cannot possibly make more than $3.00 out of the stuff coming from the lamp before it dies.[13] Now it will take $2.00 to pay proper dividends on the capital invested leaving a dollar for water. Now if you sell them to the Faure Battery Coy at $1.15 you will make sixty five cents, without any investment, before the lamp dies or in the case of a Isolated plant you make fifty cents and have no capital invested so taking all things into consideration if you can sell lamps in London for $1.15 you would make just as much as you would to invest the money in a plant necessary to work the lamps substracting the usual interest on the money invested before determining the profit from the sale of light through the lamp before it dies. I therefore see no objection to selling our lamps to the Faure Co if they will call it the Edison Lamp as at any time by shutting down on them 60 or 90 days we can deprive them of all lamps by natural breakage

Dr Muirhead.[14] This gentleman was over here snooping round but gave himself away after 15 minutes conversation so we kept closed mouths. He got no information from us but told us his firm (C. M. & Co.) have an order to make 10 000 lamps for the Lane Fox people for India and asserted that they could be made for $2.50 each but not under[15]

You will find by looking carefully through the Journal of the Chemical Society of New Castle on Tyne that late in 1878 Swan or some one for him read a paper before that Society on a Incandescent Electric Lamp regarding the smoke upon the sides of the glass walls; and in this paper he states that he used a pencil of carbon which of course is a give away on his 20 years business.[16] You will notice in his lecture before the Society of Arts in the latter part of 80 or begining of 81 he showed that he knew nothing about the value of high resistance or a flexible filament of carbon.[17] He stated that he did not believe in my method of working in multiple arc & that by the use of a very thin carbon he got higher economy. Now you know that that has nothing to do with economy. If you make two lamps one of one ohm and one of 100 ohms of the same radiating surface and adjust the volts so that each will give [16?][a] candles the economy of both will be the same leaving out conduction to the clamps &c but the current in the one ohm lamp will be one hundred times greater than in the 100 ohm lamp   Therefore when he says that economy is gained by a weaker [-- ~~passed~~?][b] current passing through the filament he shows that he is un-acquainted with the first fundamental principles. He did not know the object of high resistance and therefore it is self evident that he never had any flexible filament until I pointed it out. Like his New Castle on Tyne papers, his lecture at the Society of Arts, & his last lecture before the British Association[18] and by an accurate comparison made [by],[a] say, Hopkinson, you will get absolute proof that notwithstanding the enormous respectability of Mr Swan he is stating things that can be disproved by his own statements. I desire to call your attention to a little fact which probably has escaped you all   Lowery wrote some time ago that Brewer & Jensen[19] were hunting up Swans patents & publications and that he did not think much of their reliability as they stated that one of Swans patents early in 80 on an Incandescent lamp was was forestalled by one of my patents in the early part or middle of 1879. B. & J. were correct. You will find in my British Patent in which I announce the method of getting the air out of platinum in vacuo consolidating the same, that there is a statement that carbon may be treated in vacuo in the same manner, that is to say brought up gradually to great incandescence while maintaining the vacuum.[20] This, I believe was the main feature in Swans Paper Carbon application.

I do not see that Lane Fox can restrict the claim for a filament of carbon of high resistance in a chamber made entirely of glass in high vacuum. The only case where he used carbon in a vacuum was with a pencil of carbon & all this is very muddy. I have not seen his patent for sometime but think the Provisional & Final Specifications are somewhat mixed up.[21]

You will probably have noticed that Swan lately stated that he plated the carbon to the clamps.[22] It seems somewhat peculiar that if that were so that he does not continue so to do. My own private opinion is that he tries to claim other peoples work & carries to extreme the idea of enormous respectability while being at heart what his compatriots would call a "bloody liar." I remember that in 1878 Upton, Batch, & myself worked three days & nights on Seimens patent to figure out how the devil he connected up his armature & we never succeeded in doing it from the patent. I do not believe that a man however much skilled in the art could wire & connect up a Seimens Armature by reference to the patent. It is nothing more than an old Seimens armature with double coils arranged like Paccinottis. If the Seimens armature had $\frac{1}{2}$ an inch of iron on each side of the cylinder protruding and the wire was wound thus:—[23]   it would be the old Siemens armature thrown out of gear with a Gramme or Paccinotti winding If you should have any trouble about Royalty the bar & plate machine would undoubtedly get round his patent and ultimately all machines (both Isolated & Central station machines) will be made that way.[24]

Regarding Fabbris European proposition your statement that he proposed to pay $40,000 on the spot was the first I or anyone else here heard of it. We never received any such offer. Insull will send you by next mail copies of acct what Fabbri did propose.[25] It was a proposition without head or tail and carried its own destruction with it. If Mr Fabbri cables us in just a way that his proposition is misunderstood he must not blame us. You will see the difference when you get Insulls copies of cables between Bailey & Fabbris proposition.

Regarding Italy we made a proposition businesslike in every way, requiring as we would from anyone else moderate guarantees for the fulfillment of the contract. Instead of replying after he received our proposition that our terms were onerous and that this and that must be altered before he could take hold of it he gets in a great huff and never answers the proposition at all. Now such a course cannot be considered businesslike or at all a fair way of treating us. I value the reputation of Drexel Morgan & Co very highly, as well as the members of the firm individually but in a matter of business, I must confess I fail to see any reason for their getting mad because we do not accept any proposition they may be pleased to make instantly.[26] It appears that Mr Fabbri sometimes allows his feelings to interfere with his business. Another thing you must understand that it is a very difficult thing for me to do a thing to which every other Director offers the most decided opposition. There would never be any trouble between Mr Fabbri or Drexel Morgan & Co & myself were the interest sought to be treated entirely my own. You will find that Mr Fabbri will have my hearty support in all matters relating to English Light and in any other country ~~whic~~ where I own ~~sole~~ singly.

Yours of Nov 10th.[27] I received a cable from G.E.G[ouraud]. asking me if I would sell a share of my English Light. Now for fear that you people over there were not working in accord I was afraid to reply directly and therefore cabled him (G.E.G.) to see you, making up my mind that I would get yours and Fabbris consent before I sold the slightest amount. I also asked Mr Wright about it since which time I have received your letter and have seen Mr Wright and now know the situation of affairs. Of course under the circumstances I would not sell (G.E.G.) any interest whatever.[28] Of course men whose sole capital is the art of deceiving the uninitiated will make money in proportion to his expertness in the art; but sooner or later he must come to grief, that is to say his "Capital"(!) becomes impaired to such an extent as to incapacitate him from making any more money & I should imagine that the Dukes[29] capacity in the above direction will soon be entirely exhausted & then as you say this Telephone bubble must burst.

I wish you would ascertain what Telephone or Construction or any other companies I am a Director in (honorable or otherwise) and cable me as I desire to withdraw as a Director as I do not propose to be a Director in a Company the management of which I have nothing to do with.

So far as I my finances are concerned if [I][a] can get my money from the Telephone I think I can pull through. I never had any intention of selling English Interests and very little even after I received Gourauds cable. I would never have sold more than $1/20$ of my interest any way. I think I shall be able to pull through here all right provided that our Coy here & Batchelor help me out with the heavy stock of Dynamos I now have on hand. By the time you get this letter I shall have 90 Dynamos completed ready to ship & am turning them out at the rate of[c] 100 a month. I can make a big Dynamo every six days & still keep up 24 "Z's" a week. I am going into the building of small Armington Engines & hope to be able to supply a dozen of these a week.

Ascertain if Gouraud has received all the money and stock from the Oriental. I think the $10,000 he sent was a part of what he received and that he is keeping the remainder back. Gouraud has not up to this writing ~~recei~~ sent me any explanation of a/cs. He has promised to do so but his letters bear out your statement entirely that he is postponing the settling day.—

<u>Yours Nov 2nd</u>. Socket. We were compelled to make the change but before it was done the subject was carefully canvassed and it was decided that the Coy should purchase all the sockets Bergmann had on hand at cost and then sell the parts back to him for what he would give for them & this arrangement has been consummated.[30] We now send out new lamps & w[h]ere old sockets are in all, new ones free of charge[d] to replace them; and in that way there being but few actually sold we can change them at a small expense and then forever afterwards we will be all right. You will see from the samples sent you that it is a vast improvement   It now allows us to use globes from the glass works that we had had here to fore to throw away at a large expense. A large body of plaster of paris is used so that the strain on it is done away with and it is impossible to break a lamp loose or put such a strain on the plaster of paris as to break it. The socket is also very much better & easier to make. You will notice that Bergmann has put up a [centre?][e] piece of wood very high so that they cannot be crossed at this point. We have now got the art of mixing the plaster of paris down to a fine point. Everything is weighed out accurately the results are always the same   Previously it was mixed by boys by guess work and was a badly botched job. All parts of the plaster of Paris are varnished so as to prevent damage from water. I am not sure that the samples ~~thus~~ sent you have been thus treated as we only did this on receipt of your letter about the climate. If you have sold 500 sockets say to the

Faure Coy you can substitute 500 of the new ones you standing the loss of the change   Now is the time to make this change Plaster of Paris is the only thing that is adapted to the purpose.[31] Every oil lamp used is fixed together in the same way & all that is required is to have the strain equalized & the proportions of the plaster correct. I am sure after you have looked the whole ground over you will agree in the wisdom of making the change now. We are resocketing all our lamps at an expense of about 1¼ cents per lamp, which is not very much.

You want to call the attention of the people to whom you sell lamps to the fact that the life is based on the _average_ and that on the first two or three days some lamps will most certainly go & so on until all the bad lamps are shaken out. Be sure you explain this as people will otherwise get the impression that because some lamps go at the start our statement as to life &c must be incorrect.

You know I have not been to the Lamp Factory lately but I ascertained that the average life was running down & had got even as low as 400 hours in fact [mons?].[e] Upton had got away off his base & was trying to get back without informing me. So finding out the state of affairs I went down to the Factory Saturday before last & on the Monday morning following started for a grand bounce of the bugs.[32] I had just 18 hours sleep that week without my boots being off & the consequence was that by Friday night I not only got him all right again and the bugs out but I got 10 p[er]. E[lectric].H[orse].P[ower]. lamps to give a better curve at 48 candles than had ever been got with the 8½ p. E.HP. lamps. For instance I put up ten lamps taken out of a lot of 475 in a night run. The first lamp went in nine hours the next lamp in 20 hours and at 34 hours there were eight still burning. I have not seen the record; others were put up which I am told which turned out equally as well. I have so arranged matters that no retrograde movement will take place again. The life of the 10 p. E.HP. lamps will now probably be about 1000 hours. These we shall dilute with 10% of the 400 hour lamps now on hand which[f] will bring the average to about 600 for the present. I shall now pay a visit to the Lamp Factory once a week if not twice a week. I propose to run that concern myself just as I want it. I have seen enough in the week I spent there to convince me that by about one months solid fighting I can bring a 12 p. E.H.P. lamp up to 600 hours life & further that I can make a 250 ohm 12 p. E.H.P. lamp last 600 hours. If I were you I would sell the Faure people only 8 candle lamps.

I have not seen Crookes description of a new lamp   I think

Lowrey stated that in talking with Crookes that he stated that he has been working on lamps but found in looking over my patents that I had forestalled him.[33]

Yours of Nov [3]rd.[34a] Crystal Palace. You will have [a?][a] great advantage over Batch as your small machines will be far more reliable as the fields are all bored out very much larger & you will have no trouble from the rubbing of the Armatures Your commutators are also better & you will have no trouble from high & low blocks. The contacts between the base connected to the wires and the commutator blocks are very much better in your machines. We have just devised a new brush holder which is nc plus ultra. It is one that you can put right on the Brush Holder Arm take it off or do anything with it without stopping the machine   We now send these brush holders with each machine two to hold the regular brushes & an extra one that can be fixed on the machine in about five seconds thus the regular brush can be taken off & fixed while you are running your machines. You will therefore not have the slightest difficulty as by using this extra relief brush you can keep your regular brushes in prime order at all times while the machines are in motion. I have given orders today to ship sixty of these to you and sixty to Gouraud. I suppose you will want a supply of brushes. We have got up a automatic machine for making these brushes. I do not know what the price of the brushes will be anyway it will be quite cheap.

Below find answers to your questions:—

First. The machines will carry 75 lamps but of course will get very hot in the course of several hours but even 85 or 90 can be run for a few minutes   We have run 90 for two hours but this is a very heavy strain. The Belts are liable to slip so you want to look out that all your belts are tight as this is a bad bug when running the machines in multiple arc as if some of the belts are loose the other machines will do all the work and it will be like putting 100 lights on a machine

Second. The fields of your machines will run 90 lights at 16 candles if all the resistance is cut out always provided your belt is tight and the bobbins keep up the speed. Taking off the four machines one at a time in rapid succession will not make the belts slip to any material degree if they are very tight in the first case   You should use a belt which is the full width of the pulley and it should if possible be as horizontal as pos permissible. If you use a countershaft and the belt moves slower on the countershaft you will have to increase its width to make up for the slowness in speed. A certain number of square inches of belt must pass a given point in a given time to convey a given power.

Third. If you wish to stop one of the engines driving four machines without opening the Dynamos it would not harm the armatures of the Dynamos but the remaining eight would supply current to those four and keep them running as motors in the same direction so that if you shut off same your engine will keep on just the same. That is a little trick that fooled them down at Goerck St the other day. They thought they had made a big discovery. So the eight machines would have to run all the lights these four machines & the engines besides. They would not become low resistance arcs from this fact but continue as motors there being a counter electro motive force which is an equivalent of resistance. The sudden withdrawal of the four machines at one blow of the Switch would not over tax the other eight electrically only it would tend to make the belts slip if they were allowed to be too loose. You will remember that at Menlo Park we never had any trouble with our engines. All our machines were in multiple[f] arc. We ran for months without trouble except that caused by the effect of the weather on the belt which might have been prevented   We connected two more machines than was required to run the whole of the lights. In case of a cross on one of them we simply ~~throw~~ cut out that machine. If we wanted to fix a machine we moved the Dynamo loosened the belt, fixed it, tightened the belt again & started up.

I see you propose to use four Dynamos with one engine mechanically separate from the others. This is all right provided your engines govern well and your speed of all the sets are correct & even if they are a little out, say one set of Dynamos run at 1160 or 70, & another at 1200 the ones running[f] the highest speed will do a little more work. I suppose if one Dynamo of the four should get wrong you will simply throw it out & fix it   If the engine gets wrong you would then throw off the Dynamos & stop the engine. If you do it in this manner I do not see how you can fail to have reliability[35]   Yours truly

Thos A Edison   I[nsull]

L (letterpress copy), NjWOE, Lbk. 9:331 (*TAED* LB009331; *TAEM* 81:111). Written by Samuel Insull. [a]Faint letterpress copy. [b]Canceled. [c]"the rate of" interlined above. [d]"free of charge" interlined above. [e]Illegible. [f]Obscured overwritten text.

1. On 9 November, the Lord Mayor of London processes to Westminster to receive the Crown's assent to his election (*OED*, s.v. "Lord Mayor (1)"). Johnson's letter is in DF (*TAED* D8133ZAQ; *TAEM* 58:696).

2. Berthoud-Borel cable, an insulated telegraph cable patented and manufactured by a French firm, could be manufactured relatively cheaply in great lengths. It consisted of copper wires "wound in cotton and bathed in parafin and rosin, the whole hermetically inclosed, in perfect contact and under pressure within a lead envelope," making it especially suited for underground use. The cable was exhibited successfully at the 1878 Paris electrical exposition. In the latter half of 1881, Edison became involved with the effort of Ernst Biedermann, a Geneva merchant associated with Edison's electric light and power interests in Switzerland, to establish the United States Underground Cable Co. for manufacturing the cable in the U.S. (U.S. Underground Cable Co. prospectus, n.d. [1881], DF [*TAED* D8122U; *TAEM* 57:753]; Dredge 1882–85, 1:333–37; TAE to S. L. M. Barlow, 26 Nov. 1880, Lbk. 6:598 [*TAED* LBoo6598; *TAEM* 80:431]; Biedermann to Insull, 16 Aug. 1881, Miller [*TAED* HM810154A; *TAEM* 86:417]; on Biedermann see Docs. 1878 n. 2, and 1962).

3. Patrick Delany (1845–1924), an Irish-born electrical engineer, invented among other things automatic and multiplex telegraph systems and a lead-covered wire designed to suppress induction currents. Delany had recently moved to South Orange, N.J. *DAB*, s.v. "Delany, Patrick Bernard."

4. Johnson, George Gouraud, and Delany jointly filed a British patent specification (Brit. Pat. 2,532 [1881]) on an anti-induction cable in June (Dredge 1882–85, 2:clxx–clxxi). Johnson wrote on 9 November that "My communications from DeLany in re to the progress of our cable sales give unpleasant impressions— It seems you have finally determined upon putting down wires in your trenches & in using the B&B wires for the purpose— It is of course significant that with all my Correspondence from New York I should learn of this only through an outside source." He promised that regardless of the commercial viability of the Delany wire he would not "be led into an opposition shop" and would forgo his interest because "money hasn't sufficient power over me to lead me to reverse my position of the past 10 years." He later agreed to arrange for Delany to give Edison a sample for testing. Johnson to TAE, 9 Nov. and 11 Dec. (p. 2) 1881, both DF (*TAED* D8133ZAQ, D8133ZBD; *TAEM* 58:696, 728).

5. In August, John Kruesi made several forms of insulated wire in metal tubes. Near the end of the month he reported that "the insulation is perfect." Edison's basic conductor cable design was of two semicircular copper rods, held apart by cardboard washers, placed in a metal tube that was then filled with an insulating compound such as asphaltum. Kruesi to Samuel Insull, 18 and 20 Aug. 1881; Kruesi to TAE, 22 Aug. 1881; all DF (*TAED* D8130P, D8130Q, D8130R; *TAEM* 58:361–62, 364); Dredge 1882–85, 1:347, 611–12, 614; Francis Jehl's tests of Kruesi's insulated conductor are recorded in N-81-04-06:165–69, Lab. (*TAED* N223:79–81; *TAEM* 41:115–17).

6. DF (*TAED* D8104ZER; *TAEM* 57:260).

7. James Hood Wright (1836–1894), a partner in Drexel, Morgan & Co., was among the original stockholders of the Edison Tube Works and a significant shareholder in the Edison Electric Light Co. *NCAB* 33:443; Doc. 2058 n. 4; Edison Electric Light Co. list of stockholders, 1 July 1882, DF (*TAED* D8224ZAP1; *TAEM* 61:45).

8. Johnson wrote (see note 6; p.1) that "the machine upon which you worked night & Day & spent some hundreds of Extra money lies on the Dock yet awaiting the Completion of its final resting place   Without the knowledge I had of it however I should have made bad mistakes. . . . Mr Fabbri is therefore not wholly dissatisfied with the Delay—in fact he agrees that it has been all conducive to better work now & consequently to Earlier success."

9. See Doc. 2180.

10. It was not shipped until mid-February. See Doc. 2228.

11. It is not known what plans Batchelor had for this machine before it was destroyed (see headnote, Doc. 2122).

12. This was probably the Edison Electric Light Co.'s *Instructions for the Installation of Isolated Plant* (1881), a seventy-six page booklet (plus appendices) "published exclusively for the private use of the agents of the Edison Company for Isolated Lighting." It included specific topics such as the dynamo and its care, switches, fixtures, fuses, and lamps; it also contained more general discussions of electric energy and resistance.

13. See Doc. 2180.

14. Alexander Muirhead (1848–1920), a noted British telegraph and electrical engineer, had visited Menlo Park in 1878. He traveled extensively as a scientific advisor and a director of Latimer Clark, Muirhead, & Co., an engineering and manufacturing firm co-founded by his father. Upon arriving in New York on 5 October for a "short visit," he wired Edison to arrange a meeting. Johnson warned Edison to "look out for him" because his partner, Latimer Clark, was an advisor to Swan. *Oxford DNB*, s.v. "Muirhead, Alexander"; *DBB*, s.v. "Muirhead, Alexander"; Doc. 1339 n. 1; Muirhead to TAE, 5 Oct. 1881; Johnson to TAE, 6 Nov. 1881 (p. 5); both DF (*TAED* D8104ZDM, D8104ZER; *TAEM* 57:225, 260).

15. Johnson wrote Edison on 11 December that Muirhead had called upon him again "yesterday and occupied my time for nearly 2 hours in trying to induce me to let him have Lamps and to generally supply him with material and advice in respect to the prosecution of some experiments." Johnson recounted that he "only got rid of him finally by giving him definitely to understand that in my judgment there had already been enough attempts at appropriating Mr. Edison's invention without his leave or license and that I did not purpose being the medium through which the same thing might occur again." He offered, however, to sell outright 10,000 lamps "much better than he could make them and at a price far below that which they would cost him. This question of selling Lamps has been carefully considered by us and we have decided that in cases where large numbers are required, making it worth while for us to give the matter attention, we will sell, but we will not occupy our time or our energies in retail business." Johnson to TAE, 11 Dec. 1881 (pp. 5–6), DF (*TAED* D8133ZBD; *TAEM* 58:728).

16. This paper and the December 1878 demonstration were reported in Proctor 1879; see also Doc. 2165 n. 2.

17. This lecture has not been found or otherwise identified but see Bowers 1982 (113–15) regarding Swan's demonstrations. Swan exhibited twenty of his lamps at the Literary and Philosophical Society of Newcastle in October 1880. He also gave a demonstration in London at the Society of Civil Engineers on 24 November and, the next day, exhib-

ited thirty-six lamps to the Society of Telegraph Engineers. Swan made a broad priority claim to having invented the incandescent lamp with a filament of carbon, on the basis of unpublished experiments, as early as January 1880. Doc. 2022 n. 6; Chirnside 1979, 99–100; Swan 1946, 28; "Incandescent Electric Lights," *Nature* 23 (1880): 104–5.

18. Swan presented a paper to Section G (Mechanical Science) of the British Association for the Advancement of Science on 5 September. The paper was identified in the Association's report of transactions of the sections at the annual meeting but was not reprinted. "On the Swan Incandescent Lamp," *Report of the British Association*, 1881, 5.

19. Grosvenor Lowrey's letter has not been found. Edward Brewer and Peter Jensen were patent agents in London whom Lemuel Serrell used in connection with Edison's British specifications. See Doc. 1033 n. 1.

20. Edison referred to his British Patent 2,402 (1879), Cat. 1321, Batchelor (*TAED* MBP017; *TAEM* 92:118). His remarks on Swan may have been prompted by the summary of Shelford Bidwell's legal opinion that solicitor Theodore Waterhouse provided to Johnson on 11 November and which Johnson forwarded to Edison the next day. Waterhouse agreed with Bidwell that Edison's fundamental British carbon lamp patent of November 1879

> will give him a monopoly of the incandescent lamp as now manufactured, provided (1) that there was no prior publication by Mr. Swan or any one else of such a nature as to make the lamp on that date no longer new, & (2) that Mr. Swan's Patent of 2nd January 1880 does not prevent Mr. Edison from exhausting the air from his lamp while the carbon is incandescent. I understand that Mr. Edison made his lamp in this way at the date of this patent of 10th Nov. 1879 but he does not claim this detail, and does not in fact in that patent allude to it. I should have supposed that he then relied upon the description of the similar process as applied to metal conductors in his previous patent of 7th June 1879, & there stated to be applicable to carbon sticks, but for the fact that in his Patent of the 2nd February 1881 he has a claim No. 4 which appears to be identical with Swan's claim of the 2nd January 1880. The fact that this claim was made after Swans is, under English law, a damaging feature in our case. I cannot however think that on this point we need feel any great apprehension. The more important question is as to the publication of some descriptions of the lamp prior to the 10th Novr 1879. Mr. Swan alleges in his Lecture at Newcastle on the 20th Octr 1880 that he exhibited such a lamp as that which Mr. Edison claims as long ago as 1878. But as yet we have his assertion only for this statement, & it does not follow that exhibition would be equivalent to publication. [Waterhouse to Johnson, 12 Nov. 1881, DF (*TAED* D8133ZAS; *TAEM* 58:706)]

In his cover letter, Johnson reviewed the main features of Swan's public claims and concluded that he

> will be wholly unable to prove prior publication—and that consequently the issue is as between your Patents and others on various features—details—in fact simply a question of how much we shall have to make Public ourselves by reason of not having it properly

protected= Take for instance the Swan Patent on heating the
carbons during process of obtaining the vacuum— You apply this
process to metals & remark that it may be applied to "sticks of Carbon"—but when you come to your Carbon Patents you make no
allusion to it—and no where do you claim it for carbon— Swan
jumps in & applies it to carbon filaments & claims it= Now this
claim can easily be upset no doubt—but only by proving in our
Patents and in other ways "Publication"= We cannot so amend or
Emend our Patents as to get in a claim on the process as applied to
your present carbon Lamp—at least such is our present opinion.
[Johnson to TAE, 12 Nov. 1881 (pp. 3–4), DF (*TAED* D8133ZAR;
TAEM 58:698)]

The Edison Electric Light Co. published a detailed polemical analysis of
Edison's and Swan's respective patent positions in the U.S. in its 27 July
1882 Bulletin (CR [*TAED* CB012; *TAEM* 96:728]). In early December,
Insull sent Johnson additional comments from Edison regarding Bidwell's statements (Insull to Johnson, 2 Dec. 1881, Lbk. 9:400 [*TAED*
LB009400; *TAEM* 81:153]). Johnson proposed in early November to research the legal soundness of Edison's lamp patents "in a most thorough
manner." In addition to Theodore Waterhouse, who had advised the
Edison telephone interests, he had retained John Henry Johnson (also
involved in the telephone patents) and Bidwell, "an Expert in Electric
lighting & in Patent Law (a member of the English Jury at Paris)" (Johnson to Insull, 1 Nov. 1881, DF [*TAED* D8133ZAL; *TAEM* 58:674]).

21. Edison was probably thinking of St. George Lane Fox's British
Patent 1,122 (1879) for an incandescent "candle," formed of a mixture
of a finely ground conducting material such as plumbago with a non-
conducting material, operated in a vacuum. The specification was filed
on 20 March 1879; Lane Fox had several other patents for carbon filaments in vacuum, but these were dated after Edison's fundamental
incandescent carbon lamp specification. Dredge 1882–85, 2:lxxviii,
1:647–58.

22. See Doc. 2165.

23. The drawing was not reproduced.

24. The conceptual similarities between the basic Siemens dynamo
and Edison's wire-wound armature had already caused conflicts in the
U.S. and German patent offices. In his 12 November letter Johnson asked
if Edison knew why William Siemens had failed to get a U.S. patent (see
Doc. 1851). Johnson also recommended a thorough review of Edison's
British dynamo patents. In the meantime, he asked if he should agree
to the royalty which Siemens had evidently requested: "some of us view
that as the wiser course whilst others oppose it— Siemens is said to
be a man who prefers a Royalty to a litigation." Edison later cabled
him: "Dont offer Siemens royalty." Johnson to TAE, 12 Nov. 1881, DF
(*TAED* D8133ZAR; *TAEM* 58:698); Johnson to TAE, 5 Dec. 1881;
TAE to Johnson, 5 Dec. 1881; LM 1:109C, 110A (*TAED* LM001109C,
LM001110A; *TAEM* 83:926–27).

In reply to Edison's statements, Johnson asked:

Am I to understand that you hold your bar and plate machine
not to be a Siemen's armature? Barker will tell you that Sir William
Thomson pronounced it to be nothing more nor less than a Siemens

armature. . . . I gather that your idea is that Siemen's patent is bad because not sufficiently described. I am afraid that that will prove to be a difficult position to maintain. He will be able to call hundreds of experts to testify to the contrary. In view of the importance of this matter and of Siemen's possible action pending our negociations here we have considered it wise not to hasten the sale or public use of the small Dynamos. Of course the particular use that we are making of the large Dynamos does not give him a chance for an injunction as we are not using it at present for profit. We do not care to do anything to risk an injunction from him at the present moment. [Johnson to TAE, 11 Dec. 1881 (pp. 12–13), DF (*TAED* D8133ZBD; *TAEM* 58:728)]

25. The intended letter to Johnson has not been found. Johnson had complained to Edison in his 6 November letter (see note 6; pp. 10–11) that "Insull tells me you all think I am 'off my base' in the matter of his [Fabbri's] proposition for Europe— Well If I am then so is Lowrey for he distinctly told me that there is nothing in Bailey's present proposition which is not also included in Fabbris." He reported that Fabbri also "proposed to pay on the spot $40,000—which would have put the Co out of debt= Insull speaks of 100,000 Francs for manf'g purposes where his proposition was 1,000,000, (million)—with 500,000 more for sending experimental plants on to the Continent &c &c."

26. Nothing is known of the proposed terms. In his 6 November letter (see note 6; p. 11), Johnson explained that Fabbri felt "you have broken a promise made to him that he should have Italy on the same terms as could be obtained from any other parties— In short I gather from it all that Fabbri claims some consideration for the guarantees of good faith which his propositions bring—which you refuse to accord—but make it simply a matter of Dollars & Cents as between him & parties who give you no guarantee either in their names or in more substantial stuff."

27. DF (*TAED* D8104ZES; *TAEM* 57:274).

28. George Gouraud inquired by cable on 9 November whether Edison would "sell me part of your interest in Drexel Morgan and Companies light contract." Edison answered that he should see Johnson, whom he instructed to talk with Gouraud and "report fully." Johnson cabled the next day that it was "imperative" for Edison to retain his full interest (Gouraud to TAE, 9 Nov. 1881; TAE to Gouraud, 9 Nov. 1881; TAE to Johnson, 9 Nov. 1881; Johnson to TAE, 10 Nov. 1881; LM 1:95A, 96A, 95B, 97B [*TAED* LM001095A, LM001096A, LM001095B, LM001097B; *TAEM* 83:919–920]). He explained in his 10 November letter (see note 27) that Fabbri had "found on investigation that if he wants to bring into this thing the character of men he has been contemplating he must not have G.E.G. and . . . that G.E.G. is not the sort of man that good men will associate with in business." Johnson provided more details of Gouraud's electric light dealings in December (Johnson to TAE, 23 Dec. 1881, DF [*TAED* D8104ZFM; *TAEM* 57:314]).

29. In his 10 November letter (see note 27) Johnson made several derisive allusions to Gouraud as "the Duke," whom he suggested had tried to mount a "coup" against Edison's British electric light interests.

30. No letter from Johnson on 2 November has been found, but he dated a letter of 8 November in a way that could have been misread. In

it, Johnson complained that he had learned from Upton of a plan to alter the design of the lamp socket "to avoid the breaking of the Plaster &c— This in my judgment is radically wrong   It is of the utmost importance that once having established a standard in anything we should not depart from it— I have sold 1000 sockets & shall have to sell several thousand more ere the Lamps I have on hand and which I am now ordering shall have been disposed of." He urged that "the proper & only way is to use on the Lamp something more substantial than Plaster— It is bad for other reasons than want of strength— It softens under certain conditions and gradually crumbles." He suggested celluloid instead. Edison at first blamed the problem on faulty manufacturing by Sigmund Bergmann, but Bergmann also urged him to find a substitute for the plaster. Johnson to TAE, 8 Nov. 1881; Bergmann to TAE, 14 Oct. 1881; both DF (*TAED* D8133ZAP, D8101E; *TAEM* 58:693, 57:13); TAE to Charles Batchelor, 13 Oct. 1881, LM 1:63 (*TAED* LM001063B; *TAEM* 83:903).

The socket was altered to accommodate a slightly modified lamp base. In the base with the bevel contact ring adopted in 1880 (see Doc. 1988), screwing the lamp into the socket placed considerable tension on the plaster of Paris insulation between the wire leads, causing frequent breakage. To correct this, one lead was connected to the screw shell and the other to a small metallic button at the bottom of the base. Screwing this base into the socket compressed the plaster, which was far less injurious. This design with the button has remained essentially unchanged since that time. Howell and Schroeder 1927 (183–86) attribute the change to the middle of 1881; Jehl 1937–41 (808) describes but does not date it. Edison executed a patent application for a similar, though not necessarily identical, design on 22 November. U.S. Pat. 317,631.

Photograph of the modified screw base adopted by late 1881. Except for different materials and altered dimensions, this is the design of bulb base used in incandescent bulbs to the present day.

31. In late October, Edison made notes for a patent application on a socket made of "wood pulp or other suitable insulating compound which will harden on drying either with or without compression— Method & means of giving the glass part a strong abutment to prevent its becoming loose therein." He never filed this application. Cat. 1147, Scraps. (*TAED* NM016:104; *TAEM* 44:328).

32. See Doc. 2183.

33. William Crookes filed a British provisional patent application on 31 August (3,799) for a lamp with a carbon filament enclosed in an exhausted glass globe, through which platinum wires were sealed. Earlier in the year, Crookes filed two specifications for increasing the density of uncarbonized filaments by partially dissolving the cellulose in the paper or cotton. He also filed a specification for electroplating the carbon to platinum wires. Dredge 1882–85, 2:cxcii–cxciii, cli, clxxiii–clxxiv, clxv; Fournier d'Albe 1923, chap. 15.

34. Doc. 2180.

35. This letter is continued in Doc. 2190.

Philadelphia Nov. 23, 1881.

My dear Edison:—

Thanks for your telegram of the 12th and your letter of the 21st. I did not reply by telegraph to the question you asked in the letter because I could not recollect at once the circumstances.[1] Even now I do not recall clearly that any special form of apparatus was talked of. I remember very well that we talked during that expedition upon electricity as a motive power and upon propelling railroad cars by it. We discussed together, as I recall it, the machines in use as electric generators and the economy of their reversal so as to develop power. I had had a Gramme machine in use since 1875 and gave it always the preference over the Siemens machine. This is my recollection of what passed between us but I am afraid it is too general to be of any service to you.

I am much obliged to you also for the things you sent over to me. I made use of them as you saw in the paper I sent, before the Academy to correct and neutralize the impression given a year ago in favor of the Maxim lamp. Everybody was very much pleased and fully convinced that you were a long way ahead of everybody else.[2] By the way, did you notice in Nature for Nov. 3, page 16, at bottom of the page, a very complimentary allusion to your system of lighting?[3]

I write now to say that I expect to come over to New York on Friday morning in the 7:35 train; and that I will come at once up to 65 Fifth Avenue to see you. I shall reach there about half past nine as I suppose. There are lots of matters I want to talk over with you and we can arrange for an hour to chat quietly together. I shall probably remain till Sunday night, possibly until Monday night.[4] I have an appointment at Drexel & Co's about one on Friday. Mr. Fabbri wanted me to go in and see Mr. Wright about the London outlook, which is fine. I saw Mr. A. J. Drexel[5] here this afternoon and he is very much interested in the success of the London Company.

Hoping to see you on Friday   I am Cordially yours

George F. Barker.

ALS, NjWOE, DF (*TAED* D8104ZFA; *TAEM* 57:290). Letterhead of International Congress of Electricians, United States Delegation.

1. Edison's telegram has not been found. He asked in the letter if Barker could "recollect my speaking to you about my Electric Rail Road when I was returning from my trip to the West? Please wire me in the morning simply 'yes' or 'no.' I want this information in connection with the Rail Road Interference with Siemens." On 9 December, Barker testified on Edison's behalf in this proceeding as to the general nature of

conversations about electric power and transmission during their trip in July and August 1878. TAE to Barker, 21 Nov. 1881, Lbk. 9:313 (*TAED* LB009313; *TAEM* 81:106); Barker's testimony, 178–85, *Edison v. Siemens v. Field,* Lit. (*TAED* QD001178; *TAEM* 46:94).

2. At Grosvenor Lowrey's suggestion, Barker planned to present a paper at the National Academy of Science's November meeting in Philadelphia "as an offset to the Maxim blast last fall in New York" (see Docs. 2022 and 2033). This paper has not been found, but Barker requested from Edison 100 lamps, several chandeliers, safety catches, switches, conductors, and "whatever else you think would help me to explain the system as exhibited" in Paris. Edison gave orders the next day for these items to be sent. Barker to TAE, 11 Nov. 1881, DF (*TAED* D8135ZCS; *TAEM* 58:1119); TAE to Electric Tube Co., TAE to Edison Lamp Co., TAE to Bergmann & Co., all 12 Nov. 1881, Lbk. 9:288, 289, 291 (*TAED* LB009288, LB009289, LB009291; *TAEM* 81:99–101).

3. An unsigned editorial notice of the forthcoming Crystal Palace exhibition stated that Edison's display had aroused "very great interest. . . . The effect produced by it in Paris was quite startling, and it is generally believed that Mr. Edison has solved the problem that he set himself, viz. to produce a light to supersede gas in our houses." "Notes," *Nature* 25 (1881): 16.

4. Barker left Edison the evening of Sunday, 27 November; nothing is known of their conversation. He planned to return to New York in two weeks at Edison's request, probably to testify in the railroad interference case (see note 1). He hoped at that time additionally "to arrange for a conference between Mr. L[owrey]. yourself & myself. I should like to meet him and there are some matters to be talked over between us." Barker also inquired whether Edison knew anything about the American Electric Light Co. (It is not clear if this was the Connecticut company formed to make electric lamps under license from Elihu Thomson and Edwin Houston, or the peripherally-related Massachusetts firm of the same name that Edison later concluded was an investment fraud.) In January 1882, Barker received from Edison a gift of two shares in the Edison Electric Light Co. and twenty-five in the European light company, which he accepted with profuse but somewhat awkward thanks "as evidence that you believe that my friendship has been sincere." Barker to TAE, 7 Dec. 1881 and 13 Jan. 1882, both DF (*TAED* D8135ZCV, D8204C; *TAEM* 58:1129, 60:51); on the American Electric Light Co., see Passer 1953, 24–25; Carlson 1991, chap. 4; and Doc. 2264.

5. An heir to the Philadelphia banking house Drexel & Co. established by his father, Anthony Drexel (1826–1893) joined with J. P. Morgan to form Drexel, Morgan, & Co. Carosso 1987, 133–39; *ANB,* s.v. "Drexel, Anthony Joseph."

–2189–

From Sherburne Eaton

[New York,] November[a] 23rd. 1881.

Mr. Edison,

The application of Col. Logan and Major Mc.Laughlin for the Isolated business in California is just received by me.[1] I will bring it up at the next meeting.

We have organized the Isolated Company as follows,[2] viz. President, Eaton; Vice President not to be filled at present; Treasurer, Fabbri; Secty., Goddard; Ex[ecutive]. Com[mittee]. Banker, Fabbri and Eaton (when Mr. Lowrey comes I will resign from the Ex. Com. if he wishes it. I would still be a member ex officio.)[b] Mr. Tracy Edson resigned as a Director and on recommendation of Mr. Adams[3] and Mr. Banker his vacancy was filled by the election of Mr. John H. Flagler[4]   My[a] compensation for acting as President of the company was fixed as follows, namely, a call was given me for one year at par on the stock left in the treasury, which will be about eighty shares of stock. How does this last matter suit you? I do not care much about it as I had about as soon work for nothing and take my chances by and by as accept a call like that. It was Banker's suggestion. However, it went on the minutes of the meeting and I am willing to let it stand.

S. B. Eaton   per Mc.G[owan].

TL, NjWOE, DF (*TAED* D8121G; *TAEM* 57:715). [a]Mistyped. [b]Closing parenthesis written by hand.

1. Within six days of Eaton's letter the company selected the application of Col. James C. Logan and Frank McLaughlin, who represented a group of investors organized by Louis Glass (see Doc. 2176). McLaughlin, Logan, and Glass were all connected to Edison's mining interests in Oroville, Calif.; according to McLaughlin, Logan (b. 1847?) was the business partner of California Governor George C. Perkins, also among the prospective investors. Ladd to TAE, 29 Nov. 1881; Glass to TAE, 26 Nov. 1881; McLaughlin to James Banker and Robert Cutting, 10 Nov. 1880; Glass to Insull, 16 Feb. 1882; all DF (*TAED* D8120ZBT, D8121H, D8033ZBH, D8220I; *TAEM* 57:662, 716; 54:527; 60:760); Logan and McLaughlin agreement with Insull, 7 Feb. 1882, Miller (*TAED* HM820156B; *TAEM* 86:451); Teisch 2001, 241; U.S. Bureau of the Census 1970, roll T9-63, p. 235.3000, image 473 [Oroville, Butte, Calif.].

2. Samuel Insull explained to an unrelated correspondent that the "demand for lighting mills factories etc." had taxed the Edison Electric Light Co.'s manufacturing abilities "to their full capacity, so that now a separate company is in formation for the relief of the Parent Company from this special branch of lighting." An article in the *New York Herald* on 25 November announced the organization of the Edison Company for Isolated Lighting and the election of officers along the lines described by Eaton's letter. It stated that Miller Moore was the general manager and that the Board of Directors consisted of Edison, Eaton, Fabbri, John Flagler, Grosvenor Lowrey, and William Meadowcroft. In May 1882, those members of the Board of Directors who were also directors of the Edison Electric Light Co. were replaced to avoid conflicts of interest in the legal relationship between the two companies; Egisto Fabbri was replaced by his brother Ernesto, Banker by Moore, and Lowrey by Goddard. At the time of its formation the Isolated Company had a capital stock of $500,000; the parent Edison Electric Light Co.

held half of this in exchange for patent licenses. The Isolated Co. was authorized to operate in areas without illuminating gas service. Insull to Charles Davis, 17 Nov. 1881, Lbk. 9:300 (*TAED* LB009300; *TAEM* 81:104); "Edison's Isolated Company," *New York Herald*, 25 Nov. 1881, Cat. 1243, item 1768, Batchelor (*TAED* MBSB41768; *TAEM* 95:41); Agreement between Edison Company for Isolated Lighting and Edison Electric Light Co., 26 Apr. 1882, Defendant's Exhibit, *Edison Electric Light Co. v. U.S. Electric Lighting Co.*, 4:2363, Lit. (*TAED* QD012E2363; *TAEM* 47:1004).

3. Edward Dean Adams (1846–1931) was a financier and engineer who served as a director of the Edison Electric Light Co. from 1882 to 1886 and of the Edison Electric Illuminating Co. of New York from 1884 to 1889. He was later instrumental in the financing and construction of the hydroelectric power station at Niagara Falls. *ANB*, s.v. "Adams, Edward Dean"; "Adams, Edward Dean," Pioneers Bio.

4. John Haldane Flagler (1836–1922) was a prominent businessman in the iron and steel industry. *DAB*, s.v. "Flagler, John Haldane."

–2190–

To Edward Johnson

[New York,] 27th Nov [188]1

My Dear Johnson,

In continuation of my letter of 23rd inst I beg to state:—[1]

Regarding the regulation of course when you throw off one Dynamo the effect would be scarcely noticeable. The engines should regulate[a] good instantly bringing it up to snuff. Your method of regulating by connecting all the fields to one main line and all the other ends of the fields to the regulator[b] in multiple arc and the other end of the regulator to the other main line is correct   The resistance of all the twelve fields would be three ohms which is equal to about thirty or forty lamps. The resistance coil in this case would have to have considerable radiating surface. The part of the regulator which you have might do but it would be better to have them heavier and make better contact. I will have Clarke figure out the radiating surface & sizes of coils you will need & Insull will start Bergmann on making them & we will ship them as early as possible but if you want to use a regulator before this reaches you, you can multiple arc the coils from six or seven of your regulators Boxes. You can ascertain the heat there would be on a coil by putting fifty lamps on a machine & putting the coil in the main line between the machine and the lamps & bring the lamps up to 16 candles. As the resistance of all the fields is but three or four ohms of course the value of a ohm for regulating is tremendously increased and therefore as you approach the sixteen candles your coils will have to be subdivided into $1/10$ of an ohm or perhaps $1/4$ of an ohm—or something like that.

You ask if the coils should be $1/12$ of a single machine regulator

No! Not so bad as that   I think that of the coils of six single machine regulators—that is all the three ohm coils multiple arced together & then connected to the next series of coils so that the whole would act as one box. Then it would answer for the reason that the present box does not get hot at all but whether the subdivision is ever enough it would be difficult to say but will ask Clark—I mean whether it will make too great a difference from one notch to ano[ther]^c For fear that you may not understand my meaning I will say suppose we had a coil of four ohms and it got too hot by multiple arcing around it another coil of four ohms the resistance would be reduced one half & the surface doubled & so on.

In answer to your second question about regulation, the stoppage of one engine with four Dynamos ceasing to supply their own fields would not alter the other fields at all but the proportion of average loss in the machines to that in the lamps would be altered and you would therefore have to bring up the field in the remaining eight

Cannot you get a photograph of about two or three feet square made of the big Dynamo?

I send you a sketch made by Clarke.[2] I think he misapprehends what you want to do as I gain the impression that your shaft & engines & four Dynamos is to be entirely separate from the remaining sets—each set being entirely separate from the other & having no connection except electrically.

The small engines ordered by you some time ago are going on very slowly the parties who are making them not having much facilities. We shall soon receive their latest model which they guarantee to run fourteen consecutive days & nights without stopping one instant. We are going to build a lot of these engines at Goerck Street[3]

If you can possibly use mercury on your Dynamos it would be a good thing—[4] It is essential for the successful running of a Dynamo (& all our men are now taught so) that before starting up the brushes should be taken off on the ends, nicely cleaned and set exactly right so that the distance from the point of contact of one brush to the point of contact of the other brush is equal on both sides of the commutator & we find when these things are done no trouble is developed in the evenings run.

In using mercury it will not do to apply it to a commutator which has never had it on when the same has been running &

is hot. If the mercury is applied in the manner known to Hammer & I believe also to your own good self when the commutator is cold & then when running it will keep cold & there will no sparking at all. But in case you cannot use the mercury a rag very slightly moistened with oil or electrotypers plumbago has been found to be a good thing for the commutator.

Another thing in this connection is that we so lace our belt that the movement of it surfices ~~which~~ to give motion to the bobbin. This smoothes the commutator & bearings.[5] Hood will understand how to do this by lacing the belt slightly out of truth.

I suppose you intend to carry out in every particular the Safety Catch business. This is growing in importance every day here. We so far are safe but the arc men are rapidly having their business ruined by the fires which they cause. In fact in the last ten days their business has come to a standstill & all the Arc men have appealed to the Board of Underwriters to have a Committee appointed

The S.S. City of [~~Worcester?~~][d] Worcester on which two Dynamos and a Hampson Engine were placed still continues to work satisfactory without any breakdown[6]

Very strange to say Mr Villiard who has just returned from Oregon was asked by Maj Eaton "What has become of those Dynamos that were put on the S.S. Columbia"[7] & he replied "They are working perfectly" Where upon the Major asked if the boat is still lighted throughout with our light "Yes" replied H Villiard "They work to perfection & have never given our people the slightest trouble." This whole affair seems to me very funny. I supposed the lights had been extinguished long ago although I can now remember having sent them about 600 lamps eight months ago.[8]

I have got the nicest thing for the Isolated Business you ever saw for notifying when the lights go beyond eight or sixteen candles as the case may be. I have been fishing for it for a long time & have always felt uneasy not to have had it. It was a far more difficult thing to get than I at first supposed It is now working "bang up" downstairs here. It is a box about 8 inches ~~high~~ wide[e] 14 inches long & 6 inches thick, provided with two binding posts & is multipled arced across[a] the line just like a lamp. Underneath the box is a vibrating bell. Inside the box is a long electro-magnet of seventy ohms resistance & in the same circuit is 70 ohms of coil resistance the total of which (140 ohms) is equal to a lamp. The magnet weighs about 20 lbs the armature about one lb. The armature is set about an inch away from the magnet & has a powerful retractile spring arranged so that it cannot vary. The whole thing is on a solid iron base. The armature lever plays between points one of which is platinized. We put this across the line. Say it is used with a machine running at 105 volts. The retractile spring is so adjusted so that when the electromotive force is brought one volt higher the circuit closes & the vibrating bell (which gets its current from one of the coils in the Box by shunting) rings. Thus if the electromotive force increases or in other words if the lights go above 16 candles the bell rings & it will do it every time. The whole apparatus can be kicked across the room & not withstanding any ill usage it will when put up again ring at the proper point. It has brought out in one evening at 65 5th Ave. the terrible irregularity of those Hampson Engines. Now with this machine if we sell a Isolated Plant to go with lamps of say 106 volts we send one of these indicators with it & all we have to do is to so regulate it that it will ring at 107 volts. Then when the bell rings the customer knows his lamps are too high & all he has to do is to move the handle of the Regulator Box & so bring down the candle power & thus stop the ringing. I congratulate myself that this is a pretty good thing for Isolated business as without it we should constantly be at variance with purchasers as to the life of our lamps. Not only is it a good thing for Isolated business but we have come to the conclusion that it is a better standard of electromotive force than a standard battery   I will send you one (as soon as [I can?]^f) adjusted for [105?]^f volts. If you like it you can use your judgement about ordering more, but before I send it & before you say anything to anybody about it I must secure the patent in England.[9] So say nothing to <u>anyone</u> until I say so.

I do not see how you got Preece   I supposed he was dead against me[10]   What is your impression about him   Do you really think that he thought Hughes microphone worked on a <u>different</u> principle to the Carbon Telephone & that after a while he found he was mistaken & that my tirade against him was too severe to permit of his acknowledging that he did not at first understand the principle involved. If not what are the facts? Are they as I supposed   write me fully on this! He is certainly a very active man but he should not make such strong statements about new things as he is compelled to eat too much boiled crow afterwards. I wonder how the meal suits his digestion this time.

About competitors referred to in one of your letters the foolishness of forming an alliance with any of them is such an enormously long subject to deal with & I think you know the old old story well enough without my giving any explanation.

If we cannot raise enough money in London to put up one station without the assistance & prestige ([---]^f) of Swan I think we had better sell out & go west. And if we can get up one station we shall not want Mr Swan as the Station will be the most influential thing in England for our purpose & ~~the~~ a^e man who tried to steal my lamp would never be able to compete with gas at eighty five cents per 1000 ft. If you could get the Gas Companies to stop making gas I should be satisfied!!. By the way in connection with this I took a report of the London Gas Companies & while I am not an accountant I worked out what would be the cost of 1000 ft of gas not in the holder but made distributed sold and wiped off the books in the event of competition with the Electric Light depriving them of one half of their total business. I forget the exact figures but I fancy I made it that now their gas costs them fifty four cents 1000 ft[11] & that it would cost 97 cts if half their business was taken away. This of course is very important & would be very useful to Mr Fabbri in his negotiations when this question would come up as people would say that we must make light for sixty cents at least to make anything & compete with the Gas Companies which would be true if all our light were sold without taking any of the Gas Coys customers away   I think there is an analysis published yearly of the reports of the London Gas Companies by a man by name of Field I think although I am not quite sure of the name.[12] You could get the book at the office of the Gas Light Journal or as Insull suggests from Mr Kingsbury. For a small sum of money I think the same man who makes this analysis could give the cost of Gas in London if the output of the Companies is reduced one fifth one quarter one third one half & three quarters. This would be very valuable.

I suppose you will have to fix up a good many of our patents for disclaimer but you will notice that from the time of the Spiral carbon carbon^e filament up to the present time all our applications have their Provisionals exactly like the Finals. And more over the patents are not voluminous but each particular thing is given a patent by itself. That Platinum patent where it speaks of getting the air out of the carbon in a vacuum[13] should be carefully handled for any disclaimer if one is necessary. You will notice that my Platinum Patents have the idea of high resistance & multiple arcing running through them. You will see one of the patents has the system patented as a whole. You might get an opinion as to whether we could hold this at law provided all the claims[a] of the system is new.

About the Faure Battery I do not go very much on this for several reasons. It appears to me that the practicable way to light Rail Road trains would be first to use gas as they does on the Penn. R.R. or in case they must have the electric light a small high speed engine with a 80 half light Dynamo in an iron box on the front of a locomotive & take steam from the boiler. A two horse power engine could be made to run 1500 revolutions a minute with say 120 or 130 lbs boiler pressure

I think the economy set down in their reports is Laboratory Economy & not obtainable in practice say 100 batteries in the hands of 20 or 30 different people[14]

I return the letters as to Faure. We cannot deal with such matters from here. They certainly belong to your office.

I would entertain a proposition to build a Dynamo of any horsepower up to 1000 but I think it would be a mistake for them to have one larger than the one that will be sent from here as a spare. If they want it for charging batteries they could get have a Dynamo on each side of a 300 h.p. engine & so get the desired result.

I would prefer that you give me the plan you think should be adopted as to financing the thing there & I would give you my opinion on it better than I can originate a plan here as I do not know the conditions as you do. Very truly Yours

Thos A Edison   I[nsull]

L (letterpress copy), NjWOE, Lbk. 9:373 (*TAED* LB009373; *TAEM* 81:134). [a]Obscured overwritten text. [b]"to the regulator" interlined above. [c]Not copied. [d]Canceled. [e]Interlined above. [f]Illegible.

1. Edison's letter is Doc. 2187, written in reply to Doc. 2180.
2. Not found.
3. These were probably Armington & Sims engines. That firm had recently announced that another of Edison's orders was delayed by preparations to move their works from Lawrence, Mass., to Providence, R.I., and by their manufacture of special tools to accommodate future large orders. Johnson had complained about the quality of British engines (see Doc. 2180) but arranged about this time to borrow three for the Crystal Palace. Armington & Sims to TAE, 12 Nov. 1881, DF (*TAED* D8129ZCO; *TAEM* 58:314).
4. See Doc. 2149.
5. Edison was referring to lateral end-play of the armature shaft.
6. The *City of Worcester* entered service between New York and New London, Conn., in late September 1881. It was the first steamer on Long Island Sound illuminated by electricity. The ship had 325 B lamps operated by Z dynamos. Edison Co. for Isolated Lighting brochure, 1 Sept.

1882 (p. 6), PPC (*TAED* CA002A; *TAEM* 96:103); Dunbaugh 1992, 264–65.

7. The new steamship *Columbia* of Henry Villard's Oregon Railway and Navigation Co. was lighted by an Edison isolated plant in 1880. See Doc. 1892.

8. These lamps were requested in mid-1880 but not sent until April 1881. See Doc. 2002; Francis Upton to TAE, 4 Apr. 1881; Philip Dyer to TAE, 16 Apr. 1881; both DF (*TAED* D8123ZAP, D8123ZAX; *TAEM* 57:839, 851).

9. Edison executed a U.S. patent application for this device on 27 September and incorporated it into a British provisional specification (1,023 [1882]) filed on 3 March. Edison also described the instrument more completely in Doc. 2201. He sent two to Johnson, with detailed calibration instructions, in mid-January. U.S. Pat. 265,776; Dredge 1882–85, 2:cclxxiii; TAE to Johnson, 16 Jan. 1882, Lbk. 11:85 (*TAED* LB011085; *TAEM* 81:264).

Patent drawing of Edison's voltage regulator mechanism for isolated plants.

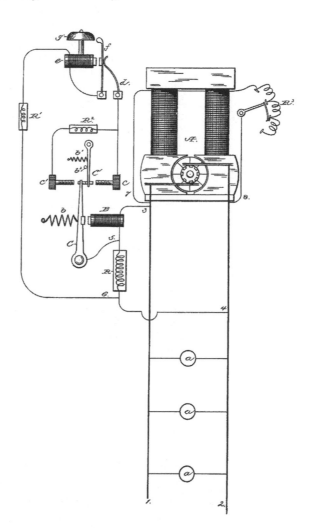

A somewhat later version of the regulating box, shown in Bergmann & Co.'s 1883 catalogue as the "Standard Pressure Indicator." Edison's 1881 regulators did not have the lamps, shown here, to indicate whether voltage in the circuit was too high or low.

10. Johnson had written Edison in late October that he had "sought out Preece—& after a 3 Hours session at Breakfast convinced him that his interests were with us— He was very frank & cordial & said that if on investigation he found the Edison system promised the most in the matter of solving the question He wanted to be identified with it= I told him we wanted him." On 12 November Johnson reported that Preece had gone to Newcastle in search of evidence concerning Swan's exhibition of carbonized paper incandescent lamps. Johnson to TAE, 22 Oct. (p. 19) and 12 Nov. 1881, both DF (*TAED* D8133ZAJ, D8133ZAR; *TAEM* 58:642, 698).

11. Gas manufacture in London reportedly cost about 37 cents per thousand cubic feet in 1882. This did not include revenue from the sale of residual products. C. M. Lungren, "Electric and Gas Illumination," *Popular Science Monthly* 21 (1882): 583; Matthews 1986, table 3.

12. John W. Field annually compiled *Field's Analysis of the Accounts of the Principal Gas Undertakings in England Scotland and Ireland.*

13. Edison's British Patent 2,402 (1879), Batchelor (*TAED* MBP017; *TAEM* 92:118).

14. At the end of December, Johnson reported that the "Faure battery concern is a fizzle; it has played out in France and has played out here. It is now denominated as the 'rotten battery' in as much as the felt rots and the batteries are so short-lived as to make their first cost an impassible barrier to their general use." Johnson to Upton, 30 Dec. 1881, Upton (*TAED* MU058; *TAEM* 95:668).

–2191–

From Francis Upton

Menlo Park, N.J., Nov. 28 1881[a]

Dear Mr. Edison:

I[b] enclose letter from Branner[1] showing that he has not received our last letters.[2]

In Oct. I received a letter from him dated July 17 saying that he should start direct for the States ~~then~~ by Oct 1, some days before the receipt of his letter.[3] I wrote to him at St. Thomas that there was no use of his going to Cuba and also to Para saying that we wanted full samples of all the fibres he could get.[4] The latter letter was short as I did not expect it to reach him.

I think it would be well to recall him by telegraph. Yours Truly

Francis R Upton.

⟨Cable Branner Para   Return. Edison[5c]

write Upton ~~to~~ [can?][d] we[e] have done so & ask him to send cost of cable— Send back Branners letter⟩[6f]

ALS, NjWOE, DF (*TAED* D8123ZFO; *TAEM* 57:1011). Letterhead of Edison Lamp Co. [a]"Menlo Park, N.J.," and "188" preprinted. [b]Obscured overwritten text. [c]Followed by dividing mark. [d]Canceled. [e]Interlined above. [f]Marginalia written by Samuel Insull.

1. John Casper Branner (1850–1922) was a botanist and geologist whom Edison dispatched to South America in late 1880. Branner spent about a year there searching for natural fibers to use as lamp filaments (see Doc. 2012); his letters from Brazil and Argentina are in Edison Electric Lamp Co.—General, (D-81-23), DF (*TAED* D8123; *TAEM* 57:756).

2. Branner was waiting to hear if he should remain in Brazil, go on to the West Indies, or return to the U.S. (Branner to Edison Lamp Co., 29 Oct. 1881, DF [*TAED* D8123ZEY; *TAEM* 57:991]). Upton had also sent to Edison Branner's previous letter, of 19 September, showing that Branner had received Upton's 11 August letter (Branner to Edison Lamp Co., 19 Sept. 1881; Upton to Edison, 19 Oct. 1881; both DF [*TAED* D8123ZDU, D8123ZEJ; *TAEM* 57:962, 980]).

3. Branner to TAE, 17 July 1881, DF (*TAED* D8123ZDB; *TAEM* 57:930).

4. These letters have not been found. Around 27 August, on the last page of a letter from Branner, Edison had written to Upton: "If you want any particular fibre for 'A' lamps you had better write Branner   Do you ever write him in answer to his letters." TAE to Upton, 27 Aug. 1881, on Branner to Upton, 2 July 1882; both DF (*TAED* D8123ZCT, D8123ZCU; *TAEM* 57:920, 923).

5. Edison cabled Branner on 1 December to "Return." LM 1:104D (*TAED* LM001104D; *TAEM* 83:924).

6. Samuel Insull reported to Upton on 1 December that he had sent the cable and was returning Branner's letters. Lbk. 9:392 (*TAED* LB009392; *TAEM* 81:150).

3   2   I

Claim Vertical Coil— Coils arranged to form arch. 2 coils plated together. ~~Vertic~~ Coil with conductor[a] of Carbon running through interior =[1]

A spiral formed of wood or woody or natural[b] fibre with the thickened ends coils & carbonized in a spiral form & secured by plating ~~above~~[c]

~~Sep~~ patent[2]

2   I

mention can all be one ckt

[A][3]

Mention in this case that separation produced by the film of an earthy oxide.[4d]

in connection with indicator of candle power or electromotive force[5]

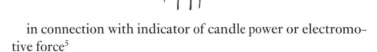

Dynamo Leakage   this   decrease tendency to crossing[6]

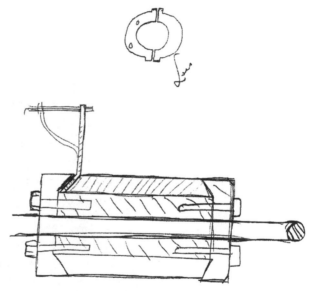

Patent. Silvered or ground[7c]

Patent   Method of making filiments of Carbon dif candle
power & Res—[8]

Witness   S. D Mott[9]

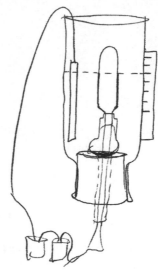

[B][10]

[Witness:] Dyer

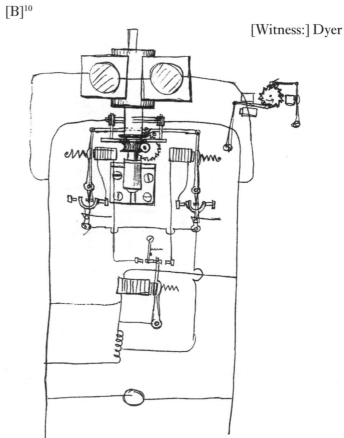

Commutator Z dynamo   Hand solder

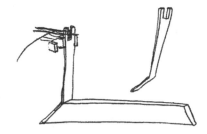

AD, NjWOE, Lab., Cat. 1147 (*TAED* NM016:120; *TAEM* 44:344). Document multiply dated; miscellaneous doodles not reproduced. [a]Obscured overwritten text. [b]"or natural" interlined above. [c]Followed by dividing mark. [d]Drawing followed by dividing mark.

1. On 25 November, Francis Upton wrote from the lamp factory that he wanted Edison "to come here to make a series of experiments on spirals. I have been thinking about the matter and am convinced that 13 or 14 per H.P. of 16 candles is possible   I wish you would make out a list of material you will need for experimenting and I will try to have it ready for you" (Upton to TAE, DF [*TAED* D8123ZFL; *TAEM* 57:1009]). Edison subsequently spent about two weeks at Menlo Park experimenting with high-resistance lamps (see Docs. 2197 and 2202). In designing such a lamp Edison drew on his 1878 "Elect[ric] Light Law" based on his observation that coiling a wire to reduce its radiating surface by a certain proportion would raise its temperature by the same ratio or proportionately reduce the amount of energy needed to maintain the original temperature (see Doc. 1577).

2. On 5 December Edison signed a patent application (Case 379) for spiral carbon filaments that would have high resistance and relatively small radiating surface. The coils in the sketch below and those marked "2" and "1" were represented in two patent drawings. Coils could be comprised of multiple carbons, either wound concentrically as in figure 2, or joined end-to-end as in figure 1. Edison's application was rejected twice and abandoned but subsequently reinstated; the patent issued in 1888. Pat. App. 379,770; Patent Application Casebook E-2537:102; Patent Application Drawings (Case Nos. 179–699); PS (*TAED* PT021102, PT023:50; *TAEM* 45:743, 868).

3. The following figure represents a high resistance filament constructed by electroplating a number of straight filament segments end-to-end to form a zigzag shape. It was the basis for a patent application that Edison executed on 13 December. U.S. Pat. 358,600.

4. Edison executed a patent application on 15 December for two filaments placed close together as shown. The ends of one filament were electroplated to those of the other. Edison used a "filling of an earthy oxide, or other suitable insulating material" to hold them apart. Near the end of the evacuation process, the filaments "are heated by an electric current to an incandescence higher than that at which they are intended to be used. This heat decomposes the material between the carbons, which is removed or partially removed from the globe with the air." Edison again did not specify the purpose of this design, which presumably

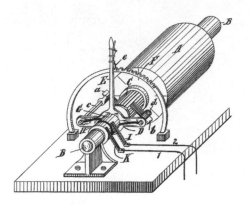

Portion of illustration from Edison's dynamo regulation patent showing mechanism for manually adjusting position of the commutator brushes.

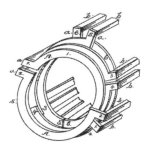

Figure from Edison's U.S. Patent 268,205 showing series of semi-circular conductors with lugs (a) to make connections between paired armature conductor bars (b).

would provide greater illumination than an ordinary lamp. U.S. Pat. 264,652.

5. This unclear sketch may depict a mechanism for manually adjusting the position of commutator brushes to regulate voltage in response to changing electrical load. Edison executed a patent application on 15 December for such means that "will not require in their operation and adjustment the attention of a person skilled in the working of electrical apparatus. This I accomplish by constructing and arranging the commutator-brushes so that they can be readily shifted to and can be conveniently and securely held in any desired position between the point of greatest generation and that of no generation, and by providing at the machine an indicator or alarm which will show in a simple manner the condition of the current" (such as that described in Doc. 2190). In this arrangement "the commutator-brushes are mounted upon arms projecting from a pivoted yoke which surrounds the armature-shaft and turns upon the axis of rotation" to positions of greater or lesser potential. U.S. Pat. 278,419.

6. This figure appears to represent semicircular conductors connecting paired armature bars. Edison executed a patent application on 1 May 1882 for an armature with these conductors instead of the solid end plates, although the stated rationale referred only to simplifying construction rather than preventing short circuits. U.S. Pat. 268,205.

7. This sketch may be related to the reusable lamps that Edison designed in May (see Doc. 2102); however, there are no related patent applications from around this time.

8. Edison executed a patent application on 13 December for a method of manufacturing lamp filaments of various specific candlepower and electrical characteristics from filament blanks of nearly the same dimensions and resistance. The device shown, similar to the patent drawing, was for "electroplating the carbon filaments for a portion of their length, preferably of copper, the non-plated portion alone becoming incandescent. Carbon filaments of a loop, arch, or horseshoe shape are plated preferably from their ends, . . . the center of the incandescing portion of the carbons being at the central point of their length and the candlepower and resistance of the carbons being dependent upon the extent of their non-plated portion." U.S. Pat. 264,653.

9. Mott signed below this drawing, but it is not clear if this should be construed as witnessing only this drawing or others as well.

10. This drawing was substantially incorporated into a patent application that Edison signed on 17 January for automating the dynamo regulation process. Like the December application discussed in note 5, this patent described a mechanism "for automatically shifting the position (relative to the neutral line) of the commutator-brushes upon the commutator-cylinder" so as to control the voltage applied to the outside circuit. However, this mechanism was to be "operated entirely and continuously by the variations of resistance in the external circuit" and to "have a continuous operation—that is to say, one not dependent upon and limited by the play of an armature-lever, but accomplished by a continuous revolution" of a worm gear and two ratchet wheels. Each wheel was acted upon by a pawl in response to an electromagnet in a shunt circuit when the current became too strong or weak. U.S. Pat. 265,779.

–2193–

Technical Note:
Incandescent Lamp

[New York,] Nov 28 '81

[A]

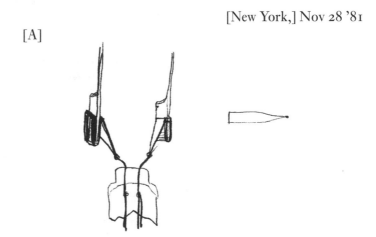

patent.[1]

~~One side[a] of the clamps shall be heavier than the other and to prevent melting by the electrical carrying~~

[B]   [C]

patent[2]

Witness   S D. Mott

X, NjWOE, Lab., Unbound Notes and Drawings (1881) (*TAED* NS81:18; *TAEM* 44:1027). Document multiply signed and dated. ªRaad"One" overwritten on "side" within canceled text.

1. There is no evidence that Edison filed a patent application related to this sketch.

2. Edison executed a patent application covering these lamps on 5 December. It was not filed until August 1882. It was promptly rejected by the Patent Office but eventually reinstated; the patent issued in 1891. The patent contained one drawing, the same as figure C; another drawing, like figure B above, was dropped. In the patent, Edison explained that between the two conventional leading-in wires "is sealed a third, which extends up between the limbs of the carbon, and is attached at the center of the arch or loop by electroplating thereto. When all these wires are properly connected to the system, the current passes through the central wire and is divided between the two limbs of the carbon, the latter thus being in multiple arc. If the central wire is disconnected and the two limbs attached to opposite wires of the system, the two are of course in series, or if one limb of the carbon is disconnected the other may be used separately." Patent Application Casebook E-2537:100; Patent Application Drawings (Case Nos. 179–699); PS (*TAED* PT021100, PT023:49; *TAEM* 45:742, 867); U.S. Pat. 454,558.

–2194–

George Van Ness[1]
Journal Orders

[New York, November 1881[2]]

Mr T. A. Edison the papers & Magazines[a] that I send to you are[3]

Weeklys[b]

    Scientific American
        "   Supplement
    American Machinist
    ~~Progress~~

Monthly[b]

    ~~Science~~
    Druggist Circular
    American Journal of Science
    ~~Builder & Woodworker~~
    ~~Philadelphia Photographer~~
    ~~Manufacturer & Builder~~
    Scientific News
    Popular Science Monthly
    ~~Self Instructor~~
    Journal of Chemistry

For Mrs Edison
Weeklys[b]

~~New York Weekly~~
Chimney Corner
Harpers Young People
American Queen
~~Fiction~~
Frank Leslie's Ill's
~~Pictorial War Record~~
Puck
Harpers Weekly
   "   Bazar
Waverly Magazine
~~Art Interchange~~

Monthlys[b]

Art Ameutuer
Demorest Magazine
~~Druggist Circular~~
~~Leslie's Ladys Magazine~~
~~Munroes Fashion Bazar~~
Nursery Monthly
Popular   "
~~Stoddards Musical~~
~~Vicks Monthly~~
Young Ladies Journal
~~Revue De La Mode~~[c]

Comes Twice a year[d]

What to Wear
Portfolio

Newark Sunday Call
   "   Daily Advertiser

~~Mrs Van Cleve Metuchen N.J.~~[4]
Monthly[b]

~~Art Ameutuer~~
~~St Nicholas~~

AD, NjWOE, DF (*TAED* D8111N; *TAEM* 57:473). [a]"& Magazines" interlined above. [b]Written outside of right brace enclosing list that follows. [c]Followed by "over" to indicate page turn. [d]"Comes Twice a year" written outside of right brace enclosing list that follows.

1. George Van Ness was a Newark dealer who had been selling magazines to Edison since at least 1878. *TAEMG1–2*, s.v. "Van Ness, George."

2. Date supplied from docket on reverse.

3. Because many of the titles in this document are either well known or self-evident, the editors have chosen to present the subscription list without individual descriptions. Information on many of these titles may be found in Mott 1938–68, vols. 2–3. Edison replied to Van Ness on 1 December and listed the magazines he wanted to discontinue. That list matches the magazines crossed out in this document. Edison also requested Van Ness to "supply Mrs Edison's papers more regularly she complains very much of your irregularity and uncertainty." On 3 December Edison directed Van Ness to send the magazines to his and Mary's residence at 72 Fifth Ave., and correspondence and bills to his office at 65 Fifth Ave. TAE to Van Ness, 1 Dec. 1881, Lbk. 9:388 (*TAED* LB009388; *TAEM* 81:148); TAE marginalia on Van Ness to TAE, 3 Dec. 1881, DF (*TAED* D8111P; *TAEM* 57:477).

4. Hattie Van Cleve, wife of Edison's employee Cornelius Van Cleve, was Mary Stilwell Edison's half-sister. Headnote, Doc. 733.

–2195–

Samuel Insull to Henry Villard

[New York,] 2nd Dec [188]1

Dear Sir,

After my interview with you yesterday I went to Menlo Park to see Mr Edison and upon my relating to him the substance of our conversation he stated that he should suspend work upon his Electric Rail Road until such time as he is in a position to pay for it himself[1]   Yours truly

Saml Insull   Private Secy

ALS (letterpress copy), NjWOE, Lbk. 9:392 (*TAED* LB009392A; *TAEM* 81:150).

1. Nothing more is known of the conversation with Insull; at this time Villard was stretching his personal and corporate financial resources to push rapid construction of the Northern Pacific Railway. On 22 November Edison had submitted to Villard an itemized bill totaling $6,186.50 for work on the electric railroad, in accord with their September contract (Doc. 2152). De Borchgrave and Cullen 2001, 320–24; Villard 1904, 2:299–302; TAE to Villard, 22 Nov. 1881, Lbk. 9:314 (*TAED* LB009314A; *TAEM* 81:107).

In February 1882, about the time that substantive electric railroad experiments resumed, Edison estimated the cost of constructing a line of unspecified length at about $25,000. In March, the Oregon & Trans-Continental Co. agreed that Villard's 1881 loan of $12,000 would be considered an outright payment should Villard or his representative consider Edison's experiments a success by 1 July 1882. At that time, however, Edison repaid the loan with interest (Cost Estimate, 25 Feb. 1882, Cat. 2174, Scraps. [*TAED* SB012ACE; *TAEM* 89:380]; TAE agreement with Oregon & Trans-Continental Co., 2 Mar. 1882; Oregon & Trans-Continental Co. to TAE, 1 July 1882; both DF [*TAED* D8249D,

D8249D1; *TAEM* 63:589, 591]). In October 1882, the Edison Electric Light Co. reported to stockholders that

> Villard has withdrawn from his arrangement with Mr. Edison and the latter has gone on with the installation at his own expense. The road was built and cars have been run over it experimentally since last Spring. The experiments, however, are not yet entirely completed to Mr. Edison's satisfaction, but as soon as he can find time from the pressure of more imperative matters, namely, those connected with the electric lighting branch of our business, he expects to go on with these experiments and perfect the railway for practical use. Until that is done no steps will probably be taken towards the formation of a company for exploiting the Edison Electric Railway and the whole subject remains substantially where it was one year ago. [Edison Electric Light Co., Board of Directors Report, 24 Oct. 1882, DF (*TAED* D8224ZBJ; *TAEM* 61:86)]

By December 1883 Edison estimated that he had spent $38,541.80 in developing his electric railway, and considered that he had a "valid claim" to recover this amount from the Edison Electric Light Co. (TAE to Eaton, 13 Dec. 1883, DF [*TAED* D8316BSG; *TAEM* 65:809]). Edison's railway accounts with Villard are in Ledger #5, Accts.; see entries indexed under "Villard, H." (*TAED* AB003:11; *TAEM* 87:384).

–2196–

From Charles Batchelor

Paris, le 14 Decembre 1888 1[a]

My dear Edison—

I have succeeded in putting all our men (that is Hennis[1], Suebel, Force, Acheson, Hipple,[2] on the payroll of the different companies from the 15th of November. In regard to Force during the time he is with me I wish you would continue to pay his wife $12 per week as you have all along done and I shall refund you the same every three months from his salary.— Also dont fail to let me know how much you have paid Mrs Suebel so that I can refund you that—[3]

No doubt you think we are a long time getting started in manufacturing but it has been very difficult indeed to find the right place— I have been in no less than 53 mills and factories within 100 miles of Paris and have often thought of splitting it up into 3 or 4 shops, but now I think I have a prospect of getting the building shown in the enclosed drawing.[4] Of course we could not buy this from our capital but I have been working on Leon and Lebey[5] two of our directors to buy it and rent it to us which I believe they will do— It is built of stone and brick and is just as good looking as it is in picture— It is outside the walls but at such a short distance that it is practically the same thing as being inside— I dont like to say now that I

shall have it but I am making strenuous efforts and the prospect looks good— We have had 2 meetings of our intended
boards and they propose to meet every night to transact business every night at 4 p.m. so the ball has fairly opened—
Porges the head and front of the concern is a very peculiar
man as probably Mr Lowery has told you, I find however that I
can get anything I want by simply making him believe that he is
the originator of the desire to have the thing I want— On the
boards when anything was proposed (no matter what) he starts
off with a loud blustering objection on general principles and
succeeds in satisfying himself in about 5 minutes that the man
who proposed it is a damn fool   afterwards when the pro-
poser insists he takes up the proposition himself and does all
you want as if it was his own   I have an amicable row over
everything ending in my getting all I want— I have done my
best to procure for Dr Moses a position in the formation of
companies for ~~g~~Germany which would be very lucrative; and
he is now under consideration by the boards— I am afraid that
Porges who was very favorable to him, has been literally talked
to death by him, and feels that he dont want to bother with it
till everything else is done towards factory etc— Mr Bailey
who is recovering[6] asked me also to tell Porges that it would
be exceedingly disagreeable to him to have to work with Moses
~~so that~~   A thing I of course would not do, but it makes it hard
for me to work for him— He wants an exclusive agency for the
whole of the German Empire with 2% interest in everything,
(whether he works or not) and all expenses and $200 per month[b]
besides— Of course when this proposition was passed round
we all with one accord put our pens through the word exclusive—[7] Lots of my lamp factory stuff I am making outside so
as to be ready when I have my factory —

Tell Kruesi to get a Frenchman and learn him all the business and I will take him   Tell him I dont want a high salary
man   Yours

Batchelor

ALS, NjWOE, DF (*TAED* D8132ZBV; *TAEM* 58:570); letterpress
copy in Cat. 1239, Batchelor (*TAED* MBLB4042; *TAEM* 93:510). Letterhead of Chas. Batchelor. a"Paris, le" and "188" preprinted. b"per
month" interlined above.

1. Charles Hennis had significant responsibility for setting up the
Edison central station in Milan in 1883. Batchelor to Hennis, 16 Jan.
1883, Cat. 1239:433; Batchelor to Hennis, 31 July 1883, Cat. 1331:116;
both Batchelor (*TAED* MBLB4433, MBLB3116; *TAEM* 93:773, 402).

2. James Hipple (c. 1855–1917) began working in 1880 as a glassblowing assistant at the Menlo Park laboratory; later he assisted William

MANUFACTURE D'ORGUES D'ALEXANDRE PÈRE & FILS — LOT DE 20000 MÈTRES DE TERRAIN
à Ivry-sur-Seine près Paris

Charles Batchelor's choice of a site near Paris for manufacturing electric light and power equipment.

Holzer in setting up the glass blowing department at the lamp factory. He went to Paris in 1881. Doc. 1926 n. 2; *TAEB* 5, App. 2; Jehl 1937–41, 898.

3. Before going to Paris, Philip Seubel arranged for Edison to send thirty dollars every month to his wife in Canton, Ohio, but Insull reportedly failed to do this before October. In February 1882, Batchelor asked Edison to advance Seubel's wife and two children first class passage to France. Seubel to TAE, 21 June 1881; Batchelor to TAE, 5 Oct. 1881 and 14 Feb. 1882; all DF (*TAED* D8135O, D8135ZCF, D8238ZAA; *TAEM* 58:900, 1087; 62:322).

4. Batchelor enclosed a printed illustration of the Alexandre Père & Fils organ manufactory situated on a 20,000 square meter lot at Ivry-sur-Seine, southeast of Paris. Purchase of the property was completed in late January 1882. *WGD*, s.v. "Ivry-Sur-Seine"; Batchelor to TAE, 23 Jan. 1882, LM 1:150C (*TAED* LM001150C; *TAEM* 83:947).

5. Georges Lebey, a member of the Paris Bourse, was a founding director of the Société Électrique Edison and the Compagnie Continentale Edison. Articles of Incorporation, both 2 Feb. 1882, both DF (*TAED* D8238Q, D8238R; *TAEM* 62:227, 252); *BDF*, s.v. "Lebey, André."

6. Joshua Bailey had been incapacitated by pleurisy since early November. Bailey to TAE, 5 Dec. 1881, LM 1:109B (*TAED* LM001109B; *TAEM* 83:926).

7. Moses complained in a letter to Edison this day that he had been invited to help form the European companies but, having obtained Edison's permission to remain, found that Porges refused the terms he proposed. He reported that Batchelor had then promised to help him obtain a position but, he claimed, had failed to speak on his behalf at the decisive meeting. Moses then applied to Edison, who did not reply directly but offered him, though Batchelor, $500 to "meet immediate wants." Batchelor denied having acted in bad faith, telling Insull later that "everybody that he [Moses] had anything to do with considered him such an infernal bore that he actually cut the position out of his own

hands, or as one or two expressed themselves to me, 'they did not wish to be worried to death' with him!" Moses to TAE, 14 and 21 Dec. 1881, both DF (*TAED* D8132ZBU, D8132ZBY; *TAEM* 58:567, 576); Batchelor to Moses, 31 Dec. 1881; Batchelor to Insull, 7 May 1882; Cat. 1239, Batchelor (*TAED* MBLB4073, MBLB4278; *TAEM* 93:531, 664).

–2197–

To Edward Johnson

New York Decr. 20. 1881

Dear Sir

I beg to confirm ~~dispatch~~ receipt[a] of cable as follows.[1]

"Vital have accurate general basis show cost not exceeding gas   cable proper method with figures mailing complete showing main reliance commercial position   am in accord Fabri."[2]

I give below very rough estimate of cost of Central Station for your personal use[3]

Central Station of one square mile 33 000 ten candle jets.

Investment

Dynamos @ 7,500—200 h.p each forced to 250	$90,000
Boilers ($12 per h.p.) say	28,000
Stacks	10,000
Piping foundations boilers, blowers Ex[haust].	
Eng[ine]. coal	8,000
Iron structure erected	8,000
Foundations	3,000
Fire proof floors	3,000
Station regulation apparatus	3,000
Meters	10,000
Erecting appliances fixing station	10,000
Mains and feeders	200,000
Total Invest plant	$373,000
Running Expenses one year	
Labor	12,000
Bad debts	1,000
Coal	16,000
Oil waste	2,500
Rent	12,000
Executive	6,000
Lamps	19,800
Depreciation	8,000
Meter Men	4,000
Taxes	3,000
Water	2,000
Total	$86,300

33,000 Lamps 3 hours daily 300 days

5 feet per hour is 148 500 000 feet at 85¢ per M. = $126,220
Gross receipts

Expenses   86,300 = Net profit $39,920
or over 10 per cent on investment.

Sold at $1= per M. would equal $148,500 giving profit of $62,200 or over 16 per cent.

This will give you a rough idea of cost and profit but it is impossible to estimate exactly when my knowledge of conditions on the other side are so slight. Yours truly

Thos A Edison   I[nsull]

P.S. Following please find copy of cable sent you in reply to one quoted above.[4]

Fifty-seven London[5] (Sent Decr 15. 81   Can only give very rough estimate at present   Investment in good portions London for every thousand feet equivalent sold be from two dollars seventy five cents to two dollars ninety cents and will pay little more than eleven percent on whole of actual investment taking a ten candle electric to equal ordinary five foot London burner. Have been working at Menlo two weeks getting high resistance lamps to admit competition London Gas   have got ten candle lamps two hundred eighty ohms, will mail detailed estimates"[6]   Thos A Edison   I[nsull]

L, NjWOE, Lbk. 9:459 (*TAED* LB009459; *TAEM* 81:169); a copy of this letter is in DF (*TAED* D8133ZBI; *TAEM*58:783). Written by Charles Mott; signed for Edison by Samuel Insull. [a]Interlined above.

1. This is the full text of Johnson's cable to Edison, received on 8 December. On 14 December, Edison replied that he was "busy collecting data   estimate comparative cost ready in few days." LM 1:112C, 115B (*TAED* LM001112C, LM001115B; *TAEM* 83:928–29).

2. This refers to a stipulation (clause 17) in the draft contract negotiated by Egisto Fabbri with financial backers of the prospective Edison company in Britain. Before the company was organized, William Thomson and Frederick Bramwell were to certify that the Edison central station system could compete economically with gas. Theodore Waterhouse to Johnson, 11 Jan. 1882; Edward Bouverie to Johnson, 25 Jan. 1882; both DF (*TAED* D8239G, D8239P; *TAEM* 62:670, 738); see also Doc. 2203.

3. Cf. Docs. 1897, 1958, 1991, and 2008; see also Doc. 2202.

4. Following is the full text of Edison's cable to Johnson on 15 December. LM 1:117C (*TAED* LM001117C; *TAEM* 83:930).

5. Cable code for Edward Johnson; see App. 4.

6. A 5 January itemized estimate typed on Edison Electric Light Co. letterhead for a "Central Station of one square mile—containing 33,000 ten candle jets," operated by twelve 200 horsepower dynamos, was presumably intended for London. The figures in that estimate are comparable to those given above by Edison; significant deviations include esti-

mates of $12,000 apiece for engines, $36,000 for boilers, and only $2,500 for smokestacks. Considerably more was allotted for coal and less for rent. Edison Electric Light Co. cost estimate, DF (*TAED* D8224111; *TAEM* 61:3).

Another set of itemized estimates for various unidentified isolated installations was also typed and dated 5 January. The prices of dynamos and equipment in that list is very similar to those in a promotional booklet published by the Edison Co. for Isolated Lighting in September 1882. Edison Electric Light Co. cost estimate, DF (*TAED* D8221A; *TAEM* 60:845); 1 Sept. 1882 circular, PPC (*TAED* CA002A; *TAEM* 96:103); see also App. 3.

–2198–

Samuel Insull to Naomi Chipman

[New York,] 21st Dec [188]1

Madam,

I am directed by Mr Edison to state that he will vacate his appartment in your house one week from this date.[1] Yours truly,

Saml Insull   Secretary

ALS (letterpress copy), NjWOE, Lbk. 9:466A (*TAED* LB009466A; *TAEM* 81:172).

1. Edison and his wife had apparently spent much of their time at Menlo Park from late November. It is not clear when they actually left the Chipman boarding house. Cf. Doc. 2245.

–2199–

To W. H. Patton[1]

[New York,] 29th Dec [188]1

Dear Sir,

The trouble as to the transmission of power[a] by electricity is that the demand is so great for machines that our Directors do not feel inclined to send out any machines simply for an experiment at their expense.[2] I talked to one or two of them about[a] the matter and they remark "why does not Mackay[3] try the experiment at his own expense as if successful the result would be more profitable to him than to us   As far as I am personally concerned I should like to try it very much for the mere science of the thing   Yours truly

Thos A Edison   I[nsull]

L (letterpress copy), NjWOE, Lbk. 9:484 (*TAED* LB009484; *TAEM* 81:177). Written by Samuel Insull. [a]Obscured overwritten text.

1. W. H. Patton was a mine superintendent (1878–1887) for the Consolidated Virginia Co., the so-called "Bonanza Firm" of the Nevada Comstock. Edison addressed this letter to him in Virginia City, Nev., at the Ophir Silver Mining Co., which the Consolidated Virginia had ac-

quired in 1878. Patton and Edison had corresponded in 1880 about using electric power generated at the Carson River in the Comstock mines. Smith 1943, 240, 213; Docs. 1949, 1957, and 1967; for a similar proposal see also Doc. 1788.

2. Patton wrote on 5 December, after having heard nothing recently from Edison about "the 'transmission of power by Electricity' business." He advised that the Brush interests were active in the area "and have made propositions in regard to the same business— Will you please advise me what your Co. proposes to do—so that I will know whether to entertain any proposition from them." Patton to TAE, 5 Dec. 1881, DF (*TAED* D8138ZAJ; *TAEM* 59:147).

3. John MacKay (1831–1902) became wealthy as the majority partner in Nevada's most lucrative Comstock mines. By 1881 he was established in New York, where he and publisher James Gordon Bennett formed the Commercial Cable Co. *DAB*, s.v. "Mackay, John William."

–2200–

From Edward Johnson

London, E.C. Dec. 29th 1881[a]

My dear Edison,

I am very sorry to say that I am met on all sides by astonishment at the price I ask for our Dynamos, namely £200. I gave Sir William Thomson a price for 3 Dynamos £180 each. He wanted it for the purpose of supplying a friend of his who had a large Factory and was about introducing Electric Light. Sir William writes to me to say that Siemens' machines have been adopted instead of ours on account of the price, namely £90 each. He does'nt say however, anything about Economic efficiency of the Siemens' machine. This seems to be a very difficult bit of information to obtain. The Paris Electric Exhibition offers no data and as far as I have been able to find none can be had anywhere. I have written Thomson a letter calling his attention to the fact that his rejection of our machines is simply based on their cost and that he gives no information whatever as to whether the difference in the economy of coal consumption will not more than compensate for the difference of cost.[1] I have asked him to accord me an interview to discuss this whole matter with him I agreeing to go up to Glasgow for the purpose. I have informed him that the Dynamo I sent him shall not be returned to me until he has either Endorsed it or condemned it, and have further asked him if he is prepared to officially pass upon the economy of our small Dynamos independent of the larger and more important experiments he is to make with our Central Station Plant later on.[2] I think I can secure his services for this purpose in which case I will have his findings printed in Circular form and will use them for the purpose of obtaining a fair price for our Dynamos. Meantime there is no doubt in my mind but that we have got to very considerably reduce our price. I find that by paying you $570 on board ship in New York that by the time we come to sell the Dynamo it costs us $620. Now add 25 per cent only as a margin of profit; it stands us in $775 equal to £155. Now if any reduction of price can be had anywhere so that I may sell these Dynamos for £150 I think I can successfully compete with Siemens and Gramme by making full use of the facts as to superior economy etc. There is no disputing the fact that a very large proportion of the people who want isolated plants have a surplus of power which they count whether rightly or wrongly as now wasted and they look more particularly to first cost than they do to running expenses. I have taken the ground that I do not care for this class of business but I am afraid that if we ignore it we will find it very difficult to secure much trade in other directions. Please carefully consider this whole matter and see if you cannot somehow reduce the first cost to me. In every other respect we are fully able to hold our own. I have the Steam Dynamo together and can be ready to run in a very few days. I propose to run it as a Steam Engine simply for some time to ease[b] up the bearings, get the knock out of the Steam Engine and generally to make it an efficient machine before applying the current. Meantime I am pushing forward the preparation necessary to receive the other Dynamo. My contract with the City is to commence Lighting on the 1st February. I want to have the second machine in place by that time and several days have now elapsed since you reported it as ready for trial but I have heard nothing from you.[3]

Preece's second Lecture at the Society of Arts for the benefit of the Juvenile members came off last night and as you will see by the notice sent you, was illuminated by the Edison Light in a very satisfactory manner. You will notice in the "Daily News" article a reference to my new chandellier.[4] It is not the Chandelier about which I have been writing to you but is a small one on the same principle. I think you will have a more glowing account of the large one as the effect will be much finer. In the small one the flowers are too large for the other parts of the fixtures thus making it appear out of proportion. Two Lamps gave out last night during the Lecture, one gave way in the carbon and the other arc'd. Am I going to have serious trouble in this respect here and at Crystal Palace? I hope not. I will have some 2,000 Lamps in operation at one time; if they are going to give way rapidly I will need a large force of men to keep replacing them which will necessarily attract considerable attention. Crystal Palace will be ready to open up the later part of next week. No one else is so far advanced. I therefore expect to be the first one to light up which of course will be a card for the Edison Light. Everybody else is at work and the exhibition promises to be a decided success. Batchelor is shipping me all your other Exhibits and I am getting them into position. What about motors? We need them both here and at Crystal Palace. You have not shown a motor on this side yet and it is exceedingly important that you should do so. I have a floor of this building[5] fitted up as a Laboratory and repair shop and ~~think of~~ am[c] arranging to drive the machinery (a few light lathes) by a motor. I also want to show one at Crystal Palace. Please send forward at once if you can do so, if not please tell me that you cannot and why. The price List of Armington & Sims Engines you sent me is rather meagre. I will, however, have some Engineer to interpret it for me. Very truly Yours,

Edwd H. Johnson

If you will let Mr Fabbri read this you will save me double work   EHJ[d]

LS, NjWOE, DF (*TAED* D8133ZBP; *TAEM* 58:807). Letterhead of Edison Electric Light System, Edward Johnson, manager. [a]"London, E.C." and "188" preprinted. [b]Obscured overwritten text. [c]Interlined above by Johnson. [d]Postscript written and signed by Johnson.

1. This correspondence between Johnson and William Thomson has not been found.

2. This refers to the stipulation proposed by London investors that Thomson attest to the economic efficiency of the Edison central station system.

3. Edison cabled on 19 December that the dynamo was "almost completed" and would be tested in two days. LM 1:121 (*TAED* LM001121A; *TAEM* 83:932).

4. The chandelier is probably like that which Johnson included in Edison's Crystal Palace exhibit (see Doc. 2226). The *London Daily News* article of 29 December did not mention Edison but gave a flattering brief description of the fixture. Preece's lecture dealt with the properties of electricity and its applications in telegraphy and telephony. According to the press reports, this was his first of two talks. "The Wonders of Electricity," *London Daily News* [29 Dec. 1881]; "Society of Arts Juvenile Lectures," *Times* (London), 29 Dec. 1881; "Wonders of Electricity," *London Chronicle,* [29 Dec. 1881]; all Cat. 1243, items 1908–10, Batchelor (*TAED* MBSB41908, MBSB41909, MBSB41910; *TAEM* 95:91–92).

5. Johnson wrote from 57 Holborn Viaduct. The building also had some office space for the British company and sleeping quarters for some of the engineers. William Hammer to William Alexander Hammer [father], 13 Nov. 1881, Ser. 1, Box 1, WJH.

[New York,] 31st Dec [188]1

My Dear Batchelor

In putting out a great number of plants as we are now doing we have found it necessary to have a Regulator for the candle power of our lamps as the parties using the light are apt to run the lamps up very high & thus cause a great many breakages. Thus the average life would be shortened great dissatisfaction caused a[nd] people get the impression that our statement as to life were not true. So I have devised an Indicator which works beautifully and I advise that hereafter all Isolated Plants shall be accompanied by one of them.[1] I will send you one with your model "Z" Dynamo provided our patents have been secured in France and other countries. It consists of an apparatus which is placed across the line just as if it were an ordinary lamp. It has a magnet with cores about eight inches long and an inch thick & is wound with wire so as to ~~make~~ give[a] the magnet a resistance of seventy ohms   In the same circuit is seventy ohms resistance, making a total resistance equal to the lamp. This magnet is provided with an armature (about four inches long and an inch thick) of iron fastened to a large sliding armature lever arranged to work the same as a relay. A large brass spiral spring made of $1/8$ in with the spiral being about $3/4$ in diameter & 4 in long is used to adjust it and is arranged so that with the proper adjustment is obtained it cannot alter. The whole thing is enclosed in a box. The big magnet acting as a relay closes the circuit in which there is a vibrating burglar alarm and which makes a great noise the circuit being a derived one closed round one of the resistances of the device enough current is obtained to ring the bell violently.

The machine is set in the following manner: Suppose that we sent out a Dynamo machine and lamps which give sixteen candles at one hundred volts. Then we take one of these boxes and put it on a circuit which is ~~adjusted~~ just 100 volts & so adjust the armature of the Regulator that by increasing the volts by one the [raise?][b] of current will attract the armature close the derived circuit and the bell will commence to ring. One volt only makes a difference of half a candle but the bell always works on this variation. Now you see if this the Regulator is put in the Lamp circuit & the machine is first started & the bell rings the[c] attendant at once knows the candle power is too high and he adjusts the field magnets ~~restance~~ resistance just enough to stop the bell ringing & then he knows his lamps are not more than sixteen candles & thereafter it is impossible for any of the lamps to go above $16^1/2$ candles without the bell ringing we find this apparatus is very reliable & brings out very conspicuously the bad governing of our small engines. when the B lamps are used the extra 70 ohms resistance is thrown out leaving only the magnet in which is the same resistance as the B lamps. So all you have to do is that when you send out a plant you must adjust the Indicating machine at your Works and send it out with it. It is so reliable that we shall use it in the Central station in the place of an Electric Dynameter for guaging the volts.[2] Bergmann is making them at $25.00 but we hope to get them much cheaper than that.

We received your cable about the model Z Dynamo.[3] This will be sent to you very shortly via Hamburg   You must not take it into France until you have got the proper authorization from the authorities as it is necessary to take very great care that nothing is done to vitiate the patents   The Paris Opera House Dynamos were ordered to be shipped to Puskas & Bailey. If I had done so my patents would have been endangered. I sent them to Geo Walker as Exhibition material ~~as Exhibition material~~. In ordering stuff from here you must be very careful about this authorization & even when you have the authority I shall always ship to Hamburg & let you get the forwarding done at your end. I am told that anything imported without the proper authorization vitiates the patent and can be taken advantage of by our competitors and that the authorities cannot make a retrospective authorization so as to make the patent good again. They cannot legally do so according to French law.

We are building a model Disc Dynamo for B lamps the same as you have at Paris and which was built at Menlo Park. We can send you the working drawings of the armature after we have thoroughly tested it if you desire.

Regarding the large Dynamo we are building six for the Illuminating Coy and could start another one as a model for you but would like to have the Coy over there send a formal order. However I think the drawings will be sufficient for you as we have got them absolutely correct so I think you can do just as well with them with explanations from us just as well as with a model Dynamo.

I spend a great portion of my time now at the Lamp Factory Upton had got badly mixed regarding the life of the lamps but now we make a curve of each days run so we know exactly what we are about and those curves show that the ten per Horsepower have an average life exceeding eight hundred hours and I am gradually increasing this. I am also working on spirals with great success.[4] I shall undoubtedly succeed in making

280 ohm A lamps 12 per horse power 600 hour life but do not promise this. This is of course sub rosa. I shall also be able in the course of six or seven weeks to materially reduce the cost of the lamps so that we can make some profit on them.

We sold 2000 resistance lamps[5] to the Western Union to be used as Resistance ion their telegraph lines at seventy five cent each.

We have had the steam on the big English machine (for spare)   We had 1300 lamps on it for several hours   We have some trouble with the brushes but I shall get them all right in a day or so & then shall ship the machine to Johnson

We are progressing very well with the Rail Road. The Locomotive is being built at Goerck St & will be finished in a week or so. Hughes is laying the track   Yours very truly

Thos A Edison   I[nsull]

L (letterpress copy), NjWOE, Lbk. 9:489 (*TAED* LB009489; *TAEM* 81:178). Written by Samuel Insull. [a]Interlined above. [b]Illegible. [c]Obscured overwritten text.

1. This is the device Edison described in Doc. 2190.

2. Cf. headnote, Doc. 2243.

3. In mid-December, Batchelor requested Edison to send a Z dynamo without wire on the field coils or armature, to be used as a model for manufacturing. Edison asked in reply if he could instead send Charles Clarke's working drawings, from which there had been "no difficulty" constructing machines; he also promised to send a model. Batchelor told him to send the drawings as well. Batchelor to TAE, 16 and 19 Dec. 1881; TAE to Batchelor, 17 Dec. 1881; LM 1:118B, 120B, 119A (*TAED* LM001118B, LM001120B, LM001119A; *TAEM* 83:931–32).

4. John Howell's 20 December note about two batches of spirals carbonized at high heat is the first lamp factory record of spiral tests. Cat. 1301, Batchelor (*TAED* MBN007:67; *TAEM* 91:360).

5. See Doc. 2085 n. 9.

–2202–

To Edward Johnson

[New York,] 31st Dec [188]1

My Dear Johnson,

Regarding your cable about 280 ohm lamps since I said what where the chances as to Fabbri making an arrangement in England I knew the economy business would come to the front and therefore I immediately went to Menlo Park and devoted myself to the production of a high resistance lamp.[1] I have made two or three of them of 280 ohms which work fairly but I think the life is not more than two hundred hours. I shall probably be able in the course of three or four weeks (the holidays having delayed me: we closed down for two weeks) to get the life up to 600 hours and 11 or 12 p. H.P. Those I have made were 16 candle lamps. I am going to try and make them 10 candles and still have 280 ohms resistance as I believe the ten candle jet giving about 16 effective candles will be the equivalent of any five foot burners in the City of London because I find that in practice in the hands of the public 5 ft of gas which should give sixteen candles does not give eight candles although with a new burner and in the hands of an Expert sixteen candles can be obtained.[2] I therefore hope and believe that I shall be able to get 19 lights of ten candles each 280 ohms resistance per electrical horsepower which means we can sell light at 85 cents and make a profit of ten to fifteen per cent on the money actually invested if I am not mistaken as to prices of material over there. But perhaps we shall not be required to sell it so low as can sell gas of 20 candlepower which gives about 12 effective is sold for £1.04   Of course you cannot expect to make a wonderfully good showing on the First Station as you will have the disadvantage of having to work up green men & there will be considerable initial cost and [------][a] expenses which being chargeable to one station instead of to a [great?][a] number which will bring the investment up but I [base?][a] my [estimate?][a] on several stations. But when it comes to fifteen or twenty stations then we shall do very much better. If we had fifteen stations the London Gas Companies could not sell gas for 85 cents but would be compelled to sell it for about $1.30 to $1.70 on account of the lessened output as I have previously explained[3]

The London Gas Companies, the public understand, pay 10% but if you look at their statement you will find that most of their Capital is in Debenture Bonds which only pay 5%. If we were allowed to bond a portion of our investment at 5% and represent the other portion by stock then the 10% or [15?]%[a] would be greatly increased.

We started some time back on making your 100 candlepower lamps. I suppose you intend to use them on the street but I should [try?][a] the effect of a mass of seven lamps in a pyramid as well. You do not say anything about meters but I have ordered fifty & Francis Jehl will go over to you with them as soon as they are ready & teach a man for you to deal with them. This meter business is very [sticking?][a]   we should know what size meters you want   we only make meters which will answer for one light up to twenty. We intend, & if it is absolutely essential for you we can make one from twenty to fifty. [When?][a] a person has 100 lights two fifty light meters can be used.[4]

We have just got a Safety Catch to be placed on the Dynamo in the Main Circuit. Perhaps you would take some.

You have not told us how you like the new socket and the tip on the lamp. We have no more trouble from this source now.[5]

It would be well for you to have in your Holborn Station a two light gas meter and to have Hammer fix it up so that you can screw different burners and then get their candle power and at the same time ascertain their cubic feet. You may learn considerable by this and it will aid you in your explanations. You should get a work called, I think, "Common Sense for Gas Users."[6] It is about $1/4$ in thick, 6 in long & $4^1/2$ in broad & has a yellow cover. It gives some very valuable "information" and is exceedingly plain. It illustrates the size of gas jets, the candle power and the amount of gas used and tells something against the gas companies although written by a gas man. It absolutely asserts that in the City of London the average amount of candle power obtained for 5 ft of sixteen candle gas does not exceed <u>seven</u> candles and I can readily believe this as you will remember our test at Bergmanns on Metropolitan Gas (21 candle power)[7]   The highest was eleven candles burning sixt feet of gas and the worst case was were one burner was using seven feet and only gave $4^1/2$ candles. This is explained by the fact that if you have a jet of with a certain opening and a certain pressure of gas and both remain relative to each other & five feet of gas pass increasing the pressure so as to pass eight feet will cause your current to act upon the flame like a Bunsen burner[8] & reduce the candle power down as low as five or six candles although the gas consumed is increased from five to eight feet. If the slit gets increased as it does after [---][a] weeks burning, and gradually gets worse, the pressure remaining constant the amount of gas will be reduced from five feet to perhaps $4^1/2$ ft but it will blow so [that][b] with[c] this Bunsen burner action the candle power will go down to 4 or 5 candles. But if it be an Iron jet the whole hole will be increased from oxidation and the pressure remaining constant it will consume more gas and the candle power will be reduced by lack of oxygen as it was decreased by increased oxygen. Now the pressure constantly varies slightly in the mains but greatly within the House. This great fall of pressure and change of pressure in the House is due to the [-----][a] permission of the Gas Companies for to the public to get their own gas fittings done by anybody. The consequence is that competition has caused the introduction of pipes incapable of carrying the gas throughout the building without great loss of pressure   Hence the jets will vary in candle power throughout the building.

Another thing the larger the flame the greater the economy as the heated body is more compact and the total surface of the jet[d] exposed to the air is small but when the jet is streaky the total surface of the jet exposed to the air is increased ~~but~~ and this not only lowers the diameter of the flame but increases the velocity of the flow and both tend to reduce the candle power. In fact the distribution of gas is in practice a very difficult matter and the talk of the gas people about candle power is all nonsense. Were it possible to make burners that would never deteorate & that the pressure in every part of a City and House was constant this Gas Distribution would be perfect but a new burner will deteorate from the very moment it is put in and in three weeks it will not give near the result first obtained. A burner taken from one house in which say ten candles were obtained from five feet of gas would if taken to another house give entirely different results as the pressure might be different in the other house and as the absolute definite relation between the size of the orifice and the pressure to obtain the best result. You can see that these conditions are utterly unattainable in practice.

It perhaps might be a good idea to keep your eye out on some square mile in London in which there is a slum near the center in which we could obtain a building cheap and if you decide on a District it might be well to hire a canvasser to obtain statistics. We can furnish you the form of books from which you can get up what is required.

Putting in this Central Station will be very much easier than I at first imagined. The men are easily broken in at laying tubes and the more I keep at it the more I am convinced that it is much easier to put in a Central Station & opperate a square mile than to put in 15 small Dynamos boilers &c & operate a mill.

I suppose in going over from one machine to another at Holborn Viaduct you will stop one dead and then start the other and have no special appliances to take the current off one & put it on the other. This would only occasion a momentary interuption but as it will occur so seldom I do not suppose you will want anything special.[9]

When you first start up it would be a good idea to have an ice box and keep ice on hand until your bearings get worn down

I wish you would have sent me a tabulated statement of prices of following:

Steam coal, Gas coal, Pea coal, Buckwheat coal & Coal Dust.
(Delivered in London & at the pit mouth & the cheapest
method of transportation)
Real Estate in London (this should be voluminous)
Gas Pipes
Raw Iron
Cast Iron
Iron Forgings
Angle Iron   the same as on the Elevated Rail Road Struc-
ture
Day laborers
Firemen
Ordinary Good Engineers
Draughtsmen
Carpenters
Masons
No 10 98% conductivity wire uncovered & covered

& price lists of everything you think would assist me in calculating   I should also like to be posted on Municipal & General
laws on Steam Engines & Boilers & erecting chimneys

I cabled you the other day to get an article from Il Nuvo
Cimento which is to be found in the British Museum. The
Magizine was for 1874.[10] My impression is that Paccinotti
published in Il Nuvo Cimento in '74 or '75 a Dynamo machine
in which the field magnet was shunted around the commutator brushes the same as with our "Z" Dynamo.[11] In fact I saw
the article the other day in the Electrician or "La Luminiere
Electrique" or Telegraphic Journal which stated that Paccinotti published such an article in 1874 in Il Nuvo Cimento[12]
Yours very truly

Thos A Edison   I[nsull]

L (letterpress copy), NjWOE, DF (*TAED* D8133ZBS; *TAEM* 58:820).
Written by Samuel Insull. ᵃIllegible. ᵇCanceled. ᶜInterlined above. ᵈ"of
the jet" interlined above.

1. Johnson cabled on 17 December: "Utmost value have few 280
lamps [Sir William] Thomson test earliest date   advance payments affected by result." Edison answered that because of the upcoming holidays they would not be ready for about four weeks. Johnson to TAE,
17 Dec. 1881; TAE to Johnson, 19 Dec. 1881; LM 1:119C, 121A (*TAED*
LM001119C, LM001121A; *TAEM* 83:931–32).

2. Johnson replied on 19 January that he did "not think well of a 10
Candle Lamp. The people in London want more light than they now
get, and although 10 Candles may be above the average of a London gas
burner and therefore a good standard for calculating cost it is not

sufficient to meet the requirements of the people who want Electric Light. They are mostly heavy consumers of gas and have adopted the best burner and in their hands gas consumption is much more carefully looked after than it is in the hands of an average consumer." In a separate letter, he also pointed out that "the B. machines are no good here— people want more light not less— The atmosphere is so heavy that it takes more light at one point than it does in N.Y." Johnson to TAE, 19 Jan. 1882 (pp. 1, 3) and 8 Jan. 1882 (p. 7), both DF (*TAED* D8239I, D8239D; *TAEM* 62:675, 659).

3. See Doc. 2190. Johnson had recently cautioned that gas producers' loss of economies of scale would be "more apparent than real" because of the increasing use of gas for heat and power. He noted, however, that these other uses would probably cause a deterioration in the quality of illuminating gas. In his 19 January reply to this document, Johnson also pointed to a trend toward increasing the brilliancy of gas street lighting by aggregating more burners in each lamp. He surmised that overall demand for gas would be "maintained on one street by just that which it loses by virtue of the Electric Light on another street. My opinion is therefore that it will not do to rely too much upon the statistics which were prepared a year or two back." Johnson to TAE, 11 Dec. 1881 (p. 27) and 19 Jan. 1882 (p. 3), both DF (*TAED* D8133ZBD, D8239I; *TAEM* 58:728, 62:675).

4. Johnson stated in his 19 January letter (see note 2; pp. 3–4) that "the absence of the meter is the only drawback to the completeness of our system. I do not want any large meters. Meters from 1 to 20 lights will be quite sufficient."

5. On the new socket and lamp base with the "tip," see Doc. 2187 n. 30.

6. Wilson 1877.

7. Edison made a series of tests of the practical illuminating power of commercial gas in September 1880; see Doc. 1990.

8. That is, draw in a greater volume of air to produce a hotter and less luminous flame.

9. Edison discussed this in more detail in Doc. 2203.

10. Pacinotti 1874.

11. Edison instructed Johnson on 9 December to "Go Brittish Museum   Translate from Nuovo Cimento Eighteen seventy four Paccinoti article on Shunt Dynamo." Johnson replied that Theodore Waterhouse would immediately attend to "your several reference in regard to the doings of Swan, Paccinotti, and others" (LM 1:113A [*TAED* LM001113A; *TAEM* 83:928]; Edison's draft (incorrectly dated 28 December) is in DF [*TAED* D8120ZCF; *TAEM* 57:686]; Johnson to TAE, 11 Dec. 1881 [p. 6], DF [*TAED* D8133ZBD; *TAEM* 58:728]). The 1874 article reportedly described a machine built by Pacinotti in 1873 and recently exhibited in Paris. According to the *Scientific American*, it was a "shunt dynamo—that is to say, the current generated is divided in parallel circuit between the fixed electro-magnet and the external resistance. This is done by means of two pairs of brushes making contact with different sections of the revolving commutator." Edison wanted the article as evidence to undermine the Siemens patent position. On 6 December he confirmed instructions already cabled to Johnson: "Dont offer Siemens royalty   Have written   Am searching." He added that he was "now looking up further information & shall hope to communicate further

with you on the subject" (Dredge 1882–85, 1:134–35; "The International Exhibition and Congress of Electricity at Paris," *Sci. Am.* 45 (1881): 377; TAE to Johnson, 6 Dec. 1881, Lbk. 9:412 [*TAED* LB009412; *TAEM* 81:155]; see also Doc. 2203). The Siemens machine with the field coil in a shunt circuit was introduced in March 1880 and is described in Dredge 1882–85 (1:284–88). The relationships among the Pacinotti, Gramme, and Siemens machines had created confusion in Great Britain for at least several years (Higgs and Brittle 1878, 65–68, 86).

12. The article to which Edison referred has not been identified.

–4– January–March 1882

Cold and snowy weather early in the new year interfered with the construction of Edison's electric distribution system in New York City, and the laying of underground cables was halted until the middle of February.[1] Anxious that his enterprise should not fall into "the rut of indolence" during the winter, he urged Charles Clarke to start planning for the next district, further uptown in the vicinity of Madison Square.[2] He also spent more time at the Edison Machine Works on Goerck Street, some thirty blocks from his office, where in February Charles Dean and his men completed the third Jumbo dynamo. The machine was slightly larger than but otherwise similar to its predecessor at the Holborn Viaduct plant in London. Edison was "absolutely satisfied" with its design but encountered problems with the steam engine that delayed its shipment in February.[3]

Edison gave much attention during the first quarter to affairs of the prospective Edison Electric Light Company, Ltd., in London. Potential investors there were becoming uneasy about the soundness of Edison's patents and the electric light's economic competitiveness with gas. Remembering their unhappy experience with Edison's telephone patent, Edward Johnson and several principal investors recommended an outside legal review. An aggrieved Edison responded that the "only advantage gained by taking out Patents at all is that you have the privilege of paying heavy fees to the British Government & have the honor of receiving a piece of parchment."[4] Quickly, however, he threw himself into an exhaustive correspondence that answered in detail the experts' queries and gave Johnson information to allay most of the concerns. He also agreed to submit five patent specifications for disclaimer, or amendment. Edison remained disappointed with the financial terms negotiated by Egisto Fabbri but ultimately accepted; the Edison Electric Light Company, Ltd., was organized in March.

Edison relied heavily on Johnson's advice even when it ran counter to his own inclinations. Acceding to Johnson's forceful suggestion, he assented to having British patent attorney Thomas Handford review and file British specifications prepared in New York. Mindful of the need for scientific allies in future court battles in Britain, Johnson also urged him to acknowledge the public congratulations of a former adversary, William Preece, telegraph-electrician of the British Post Office; Edison did so in a cordial letter, which drew an equally amicable reply.[5] With reluctance, Edison additionally agreed to a royalty in Britain to the Siemens interests there, who claimed to have patented the basic design of the armature in his dynamo.

Johnson was meanwhile gaining recognition as Edison's man-on-the-scene in London. A newspaper there described Edison's Crystal Palace exhibit as "the wonder of the show, and his representative is certainly the prince of all showmen. There is but one Edison, and Johnson is his prophet."[6] When he was not showing Edison's display to reporters or dignitaries, Johnson was busy supervising the construction of the demonstration central station at Holborn Viaduct. Assisted by William Hammer, John Hood, and eventually Francis Jehl, he laid conductors along a half mile of the Viaduct. The district ultimately encompassed 164 street lights and nearly 500 indoor lights at dozens of buildings, including the Post Office and City Temple.[7]

Edison continued to rely on familiar associates to administer his affairs in the United States, with one notable change. He hired patent attorney Richard Dyer to take charge of his patent affairs when his previous attorney, Zenas Wilber, and George Dyer (Richard's father) dissolved their partnership.[8] Samuel Insull, who had worked for Edison a full year in February, self-consciously shouldered increased responsibilities. He began to routinize correspondence by stamping outgoing and incoming letters, and he wrote and signed letters in Edison's name. Apparently reacting to Edison's growing trust, young Insull also took it upon himself to ensure that his employer undertook no new commitments without "my disputing every point & bringing forward every imaginable difficulty."[9]

During discussions of the proposed contract with the British company, he observed that not only Edison but "such old business men as Mr Fabbri and Mr Lowrey not only listened to but seemed to encourage my comments on various points."[10]

Insull's organizational skill and knowledge of business proved crucial early in 1882, when Edison spent much time away from his office. Edison's whereabouts at this time are hard to reconstruct fully. He and his wife had given up their New York lodgings at the end of 1881, and Edison made frequent trips to Menlo Park. He also traveled briefly to Washington, D.C. in February for an unspecified event for which he dressed, according to Insull, like "a regular masher."[11] The day he left, Mary Edison planned to attend a ball in New York with a party that included Insull.[12] She then went to South Carolina, presumably with her children, while Edison returned north. Mary was plainly unwell and her doctor had advised an extended vacation. Edison himself badly needed a respite from work, Insull thought, and he left, exhausted, on 1 March to join his family in Florida. He put Insull "in full charge of everything" and, uncharacteristically, did not communicate with him until his return trip at the end of the month.[13]

Edison found time at Menlo Park to return to inventing. He erected a large stamping mill outside the laboratory in February. About the same time Charles Hughes resumed trials on the electric railroad there. Hughes sought to overcome the difficulties of running a direct current motor under highly variable loads. Much of his work involved matching mechanical and electrical components, which had a more complex relationship under these conditions than in a dynamo running at constant speed. Edison also returned to the problem of regulating dynamo output. Although the immediate stimulus for doing so is not clear, he executed five patent applications on this subject on 10 February and another five at the end of the month.[14] These were among eighteen applications from February and March that resulted in U.S. patents. He also supplied substantial material about his dynamos for an enlarged English edition of a major German text in electrical engineering.[15] Edison stayed in close contact with the Menlo Park lamp factory, where he oversaw experiments with high candlepower street lamps and various filaments and coatings.

The Lamp Works having largely exhausted its credit, factory manager Francis Upton began the year with a manufacturing hiatus to reduce expenses. When a pieceworker devised a process to greatly increase his output, William Holzer, soon to be the superintendent of manufacturing, reduced his pay rate. Edison overruled this decision with the remark that "we must carry out all our promises if it busts the Co."[16] At the end of March, the factory closed in preparation for the move to Harrison, New Jersey, where Edison expected to lower labor costs and double capacity.

Formation of the Compagnie Continentale Edison and two related companies on 2 February gave Charles Batchelor the means to continue setting up a factory for manufacturing lamps and electrical equipment at Ivry-sur-Seine, outside Paris. He complained acidly that the "lazy, slow and very bad workmen" in France made for "very difficult work."[17] Elsewhere in Europe, Joshua Bailey and Theodore Puskas negotiated for isolated plants and utility companies in smaller cities not covered by the Compagnie Continentale, including Strasbourg and Milan, and began to plan for a factory in Spain.[18]

To support his expanding manufacturing enterprises, Edison continued to seek money from the liquidation of his English telephone interests. He also looked forward to the proceeds of organizing the English electric light company, and borrowed $10,000 from Egisto Fabbri in late January against these expected payments (which he did not receive until April).[19] He agreed to loan his brother Pitt $125 about the same time but Insull delayed sending the check because Edison was "pinched for funds." Edison promised his brother another $250 in mid-February.[20]

1. The Edison Electric Light Co. began reporting the progress of the district and Pearl St. station in its Bulletins about the time that work resumed. See Edison Electric Co. Bulletins 3–6, 24 Feb., 17 and 27 Mar. 1882, all CR (*TAED* CB003; CB004, CB005, CB006; *TAEM* 96:674, 676, 681, 688).

2. Doc. 2223. A canvass of the proposed new district between 24th and 38th Sts. and 8th and Madison Aves. was completed by June. It showed a total of 41,000 gas lights in houses, 500 in various buildings (including hotels and theaters), 220 pumps, and 2,284 sewing machines. A contemporary article noted that "buildings where the most sewing-machines are found contain the fewest gas jets." "The Electric Light in Houses," *Harpers Weekly* 26 (1882): 394; in Cat. 1018:11H, Scraps. (*TAED* SM018039a; *TAEM* 24:253).

3. Doc. 2215.

4. Doc. 2203.

5. Docs. 2217 and 2225.

6. "The Electrical Exhibition at the Crystal Palace," *Daily News* (London), 8 Apr. 1882, 6.

7. Edison Electric Light Co. Bulletin 9:8, CR (*TAED* CB009; *TAEM* 96:706).

8. Dyer drafted Edison's foreign specifications as well, then sent them to agents abroad to revise and file them. This was a particularly significant practice for patents in Britain and its colonies. Thomas Handford to Theodore Waterhouse, 3 Mar. 1882; Waterhouse to Johnson, 4 Mar. 1882; DF (*TAED* D8248Y, D8248X; *TAEM* 63:432, 430); Dyer to Johnson, 1 Apr. 1882; Dyer to Handford, 3 Apr. 1882, Lbk. 12:6, 19 (*TAED* LB012006, LB012019; *TAEM* 81:498, 505).

9. Insull to Johnson, 20 Feb. 1882, LM 3:54 (*TAED* LM003054; *TAEM* 84:57).

10. Insull to Johnson, 8 Jan. 1882, LM 3:11 (*TAED* LM003011; *TAEM* 84:12).

11. Insull to Johnson, 22 Feb. 1882, LM 3:64 (*TAED* LM003064; *TAEM* 84:67).

12. Mary had four tickets to the annual Arion Ball. She and Insull were evidently joined by Sherburne Eaton and his wife. Sarah Guernsey to Mary Edison, undated Jan. 1882, DF (*TAED* D8214D; *TAEM* 60:700); Insull to Edward Johnson, 20 Feb. 1882, LM 3:54 (*TAED* LM003054; *TAEM* 84:57).

13. Insull did not even know when to expect Edison's return. He wrote Charles Batchelor on 25 March that he had "not heard one word from him since he left on the 1st of this month—with the exception of a short telegram from his wife," which is Doc. 2241. Insull to Batchelor, 7 and 25 Mar. 1882, Lbk. 11:415, 501 (*TAED* LB011415, LB011501; *TAEM* 81:429, 469).

14. See headnote, Doc. 2242.

15. Schellen 1884.

16. TAE marginalia on Upton to TAE, 29 Mar. 1882, DF (*TAED* D8230ZAC; *TAEM* 61:741).

17. Batchelor to TAE, 28 Mar. 1882, DF (*TAED* D8238ZAN; *TAEM* 62:358); Batchelor to Upton, 3 Apr. 1882, Cat. 1239:194, Batchelor (*TAED* MBLB4194; *TAEM* 93:611).

18. For overviews of these plans see Bailey and Puskas to TAE, 26 and 29 Mar. 1882, both DF (*TAED* D8238ZAL, D8238ZAP; *TAEM* 62:353, 361).

19. Ledger 8:218, Accts. (*TAED* AB004:93; *TAEM* 88:95); TAE to Fabbri, 23 Jan. 1882, Miller (*TAED* HM820156A; *TAEM* 86:450); Fabbri to TAE, 23 Jan. 1882, DF (*TAED* D8239K1; *TAEM* 62:704).

20. Insull to Pitt Edison, 19 Jan. and 1 Feb. 1882, Lbk. 11:127A, 193A (*TAED* LB011127A, LB011193A; *TAEM* 81:290, 314).

EXAMINATION OF EDISON'S BRITISH ELECTRIC LIGHTING PATENTS   Doc. 2203

Edison had always relied on his U.S. patent attorneys and their London agents to prepare and file his patent specifications in Great Britain. In the latter part of 1881, with negotiations moving ahead to form an Edison electric lighting company there, patents covering his lighting system were subjected to systematic review by outside legal experts for the first time. Frederick Bramwell, an eminent barrister, was retained for this purpose at the suggestion of William Preece. Bramwell prepared a report in November on the soundness of Edison's lamp patents; while he moved on to examine other parts of the system, his conclusions were vetted by another expert, John Henry Johnson.[1] Edward Johnson relayed these opinions to Edison with his own commentary.[2] Other authorities were recruited later. Their collective findings gave weight to Edward Johnson's argument that Edison would be better served in the future by having new specifications carefully reviewed by an expert in the intricacies of the British system.[3]

Doc. 2203 is the beginning of Edison's response to this process.[4] Despite his skepticism of British patent law (he once sarcastically called it a "beautiful system . . . invented I think by King Canute"),[5] Edison gave the matter his careful attention. He remembered his unhappy experiences with telephone patents, as did Johnson and several prospective organizers of the new lighting company who had been principals in the now-defunct Edison Telephone Co. of London.

Edison faced the grave possibility that his more recent lamp specifications could be read as having been anticipated by his earlier ones; the invalidation of a single claim for lack of novelty (or any other reason) would nullify the entire patent. Other concerns centered on the specificity of his patent claims. The formal claims constituted the legally binding portion of a specification. A specification without claims would be interpreted to apply to everything described in it but, as Johnson explained, when one or more claims was present, only the features specifically enumerated there would receive patent protection, the remainder of the text being merely descriptive.[6] This presented a serious problem in Edison's carbon lamp specifications, which included essential processes and principles that he had patented in connection with his early platinum-burner lamps but did not claim particularly for the carbon filament. Edison eventually consented to have five lamp specifications submitted for disclaimer, an amendment process in which claims could be restricted or excised.[7]

Edison's misapprehension of the narrow function of patent claims also informed his effort to invalidate the basic Siemens patent for a longitudinally-wound dynamo armature. He had hoped that his large direct-driven steam dynamos, like those being readied at the Holborn Viaduct demonstration central station, would be clear of the Siemens interests. The London experts agreed unanimously, however, that the patent's carefully contrived claims could withstand any argument that might be made against their originality. Rather than face a long legal fight, Edison later consented to pay a royalty on the design.[8]

1. Edward Johnson to TAE, 12 Nov. 1881, enclosing Theodore Waterhouse to Johnson, 12 Nov. 1881; Waterhouse to Egisto Fabbri, 16 Dec. 1881; all DF (*TAED* D8133ZAR, D8133ZAS, D8133ZBF; *TAEM* 58:698, 706, 770).

2. See Johnson to TAE, 15 and 19 Nov. and 11 Dec. 1881, DF (*TAED* D8133ZAT, D8133ZAU, D8133ZBD; *TAEM* 58:710, 712, 728).

3. Johnson had excoriated Edison for continuing to take out patents "through the present channel. . . . Your continuance of the old worse than foolish method of filing your English Patents gives the impression that you are careless of English interests. We are today compelled in our negotiations to face the drawback of extreme weakness in our Patents at a cost to you of millions, and yet you continue right along in the same old beaten path. No practical improvement can be made by an oversight only had in New York." Johnson to TAE, 11 Dec. 1881 (pp.10–11), DF (*TAED* D8133ZBD; *TAEM* 58:728).

4. See also Docs. 2221 and 2228.

5. Doc. 1600.

6. Davenport 1979, 31; Johnson to TAE, 16 Jan. 1882, Ser. 1, Box 1, Folder 2, WJH.

7. See Docs. 2221 and 2285.

8. Johnson's précis of this reasoning is quoted in Doc. 2203 n. 27; see also Docs. 2099, 2103, and 2226.

–2203–

To Edward Johnson

[New York,] [2nd][a] Jan. [188]1[2][1]

My Dear Johnson,

Your favour of Dec 11th comes duly to hand.[2]

Big Dynamo. It is a great pity that Batchelor has not informed me of the fact that the bushings in the commutators brush holder [--- ----][a] had been carbonised.[3] We will increase the insulation. The carbonisation could not have been due to the heating of the Journals as we had no trouble about that after the first three or four days running. The insulation must have been very thin and [----][a] and oil produced a surface conduction which tended to carbonise them.

I do not know whether it is a good plan to run both Dynamos at once.[4] I am under the impression that it would be best to connect both dynamos up in such a manner that one only should be doing the work at full speed while the other could be running say 30 revolutions a minute with the wires disconnected and if the one doing the work should break down

I think the machine could be disconnected and the wires connected to the other and full speed obtained in something less than 20 seconds, if your man were properly trained. This would be such a short stoppage of light that no complaint would arise and it would not be liable to occur very often any way. The reason why the spare Dynamo should be run slowly (80 revolutions) is so that you can start up quickly having its cylinder all hot and without any fear of [----][5a] You can rely on Hood[6] more than you can on Porter. Porter set the valves on our Engine here and we had to alter them.[7]

I have not heard from Batchelor what he is going to do with his big Dynamo in fact I get no information at all from him. I suppose he cant find time to satisfy my curiosity by writing.[8]

We are having the chalks made for your telephones. Insull has sent you some more of those Instruction Books[9]

I believe in selling lamps [----][a] you can in large quantities a thousand or upwards but you must be careful not to sell them for foreign use as even I have not got patents[10]   I have contracts for my interests such as my contracts with Gouraud.[11]

Patents.[12] In my subsequent patent as to getting the air out of the carbon of course you must read the claim by the specification & when they are read in that light I do not see that they are too broad.[13] You say you cannot bring your patent experts to the opinion that the claim for a filament of carbon of high resistance is good. Can you bring them to the point of rendering an opinion as to whether a filamentar[y] carbon for giving light by incandescence placed in a sealed glass chamber from which the air has been [removed][a] would be a good claim. By leaving out high resistance that is to say disclaiming high resistance you would then have a filament of carbon in connection with the rest of the claim. In the patent for a paper Carbon filed, sometime in December you will notice that I stated that the carbon is flexible.[14] Now there is one point about the English law that I am not certain about i.e. I understand that it is not necessary to make claims that you are entitled to everything in the specification as long as it conforms to the law and is new & novel & useful. Grammes patent has no claims. If this is so then the paper carbon patent would be held to cover a flexible carbon.[15] You say that the experts say that the filament of carbon of high resistance is anticipated in detail in so many directions that there is nothing [except][a] the degree pure & simple for [----][b] to lay hold of. Now I should like to know where a filament of carbon of high or low resistance has ever been used before my patent. I would also like to know whether a filament of carbon of high resistance has been used. Also whether a filament of carbon has been used in high vacuo in a chamber made entirely of glass through which platinum wires are sealed. I should also like to know if a claim taken not in its broad sense but as a combination claim is not good. Combination claims are good even if all the parts are old. You will notice that our claim is really a combination claim. I think abandoning all else and seeking only to make good a claim for carbonised fibre is a mistake. While carbonized fibre is the best and would probably hold it would let too many people in.

Memo. Be careful and dont disclaim the coil carbon for diminishing the radiating surface as it is on this line that the high resistance lamp will be obtained.[16] That I believe can be held against everybody. I think that there is no immediate hurry about bringing any suits; the longer you are at these patents the more thoroughly the experts will get to understand our system and the better the patents will appear. You know that Betts[17] & Dickinson[18] both said at first that we could not claim a difference in degree but after a while when they became more familiar with the results produced by this change in degree their opinion altered and we now have strong opinions from them that there is not a shadow of a doubt but ~~what~~ that the Court will sustain the patent   what I want to know where and by whom before my patent was a "filamentary carbon for giving light by electrical incandescence ~~used~~ in a glass chamber from which the air has been exhausted" used.[19] Now I maintain that if no reference can be found [---]ᶜ were any person did this before and that I was consequently the first to produce it and it is acknowledged to be a useful and great advance in the art of electric lighting then I must be entitled to it. That is to say that I am entitled to a lamp for giving light by incandescence consisting of a filament of carbon in high vacuo maintained under conditions that said vacuo is stable. Filaments of carbon might have been produced before: vacuums have been produced before but who has used or combined in a vacuum a filament of carbon and I think that the November Patent and the Coiled Carbon is clear and explicit and not complicated. If your people are afraid of the words high resistance why not make the claim thus:—

An Electric lamp for giving light by incandescence consisting of a filament of carbon

The second, third, & fourth claims to be the same.

If this patent will not hold as thus amended then it seems to me to be absolutely useless to take out English patents.

I should like to ask if the second patent wherein paper carbon is spoken of and claimed is not good or is it rendered negatory by the patent of Nov 10th speaking of fibrous material such as paper thread wood &c and if so you would probably have to disclaim claim 1st.[20] The third claim ought to be good and is important on account of the broadened ends. The fourth claim does not amount to anything   The second claim might be good in so much as it describes a method of carbonizing filaments in a <u>definite</u> shape. You are certainly wrong about this fibre claim business   Yours would be an excessively narrow view of the matter and could be very readily got around You say that the process as applied to Swan becomes public property when speaking of Swans patent. It may be very clear to you but it is not clear to me why getting air out of carbon by means of Electrical Incandescence in vacuo is thrown open to the public by the action of a subsequent inventor when another inventor has shown it and claimed it in a patent several months prior[21]   Do I understand that the June 79 patent in which a process is described of eliminating the air from carbon by electrical Incandescence in vacuo is public property and invalid. If so is it by reason of prior publication of some other inventor or why?

You say that I clearly set forth the value of high resistance in patent no 4 as to anticipate myself in patent no 5?[22] Hence, is Patent no 4 good? If so will not that patent hold a high resistance lamp?

Regarding taking out those patents you know that up to June 1880 Serrell prepared these patents[23] and they were sent to Brewer & Jensen who filed them just as they were received from Serrell.[24] After that Wilbur prepared the patents sent them to Serrell, who merely ~~kept~~ copied them and then forwarded them to Brewer & Jensen who filed them   So Serrell is to blame previously to June 80 & Wilbur after. You must consider that owing to the enormous number of applications for patents that to have got the patents in any good kind of shape it would have taken the time of one man continuously. It would have had to have been done on this side because I fail to see how an^d imperfect application from this side could have been made perfect by solicitors on the other side who could have no knowledge of the system. However hereafter we will send forward the applications through any person whom you may designate. No improvement however, can be expected in the old way without a man is hired to attend expressly to attend to^e that business.

Can you tell me how to make an inch of platinum wire of a certain size any greater resistance than it naturally is. Is not the resistance of a definite piece of platinum wire always the same. Hence how could one claim a platinum loop of wire of high resistance. Now it is a very easy matter to make an inch of carbon wire of any kind of resistance you want by selecting out a carbonizing material whose cells are either large or small hence it is possible to have a claim on a carbon wire loop of high resistance   I merely make this little note just as it comes to my mind, as you may possibly be able to make use of this fact. When you speak of my previous patents detailing the necessity of high resistance you are correct. The method however adopted for obtaining the high resistance was not by specifically increasing the resistance of the material or conductor employed but by employing greater length of it, while in the carbon patent the claim is for a peculiar kind of carbon that is to say a filament of carbon aggregated together in such a manner that it shall have high resistance.

You say "Had your patent for the application of the process not been so religiously restricted to metalic conductors &c &c." I ask in this relation again as to the necessity of making claims in an English Patent.

I will see Mr Fabbri about engaging young Dyer[25] to take charge exclusively to go over the whole ground of English Patents & I strongly advise that no action about disclaiming any patents before we have gone over the whole business because with hasty action we might disclaim some things which have a value & which it might be afterward shown it is not necessary nor vital to disclaim.

I do hold that my bar & plate armature is not an infringement of the Seimens machine or his patent[26]   His patent as worded cannot possibly cover it as worded. I cannot possibly see how Seimens in view of Grammes patent can bring an action against any party until he has disclaimed certain things shown by Paccinotti & Gramme—Seimens using the Gramme & Paccinotti method of winding and connecting and taking off the current.[27] All the difference between the Seimens method of winding & Grammes & Paccinottis (especially Gramme) is that he winds with the end like his old armature and Gramme passes it underneath—the face winding is the same.[28] All that Seimens can claim would be the end winding and as this was done in his old machine I do not see where the invention comes in. It is merely employing the Gramme winding on his old machine free to the public.

It might be a good idea to hire an expert to report favorably on our patents bringing up all the arguments he can to uphold the patents and between the two sides I am sure it would advance the understanding of the whole much better.

I think you will find the life of the lamps all right providing you arrange to keep them at the proper candle power and not allow them to go above that is to say keep them at sixteen candles. Lamps you get hereafter will have very long life   All you have to do is to keep them down to their proper candle power and you will have no material trouble from breaking. Of course the first four or five nights all the weak lamps will be shaken out & then your breakages will diminish and gradually get down to a mere nothing.

I see in the contract submitted by Mr Fabbri that we are to compete with gas at the same price giving an equal amount of light. This will be all right as I have not the slightest doubt but that ten candles per 5 ft burner is the average of London gas. Most works on London gas give it lower. I base my estimate on the ten candle burner being equivalent of a 5 ft London jet in actual practice. You know I have always stated that we can compete with gas and that our only difficult place was London & that it is essential in competing that we should have a respectable number of stations because it cannot be expected that with a single station we can go and compete on price only, where so much general expense would have to be charged to so small an output. My figures are for a dozen or so stations. Of course it does not follow that we would have to compete on price. It might be possible that even one station would pay in localities where the quality and purity of the light would have a commercial value of fifteen or twenty cents over that of gas. Of course I have said nothing about power which is a considerable factor in helping us to compete.

In your Crystal Palace Exhibition have a good man to attend to the commutators, keeping the commutators clean, the ends of the brushes in nice order, setting them accurately, using light pressure on the brushes &c and you will never have trouble from these sources. You should employ an extra brush holder which can be slipped on and off while the dynamo is running & thus have an opportunity of setting brushes right when they are sparking without stopping the machine. we foind these extra brush holders of great value especially where we have several machines for if a brush gets sparking all we have to do is to put on the extra brush holder take the other one off and fix it and put it back again.

We have no trouble in gradually moving the Dynamo and so slackening the belt.[29] We always did it at Menlo and never had any trouble. If it is done gently it will not fly round and do mischief as you imagine Friction clutches are very unreliable things without you have a good one Of course all your regulating resistance has gone long ago.

I do not know what you propose to use to keep your candle power constant. You should use something at any rate. We will send you a bell indicator for for indicating when the candle power exceeds sixteen candles which you can use but I fear it will not reach you in time & you will have to use a galvanometer or something else which Hammer can fix up for you.

Of course you never want to open a field magnet with a machine running ~~with~~ when[d] the bobbin is connected across with the main wires as it would then burn the bobbin up. The bobbin should always be open first. By taking off the field the bobbin would not run as a motor because it has nothing to work against and it then acts as a dead cross circuit[30] but if you take off the belts and leave the bobbin & field undisturbed it will then act as a motor.

We will try & get a photograph taken of your new machine.

I do not look with favor upon using gas engines if it can possibly be avoided as it will prove a source of revenue to prolong the fight with the gas companies.[31] The little Armington & Sims engines are above criticism—cheap, reliable, and perfect in every way.

About the commutator brushes on your English machine if great care is taken with the mercury and the brushes are ground properly and the burs all taken off the [---- of?][a] the wires where they come in contact with the commutator you will have no trouble about getting rid of the sparking. We ran [--][a] hours with 1000 lights with scarcely any more spark than there was in a regular "Z" machine. The machine I have here now gives us some [trouble?][a] with the sparking for the reason that it is of such [------][a] low resistance that the short circuiting spark between one bar and another is considerably more than in your other machine This sparking always increases with the diminution of the internal resistance that is to say as the machine gets better You cannot always [have?][a] good things for nothing and beside I put 1360 full lights on this machine the first time we tried it and I do not propose to take off ~~the~~ a light until we take the current off easily & nicely. 1360 lights are altogether too much for our boiler and engine and we are only enabled to run about five minutes before the steam gets down so that the engine slows right down but during these five minutes the commutator brushes are lively and if I can take this current off nicely you certainly will not have much trouble more especially as my impression is that your engine will not run more 900 or 1000 as a regular thing. Hammer was here when your machine was run for 7½ hours and attended to the commutators and his statements are correct that there was not the slightest difficulty in taking the current off.[32]

We put a Safety Catch in every branch and we are thinking of putting it in both sides of the branch as the boys will make joints and run wires down the sides of gas pipes and so cross them. We had a case the other day where the gas pipes being used somewhere in the building. The positive wire got connected with the gas pipe and in another case the negative wire got connected and it so happened there was no safety catch on the negative wire at one place or a safety catch on the positive wire at another place. We had a bad cross and burnt the moulding.

The New York Board of Underwriters have practically adopted our rules   I will send you a copy as soon as they are published.[33]

I feel convinced that the size of wires used to run down to the lamps should be increased over and above what has been our practice. With the present wires they being small when they come in contact the arc is quicker than the Safety Catch and the wire is fused; while if it was larger its greater conductivity of heat would give the Safety Catch time to fuse and this arc & fusing of the copper would be avoided. In fact I believe the wires that run to the lamps or down a chandelier or in places where there is a chance of a cross—should be very large in fact it might be larger than the wires which are led down from the mains. You will probably see my point.

Our first electromotive force regulator showed the fall of the electromotive force by ringing another bell but as it required a pretty fine adjustment we abandoned it. The main object of this indicator is to prevent quarrelling with our customers about the life of the lamps for by the use of this instrument our words as to life of lamps will be found more than true; whereas without it Customers are of course liable and would, and do, undoubtedly run the lamps at a very high candle power.

I have read the Society of Arts lecture of Preece[34] and think that in face of the prejudice of English Journals & Scientists he shows great great independence of mind and does himself great credit to express himself so clearly and decidedly especially considering his earlier battle for utterances[d] the subject of Incandescent Lighting.[35] I will send you in a day or so an autograph letter addressed to Preece which you must use your discretion as to delivery to him.[36]

Do not fail to remember the fact that there are 8 or 9 things in our system of general distribution any one of which if maintained will give us the commercial monopoly of the general distribution business and the patents which are now coming out and which will come out in the course of the next year will be just as important in maintaining the monopoly as the patents which have been already issued.

I noticed that the Telegraphic Journal gave us a very handsome send off[37]   I do not remember anything in the Electrician.

I have read the contract brought over by Fabbri and am favorably impressed. Details will require [altering?][a] but nothing material.[38]

4th Jan 1881 2[f]

I have definitely arranged to have Dick Dyer go through our English Patents make a thorough analysis for use in disclaiming also for use in taking out future patents

I notice that in the Patent of 15th Dec 1879 that in the first paragraph in the Provisional the following words occur. "In a former application made by me for letters patent in Great Britain an improvement in Electric lamps is set forth wherein a filament of carbon is enclosed in a glass bulb and the atmosphere removed as nearly as possible and the carbon is brought to incandescence by an electric current to <u>form</u> the lamp" What does this mean? Does it mean that the lamp is not formed until the carbon has been brought to incandescence or not!

I have been thinking over the question of high resistance and make a few more notes as follows:

When we say "a filament of carbon of high resistance" it ~~doesnt~~ is[d] not altogether a question of degree of resistance but the words high resistance are indicative of the quality of the carbon. In the specification I speak of carbonizing vegetable matter and as all vegetable matter is formed of cells a filament of this kind of carbon will necessarily be of high resistance and the word high resistance will in their turn be indicative of the quality of the carbon.[39] If we were to make a filament <u>artificially</u> it would be of low resistance hence the words "high resistance"means a filament of carbon obtained in a <u>natural</u> state whereby it will offer high resistance so you see it is not altogether a question of <u>degree</u>—that might be left out—it is a question of <u>nature</u>. You will notice if I remember rightly, in the specification that I state that carbon produced from organized[40] matter has this essential <u>quality</u> of <u>high resistance</u>. If we were to use all kinds of carbon, one might be low resistance carbon whereas the other might be high resistance carbon yet both might be filaments and both of the same size and length Hence the words "Carbon of high resistance" would be indicative of special natural quality of the carbon and it is this special <u>quality</u> of the carbon which I desire to claim that is to say high resistance carbon. Do you twig![41] I give this simply as an idea. You may be able to say something of it.

I cannot understand this English Patent business— Swans patents you say are no good; Lane Foxes patents are no good; Bells telephone patents were no good; the United Telephone Coy (judging from its policy of compromise) do not seem to think my Telephone patents are any of them good— Really it would not seem to matter much whether the applications are drawn by American or English Lawyers. In fact I am begining to think that the only advantage gained by taking out Patents at all is that you have the privilege of paying heavy fees to the British Government & have the honor of receiving a piece of parchment the letter press on which would seem to be worth but little more than the paper it is printed on—if this is not the case well then some of your lawyers are very wide of the mark in the opinions they give   Very truly Yours

Thos A Edison   I[nsull]

L (letterpress copy), NjWOE, Lbk. 11:1 (*TAED* LB011001; *TAEM* 81:223). Written by Samuel Insull. "Per S.S. Gallia" written to left of dateline to indicate ship carrying letter. [a]Illegible. [b]Faint copy. [c]Canceled. [d]Interlined above. [e]"attend to" interlined above. [f]"2" underlined and circled.

1. Date from docket notation.

2. Johnson's letter to Edison of this date was a thirty-three page answer to Docs. 2187 and 2190. In making his reply Edison closely followed Johnson's letter, the most germane points of which are noted below. Johnson to TAE, 11 Dec. 1881, DF (*TAED* D8133ZBD; *TAEM* 58:728).

3. Johnson wrote in his December letter (pp. 2–3; see note 2) that "Reports are reaching me from Paris from the boys now arriving from there to the effect that the continued use of the Dynamo has developed some evils of which you of course could not be aware by a few hours of running. For instance I am told that the Insulations in the Commutator extensions become thoroughly carbonised from heating. I take it that this heating is from friction of the bearings and not from the Electric current."

4. In his 11 December letter (p. 3; see note 2), Johnson proposed "to run the two [dynamos] taking therefrom only the capacity of one so that

in case of the necessity of stopping I can draw upon the remaining one for its full capacity for a brief time and in this way, and this way only, can I hope to secure that which I consider of very great importance, namely, 'reliability.'"

5. Johnson later explained that he wished to run both dynamos at half capacity during peak evening hours, switching from one to another only in case of accident and "would not like to have a stoppage of my lights even for so short a space of time as 20 seconds   it would throw everybody into Confusion and would create a nervousness which would be a serious detriment to us." Edison subsequently adopted the practice of keeping a spare machine (called a "relay dynamo") in motion at his New York central station. Johnson to TAE, 19 Jan. 1882 (pp. 4–5), DF (*TAED* D8239I; *TAEM* 62:675); Charles Clarke memorandum, 14 Dec. 1931, enclosed with Clarke to C. J. Leephart, 15 Dec. 1931, acc. no. 30.1507.1, MiDbEI(H).

6. John Hood, a machinist, began working for Edison in early 1878. He ran at least some of the steam engines at the Menlo Park laboratory and machine shop, including the Porter-Allen engine on the experimental direct-connected dynamo at the beginning of 1881. He was in London to assist Johnson by the end of that year. *TAEB* 4:135; 5 App. 2; Johnson to TAE, 16 Dec. 1881, LM 1:118A (*TAED* LM001118A; *TAEM* 83:931).

7. Johnson wrote (pp. 3–4; see note 2) that he was awaiting unspecified modifications to the valves and governor from Charles Porter.

8. Johnson reported in his 11 December letter (p. 4; see note 2) that he did "not know what Batchelor proposes to do with his Dynamo. It is very difficult for me to get anything out of him. He is doubtless (like myself) fully employed in attending to his own business and has no time to waste in answering idle questions and having decided not to take the Dynamos from him of course it is an idle question for me to ask him what he proposes to do with it."

9. These have not been found. Johnson had received one copy and asked for more in his 11 December letter (p. 4; see note 2).

10. See Doc. 2187 n. 15.

11. See Docs. 1978 n. 1, 1920, and 2161.

12. See Doc. 2210.

13. Johnson had complained in his 11 December letter (pp. 6–7; see note 2) that in the case of British Patents 2,402 of 1879 and 578 of 1880 (Cat. 1321, Batchelor [*TAED* MBP017, MBP023; *TAEM* 92:118, 149]), Edison's broad claims did not distinguish sufficiently between the original process and the subsequent improvement. The first specification concerned Edison's process for removing occluded gases from the wires of his metallic lamp; the second referred in passing to making lead-in wires and clamps from platinoid metals "prepared and treated by heat in vacuum."

14. A flexible carbon filament was described in the text of Edison's British Patent 4,576 of 1879 (Cat. 1321, Batchelor [*TAED* MBP019; *TAEM* 92:134]) but not included in the four claims. The first claim was for "an electric lamp for giving light by incandescence consisting of a filament of carbon of high resistance, made as described and secured to metallic wires, as set forth."

15. Edison's understanding of the law was essentially correct at this

time; see headnote above. Johnson explained in reply that "in the case of your Platinum Patent wherein you first specify the utility of high resistance you were under the impression that high resistance can be held by this Patent   Now it can be held by this Patent for Platinum but for Platinum only." Johnson to TAE, 16 Jan. 1882, Ser. 1, Box 1, Folder 2, WJH.

16. Edison's third claim in his British Patent 4,576 of 1879 (Cat. 1321, Batchelor [*TAED* MBP019; *TAEM* 92:134]) was for a "coiled carbon filament or strip arranged in such a manner that only a portion of the surface of such carbon conductor shall radiate light." This was Edison's first British specification for the incandescent carbon lamp.

17. Patent attorney Frederic Betts (b. 1843) acted as counsel for a number of major entities, including the Edison Electric Light Co., the City of New York, and the Westinghouse Air Brake Co. He also served on the law faculty of Yale University from 1873 until 1884. *NCAB* 2:38.

18. Edison presumably meant New York patent attorney and technical expert Edward Dickerson, Sr. (1824–1889), whom the Edison Electric Light Co. retained in 1879. Doc. 1011 n. 3; Grosvenor Lowrey to TAE, 31 Oct. 1878, DF (*TAED* D7820ZBD; *TAEM* 18:56).

19. Edison quoted loosely from the second claim of his British Patent 4,576 (1879).

20. Edison's British Patent 5,127 of 1879 (Cat. 1321, Batchelor [*TAED* MBP020; *TAEM* 92:138]) contained four claims. The first was simply for "an electric lamp formed of carbonized paper." The second was for the process of heating a paper filament in a mold "to drive off the volatile portions"; the third for filaments having broadened ends for the clamps; and the last for the spring clamps themselves.

21. Edison referred to Joseph Swan's British Patent 18 (1880), his first in incandescent electric lighting. The patent was not for the lamp itself but for the process of evacuating the globe to prevent "wasting" of the carbon. Swan 1946, 23; Dredge 1882–85, 2:xcvii.

22. Edison's British Patents 2,402 of 1879 and 4,576 of 1879 (Cat. 1321, Batchelor [*TAED* MBP017, MBP019; *TAEM* 92:118, 134]) were his fourth and fifth specifications for electric lamps. The former explained the significance of high resistance lamps to an economical and efficient system.

23. In Doc. 1828 n. 1 the editors stated that Zenas Wilber took charge from Lemuel Serrell of all of Edison's new patent applications in January 1880; Serrell apparently remained involved with at least some new applications through June.

24. See headnote above.

25. Patent attorney Richard Dyer (1853?–1914), the son of George and brother of Philip Dyer, practiced with Henry Seely. He had been involved with Edison's affairs for some time and became Edison's principal patent attorney when his father and Zenas Wilber dissolved their partnership on 1 February 1882. Doc. 1270 n. 16; George Dyer to TAE, 30 Jan. 1882, DF (*TAED* D8248K; *TAEM* 63:382).

26. The dynamo specification in question (Brit. Pat. 2,006 [1873]) was taken out jointly by Charles William Siemens, his brother Werner, and Hefner von Alteneck. Dredge 1882–85, 2:iv–v; see also Doc. 2187 esp. n. 24.

27. Johnson pointed out in reply that Edison was again mistaken. Siemens

may describe what he pleases but if he does not claim it but does claim something that is new then his Patent stands good and needs no Disclaimer. What he does claim is not the longitudinal winding simply but the longitudinal winding over the entire surface of the cylinder, or rather more accurately, a number of coils consisting of 8 or more. Now his old armature was a longitudinally wound iron cylinder but wound with one coil and Gramme and Paccinotti had already published the means of obtaining a constant current by continuous winding so that this too was old but Siemens' claim is a combination claim for these old things namely, longitudinal winding over an iron core and a great number of coils to get constant current. . . . We have been all over the ground very thoroughly and invariably come to the same conclusion, namely, that Siemens old Armature made public longitudinal winding both over a shell and over an iron cylinder and that Paccinotti made public the principle of distributing the coils throughout the annular space but that Siemens was the first to combine the two and as this is all he claims, his Patent is good.

Johnson stated that he and John Hopkinson believed that Edison's large steam dynamo, while in other respects unique, would infringe the Siemens patent. Johnson to TAE, 16 Jan. 1882 (pp. 4–6), Ser. 1, Box 1, Folder 2, WJH.

28. In the Gramme armature, the wire was wound around a metal ring and returned to the starting point through the inside of ("underneath") the ring. In the basic Siemens drum armature patented in 1873, the wire was wound longitudinally over the outside of a cylinder, each loop being completed across the cylinder's ends. This pattern was adapted from the older Siemens shuttle armature, devised in 1856, which employed a figure like a weaver's shuttle with an H-shaped cross-section instead of a cylinder. King 1962, 369–75; Dredge 1882–85, 2:iv–v.

29. In most Edison isolated installations, belt tension was adjusted by using a horizontal jackscrew to slide the dynamo bed plate along metal rails (Charles Clarke to James Bishop, 18 Jan. 1933, acc. no. 30.415.1, MiDbEI(H)). Johnson wrote (pp. 20–21; see note 2) that he preferred to have a friction clutch to loosen a drive belt rather than move the dynamo itself because "in a compact mass of Dynamos and so many belts it strikes me that it is not a wise thing to risk the sudden release of a belt travelling at such a high speed. It is too liable to fly round and do general mischief."

30. That is, because the armature would not be moving in a magnetic field, it would not develop a counter electromotive force against the current of the other dynamos in the outside line. Edison included a switch for breaking the armature circuit before the field circuit in a patent application filed in February 1883. To prevent this happening in isolated plants with more than one dynamo, Edison arranged the switch so that it opened the main circuit while leaving the field energized by current from the other machines. U.S. Pat. 280,727; see headnote, Doc. 2126.

31. Johnson had complained (pp. 22–23; see note 2) that he could not find high speed steam engines in England and that ordinary engines required expensive countershafts. He promised to send Edison a small gas engine and to arrange for control of pertinent patents in the U.S.

32. Johnson noted (pp. 23–24; see note 2) that William "Hammer feels confident of his ability to manipulate the mercury business as satisfactorily as was done in your shop. I cannot honestly say that I share in his confidence"; see Doc. 2149.

33. The New York Board of Fire Underwriters adopted on 12 January five safety rules developed in consultation with the Edison Electric Light Co., which expected them to be adopted throughout the United States (Edison Electric Light Co. Bulletin 1:1, 26 Jan. 1882, CR [*TAED* CB001; *TAEM* 96:668]; "Electric Lighting," *American Architect and Building News* 11 [1882]: 321). Among them was a requirement for an automatic circuit breaking device where a large conductor met smaller conductors within a building. The guidelines are quoted in Preece 1881 (102–3).

34. Preece 1881.

35. See Doc. 1825 esp. n. 13.

36. Edison's letter is Doc. 2217.

37. A flattering biographical article appeared in a special 5 November issue of the *Telegraphic Journal and Electrical Review* devoted to the Paris Exposition. It concluded with the promise that should Edison "fulfil his intention of coming to England, he will meet with a cordial reception from English electricians." "Thomas Alva Edison," *Teleg. J. and Elec. Rev.* 9 (1881): 431–33; "The Paris Electrical Exhibition: Foreign Sections. Edison," ibid., 436–45; both Cat. 1243, items 1745–46; Batchelor (*TAED* MBSB41745, MBSB41746; *TAEM* 95:24, 26).

38. A few days later Insull sent Johnson a list of items Edison wished to alter in the proposed agreement. Insull to Johnson, 8 Jan. 1882, LM 3:11 (*TAED* LM003011; *TAEM* 84:12).

39. Edison made several references to carbonized "fibrous material" in his British Patent 4,576 (1879) but did not elaborate on the special qualities of vegetable matter for making high resistance filaments until the September 1880 provisional specification. British Pat. 3,765 (1880), Cat. 1321, Batchelor (*TAED* MBP027; *TAEM* 92:172).

40. That is, organic. *OED*, s.v. "Organic" 1.

41. To perceive or to understand. *OED*, s.v. "Twig" 1b, 2.

–2204–

From John Ott

Menlo Park, N.J., Jan 5 1882[a]

Mr Edison

I burnt several Carbons in sand and quarts but find it a failure, as when it comes out of the fiurnice, it resembles that of old Aincent Pottry and is nessiary to be broken on the anvil.[1]

I tried other substances. I shall give you the folling table below

Kaolin one part Wood five parts, makes a verry hard Carbon smooth and close but resistance high.[2]

Kaolin and Wood Res 525, 470, 415, 480, 400, 540, 500. You will see the resistance are not equal. in wood carbon they are more equal. Res 345, 380, 405, 440,[b] 340, 345, 360, 345, 495, 365, 440. This lot is lower not eaven.

I sent model to [Dreon?][c] & Co[3] they say they canot let us have them sooner than two weeks as it takes that long to dry

John. F. Ott.

⟨Keep on experimenting & send in Daily reports addressed to 65 Fifth Ave⟩[4]

ALS, NjWOE, DF (*TAED* D8232A; *TAEM* 61:956). Letterhead of Edison Electric Lamp Co. [a]"Menlo Park, N.J.," and "188" preprinted. [b]Followed by "over" as page turn. [c]Illegible.

1. These experiments apparently concerned the longstanding problem of allowing filaments to shrink while still retaining their shape during carbonization (see, e.g., Docs. 1961, 1966, and 1973). John Howell recalled that at some unspecified time the lamp factory began to carbonize filaments in iron boxes packed with "peat moss which was first roasted. This shrunk the same amount as the bamboo and kept the fibers in shape." During the first (of two) firings the temperature was brought up slowly "for if we went too fast the tar which was distilled out of the bamboo and peat would come too fast and stick all the carbons together." During this heating "the mass of peat and fibers shrunk away from the box leaving an empty space all around." Jehl 1937–41, 811.

2. John Howell noted spirals carbonized in wood charcoal and in a mixture of charcoal and kaolin on 11 January; no other records are extant. Cat. 1301, Batchelor (*TAED* MBN007:67; *TAEM* 91:360).

3. Unidentified.

4. This marginal note is the basis for Edison's one-sentence reply to Ott on 9 January (Lbk. 11:43 [*TAED* LB011043A; *TAEM* 81:255]). The few extant reports by Ott from this period appear to concern tests similar to those described in this document. One refers to wood spirals, suggesting that Ott was continuing the experiments on spiral filaments that Edison had been making at the end of 1881 (Ott to TAE, 14 and 18 Jan. 1882, both DF [*TAED* D8232F, D8232G; *TAEM* 61:962–63]; see also Doc. 2212).

–2205–

To French Minister of Posts and Telegraphs[1]

[New York,] Jany 7 [188]2

Sir

I have the pleasure to acknowledge the receipt of your esteemed letter of the 17th December 1881 announcing that I have been appointed an officer of the National Order of the Legion of Honor, and I have since received through our Secretary of State the Diploma and Insigna of the order[2]

I beg to thank you for your extreme courtesy in sending me a notification of the very great honor awarded me   I am Sir Your Obedient Servant

Thomas A Edison.

LS (letterpress copy), NjWOE, Lbk. 11:41 (*TAED* LB011041; *TAEM* 81:254). Written by Charles Mott.

1. Louis Adophe Cochery (1819–1900) was French Minister of Posts and Telegraphs from 1879 to 1885. *Ency. Brit.* 1911, s.v. "Cochery, Louis Adolphe."

2. The French National Order of the Legion of Honor was granted to exhibitors who made a significant contribution to the success of the Paris Exposition. The French Minister of Foreign Affairs forwarded the award to the U.S. State Department, which informed Edison that he was entitled "to wear a round red rosette in your button hole for every day use, while the big cross hitches on your coat for full dress." Secretary of State Frederick Frelinghuysen sent the diploma (displayed at Glenmont, Edison's West Orange home) and decoration on 7 January. In 1878 Edison had earned a lesser award, Chevalier of the Legion of Honor, at the Paris Universal Exposition. Cochery to TAE, 17 Dec. 1881; Adee to TAE, 6 Jan. 1882; Frelinghuysen to TAE, 7 Jan. 1882; all DF (*TAED* D8135ZCW, D8215A, D8215B; *TAEM* 58:1132, 60:732, 734); Doc. 1519 n. 1.

–2206–

To Christian Herter[1]

[New York,] 9th Jan [188]2

My Dear Sir,

I duly received the set of Emersons works which you were good enough to send me.[2] I am reading them with very great pleasure

I should have acknowledged them long ere this but the package came without anything to show who sent them and it was only by accident that some days after their receipt my attention was called to them   Very truly Yours

Thos A Edison

LS (letterpress copy), NjWOE, Lbk. 11:39 (*TAED* LB011039A; *TAEM* 81:252). Written by Samuel Insull.

1. Christian Herter (1840–1883) headed the interior decorating firm of Herter Brothers. His last major project was the Vanderbilt family residences in Manhattan, where an Edison lighting plant was installed in 1881 (see Doc. 2108 n. 2). He had worked in the home of Darius Ogden Mills, who was later a director of the Edison Electric Light Co. and Edison Electric Illuminating Co. of New York. Herter Brothers continued to operate after Herter's death and later worked on Glenmont, Edison's home in Llewllyn Park, N.J. *DAB,* s.v. "Herter, Christian"; *ANB,* s.v. "Mills, Darius Ogden."

2. The editors have found no further information about this gift.

To Edward Johnson

[New York,] Jany 9. 82

Fiftyseven London[1]

Only trouble is with Engine[2]   Dynamo will run all lights
that engine can without heating   you can run eight hundred
fifty easily if you can keep boiler pressure up will carry thirteen hundred   Dynamo will stand anything after you acquire
knack attending brushes[3]   we run continuously eight hundred fifty   shall we run two or three nights longer and ship
will knock of engine cause any trouble[4]   can send small bar
dynamo ten days   Francis leaves in thirty days with supply
meters[5] you and Batchelor will instruct your men and return
use "Z" dynamo as motor will ship smaller one this week[6]
Fabbri cables to morrow concerning contract[7]

L (telegram, copy), NjWOE, LM 1:136A (*TAED* LM001136A; *TAEM*
83:940). Written by John Randolph.

1. Cable code for Edward Johnson; see App. 4.
2. Samuel Insull wrote Johnson the previous day that he was "inclined to think that the Porter Allen engines will give a very great deal of
trouble. Porters reputation with Edison is 'petering' out & he (E) would
like to have an Armington & Sims Engine to send over with the new machine. Edison went to the Works early this morning & he is now testing
the Dynamo after having made some alterations on the engine." Insull
to Johnson, 8 Jan. 1882, LM 3:11 (*TAED* LM003011; *TAEM* 84:12).
3. For Edison's detailed instructions about setting up and maintaining the commutator brushes see Docs. 2149 and 2190.
4. Johnson cabled back on 11 January to "get knock out if possible
ship not later than middle of next week." Johnson to TAE, 11 Jan. 1882,
LM 1:137 (*TAED* LM001137B; *TAEM* 83:940).
5. Jehl sailed with the meters on the *Arizona* from New York to Liverpool on 14 February. TAE to Johnson, 14 Feb. 1882, Lbk. 11:133 (*TAED*
LB011256; *TAEM* 81:333).
6. Edison shipped two motors on 20 January and a total of ten dynamos on 11 and 13 January. TAE to Johnson, 23 Jan. 1882, LM 1:151B
(*TAED* LM001151B; *TAEM* 83:947).
7. Egisto Fabbri's message has not been found. Johnson stated in his
11 January cable (see note 4): "Fabbris cables receiving our attention."

*From Charles
Batchelor*

Paris, le Jan 9. 1882.[a]

My dear Edison,

We have just started our Strasbourg plant and it is working
nicely in the Gare or R.R. station there—[1] The first night was
eventful as usual; that is that although we had only on 35 lights
after running 4 hours a man noticed fire coming out of armature at back side. I had Suebel there and after he stopped he investigated it and found that $\frac{1}{2}$ inch of a top layer and $\frac{1}{2}$ inch

of a bottom layer had burned right away and he had to patch in again so much— This was defective insulation between the two— Of course no one knew anything about this but ourselves as it never stopped— I am very much afraid that they do not give these machines a 5 hours trial with a load on, or <u>you would certainly see these defects</u>; remembering the many nights I had to put in at the Opera House, doctoring them up, I always feel scared till I hear that they have run two or three days with a load on—[2] Please make more rigorous trials. Yours

Chas Batchelor

⟨Do not have any trouble here machines are run four hours with 80 lights on for test Strasbourg machine probably one of the first ones made which were defective⟩[3b]

ALS, NjWOE, DF (*TAED* D8238H; *TAEM* 62:212). A letterpress copy is in Cat. 1239:96, Batchelor (*TAED* MBLB4096; *TAEM* 93:544). [a]"Paris, le" preprinted. [b]Marginalia written by Samuel Insull.

1. In early October 1881 Theodore Puskas and Joshua Bailey secured orders for the railway stations at Strasbourg, Mulhouse, and Metz through a junior administrator of the German telegraphs who had attended the Paris Electrical Exposition and had "studied thoroughly" Edison's lighting system there. Bailey believed that successful installations in these stations would lead to more contracts in Germany, including a 3,000 lamp installation for the new university campus in Strasbourg. The initial installation at Strasbourg consisted of a Z dynamo and sixty 16 candlepower lamps for illuminating a portion of the station. It began operation on 5 January 1882, and four days later Philip Seubel reported that the station director was "so delighted . . . with the new light that he arranged a great Dinner. . . . The RR officials would not get tired to fill up their glasses with champagne and drink the health of 'Herr' Edison, of which fact I make mention upon the special request of the directors." Batchelor reported in February that the "Strasbourg plant never missed an hour since we patched it and it runs from sunset to sunrise." The success of this small installation led to negotiations in April for lighting the entire station; the plant was expanded to 1200 lamps in March or April 1883. Bailey and Puskas to TAE, 5 Oct. 1881, LM 1:53B (*TAED* LM001053B; *TAEM* 83:898); TAE to Edison Machine Works, 6 and 7 Oct. 1881; Bailey to TAE, 7 Oct. 1881; Seubel to TAE, 9 Jan. 1882 (misdated 1881); Batchelor to TAE, 11 Feb. 1882 and 23 Jan. 1883; all DF (*TAED* D8129ZBR, D8129ZBT, D8238J1, D8132ZAV, D8238X, D8337H; *TAEM* 58:288, 291, 419, 501; 62:315; 67:605); Batchelor to Eaton, 3 Dec. 1881, Cat. 1239:27, Batchelor (*TAED* MBLB4027; *TAEM* 93:504); Edison Electric Light Co. Bulletins 3, 7:6, 17:11, and 18:18; 24 Feb. and 17 Apr. 1882, 6 Apr. and 31 May 1883; all CR (*TAED* CB003, CB007, CB017, CB018; *TAEM* 96:674, 692, 809, 827).

2. For Batchelor's accounts of similar troubles at the Paris Opera House see Doc. 2179 and Batchelor to Seubel, 9 Jan. 1882, Cat. 1239:97, Batchelor (*TAED* MBLB4097; *TAEM* 93:545).

3. Insull's marginalia, probably dictated by Edison, was the basis of

Edison's reply. TAE to Batchelor, 26 Jan. 1882, Lbk. 11:159 (*TAED* LB011159A; *TAEM* 81:303).

–2209–

From Francis Upton

Menlo Park, N.J., Jan. 9, 1882[a]

Dear Mr. Edison:

The factory started off this morning quite promptly.

All our men have come back so we have lost nothing by stopping.[1]

Holzer is getting ready a number of experiments for you.

Branner has been here.[2] I have sent his samples to E. Newark where I am to meet him as soon as they reach there to examine them with Bradley[3] and Branner.

Branner is very anxious to make an exhaustive report, but wants pay for his time while working on it. He says he can make a very interesting and full account of his journeys.

He wants $100 a month for three months[4] and money for cuts and printing. Do you want anything done in the matter or only have him tell where he got his fibres and how we should go to work to get any more.

He is very anxious to gain a little fame.

The new boxes are on on one line of pumps[5] and seem to work well. I have been working all day on them getting them in perfect order, so that mistakes will not likely be made. Yours Truly

Francis R. Upton

The Corning Works write that they make nothing but lead glass.[6] Glass with lime instead of lead is very brittle.

⟨Only want to know where Branner got his fibres from & how we are to go to work to get more   Dont care anything about an account of [his][b] journey⟩[7]

ALS, NjWOE, DF (*TAED* D8230C; *TAEM* 61:721). Letterhead of Edison Lamp Co. [a]"Menlo Park, N.J.," and "188" preprinted. [b]Illegible.

1. The factory was shut for two weeks over the Christmas and New Year's holidays (see Doc. 2202), presumably to alleviate overproduction; cf. Doc. 2159. While the layoff would also have reduced costs briefly, the factory continued to suffer cash-flow problems. By the end of 1881 it owed $18,000, mostly in short-term debts, and carried a $29,400 inventory of about 84,000 lamps. Philip Dyer to TAE, 20 Dec. 1881, DF (*TAED* D8124ZAV; *TAEM* 57:1079); Edison Lamp Co. to Frank McLaughlin, 20 Dec. 1881, Upton (*TAED* MU057; *TAEM* 95:666).

2. After Edison recalled him from Brazil (see Doc. 2191), John Branner returned to New York before 3 January. George Phelps to TAE, 3 Jan. 1882, DF (*TAED* D8230B; *TAEM* 61:721).

3. James Bradley, a longtime Edison employee, was getting the new lamp factory ready in Harrison. Doc. 1080 n. 4; *TAEMG2*, s.v. "Bradley, James J."

4. Calvin Goddard notified Edison at the beginning of February that the Edison Electric Light Co. was still paying Branner a monthly one hundred dollar salary, and wondered if it should continue to do so. Goddard to TAE, 2 Feb. 1882, DF (*TAED* D8230T; *TAEM* 61:736).

5. Nothing is known of these apparatus.

6. The Corning Glass Works supplied tubes and globes to the lamp factory from the beginning of production there. Doc. 1971 and headnote, Doc. 2098.

7. The *New York Times* published a lengthy interview with Branner about his travels, portions of which it later retracted. "A Brooklynite in Brazile," *New York Times*, 23 Jan. 1882, 5; "The Travels of Mr. Branner," ibid., 27 Jan. 1882, 8.

–2210–

From Lemuel Serrell

New York, Jan 11, *1882*[a]

My Dear Sir

Messrs Fabbri & Lowrey desire patents in South Australia & Tasmania: They do not think it safe to put more than one English patent in one patent there, but desire you to look over the English patents and designate which ones you consider absolutely necessary for the protection of the Light inventions that are now employed

Please indicate which of these English patents must be protected there and inform me at your early convenience and oblige. Yours truly

Lemuel W. Serrell

ADDENDUM[1b]

[New York, c. January 12, 1882]

2402   of 1879 — 1st claim if [c] broadened to include carbon
                 filaments, 16th claim

45776 of 1879 — 2d claim

5127   of 1879 — 1st & 3d claims

33   " 1880 ———

578   "   "   — 1st, 2d, 3d, 4th [--][d] 7th 9th, 10th, 11th,
                 12th, 13th 14th, 15th, 16th and 17th claims

602   of 1880 ———

1385   "   "   1st and 2d claims

367765 "   "   1st, 4th, 5th, 6th, 7th, 9th 10th, 11th,
                 12th, 19th, 20th, 21st 27th.

3880   of 1880 — 1st, 2d.

3964   " 1880 — 1st, 5th, 22d, 23d 24th, 25th.

4391	"	1880 ———	
539	"	1881 — 1st, 6th, 7th	
562	"	" — 4th, 5th	
768	"	" ———	
1240	"	" ———	
1783	"	" ———	provisional should be made to correspond with complete.
1802	of 1881 ———		

Should also include <u>Meter</u>—described but not claimed in Eng pat.[e] 4226 of '78   Conductors—U.S. Appl 307.[2]

ALS, NjWOE, DF (*TAED* D8248C; *TAEM* 63:372). Letterhead of Lemuel Serrell. [a]"*New York*," and "*18*" preprinted. [b]Addendum is a D written by Edison. [c]Obscured overwritten text. [d]Canceled. [e]"Eng pat." interlined above.

1. For the sake of simplicity, the editors have chosen not to explicate the numerous individual patents and claims in Edison's list below, which he prepared without further comment. Edison analyzed the scope and importance of several in Doc. 2203 (see also headnote). His British patents prior to 1881 are in Cat. 1321, Batchelor (*TAED* MBP; *TAEM* 92:3); facsimiles of those from 1881–1883 will be published on the Edison Papers website (http://edison.rutgers.edu).

2. This application issued in December 1881 as U.S. Patent 251,552.

–2211–

To Edward Johnson

[New York,] 15th Jany [188]2

My Dear Johnson,

Your letter 23rd December 1881.[1]

As to the fact that the gas engine proves the economy of our system I would not be quite so sure of this for the reason that if you take into consideration the interest and depreciation the Dynamo and its appliances and the life of the lamp and their cost you would find that that it would not turn out to be true. If you attempt to make it true by increasing the economy of the lamps you will get [left?][a] by the lessened life of the same

You are quite right in stating that [one of?][a] the main factors (in fact the principal factor) in the economical introduction of our system is the high resistance lamp[2]   The real factor is the greater number per horse power providing the life is always kept the same or is not diminished by the increased economy This reduces the investment in Dynamo steam engines & boilers in proportion to the increased number per horse power

In the Printers Proof you sent me you say "of course it is a well known law that a given amount of heat acting upon two bodies one a large and one a smaller one will produce results in proportion to the ratio of their size." You should not have used the word <u>heat</u> but should use the <u>energy</u> instead

As to ~~Tt~~he words "out of the globe and leaving the carbon in a atmosphere of such exceeding rariety[b] as to preclude the radiation of any material degree of heat from the walls of the globe" you must take out the word radiation and put in the following: "As to preclude the carrying by convexion of any material degree of heat"   The reason for making this alteration is that when the globe is perfectly exhausted all the light and heat is radiated to the globe and through it. In fact in a perfect lamp all the light is radiant while if there be air in the globe heat is carried through the globe by convexion currents of the air.

You speak of others obtaining a thin carbon. Where do you find a reference to other people using a thin carbon?[3]

You say "this is done by sealing them in a vessel containing nitrogen and gases and electrically energising them. You should say "sealing them in a vessel containing hydrocarbon

In connection with this building up process you might mention that this form of carbon thus deposited is amorphious in its structure and does not withstand the carrying action of the Electric Current in the vacuum as well as naturally arranged carbon.

You will find in a later patent (2492 of 81) a claim for copper wires soldered to the platina wires within the vacuum so as to save platinum.[4] You will also find I think several patents on the method of carbonising the fibre[5]

Your letter Dec 29th.[6] I think you will find upon making further enquiries about the price of the[c] Seimens machine as compared to our machine that ours at the price you gave will in all cases be found cheaper than Seimens when you take into consideration the amount of horsepower applied to the amount of light on the same number of lamps of the same candle power. Seimens machines according to Hopkinsons tests when the value of his standard cell is connected is only 83% machine scientific efficiency[7] and with incandescent lamps I do not believe he will get more than 55 or 60% [ofs?][d] the lamps I even doubt if he got this much. Their machines besides are not nearly as solid and substantial as ours. We can of course skim our machines down and make them ~~appear~~ cheaper. I think I could build the present machine or rather alter it so that it would only cost $350 but it would reduce the economy and increase the subsequent troubles of taking care of it. We cannot possibly [reduce?][d] our price because we only made $15 each on the first 100 machines the stock itself without any labor costs $[310?].[d] By making the field magnets one third as long & by loosing a horse power in the field magnets making a shallow commutator with only a few blocks and the Iron cores of thick plates—this would enormously reduce the price but it appears to me that I would sell the present Dynamo and not the inferior one even if I did not sell as many as in the long run we shall come out ahead.

I do not believe you will have any trouble with your lamps; to be sure the lamps which you have are not so good as those turned out now but they certainly have over a 600 hour life. You must remember that if you put up 100 lamps and run them at sixteen candles for 6 hours there will be two or three go in the first night (sometimes as many as four the next night two, the next two or three nights after probably one and then they will run along for five or six days without one going. The weak lamps are all shoot out the first three or four nights. I will bet my life that if you keep these lamps at sixteen candles and keep a time record they will last over 600 hours average. Except possibly the "B" lamps which may possibly not go over 500 hours but I suggest that you order with new sockets 2000 of our new lamps and if you keep them at 16 candles and they last [less?][d] [th?]an[e] average of 800 hours I will agree to eat them. In running our old curves considerable number of lamps arced but of course they were at 48 candles while the new lamps scarcely ever are at 48 candles thus showing a great improvement in this respect

⟨Note. Upton has just come in while I am writing this from my notes. He says that not one single lamp has arced this year Insull⟩[f]

About Motors. You can put half a dozen regular "Z" Dynamos (machines for "A" lamps anywhere on your main conductors and putting in ten or fifteen ohms resistance in the circuit with the Dynamo: place it across the conductor the same as a lamp   Then close the circuit and the Dynamo will start off and then you can keep plugging out the resistance until you have got it all off and the Dynamo will run fr 1500 or 1800 revolutions a minute and only absorb one half a horsepower so if you have six of these Z Dynamos used as motors only to run themselves you will use only about three horsepower but to make it more effective you should run a belt from one or two of them up to a pulley in Journals connected to the wall or on the floor   This will give the idea of motion. You can also after the Dynamo has been started and is running and all the resistance is out rig up some scantling[8] so that persons can try and stop the Dynamo but do not rig the scantling so powerful as to enable them to reduce the velocity below 600 or 800 revolutions and then you will only absorb 76 or 7 house power out of that one particular Dynamo.

My impression is that you can get the "Z" Dynamo field magnets [and?][d] base and boxes[9] contracted for in England much cheaper than we can get them here on account of the low price of material if so we could send you the Switch Armature[10] with shaft complete

We shall send you by next steamer a 15 "A" Light machine. The speed I think necessary to get the proper electro motive force up is 2000 revolutions per minute. It will work twenty five lights without materially heating but we call it a 15 Light machine. Make use of this as a motor and it will give about two horse power. As this machine will run 40 B lamps and can be made for probably $225 or there abouts you might compare this with the capacity of a Seimens Machine at £90.

The Armington & Sims (owing to their being in a state of transition from Lawrence to Providence) engines which you ordered have not come to hand   The price of Engine complete for running one "Z" Dynamo very economically and two Dynamos pretty economically is $550 but the price list[g] we sent you would would be quite sufficient when translated to you by an Engineer. The engine required depend entirely on the Boiler pressure. If with 80 lbs a certain engine will only run two Dynamos with a certain economy you can increase the Boiler pressure so that it will run two Dynamos with the same economy as it will run one with a lower boiler pressure

Your letter 31st Dec. I am sorry to say that in your arguments with Bidwell[11] you have made a mistake.[12] You are generally right but in one of the details you are wrong   You state that one lamp of 125 ohms has one Ampere[13] of current passing through it. If you take one "A" Carbon there will be one Ampere passing through it and it will give say 16 candles. If you cut it in two parts you will have two B lamps and if these two parts are multipled arced you will have two amperes passing through them. (i.e.) an ampere through each hence we get 16 candles but have two amperes yct the energy [in?][a] foot lbs are the same. If we divide these two parts into 4 we will have 4 parts each giving 4 candles total 16 candles. But there will be four amperes passing. If say the A lamp is 100 ohms and you split it up with 100 parts and multiple arc all these parts between your plates there will be 100 amperes passing from plate to plate and the total candle power will be 16 candles but the total resistance will only be one one hundredth of one[b] ohm and not one ohm as you say because each part being one ohm multiple arcing 100 carbons of one ohm each would reduce the resistance to $1/100$ of one ohm. Hence if we leave out the loss of heat from conduction to the plates and say the loss is no more for the 100 peices than for one peice in an "A" lamp. You would have 16 candles and get the same number per horsepower as if it was a continuous fibre. You see that if you have a Carbon that gives one Ampere and sixteen candles you may subdivide it into a thousand peices and each will have passing through it an Ampere   Take for instance the case of 16 candle B lamps in place of 10 candle A lamps; there is exactly the same economy and number per horsepower but one has a resistance of say 100 ohms while the "B" 16 candle lamps have a resistance of 25 ohms but the radiating surface and the number of foot lbs of energy radiated is exactly the same. So you see if we had a one ohm conductor and 100 "A" lamps 16 candles the aggregate resistance of the lamps would be one ohm hence if 10 horse power was used it would be divided over 2 ohms—5 horse power would be lost in the Conductor and 5 horse in the lamps. If we should place 100 16 candle B. lamps instead of A lamps the aggregate [revolutions?][h] resistance[g] would only be $25/100$ of an ohm or $1/4$ of an ohm and as the conductor is one ohm there would be 8 horsepower lost in the conductor and 2 in the lamps. Of course if five horsepower is required to bring the lamps up to proper candle power it will require the same horse power for the B 16 hence we must increase the horsepower to 25 horsepower of which 20 horsepower will be lost on the conductor and five in the lamps and so on ad infinitum but if we put four times the mass of copper in the conductor it would reduce its resistance to a quarter of an ohm and as the resistance of the 100 B 16 candle lamps is $1/4$ of an ohm the loss of energy would be the same as in the first instance with the A lamps.[14]

To sum up:

The amount of copper necessary to use in the Conductors is in direct proportion to the resistance of the lamp under like conditions. If I spend in copper $100,000 for 20,000 100 ohm lamps I will have to spend $200,000 for 20,000 50 ohm lamps all giving the same candle power.

Roughly to light the City of London each additional ohm of resistance in the lamp will effect a saving of $57 142

If Bidwell took your Ampere illustration in he is not well up on Amperes

You were all right only you should have said each column or subdivision uses one ampere and not one ampere for the whole.

I can see from the printed matter regarding the lamp that you are finding out a great many things that you did not know when you left. The printed matter is astonishingly accurate considering what little you have had to do with the energy portions of the scheme and with the corrections made I endorse it.

I want to call your attention to another little funny thing regarding the economy of our lamp (that is to say a high resistance carbon) i.e that those bodies which are good conductors of heat are also good conductors of electricity and vice versa. Now we have determined that the loss of energy by conduction down the clamps ~~is~~ is 256 foot pounds or one eighth of the total energy that it takes per lamp and this loss increases just in proportion ~~just~~ as the length of the carbon decreases in its resistance. The loss by conduction when a platinum loop is used is so great that it does not get to full incandescence for half an inch above the clamps—an eighth of an inch above the clamps it is not even visibly red so you see there is another gain in economy by using high resistance carbons.

I have handed your letter of 23rd of Dec on Telephone to Insull to go into it when he can get a few spare moments[15]   I am much obliged to you for the information contained therein and with what we already know it will doubtless make matters clear. Gouraud wrote me again in regard to Glasgow Commission to Moore[16] & Stoddard[17] and also about the sale of United Shares by which I lost $25 000 but he simply reiterated his former statements so I replied that I had no wish to carry on an interminable correspondence and should therefore add nothing to what I had already said in former letters[18]   So that is how the matter stands. Telegraph me the moment Gouraud gets my remaining United Shares quoting current prices of same.

I think the contract submitted by Mr Fabbri is all right with the exceptions of the alterations cabled.[19] Of course not getting a large amount down hurt me as well as the purchasers in England but they of course cannot understand how the extra payment of money to me can tend to their benefit although it certainly would tcn times over. I suppose however that in the long run it will be better for us to have taken less money as it must add to the value of our shares.

It is not much use to say anything here about your big machine. We have got the brush business down fine excessively convenient and you will have no trouble from this as the current can be taken off a great deal stronger than the engine can run. Our main trouble is the Porter Engine. After we got disgusted with it we brought Porter over and he became disgusted with it and now he is rebuilding one of the Engines, and promises to bring it here by Tuesday, and says it will run silent and give no trouble and Mr Fabbri thinks we had better wait for this.[20] We will send you a full letter regarding the workings of the machine when we ship it. Yours very truly

T A Edison

LS (letterpress copy), NjWOE, Lbk. 11:92 (*TAED* LB011092; *TAEM* 81:271). Written by Samuel Insull. [a]Faint. [b]Obscured overwritten text. [c]"price of the" interlined above. [d]Illegible. [e]Edge of original not copied. [f]Marginalia written by Insull. [g]Interlined above. [h]Canceled.

1. Edison referred to one of three letters from Johnson on this date. The others concerned an arc light patent and details about telephone business arrangements in Britain, its colonies, and Europe. Johnson enclosed with his letter a printer's proof of the first section, pertaining to the lamp, of a booklet on "The Edison Electric Light." He noted that he had completed a second section (of eight) in manuscript but "I have to revise and re-write a great deal of this stuff before it becomes perfectly clear to my own mind, and therefore, before I can make it clear to others, or in other words, I am educating myself as I go along. If my argument and deductions are not perfectly sound you must devote a little time to correcting them." The proof contains handwritten emendations by Johnson and marginal notes by Edison. Johnson to TAE, with enclosure, 23 Dec. 1881, DF (*TAED* D8133ZBM; *TAEM* 58:788); Johnson's other letters to Edison of that date are DF (*TAED* D8104ZFM, D8133ZBL; *TAEM* 57:314, 58:786).

2. Johnson argued in his letter (see note 1) that increasing lamp resistance would permit greater resistance in the conductors and consequently would enlarge the service area for a given amount of copper. The cost of distribution was "the question upon which our ability to supply Electric Light at the cost, or less than the cost of gas, turns" because the cost of producing electric light was less than that of gas light: "therefore, the only question is, will the investment necessary to effect a practical distribution of the Light be such as to eat up the margin of difference in cost? I tell our people that this is the only question yet to be determined and it is to be determined solely by the resistance in a Lamp and that, therefore, any advance in the direction of increasing this resistance enables us the better to compete with gas."

3. In the printer's proof of "The Edison Electric Light" (p. 3; see note 1) Johnson countered the argument "that Mr. Edison's claim for high resistance in *combination* with a slender filament is anticipated by the recorded efforts of others to obtain *thin* carbons, in order to obtain an economical conversion of energy into light. . . . There is, in fact, no indication that in searching for *thin* carbons they were necessarily searching for *thin* carbons of *high resistance;* on the contrary, all the evidence points in the opposite direction, viz., thin carbons and *low resistance.*"

4. See headnote, Doc. 2098.

5. Brit. Pats. 5,127 (1879) and 3,765 (1880), Cat. 1321, Batchelor (*TAED* MBP020, MBP027; *TAEM* 92:138, 172). Edison's British Patent 578 of 1880 (Cat. 1321, Batchelor [*TAED* MBP023; *TAEM* 92:149]) also dealt with substituting copper for part of the platinum lead-in wires.

6. Doc. 2200.

7. In 1879 Francis Upton calculated this efficiency for the small Siemens machine based on the experiment described in Hopkinson 1879. Because of questionable battery voltage in the test arrangement, Edison disputed the higher efficiency that John Hopkinson had ascribed to the machine. See Doc. 1772 esp. n. 7.

8. A small beam (specifically less than five inches square) used for braces, ties, studs, or similar construction. *OED*, s.v. "Scantling" 7; Knight 1881, "Scantling."

9. That is, journal boxes.

10. Edison referred to the standard interchangeable armatures with which the Z dynamo could run either 16 or 8 candlepower lamps. See headnote, Doc. 2126.

11. Shelford Bidwell (1848–1909) was a barrister whom William Preece had recommended to review Edison's electric lighting patents in Britain. He reported on Edison's claims for the lamp in November. Bidwell had been a juror at the Paris Exposition and belonged to the Society of Telegraph Engineers; he was elected a Fellow of the Royal Society in 1886. Theodore Waterhouse to Johnson, 12 Nov. 1881; Waterhouse to Egisto Fabbri, 16 Dec. 1881; both DF (*TAED* D8133ZAS, D8133ZBF; *TAEM* 58:706, 770); "Congrès International des Électriciens, Liste Des Membres du Jury," *La Lumière Électrique* 4 (1881): 417–420; Obituary, *Engineering* 88: (1909), 864.

12. Johnson reported on 31 December that he had "considerable trouble in convincing Bidwell & others that Resistance as such was not a factor in the Economical conversion of a given unit of Energy active on a given carbon surface— That is to say that with a given surface & a given current active on it the same light would be obtained whether the resistance was one ohm or a thousand." He described a thought experiment he used to illustrate this point. In this hypothetical situation, Johnson started with a 6-inch Edison filament of 125 ohms resistance in which the passage of a 1 ampere current would produce 16 candlepower. He divided the filament into 125 equal segments, each having resistance of 1 ohm. Johnson concluded that each segment would have "$^1/_{125}$ of the total length of carbon yielding $^1/_{125}$th of the total Line of Light—or $^1/_{125}$th of 16 candles= and measuring $^1/_{125}$th of the total Res, or 1 ohm." Johnson then imagined standing the segments on end, one next to the other, and placing the entire group between conductive metal plates at top and bottom. He reasoned that with the passage of the same 1 ampere current each piece "must yield its $^1/_{125}$ of the total Light= & the whole No. of columns must use at up 1 ampere & yield 16 candles of Light— But the resistance is changed from 125 ohms to $^1/_{125}$th of one ohm" for the entire circuit. Johnson to TAE, 31 Dec. 1881, DF (*TAED* D8133ZBQ; *TAEM* 58:812).

13. The International Congress of Electricians at Paris adopted a number of absolute definitions of electrical units in September. Among them was the ampere, the amount of current which flows under one volt through one ohm, previously known as a weber. William Thomson

played crucial roles both in defining the ampere and in securing its adoption by the International Congress. This appears to be Edison's first use of the term. "Electrical Measures," *Sci. Am.* 45 (1881): 295; Tunbridge 1992, 34–40.

14. Fearing that Johnson may not have understood this explanation, two weeks later Edison framed it as an analogy with battery current. TAE to Johnson, 27 Jan. 1882, Lbk. 11:164 (*TAED* LB011164; *TAEM* 81:305).

15. DF (*TAED* D8104ZFM; *TAEM* 57:314).

16. Michael Moore, an American associate of George Gouraud, was an organizer of the Edison Telephone Co. of Glasgow. Doc. 1854 n. 5.

17. A partner of Michael Moore in the Edison Telephone Co. of Glasgow, but otherwise unidentified. TAE to George Gouraud, 5 June 1881, Lbk. 8:306 (*TAED* LB008306; *TAEM* 80:953).

18. Edison had disputed with Gouraud since June his liability for funds due to Moore and Stoddard (see Doc. 2107 n. 5) and the contested sale of Edison's United Telephone Co. shares (see Doc. 2079). Gouraud had advanced him $10,000 in November. In mid-December, Gouraud repeated his understanding of these transactions, offering to absorb the loss from overselling the United shares. Edison replied only that he had nothing new to say in either matter (Gouraud to TAE, 10 Nov. and 14 Dec. 1881, both DF [*TAED* D8104ZET, D8149ZAE; *TAEM* 57:279, 59:983]; TAE to Gouraud, 4 Jan. 1882, Lbk. 11:28 [*TAED* LB011028; *TAEM* 81:249]). A disagreement with the United Co. over liability for the purchase of 500 telephones further delayed the settlement, which did not occur until April (Gouraud to TAE, 11 Mar. 1882, DF [*TAED* D8256K; *TAEM* 63:862]; TAE to Gouraud, 21 Apr. 1882, Lbk. 12:139 [*TAED* LB012139; *TAEM* 81:577]).

19. Edison had wired Johnson on 9 January that Egisto Fabbri would cable the next day about the English contract but this message (evidently addressed to Edward Bouverie) has not been found. Solicitor Theodore Waterhouse foresaw few difficulties with the proposed alterations, except regarding the provision that William Thomson and engineering expert Frederick Bramwell attest to the system's economic competitiveness (clause 17), which Bouverie and another principal investor regarded as "fundamental." This requirement was eliminated; in return, Edison waived part of the cash advance. TAE to Johnson, 9 Jan. 1882; Insull to Johnson, 17 Jan. 1882; LM 1:136A, 3:20 (*TAED* LM001136A, LM003020; TAEM 83:940, 84:21); Waterhouse to Johnson, 11 Jan. 1882; Johnson to TAE, 1 Feb. 1882 (pp. 3–4); both DF (*TAED* D8239G, D8239Q; TAEM 62:670, 741).

20. Edison cabled Johnson the next day (Monday) that he could "ship dynamo any time but engine not satisfactory myself or Porter who promises another Wednesday   Fabbri and self think better wait if you willing." Johnson replied: "Dont ship till OK but waste no time parties hesitating." A day or two after this, the engine was returned to Charles Porter's shop in Philadelphia; Edison expected it back on 1 February. Porter later blamed the problems on faulty supervision and poor workmanship at the Southwark Foundry, which built the engines. TAE to Johnson, 16 Jan. and 1 Feb. 1882; Johnson to TAE, 17 Jan. 1882; LM 1:142C, 153B, 144A (*TAED* LM001142C, LM001153B, LM001144A; *TAEM* 83:943, 948, 944); Porter 1908, 309–10; see also Doc. 2006 n. 2.

From Francis Upton

Menlo Park, N.J., Jan. 17, 1882.[a]

Dear Mr. Edison:

The spiral lamps show about eleven per horse power, but are all very short lived judgeing from the blue in them at sixteen candles and that four of ten were gone at noon today.

760 minutes
1145
1415
1795

giving 16 candles. John Ott is trying a large number of experiments and is getting very good shaped spirals[1]

Howell has had a lamp on the guage today with a piece of charcoal in a bulb connected with it[2]

The vacuum has been[b] steadily rising with the lamp burning at about 48 candles. It rose from $1/30,000$ at 11 A.M. to $1/130,000$ at 4 P.M.[3]

Holzer is going to make a number more lamps for him and we will soon have a life test.

Do you want the lamps with a double chamber run through and tested for life, or shall we wait for you to come out here?[4]

How about the names on our paper as asked the other day? Shall we keep the names and add that of Maj MacLaughlin?[5]

Welsh is doing fairly running all the forms in boxes packed with wood charcoal. We have not cleaned the forms so far this week and have had no sticking.

The carbons are very good resistance and excellent body and color. Lawson does not want to take charge and experiment with Welsh here; as he says he could spoil any experiment if he so wished to, especially as he Lawson[c] would have to use the men that are under Welsh now.[6] He wants us to build our furnace at Newark and let him start there with full sweep. I will write you fully about this if you desire and where and how we want to build the furnace.

I am running through the new lamps as fast as they come and putting in old lamps enough to keep the socket room busy. The old lamps are nearly all marked and are mostly the N.Y. Volts or high so that they do not injure the Volts we use for the England the Continent or Isolated.[7]

There are a great many good lamps among the old, I feel sure that they would average 600 hours at 16 candles. Those that tend to leak or play out have had a chance in the long time that they have been stored.

We are now averaging about 800 lamps a day from the

pumps. The curves are about the same as before and there is a very small number of arcs

Only one lamp in twenty five at 16 candles has given out in the past eight days. Howell is making out the time sheet for them and will send it to you tomorrow.[8] Yours Truly

Francis R. Upton.

ALS, NjWOE, DF (*TAED* D8230K; *TAEM* 61:728). Letterhead of Edison Lamp Co. [a]"Menlo Park, N.J.," and "188" preprinted. [b]Obscured overwritten text. [c]Interlined above.

John Howell's drawing of a lamp with a side chamber of thickened glass containing "cocoanut charcoal about size of pea."

1. See Doc. 2204 esp. n. 4. On 23 January John Howell recorded life tests of lamps with spirals of different substances coated with vegetable oils. Cat. 1302, Batchelor (*TAED* MBN008:3; *TAEM* 91:368).

2. John Howell was experimenting with absorbent charcoal as early as 6 January. His sketches of that date relate to a modified lamp design (but see also Doc. 2219). According to Upton, Howell found that "lamps with a piece of charcoal near the inside part did not show any longer life than the usual run of lamps" (Cat. 1301, orders 595–96, Batchelor [*TAED* MBN007:69–70; *TAEM* 91:362–63]; Upton to TAE, 20 Jan. 1882, DF [*TAED* D8230L; *TAEM* 61:731]). The next day, however, Edison or Howell drew two lamps, each with bits of absorbent material in a small tube joined to the bulb base (Cat. 1148, Lab. [*TAED* NM017:39, 43; *TAEM* 44:399, 403]).

Two weeks later, Edison executed a patent application (with two drawings similar to his own) for this arrangement, which was intended to absorb gases released from the lead-in wires. Because gases would continue to be released during normal operation,

Edison's 21 January drawings of lamps designed for continuous absorption of gases released from the metal conductors during use.

the lamp as commercially sold and used must be provided with means ever present for preventing the exuding gases from affecting the quality of the vacuum. I accomplish this object by permanently providing the lamp with a material which will absorb completely the occluded gases as they are given off. For this purpose I use a piece of charcoal, preferably a dense cocoanut charcoal. . . . This charcoal must be removed as far as possible from the light, in order that it may not be heated thereby . . . which heating would decrease its capacity for absorption of gases. To do this I inclose the charcoal in a glass tube having a closed outer end, and connected at its inner end to the chamber of the lamp. . . . After the lamp is sealed this tube is heated at its junction with the globe and bent down against the neck of the lamp, it being included in and covered partly or wholly by the molded base. [U.S. Pat. 401,646]

3. It is not clear what units were used to rule the factory's vacuum gauges. Lamps were routinely exhausted to a pressure of about one millionth of an atmosphere. See Docs. 1816 n. 3 and 1873.

4. The "double chamber" lamps may have been like the sketch John Howell made on 6 January of a "Double Lamp one opening into the other" (Cat. 1301, order 597, Batchelor [*TAED* MBN007:70; *TAEM* 91:363]). At the same time, he was also designing an experimental lamp with a small attached bulb or tube to contain a desiccant (see note 2).

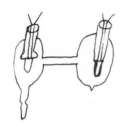

John Howell's 6 January sketch of a double lamp.

5. Edison sold Frank McLaughlin a five percent share in the lamp company in December (leaving himself a 70% stake). The lamp company's old letterhead carried the names of the four partners (Edison, Upton, Charles Batchelor, and Edward Johnson); it was using new letterhead without names by early February. TAE to Edison Lamp Co., 9 Dec. 1881; Edison Lamp Co. to Insull, 20 Jan. and 6 Feb. 1882; all DF (*TAED* D8123ZFT, D8230M, D8230U; *TAEM* 57:1015; 61:731, 736).

6. Welsh was discharged a few weeks later. Upton subsequently testified that Welsh had failed to give accurate reports of experiments and "this lack of truthfulness had become a byword in our factory." An assistant, Robert White, was let go at the same time because "we did not think it prudent to discharge Welsh and leave his intimate friend behind." Upton's testimony, 3256–57, *Edison Electric Light Co. v. U.S. Electric Lighting Co.*, Complainant's Rebuttal, Vol. 5, Lit. (*TAED* QD012F:133; *TAEM* 48:135).

7. Lamps selected for New York gave their rated candlepower at 110 volts. Those for England did so at 106–108 volts and were said to operate on "London volts." TAE to Edward Johnson, 16 Jan. 1882, Lbk. 11:85 (*TAED* LB011085; *TAEM* 81:264).

8. Howell to TAE, 18 Jan. 1882, DF (*TAED* D8232J; *TAEM* 61:968).

–2213–

From Leslie Ward

Newark Jan 18th 82

Dear Sir.—

My experience in treating your wife has convinced me that her uterine troubles yield much more readily to treatment when her nervous system is in a fair condition.—[1] I have noticed during the past few weeks that she seems very nervous and despondent and thinks that she will never recover &c.— I believe that an entire change would be of benefit and if you could take or send her to Europe for a few months she might return improved in health and be better pleased with her surroundings here—[2]

She seems so changed physically and mentally of late that something ought to be done and I am suggest nothing better that the above— Yours truly

Leslie D Ward[3]

⟨Ansd   I am going take her off on a trip soon! Etc⟩[4]

ALS, NjWOE, DF (*TAED* D8214C; *TAEM* 60:698).

1. Useful overviews of the contemporary diagnosis and treatment of women include Gijswijt-Hofstra and Porter 2001, Veith 1965, and Micale 1995, part 1.

2. Little is known of Mary Edison's health. She seems to have suffered at least one prior period of acute anxiety, in 1878. Shortly before Christmas 1881 Mary threw a party for New York friends in Menlo Park, but the editors have found nothing more of her activities at this time.

See Doc. 1394; Samuel Insull to Pennsylvania Railroad, 17 Dec. 1881, Lbk. 9:451 (*TAED* LB009451; *TAEM* 81:165).

3. Newark physician Leslie Ward (1845–1910), the Edison family doctor since at least 1878, was a co-founder of the Prudential Insurance Co. and one of its executives. He retired from practice in 1884 to devote his time to that company. Edison allotted him one stock share in the Edison Electric Light Co. in August 1882. Doc. 1394 n. 2; *NCAB* 25:103; Obituary, *New York Times*, 14 July 1910, 7; Carr 1975, 18–20, 25–27, 36–37, 39, 67, 73; Ward to Samuel Insull, 28 Aug. 1882, DF (*TAED* D8204ZFZ; *TAEM* 60:271).

4. Edison's response has not been found. Mary arrived in Detroit this same day. One month later she was in South Carolina. Mary Edison to TAE, 18 Jan. 1882, DF (*TAED* D8214B; *TAEM* 60:697); see Docs. 2233 and 2234.

–2214–

From Charles Batchelor

Paris, le Jan 20th ~~1881~~1882[a]

My dear Edison,

Your regulator idea is good and am glad to hear you are sending me one—[1] You may be sure I shall take all precautions necessary for the bringing in of the model etc—

I have now opened an office or rather reshipping place at Hambourg and have got Force and another young man there fitting it up—[2] Hamburg is a free port— I shall send Force to Finland very shortly to put up 5 plants there—[3] Acheson is at present putting up a plant in Milan Italy[4] and Suebel is doing good work in Germany—[5] I have an order for lamps for Krupp's Works at Essen[6] and I am going to send Seubel there to put them on the machines as they want to use the machines they have been using for Arc lights— How about your Arc-light? Is there anything new on it?[7] Yours

Batchelor

ALS, NjWOE, DF (*TAED* D8238K; *TAEM* 62:218). [a]"Paris, le" and "188" preprinted.

1. Edison described this device in Doc. 2201.

2. Martin Force and the Hungarian engineer Etienne de Fodor (born Istvan von Fodor) opened an office in Hamburg, a duty-free port, a few days before this letter. Batchelor to Force, 18 Jan. 1882, Cat. 1239:118, Batchelor (*TAED* MBLB4118; *TAEM* 93:558); "de Fodor, Etienne," Pioneers Bio.; see also Myllyntaus 1991, 302.

3. On 3 December Batchelor ordered six fully equipped lighting plants for Russia, which arrived in Hamburg on 7 January. One was to be installed at a naval base in St. Petersburg and the remaining five, each 60-lamp Z dynamos, were for the cotton mill of Finlayson & Co. at Tammerfors, Finland, an industrial city now called Tampere. (The mill's manager was the father of Edison associate Charles Nottbeck.) Since

Finland was at this time a province of Russia, the plants were forwarded from Hamburg through St. Petersburg to Tammerfors, where Force went in March. Force returned to New York in April, after his wife became seriously ill, leaving de Fodor to manage their installation. Batchelor to Eaton, 3 Dec. 1881; Batchelor to Puskas & Nottbeck, 9 Jan. 1882; Batchelor to de Fodor, 1 Dec. 1882; Cat. 1239:27, 98, 382; Batchelor (*TAED* MBLB4027, MBLB4098, MBLB4382; *TAEM* 93:504, 546, 738); Edison Electric Light Co. Bulletin 2, 7 Feb. 1882, CR (*TAED* CB002; *TAEM* 96:672); *WGD*, s.v. "Tampere"; Batchelor to TAE, 11 Feb. and 13 Apr. 1882, both DF (*TAED* D8238X, D8238ZAT1; *TAEM* 62:315, 376); TAE to Batchelor, 3 Apr. 1882, LM 1:181D (*TAED* LM001181D; *TAEM* 83:962); Batchelor to Force, 10 Jan. and 7 Feb. 1882; Force to Batchelor, 28 Jan. 1882 and undated Feb. 1882; all Force; Myllyntaus 1991, 302.

Batchelor asked Edison to inform him of changes to equipment shipped to Europe, such as the new style socket sent for Finland, because "the first intimation of this change I get from Finland [is] from parties who wish to return the lamps that I sent from Paris as they dont fit." Batchelor told Edison at the end of March that the Tammerfors and St. Petersburg plants both "continue to give excellent satisfaction." The success of these installations led to orders for ten more by the end of March 1882. Batchelor to Eaton, 15 Dec. 1882; Cat. 1239:407; Batchelor (*TAED* MBLB4407; *TAEM* 93:758); Batchelor to TAE, 7 Feb. and 28 Mar. 1882; Bailey and Puskas to Eaton, 28 Mar. 1882; all DF (*TAED* D8238T, D8238ZAN, D8238ZAP; *TAEM* 62:309, 358, 361); see also App. 2.

4. Edward Acheson installed an isolated plant at the Teatro alla Scala opera house that illuminated the foyer with 90 lights in February. In 1881, the theater premiered *Excelsior*, an extravagant allegorical ballet of Progress in which electric lighting figured prominently, to wide acclaim. Edison Electric Light Co. Bulletin 4:6, 24 Feb. 1882, CR (*TAED* CB004; *TAEM* 96:676); Sinclair 1989; *NGD*, s.v. "Manzotti, Luigi."

5. After Philip Seubel installed the Strasbourg railroad station lighting plant, Batchelor directed him to demonstrate Edison's lamp at a factory in Kreuznach that employed Siemens dynamos. Seubel was placed on the payroll of the new Société Électrique Edison, formed to exploit Edison's electrical power and light patents in France, in February. Batchelor to Seubel, 10 Jan. and 6 Feb. 1882; Cat. 1239:102, 30, 317; Batchelor (*TAED* MBLB4102, MBLB4030; *TAEM* 93:548, 506).

6. The eponymous steel and armaments firm controlled by Alfred Krupp operated a vast mill complex in the heavily industrialized Ruhr city of Essen. Tenfelde 2005, 15–24.

7. In May 1881, Edison had a measured drawing made of a combination incandescent and arc lamp which could fit into an existing Bergmann socket. In November, he had executed five patent applications for arc lamps. Two of these, including one expressly for arc and incandescent lamps in the same circuit, were rejected. Two others, pertaining to regulation of carbons in arc lamps, were allowed in 1882. The fifth, similar to the latter two, issued in April 1884. Batchelor and Ott drawing, 11 May 1881, Oversize Notes and Drawings, Lab. (*TAED* NS7986BAG; *TAEM* 45:7); Patent Application Casebook E-2537:74, 82; Patent Application Drawings, Cases 370 and 373; PS (*TAED* PT021074,

PT021082, PT023:46–47; *TAEM* 45:740–41, 864–65); U.S. Pats. 263,138; 265,775; and 297,580 see also Docs. 2216 and 2229.

–2215–

Samuel Insull to
Charles Batchelor

[New York,] 20th Jan [188]2

My Dear Batchelor,

I am almost ashamed to write you about amount received from Goddard ($186.25) & from you ($25).[1] It appears when I got them I paid them into my Bank account & omitted to enter them in my personal Cash Book & as I did not have my Bank Book made out till the end of the year I overlooked[a] the matter entirely and did not discover that I had in my possession money belonging to you until I compared my cash Book & Bank Book the other day when I found I had $211.25 belonging to some one else. Your letter showed me whose it was so today I have sent to Drexel Morgan & Co a check for this amount ($211.25) & doubtless they will advise you of it

I must claim your very kind indulgence for this big blunder on my part & trust you will excuse me.

Isolated Stock.[2] Edison arranged to pay for his Isolated Stock in Dynamos so I suggested that he should subscribe for yours & your wifes in the same way as you having a 10% interest in the Machine Works that would give you 10% of the Isolated Stock (between 50 & 60 shares) so that is how we arranged matters. As calls are made we put in a bill for "Z" Dynamos

T.A.E. has paid the Lamp Co assessments & Machine Works assessments for your account.[3] When either of these want money on your account they notify me & I send them Edisons check

I was notified by Eaton that you wanted Edison to pay for some E.E.L. Coy of Europe Bond & Edison also received your letter on this subject.[4] Edison is short of money just now but immediately he is in funds I will make the payment

I have been hoping to send you a complete statement of your account but have been waiting the receipt of Canadian Telephone Royalties so as to close up everything for 1881   But the G&S are slow in sending this in so I will go at it Sunday and make it up without that.[5] We have paid out I fancy so far as memory serves without referring to the Books just about as much as we have received.

Now for a little general information.

Central Station. The structure Boilers &c are all in their place ready to receive the Dynamos. These we are now going ahead on. The second Dynamo which we are going to send to Johnson is a splendid machine but the engine that Porter put on it disgusted Edison. The workmanship was terrible. It knocked terribly was greatly out of ~~align~~ alinement & was not at all up to the work which Porter said it would do. Directly 1200 lamps were put on steam ran down & to run ~~9~~800 or 900 lights was almost impossible. There was a terrible loss of power which Edison got over to a great extent by putting rings in the head of the cylinder & thus lessening the clearance. Anyway the Engine,[a] whatever Doctoring it received, was not by any means equal to the work required so Porter came on and he also saw it would not do to send it to London so he is re-building thoroughly another engine which we shall have in a day or so. Immediately it was decided to have a new engine Edison decided to try an experiment as to the maximum capacity of the Dynamo   He brought up the Boiler pressure to 158 pounds put on 1340 lamps and brought the volts up to 117 the lamps being 100 volt lamps these calculated as equal to 1650 lamps   there was no trouble with the Dynamo nor was there any trouble with the commutator (Edison has immensely improved his commutator & brush arrangement after considerably further experimenting) but after running three minutes the [~~eup?~~][b] cuppling broke from incapacity to stand such an enormous strain it being made of cast iron—future ones will be made of steel.[6] Edison saw the engine must have been developing 220 or 230 horsepower & he feels confident the dynamo would stand taking 2000 lights off. I think the internal resistance of the armature is $^{38}/_{10,000}$ but I am not quite sure.

Dean has got up a first class winding arrangement for putting on the paper insulation which does the work very quickly. He has also got up a machine for finishing off the copper plates [fore?][c] the armature in less than three days whereas it took over ten men to do the work in about two weeks previously Edison is just absolutely satisfied with the Dynamo. I rather fancy however that his feelings as to Porters Engines are not quite so assuring. All the Engines Porter has made have to go back to Phila again to be thoroughly overhauled.

The Lamps now being turned out Edison tells mc are the best hc ever made. I will get some curves sent you the next time I am out there probably Wednesday next.

Kruesi has about $^2/_3$ or $^3/_4$ of the mains laid. The weather gives him considerable trouble and as we have had so much rain here of late he is often obliged to stop work

The Isolated Business promises extremely well. I believe Eaton has written you as to this.

Parent Stock[7] is away down. Two shares sold yesterday at 575 European 60–70 Ore milling nothing doing but about 80 nominally. Isolated 20–25 premium but people are not selling. Illuminating plenty of sellers but no buyers nominally offered at par or 90, but only about 50 bid for it. Very Sincerely Yours
Saml Insull

Later   Since writing above yours of 5th Jan'y has come to hand and I am much obliged for your thinking of me in the matter.[8] My cable address is "Insull New York"

I have entered "Index" in Code Book[9]   Please enter following in yours:—Indian: Put [---][c] European Light shares in name of [---][c]

Your drawings of Central Station Dynamo are ready.[10] I do not think it advisable that they be sent you until Johnsons new machine is shipped when I will get Edison to go through & give me a letter of explanation   I will ask Edison whether drawings of special machine should not be made & sent you[11]   S. Insull

P.S. If in the above you find any bulls in description remember I am not an Electrician or a machinist but all I do is to try & reproduce what I have heard Edison say   I.

ALS (letterpress copy), NjWOE, LM 3:32 (*TAED* LM003032; *TAEM* 84:32). "PERSONAL" stamped on first page. [a]Obscured overwritten text. [b]Canceled. [c]Illegible.

1. This matter dated back to August 1881, when Batchelor asked Insull to deposit on his behalf a check from Philip Seubel. He also told Insull to expect a check for $186.25 in his favor from Calvin Goddard. Batchelor inquired about these transactions in September, October, and most recently, early December. Batchelor to Insull, 6 Aug., 15 Sept., 6 Oct.1881, all DF (*TAED* D8135ZAI, D8135ZBR, D8135ZCH; *TAEM* 58:960, 1066, 1090); Batchelor to Insull, 3 Dec. 1881, Cat. 1239:24, Batchelor (*TAED* MBLB4024; *TAEM* 93:503).

2. Batchelor inquired in early December if Edison had made any arrangements for him to receive stock in the new Edison Co. for Isolated Lighting. Batchelor to Insull, 3 Dec. 1881, Cat. 1239, Batchelor (*TAED* MBLB4024; *TAEM* 93:503).

3. The Edison Lamp Co. made frequent assessments on its partners in 1881. When it acted to raise another $10,000 for immediate expenses in November, Philip Dyer asked Edison to remit on behalf of Batchelor and Edward Johnson; Batchelor's share was $1,000 (Philip Dyer to TAE, 2 and 5 Nov. 1881, both DF [*TAED* D8124ZAN, D8124ZAO, D8124ZAP; *TAEM* 57:1070–71]); Cat. 1318:5, Batchelor [*TAED* MBA001:7; *TAEM* 93:855]). Edison also paid $2,500 to the Edison Machine Works on Batchelor's account in November, presumably to cover an assessment by that firm. Charles Rocap to TAE, 9 Nov. 1881, DF (*TAED* D8129ZCM; *TAEM* 58:313).

4. This letter has not been found. In November, Batchelor pointed

out that the Paris investors were reluctant to pay him a salary or even meet his living expenses. He told Edison he trusted "that at that end you have been able to get something for our labor." Batchelor was required to buy shares in the various new Edison companies in France in order to serve on their boards. To do so, in early 1882 he considered selling some of his Edison stocks in the U.S. Insull advised him to wait until about mid-March, when Edison expected to receive cash from Britain. Batchelor to TAE, 14 Nov. 1881, DF (*TAED* D8132ZBL; *TAEM* 58:532); Insull to Batchelor, 7 Mar. 1882, Lbk. 11:415 (*TAED* LBo11415; *TAEM* 81:429).

5. The Gold & Stock Telegraph Co. acquired Canadian rights to Edison's telephone inventions in 1878 (see Doc. 1417). Insull's statement of this account has not been found but Batchelor was credited at the end of 1881 with 20 percent of Edison's $2,000 royalty for the year (Cat. 1318:80, Batchelor [*TAED* MBA001:32; *TAEM* 93:855]). Records of Edison's transactions with Batchelor and other associates prior to this time, as well as those of Gold & Stock royalties, are in Private Ledger #2, Accts. (*TAED* AB006; *TAEM* 88:163).

6. Cf. Doc. 2149 n. 11. Charles Porter replaced the "worn and broken coupling" with a sturdier design, a description and sketch of which he sent Edison a few days later. Porter to TAE, 21 and 23 Jan. 1882, both DF (*TAED* D8233E, D8233G; *TAEM* 61:1006–7).

7. That is, shares of the Edison Electric Light Co.

8. Batchelor reported "quite a demand here for Edison Electric Light Co of Europe stock." He suggested that if Insull would provide a private cable address "I probably can help you to turn over something— What I should do would be to cable you—can you procure so many shares for so much and if your answer is in the affirmative I will cable you the money and you can remit the shares to me   I do not care to bother about this business myself but the people here are my friends and I should like you to have any benefit if you can get them for lower than they want them." Batchelor to Insull, 5 Jan. 1882, Cat. 1239:86, Batchelor (*TAED* MBLB4086; *TAEM* 93:540).

9. Batchelor took with him to France in July 1881 a book of cable codes. By this time he had registered at least several dozen additions for specific equipment orders or financial transactions. He defined "Index" in his 5 January letter as "Can you get ____ for ____." Batchelor to TAE, 30 July 1881, DF (*TAED* D8135ZAE; *TAEM* 58:945); Batchelor to Insull, Cat. 1239:86, 5 January 1882, Batchelor (*TAED* MBLB4086; *TAEM* 93:540); see App. 4 and Batchelor's correspondence with Insull in Cat. 1239, Batchelor (*TAED* MBLB4; *TAEM* 93:470); Exhibitions— Paris Electrical Exhibition (D-81-35); and Electric Light—Foreign— Europe (D-82-38); both DF (*TAED* D8135, D8238; *TAEM* 58:870, 62:198).

10. These have not been found, but presumably were those offered by Edison in Doc. 2201. The drawings had not been sent by mid-April, when Edison explained that successive design changes necessitated a new set of drawings, which he expected to have in two weeks. He also promised to send drawings of the new 250 light machine, the first of which was being tested. TAE to Batchelor, 19 Apr. 1882, Lbk. 12:105 (*TAED* LBo12105; *TAEM* 81:557).

11. Batchelor subsequently requested drawings of a punch press used

at the Edison Machine Works. Anticipating that special tools and machines could be made more cheaply in France, Insull asked for complete drawings from the Machine Works in late February, in advance of Batchelor's request. They were not yet ready in late March. Insull to Batchelor, 24 Mar. 1882, Lbk. 11:484 (*TAED* LBo11484; *TAEM* 81:460).

-2216-

To Edward Johnson

[New York,] 23rd Jany [188]2

My Dear Johnson,

Your letter of 6th inst came duly to hand.[1]

Regarding the test by Sir Wm Thomson I do not think silk winding is so good as wire winding.[2] On the big Dynamo we use $^1/_{16}$ in pianoforte wire for winding, the centrifugal force being so very great   There is no trouble with the Dynamos now as we bore the hole in the field out very much larger. The trouble in all Batchelors machines was that the fields were too small and the momentum of the parts swelled the armatures and caused them to catch.[3] Evens the machines you have could have a space of $^1/_8$ of an inch more than they now have & still the field magnet would be sufficiently powerful to bring them up to the required volts. To give me proper data to ascertain about the Seimens machine please give me the internal resistance of the field magnets the number of volts across the terminals of the machine when the full load of lamps are on. I should also like to have the number of volts across the terminals when the external resistance is fifteen times greater than that of the internal also ten times and five times. I would also like to know the diameter of the armature and its length and the velocity at which the tests are taken also the amount of resistance in the field magnets and the variation it makes in the number of volts when the machine is worked ten to one.

Do you know that I have the impression that we were the first in England to multiple arc the field across the line containing a resistance for regulating the strength of the field and I am quite sure that we were the first to regulate the field of force magnets by putting in the field of force circuit[a] an adjustable resistance. Please look through the patents for this point.[4] I think before long the boot will be on the other foot and a good many of these fellows will be infringing this patent Seimens for instance.

You spoke about the greater economy of our machine being an important advantage over the Seimens machine when the prices are mentioned   But is there not another element [~~conserned?~~][b] concerned which is that our machine will run more lights. Upton is now making your hundred candle power lamps He is also going to make a lot of 150 candle power lamps of pretty good economy and 600 hours life for use against the arc men here. For instance we get into a factory where they have already bought a arc machine which gives a dozen or fifteen arc lamps and which is the property of the concern where it runs we take out the arc lamps and substitute a 150 candle[c] Incandescent Lamp requiring the same horse power as the arc lamp the whole of them being worked in series so that no changes are made the whole being perfectly steady and giving a large ball of fire and no light being lost by a ground glass globe Our idea is that they will much prefer to run these lamps than the arc lamps. Such being the case we will get the future sale of these lamps which if they last 600 hours are far more economical than the arc lights for the reason that it costs a cent and a half an hour for the Carbons which will be nine dollars for 600 hours whereas we can afford to sell these lamps for three or four dollars. The 8½ p[er]. E[lectric]. h[orse]. p[ower]. lamps giving 6 p. I[ndicated]. h.p. Armington and Sims small engines you can guarantee with perfect safety   Their[a] average life will be 600 hours providing they are run at the normal candle power of Sixteen candles. In fact these lamps will last twice as long but it is best to say that they will last 600 hours because our statements will more than come true and it will establish a confidence in the same ratio as a loss of confidence begot by the proverbial and persistent liars who infringe our rights. we are going to test the Disc Dynamo in a few days and if everything is satisfactory will ship it at once to you for Sir William Thomsons test[5]

Our last test of the large Dynamo showed about 6¾ lamps sixteen candles each per i.h.p. undoubtedly it gives seven p.i.h.p. as when we first started we only got about five p.I.h.p. but finding out[d] errors in the Indicator and other places have gradually brought the economy up   It is extremely difficult to indicate these powerful high speed engines and get anything like a proper & true Indication as there are so many sources of error. When we first started there was a loss of 45 h.p. now there was no place so heated nor was the heat of all the aggregate heat of all the parts sufficient to account for more than a loss of ten h.p. hence we knew the error must be in the Indicator and by gradually finding out the errors we have reached higher economy. We had decided after running the Porter Engine on and off for a week and the last three nights all night long to accept Mr Porters offer of a new Engine that is to say the same kind of an engine thoroughly over hauled and everything put in alinement. After I determined to take the Engine off I thought I would try an experiment with the Dynamo. So we brought the boiler pressure up to 158 lbs ~~pre~~ put on 1350 lights and brought the volts up to 117 the lamps being 100 volt lamps and calculated that we had the <u>equivalent</u> of 1650 lights on the machine. There was no trouble at all with the Dynamo nor was there with the commutator but after running three minutes the cuppling broke from pure incapacity to stand such an enormous strain. The engine must have been developing between 220 and 230 h.p. This cuppling that broke ~~must have been~~ was made of cast iron. We shall put in steel cupplings with the new engine. You will be highly pleased with the new brushes and commutator arrangement   We spent $1500 in experimenting on brushes as we found that the sparking increased as the internal resistance of the machine decreased   You will have no trouble on this score. I will write full instructions about these matters when the machine is ready for shipment.

I understand from Barker that his test of Swan and Lane Foxe's lamps at Paris Exposition that they measured only 27 ohms hot when giving 16 candles Maxims averaged 52 and ours 148.[6] From this I imagine that both Swan and Lane Fox tried to even up their botch work by the deposit of carbon from the hydrocarbon like Maxim. If this is true Swan does not seem to succeed very well at the Savoy Theatre.[7] These fellows will find out the longer they keep at it how extremely difficult it is to make Electric Lamps by the thousand with long life and even candle power.

Down at the Lamp Factory we are gradually narrowing down the variation in the volts. Our output in December all came within seven volts of each other and I believe that we shall be able in the course of six months be able to make all our lamps of the same volts and candle power. Upton has sent you a life test of twenty seven lamps put up at 16 candles, after I had got the Factory on its feet again making first class lamps. You will see that those have remarkable life and we calculate that they will last about twelve or thirteen hundred hours on an average   I have not succeeded very well with the Spiral High Resistance lamps. There is no difficulty in making the lamps 250 ohms 11 p.E.h.p. but they have a very brief life and it will take some time to bring the life up to the proper standard. This is perhaps a good thing as the [extreme?][e] difficulty is one of our best precautions. The present lamp is childs play as compared with the lamp I am now working on

It may possibly interest you to know that the Maxim Light has been taken out of the Jewellers Store in Eighth Avenue out of the Cafe of the Hoffman Hoffman House[8] the Equitable Bldg,[9] Fisk & Hatch,[10] Hatch & Foote[11] & abandoned at Cincinnatti[12] leaving them only Caswell & Hazards Drug Store[13] and the Penn. R.R. Ferry Boat.[14] It may also interest you to know that we received a letter from the Penn. R.R. People asking us for an estimate to light up all the Ferry Boats[15] making the specification almost identical with the Maxim Plant now in one of the Boats but saying nothing about the life of the lamps nor the power which conclusively proves that this asking us to bid is merely a blind to cover up a scheme as we are informed that Thompson[16] is interested in the Maxim Light. We are going to bid notwithstanding giving them the regular prices and render their little scheme nugatory by writing them a letter calling attention to the fact that the life of the lamps and the economy of the same have been left out of the Specifications

We enclose you list of Isolated Plants installed to date.[17] I also enclose you several copies of the last rules adopted by the Board of Underwriters which rules were suggested by us and adopted by them after consultation with all the Light Companies and arguments before the National Board of Fire Underwriters.[18] We received the copies of the Morning Advertiser containing your Article[19] which is first class in every respect and has been very favorably commented on by Mr Fabbri Lowrey and others.

I do not know whether I shall see D'Oyle Carte[20] but Oscar Wilde[21] to whom D'Oyle Carte is agent has expressed a desire to call on me. So perhaps I may also see the latter named gentleman

We have sent you some copies of Du Moncels Articles & Howells Paper from Van Nostrands Magazine by Wells Fargo Express. Van Nostrand is now printing a small Hand Book forming one of their Science Series containing Du Moncels Article Preeces Lecture & Howells Test with an article by Seimens besides[22]   Please say how many of these you want. Howells Test was made with the Bar & Disc machine[23] which Batchelor has   Thompson will be pleased with this. I will with the aid of Insull try and get up a letter on Preece & enclose it to you for delivery[24]

I cabled you that in the new Large Volumes of my Telephone Interferences with which you are familiar there is an exhibit 7-12-1 dated July and the other August relating to the Phonograph[25] showing that I had the thing in my mind when

I filed the Provisional Specification on the patent you are now litigating. I hope you have the Volume. Batchelor had it at the Exposition so I will not send you one. I think you will also find in the back part of that Book or in the smaller volume all your lectures and the necessary material you desire on that subject. The idea came of course from that Morse Operator with the spiral on the Disc for recording Morse Characters and reproducing the same from the indentations which were made— long previous to the filing of the English Specification[26]

Could you not hire a boy and station him at some point were he could see the nightly burning of Swans lamps to ascertain their life. Yours truly

Thos A Edison   I[nsull]

L (letterpress copy), NjWOE, Lbk. 11:142 (*TAED* LB011142; *TAEM* 81:292). Written by Samuel Insull. [a]Obscured overwritten text. [b]Canceled. [c]Interlined above. [d]"finding out" interlined above. [e]Illegible.

1. DF (*TAED* D8239B; *TAEM* 62:626). Johnson's reply to the present document is Doc. 2226.

2. Before making a comparative test of an Edison dynamo, William Thomson planned to wind the armature with silk thread to hold the bars in place against centrifugal force. Thomson and Johnson believed that the centrifugal expansion of a bar armature may have caused the destruction of one of Charles Batchelor's dynamos. Johnson to TAE, 6 Jan. 1882; Thomson to Johnson, 2 Jan. 1882; both DF (*TAED* D8239B, D8239C; *TAEM* 62:626, 654).

3. See Doc. 2147.

4. Both elements are contained in Edison's British Patent 602 of 1880 (Batchelor [*TAED* MBP025; *TAEM* 92:158]).

5. See Doc. 2228.

6. No correspondence on this subject from George Barker has been found. The test results reported in Crookes 1882 (212), in which Barker participated, are somewhat different. At 16 candlepower, the resistance of the Edison lamp is given as 137.4 ohms, of Swan lamps 32.78 ohms, of Lane-Fox lamps 27.40 ohms, and of Maxim lamps 41.11 ohms.

7. Johnson reported in his 6 January letter (see note 1) that the stage and auditorium of London's Savoy Theatre were lit by Swan's lamps. A detailed report from *Engineering* on the installation is found in an Edison scrapbook. "The Lighting of the Savoy Theatre," *Engineering*, 3 Mar. 1882, Cat. 1018, Scraps. (*TAED* SM018062a; *TAEM* 24:265).

8. The Hoffman House Hotel at 1111 Broadway, in Madison Square, was synonymous with its barroom, where many prominent New York men routinely gathered. *Rand's* 1881, 258; Batterberry and Batterberry 1999, 144–45; "Vanished is Hoffman House Bar," *Washington Post*, 27 May 1906, SM4.

9. The headquarters building of the Equitable Life Assurance Society at 120 Broadway was home to the United States Electric Lighting Co., and the site of the first commercial use of Hiram Maxim's incandescent lamp. Doc. 1617 n. 4; Friedel and Israel 1987, 193–94.

10. The banking house of Fisk and Hatch was founded in 1862 by Alfrederick Smith Hatch and Harvey Fisk. The firm was noted for its sale of government securities and railroad bonds. Located in 1882 at 5 Nassau St., the partnership became an early customer of the Pearl St. station; it dissolved in 1885. *Wilson's Business Directory* 1879, 38; *ANB*, s.v. "Hatch, Alfrederick Smith"; *NCAB* 11:261; Carosso 1970, 17, 25, 27.

11. Hatch & Foote, a banking and brokerage partnership, was formed in 1867 after the reorganization of Fisk & Hatch, in which Daniel B. Hatch and Charles B. Foote were clerks. The firm was located at 12 Wall St.; it failed in 1900. "Death of Charles B. Foote," *New York Times*, 21 Sept. 1900, 1; *ANB*, s.v. "Hatch, Alfrederick Smith"; *Wilson's Business Directory* 1879, 38.

12. Nothing is known of this installation.

13. Caswell, Hazard & Co., originally of Newport, R.I., was a druggist and dealer in toiletries and surgical instruments. Its New York shops were at 1099 Broadway and 672 Sixth Ave. *Trow's* 1883, 260; "Caswell, Massey & Co.," *New York Times*, 7 June 1885, 5; "End of a Long Litigation," ibid., 26 Feb.1887, 5; "107 Years Old", ibid., 16 Apr. 1887, 5.

14. See Doc. 2136.

15. This letter has not been found, nor has any other correspondence with Edison on this subject, but see Doc. 2136.

16. Frank Thomson (1841–1899) was second vice-president of the Pennsylvania; he was later its president. *DAB*, s.v. "Thomson, Frank."

17. The enclosure has not been found. The Edison Electric Light Co.'s bulletins from this period identify and describe many isolated installations. Edison Electric Light Co. Bulletins, CR (*TAED* CB000; *TAEM* 96:667).

18. The enclosure has not been found, but see Doc. 2203 n. 33.

19. Johnson sent this article in his letter of 6 January (see note 1). He wrote it as a 4 January letter to the editor of the *Morning Advertiser* in which he attempted to "dispel the halo of mystery with which electrical action and effects are surrounded, and to plant in its stead a sense of the mastery over the power which it is perfectly easy for the general public to possess." He discussed the risks of electrocution and fire in arc and incandescent lighting systems. Johnson was also named as the source of similar information supplied by an unidentified correspondent to the *Daily Free Press.* "The Alleged Danger of Electric Lighting," *Morning Advertiser,* 5 Jan. 1882; *Daily Free Press,* 3 Jan. 1882; Cat. 1243, items 1963 and 1958, Batchelor (*TAED* MBSB41963, MBSB41958; *TAEM* 95:111, 110).

20. Richard D'Oyly Carte (1844–1901), theater impresario and manager, completed the Savoy Theatre in late 1881. He promoted English operas, especially those of William Gilbert and Arthur Sullivan, with whom he had a famous collaboration. He had solicited Edison in November 1878 to make an English lecture tour on electric lighting. He arrived in New York on 11 January, ostensibly for a vacation. *Oxford DNB*, s.v. "Carte, Richard D'Oyly"; D'Oyly Carte to TAE, 26 Nov. 1878, DF (*TAED* D7802ZZMD; *TAEM* 16:486); "The Oily D'Oyly Carte," *Atlanta Constitution*, 14 Jan. 1882, 5.

21. Dublin-born Oscar Wilde (1856–1900), noted wit and esthete, began a sensational tour of the U.S. and Canada in early January 1882. There is no evidence of his meeting or corresponding with Edison. *Oxford DNB*, s.v. "Wilde, Oscar O'Flahertie Wills."

22. Van Nostrand published *Incandescent Electric Lights* in 1882; a copy of the second edition, issued the same year, is in Series 3, Box 44, Folder 5, WJH. It was a compilation of du Moncel 1881 with an abridgement of the subsequent discussion (5–43), Preece 1881 (45–81), Siemens 1880 (143–87), and several brief unsigned articles. Added to the second edition were Jehl 1882a (85–103) and Howell 1882 (105–42), which was Howell's master's thesis at the Stevens Institute.

23. That is, a standard Edison armature with longitudinal induction bars connected through end plates or disks.

24. Doc. 2217.

25. Edison instructed him to "see exhibit hundred fifteen exhibit twelve large photolithographic volume in telephone interference Batchelor has it in Paris   Also exhibit seven twelve." Exhibit 115-12 included a drawing by Edison, dated 17 August 1877, of a method to "record the talking by undulating." This instrument used a telephone receiver to make indentations on a moving strip of paper. Exhibit 7-12, dated 11 July 1877, is Doc. 964. One proposed device shown there used rotating wheels with teeth of specific pitch and height to produce vibrations in a slender spring corresponding to phonic components of human speech. TAE to Johnson, 20 Jan. 1882, LM 1:148C (*TAED* LM001148C; *TAEM* 83:946); TI2 (*TAED* TI2253; TAEM 11:425).

26. Edison's embossing telegraph recorder and repeater is Doc. 857.

–2217–

To William Preece[1]

[New York,] January 23 [188]2

My Dear Mr Preece.

It was with very great pleasure indeed that I read your lecture on Electric Lighting at the Paris Exposition, and more especially your remarks upon my system as exhibited at Paris

I must thank you most heartily for your very favorable endorsement of my system and congratulate you on expressing your views in so independent and forcible a manner.[2]

Your lecture will be reprinted in Van Nostrands Magazine for February,[3] copies of which I will send you when published. Arrangements have also been made for its early publication in conjunction with Count Du Moncels article from "La Lumiere Electrique" and several other articles, in book form as one of "Van Nostrands Science Series"[4]

I can scarcely close this letter without referring to our past troubles, in which I am very glad to say I think we were both mistaken, and to repeat my gratification at having your approval and support in respect to this larger subject.[5] I am my Dear Mr Preece   Very truly yours

Thomas A Edison.

ALS (letterpress copy), NjWOE, Lbk.7:329 (*TAED* LB007329; *TAEM* 80:555).

1. Samuel Insull sent this letter to Johnson for him to deliver to William Preece. Insull to Johnson, 23 Jan. 1882, Lbk. 7:325 (*TAED* LB007325; *TAEM* 80:552).

2. Preece 1881 (103) stated that "the completeness of Mr. Edison's exhibit was certainly the most noteworthy object in the exhibition." His description of the system occupied more than a half page of the article.

3. "Electric Lighting at the Paris Exhibition," with discussion, *Van Nostrand's Engineering Magazine* 26 (1882): 151–63.

4. See Doc. 2216 n. 22.

5. The "past troubles" were the 1878 microphone controversy (see *TAEB* 4, passim), to which Preece did not refer directly. He stated in conclusion (p. 103) that "Mr. Edison's system has been worked out in detail, with a thoroughness and a mastery of the subject that can extract nothing but eulogy from his bitterest opponents. Many unkind things have been said of Mr. Edison and his promises; perhaps no one has been severer in this direction than myself. It is some gratification for me to be able to announce my belief that he has at last solved the problem that he set himself to solve, and to be able to describe to the Society the way in which he has solved it."

–2218–

From Otto Moses[1]

Paris, Jan. 23. 82

My dear Mr. Edison,

I have just returned from a trip to London, or rather to the Exposition, or I might with truth say to the Edison Exhibit at the Crystal Palace; for with the exception of the Weston lamps,[2] run by the E. Light and Power Generator Company,[3] there was nothing of any importance besides your dynamos and 250 of your lights. Last Tuesday, the day I arrived, Johnson had invited about 125 guests, engineers and press, to witness the first display. The dinner was to be served at 7, and at 6.45 the safety plug[4] blew out of the Roby Engine![5] However, the feast was not delayed. By the light of the customary lines of gas jets, the dinner was gradually consumed, everyone enjoying the meal but those of us who were waiting for the repair of the damage. Of course when the accident happened, we wanted as soon as possible to be able to find out the extent of the damage and the possibility of starting up again that night. So as soon as the clouds of steam had disappeared, cold water was poured over the boiler, after drawing the fire, and as soon as a man could get in, the field was surveyed.

In the mean time Johnson entertained the guests. The last course had been served and segars and coffee [~~bourght?~~][a] brought[b] when a messenger announced that in fifteen minutes the light would be turned on. this Johnson communicated in a cool little speech, and asked for a little more patience. Then when the moment came, Hammer commenced a little fire-works with the switches, which 'took' immensely. Then speeches began in which the Chairman & Col. Gouraud made capital ~~speeches~~ hits, the Colonel relating in a most entertaining way how you recommended Johnson for his mission, and how fortunate all were in having him again in England.[6]

The illumination was very pretty, a chandelier effect being most striking where[b] a few lights, by reflection from a white screen and passing through chains of facetted glass, seemed to come from ten times their numbers.

I examined the E.E.L. System building 57 Holborn Viaduct, and was very much pleased with the whole work. On Thursday night there was a grand display from 400 lights in the offices &c. Wires are laid for all the lights the two machines will carry. Hood has his dynamo in tip top order everything is compact and handy.

Col. Gouraud seems to be working everything and everybody with his usual ability and Johnson is 'all right.' Faithfully Yours

Otto A Moses

ALS, NjWOE, DF (*TAED* D8242B; *TAEM* 63:140). [a]Canceled. [b]Interlined above.

1. Samuel Insull wrote on this document: "Edison thought you might like to see this." It is not known to whom he gave it.

2. Edward Weston (1850–1936), an English-born inventor working in Newark, headed the Weston Dynamo Electric Machine Co., one of the largest U.S. manufacturers of electroplating dynamos. By this time Weston had been working in arc lighting for several years. His lamp took numerous forms, all distinguished by the short length of arc which required comparatively low voltage. *ANB*, s.v. "Weston, Edward"; Woodbury 1949, 87–106; Prescott 1884, 471–516; see also Doc. 1832 n. 1.

3. The Electric Light and Power Generator Co., a relatively new concern, had one of the three franchises for experimental street lighting in the City of London. Later named the Maxim-Weston Electric Light Co., in early 1882 it became the first to install Weston arc lamps in England. Bourne 1996, 82—83.

4. A fusible plug in the boiler that was designed to melt above a certain temperature. Knight 1881, s.v. "Safety-plug."

5. Beginning in 1854, Robey & Co., of Lincoln, England, manufactured stationary and portable steam engines for a variety of agricultural and industrial uses. The company began to illuminate its factories with Edison lamps about this time. Website of the Robey Trust (http://www.therobeytrust.co.uk/index.htm); Southworth 1986, passim; "Faits Divers Éclairage électrique," *La Lumière Électrique* 5 (1882): 288; Robey & Co. to Arnold White, 11 July 1882, DF (*TAED* D8239ZDF; *TAEM* 62:1015).

6. Johnson sent Edison a printed invitation to the opening affair with

the handwritten admonition to "be sure to put in an appearance   All these people expect to see you." He provided a detailed account of the evening's events a few days afterward. The display consisted of more than 200 lamps run by four dynamos from a single steam engine; the other two engines were not yet installed. (Johnson to TAE, 16 and 19 Jan. [pp. 6–12], both DF [*TAED* D8242A, D8239I; *TAEM* 63:139, 62:675]). He also promised to send newspaper accounts, which probably are those found in a scrapbook ("The Electric Exhibition at the Crystal Palace," *Daily Chronicle* (London); "The Edison Light at the Crystal Palace," *Standard* [London]; "Edison Light at the Crystal Palace," *Daily News;* "Electric Lighting," *Daily Telegraph;* all 18 Jan. 1882; Cat. 1327, items 1982, 1892, 1995, 2000, Batchelor [*TAED* MBSB51982, MBSB51992, MBSB51995, MBSB52000; *TAEM* 95:117, 121–23]).

–2219–

*Memorandum:
Incandescent Lamp
Patent*

Menlo Park, N.J., January 27 1882[a]

Take out this patent U.S.[1]

Object ~~process~~ of ~~Exhausting the~~ manufacturing Electric Lamps

Invention consists in ~~first Exhausting the air from the containing chamber by means of charcoal~~ process of completing an incandescing Electric Lamp

first by heating the chamber Externally combining therewith a chamber containing an absorbant of gases such as Charcoal, heating the charcoal to drive out all the gases possible

becomes the charcoal chamber is connected to the lamp— connecting the charcoal chamber to the lamp while both are hot. The use of a spark gauge to ascertain the state of the vacuum.

When sufficiently high the ~~carb~~ Lamp is connected to the Electric Circuit & gradually heated by the current throwing out air which is absorbed by the charcoal. ~~When~~ the lamp ~~comes up~~ is ~~brilliant~~ brought up beyond the point where it is to be burned regularly, it is allowed to burn~~ed~~ for some time, then is sealed off at .X.

The tube C is then affixed to another lamp and the same process takes place. The tube C & rubber joint d might be dispensed with and the mouth of the charcoal chamber E extended so as to be sealed on each time to a lamp

By this process Vacuum pump can be dispensed with although in some instances a steam operated vacuum pump may[b] be used with economy to ~~obtain~~ partially Exhaust the globe=[2]

I am aware that charcoal has been used with a mercury pump (see my other patent) but the operation of Electrical treatment of the incandescent conduct was done while the Lamp was connected to the pump—[3] The main object of this invention is to save the Expense of a ~~vast~~ great number of pumps necessarily used on account of the necessity of slowly treating the incandescing conductor

Dick=[4] Want to get a process patent on this.[5]

Dewar of Scotland[6] got a vacuum in a chamber by two stopcocks thus[7]

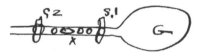

S S are stop cocks X the charcoal. he heated the charcoal with S1 closed   this threw out the air   he closed S2 & opened S1 the charcl absorbed air from G   he then closed S1 & opened S2 & heated charcoal again but ~~I dont claim~~ he never could get a high vacuum this way as the space containing the charcoal in manipulating the cox would always keep the vacuum down a little.[c]

Our device is different. besides As with a lamp & Electrical heating of the conductor[b] is another point etc.

~~Other~~ You can mention drying substances such as phosphoric anhydride may be used in connection with the charcoal to ~~dry the~~ absorb the aquious Vaper in the Lamp.

T A Edison

ADS, NjWOE, Lab., Cat. 1148 (*TAED* NM017:65; *TAEM* 44:425). Letterhead of Edison Lamp Co. [a]"Menlo Park, N.J.," preprinted. [b]Obscured overwritten text. [c]Followed by dividing mark.

1. Edison executed a patent application based on this document on 28 February 1882. The patent closely followed the main points of the memorandum except as indicated in notes 2 and 3 below. It issued in May 1883 as U.S. Pat. 278,416.

2. Edison's patent specification stated that the process of absorbing of gases by charcoal might have to be repeated to attain sufficient evacuation. In the patent drawing, **C** is the chamber filled with charcoal **D**, connected to the lamb globe **A** through rubber fitting **c**. U.S. Pat. 278,416.

The patent specification pointed out another expedient. Prior to connecting it with the chamber containing the charcoal, the lamp "may be filled with an atmosphere of hydrochloric-acid gas, hydrobromic-acid gas, or ammoniacal gas, so as to displace as far as possible the air therein. The charcoal absorbs this gas without losing very greatly its capacity to absorb air, and a better vacuum may by its use be obtained." In a lamp patent resulting from related research about this time (see Doc. 2212 n. 2) Edison noted that chlorine or bromine compounds placed in the lamp globe would, when heated, combine with hydrogen to produce hydrochloric acid or hydrobromic acid gas. Charcoal would absorb these gases without impairing its ability to take up others. U.S. Pat. 401,646.

3. Edison stated in the specification that "in my Patent No. 248,428 I describe the use of heated charcoal in connection with a mercury

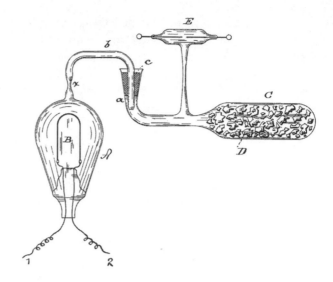

February patent drawing for chemical exhaustion of lamps without aid of mercury pumps.

vacuum-pump for exhausting incandescing electric lamps; but in that instance the charcoal was used as an auxiliary to the mercury-pump, while by my present invention I dispense altogether with such pumps."

4. Richard Dyer.

5. That is, a class of patents for an art or process for accomplishing a particular result, as distinct from patents for designs, machines, manufactures, or compositions of matter. Prindle 1908, 20–21.

6. James Dewar (1842–1923) was a professor of natural philosophy at Cambridge and of chemistry at the Royal Institution, where he experimented prolifically in spectroscopy and cryogenics. Dewar and George Liveing jointly determined the absorption spectra of many elements, particularly of metallic vapors. *DSB*, s.v. "Dewar, James."

7. Edison may have been referring to Dewar and Tait 1874 or Dewar and Tait 1875. The former brief paper described (but did not illustrate) the use of coconut charcoal to absorb residual gases in an evacuated tube. The authors noted that heating the charcoal would drive the gases out again, and that the process could be repeated indefinitely. The second paper was published only as an abstract of a demonstration to the Royal Society of Edinburgh, again without illustrations. According to a further description contained in Dewar 1927 (1:127–28), it concerned using a piece of charcoal that was allowed to cool only after the vessel was fully evacuated and sealed, so as to absorb residual gases.

–2220–

Samuel Insull to George Gouraud

[New York,] 30th January [188]2

My Dear Col,

I want to get full information about those elevators that are in use in Stewarts Building 11 Queen Victoria Street.[1] Is there a patent on it in England & if so is it patented in the United States. If Patented in England could you send a copy of the English Patent? Also can you let me have Stewarts opinion as to the working of the Elevator.[2]

My object in asking for this information is that I thought this style of Elevator would be adapted for use in connection with Mr Edisons Electric Rail Road System & on my describing the way it runs to Mr Edison he thought it would suit especially on Elevated Roads run by electricity[3]

I suppose you are anxiously awaiting the arrival of Johnsons second Dynamo & we are just as anxious to get rid of it. It has been standing in our shop at Goerck Street for some weeks already for shipment except the High Speed Porter Allen Engine. Mr Edison was not satisfied with it and ordered a new one. This was promised a fortnight ago & has again been promised for tomorrow certain.[4] The machine tests splendidly & will carry with perfect ease 1650 lights of 16 candles—considerably more than the Engine attached to it will supply Power for, which is very gratifying as it will enable us to be thoroughly sure of its reliability.

The Central station work here is progressing as well as can be expected. But the weather this winter has been so changeable that we have been delayed considerably in laying the mains.[5] As soon as the weather becomes settled the work will be very speedily concluded & then it will but a short time before we are able to start up.

Is there any prospects of old telephone matters being settled up.[6] Mr Edison wants pretty badly our expenditure being [$10,000.?][a] a week which you may imagine comes very heavy on Mr Edison   Very truly yours

Saml Insull

ALS (letterpress copy), NjWOE, Lbk. 7:331 (*TAED* LB007331; *TAEM* 80:556). [a]Illegible.

1. This building previously housed the headquarters of the Edison Telephone Co. of London. Queen Victoria St. was relatively quite new, having been opened in 1871. An 1891 handbook described it as "one of the finest in the City. . . . lined on both sides with large, lofty, solidly built and ornamental buildings." Jehl 1937–41, 1033; Wheatley 1891, s.v. "Queen Victoria Street."

2. The editors have not identified Stewart, the presumptive owner of the building. Gouraud replied on 15 February and promised to get Stewart's opinion of the elevator. On 8 March, Gouraud's assistant F. R. Grigg sent Insull comments from Stewart and the elevator's unnamed inventor. Grigg said that "Mr. Stewart speaks very highly of the apparatus, stating that he considers it the most perfect out for elevating a continuous stream of passengers without their having to wait any time." This "cyclic elevator" (also called a paternoster lift) used a chain to move

twelve passenger cars in a continuous loop. The inventor told Grigg that it had been installed in three other locations and that he was in negotiations to install it in the planned London underground railway. He estimated the cost of installing it on an elevated railway at about £775 per elevator. Gouraud to Insull, 15 Feb. 1882; Grigg to Insull, 8 March 1882; both DF (*TAED* D8204R, D8204ZAC; *TAEM* 60:71, 86); *OED*, s.v. "paternoster, n. 6."

3. Insull wrote to Johnson on 4 February about Edison's wish to drive a similar device by electricity for stations of the New York Elevated railroad. Johnson replied with detailed comments about the elevator's operation (without indicating its motive power) and the likely problems of adapting an electric motor to its operation. Insull to Johnson, 4 Feb. 1882, Cat. 2174, Scraps. (*TAED* SB012ABY; *TAEM* 89:369); Johnson to TAE, 19 Feb. 1882, DF (*TAED* D8239V; *TAEM* 62:766).

4. Johnson cabled Edison on 30 January that "dynamo delay serious when will you ship." In a 1 February letter, he explained that he was "to have lighted up the Holborn Viaduct today as per my Contract with the City." Edison had expected the engine back from Philadelphia about this time, but it was not until 10 February that he reported that the engine and dynamo "worked perfectly" and were being prepared for shipment. Edison asked if Johnson wanted them shipped immediately via Liverpool or in ten more days directly to London. Johnson responded that he was "damaged by delay already" and would arrange for transhipment through Liverpool. The engine and dynamo sailed on 14 February with Francis Jehl; the piston was sent later. Johnson to TAE, 30 Jan. and 11 Feb. 1882; TAE to Johnson, 1, 10, and 15 Feb. 1882; LM 1:153A, 158A, 153B, 156D, 157C, 159A (*TAED* LM001153A, LM001158A, LM001153B, LM001156D, LM001157C, LM001159A; *TAEM* 83:948, 950–951); TAE to Johnson, 14 Feb. 1882, Lbk. 11:256 (*TAED* LB011256; *TAEM* 81:333); Johnson to TAE, 1 Feb. 1882, DF (*TAED* D8239Q; *TAEM* 62:741); see Doc. 2228.

5. By mid-November 1881, the Edison Electric Illuminating Co. had laid about 3 miles of the approximately 13 to 15 miles of underground mains needed for the Pearl Street station. Bad weather suspended the work in January but by the end of February about 7½ miles of mains were laid; another 3 miles were set down in March. Installation of the mains was completed in July 1882. "The Distribution of Light and Heat in New York City," *Sci. Am.* 1881 (45): 319; Jehl 1937–41, 1041; Edison Electric Light Company Bulletin 3:3, 24 Feb. 1882, CR (*TAED* CB003; *TAEM* 96:674); "Progress of Electric Lighting in New York City," *Sci. Am.* 1882 (46): 281.

6. Gouraud replied on 15 February that he was still working to resolve outstanding telephone matters. He expected to "know in a few days precisely how it stands— Of course you will explain to Mr. Edison that this is a matter that is not in my hands; but everything that lies in my power to get it settled you may rest assured will be done." Gouraud to Insull, 15 Feb. 1882, DF (*TAED* D8204R; *TAEM* 60:71).

[Menlo Park,][1] 3rd Feby 1882

My Dear Johnson,

Yours of 20th Jany[2] came to hand just as I was leaving 65 5th Ave this evening.

I might twit you on a number of your comments were it not that the several matters relating to the contract are now closed up as today Mr Fabbri received Mr Waterhouse's[3] cable that the English Contract is closed[4]

As to Arnold White[5] & my opinion of him it will be time enough for you to deplore your inability to change your views on a given subject because forsooth I hold opposite views when you are asked to.[6] I do not see that you have any right to say that my information is filtered through Gouraud. Simply because he & myself are of the same opinion on one subject is no proof that my opinion is colored by his. My opinion of A.W. was formed when G.E.G. thought him a paragon of perfection and I have never seen any reason to think contrary wise to what I then thought.[a] Of course it is not of the slightest importance to me as to as to what that opinion may be so far as his employers are concerned. I do not think it would be advisable for Mr Edison to oppose his appointment.

As to that little talk I had with Mr Fabbri & Mr Wright about money matters I am perfectly prepared to stand by everything I then said. Mr Wright made certain insinuations which I do not think he would care to be reminded of nor do I think he would for a moment think he would now think of repeating them & what I said in reply was not one tenth as strong as what Mr Edison had instructed me to say

As to that French prosposition of course it is very annoying that Mr Fabbris proposal was not put at all in the light it should have been. But the fault for this error lies with Mr Fabbri's friends so I do not now even see why he should feel in the slightest degree sore about it so far as Mr Edison and his associates are concerned.[7]

Mr Edison has taken Dick Dyer into his personal employ to look after patent work. He has a desk in our room & will attend to the preparing of all cases. While you are in England English cases will be sent to you for you to put in the hands of a Patent Solicitor you may sellect. Dyer has just completed a Digest of Edisons English Patents a copy of which I hope to send you by next mail.[8] He will now go thoroughly into Bidwells last report[9] & your long letters relating thereto: so soon as he has finished Mr Edison will write you on the subject. I think that in the future Edisons Patent business will receive much better attention as this last move seems to promise extremely good results.

You will most certainly have to pay for your new issue Parent Stock. Batchelor & Upton will have to do the same. Of course if there is a chance of getting it for you on better terms I shall do so but I dont think there is.

Bergmann & Co (note the Co this time)[10] seem to be doing extremely well. We give them a good many orders on foreign account & the Isolated Co come up to time pretty well. I gave them orders for about $750 of sundries this a.m. & shall give them orders for $1500 or $2000 more on Monday morning when I return to town.

Of course this same activity with B & Co means increased business for Lamp Co. what with your order for 2000 lamps & orders from Batch I shall have given them orders for about 9000 lamps this week[b] by the time Saturday night (tomorrow) arrives.

We have had quite a good deal of snow lately which has entirely stopped Kruesi from laying his mains   This is a great nuisance & must naturally cause considerably ~~great~~ longer delay in starting up the first District

The Electric Locomotive for the new Road is now out here & will be running in a week.[11] Hughes has one mile of the Road actually completed; about another mile requires the metals to be spiked and the remainder is all graded.[12] As soon as the weather breaks he will be able to finish up very rapidly as he will be able to use the Locomotive for drawing ballast R.R. ties & iron which of course means quick work.

Edison has had a Stamp mill erected behind the machine shop out here for use in Ore Milling Experiments[13]   It will be working very shortly.

The Lamp Factory claims a great deal of Edisons time now. He has been out here a great deal of late & told me a few minutes ago that he thinks he has struck the radical bug in the lamp. Anyway he is here right along watching matters extremely closely & of course great good must come of it. They have one of your hundred candle power lamps up in the Photometer Room just now. It gives a beautiful soft light & will be very taking I am sure.

The High Resistance Special Carbon experiments are going on here right along.[14]

Edison proposes to move the Factory to Newark by April 1st. He says if a big rush was to come on in orders for Lamps it would throw matters here into great confusion where as if he gets to Newark the difficulty to double the present capacity would be very slight. Upton had an offer yesterday of $5000 more for the Newark property than the Lamp Co paid for it. This offer came from the man who sold the property to us.[15]

Edison was very pleased with your Article on the Fire Insurance & Protection from Fire Question which appeared in the Morning Advertiser as you will have learned from letters from him personally.[16] He was also of course delighted with the result of the Edinburgh carbon Transmitter trial.[17]

I had my first Sleigh Ride the other evening. I went out to Roselle[18] with Moore.[19] He has a splendid turn out & I enjoyed myself immensely. Coming out here tonight I stopped off at Rahway[20] & hired^c a sleigh & drove myself over sending the turnout back by one of the Lamp Factory Boys.

But there it is just 4 A.M. & I think I will save sleigh riding talk for Mrs E.H.J. when I have a little chat with her through the medium of pen ink & paper on Sunday. So good morning my dear fellow. I will see if I cannot find a Board with a soft side to it & take a nap   Yours as ever

Saml Insull

I said in my last letter I hoped to send you a long letter from the old man by the same mail. You will understand that this is now waiting pending Dyers completion of his Analysis of Bidwells report.[21]

ALS (letterpress copy), NjWOE, LM 3:47A (*TAED* LM003047A; *TAEM* 84:49). ^a"to what I then thought" interlined above. ^b"this week" interlined above. ^cObscured overwritten text.

1. The place is evident from the contents of the document.

2. Not found.

3. Theodore Waterhouse, a London patent solicitor, had advised the former Edison Telephone Co. of London, of which he had also been a director. Docs. 1799 n. 5 and 1825.

4. This cable has not been found. The agreement to organize the Edison Electric Light Co., Ltd., was dated 18 February, when Edison signed and returned it. The company was to be formed within two years by Edward Bouverie and associates with a capital of £1,000,000, for the purpose of acquiring Edison's patents. Drexel, Morgan & Co. agreed to sell Edison's patents for £30,000 and also to hold five percent of the shares. The company was incorporated on 15 March 1882. TAE agreement with Drexel, Morgan & Co., Grosvenor Lowrey, Fabbri, and Bouverie, 18 Feb. 1882; TAE agreement with Edison Electric Light Co., Ltd., 31 Mar. 1882; both CR (*TAED* CF001AAE1, CF001AAF; *TAEM* 97:295, 301); Johnson to TAE and TAE to Johnson, both 20 Feb. 1882, LM 1:161C, 161D (*TAED* LM001161C, LM001161D; *TAEM* 83:952).

5. Arnold White (1848–1925), a former coffee planter in Ceylon, had been manager of the Edison Telephone Co. of London from 1879 to

1880. He became manager of the Edison Electric Light Co., Ltd., when it was formed in March. *Oxford DNB*, s.v. "White, Arnold Henry."

6. Insull had stated to Johnson on several occasions his strong opposition to White becoming secretary of the new electric light company, and believed Edison would take the same view. He referred to White's 1880 attempt to end a dispute between Edison and the board of the Edison Telephone Co. of London, which George Gouraud denounced in Doc. 1989. Because White's alleged effort to "dish" Edison and his associates "was so clear that I do not think it can be explained away," Insull believed he was "entirely the wrong man to hold the position." He later expressed his surprise to Johnson "that you above all others should endorse A.W. You talk about G.E.G. [George Gouraud] & his modus operandi   why do you not give the same mead of justice in A.W.s case!!" Insull to Johnson, 8 and 17 Jan. 1882, LM 3:11, 20 (*TAED* LM003011, LM003020; *TAEM* 84:12, 21).

7. The incident to which Insull referred is not clear. He explained that he did not understand what had happened with the negotiations for Europe but had the impression that "Fabbri admitted that his proposition was not put properly before the European directors. The trouble was that Mr. Wright would not discuss the matter but would simply take 'Yes' or 'No'   we had therefore to read the proposal according to our own lights & if Edison was mistaken in the view he took he (F) must not blame him (E) but those whose duty it was to present the matter here. However I fancy Mr. Fabbris inclination is to say no more about it as it can do no good & might produce ill-feeling." Insull to Johnson, 8 Jan. 1882, LM 3:11 (*TAED* LM003011; *TAEM* 84:12).

8. Edison reportedly mailed the digest with Doc. 2228 but it has not been found.

9. Probably Shelford Bidwell's comprehensive report on all Edison electric lighting patents, which Johnson forwarded on 16 January but has not been found. Johnson to TAE, 16 Jan. 1882, Ser. 1, Box 1, Folder 2, WJH.

10. Sigmund Bergmann had been using the letterhead of Bergmann & Co. since at least October 1881. Bergmann & Co. to Francis Upton, 12 Oct. 1881, DF (*TAED* D8101D; *TAEM* 57:12).

11. See Doc. 2237.

12. For Edison's description of his method of constructing the electric railway track see TAE to Théodore Turrettini, 27 Oct. 1882, Lbk. 14:345 (*TAED* LB014345; *TAEM* 81:966).

13. According to a report prepared by Sherburne Eaton, the mill was for crushing gold- and silver-bearing ores in connection with a new process for recovering the metals. Edison finished a patent application in June and another in August for separators particularly adapted to working gold-bearing ores; Edison described one as especially suited to the streams of material produced by hydraulic mining. Eaton report to Edison Ore Milling Co., 17 Jan. 1882, CR (*TAED* CG001AAI2; *TAEM* 97:416); U.S. Pats. 400,317 and 263,131.

14. Nothing specific is known of the experiments at this time, but see Doc. 2216.

15. When Francis Upton reported this offer to Edison he noted that they could probably get $10,000 more than they had paid, "so you see we bought at a bargain." Upton to TAE, 2 Feb. 1882, DF (*TAED* D8230S; *TAEM* 61:735).

16. See Doc. 2216.

17. Johnson cabled Edison the favorable outcome of this case on 2 February (LM 1:153D [*TAED* LM001153D; *TAEM* 83:948]). The suit had been brought in Scotland by the United Telephone Co. alleging infringement of Edison's fundamental telephone patent and two others of Alexander Bell. This was the first legal test of Edison's fundamental British telephone patent since it was amended in early 1880, in part to correct technical flaws (especially pertaining to the phonograph) that were at issue in this case (see Docs. 1237, 1272, 1318, 1392 n. 4, 1825 n. 5, 1870, 1903; and Johnson to TAE, 6 and 29 Jan. 1882, both DF [*TAED* D8239B, D8239N; *TAEM* 62:626, 709]). The judge decided generally that Edison's patent was valid and specifically that it applied to the transmitter in question. He also upheld the Bell patents (Opinion of John McLaren, 1 Feb. 1882; DF [*TAED* D8248M; *TAEM* 63:384]).

18. The town of Roselle, N.J., is located about two miles west of Elizabeth. *Ency. NJ*, s.v. "Roselle"; see Docs. 2336 and 2343; Insull to Ladd, 8 March 1883, Lbk. 15:431 (*TAED* LB015431; *TAEM* 82:215).

19. Miller Moore (1842–1930), a resident of Roselle, was an engineer and businessman. From 1874 to 1880, he was secretary and general manager of the Locomotive Engine Safety Machine Co., in New York. He became head of the Edison Electric Light Co.'s Bureau of Isolated Lighting in April 1881 and, a year later, was appointed manager of the Edison Co. for Isolated Lighting and had responsibility for setting up the demonstration village plant in Roselle. He managed the Isolated Co. until 1883, when he became a general agent for Armington & Sims, where he continued to work closely with Edison companies. Moore agreement with Edison Electric Light Co., 30 Apr. 1881, DF (*TAED* D8126X1; *TAEM* 227:513); Jehl 1937–41, 930–31, 1093–94; "Moore, Miller F.," Pioneers Bio.

20. Rahway is a city on the Pennsylvania Railroad line about five miles northeast of Menlo Park. *Ency. NJ*, s.v. "Rahway."

21. Dyer's report has not been found. Edison sent it to Johnson on 6 March with sixteen pages of his own specific comments on Bidwell's suggestions for patent disclaimers. TAE to Johnson, 6 Mar. 1882, Lbk. 11:385 (*TAED* LB011385; *TAEM* 81:409).

–2222–

Memorandum to Richard Dyer

[Menlo Park, c. February 3, 1882[1]]

Dyer

has the ring dynamo been taken out in England[2]

⟨Engl 30   U.S. 335⟩[3a]

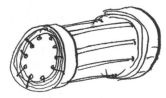

has the worm & worm wheel ~~ad~~Automatic adjustment of the Swinging brush holder magnet on each side with ratchet & pawl[b] been taken out[4]

Both are important—

You have given me no report as to whether I would infringe Brush or if there is any valuable difference between Brush & this^c

⟨Patent 224,511⟩^a

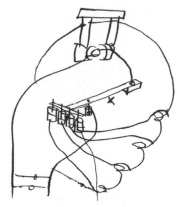

⟨S. D. Mott   Feby 8th 1882⟩^{5d}

X [are?]^e Expansion strip heated by main current   please advise=

E[dison]

⟨Brush's pat. 224,511 does not describe the thermal device with sufficient clearness to prevent obtaining claims on above which while ~~never~~ specific in statement will cover the principle⟩^{6a}

ADS, NjWOE, Lab., Cat. 1148 (*TAED* NM017:83–84; *TAEM* 44:442–43). Commas added to patent numbers for clarity. ^aMarginalia written by Richard Dyer. ^b"magnet . . . & pawl" interlined above. ^cFollowed by "over" and pointer to indicate page turn. ^dMarginalia written by Samuel Mott. ^eCanceled.

1. Richard Dyer replied to this document on 4 February. DF (*TAED* D8248N; *TAEM* 63:394); a letterpress copy is in Lbk. 11:214 (*TAED* LB011214; *TAEM* 81:317).

2. Edison referred to the method of connecting induction bar pairs through concentric end rings, as shown in Doc. 2102.

3. Dyer replied (see note 1) that a British specification had been filed in July 1881. That patent (Brit. Pat. 2,954 [1881]) applied to a disk dynamo but included "concentric insulated rings" for connecting induction bars (including more generally those of a bar armature) at each end of the armature, as described in Edison's Case 335. The U.S. application contained a figure (number 1) resembling the one in this document; it issued in December 1881 as U.S. Patent 251,537. Dredge 1882–85, 2:clxxvii–clxxviii.

4. Dyer responded (see note 1) that this mechanism had not yet been included in a British patent but could be incorporated in a forthcoming case after George Seely obtained a copy of the relevant U.S. speci-

fication. (Edison had considered using brush position to regulate dynamo output in late 1880 but did not attempt to patent it then; see Doc. 2036 esp. n. 5.) He had executed an application in January 1882 for a mechanism to regulate dynamo output voltage by rotating commutator brushes relative to the commutator axis. This was among the last cases prepared by Zenas Wilber and was not filed until August; it issued in October 1882 as U.S. Pat. 265,779. The British specification was filed on 9 March as British Patent 1,142 (1882). This regulation method was also an important part of U.S. Patent 379,771; see also Dredge 1882–85, 2:cclxxvi.

5. Dyer reported in his letter (see note 1) that he gave this sketch to Samuel Mott to prepare the patent drawing. The drawing that accompanied the specification was ultimately made by another draftsman, Edward Rowland (U.S. Pat. 265,780).

6. Charles Brush filed a patent application in November 1879 for regulating dynamo output by placing a variable resistance in series with the field magnets to alter the strength of the magnetic field. This was placed in interference with an Edison application. Brush's application issued in February 1880 as U.S. Patent 224,511. Edison divided his application into two parts, one of which proceeded to issue. He refiled the remaining portion in May 1880; three months later it was placed in interference with the Brush patent and another application. Patent Application Casebook E-2536:88, PS (*TAED* PT020088; *TAEM* 45:706); Wilber's testimony, pp. 64–66, *Keith v. Edison v. Brush*, Lit. (*TAED* QD002:33–34; *TAEM* 46:146–47).

The Brush patent described methods to vary the field circuit resistance manually or automatically. In the latter case, an electromagnet responding to the strength of current in the main circuit would act upon a lever to place more or less resistance in the circuit. In his 4 February reply to Edison (see note 1), Dyer reported that the Brush patent "speaks generally of a thermal arrangement" for regulation. In fact, Brush's patent did not contain means to alter field circuit resistance through thermal effects but Dyer may have assumed that his broad claim for automatic control of the field resistance could be construed to cover Edison's arrangement. He concluded, however, that Brush's claims or specification did not have "sufficient definiteness to prevent you from obtaining claims, which, while somewhat specific in statement, would cover essential features."

Edison executed an application on 28 February for the device sketched in this document. It consisted of several discrete resistances

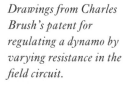

Drawings from Charles Brush's patent for regulating a dynamo by varying resistance in the field circuit.

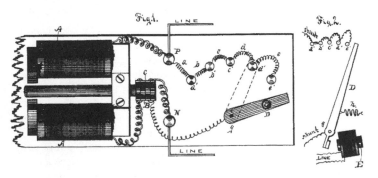

placed in series with the field magnet and a device which successively shorted out these resistances as the dynamo's external load increased. This instrument consisted of "a double strip composed of two metals of different degrees of expansion" placed in the main circuit which expanded or contracted as current through it increased or decreased. At the end of the strip was an arm which "carries a number of contact-points, each dipping into a mercury-cup, the contact-points being of different lengths, or the mercury of different heights, in order that successive contacts may be made and broken." Each mercury cup formed a shunt around a resistance so that as the bimetallic strip expanded it successively short-circuited the resistances, thus lowering the resistance in series with the field magnet and increasing the field current. This application was not filed until August; it issued in October 1882. U.S. Pat. 265,780.

-2223-

To Charles Clarke

[Menlo Park,] Feby 6/82

Clarke.

Bergman promised to have those sounders finished Saturday.[1] I [to]ld[a] Andrews[2] how to test them [al]so[a] Bergmann please give them [your?][a] assistance. I consider [it][a] a very important thing for [th]e[a] success of the Isolated. [It?][a] does not matter if the [vari]ations[a] in practice are [3?][a] or 4 candle power or [-][a] times for a short period   [The?][a] variations are from 3 to 10 candles; it will not be on the [w]hole[a] of the lamps and [t]hey[a] have an average life [over][a] 600 hours to compensate[b] [f]or[a] these irregularities—only [w]ant[a] a general regulation [so?][a] our words as to 600 hours [w]ill[a] come true in practice.

How about Claudius going ahead with the second Central station[3]   has arrangement been made regarding the style of board and arrangement at Laboratory   will have anything done you desire and that quickly. We must not fall into the rut of indolence because one department is unavoidably delayed

Edison.

ALS, Hodgdon (*TAED* B017AA). [a]Left edge of document not photocopied; original document not available. [b]Obscured overwritten text.

1. The sounders were evidently for the voltage indicator described in Docs. 2190 and 2201.

2. William Symes Andrews (1847–1929) was a machinist and mechanic who began working for Edison at Menlo Park in November 1879. He took charge of the Testing Room at the Edison Machine Works in late 1882 or early 1883. He subsequently served as chief engineer for the Edison Construction Department, then worked for General Electric from 1894 until his death. "Andrews, William Symes," Pioneers Bio.; Obituary, *New York Times*, 2 July 1929, 22; *TAEB* 5: chap. 4 introduction and App. 2; Andrews test report, 16 Feb. 1882, DF (*TAED* D8235E; *TAEM* 62:31); Jehl 1937–41, 684.

3. In June, Sherburne Eaton told Edison that Charles Clarke wanted "to begin with his preliminary work in the second district, namely Madison Square district. He wants to make correct measurements and to plan the entire district thereby facilitating and cheapening the laying of conductors." Edison agreed that Clarke should start making preparations which "will enable us to keep our men busy" (Eaton to TAE, 14 June 1882, DF [*TAED* D8226X; *TAEM* 61:277]; TAE to Eaton, 16 June 1882, Lbk. 7:482 [*TAED* LB007482; *TAEM* 80:620]; see also Doc. 2275 n. 4). Data for 2,400 buildings in the area between Twenty-fourth and Thirty-fourth Sts. were tabulated by mid-October 1882, approximately the time that the Edison Electric Light Co. printed a three-page set of "General Instructions to Canvassers." The district included about 41,000 gas lamps, 2,300 sewing machines, and four dozen steam engines. There are no extant records of Claudius's work on this project until February 1883, when he evidently began testing a model of the distribution network like the layout he had built at Menlo Park for the Pearl St. district ("The Electric Light," *Operator* 13 (1882): 465; N-82-08-28:1–11, Lab. [*TAED* N231:1–6; *TAEM* 41:450–55]; Edison Electric Light Co. pamphlet, c. 1882–83, CR [*TAED* CD003; *TAEM* 97:190]; see also Doc. 2028 n. 2; Edison Electric Illuminating Co. memorandum, 12 Oct. 1882, DF [*TAED* D8223H1; *TAEM* 60:1031]). In 1887 the Edison Machine Works, having absorbed the Electric Tube Co., prepared a cost estimate of the conductors for the second central station district. The generating station, eventually located on West 26th St., did not begin operating until 1888 (Edison Machine Works cost estimate, 12 May 1887, DF [*TAED* D8733AAF; *TAEM* 119:890]; Martin 1922, 87–88, 172).

–2224–

Memorandum to Richard Dyer

[Menlo Park,] FEBRUARY 6, 1882.

Dick—

Please write up the specifications for this patent, and keep it until I get in.[1]

Method of deriving two independent circuits from a dynamo or magnetic electric machine, each of which has a different electromotive force regulatably independent of each other. Such extra circuit useful for working the field of force magnets or the field of force magnets multiple-arc'd across a circuit containing lamps requiring lower volts or emf.

X is the bobbin; C and C¹ are the regular brushes; 1 and 2 the regular circuit across which lamps requiring the highest volts are placed. a b are extra brushes one above the centre the other below the centre; say several blocks[2] to the right and left of C.[3] These brushes are connected to an arm, and are swung around by the handle, the brushes being pivoted as well, the handle may be worked so the brushes a c are brought in line with C, or by putting handle at angle separated when a and b are furthest from the centre or line or block upon which C rests, there is the greatest electromotive force; these two brushes a b are connected together, forming one pole of the second circuit, while C forms the other pole.

a and b when connected together do not short circuit the wire on the machine as both sides of the bobbin, are sending currents in the same direction. Fig. 2 shows the two arms on separate swings, so they may be brought to or from the centre independently.

T.A.E.

PD (transcript), NjWOE, Lit., *Mather v. Edison v. Scribner*, Edison's Exhibit G (*TAED* QD003052B; *TAEM* 46:204).

1. This memorandum served as the basis for a patent application that Edison executed on 28 February. It described means of regulating two or more independent main circuits operating at different voltages. One of these circuits operated directly from the dynamo armature while the other was in parallel with the field circuit. The field circuit and second main circuit got their power from another set of commutator brushes whose position on the commutator could be mechanically adjusted to vary the voltage. During a subsequent patent interference, John Ott testified that he tested both manual and automatic means for adjusting the secondary brush position in late May and June 1882. Edison testified that he had used this method successfully for about two months on a Z dynamo for lighting his Menlo Park house. The February application

Of the four drawings in Edison's U.S. Patent 379,771, this one corresponds most closely to the description provided to attorney Richard Dyer. The extra brushes for the secondary circuit are both on the same side of the commutator.

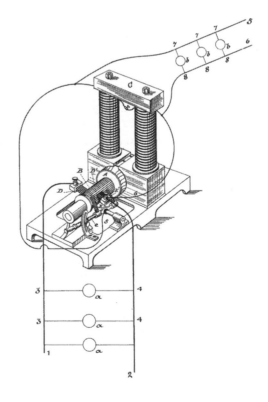

was among those withheld by Zenas Wilber and did not reach the Patent Office until August (see Doc. 2323). It became the subject of an interference with two other applications for voltage regulation; it finally issued in March 1888 as U.S. Patent 379,771. Testimony of Ott (3 Oct. 1883), Dyer (5 Nov. 1883), and TAE (6 Nov. 1883), pp. 1–19, 37–49; *Mather v. Edison v. Scribner;* Lit. (*TAED* QD003001, QD003037, QD003041; *TAEM* 46:178, 196–98).

2. That is, commutator sections.

3. No drawing or sketch has been found to accompany this letter and the alphabetic designations for various components are different in the patent drawings.

–2225–

From William Preece

London   Feby 10. 82

My dear Edison

I have received your letter of Jany 23rd[1] with a great deal of pleasure.

I have never ceased to regret our estrangement and I am glad to think that bygones can be bygones.

I wish you every possible success in your present grand work, and I congratulate you sincerely on what you have done already. Yours very truly,

W. H. Preece

ALS, NjWOE, DF (*TAED* D8204P; *TAEM* 60:67).

1. Doc. 2217.

–2226–

From Edward Johnson[1]

London, E.C. 13th Feby. 1882[a]

My dear Edison,

I have 8 machines in operation at the Crystal Palace. Two of them (One particularly) get quite warm with the load they are now carrying, namely 64 lamps each, the others remain perfectly cool. I presume these two are of the 5 that were first shipped with my big dynamo, and that they are not the same as those subsequently shipped. I am going to take them out and put two others in their place out of those you have just sent me. I want this plant to run 70 lamps each for 5 hours every night. I know that 6 of the 8 machines will do it without the slightest difficulty and I want to get 12 equally as good. I am endeavouring to get you the data about the Siemen's machine. I don't know just how I will do it but I will try and find a way. I am conducting a siege against Siemens with a view of bringing about an amicable arrangement of the armature question. We have decided that if he is reasonable, we will adjust this matter with him by paying him a royalty, but that if he is unreasonable we will fight him. The first approach has been made by Dr Hopkinson as our Plenipotentiary and with satisfactory results. He finds that Siemens is prepared to deal with us in a proper spirit and thinks that we can make a very easy arrangement with him. At all events he is bringing Siemen's to the Viaduct to see me and to see the machines in operation. I think it not unlikely that we shall make an arrangement with him either by the payment of a direct royalty or by a contract for the manufacture of our dynamos or at least of the large heavy iron parts, you supplying the armatures &c.[2] I will have the subject of regulating the field by an adjustable resistance particularly investigated but I am of the impression that when we were on that subject before we found that your patent ~~failed us too.~~ was good—[3b] There is no doubt in my mind at all that they will all want ~~too~~ use this sooner or later (Siemen's among the rest.) I have called attention to this and to its bearing on our present negotiations with Siemen's. Of course I have not lost sight of the fact that our machines will run more lights tha~~t~~n the Siemen's machines   I have asked Sir Wm Thomson in his experiments with our machines and with the Siemen's machine to let me know exactly what the difference is. It is the absence of accurate knowledge of the effective capacity of the two machines which prevents me from stating the facts in my circular and thus justify the price. I shall be very glad indeed to get some of your high candle power incandescent lamps. They would be of extraordinary value in the Crystal Palace Exhibition. As soon as you get any, please dont lose a moment in sending a few on to me. I have been a little astray in regard to the capacity of the big machine. I understood all along that of your 10 per Electrical HP lamps we could get 8 per indicated H.P. from this machine. You now tell me 6¾. I am afraid this will be a disappointment to our people.

You said you were going to send me some new brushes for the little dynamos. They have not come to hand—did you ever send them? No, the Savoy Theatre is not a success; the amount off power that is expended there for keeping up 1200 lamps (which do not average more than 5 or 6 candles) is something prodigious. I hope the 2000 new lamps I ordered will be first quality in every respect. I shall put them in operation in Crystal Palace, although to do so I will have to change the old sockets for the[c] new. I am glad to have your information about the Maxim light which will assist me materially in answering questions which I am frequently asked. They have not done much here, in fact the Manager of the Company which controls the Maxim lamp openly denounces it as a complete fraud. It is impracticable to have any surveillance of the Swan lamps. The only place where it would be at all possible would be at the Savoy Theatre, and that of course is out of the question since it could be only accomplished by the connivance of the manager who is himself in America. We have 500 & some odd lamps now in nightly operation at the Crystal Palace. No one else has yet shown over 50 incandescent lamps. Our display is in fact the one object of interest at Crystal Palace so far. Our Exhibition will be the most complete ever made by anyone. I have arranged your Exhibits in the room which is to be my headquarters in the order of their invention and classification; the first section, being a space entirely across one end of the room, is devoted to your contributions to type printing telegraphs; the next section to your duplex and quadruplex telegraphs, the next section to your messenger & general call service telegraphs; to one side of the centre stand is your electrical pen. The central stand is a large square table immediately under the Chandelier[4] and about 10 ft square, in the centre of which is erected an octagonal structure upon which are placed various samples, pictures &c &c. Around this are show-cases in which are placed your various miscellaneous inventions and the experimental apparatus which shew the steps by which your telephone and other inventions were led up to. Surrounding this is a complete demonstration of all your telephones and your various scientific apparatus including the ~t~Tas~o~im~o~eter, the pressure relay, the motograph relay &c &c arranged and classified in their proper order. The stage is occupied by your Phonograph, Musical Telephone, & Loud Speaking Telephone,[5] which will be in constant operation and in first class condition throughout the Exhibition. Above and surrounding all is your last contribution namely the Electric Light. To one side in an alcove formed[d] by a closed doorway we are placing ~the~ a[c] Z. dynamo to run as a motor and mounted on top of it will be a little 15 light dynamo driven by it and from this small dynamo we will supply a few lamps. Of course the current to drive the motor will be taken from the main system. I understand that some of the other Companies are expressing some disgust with the Exhibition saying they wish the damned thing had never been thought of. The fact is we have by being in advance of everybody else and by the completeness of the work we have done, as well as the magnificence of our display, stolen all the thunder the Exhibition has to give. ~~Those~~ Others may come after us but they are simply dropping along in our footsteps and there are but few crumbs of comfort for them to pick up. Good by   Very truly yours

Edwd. H. Johnson

P.S. I understand there is a movement on foot to formally invite you to come to this country. What do you say? Suppose I get the Holborn Viaduct and Crystal Palace in full blaze with about 2000 lights and everything working perfectly smooth, don't you think you could afford to take a month and come over here? It would be money in your pocket; the fact of your presence here I am quite sure would give a stimulus to your various enterprises which would nett you a very considerable enhance of values on your shares. They certainly could run things for 30 days in New York without you. I should like to have you here in April. The Crystal Palace Company have arranged for a series of popular lectures on Electricity. The first four are to be given by Sylvanus Thompson,[6] after which Siemens is to give one on his light, Swan on his, Lane-Fox on his, and I am asked to give one on yours. They will not permit any of us to employ a professional lecturer for the purpose. If I go in for it (as I suspect I must) you may rely on its being properly done although it will not probably be as scientific or technical as some of the others. These lectures are to be held in the Concert Room which is of course lighted by your lamps. It is now conceded on all sides that we displayed the best judgment in the selection of our site— Good bye— E. H. Johnson

LS, NjWOE, DF (*TAED* D8239U; *TAEM* 62:757). Letterhead of Edison Electric Light System. Edw. H. Johnson, manager. [a]"London, E.C." and "188" preprinted. [b]"was good—" interlined above. [c]Interlined above. [d]"d" interlined above.

1. This letter refers generally to Doc. 2216.
2. After meeting with Charles William Siemens (and John Hopkinson and Frederick Bramwell) about a week later, Johnson reported "that we will either pay Siemens a royalty, make an arrangement with him for the manufacture or secure him as a Shareholder or all three." He urged reaching a settlement because the dynamo "is our only weak point—& as the Patent hasn't long to run—we shall probably never pay him very much." (British patents at this time ran for a maximum of fourteen years [Davenport 1979, 32].) In early March, Johnson consented to a royalty of 6% of the gross value of each dynamo manufactured but made no admission of infringement. He told Edison that although "you might ultimately be able to avoid him," the arrangement "will do you far more good than it will cost. It gives you a high moral ground to stand on in asserting your rights with others." Edison was not wholly persuaded and cautioned Johnson that Siemens's failure thus far to take legal action

against infringers could be a tacit acknowledgment of the weakness of his patents. Johnson to TAE, 19 and 26 Feb., 4 Mar. 1882, all DF (*TAED* D8239V, D8248Q, D8239ZAF; *TAEM* 62:766, 63:398, 62:792); TAE to Johnson, 6 Mar. 1882 (pp. 13–14), Lbk. 11:385 (*TAED* LB011385; *TAEM* 81:409); see also Johnson's correspondence with Siemens Bros. in Electric Light—Foreign—U.K.—General (D-82-39), DF (*TAED* D8239; *TAEM* 62:625).

3. Edison's British Patent 602 (1880), Cat. 1321, Batchelor (*TAED* MBP025; *TAEM* 92:158).

4. This large chandelier is prominently visible in an illustration of Edison's exhibit room published by the *Illustrated News* of London.

5. That is, the electromotograph telephone.

6. Silvanus Phillips Thompson (1851–1916) was professor of physics at University College in Bristol and a delegate to the recent electrical congress in Paris. He was elected to the Society of Telegraph Engineers in 1882 and helped draft the Electrical Lighting Act of that year. Thompson was known for the wide range of his interests and his commitment to technical education in Britain. His classic 1884 work on *Dynamo Elec-*

Edison's exhibit room at the Crystal Palace, illustrated in a London newspaper; the floral chandelier commissioned by Edward Johnson is in the foreground.

THE EDISON EXHIBIT

tric Machinery went through numerous editions, including Thompson 1902. Thompson lectured on successive Wednesday evenings from 22 February to 15 March on electric currents, electric arc lights, and electric incandescent lights. *Oxford DNB*, s.v. "Thompson, Silvanus Phillips"; "Crystal Palace Electrical Exhibition," *Electrician* 8 (1882): 227.

–2227–

From John Ott

Menlo Park, N.J., Feb 13 1882[a]

Mr Edison

I made the conections in Locomotive,[1] and find several weak points. There wants to be a reatone[2] between the Armiture sirkit enableing the starter to throw in resistance while starting then turning out as speed is obetained.[3] Also the reverse switch wants altering. It wants to brake with a spring so as to prevent sparking.[4]

I also conected the field Magnet in multiple ark and inserted ten ohms resistance, and it started quite easy.[5]

Mr Vancleaf[6] and my self eased all the barings, one Man being able to start quite easey on the leavel.

I also find the present belt tightner of litle or no use as it is on the rong side or rong belt, if belt is mad luse enough to let Armiture slip the tightner will not take up the slak.[7] Mr Hues[8] said he would not do enything for a few days on the railroad as the ground is so soft and tracks crucket by the frost

I should like to know what to go on with next[9]

John. F. Ott.

ALS, NjWOE, Cat. 2174, Scraps. (*TAED* SB012ABZ; *TAEM* 89:370). [a]"Menlo Park, N.J.," and "188" preprinted.

1. The Menlo Park staff working on the electric railway received a new motor from the Edison Machine Works on the previous day. Edison had two new electric locomotives built during the first half of 1882, one for passenger service and one for freight. This motor was for the passenger locomotive. N-82-03-12:1, Lab. (*TAED* N249:1; *TAEM* 41:861); see Doc. 2181.

2. Ott likely meant "rheotome," a device which periodically interrupts an electrical current. Knight 1881, s.v. "Rheotome."

3. Upon starting, a motor is susceptible to a high inrush current. When a motor is at rest there is no back electromotive force (generated by the armature's rotation within the motor's magnetic field); when it is connected to an electrical power source the initial current is limited only by the resistance of the armature. This current may be several times greater than its normal current at full load and may cause serious sparking or flashing at the commutator, or damage the insulation. A resistance in series with the armature is necessary to limit starting current to an acceptable level. The starting resistance is typically removed after the motor has been started. On the same day as Ott's letter to Edison, Charles

Hughes recorded that he had burned out the motor's armature as a result of a short circuit produced by the previous day's tests. Pender 1922, 215–16; N-82-03-12:3, Lab. (*TAED* N249:2; *TAEM* 41:862).

4. The reverser switch was used to start the locomotive and to invert the direction of the current to the motor, enabling it to run in reverse. A notebook entry by Ott dated March 1881 showed a sketch of a circuit breaker which used a spring-loaded contact. Hughes was working on a reverser at the end of April. Jehl 1937–41, 581, 585, 586; Ott notebook, p. 36, Hummel (*TAED* X128B:36); Hughes to TAE, 25 April 1882, Cat. 2174, Scraps. (*TAED* SB012BAM; *TAEM* 89:420).

5. On the previous day Charles Hughes recorded that he installed two field coils, of 1.7 and 10 ohms resistance each, in parallel: "The larger coil to be used for starting and the 10. ohm coil after getting under way and the large coil cut out entirely. Worked fairly but took too much power and heated the magnets and made 3 Dynamos in multiple arc heat up very much," to the point of melting solder. On 17 February, he described a tradeoff between starting the motor smoothly and maintaining its ability to pull a heavy load: "Put a 'B' armature of .035 ohm resistance into motor, and could not start it while using the 10 ohm coil around field magnet. Used the large coil of 1.7 ohms and started at high rate of speed but found that we lacked power to start heavily loaded car." N-82-03-12:1–5, Lab. (*TAED* N249:1–3; *TAEM* 41:861–63).

6. Cornelius Van Cleve (b. 1842?) began working as a laboratory assistant at Menlo Park in June 1880. Nicknamed "Neal," he was the husband of Hattie (Harriet) Van Cleve, Mary Stilwell Edison's half sister. He participated in much of the work with Charles Hughes on the electric locomotive from February to April and in November and December 1882, co-signing most of the notebook entries in this period. U.S. Bureau of the Census 1970, roll T9-790, p. 291.4000, image 0424 [East New Brunswick, Middlesex, N.J.]; N-82-03-12; Lab. (*TAED* N249; *TAEM* 41:861); *TAEB* 5, App. 2.

7. Charles Hughes noted this same difficulty the previous day. On 15 February he replaced the armature countershaft pulley with a Mason clutch, a form of friction clutch patented by locomotive builder William Mason in 1862 (N-82-03-12:1, 5; Lab. [*TAED* N249:1, 3; *TAEM* 41:861, 863]; "Mason's Patent Frictional Clutch," *Sci. Am.* 8 [1863]: 128). Francis Jehl later described the utility of the experimental belt tension mechanism: "The leather belt arrangement allowed us to start the motor . . . and then as it gained speed and the pulley lever was gradually pressed down, we could tighten the belt, start the locomotive and regulate its speed. Slipping occurred, of course, and it was soon learned that the operator had to exercise judgment in throwing the power onto the driving wheels to prevent damage to the armature" (Jehl 1937–41, 585).

8. Charles Hughes.

9. Edison's reply to Ott has not been found. Ott apparently stopped working on the electric locomotive soon afterward and turned his attention to elements of Edison's electric lighting system, including voltage regulation, dynamo construction, lamps, and meters. Ott notebook, pp. 24–25, 30; Hummel (*TAED* X128B:25–26, 31); N-82-05-26, Lab. (*TAED* N204; *TAEM* 40:575).

Hughes continued work on the electric locomotive until the end of April and resumed work again in November and December. Major technical problems which Hughes tried to solve included excessive heating of the motor armature, insufficient power, inefficient mechanical power transmission, low voltage at the far end of the track, and short circuits in the tracks. N-82-03-12, Lab. (*TAED* N249; *TAEM* 41:861).

–2228–

To Edward Johnson

New York, 15th Feb *1882*[1a]

My Dear Johnson,

Your letter 16th Jany:[2] I now understand the value of claims in English Patents but what you say does not in any affect my arguments at all   I only asked about it as a matter upon which I desired information more particularly with reference to the Gramme patent.[3]

I send you some notes of Dyers relating more particularly occluded gas taken out of carbon in vacuuo.[4] You will remember that I used platinum clamps with platinum wire running through glass to connect with the carbon and in a subsequent patents you will find that I made special devices to more effectively get the occluded gases out of the clamps than was possible by mere heating.[5] Now you will see that when I bring up one of [my?][b] filaments in vacuuo I heat the carbon to a higher incandescence for the sake of heating the clamps and get out the occluded gases which is especially necessary where metalic plated clamps are used as plated metal has an immense quantity of occluded gases   So you see I am still carrying out my first patent in so getting out these gases and the others are infringing this patent as the act of heating up the carbon must necessarily throw out gases from the clamps which they all use and this statement is backed up by my subsequent patent   Dyer gives expression to this view in his above referred paper.

With reference to my Patent No 5 as to my using carbon having a specific high resistance I still maintain what I before stated and Dyer has made an argument on this subject which I send you herewith. Of course if the patent can be sustained for a carbon filament having a high resistance as compared to previous statements in this line all right but if not the patent will admit of the other construction. If you will read it again carefully in the light of Dyers argument you will see that it fairly bristles with allusions to the specific kind of carbon of high resistance and that making this kind of carbon into a filament gives a filament composed of carbon of high resistance and that the coiling of the carbon was for an entirely different object; that is to say if the carbon was of very high resistance and made straight there would be such a small mass of matter and such a small amount of energy at any one given second that the slightest change in the strength of the current or at the commutator brushes would cause the loop to show it by vibration of light. That is its mass being so small it would loose its heat almost instantly while if the carbon was made a little more homogeneous i.e made of a little lower resistance of[c] a greater length to compensate for this the coiling of the same would keep the radiating surface down and the specific heat of the whole being then raised by increasing the mass it would not be so sensitive to slight changes in the current. You will see that this is clearly set out in the Provisional as well as the final Specification.[6]

It is quite impossible for me to send Dyer over to you as he is attending to all the foreign business & is far more valuable to me here than he would be in England. He is now preparing Foreign Cases which will be sent to you to be filed as a communication from me through the channels you think best. No patent will be sent without pretty good examination of all previous patents by Dyer who has gone over the whole of them and has made a Digest for reference a copy of which is enclosed herewith as it may be useful to you.[7]

Regarding the Radial Bar Disc Dynamo I have the one which was intended for the Paris Exposition just finished. I have not tried it yet. They need not get scared on the Dynamo question.[8] I cabled you several days ago to know whether the Gramme patent in view of the Paccinotti and Elias machines is valid in England[9]

Please glance your eye on figures 7 & 8 in the English Patent showing the radial bar disc Dynamo which you speak about.[10] Say what you think about Seimens claim for winding over the whole of the cylinder. In this patent I neither go over the ends of the cylinder like Seimens or through the cylinder like Gramme. I enclose you Dyers report to me as to the Seimens patent.[11] I think you will find that he has made a closer analysis than Hopkinson & that he throws new light upon the subject. I am strongly of the impression that I remember reading a number of years ago of the old Seimens armature which was covered with several coils. I think it was in an Electro motor. Do not lose sight of the fact that an Electro motor is reversible & this becomes a Dynamo machine & that the English Patents contain thousands & thousands of such motors and that it would be a mistake to examine magneto or Dynamo machines only. In England it will only be a difference in the manufacturing cost of the machine no matter what point I am betting on as I have so many ways of accomplishing the same result with means equally efficient and only differing as to cost.

Dyer is going into Bidwells final report and I will send his comments on it as soon as possible[12]

Did you find that little squib about a lecture delivered by some person I think named Fleming on the Swan lamp published in the Journal of the Chemical Society in January February March or April 1880 which I told you about.[13] You should get a copy of the Proceedings of the Newcastle on Tyne Chemical Society in which perhaps the full paper is to be found. If the paper is as I think it is it might be as well to show it to Sir William Thompson as well as it shows Swans lecture in which he clearly sets forth that he did not know the object of high resistance lamps but merely expressed the idea that what was[d] wanted was carbon made out of paper.

We have tested the bar and little end Disc "Z" Dynamo— the same style as Batchelor had but with end discs a little [sma? thicker]e thinner. It was intended to run 150 B lamps but it is so very economical and is of such very low resistance that we actually over 300 B lights at 9 or 10 candle power on it and it was only 158°[f] degrees Fahrenheit after running one hour and a half. I propose to send this armature which will [--]e fit any of your machines by [--]e the next steamer[14]   You can then have Hopkinson test it as Howell did whose paper I sent you and also Sir William Thompson and I think it will be found that when fifty horse power is taken out of the machine it will be a 97% machine of which 91 or 92% is available outside of it.[15]

In case you might order those 200 light machines for Covent Garden[16] Dean is going ahead to make three or four different sizes to strike the size. We are going to wind these with wire as the connections of the bar and Disc machines are somewhat expensive in the absence of special tools the making of which can only be warranted on our receiving large orders

In a few weeks we expect to start putting together six large Dynamos simultaneously for the Down Town Station

Francis Jehl as I have already written you left yesterday on the Arizona with 20 Twenty Light meters and 5 Fifty Light meters.[17] These meters are remarkably accurate   They beat anything ever got up. They are cheap & reliable. You will have to provide him with a pair of accurate assay balance scales and some sundries. Also a good man for him to instruct in the work

I think a German Chemist would be as good a man as could be got and I think you will find many of them in London who will work for a moderate salary

The Dynamo that went yesterday will stand all the lamps you can put on it that the Engine will take care of or the brush man can handle   Francis will show you some of the knacks about these new brushes—they are very nice the whole arrangement is very handy and brushes may be put on and taken off with the greatest ease when the machine is running. All that it requires is a little practice in setting the brushes. Francis will make your amalgams when he arrives as we have found that zinc amalgam is preferable to copper. Experience will dictate to you whether to run both machines connected together at full velocity and allow the whole load to be distributed over two machines or whether to run one machine at full capacity and run the other slowly ready to throw over in case of any accident to the first machine. I think the latter is preferable as with proper arrangement it would only make a slight dip not more than five seconds and that only when an accident occurs which should not occur more than once in two or three weeks. It may be possible that you can so arrange it when an accident occurs and you are compelled to stop the first machine that you can speed up the second machine and throw over without making a dip in the light. That is supposing one machine was running full speed the other one slowly before you stop the first machine speed up the second machine connect it aceross the line let it take half the load then disconnect the first machine and thus it will take the whole load. This would only make a slight variation in the candle power which could be instantly brought back by the regulator. I do not think it would make more than three or four candle power difference in [ᴱᴸ?]ᶜ the light. In this way you would only require one Engineer and the whole attention could be paid to one machine. I do not know whether I mentioned it before but I think it would be a good idea to have a flexible hose to run to the machine with cold water so that it can be put on any bearings. In our new central station machines we are having water jackets put on all of our bearings. Every three or four days it would be a good idea before starting on the regular run to run the machine slowly and with a coarse cut file smooth off the commutator and reamalgamate it. In this way you will wear your commutator down very ~~slowly~~ smoothly. In our Central Station machines we are arranging the commutator strips so that they can be screwed and unscrewed so that the whole commutator can be changed by screwing in new strips and made entirely new in about three hours and as we can have machines idle at all times in the Central Station we shall not have any trouble about the commutator business.

The reason why this last machine sent you will spark more with a quick load[18] is because it is of such an exceedingly low internal resistance in fact it is a 200 horse power Dynamo. Hence when two bars are short circuited by the Commutator Brush it means a very great deal. It is almost equal to three of the Paris machines multipled arc'd together the resistance of the Paris bcing tcn thousandths while this is only thirty eight ten thousandths or eight ten thousandths more than three Paris machines multipled arc'd together.

The Piston was not sent on the Arizona as it made a little knock and Porter when we took the Engine down took it to Philadelphia to fix it and promised to send it to you by express it will doubtless reach you before you need it. Before testing the machine the last time we had our chimney fixed and while the Engine was indicating 250 h.p. the Boiler which is the same as yours actually ran up from 125 to 138 lbs boiler pressure whereas previously with a small chimney it would run down from 135 to 90 lbs inside of two minutes. If you have a good draught you can get two hundred horse power out of these Porter Engines   You will have plenty of volts to regulate on when you have [---]^c 1200 lights on & when you have a man to keep a constant electromotive force you will find that the breakage of lamps will be excessively small.

We have made a number of experiments at the Lamp Factory and have ascertained that 16 candle lamps giving $5^{36}/_{100}$ horse power in the current will last 35 000 hours that if the economy be increased to $8\frac{1}{2}$ per horse power in the current the life reduced away down to 2200 hours & if you take 10 p.h.p. in the current life is 800 hours providing the electro motive force is kept constant. So you see if in certain places economy is no object lamps can be made that will last almost indefinitely.

The lamps we are now making and which I have written you about and am sending you some are ten per horse power in the current   I do not know whether I have mentioned before what we did with some of them   If I have what I give below is the latest. 26 of them were set up in the Lamp Factory, where a man sets night and day & keeps a constant electromotive force— [----]^c Last^g Saturday night, the average of all the lamps was 1220 hours twenty were still running. We made a number of 100 candlepower lamps but found the clamps were not thick enough to carry the current at the point of contact between the copper and the clamps. They gave a very fine light. I hope in a few days to have ~~sufficiently~~ some fibres with sufficiently thick ends.

I have not had good success with the method of getting high resistance by coiling owing to the fact that the fibre filament being excessively fine the whole of the surface was exposed to radiation except when each convolution almost touch the other but I am now working on a new line which promises to give us exactly what we want.

Yours letter Jany 29th.[19] Letters to Preece[20] & Sir John Lubbock[21] were sent you some time back

We have shipped Dynamos to you as follows:—

Oct 4th	3 "Z" machines
" "	1 "C" (Central Station)[22]
Oct 25th	
Nov 2nd[h]	20 Z machines
Jany 11th	4 " "
13	6 " "
19	2 E " for motors
Feb 13	1 C. (Central Station)

We have shipped to Col G. E. Gouraud for <u>his own account</u> the following:—

| Nov 4th | 3 Z Dynamos |
| " 10th & 11th | 17 " " |

I think this will give you the information you desire

I see from Mr. Bouveries[23] letter that he thinks I am like all the rest of the Inventors that I am flighty and have no idea of practical financial questions &c &c.[24] He might consult with Mr. Pierpont Morgan as to ~~whether~~ that point when he arrives in London[25] He should put the square question to Pierpont Morgan as to whether I have ideas of business and whether I can conduct several large manufacturing establishments besides superintending the installation of the Light here, having no small say in the policy of the various Companies & conducting experiments on half a dozen widely different subjects all at the same time. Drexel Morgan & Co do not hold such views as Mr Bouverie or they would certainly not have put into the Electric Light the amount of money they have simply on my statements as to ultimate success. They do not usually go into business on the wild assertions of a flighty inventor   Very truly yours

T A Edison

Written in great haste to catch mail   will write in a day or so   Insull

LS (letterpress copy), NjWOE, Lbk. 11:276 (*TAED* LB011276; *TAEM* 81:351). Written by Samuel Insull; stamped Thomas A. Edison to Edward H. Johnson. ᵃ"*New York*," and "*188*" from handstamp. ᵇIllegible. ᶜObscured overwritten text. ᵈInterlined below. ᵉCanceled. ᶠDegree symbol canceled. ᵍInterlined above. ʰ"Oct 25th" and "Nov 2nd" enclosed by brace at right.

1. At the end of January Samuel Insull began using a handstamp to inscribe "FROM THOMAS A. EDISON, NO. 65 FIFTH AVENUE" and the dateline on some outgoing letters. At the same time, he also began stamping "TO EDWARD H. JOHNSON, LONDON, ENGLAND"; later he adopted a similar stamp for Charles Batchelor. TAE to Johnson, 27 Jan. 1882, Lbk. 11:164 (*TAED* LB011164; *TAEM* 81:305).

2. Johnson's twenty-four page reply to Docs. 2202 and 2203 is in Series 1, Box 1, Folder 2, WJH.

3. See Doc. 2203 esp. nn. 15 and 26.

4. Richard Dyer's notes have not been found.

5. Edison's British Patent 562 (1881) pertained to heating the clamps by focusing the beam of an arc light on them through a mirror and lens (Dredge 1882–85, 2:cxliv). This was similar to his U.S. Pat. 248,428; see also U.S. Patent 265,777.

6. Edison stated in both forms of this 1879 patent that when rolling carbon compounds into slender filaments, "to make the light insensitive to variations of the current a considerable mass of matter should be used in order that the specific heat of the lamp may be increased, so that it takes a long time to reach its full brilliancy and also to die away slowly. To do this it is better to have the carbon as homogeneous as possible, and obtain the requisite resistance by employing a filament several inches long and winding the same in a spiral form so that the external radiating surface shall be small." Brit. Pat. 4,576 (1879), Batchelor (*TAED* MBP019; *TAEM* 92:134).

7. The patent digest has not been found but Johnson acknowledged it as "very valuable." In early March, he sent Edison a summary of Shelford Bidwell's comments on it. In his 16 January letter (p. 15, see note 2), Johnson had urged Edison to send Dyer to London to attend to patent business. If he were "coached thoroughly," Johnson argued, Dyer "could come over here and in a fortnight accomplish more for you . . . than he can do in New York in a year. Think it over and you will agree with me." Johnson to TAE, 26 Feb. and 4 Mar. 1882, both DF (*TAED* D8248Q, D8239ZAF; *TAEM* 63:398, 62:792).

8. In his 16 January letter (pp. 15–16; see note 2), Johnson stated that "our people have been so much exercised about this Patent of Seimen's that I felt they needed a little backbone, but I had none to spare until the arrival of your specification from Brewer & Jensen for your Disc Dynamo Patent. This gave me a hint and I used it. I gave Waterhouse to understand that this was a Baby of yours from which you expect great

things; that you believe it to be the foundation of the Dynamo for the future but that you are not hurrying it at all inasmuch as you are very anxious to keep it perfectly quiet until you Patent is secured." He showed the specification to Shelford Bidwell, who reportedly declared it "most extraordinary" and a "new departure in the right direction."

9. Edison made this inquiry at the end of a brief cable to Johnson about another matter on 6 February (LM 1:154C [*TAED* LM001154C; *TAEM* 83:949]). A man named Elias in Amsterdam invented an electromotor with a ring armature in 1842. The ring was divided into six sections, each wound with a single layer of wire, in a manner similar to that later adopted independently by Pacinotti and Gramme. The ring rotated within a larger wire-wound ring which served as the field magnets. The Elias machine was exhibited at the 1881 International Exposition in Paris (King 1962, 371; Dredge 1882–85, 1:102–5; Prescott 1884, 701–3).

10. These figures in Edison's British Patent 2,052 (1882) on dynamos corresponded to the two drawings in his U.S. Patent 264,646; see Doc. 2082 esp. n. 2.

11. Dyer's seventeen page analysis raised several questions about the validity of the 1873 Siemens patent. He found its relationship to Edison's small wire-wound armature ambiguous, yet concluded definitively that it could not be interpreted to cover the Edison large bar and disk armature. Dyer to TAE, 14 Feb. 1882, DF (*TAED* LB011257; *TAEM* 81:334).

12. Bidwell's report has not been found, nor has Dyer's gloss on it. Edison sent the latter in early March with sixteen pages of his own comments. TAE to Johnson, 6 Mar. 1882, Lbk. 11:385 (*TAED* LB011385; *TAEM* 81:409).

13. This report has not been found but see Doc. 2165 n. 2.

14. When tested on 17 February in an L dynamo, this armature induced enough current for 600 A lamps. Clarke "Electrical Experiments & Tests" notebook, acc. 1630, box 30, folder 3, EP&RI (*TAED* X001K1).

15. That is, 97% of the mechanical energy at the armature pulley would be converted to electricity; somewhat less could be drawn off to perform useful work, the remainder being lost internally as heat.

16. The Italian Royal Opera at Covent Garden, London, was one of Britain's foremost theaters (*London Ency.*, s.v. "Covent Garden [Royal Opera House]"). At the prompting of Egisto Fabbri, company officials approached Johnson about lighting the theater before the coming opera season in April. Johnson promised Edison that he could do a more successful installation than the Swan lights at the Savoy; he planned to submit a cost estimate about this time (Johnson to TAE, 19 Jan. [pp. 20–21] and 19 Feb. 1882, both DF [*TAED* D8239I, D8239V; *TAEM* 62:675, 766]; "Royal Italian Opera", *Times* (London), 19 Apr. 1882, 12).

17. Jehl was to set up the meters and train someone in their operation, then go to Paris and do the same there. TAE to Johnson, 14 Feb. 1882, Lbk. 11:256 (*TAED* LB011256; *TAEM* 81:333).

18. That is, strong or full of energy; a heavy load. *OED*, s.v. "quick."

19. DF (*TAED* D8239N; *TAEM* 62:709).

20. Doc. 2217.

21. Sir John Lubbock (1834–1913), a major shareholder in the Edison Electric Light Co., Ltd., had played a similar role in the Edison Telephone Co. of London. He was a prominent banker and member of

Parliament, as well as a noted entomologist and member of the Royal Society (Doc. 1741 esp. n. 2). Lubbock had promised in December to aid in forming an Edison lighting company in Great Britain. Samuel Insull sent Edison's handwritten reply to Lubbock and the letter to William Preece (Doc. 2217) to Johnson for him to deliver (Lubbock to TAE, 8 Dec. 1881, DF [*TAED* D8133ZBC; *TAEM* 58:726]; TAE to Lubbock, 23 Jan. 1882; Insull to Johnson, 23 Jan. 1882; Lbk. 7:328, 325 [*TAED* LB007328, LB007325; *TAEM* 80:554, 552]).

22. The first London Jumbo was packed on this date and sailed on 6 October. See Doc. 2162.

23. Edward Pleydell Bouverie (1818–1889), the founding chairman of the Edison Electric Light Co., Ltd., had played a similar role in the Edison Telephone Co. of London. Doc. 1765 n. 2; *Oxford DNB*, s.v. "Bouverie, Edward Pleydell."

24. In a letter to Johnson, a copy of which was sent to Edison, Edward Bouverie said that he did "not wonder that Mr Edison is disposed to growl at the caution which has been displayed by people on our side. . . . Like all creative and poetic minds he sees no difficulties where men of an ordinary understanding require to make their ground good. This is one of the distinctive qualities of genius, their flight is so high & strong that they are apt to forget they may fly too near the sun & have their wings melted. This, I suppose is the true meaning of the fable about Phaeton & explains Mr Edisons own pecuniary straights." Bouverie to Johnson, 25 Jan. 1882, DF (*TAED* D8239P; *TAEM* 62:738).

25. John Pierpont Morgan, head of Drexel, Morgan, & Co. in New York, was traveling in Egypt; he arrived in London in early March (*ANB*, s.v. "Morgan, John Pierpont"; Strouse 1999, 199–205). Morgan was among the earliest and strongest financial supporters of Edison's electric light research (see e.g., Docs. 1586, 1595, and 1607).

–2229–

To George Gouraud

New York, Feby 18 *1882*[a]

Dear Sir

I beg to confirm cable as follows

To you Feby 18, 82   Liquify   cannot approve sale dynamos domestic company   Written[1b]

You will understand from the above that I do not approve of the proposed company to purchase the arc light referred to in yours of 28th Jany.[2] I do not believe to the particular Arc Light you refer to and furthermore I do not see the necessity for an Arc Light when we have an Incandescent Light which is capable of giving any amount of candle power required as you will shortly see from the Light given by the Street Lamps which I am going to send over for use on Holborn Viaduct.

As to selling the plants you obtained from me to the Domestic Electric Lighting Company[3] I would point out to you that my Electric Light Patents are not owned by me for England & I am therefore unable to give my permission for you to sell apparatus which they cover in England.[4] I might also add that I was under the impression that you bought the plants in order to exploit my Light in the countries which are covered by your contract with me.[5] Success in these countries depends entirely on prompt action so as to occupy the field before others can get well established and I should think our mutual interests would be best served by the immediate installation of the twenty plants you have on hand in twenty different cities of the various countries referred to in our contract.

I shall await with pleasure the receipt of the "analysis of comparative prices" you promise me[6] & will examine it carefully & then write you my views on the subject   Very truly yours

Thos A Edison   I[nsull]

L (letterpress copy), NjWOE, Lbk. 11:319 (*TAED* LB011319; *TAEM* 81:379). Body of document written by Samuel Insull; handstamp of Thomas A. Edison. ᵃ"*New York*," and "*188*" from handstamp. ᵇSalutation and text up to this point written by Charles Mott.

1. This is the full text of Edison's cable to Gouraud on this date. "Liquify" was a prearranged code for Edison to signify disapproval of Gouraud's plan to purchase the Solaire lamp (see note 2 below). The remainder of the message is the subject of note 3 below. LM 1:160B (*TAED* LM001160B; *TAEM* 83:952).

2. Gouraud had summarized his plan to form a company for the Solaire lamp, which he described as a "semi-incandescent" arc light. The patent had been brought to his attention by Otto Moses, whom Gouraud thought should purchase it because "all the Arc Companies are working incandescent lamps of some sort . . . [and] it seems to me that the development of your system will involve, to some extent, the desirability of an Arc light to work in conjunction with your incandescent system where an arc system might be required." Gouraud to TAE, 28 Jan. 1882, DF (*TAED* D8239M; *TAEM* 62:707).

3. Samuel Insull informed Johnson of newspaper reports indicating that Gouraud had organized this company for the "miniature Central Station business," by which he apparently meant isolated plants in theaters and hotels. It was not specifically an Edison company but would reportedly "give preference to the Edison System." Insull expressed his own strong disapproval of the arrangement, presumably mirroring Edison's opinion, and accused Gouraud of failing to disclose his true relationship with the company. Insull to Edward Johnson, 24 Mar. 1882, Lbk. 11:486 (*TAED* LB011486; *TAEM* 81:462).

4. Drexel, Morgan & Co. owned Edison's British patents (see Doc. 1649). Gouraud reported that he could not sell dynamos at the price he had paid Edison. He proposed selling the machines already on hand for "carrying out some orders which the Domestic Electric Lighting Company have for lighting large buildings &c." He reasoned that "as those contracts . . . have to be carried out by some one's dynamos I have taken such steps that they [the Domestic Co.] will take those that I have if there is no objection to their being used here." Gouraud to TAE, 28 Jan. 1882, DF (*TAED* D8239L; *TAEM* 62:705).

5. Edison was referring to the two agreements with Gouraud signed on 18 August 1880 (see Doc. 1978) concerning the commercial development of electric light and power patents in regions outside of Great Britain and the major nations of continental Europe.

6. Gouraud made this pledge in his letter about the dynamos (see note 4).

–2230–

Notebook Entry:
Pyromagnetic
Generator

[New York,] Feby 20 1882

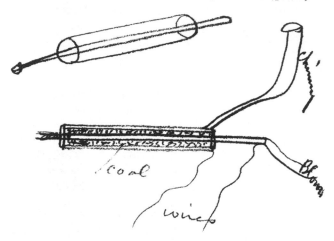

Direct Conversion from Coal.[1]

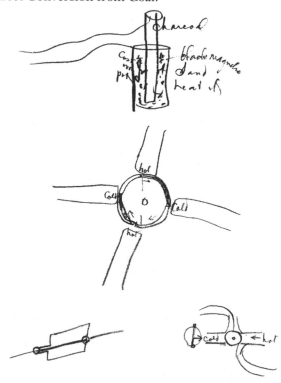

[A][2]

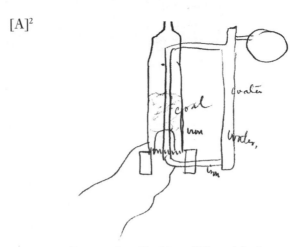

Direct Conversion Coal into E[lectricity]

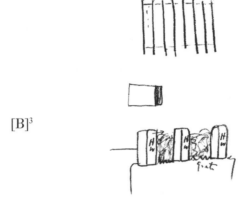

[B][3]

X, NjWOE, Lab., N-80-10-01:99, 98, 101–3 (*TAED* N304:59, 58, 60–61; *TAEM* 41:1109–12). Document multiply signed and dated.

1. Figure label at left is "Cast iron pot." This is the first extant evidence of Edison's investigation into producing electricity directly from the combustion of coal. He evidently did not pursue it until he was back at Menlo Park in May. Martin Force experimented then with shaping and heating finely ground carbon under pressure (N-82-05-15:9–13, Lab. [*TAED* N203:5–7; *TAEM* 40:477–79]). At the end of May, Edison executed a patent application embodying some of the ideas represented here. The application covered a generating cell consisting of an iron vessel heated from the outside in a furnace. The vessel would contain a fusible oxide of some kind, into which was immersed an electrode of pressed carbon. He explained that

> The heat of the furnace fuses the oxide at the same time that it
> raises the carbon to a temperature at which it combines rapidly with
> oxygen. A reduction of the oxide takes place, the oxygen combining
> with the carbon and forming carbon monoxide . . . while the metal
> or metalloid or other product is carried to the other electrode, being
> deposited upon the walls of the containing vessel or pot. During the

oxidation of the carbon an electric current flows through the circuit of the cell. [U.S. Pat. 460,122]

The patent issued in 1891.

In October 1882 Charles Batchelor told Edison that a Dr. Beard in Paris had published a description of a similar process for "direct production of Electricity by combustion." Batchelor obtained four samples of his briquettes and found that he obtained an ampere of current from each of two he burned. He promised to forward the other two to Edison for his study. Batchelor concluded, "It looks very small at present, but six months of our experimenting might show it a much better thing." Batchelor to TAE, 27 Oct. 1882, DF (*TAED* D8238ZEB; *TAEM* 62:549).

Edison returned to the subject in September 1883, when he drafted a patent application for a generating cell similar to that shown here. He conducted a long series of experiments the following summer on the "direct oxidation" of coal into electricity. He reportedly obtained a strong current but, after an accident blew out the laboratory windows, decided the process was too dangerous to exhibit at the 1884 Electrical Exhibition in Philadelphia. Cat. 1149, Scraps. (*TAED* NM018:56–60; *TAEM* 44:585–89); U.S. Pat. 490,953; N-82-05-15:104–81, Lab. (*TAED* N203:42–79; *TAEM* 40:514–51); "Producing Electricity Directly from Coal," *Operator* 15 (1884): 181; Israel 1998, 233–34.

2. The water pipes are presumably to carry off excess heat for other uses, which Edison expected would increase the overall efficiency of the conversion process.

3. Figure label is "grate."

–2231–

To Theodore Waterhouse

[New York,] 21st Feby [188]2

My Dear Sir,

I duly received your favor of 4th inst enclosing a copy of the Judgment in the Scotch Telephone case which I read with great pleasure.[1]

It is extremely gratifying to me that my invention of the Carbon Transmitter should have been recognized as covering the microphone and that my Patent for Great Britain for the same should have been declared valid

Considering the great controversy which has taken place on this Carbon question and the many doubts so often expressed as to the validity of my Patents the result obtained gives me still greater confidence in my belief that my Electric Light Patents (which are now passing through the same criticism as did my Telephone Patents some two or three years back) will be eventually be proved to hold the monopoly of Electric Lighting by Incandescence

I need hardly say that I fully reciprocate the very cordial feeling that you say exists among the intending shareholders in the proposed Light Co that the partnership between the A. & B. shares should be of the most harmonious and friendly character.[2] I am very pleased that in this new & greater enterprise I should have associated with me many of the same gentlemen who exploited my Telephone in England and feel sure the connection will prove advantageous to us all[3]   Very truly yours

Thomas A Edison

LS (letterpress copy), NjWOE, Lbk. 11:334 (*TAED* LBo11334; *TAEM* 81:387). Written by Samuel Insull.

1. Waterhouse's 4 February letter is in DF (*TAED* D8256C; *TAEM* 63:848); see Doc. 2221 n. 17.

2. Waterhouse indicated in his letter (see note 1) that other prospective investors shared his hope "that the partnership between the A & B shares should be of the most harmonious & friendly character." The A shares of the Edison Electric Light Co., Ltd., were for subscription to investors; Edison was to be given B shares, which had one-half the voting power and less favorable terms for dividends; Edison received "one fully paid-up B share for every £10 of capital actually paid up or credited as paid up in respect of A shares." This would amount to 5,000 B shares (£10 at par) when the A shares were fully subscribed. TAE agreement with Drexel, Morgan & Co., Grosvenor Lowrey, Fabbri, and Bouverie, 18 Feb. 1882, CR (*TAED* CFoo1AAE1; *TAEM* 97:295).

3. Among the directors of the new company were three who had invested in the Edison Telephone Co. of London: Edward Bouverie, Sir John Lubbock, and (Viscount) Thomas Francis Anson. The stockholders as of June 1883 also included Waterhouse and additional telephone investors. Articles of Association (p. 9), 15 Mar. 1882; list of stockholders, 13 June 1883; both CR (*TAED* CFoo1AAE, CFoo1AAO; *TAEM* 97:277, 347); see Doc. 1799.

-2232-

From José Navarro

New York, Feby 21st 1882[a]

Dear Sir:

As this Company[1] has commenced operations under the verbal understanding with you that you will furnish it with instruments lamps machinery fixtures supplies &c at the same price you charge to the "Illuminating" & "Isolated" Companyies & there is no written document to that effect I beg you to address in an official letter stating the terms of our agreement.—[2] I am, dear Sir   very Truly Yours

J. F. Navarro[3]   Treasurer   pr F. M[4]

L, NjWOE, DF (*TAED* D8237H; *TAEM* 62:154). Letterhead of Edison Spanish Colonial Light Co. [a]"New York," and "188" preprinted.

1. The Edison Spanish Colonial Light Co. was incorporated on 10 January 1882 by Edison, José Navarro, Grosvenor Lowrey, George Soren,

and Fausto Mora to exploit Edison's electric light and power patents in Cuba and Puerto Rico. Three days later Edison was elected president and Navarro treasurer. This company was the successor to the Edison Electric Light Co. of Cuba & Porto Rico incorporated by these men on 9 May 1881; the same day they also incorporated a local company, the Edison Electric Light Co. of Havana. In Doc. 1571 Lowrey had noted the importance of obtaining local concessions in Cuba which he thought could be done better from New York by persons with Cuban connections. Certificates of Incorporation, Edison Electric Light Co. of Cuba & Porto Rico and Edison Electric Light Co. of Havana, both 9 May 1881; Edison Spanish Colonial Electric Light Co., 10 Jan. 1882; all NNNCC–Ar (*TAED* X119KA, X119MA, X119OA); Edison Spanish Colonial Electric Light Co. minutes, 13 Jan. 1882; Navarro to TAE, 5 Mar. 1882; Mora to TAE, 1 and 8 June 1882; Lowrey to TAE, 11 Jan. 1884; all DF (*TAED* D8237E, D8237J, D8237P, D8237Q, D8434E; *TAEM* 62:149, 156, 162, 163; 73:762); TAE agreement with Navarro, Lowrey, Soren, and Edison Spanish Colonial Light Co.; 9 Feb. 1882; TAE and others agreement with Lowrey, 3 May 1882; both Miller (*TAED* HM820157, HM820163; *TAEM* 86:452, 86:468).

2. No such letter has been found.

3. New York financier José Francisco de Navarro (1823–1909), an original stockholder in the Edison Electric Light Co., had acquired rights to Edison's electric light and power patents for Cuba in 1880. Docs. 1571 and 1920 n. 2; Obituary, *New York Times*, 4 Feb. 1909, 9.

4. Probably Fausto Mora (c. 1836–1914), who was likely the son of the partner in the firm of Mora, De Navarro & Co. Since 1878 Mora had worked with Navarro to exploit the Edison light and power system in Cuba. His brother was photographer José Mora, for whom Edison sat for a portrait in 1879. Previously the editors had conjectured that Navarro's partner was Joseph Mora but his birth date of 1844 would preclude this because the firm was founded in 1855. U.S. Customs Service 1962, reel 153, list 542, 23 June 1855; "Mora Left $170,435 Here," *New York Times*, 8 Feb. 1914, 13; "The Mora Paintings," *New York Times*, 18 Feb. 1890, 4; Docs. 1790 n. 17 and 1604 (n. 3 identifies Joseph Mora); Navarro to TAE, 19 Feb. 1881, DF (*TAED* D8131E; *TAEM* 58:387); *NCAB* 15:246.

–2233–

Mary Edison to Samuel Insull

Aiken SC   Feb 23 1882.[a] 10:27

Arrived here yesterday   send Letters to Highland Park Hotel[1]

Mrs Edison

L (telegram), NjWOE, DF (*TAED* D8214G; *TAEM* 60:703). Written on Western Union Telegraph Co. message form. [a]"1882." preprinted.

1. The Highland Park Hotel, established around 1871, was a large winter resort in Aiken, S.C. The region promoted its "health restoring influences" for visitors seeking "recuperation, rest, or pleasure"; the clear and relatively dry air particularly attracted tuberculosis patients. Situated on the South Carolina Rail Road, Aiken was about thirty-one hours from New York. *Aiken* 1883, 5, 7, 30.

–2234–

To Edward Johnson

Fiftyseven London[1]

Taking family Florida two weeks[2] cannot get working drawings covent garden special machine until return   Hence beware time contract   Other matters proceed as usual.

L (telegram, copy), NjWOE, LM 1:165D (*TAED* LM001165D; *TAEM* 83:954). Written by Charles Mott.

1. Cable code for Edward Johnson; see App. 4.
2. The details of this trip are unclear. Edison arrived in Washington, D.C. on 20 February. The next night he attended an unknown formal event, for which he evidently purchased a new dress suit. Samuel Insull reported that he "looked a regular masher." Edison returned north on 22 February (TAE to Insull, 20 Feb. 1882; Cavanagh, Sandford & Co. to Insull, 16 Feb. 1882; both DF [*TAED* D8204T, D8204S; *TAEM* 60:74, 73]; Insull to Johnson, 22 Feb. 1882, LM 3:64 [*TAED* LM003064; *TAEM* 84:67]). That same day Mary Edison, who attended a ball in New York on 20 February, arrived in South Carolina (see Doc. 2233). Edison left for Florida about the first of March. In advance of the trip, Richard Dyer wrote his father that Edison was "going for his health" (Insull to Edward Johnson, 20 Feb. 1882; LM 3:54 [*TAED* LM003054; *TAEM* 84:57]; Insull to Charles Batchelor, 25 Mar. 1882; Richard Dyer to George Dyer, 18 Feb. 1882; Lbk. 11:501, 315 [*TAED* LB011501, LB011315; *TAEM* 81:469, 375]; see also Israel 1998, 231).

–2235–

From Edward Acheson

Paris, le 26th Feb. 1882[a]

Dear Sir

Four days ago I returned from Milan Italy, where I was sent, by Mr Batchelor to install a plant. Mr Shepherd the Company's Agent[1] succeeded in obtaining the entry to the Scala Theater, in the "Redotto" or drawing room of which are three chandeliers containing in all 292 gas jets.[2] I replaced each of the jets with a "B" lamps. I put one "A" lamp in the private box of one Sig. Bussie, the C̶h̶ Cashier of the Bank of Italy,[3] also one "A" in the Courtyard of the Royal entrence to the theater, also one "A" and 2 "B" in the engine room. It was a most perfect success. Several gentlemen who were at the Paris Exposition and in London, agreed it was the finest exhibit of electric lighting they had yet seen. Even the agent of Seamen's[4] agreed to this.

The people were wild over it. At the time I left Milan there were excilent prospect for the formation of a Company of five or six million frances for the opperation of Italy entire. It is I think the desire of the Paris Company to divide Italy, giving to one Major Garbin the southern portion.[5]

The company here were not sure of Mr Shepherd as he was

unknown to them. Having been in Milan for one month I am free to say, Mr Shepherd has a <u>very</u> wide circle of influenetial friends, is greatly liked and stands strong in Italy. His[b] friends reach out pretty much all over Italy.

Knowing you can roll the ball in either direction, I write in the interest of both Mr Shepherd and myself. I have already talked with Mr Batchelor of this and told him that the leaders of this Milanese Company hold out to me the Cheif Engineer ship of their Company, and that I left Milan with the understanding that in case they were successful in their treaty with the Paris Company I would after the Completion of my duties at the factory here, which probably will[c] be in five months, return to Milan permanently[6]

It is not my intention to remain in France. Indeed I may not be wanted. There is nothing for an American in France. The French know it all. I am now doing what you in the shop at Menlo Park engaged me to do, namely assist in getting into opperation a factory in France   after this is done, which won't be long, I will consider my work in France completed, and I shall then look for other fields, where the grass grows longer and is more easily gathered.

Every thing is tending toward an early starting of lamp factory. The buildings at Ivry are well adopted for our work.

I was rather uncertain wether to write you or not, as in case Mr Batchelor heard of it he might think it unjust to him. I hope this will not occur as I have the kindest feeling toward him, and am determined to render him all assistance possible. I was induced to write you, knowing you[b] rejoice in the success of your subjects and the gratification of their ambition.

You may consider Italy a most promising field, for the Edison light. Yours very Respectfuly

E. G. Acheson

P.S. I may add that Mr Batchelor is satisfied that I go into the engineering dept after the factory is going. As Hipple will then run it. Acheson

ALS, NjWOE, DF (*TAED* D8238ZAD; *TAEM* 62:327). Letterhead of Société Électrique Edison. [a]"Paris, le" and "188" preprinted. [b]Obscured overwritten text. [c]Interlined above.

1. James Shepherd, Edison's agent in Milan, superintended the installation at the Teatro alla Scala. Szymanowitz 1971, 81.

2. The plant ran for about twenty nights during the theater's winter season, after which it was moved to the nearby Cafe Biffi. At the end of March Bailey and Puskas wrote Sherburne Eaton that "a great success has been made at Milan in the Scala Theatre, where the light was shown

during nearly two months, afterwards in street lighting, and now in the great cafe Buffi of Milan, exciting greater enthusiasm than ever." Francis Jehl reported a few weeks later that "the Edison plant works very well here, considering the difficulties under which they labor. The 'Cafe Buffi' which has the light is one of the principle ones in town, and it is said that since he has had it, he does double the business that he done before. The cafe is situated in the middle of the 'galleries' [Galleria Vittorio Emanuele] which is something like the 'Crystal Palace' in London only that the sides are the houses of the street while a roof of glass extends from one side of the street to the other. The whole town comes here in the evening, so that the light could not be exhibited better." By the fall, Edison's representatives in Milan were planning to light the Teatro alla Scala for the 1883 season from the planned central station (Bailey and Puskas to Eaton, 29 Mar. 1882; Jehl to TAE, 14 April 1882; Colombo to TAE, 10 Oct. 1882; all DF [*TAED* D8238ZAP, D8238ZAU, D8238ZDT; *TAEM* 62:361, 378, 540]; Acheson 1965, 23). Italian-language clippings about lighting the Teatro alla Scala are in Cat. 1327, Batchelor (*TAED* MBSB5; *TAEM* 95:114).

3. Felice Buzzi, an agricultural landowner, cofounded the Comitato per le Applicazioni dell'Elettricita. Scalfari 1963, 11.

4. The Siemens agent at Milan has not been identified.

5. Elsewhere referred to as Major Garbi, this may be Alexander Garbi of Florence, to whom Egisto Fabbri bequeathed a modest annual sum on the basis of an unknown association between the two men. Joshua Bailey traveled to Florence on 27 March to close negotiations with him. Bailey and Puskas cabled Edison on 1 April that they had arranged for him to represent the Edison interests in Italy, and Insull informed Fabbri of this two days later. However, the formation of the Comitato per le Applicazioni dell'Elettricita was delayed until July. Capitalized at 3 million francs, this syndicate was responsible for manufacturing equipment and operating central stations. "Will of Ernesto [Egisto] P. Fabbri," *New York Times*, 27 Dec. 1894, 14; Bailey and Puskas to Insull, 26 Mar. 1882; Bailey and Puskas to Eaton, 29 Mar. 1882; Bailey and Puskas to Eaton, 17 July 1882; all DF (*TAED* D8238ZAL, D8238ZAP, D8238ZCC; *TAEM* 62:353, 361, 454); Bailey and Puskas to TAE, 1 Apr. 1882, LM 1:180D (*TAED* LM001180D; *TAEM* 83:962); Insull to Fabbri, 3 Apr. 1882, Lbk. 12:10 (*TAED* LB012010; *TAEM* 81:501).

6. Samuel Insull replied that Edison was in Florida when this letter arrived but had since written to Batchelor "in such a manner as will no doubt obtain the object you desire." On the same date Edison wrote Batchelor asking his views on bringing Acheson back to New York for experience at the Pearl St. station before returning him to Europe. Though Acheson subsequently asked to become an Edison lighting agent in Pennsylvania, he remained in Europe and installed isolated plants in several cities. He was the Compagnie Continentale's chief engineer at Milan in December 1882 when he resigned in a salary dispute and was succeeded by John Lieb; Acheson worked briefly for Edison again in 1884. Insull to Acheson, 5 Apr. 1882; TAE to Batchelor, 5 Apr. 1882; TAE to Colombo, 12 Oct. 1882; Lbk. 12:36, 37; 14:277 (*TAED* LB012036, LB012037, LB014277; *TAEM* 81:516, 517, 939); Bailey to TAE, 13 Dec. 1882, DF (*TAED* D8238ZEY; *TAEM* 62:590); Szymanowitz 1971, 79–101.

From Edward Johnson

London, E.C. Feby 2~~9~~6th 1882[a]

My dear Edison,

You will see by the papers that Crystal Palace Electrical Exhibition is now formally opened.[1] Lane–Fox, Swan & Maxim are making prodigious effort to rival your Exhibit and last night Maxim had a Chandelier with 96 Lamps on it ~~and~~ a very gorgeous Crystal affair but producing such a blaze of Light as to offend the Eye; it is consequently a failure. He has some other smaller Chandeliers however which are more satisfactory but they are all on the lines of Gas Chandeliers nothing new or particularly novel. Swan is exhibiting one or two long sweeping stem arrangements like he had at Paris—nothing new. The Brush Company are however exhibiting a Crystal Chandelier with Lane–Fox Lamps the bulb of which is white porcelain. The effect is very beautiful as the Lamp is simply a white bulb of light. The Arc Lights throughout the building were very generally in operation and altogether the Palace now begins to look like a complete affair. There were upwards of 25,000 or 30,000 people at the Palace yesterday and altogether the thing may be said to be fairly launched and a creditable display. The Duke and Duchess of Edinburgh[2] were the Royal Visitors on the occasion and were Entertained at a private dinner to which a select few were invited myself among the number. The party arrived at the Palace at $\frac{1}{2}$ past 6. They passed first through the Concert Room where your lights received their hearty admiration. They then passed down through the various Exhibits at the South end of the Palace, took a view of the entire length of the Palace from the Clock Tower and returned through Siemens' arc light Exhibit to the Swan Exhibit in the Picture Gallery and were there detained 5 minutes waiting in vain for Swan to get his Lights in operation and from this abortive attempt they came immediately into the Entertainment Court where I with white Kids and Swallowtail awaited their arrival, the doors having been kept closed throughout to keep the general public out but a large number of privileged persons were already admitted by Card. Among these were Messrs MacLoughlin[3] Logan,[4] Francis &c late arrivals from America who will doubtless report to you their impressions[5] Receiving Royalty is a new role for me and I had no advice but was informed afterwards that I did the thing in a creditable manner. At all events their Royal Highnesses were so interested in what they saw and so pleased with the beauty and taste of the entire display as well as with the completeness of all your work that they remained so long in your rooms as to preclude their visiting the North end of the Palace so that they went directly from the Entertainment Court to the Dining Room thus omitting entirely to visit Maxim, Lane–Fox, Brush &c which was of course a sore disappointment to these people. At the Dinner Table the Duke referred only once to what he had seen and then in terms of admiration and great animation at the completeness and beauty of Mr. Edison's work. The conversation on this subject was carried on between his highness and Professor Spottiswoode[6] across the Table and was therefore distinctly heard by Messrs Swan, Siemens, Brush, Maxim or their representatives who were guests at the Table. I was assured on all hands that we had scored heavy, that in point of fact the Royal party which was composed of a large number of other distinguished persons besides the Duke and Duchess, would leave the Palace with but one impression; namely, that the only thing shown them worthy of their special attention and admiration was Mr Edison's Exhibit. The Chairman of the Crystal Palace Company[7] whispered in my ear as the Royal party were leaving that he had something exceedingly important to communicate to me but that it was impossible for him to do so there and he wished to see me early this week. He said that the Duke was full of Edison and could talk of nothing else. Of course the Chairman's place at the table was between the Duke and Duchess and he consequently knows their minds. I am somewhat curious to know what he has to communicate— will advise you in my next letter. I will leave to the papers and to your occasional correspondents McLoughlin, Francis and others to give you a better idea of the complete success of the evening to us and its consequent failure to other Exhibitors. I shall now have to spend a considerable portion of my time at the Palace. The various Corporations &c throughout the Country are sending Deputations to the Palace to investigate the subject of Electric Lighting   It is absolutely necessary that someone capable of properly setting forth the merits of your system should be present. In the absence of any other I shall have to do this work myself. Fortunately it is night work as the Exhibition is only of interest in the evenings. I shall therefore be able to give it considerable attention. Very truly Yours,

Edwd H. Johnson

The papers of course represent that all the Exhibits were "viewed" by the Royal party—but in point of fact such was not the case— The Duchess asked me whether it was likely you would be coming to England soon— EHJ[b]

Edison isolated lighting installation at the Crystal Palace Exposition.

LS, NjWOE, DF (*TAED* D8239ZAE; *TAEM* 62:784). Letterhead of Edison Electric Light System, Edwd. Johnson, manager. a"London, E.C." and "188" preprinted. bPostscript written and initialed by Johnson.

1. Press descriptions of the Crystal Palace exhibition include "The Electrical Exhibition at the Crystal Palace. Visit of the Duke and Duchess of Edinburgh," *London Daily News,* 27 Feb. 1882; and "The International Electrical Exhibition," *Times* (London), 27 Feb. 1882; both Cat. 1327, items 2079, 2082, Batchelor (*TAED* MBSB52079, MBSB52082; *TAEM* 95:149, 151); see also "Electricity at the Crystal Palace. II. Edison's Electric Light," *Nature,* 9 Mar. 1882, Cat. 1327, item 2112, Batchelor (*TAED* MBSB52112; *TAEM* 95:158) and excerpts in Edison Electric Light Co. Bulletin 5, 17 Mar. 1882, CR (*TAED* CB005; *TAEM* 96:681).

2. Prince Alfred (1844–1900), second son of Queen Victoria and Prince Albert, was Duke of Edinburgh and Duke of Saxe-Coburg and Gotha. He was also a naval commander. He married the Grand Duchess Marie Alexandrovna (b. 1853), the only daughter of Alexander II of Russia, in 1874. *Oxford DNB,* s.v. "Alfred, Prince"; *Dod's Peerage* 1881, 88.

3. Frank McLaughlin.

4. Col. James C. Logan.

5. Francis Jehl arrived in London two days earlier. He sent Edison a shorter but substantially similar account, adding that the Duke and Duchess spent an entire hour at the Edison exhibit. Jehl to TAE, 26 Feb. 1882, DF (*TAED* D8239ZAD; *TAEM* 62:781).

6. William Spottiswoode (1825–1883), mathematician and physicist, was president of the Royal Society from 1878 until his death. Spottiswoode was famed for the clarity of his public lectures at the Royal Institution and was also an accomplished linguist. *Oxford DNB,* s.v. "Spottiswoode, William."

7. Mungo McGeorge (1819?–1897), a wholesale warehouseman, held this position during the exhibition. Obituary, *Times* (London), 22 Oct. 1897, 1; "The Proposed Electrical Exhibition," ibid., 25 Oct. 1881, 11; "Failure of the Late Chairman of the Crystal Palace Company," ibid., 13 Aug. 1885, 12.

*Memorandum of
Conversation: Electric
Railway*[1]

when are you going to fit up a train on the elevated lighted by electricity[a]

Isolated dept has that & they could do it any time—[3] dont think there is much money in it   Gould[4] dont pay anything=[b]

Will have 3 miles Electric RR running at Menlo in 4 weeks tried new locomotive tonight[5b]

you are mistaken about Gould not <u>paying</u> anything— ~~lets~~ when you are ready with that engine let me know and give Gould & myself a private exhibition— It may be very profitable to <u>all</u> 3 of us—[a]

The biggest thing financially[c] outside of Electric R[6b]

Dont forget to let me know when you are entirely ready so I can take Gould out to see it— <u>we want it</u>— I will take care of the rest—[a]

All right   Can take you over the road at 50 miles per hour in a 10 hp Locomotive[d] ~~& on Elevated is the use of~~   Electricity taken from the rail to run the London Continuous Elevator system at all the stations of Co—[7b]

I want the first chance at it for our roads. Siemens has been here and over the roads with this view—but we can make more money by the Edison Company[a]

Lowrey & Fabbri will tell you that his Electric R is a mere toy & farce[8] whereas it requires a 100 hpow Electric[d] Loco for Elevated & one that will not get out of order[e]

Can you haul 4 cars 100 passengers–each and start them out quick from a station—[a]

Can give you 100 indicated hp when running & 250 indicated hp[f] to start= Electric Locos are very powerful in starting   I propose to use small Extra motor to work[g] vacuum pump for brakes so no change in Cars necessary—[b]

What will it weigh?[a]

800 lbs for small motor= Big Loco weigh ~~about~~ little less than yours[9b]

D, NjWOE, DF (*TAED* D8232P1; *TAEM* 61:966–67, 969). Order of text altered to reflect order in which it was written. [a]Paragraph written by interlocutor. [b]Paragraph written by Edison. [c]Interlined above. [d]Interlined below. [e]Paragraph written by Edison and followed by dividing mark to indicate change in text order. [f]"indicated hp" indicated by ditto marks. [g]Obscured overwritten text.

1. This document is an exchange of written comments between Edison and an unidentified person. The editors believe, based on context and a comparison of handwriting, that this was Robert Macy Gallaway

(1837–1917). An incorporator of the Edison Electric Illuminating Co. of New York, Gallaway was president of the Manhattan Railway Co. from April to November 1881, when Jay Gould succeeded him. Gallaway became vice president; he remained the company's executive officer and a member of the board of directors (Obituary, *New York Times,* 14 Nov. 1917, 15; "New-York," ibid., 17 Apr. 1881, 7; "Jay Gould's Lunch Party," ibid., 10 Nov. 1881, 3; Edison Electric Illuminating Co. of New York, Certificate of Incorporation, 16 Dec. 1880, NNNCC-Ar [*TAED* X119JA]). This placed him in a favorable position to act as an intermediary between Edison and Gould. In May 1882, he sought a demonstration of Edison's electric locomotive (Gallaway to TAE, 17 May 1882, DF [*TAED* D8249O; *TAEM* 63:599]).

2. The particular date, place, and other circumstances concerning the authorship of this document are unclear. However, it was written on the back of two lamp test reports dated 18 January 1882, thus fixing the earliest possible date of its creation. Testing of the electric locomotive first occurred in mid-February. Marshall to TAE, and Howell to TAE; both 18 Jan. 1882; both DF (*TAED* D8232H, D8232J; *TAEM* 61:965, 968); Insull to Johnson, 3 Feb. 1882, Lbk. 3:693 (*TAED* LM003047A; *TAEM* 84:49); TAE to R. M. Hughes, 16 Feb. 1882, Lbk. 11:305 (*TAED* LB011305A; *TAEM* 81:371).

3. Edison had contracted with Henry Villard in September 1881 to develop an electric locomotive. Villard released Edison from this contract in December 1881, presumably leaving him free to market his system to other railroads. See Docs. 2152 and 2195.

4. The financier Jay Gould (1836–1892), in concert with affiliated capitalists, monopolized New York City's elevated railroad system in 1881. He acquired control of the Metropolitan Elevated Railroad and the Manhattan Elevated Railway Co., then arranged to merge the New-York Elevated into the latter. Gould acquired an interest in Edison's quadruplex and automatic telegraph inventions in the 1870s. Grodinsky 1957; Klein 1986; *ANB,* s.v. "Gould, Jay"; Docs. 522 and 526; Israel 1998, 101–104.

5. Charles Hughes first tried the locomotive motor on 12 February but was unsatisfied with the results. It started the car only with difficulty and created excessive heat in the three dynamos supplying the current. Hughes worked on improving and modifying the locomotive over the next two weeks. On 26 February he gave a demonstration of the locomotive and passenger car, carrying some twenty passengers about six miles; one of the passengers was Henry Villard's engineer, John C. Henderson, who had played an important role in the installation of the isolated lighting plant on the steamship *Columbia* in 1880. After further improvements to the locomotive, track, and dynamos, Hughes carried thirty-five people in the passenger car on 18 March and pronounced the locomotive "in good shape." He continued to work on this and the freight locomotive through April. In May, Edison claimed that the passenger locomotive worked "splendidly," producing 15 horsepower to pull 120 people in four cars at twenty miles per hour. Hughes returned to this work in November and December. N-82-03-12, passim, Lab. (*TAED* N249; *TAEM* 41:861); Doc. 1892; TAE to Thomas B. A. David, 11 May 1882, Lbk. 12:284 (*TAED* LB012284; *TAEM* 81:652); Jehl 1937–41, 565–96.

Edison's electric railroad at Menlo Park, with one of his locomotives, in the late spring or summer of 1882. The engineer may be Charles Hughes.

6. It is possible that Edison meant to write L for "light" instead of R for "railroad."

7. Edison was referring to the cyclic elevator system discussed in Doc. 2220.

8. Werner Siemens developed the first practical electric locomotive and exhibited it at the Berlin International Exhibition of 1879. He then built a two-axle electric tram operated from overhead conductors. This vehicle ran over a 500 yard track at the Paris exposition in 1881, where Grosvenor Lowrey and Egisto Fabbri would have seen it. "Electrical Railway," *Sci. Am.* 42 (1880): 137; "The Electric Railway at West End, Near Berlin," *Sci. Am.* 47 (1882): 134; Guillemin 1891, 793–94.

Edison was in interference proceedings in late 1881 and 1882 with Siemens and Stephen Dudley Field over their respective electric railway patents. Edison and Field merged their electric traction interests and formed the Electric Railway Company of the U.S. in 1883. This company exhibited an electric locomotive at railway expositions held at Chicago and Louisville in 1883. In November 1882, Joshua Bailey negotiated a consolidation of the Edison and Siemens dynamo and electric railway patents but Edison balked (see Doc. 2379). *Edison v. Siemens v. Field,* Lit. (*TAED* QD001; *TAEM* 46:5); *DAB,* s.v. "Field, Stephen

The electric tram of Werner Siemens at the Paris electrical exposition of 1881.

Dudley"; Eaton to TAE, 9 Aug. 1882, DF (*TAED* D8226ZBA; *TAEM* 61:368); TAE to Bailey, 11 Dec. 1882, Lbk. 15:29 (*TAED* LB015029; *TAEM* 82:41).

9. That is, the steam locomotive in use on the Manhattan Elevated Railroad.

CENTRAL STATION "JUMBO" DYNAMO
Doc. 2238

Even before the "C" dynamo for the Paris Exposition was completed,[1] Charles Clarke began planning a still larger version. The new one had a dozen horizontal magnet coils and an armature of the same diameter but substantially longer (at $42\frac{1}{2}$ inches). It had 106 induction bars connected to 53 commutator blocks and a resistance of 0.0049 ohm. Driven by a 125 horsepower Porter-Allen engine, it was capable of operating 1,000 lamps at 103 volts and was intended for Edison's demonstration central station in London.[2] It was shipped in early October despite problems with the connection between engine and dynamo, and began operating on 12 January 1882.[3] In February the Edison Machine Works shipped a larger "C" (designated No. 3) to Holborn. Its armature was $46\frac{1}{2}$ inches with 98 induction bars and a resistance of 0.0039 ohm. Eight field coils were situated in a plane above the armature and four below it. The topmost four had the 8 inch diameter used in previous machines; the remaining eight were enlarged to

One of the Jumbo dynamos installed at Holborn Viaduct, London.

9 inches.[4] They were apparently connected in two groups of six coils in series, though subsequent practice was to use three multiples of four coils. The asymmetrical coil placement probably contributed to the heavy commutator sparking that plagued all Edison's dynamos of this type. This machine began working on 8 April.[5]

The latter Holborn "C" served as the prototype for central station dynamos built by the Machine Works until about 1884. These included a group of six constructed for the Pearl Street station in the spring of 1882. Fabrication of these machines probably began in March, but the design was essentially fixed in February with completion of the Holborn dynamo.[6] Specific details, such as the number of armature and commutator bars, varied somewhat but the machines were essentially the same.[7] The machines were tested in April and May on 1,400 lamps, though they were conservatively rated for 1,200 lamps to compensate for voltage drop through the distribution system. Edison considered them "far superior" to the Holborn dynamos.[8] Number nine was installed in June and, in September, became the first in regular operation at Pearl Street. It initially ran with a Porter-Allen engine, which Edison replaced with an Armington engine at the end of 1882.[9] The armature shaft was driven through an Oldham coupling, which permitted a small degree of misalignment between the shafts. As in all Edison's large central station dynamos, electrical connections in the armature were made between gold-plated surfaces that were bolted, rather than soldered, together.[10] Edison decided to keep one of these machines in motion as a spare at Pearl Street in case one of the others should suddenly fail.[11]

Number nine was the only dynamo to survive the January 1890 fire that partially destroyed the Pearl Street station, and it briefly returned to service. It was displayed at the 1893 Columbian Exposition in Chicago and, in 1904, at the Louisiana Purchase Exposition in St. Louis.[12] During the First World War it was in storage in Hoboken, N.J., where vandals stripped away the copper wiring. The machine was briefly exhibited (though not operated) in New York before Henry Ford acquired it in 1930 and had it restored for his Greenfield Village museum in Dearborn, Mich.[13]

Charles Clarke later recalled that the Machine Works produced a total of 23 "C" dynamos, including the one for Paris and the first two for London. Another went to Holborn at the end of 1882; the Pearl Street station eventually employed eight; ten (including one from Holborn) went to Milan, Italy; and two to Santiago, Chile.[14]

All these machines were formally designated "C" dynamos

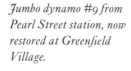

Jumbo dynamo #9 from Pearl Street station, now restored at Greenfield Village.

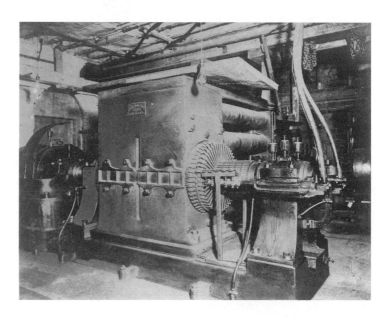

but came to be called "Jumbos" by May 1882.[15] P. T. Barnum's famous elephant of that name arrived in New York just a month earlier, creating a sensation in the newspapers. William Hammer took credit for applying the nickname, supposedly because the Holborn machines were shipped aboard the *Assyrian Monarch,* the vessel that carried the elephant. The Holborn station was completed before the animal reached New York, however, and there is no evidence that any of the early dynamos traveled on that ship.[16]

1. See headnote, Doc. 2122.

2. Clarke 1904, 39–40; "The Installation of the Edison Light," *Electrician* 8 (1882): 202–3. Clarke's notes and calculations for the first London machine, as well as loose undated entries which may pertain to it, are in N-81-07-11:15–45, N-81-02-20, Lab. (*TAED* N220:4–19, N214:107; *TAEM* 41:6–21, 40:1052); his notes for central station dynamo no. 4 follow in N-81-07-11:57–65 (*TAED* N220:25–29; *TAEM* 41:27–35).

3. See Docs. 2162, 2169, and 2228; Clarke 1904, 44.

4. Clarke began designing the second Holborn machine in October (see Doc. 2167 n. 1). The dynamos are described generally in Dredge 1882–85, 1:265–70. Originally driven by Porter-Allen engines, the Holborn machines evidently received Armington & Sims engines in 1883. Clarke 1904, 40–41; see Docs. 2396 and 2403.

5. William Hammer notebook 8:44, Ser. 1, Box 13, Folder 2, WJH; Clarke 1904, 44, 50; Clarke to James Bishop, 8 Jan. 1931, acc. no. 30.1507.1, MiDbEI(H).

6. Edison expected to begin building the machines in March (see Doc. 2228). Progress of their construction and installation was noted in the Edison Electric Light Co. Bulletins, CR (*TAED* CB; *TAEM* 96:667); for Edison's expected schedule see Doc. 2228.

7. Most evidently had 49 commutator bars; number nine had 51, and number seven had 48 ("Electrical Experiments & Tests" notebook, acc. 1630, Box 30, Folder 1, EP&RI [*TAED* X001K2]). Charles Clarke's extensive correspondence with James Bishop concerning Henry Ford's restoration of the dynamo in Doc. 2238 is a rich source of detailed information about this machine's design (acc. no. 30.1507.1, MiDbEI(H)).

8. Clarke 1904, 41; Lieb 1904, 65; TAE to Johnson, 7 May 1882, LM 1:206D (*TAED* LM001206D; *TAEM* 83:975). Tests of several of these machines in April and May are reported in "Electrical Experiments & Tests" notebook, acc. no. 1630, Box 30, Folder 1, EP&RI [*TAED* X001K2].

9. Clarke 1904, 46, 50, 47; Clarke "Electrical Experiments & Tests" notebook, acc. no. 1630, Box 30, Folder 1, EP&RI (*TAED* X001K2).

10. Clarke to Bishop, 26 Oct. 1931 and 3 Sept. 1932; both acc. no. 30.1507.1, MiDbEI(H).

11. Charles Clarke memorandum, 14 Dec. 1931, enclosed with Clarke to C. J. Leephart, 15 Dec. 1931; acc. no. 30.1507.1, MiDbEI(H).

12. Clarke 1904, 50–51.

13. Clarke to Bishop, 20 Oct. 1938, acc. no. 30.1507.1, MiDbEI(H).

14. Clarke 1904, 45, 53–55.

15. Edison to Clarke, 11 May 1882, Lbk. 12:285 (*TAED* LB012285; *TAEM* 81:653).

16. Harding 2000, 41–62; William Hammer 1904, 452. The *Assyrian Monarch* was scheduled to leave New York on 6 October, about the time the first Holborn machine went, but was delayed to the middle of the month. Hammer's confusion also may be related to the third Holborn machine having been sent on the *Grecian Monarch* in early November 1882. "Shipping News," *New York Herald,* 14 Sept. 1881, 10; "Shipping News," ibid., 15 Oct. 1881, 10; Edison Electric Light Co., Ltd., to TAE, 31 Oct. 1882; TAE to Edison Electric Light Co., Ltd., 31 Oct. 1882; LM 1:126A, 126B (*TAED* LM001262A, LM001262B; *TAEM* 83:1003).

The large Paris dynamo went aboard the *Canada* on 7 September 1881; the first Holborn machine was scheduled for the *Greece* on 5 or 6 October; and the most recent Holborn Jumbo sailed on the *Arizona* on 14 February 1882. See headnote, Doc. 2122, and Docs. 2162, 2207 n. 5, and 2220 n. 4.

–2238–

*Production Model:
Central Station
Dynamo*[1]

[New York, February 1882][2]

M (historic drawing) (est. 4.27 × 2.67 × 1.98 meters; 27,425 kg), Dredge 1882–85, 2:340.

1. See headnote above.
2. The dynamo shown here was Jumbo No. 9, installed at Pearl St.

–2239–

To Edward Johnson

New York, 7th Mar *1882*[a]

My Dear Johnson,

I can probably arrange to go over to England in the course of three or four months, i.e when we have got the Central Station fairly started here and provided the unexpected does not arise in the meantime.

The bar armature sent you will prove far superior to the one that Batchelor had and will give an economy of about 96% of which about 90% will be available outside when 12 h.p. is being taken out[b] off it.[1] It is the most perfect machine we ever made and has only about ~~that~~ eighteen thousandths resistance with only the same size as the regular B armature and about the same radiating surface. You can run[c] easily 250 B lamps on it and I think it will carry 115 to 125 of the 16 candle B lamps without any difficulty. It will greatly astonish Sir William Thomson when he comes to test it. You can put it in any one of the Z Dynamos and there will be quarter of an inch clear space that you can look through   Get Sir William to make a test of it in the same way that Howell made his test both as to scientific and commercial ~~econo~~ efficiency and I feel convinced that if he makes this test he will have to admit that my machine is the most perfect of all machines yet devised and leaves nothing to be desired or possible

These machines are rather costly or rather the armature is. But they can be made quite reasonably providing 100 or 150 were ordered to warrant us in making a lot of special tools to enable us to turn out the work cheaply

You speak in one of your letters about some of the machines heating and others not.[2] You know that if the brushes are set exactly on the neutral point [ie?][d] the [ends of?][d] the brushes on the commutator block which is exactly opposite the centre of the field piece all the machines will have the same electro motive force providing all the fields are arranged to get the same current. But if say of five machines and four of them are set right and the fifth one has its brushes set say several blocks removed from the centre in the direction of rotation it will diminish the electro motive force of that machine and it will become less heated than the others. If you have five machines and four of them have the points of the brushes set up more towards the opening while the fifth one is set exactly right the fifth one will give the highest electromotive force and do the greatest amount of work and become heated. If the four machines had their brushes set away up so as to be on a line with the block at the opening of the field there would be no current at all given from these machines and the fifth one would have to do all the work so you can see how a slight change in the position of the brushes will cause a machine to carry an abnormal load and doubtless this is the cause of your trouble

Please let me have your views as to the disposition of the Exhibit now at Crystal Palace. Gouraud I fancy wants to make a permanent exhibition of the things there. I want you to tell me what you think about it and where they should go after the exhibition closes.[3]

Swan promised Barker some Swan lamps and said he was to get them from Montgomery the Swan Coy's agent here.[4] Barker cannot get them and wants you to see what you can do. He wants about half a dozen

Can you get from the United Telephone Coy half a dozen or a dozen of the Motograph Telephones made for the Edison Co by Bergmann[5] They do not use them and I think they ought to let ~~them~~ you have them for me. Barker wants six of these Yours very truly

Thos A Edison I[nsull]

L (letterpress copy), NjWOE, Lbk. 11:401 (*TAED* LB011401; *TAEM* 81:425). Written by Samuel Insull; handstamp of Thomas A. Edison to Edward H. Johnson. ª"*New York*," and "*188*" from handstamp. ᵇInterlined above. ᶜObscured overwritten text. ᵈIllegible.

1. This armature, also described in Doc. 2228, was shipped on or before 24 February. TAE to Edward Johnson, 24 Feb. 1882, LM 1:163C (*TAED* LM001163C; *TAEM* 83:953).

2. See Doc. 2226.

3. The editors have not found a reply from Edison on this point, but see Doc. 2292.

4. Thomas Montgomery was connected with the Brush Electric Light Co. in the U.S.; he and Henry Edmunds were trying to organize companies in New England to use Swan and Brush lights. Eaton to TAE, 24 May 1882, DF (*TAED* D8226H; *TAEM* 61:175); "Anglo-American Brush Electric Light Corp.," *Times* (London), 14 Dec. 1880, 13; "Miscellaneous City News. Storing the Electric Current," *New York Times,* 20 Dec. 1882, 8; Tritton 1993, 53–54, 82.

5. See Docs. 1784 and 1813 n. 3.

–2240–

From Charles Porter

PHILADELPHIA, Mar 8th 1882ª

My Dear Sir:

I have been invited to prepare for the meeting of the Am. Inst. of Mechanical Engineers, which meets at the Hall of the Franklin Institute[1] in this City on the 12th of next month, a paper on the application of my engine to your Dynamos. I have been so occupied that I did not attend to the matter, but on the request being somewhat urgently repeated, I have thought if you were agreeable I would endeavor to comply. Should want to make thorough work of it if I do it at all, and if you feel inclined to have it done I will go to N.Y. and see you about it.[2]

Three or four more engines are about ready. Very Truly Yours

Chas. T. Porter.

ALS, NjWOE, DF (*TAED* D8212E; *TAEM* 60:631). Letterhead of Southwark Foundry and Machine Co., C. T. Porter, Vice-President. ["PHILADELPHIA," and "188" preprinted.]

1. Founded as a Philadelphia mechanic's institute, the Franklin Institute by this time was a nationally prominent center for science and technology. It still occupied its original building (c. 1825) on Seventh St. in downtown Philadelphia. Sinclair 1974, passim, 47–48.

2. Porter asked Edison to meet him at the Goerck St. works on 4 April. The resulting paper was published as Edison and Porter 1882; Porter apparently also made arrangements for publication in the *American Machinist* (Porter to TAE, 3 Apr. 1882; American Machinist to TAE, 25 Apr. 1882; both DF [*TAED* D8233Y, D8207S; *TAEM* 61:1024, 60:537]). The paper gave detailed descriptions and specifications of the Edison Jumbo dynamo, the Porter-Allen engine, and the coupling between them. It also presented engine indicator diagrams made under various loads, interpreting them to show the dynamo's efficiency and number of lamps operated per mechanical horsepower.

–2241–

Telegrams: Samuel Insull to/from Mary Edison

March 13, 1882[a]
[New York]

Mrs T. A. Edison
When will you probably arrive here   No necessity your hurrying on account of business

S. Insull

Green Cove Fla[1]   12:20 p[m]
Mr Saml Insull
We will remain here until Wednesday[2]

Mrs T A. Edison

L (telegram, copy) and L (telegram), NjWOE, Lbk. 11:432 (*TAED* LBo11432; *TAEM* 81:436) and DF (*TAED* D8214I; *TAEM* 60:705). First message written by Charles Mott; second message written on Western Union Telegraph Co. message form. [a]Date from document, form altered.

1. Green Cove Springs was a small health resort about twenty-five miles up the St. John's River from Jacksonville. It was a winter haven for tuberculosis patients and the site of warm sulphur springs. The area reportedly also offered opportunities for hunting and fishing, and Edison arranged to have several birds mounted and shipped to him. *WGD*, s.v. "Green Cove Springs"; Walton 1873, 185; Hoyt & Dickinson to

TAE, both 8 Apr. 1882, DF (*TAED* D8204ZAS, D8204ZAT; *TAEM* 60:111–12).

2. The family's whereabouts after Wednesday, 15 March, are unknown. Two days later Insull wrote Johnson that Edison had been away "two weeks & I very much hope he will not return for another two weeks at least"; the next week he told Charles Batchelor that Edison was still away and he had "not heard one word from him since he left on the 1st of this month—with the exception of a short telegram from his wife." During this time Insull carried on correspondence in Edison's name; Edison apparently did not return until 28 March, when he wired from Newark that he would reach Menlo Park that night. Insull to Johnson, 17 Mar. 1882; Insull to Batchelor, 25 Mar. 1882 Lbk. 11:459, 501 (*TAED* LBo11459, LBo11501; *TAEM* 81:450, 469); TAE to Insull, 28 Mar. 1882, DF (*TAED* D8244G; *TAEM* 63:188).

DYNAMO VOLTAGE REGULATION   Doc. 2242

Edison had devised many methods to regulate dynamo voltage from 1878 to late 1880 in order to provide the intended voltage to the lamps and to eliminate fluctuations in their intensity. These designs focused on compensating for variations in load and dynamo speed, and were appropriate for isolated plants. In 1882 Edison renewed work on dynamo regulation, presumably to devise an automatic regulation method for central stations.[1] He executed a total of twenty-two U.S. patent applications on this subject between January and July, ten of them in February and seven more in May. However, these applications were not filed at the Patent Office until August. Edison's patent attorney Richard Dyer turned them over to Zenas Wilber, patent attorney for the Edison Electric Light Company, who kept the fees but did not file the papers. When Dyer began acting as patent attorney for the company in August 1882 he promptly filed the applications withheld by Wilber.[2]

Edison's general method of regulating dynamo voltage was to change the amount of current through the field coils. He devised several specific means to do this. The simplest was to use resistances in series with the field coils, each of which could be switched in or out by relays responding to changes in load current. Several designs placed a motor in series with the field coil. The motor would act as a variable resistance by producing a counter-electromotive force—essentially an opposing current—as its speed increased with the load on the dynamo. Several other approaches regulated simultaneously for changes in armature speed and load current. A final method was to use several windings on the field magnet, each of which responded to changes in distinct parts of a plant's load, such as different feeders in a central station district.[3]

By the beginning of June 1882 Edison had concluded that the best way to regulate voltage was to use an electromagnet to control a mechanism which compensated for both variations in the number of lamps and speed of the engine driving the dynamo.[4] However, by July 1882 Edison had still not settled on the best specific means for voltage regulation. He wrote William Hammer, "We have 3 different varieties of automatic regulators to go with Isolated plants working at Menlo going through the Evolution test & I think one of them will prove reliable."[5] Edison concluded that automatic regulation methods were not reliable enough for central stations. The first central stations used manually switched resistances in the field circuit to compensate for variations of 1 or 2 volts above or below nominal voltage, as indicated by red and blue lamps, respectively. The total resistance which could be switched into the field circuit was about 2 ohms. As late as 1884, one former Edison employee recalled, the staff at Pearl Street included two "regulators" who each worked twelve-hour shifts, seven days a week: "Their regulating standard was a Thomson galvanometer, and their job was to sit at the desk on which the galvanometer was mounted and note the fluctuations from zero and operate the dynamo rheostat accordingly. This was a monotonous job from which about the only digression was serving the infrequent customer who entered the station for lamp renewals."[6]

1. For Edison's prior work on regulation see headnote, Doc. 2036. In October 1882 (see Doc. 2345) he estimated that he had tested about thirty different kinds of regulation schemes before finding one he thought workable for variations in both load and engine speed.

2. See Docs. 2224 n. 1 and 2323.

3. Details of work on automatic dynamo regulation between September 1881 and October 1882 are in N-82-06-08, N-82-05-26, PN-82-04-01; Cat. 1147, Cat. 1148; Lab. (*TAED* N197, N204, NP016, NM016, NM017; *TAEM* 40:332, 574; 44:10, 224, 360); and Ott Notebook, Hummel (*TAED* X128B).

4. This design philosophy enabled Edison to make broader claims in two dynamo regulation patents filed in August 1882. Dyer to Wilber, 5 June 1882, Lbk. 7:403B (*TAED* LB007403B; *TAEM* 80:584); U.S. Pats. 264,660 and 265,779.

5. Doc. 2322. In 1883, Bergmann & Co. had in its catalogue an automatic regulator for Z, K, and L isolated plant dynamos. Bergmann & Co. circular, [1883], CR (*TAED* CA002C [image 77]; *TAEM* 96:185).

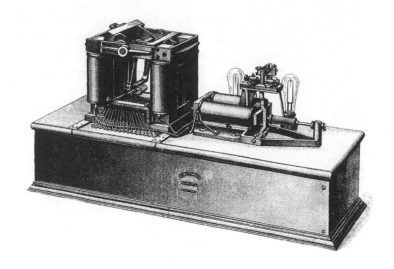

The automatic isolated plant regulator manufactured by Bergmann & Co. The lamps provided a visual check of its operation: one would light when voltage rose too high, the other when it fell too low.

6. Jehl 1937–41, 1053, 1083; Charles Clarke to James Bishop, 8 and 28 Jan. 1931, acc. no. 30.1507.1, MiDbEI(H).

–2242–

Technical Note:
Voltage Regulation

[Green Cove Springs, Fla.,] Mar 13 82

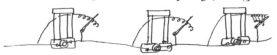

Patent several Dynamos worked Either as generators or motors all[a] in series   Each having its own field multipled arc'd across its armature with an adjustable resistance in the field so that EMF on each can be adjusted seperately—[1]

Reg field   Isolated.[2]

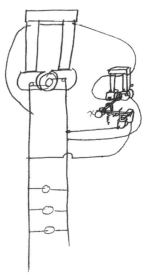

X is very large or it is geared up to brake wheel, so as to make the action exceedingly sensitive.[3]

D[itt]o[4]

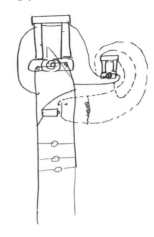

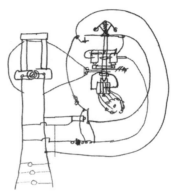

Details 2nd patent Reduc[in]g Emf[5]

[A][6]

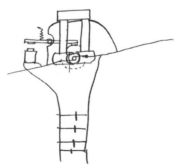

[B]⁷

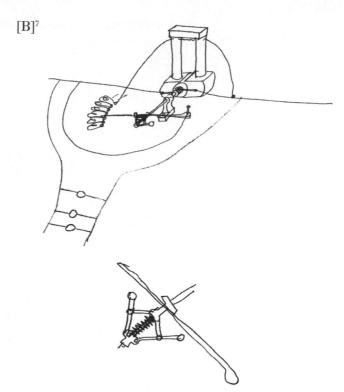

Details 2nd Emf patent

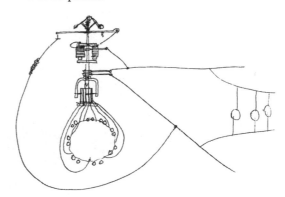

2nd patent Emf (Details)

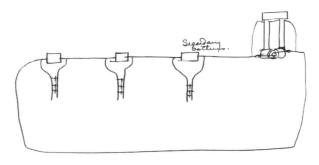

X, NjWOE, Lab., Cat. 1148 (*TAED* NM017:100, 102, 101, 104, 103; *TAEM* 44:460, 462, 461, 464, 463). Document multiply dated. ªObscured overwritten text.

1. Dynamos connected in series produce a higher voltage on the line than if connected in parallel. Edison may have been thinking of using higher voltage for more efficient transmission of current across relatively large distances, such as from generators driven by the flow of the St. John's River, in the vicinity of Green Cove Springs, or the nearby springs. While traveling in the West in August 1878, just prior to starting intense research in electric light and power, he had entertained the possibility of carrying power from the Platte River to mines hundreds of miles away. A high voltage system would also be suited to operating arc lights (*TAEB* 4:374–75; "A Great Triumph," *New York Mail*, 10 Sept. 1878, Cat. 1241, item 878, Batchelor [*TAED* MBSB20878X; *TAEM* 94:349]). Edison returned to the subject of voltage reduction in May (see Doc. 2276).

2. The drawing below shows a motor-driven centrifugal governor controlling a variable resistance in the dynamo field circuit. This arrangement is similar to one covered in a patent application that Edison executed on 1 May. U.S. Pat. 264,665.

3. This is a variation on the previous device. The motor's speed was governed by a friction brake controlled by an electromagnet in the line circuit. The brake would be lifted as line voltage increased, allowing the motor to turn faster and apply a greater counter-electromotive force. This device is similar to one shown in Doc. 2276 and was embodied in a patent application that Edison executed in May. U.S. Pat. 264,667.

4. Edison executed a patent application on 22 May incorporating this device. The motor, at right, applied a counter-electromotive force to the dynamo field circuit. Its speed was governed by a "vibrating circuit-controller" (an electromagnetic relay); in response to high voltage on the main line, the relay would close the motor field circuit, accelerating the motor. A shunt with resistance provided that the motor would not stop completely even when the controller opened the circuit. U.S. Pat. 264,672.

5. No prior patent application covering the ideas in this document has been found.

6. In this variation on the use of a motor to provide counter-electromotive force, the motor appears to be in the main circuit.

7. Edison made a variation on this device, in which a centrifugal governor controlled a variable resistance in the dynamo field circuit, the subject of a patent application executed on 1 May. U.S. Pat. 264,665.

PEARL STREET CENTRAL STATION   Doc. 2243

Edison had decided in 1880 to build his first commercial electric lighting system at the center of one of the world's great capital markets, New York's financial district in lower Manhattan. The boundaries of the distribution system he built in 1881 and 1882 encompassed banks, brokerages, offices, and

Exterior of Pearl Street
central station buildings.

newspaper publishers, just blocks away from other important institutions such as the Western Union Telegraph Company building and City Hall.[1] The central station itself occupied two adjoining buildings on Manhattan's Lower East Side, near the new Brooklyn Bridge. The mechanical and electrical portion of the plant was located in 257 Pearl Street. The companion building at 255 Pearl Street was used for office, storage, and sleeping spaces; two Jumbo dynamos were installed in the basement in the spring of 1884. The combined lot was 50 feet wide and about 100 feet deep. John Lieb later wrote that 257 "was originally erected for commercial purposes, and as it was incapable of sustaining the weight of the engines and dynamos planned to be installed on the second floor, the old flooring was torn out, and a floor of heavy girders supported by stiff columns was substituted. This heavy construction, not unlike the

Dynamo Room.

supporting structure of the elevated railroad, was erected so as to be independent of the building walls, and occupied the full width of the building . . . and about three-quarters of its depth."[2] A detailed description of the plant in the 26 August 1882 *Scientific American* reported that the ironwork included "pillars planted on heavy plates resting on three feet of solid concrete."[3] Four 240 horsepower Babcock & Wilcox boilers occupied the basement. Above them on the new iron frame rested the six engine and dynamo assemblies, each unit weighing about thirty tons and rated for 1,200 lamps.[4] Not shown in the drawing were a traveling crane and hoists running the length of the room immediately above the dynamos. On the third floor, copper wire resistances wound on large wooden frames were used for manual regulation of the dynamo fields. An automatic indicator lighted a red lamp when the voltage rose too high, and a blue lamp when it dropped; the indicator was calibrated every few days against a reflecting galvanometer at the Edison Machine Works. Separate mechanisms, the so-called "cheese-knife" regulators, were available to manage the voltage on individual feeders.[5] Ashes and coal were transported by a 20 hp. engine to and from a vault under the sidewalk at the front of the building.[6] The top floor housed a battery of 1,000 lamps used to test dynamos removed from the main circuit for inspection or repair.[7]

The large coils of copper wire were used to regulate dynamo voltage by manually varying the field resistance.

Construction of the station began in late summer 1881 and the Edison Electric Light Co. published progress reports in its Bulletins over the next year.[8] On 22 September Edison, John Kreusi, and their work crews began excavating at the corner of Peck Slip and South Street for the network of underground conductors.[9] Work stopped during the winter because of frozen ground, but resumed in late February 1882. By the first of April the crews had installed about two-thirds of the roughly 80,000 feet of conductors, although they had not yet connected the feeders to the mains. By mid-April 1882 the central station structure (including boilers) was largely complete. The mechanical and electrical equipment was not yet set up, although assembly of the dynamos was well under way at the Edison Machine Works and the finished Porter-Allen engines were on hand. The first engine and generator assembly was tested on 5 July.[10] By late July crews had completed the installation of the underground feeders and mains except for connecting to individual buildings (946 in all). As they did this, a number of buildings were lighted to test the engines and dynamos.[11] After the station officially began service on 4 September, it remained in operation until a fire partially destroyed it in January 1890.[12] It was reconstructed and operated until 1894. The buildings were sold soon afterward and no longer exist.[13]

The Edison Electric Illuminating Co. paid about $300,000 to acquire the Pearl Street properties and construct the central station and distribution system. Administrative expenses, in-

One thousand incandescent lamps installed on the top floor could be used to test individual dynamos off-line.

terest, canvassing, and patent license fees to the Edison Electric Light Co. brought the total cost chargeable to the first district to more than a half million dollars.[14] The company did not charge customers for current until early 1883.[15]

1. Edison had the area of his proposed first central station canvassed in 1880 to tabulate potential demand for electric light and power (see Doc. 1995). At that time, there were reportedly about 18,000 gas jets, 90 elevators, and 80 sewing machines there ("The Electric Light in Houses," *Harpers Weekly* 26 [1882]: 394; in Cat. 1018:11H, Scraps. [*TAED* SM018039a; *TAEM* 24:253]; see also Doc. 2338 n. 5). Partial lists of central station customers as of April and October 1883 are in Edison Electric Light Co. Bulletins 17:3 and 20:3 (*TAED* CB017, CB020; *TAEM* 96:809, 867); a list of the first year customers itemized by type of business is in Jones 1940, 183–87. For a discussion of Edison's choice of location see Bazerman 1999, 228–32.

2. Clarke 1904, 50; Lieb 1904, 61; the map in Jehl 1937–41 (1043) represents slightly different depths for the two lots.

3. "The Edison Electric Lighting Station," *Sci. Am.* 47 (1882): 127–30.

4. Charles Dean's itemization of the weight of the machine's components was published in Edison Electric Light Co. Bulletin 10:7–8, 5 June 1882, CR (*TAED* CB010; *TAEM* 96:714). In March 1883, Edison

executed an unusually long application for a broad patent on the form and regulation of dynamos and high-speed steam engines in large central stations, emphasizing the economic advantages of his arrangements embodied in the Pearl St. station. He received U.S. Patent 281,351 in July 1883.

5. See Docs. 2269 and 2272.

6. *Edisonia* 1904, 61–73; Clarke 1904, 43; Jehl 1937–41, 1049–53.

7. "The Edison Electric Lighting Station," *Sci. Am.* 47 (1882): 127, 130. Julius Hornig made an incomplete estimate of the cost to equip the station on 5 January (DF [*TAED* D8223A; *TAEM* 60:1020]). More complete cost estimates by Hornig, as well as mechanical and structural specifications of the plant, are in Series 2, Box 23, Folder 1, WJH; see also the Edison Electric Light Co.'s comprehensive 3 August estimate of construction and operation costs of a central station district and plant (DF [*TAED* D8227ZAT; *TAEM* 61:525]).

8. See Doc. 2145; Edison Electric Light Co. Bulletins, CR (*TAED* CB000; *TAEM* 96:667).

9. Hammer memorandum, 22 Sept. 1882, Ser. 2, Box 23, Folder 1, WJH. During the winter and spring, the New York Steam Co. built its own underground network of pipes for a central station steam heating system in an area slightly overlapping the Pearl St. district. The simultaneous construction resulted in damage to Edison's conductors on at least one occasion. "The Distribution of Light and Heat in New York City," *Sci. Am.* 45 (1881): 319–20; Sherburne Eaton to TAE, 2 June 1882, DF (*TAED* D8226O; *TAEM* 61:227; see App. 1.B.18).

10. Edison Electric Light Co. Bulletins 7:1–3, 17 Apr. 1882; 12:2, 27 July 1882; CR (*TAED* CB007, CB012; *TAEM* 96:692, 728); Martin 1922, 47.

11. Edison Electric Light Co. Bulletins 4:9–10, 24 Feb. 1882; 9:1–2, 15 May 1882; 12:1, 27 July 1882; 13:1, 28 Aug. 1882; all CR (*TAED* CB004, CB009, CB012, CB013; *TAEM* 96:676, 706, 728, 738); see also Doc. 2220 n. 5. During the summer, the company sent a circular letter to every property owner and tenant in the district, advising that connecting each building to the mains would take only a few hours and would be done without charge "subject to the terms of our contract with you, which is, that if you ultimately decide to adopt our light permanently, you will pay the expense of making the connection and the expense heretofore incurred in wiring your building; but if you decide not to adopt our light, no charge whatever is to be made against you." "Edison's First District Completed," *Operator* 13 (1882): 320.

12. *Edisonia* 1904, 73; Jehl 1937–41, 1069–72.

13. Jehl 1937–41, 1080; Lieb 1904, 73.

14. See Doc. 2289.

15. Jones 1940, 209, 212.

Drawings: Pearl Street
Central Station[1]

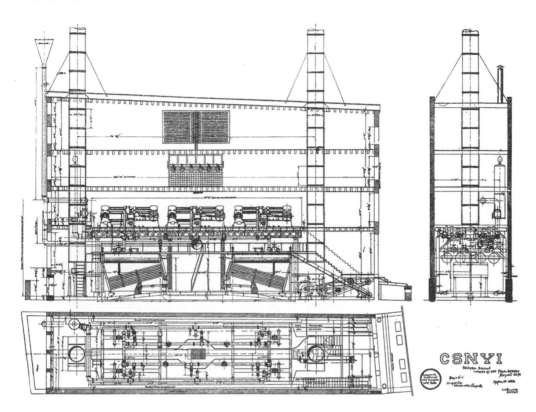

CSNYI[2]   CENTRAL STATION LOCATED AT 257 PEARL
STREET NEW YORK CITY

CAPACITY OF THE STATION ONE THOUSAND HORSE
POWER[a]

SCALE $^{1}/4'' = 1'$

H. M. BYLLESBY[3]   ASSISTANT

J. L. HORNIG   CONSTRUCTING ENGINEER.

M (historic drawings) (est. 7.6 m × 30.5 × 27.1), *Edisonia* 1904, leaf facing p. 64. A copy is in MiDbEI, EP&RI (*TAED* X001H1AL). [a]"CAPACITY . . . POWER" circled.

1. See headnote above.

2. Probably Central Station New York Illuminating or Central Station New York I.

3. Henry Byllesby (1859–1924) began working as a draftsman for the Edison Co. for Isolated Lighting in June 1881. A few months later, under

the supervision of Charles Clarke and Julius Hornig, he prepared drawings for the Pearl St. and Santiago, Chile, central stations. In November
1882 the Edison Electric Light Co. sent him to Canada to install dynamos in textile mills. Byllesby also supervised the installation of Edison exhibits at Louisville, St. Louis, and New Orleans in 1884 and 1885.
After leaving Edison's employ he became Vice President and General
Manager of Westinghouse Electric Co. "Byllesby, Henry Marison,"
Pioneers Bio.; *NCAB* 15:310–16.

–2244–

*Samuel Insull to
Edward Johnson*

[New York,] 21st Mar [188]2

My Dear Johnson,

Please sign the enclosed assignment in the presence of two
witnesses & return it to me so that I can have it recorded in the
Patent Office. All things taken out by Bergmann are treated in
this same manner.[1] I arranged with Edison some time back
that all devices got up by Bergmann should be patented the
consideration being that half interest should be assigned to
you   I take care that all cases filed in U.S. Patent Office by
Bergmann are treated this way. Please return enclosed as soon
as possible. There is <u>no</u> necessity for it to have a consular
certificate but simply the signatures of two witnesses

I received that socket from you and as Edison was away
handed it over to Bergmann who said that for the sake of your
reputation he would get up a better looking affair! However
when Edison gets back I will submit what you sent to him.

I also note what you say about Insulite[2] especially your pon-
derous joke on my name— Is the atmosphere affecting you—
You should send us a specimen of this new insulation

Stocks here are pretty low just now. For Illuminating there
seems really no market in fact I do not suppose there has been
a transaction in it ~~since~~ for months. The same may be said of
Ore Milling. As to European the last I heard of it was that it
was offered at 55 & no purchasers. Why it keeps so low seems
very surprising. The Parent stock can be bought at from 350 to
375 but I do not suppose you could sell any to net over 300. As
to Isolated there is some enquiry for it but there is none offering for sale—in fact no price at all is named. People seem to
think it a good thing to keep in which belief I think they can
with confidence repose considering that the Isolated Co is doing quite a solid business.

Do you know that Holzer has been appointed Superintendent of the Lamp Factory under Upton & he has had given

him 3% interest in the business.[3] This 3% will of course be taken off Edisons share but the Lamp Co had to carry Holzers share of ~~Holzers~~ assessments which are to be cleared off by his being credited with salary ~~at~~ from[a] a date to be arranged & his share of[b] first profits are also to go towards paying the assessment but it will end in his having a 3% in the Lamp Factory costing him nothing. He talks bigger than ever about what <u>he</u> has done & the improvements <u>he</u> has made[4] whereas he knows no more about practical lamp manufacture than does ~~the~~ Barney the night watchman   I begin to think from what I see of others that to make money with Edison you must blow your own trumpet so loud that Edison will give you a big slice to quiet you in other words fancy yourself a hell of a fellow & the Great Mogul will do likewise. What do you say   Take Jim Adams,[5] Holzer, Moses, ~~& a~~ Griffin[6] &c &c for instance.

Things proceed here pretty well   Kruesis work goes along nicely & quietly. The Dynamos at Goerck St. go along all[c] right but will in my opinion cost a great deal more than they would had we any really good style of Superintendence there.[7] The old man begins to think that the best thing he can do is to sell it to the E.E.L.Co. It is a very fine property & would yield very large profits had we a man there who attended to his work properly.

The central station begins to look quite nice. The structure to carry the Dynamos the boilers the coal conveyors &c are all in their place & if the Dynamos were ready & the tubes laid it would not be very long before we should be in a position to start up.

The Oregon Steamship & Navigation Coy[8] are building a new boat which will be fitted up in palatial style & will be lighted by Edison Lights.[9] We are just sending out a new plant for the S.S. Columbia & the Coys Docks at Portland Oregon & San Francisco & Villiards country house just outside N.Y.[10] will all be lighted up shortly by Edison Light.

The Rail Road at Menlo Park is fast approaching completion. We can now run over about 1 3/4 miles of it and can obtain any desired speed by putting in different forms of Armatures. At 25 miles an hour on Saturday we took 39 people a four ton locomotive and a Car—a cross between a two horse & a Bobtail Car—[11] The armature of the motor was as cool as possible when we got through and so little power was consumed that at the Dynamo Room by the machine shop they could tell whether we were running or not by looking at the Commutator Brushes of the Dynamos. A slow moving Freight Locomotive & ten freight cars each to carry four tons are now being built. As it is I fancy we must now have more to show than Siemens has. What do you say?

Edison has had a fine stamp mill erected in a new shed at the back of the machine shop for the purpose of ~~working~~ crushing Sulphurets & other ores to be worked in connection with his Ore milling processes   I think he will work on this a great deal of his time so soon as he gets back to Menlo

I must close now as I have a good many letters to write for tomorrows steamer

I fancy Edison will be back about 1st of next month   Yours as ever

Saml Insull

Of course I have done nothing so far with your European stock.

ALS (letterpress copy), NjWOE, LM 3:69–71, 73, 72, 74 (*TAED* LM003069; *TAEM* 84:71–73, 75, 74, 76). [a]Interlined above. [b]"his share of" interlined above. [c]Obscured overwritten text.

1. Johnson assigned to Sigmund Bergmann a one-half interest in two patents for which he had applied in November 1881. Both dealt with connections in a household circuit; both issued in September 1882. U.S. Pats. 264,298 and 264,299.

2. John Ambrose Fleming patented an electrical insulating compound of this name consisting of impregnated wood, sawdust, cotton waste, and paper pulp. It was reportedly moderate in cost, and a company was organized in April for its manufacture. Fleming gave Johnson a sample to take to New York in June. During the summer, it was used in the installation of Edison lights at the Post Office in London. *OED*, s.v. "insulite"; "Insulite," *Electrician* 8 (1882): 396; DF (*TAED* D8239ZCK; *TAEM* 62:991); "The Electric Light at the Post Office," *Times* (London), 22 Aug. 1882, 5.

3. William Holzer became superintendent of manufacturing (see Doc. 2260; Jehl 1937–41, 815). His three percent share was represented in the incorporation papers of the Edison Lamp Co. in March 1883 (DF [*TAED* D8332ZDW; *TAEM* 67:294]).

4. Holzer had applied for a patent in November 1881 on a method of sealing the platinum lead-in wires into the glass globe. It issued in September 1882 as U.S. Patent 264,698 and was assigned to the Edison Electric Light Co. See also Richard Dyer to Zenas Wilber, 10 June 1882, Lbk. 7:435 (*TAED* LB007435; *TAEM* 80:600).

5. James Adams (d. 1879) became an experimental assistant to Edison in 1874. As he did for other principal assistants, Edison had assigned Adams an interest in patents resulting from inventions on which he worked (see Docs. 1652, 637, 1345, and 1717 n. 2). Edison also gave him thirty shares of Edison Electric Light Co. stock in January 1879, apparently as a gift (see Doc. 1668).

6. Stockton Griffin was Edison's personal secretary from 1878 until he resigned in early 1881. Edison gave him twenty shares of Edison Electric Light Co. stock in January 1879. Docs. 1322 n. 1, and 1668.

7. Charles Dean held this position.

8. Henry Villard organized this holding company in 1879 to consolidate his Pacific Coast rail and steamship lines. *ANB*, s.v. "Villard, Henry."

9. The *Queen of the Pacific* was equipped by June with two Z dynamos and about 250 B lamps. Edison Electric Light Co. Bulletin 11:9–10, 27 June 1882, CR (*TAED* CB011; *TAEM* 96:720).

10. Villard purchased Thorwood, an eighty-acre estate on the Hudson River at Dobbs Ferry, N.Y., in 1879. Buss 1978 [1977], 94.

11. A popular name for a small tram vehicle pulled by one horse. *OED*, s.v. "bob-tail."

–2245–

To M. Russell[1]

[New York,] March 23 [188]2

Dear Sir

Referring to yours of March 19, 82[2] I beg to reply that I shall not require rooms as I am about to remove from the City

Yours Truly

Thos A Edison   I[nsull]

L (letterpress copy), NjWOE, Lbk. 11:478 (*TAED* LB011478; *TAEM* 81:457). Written by Charles Mott; signed for Edison by Samuel Insull.

1. Insull addressed this letter to 14 and 16 West Seventeenth St. The only occupant with this name that has been identified for this address was Margaret Russell, a widow. With Edison in Florida, Insull apparently acted on his own in declining Russell's inquiry. *Trow's* 1881, 1343.

2. Not found.

–2246–

William Meadowcroft[1]
Memorandum: Ore Milling

[New York, March 27, 1882[2]]
Edison Ore Separator[a]
Investment for One Separator[3b]

Separator . . . . (deposit)	$600
Dynamo	350
8 Horse power "Baxter" Engine & Boiler . .	800
Sheds . . . . say	500
Sand Elevator, sieves &c	200
Dryer	250
Bags for transporting the iron—	
$160 per 1000	
1000 Bags will hold 75 tons—[c]	
Incidentals . . . . say	250[b]

Running expenses per diem
of one Edison Separator for 1 Day[4b]

Coal	54
Wood for dryer—½ Cord—	2.00
Superintendent	2.50
Engineer	1.75
10 Laborers . . . . @ 1.50	15.00
Man & Team for hauling sand	3.50
Renewal of sieves—	.15
Interest on investment @ 6%	.50
Total daily running expenses—	$25.94[b]

The above is the average daily expense of our working at Quonocontaug, Rhode Island.[5] At this expense we have run through the separator on 60 to 65 tons of sand per day with an average output of 9 tons of iron ore— We have, however, made as much as 12 tons of iron per day. The above plant and daily expenses will cover a run of 65 tons of <u>sand</u> per day through the separator. The amount of <u>iron</u> obtained depends upon the per centage contained in the sand— The best of our sand at Rhode Island only contains about 20 per cent of iron and our output has been as high as 12 tons of iron per day—

We have sold our product so far to[d] the Poughkeepsie Iron & Steel Co Poughkeepsie, N.Y. at $10 per ton—[6]

The quality of the magnetic iron after separation is very fine— The analysis shows out of 100 parts:

Pure metallic Iron	71.42
Oxygen with the Iron	27.22

AD, NjWOE, Miller (*TAED* HM820161; *TAEM* 86:463). Multiple dashes across columns transcribed as multiple periods. [a]"Edison Ore Separator" written by Samuel Insull. [b]Followed by dividing mark. [c]"Bags . . . 75 tons—" enclosed by brace. [d]Obscured overwritten text.

1. William Meadowcroft (1853–1937) was secretary to Sherburne Eaton and had authority over the Edison ore milling installations in Rhode Island and Long Island. Meadowcroft remained connected with Edison enterprises for many years and, in 1910, became his "assistant and confidential secretary," a position he held until Edison's death. "Meadowcroft, William, Henry," Pioneers Bio.

2. Date taken from canceled docket note.

3. With the Edison Ore Milling Co.'s sole customer having withdrawn from the market in recent days (see note 6), this document was evidently intended as a prospectus for potential operators of an ore separating plant. The company agreed in January on a policy to lease separators with a $600 deposit. A further requirement, not listed below, was a royalty of twenty-five cents per long ton with a guaranteed minimum of $300 for the first year. Edison Ore Milling Co. report, 17 Jan. 1882, CR (*TAED* CG001AAI2; *TAEM* 97:416).

4. These figures are roughly comparable to those listed in a November memorandum by M. R. Conley, who became superintendent of the Rhode Island operation at that time. Conley also identified a number of capital improvements to be made. Conley cost estimate, 12 Nov. 1881; Edison Ore Milling Co. agreement with Conley, 14 Nov. 1881; both DF (*TAED* D8139U, D8139U1; *TAEM* 59:207, 211).

5. See Doc. 2175.

6. After receiving samples of concentrated Rhode Island ore, the Poughkeepsie company agreed in February 1882 to purchase between 125 and 150 long tons at this price. The firm rescinded its order prior to 28 March because of mechanical and financial difficulties. Edison Ore Milling Co. agreement with Poughkeepsie Iron & Steel Co., 10 Feb. 1882, DF (*TAED* D8246E; *TAEM* 63:275); Sherburne Eaton to M. R. Conley, 28 Mar. 1882, LM 5:150 (*TAED* LM005150; *TAEM* 84:271); see also Doc. 2393.

–2247–

From Charles Batchelor

Ivry-sur-Seine le 28 March 1882[1a]

Sir

Messrs. Finleyson & Co of Tammerford are very anxious to install a large dynamo for 1000 lights[2]   Can you let us have a price for it?

The light in St Petersbourg & Tammerford continue to give excellent satisfaction—

I wish you would please send me full drawings of the 200 light machine.

I cabled you yesterday the following:—

Abdomen health good he is life of business[3]   contracts will be sent   do you want all patents taken in your name.[4]

If I thought that Bailey did not write to you regularly I should have written you oftener— There is no doubt that Bailey has done, and is now doing, all the work of the companies, and the loss of him at present would set the project back at least a year— I do not think he is satisfied, as he feels that this work is not appreciated at home, and I am afraid he will leave it altogether, if he should do so, at present, it would be very disastrous. I have great difficulties in the factory with the workmen. they are lazy, slow and very bad workmen indeed   of 25 workmen hired in the morning sometimes they are all gone before night. this of course takes up a great deal of time & if I loose Bailys help in the other two companies I am sure we shall do nothing for the first year.

I should judge from letters received that the European Co is not quite satisfied with what is being done here but I think you will allow that our orders have been good considering the short time we have been started and the difficulties we have had to overcome. I do not think the contracts are understood over there as I had a letter from Upton in which he says that all the French Companies are frauds and he feels compelled to tell his friends who he had persuaded to put money into stock & bonds of the European Co that their stock was <u>absolutely worthless</u>.[5] I think Mr Lowery ought to have explained the thing better than that.

Will you tell me what is <u>your</u> private opinion of our arrangement in Europe? Yours faithfully

Chas Batchelor

LS, NjWOE, DF (*TAED* D8238ZAN; *TAEM* 62:358). Letterhead of Société Industrielle et Commerciale Edison. ª"Ivry-sur-Seine le" preprinted.

1. This letter was among the first stamped with the date received, and blank lines for the reply date and file number. It was received on 10 April.

2. By the end of the year this large cotton mill operated one E, two K, and two Z dynamos, all by waterpower. Batchelor to Sherburne Eaton, 15 Dec. 1882, Cat. 1239:407, Batchelor (*TAED* MBLB4407; *TAEM* 93:758).

3. "Abdomen" was the cable code for Joshua Bailey (see App. 4). Samuel Insull had wired Batchelor to "Cable us ac[coun]t condition Baileys health likelihood his attending business   hear nothing from him   is he permanently disabled." Otto Moses had reportedly indicated that Bailey was "too sick to attend to business" and that the affairs of the European Edison company were consequently neglected. Insull to Batchelor, 23 Mar. 1882, LM 1:176D (*TAED* LM001176D; *TAEM* 83:960); Insull to Batchelor, 25 Mar. 1882, Lbk. 11:496 (*TAED* LB011496; *TAEM* 81:467).

4. This is the full text of Batchelor's cable to Edison on 27 March. LM 1:177D (*TAED* LM001177D; *TAEM* 83:960).

5. Batchelor was apparently referring to a 12 March letter from Upton, to which he promised to respond. Neither letter has been found. Batchelor to Upton, 3 Apr. 1882, Cat. 1239:194, Batchelor (*TAED* MBLB4194; *TAEM* 93:611).

–2248–

From Francis Jehl

London, E.C. March 28th 1882[1a]

My Dear Sir

It is about time no doubt that you would like to know, what progress there has been made, and also the rapidity in meter work. Then also how soon I shall be able to go to Batch. Well! The first two weeks after my arrival, was spent in geting a room in order, and also in procuring such chemicals, and apparatuses as are necessary in meter manipulations. The balance that I have is a very fine one, and almost as good as the one you have in the Goerck St laboratory, it was made by Oertling,[2] the price being only 20.£. The Oertling folks are the ones that standardized[b] all the governments weights, and have that reputation in England, what Beacker[3] has in America. The young man they gave me to instruct in the meter line, was employed before he came into Edison Co in some chemists place, and he understands well how to weight, and take care of the balance. That's one comfort— I have had him all last week engaged on the zinc plates, the manner we prepare them, and also in making the Standard solution, also had him put a meter together, and which is now at Crystal Palace registering twenty four lights. As far as I have gone with him, I think he understands what I have told him, and now remains the electrical part, such as making compensation resistance, and also in making shunts. There is a great deficiency in this respect, the youths over here seem to lack electrical knowledge, some go under the impression that all that is necessary for electricians to know, is how to fix up batteries, and hang some lamps up &c   I dont doubt, however; that I shall soon be able to teach him all that is necessary in a few weeks. Please let me know if you intend to have meters made in England? I dont think it is necessary, in fact one would have to wait a long time, as everything is done at a very slow rate. (except eating)

I saw Batch, last week, he came over to England for the purpose of buying machinery, and stop'd a day with us; and from what he tell's me, the sooner I can get to him the better, as no doubt I can assist him a good deal in his place which he is now fiting up.

I saw our Hungarian prince (Mr Puskas) at the Crystal Palace the other night, and he is greatly pleased with the Edison display, and seems to be very anxious for the Paris place to be started and in working order, but the difficulties they have to contend with is surprising, the french law not allowing them to bring anything into the country; they being compelled to manufacture everything them self and can only use[c] foreign drawings, hardly anything else. Talking of your exhibit at the palace, I need not go into the details, for those are nearly all justly illustrated in the papers you receive, but I can say this, judging with impartiality, not being biased by any interests, that the "Edison exhibits" is the finest there is. One never see's any fluctuation in the light, something which is always seen at the Swan show. In the British Electric Co's[4] place there is a machine crossed almost every night, and yet these facts are never know to the public, they manage somehow to keep such things out of the papers, they have their machines next to our, so we can always tell[c] when something happens. No machines crossed yet at our place. The maxim lamps are burning now and then, sometimes I see them all ablaze and then other times they are stop'd, what the trouble is I dont know. One always finds the Edison lamps the first to be lit, and the last to stop. Billy Gladstone[5] spent about fourty five minutes at the Edison exhibit the[d] other night, and went through everything, and he is one of those duffers ~~that~~ who[c] understand what you tell him. ~~and~~ While Johnson was explaining the system to him, one of Swan's men manage'd to get into our court, and did his best to get the Major that was taking the party around, to have him take Gladstone and his party into the Swan exhibit, but it would not wash   Then when he saw that it was of no avail, he got one of his lamp and ~~waited~~ wanted to have it presented to Gladstone but we gave him some of our lamps and the major told this Swan agent that he would do no such thing— The child then took his departure—.

I visited the South Kensington Museum yesterday evening, and there in the Patent office Museum, saw those instruments you gave to them some time ago, the tasimeter electric pen, chalk telephone,[6] stock instruments Phonograph,[7] and that paper carbon lamp, that burned 1390 hours. All these are on one shelf and your photograph in the center. Next to your lamp is a Lane Fox lamp, onc I am sure that never had a current passed through it, the carbon is of a shape similar to our "A" lamp ones, and is inserted into the glass glob ~~and~~ with black sealing wax, showing that it could not hold a vacuum. In front of this lamp is a small placard bearing the following inscription. "Incandescent Electric lamp   St George Lane Fox letters Patent A.D 1880 no 3494"   Now perhaps you will remember when you gave your lamp away to this museum you wrote the following words, ("This lamp was lighted 1390 hours and gave a light of 13 candles   Thomas A Edison Menlo Park Sept 14 1880")

Now looking at these two lamps, it seems to me that the object of the Lane Fox lamp there, is to try and deceive the strangers that visite the place, by making both lamps look as if there is no priority in either of them, as both bear the date 1880.

(Of course I might be wrong in this conception, and very unjust, to doubt anything, as the folks that reign over here are always ~~actuate~~ actuated[c] by honesty and purity of intention and would under no consideration defraud you.) Hoping you are well   I remain your   Obedient servant

Francis Jehl

ALS, NjWOE, DF (*TAED* D8239ZAN; *TAEM* 62:811). Letterhead of Edison Electric Light System. [a]"London, E.C." and "188" preprinted. [b]"ard" interlined above. [c]Interlined above. [d]Obscured overwritten text.

1. Stamped with date received (10 April) and blank lines for reply date and file number.

2. Ludwig Oertling (1818–1893) began manufacturing fine chemical balances in London under his own name in 1847. The British government used an Oertling vacuum balance to compare weights against the national standard from 1872 to 1892. *Oxford DNB*, s.v. "Oertling, Ludwig"; Turner 1983, 65, 67.

3. Becker & Sons, established by Christopher Becker (1806–1890), a noted Dutch maker of precision balances, had a factory in New Rochelle, N.Y., and offices in New York City. Edison had at least one Becker balance at his Menlo Park laboratory. Doc. 1665 n. 8; "Mr. Becker's Wife Missing," *New York Times*, 7 May 1886, 3.

4. The British Electric Light Co. was an arc lighting company that entered the incandescent business about this time and shortly after began to manufacture Edison lamps under a licensing agreement. Bright 1972 [1949], 106, 108 n. 52.

5. William Henry Gladstone (1840–1891), oldest son of William Gladstone, the British prime minister, was a member of Parliament. The senior Gladstone reportedly visited on 20 March with a party that presumably included his son. *Oxford DNB*, s.v. "Gladstone, William Ewart"; undated excerpt from the *Chronicle* (London) in Edison Electric Light Co. Bulletin 8:10, 27 Apr. 1882, CR (*TAED* CB008:6; *TAEM* 96:698).

6. The electromotograph telephone receiver.

7. These pieces were on a list of items that the director of the British Patent Museum took from Menlo Park in October 1880. Stuart Wortley to TAE, 21 Sept. 1880; Wortley to H. G. Lack, 16 Oct. 1880; both DF (*TAED* D8004ZFN, D8004ZGA; *TAEM* 53:226, 242).

–2249–

From Joshua Bailey and Theodore Puskas

Paris, le 29 Mar. 1882[1a]

Dear Sir:

We have not made reports the last week owing to the great pressure of business consequent on the organization of the companies. The necessity for attention to this was so pressing that we do not doubt you will excuse the delay in considering the work done. Also the stenographer heretofore employed having been ill for some weeks it was beyond the strength of our Mr Bailey who was in convalescence to do other work than that above referred to

The several Companies have been organized for work[b] for work[c] on following basis.[2]

The Committee of Management of the Manf'g Co. is composed of Mess Batchelor, Puskas Chatard[3] & Favier,[4] Mr. Batchelor being the Director delegated & having the signature for all matters, and having been voted full power by the Directors.

The Committee of the "Continentale" is composed of MM. Bailey Raud[5] & Magnan[6] with the two first as active directors in office.

The Committee of the small plants is, Bailey, Chatard, Serurier[7] and Rau, and all the persons named take active part; Mr Serrurier mainly employed in outside visits. The special direction of correspondence and [reception?][d] of persons calling, as also the carrying on of negotiations has been with Bailey of necessity as the only one acquainted with various matters in train, and also with the conditions in which negotiations could be made. All the gentlemen above named are taking a more or less active part in the work, and especially MM. Rau, Chatard and Serrurier, gradually relieving Bailey to an extent. The increase of business relations is however so great that there is more than sufficient for all.

The attempt to bring Mr. Wirtz[8] in as President failed owing to the financial crash.[9] He not being a business man was frightened by the responsibility of connection with a share company, and this feeling was increased by the fact that one of his associates in the Institute was president of the Boutoux society[10] which went under in the crash. The societies were therefore organized with Mr. Porges as president, of the three, and with

Favier as V.P. of Manf'g Co.
Bailey "   "   Continentale &
Leon "   "   Electrique.

The principal matters now in hand are,

(1)   An important negotiation is on foot with parties in Germany since some weeks, involving the starting of a factory and a company for exploitation: the factory with capital of 1,500,000 marks, and exp[loitation] with 3,000,000 marks.[11] The manf'g Co. is on same basis as the Paris Co. and the terms proposed for the Exploitation is 20% com. on what comes to us. The latter would make installations of large dynamos for demonstrations for contracts with cities.— We have two plants at Frankfort, which will be put up in few days; another

will be shown at Hamburg also about April 1. The lamps have been shown on Railway trains at Frankfort with great success; also at Stuttgart where important installations are in course of being made; at Berlin we shall vote this afternoon authorizing an expenditure of 100,000 fcs for a trial installation, in a building that will give 3000 lamps.

The Strasbourg depot is great success, and we are treating for putting in the whole station. We have many other important matters in train in Germany. Bailey went to Frankfort a few days ago, and Rau leaves for Berlin tomorrow night to follow up the affair in consideration.

Italy. A great success has been made at Milan in the Scala Theatre, where the light was shown during nearly two months, afterwards in street lighting, and now in the great cafe Baffi of Milan, exciting greater enthusiasm than ever.— We have propositions for formation of company for Italy with capital of 10,000,000 fcs. We have installations engaged at Rome, Ferrare and other points, as soon as machines arrive. We have also many propositions for agencies. Bailey leaves for Florence day after tomorrow to meet Maj. Garbi, and to organize the various matters in train for Italy. Maj Garbi was indisposed towards us by the fact that an exhibition was made at Milan, commenced while we were in the differences with our Paris friends. We have had personal correspondence with him, and all parties here being of accord in desiring to make him representative in Italy hope that by personal interview the difficulty will be removed.

Spain. We are about closing contracts giving option for establishing factory, & for exploitation Co, with same parties we wrote about in previous letters.[12] The old negotiation was broken up by the fact that under our statutes the business could not be realized as proposed. We are in doubt whether under Spanish law patents will not be lost if we import goods into Spain before manf'g, and have ordered to Genoa the plant you shipped us to Barcelona.

Russia. Very successful exhibitions have been made at St. Petersburg &ᵉ at Tammerford in Finland. Ten plants have been sold there and are in course of installation by Mr. Nottbeck[13] who has Force with him. We are organizing for Southern Russia & Polish Russia.

Austria–Hungary. Mr. Francis Puskas[14] has ordered 2 plants for Buda-Pesth. We have agreed to take part in the Electricty exposition at Vienna in Sept. and have taken a large space.[15] The Emperor's Room at Palace of Exhibition will be lighted by us. We have propositions under consideration for lighting the Parliament house at Vienna also a Theatre at Brunn etc. etc.— One of directors your company will go to Vienna in couple of weeks.

~~Denmar~~ We have orders for a great factory at Styria and shall make installation in two weeks.

Belgium. We are now considering lighting of Chamber of Deputies at Brussels; a great factory at Lys of 2400 lamps, another at Gand,[16] & several minor affairs.

Denmark. We are in correspondence with parties but have done nothing definite

Holland. We have about closed for an installation at Amsterdam, in the great café & Concert Garden of Krasnapolsky,[17] one of the great fortunes of Holland, who proposes to ally himself with us. In Rotterdam also we are in course of making arrangements for showing light.

Roumania. We have an important business on foot for Bucherest. This is a rapid review—a birds eye view only of what we are doing.

You will readily understand what labour is involved in organizing such extensive operations and you will pardon the seeming neglect of advising you, under the circumstances referred to at beginning of this letter. We suggest that you use for your circulars only such matters as here referred as done. In the course of April we can count on being better organized and will keep you regularly informed of progress. Will write you ~~bu~~ by[e] Monday mail[e] regarding Founders shares & other matters of importance before leaving for Italy.

We remain dear Sir, Yours very truly

Puskas & Bailey

ALS, NjWOE, DF (*TAED* D8238ZAP; *TAEM* 62:361). Written by Joshua Bailey; letterhead of Continental Edison Co. [a]"*Paris, le*" and "188" preprinted. [b]"for work" badly smeared. [c]"for work" interlined above. [d]Illegible. [e]Interlined above.

1. Stamped with date received (6 April) and blank lines for reply date and file number.

2. The Compagnie Continentale Edison (holding company), the Société Industrielle et Commerciale Edison (for manufacturing), and the Société Éléctrique Edison (for isolated lighting) were all incorporated on 2 February 1882. See Doc. 2182 n. 2.

3. Alfred Chatard, a civil engineer, was among the founding investors in the Société du Téléphone Edison in 1878. He became one of the original directors of the Société Éléctrique Edison in February. See Doc. 1449 n. 1; Société Éléctrique Edison articles of incorporation, 2 Feb. 1882, DF (*TAED* D8238Q; *TAEM* 62:227).

4. Possibly Paul-André Favier (1837–1889), a former army officer and, since 1872, a full-time inventor. Favier devised several forms of explosives and methods of handling them. *DBF,* s.v. "Favier (Paul-André-Arthur-Marie-Auguste)."

5. Louis Rau (1841–1923) was a French businessman who for many years played a major role in the commercial development of Edison's electric light in France. He was also responsible for organizing Edison interests in Germany in 1883. "Rau, Louis," Pioneers Bio.

6. Probably Claude Magnin, one of the original shareholders in the Compagnie Continentale. Compagnie Continentale Edison articles of incorporation (p. 15), 2 Feb. 1882; Banque Centrale du Commerce to TAE, 9 Jan. 1882; both DF (*TAED* D8238R, D8238J; *TAEM* 62:252, 215).

7. Charles-François-Maurice Sérurier (1819–1887) was a French count and an investor and manager in the Compagnie Continentale Edison. Révérend 1974, 135–36; Batchelor to Serrurier, 20 April 1882, Cat. 1239:244 (*TAED* MBLB4244; *TAEM* 93:647).

8. Charles-Adolphe Wurtz (1817–1884) was a distinguished academic chemist. Early in his career Wurtz was professor at the agricultural Institute of Versailles; he became professor of organic chemistry at the Sorbonne in 1875. Wurtz also served as mayor of the seventh arrondissement of Paris and, since 1881, as a member of the French Senate. He is listed among the original shareholders of Edison's Parisian companies. *DSB,* s.v. "Wurtz, Charles-Adolphe"; *Gde. Ency.,* s.v. "Wurtz (Charles-Adolphe)"; Banque Centrale du Commerce to TAE, 9 Jan. 1882, DF (*TAED* D8238J; *TAEM* 62:215).

9. The Union générale bank grew rapidly from 1878 to 1882 on the basis of promoting large industrial and financial enterprises in Austria-Hungary and France. After one of its subsidiaries failed on 19 January 1882, however, the Union générale itself collapsed in a matter of weeks, provoking a severe financial crisis that led to one of France's longest depressions. Fox 1996, 185; Cameron 1961, 198–99.

10. Probably a financial syndicate of Eugène Bontoux (1820–1905), sometimes spelled Boutoux, the engineer and railroad promoter who founded the Union générale in 1878 and, as its director, was held criminally responsible when the bank collapsed. *DBF,* s.v. "Bontoux (Paul-Eugène)"; Byrnes 1950, 130, 133.

11. Bailey concluded negotiations by late April for the formation of German manufacturing and lighting companies on terms similar to those given here. Bailey and Theodore Puskas to TAE, 27 Apr. 1882, LM 1:200B (*TAED* LM001200B; *TAEM* 83:972).

12. Bailey wrote a few days earlier that "a plan of contract for the establishment of a factory and an exploitation Co. in Spain has been agreed on" but not yet signed. Bailey and Puskas to Samuel Insull, 26 Mar. 1882, DF (*TAED* D8238ZAL; *TAEM* 62:353).

13. Charles Nottbeck (1848–1904; born Carl Nottbeck) was a Finnish electrical engineer from Tampere or Tammersfors, the site of the first Edison lighting installation in Finland. After completing engineering training at Russian and Swiss universities, he traveled to the United States to work with Edison. He helped to commercialize Edison's telephone in Russia in 1880 and subsequently entered into a partnership with Theodore Puskas to introduce Edison's electric lighting system there. He later became the general agent for Edison lighting equipment in Russia and Finland. Minutes of Edison Telephone Co. of Europe, Ltd., 14 Jan. 1881; Batchelor to TAE, 7 Feb.,

16 and 28 Mar. 1882; Bailey and Puskas to Eaton, 29 Mar. 1882; all DF (*TAED* D8148M1, D8238T, D8238ZAI, D8238ZAN, D8238ZAP; *TAEM* 59:753, 62:309, 351, 358, 361); Myllyntaus 1991, 302.

14. Francis (Ferenc) Puskas (b. 1848), Theodore's brother, lived in Budapest. Gábor 1993, 34; Doc. 1751.

15. An International Electrical Exhibition at Vienna in September was announced a few weeks before this. It was postponed a year on account of a similar event scheduled for Munich at about the same time. "An Electrical Exhibition at Vienna," *Electrician* 8 (1882): 241; "Electrical Exhibition at Vienna," ibid., 10 (1882): 74.

16. The French name for Gent. *WGD*, s.v. "Gent."

17. The Polish-born entrepreneur and hotelier, Adolf Wilhelm Krasnapolsky (1834–1912) was an original investor in N.V. Nederlandsche Electriciteit-Maatschappij, the company founded in June 1882 to represent the Dutch interests of Compagnie Continentale Edison. One attraction at Kasnapolsky's Amsterdam hotel was the café-restaurant and wintergarden constructed in 1880. The Grand Café had a comparative display of gas lights and arc lights, and, by July 1882, about sixty Edison A lamps. Edward Acheson supervised installation of the Edison system. Hartmann 2001, 38; Heerding 1986, 114–15; Vreeken and Wouthuysen 1987, 99–102; "Electric Light in Amsterdam," *Electrician* 9 (1882): 194; Compagnie Continentale Edison, Bulletins 2–3, 10 Oct. 1882 (p. 22) and 15 Feb. 1883 (p. 15), PPC (*TAED* CA007B, CA007C; *TAEM* 96:336, 349).

–2250–

From Sherburne Eaton

New York March 29th. 188[2][a]

Dear Sir:—

Please find hereto annexed a copy of a resolution adopted by this company March 2nd. 1882. Pursuant to that resolution will you kindly inform us what your views are with reference to a profit to yourself as manufacturer, in order that we may confer intelligently with you on the subject matter covered by the resolution. Pursuant also to the policy of the company adopted in this resolution, will you kindly furnish us with a list of manufactures (exclusive of lamps) heretofore furnished by you to others than this company and its authorized licensees. Permit us to call your attention to the second clause of the resolution which provides that the license given to you to manufacture under the patents of the company is limited to such orders or transactions as shall be submitted to the company before acceptance or execution.[1]

Our Vice President, Mr Eaton, will be happy to confer with you on this matter at your early convenience. Very truly yours, The Edison Electric Light Company by

S B Eaton   V.P.

ENCLOSURE[b]

[New York, March 2, 1882[2]]

RESOLVED. That in the judgment of this committee it is the true interest of this company and of Mr Edison to encourage the adoption of the Edison system of lighting in this and foreign countries by extending to purchasers during the present year the inducement of as low a price as possible, and that to that end this company will participate with Mr Edison in a mutual sacrifice of immediate profit and that to the degree that Mr. Edison as a manufacturer is willing to reduce his profit the Company will reduce its claims for royalty, and that the Vice President be authorized to arrange with Mr. Edison in lieu of royalty[c] upon machines and other patented articles manufactured for foreign countries for a participation in his profit of not more than 33 1–3 per cent and that in the meantime the Vice President be authorized to arrange with purchasers for foreign countries for the privilege of purchasing at Mr. Edison's present prices;[3] and it was further

RESOLVED. That the effect of the foregoing resolution shall cease on January 1st. 1883, and shall not apply to any orders or transactions except such as shall be submitted to the Company before acceptance or execution by Mr. Edison, in order that the Company may protect itself against having its own orders delayed or neglected in favor of foreign orders.

TLS, NjWOE, DF (*TAED* D8224K; *TAEM* 61:16). Letterhead of Edison Electric Light Co. [a]"New York" and "188" preprinted. [b]Enclosure is a TD. [c]Mistyped.

1. Cf. Docs. 2106 and 2164.

2. Date taken from document.

3. At its 5 January meeting the company had resolved that because Edison's inventory of dynamos and material so exceeded its own ability to absorb them, Edison should be permitted to manufacture equipment for England without paying royalties for one year, so long as this privilege did not interfere with completion of its own orders. Goddard to TAE, 13 Jan. 1882, DF (*TAED* D8224C; *TAEM* 61:6).

5

April–June 1882

Two years after moving to New York City to bring his electric lighting system into commercial operation, Edison stepped back from the press of technical, organizational, and financial matters that had consumed his attention. Not long after vacationing with his family in Florida, Edison fell ill with "a terrible cold" and remained in bed at the Everett House (having given up the rented house across the street from his Fifth Avenue office at the end of 1881) from about 10 April through the end of the month. His secretary, Samuel Insull, ran his affairs, as he had during the vacation, noting that "Mrs E refuses to let anyone talk business to him."[1] Scarcely a week after resuming limited activities, Edison withdrew to Menlo Park with Insull and patent attorney Richard Dyer.

Edison enjoyed a productive springtime and early summer at Menlo Park, where he seems to have rediscovered the pleasure of allowing his mind to follow its own course. He planned to try a "great many things here that its difficult to do" at the Edison Machine Works, the closest thing to a laboratory he had in New York.[2] He quickly turned to experimenting and, in a return to his former practice, to recording in notebooks a wide variety of ideas just as they occurred to him. Most of these related to his electric light, but were not necessarily prompted by immediate engineering or production problems. He conceptualized a high voltage (300 or more volts) direct current distribution network; what he came to call the "village system" would require less copper in the main conductors than the 110 volt system being built in New York, and promised economy in areas of relatively light population density. He also explored solutions to the baffling phenomenon of "electrical carrying" (or "carbon carrying") in his lamps, attempted to design more efficient dynamos, investigated the use of processed cellulose in lamp filaments, and began to evaluate storage batteries for electric distribution systems. He fused a number of these ideas into an expansive draft caveat (Doc. 2307) sometime in the second half of June. He also executed at least three dozen U.S. patent applications, thirty-one of which were successfully issued. For the first time, he requested regular updates on the status of his applications from patent attorney Zenas Wilber, whom the Edison Electric Light Co. still retained.[3] Returning to the subject of electric railroads, he purchased a rail car that he planned to equip with motors and test on the new experimental railroad that Charles Hughes completed in April, about the time an electric freight locomotive was finished. In mid-June, he wished to hire two instrument makers for the laboratory, which he described as "very bare."[4]

Edison went to New York on occasion but generally relied on a few intimate assistants to convey essential information and carry out his decisions from Menlo Park. Insull handled a wide variety of matters in his correspondence from there. So, too, did Richard Dyer, a new member of the inner circle. Edison employed Dyer directly and treated him as an associate far more than either of his two predecessors, giving him significant responsibility for the first comprehensive review of his patents and applications. Sherburne Eaton, vice president (and effectively the chief executive) of the Edison Electric Light Company, sent frequent memoranda about the company's affairs in New York and its dealings with satellite Edison companies and investor syndicates throughout the United States, Great Britain, and continental Europe.[5] Francis Upton, head of the lamp factory (then being relocated to Harrison, near Newark), planned to visit Edison each Friday morning.[6] Edison asked Charles Clarke, chief engineer of the Edison Electric Light Company, to "Please keep me posted by brief notes of the progress of things."[7] Clarke's responsibilities included overseeing construction of the New York central station on Pearl Street and developing the new K dynamo for isolated plants; he also made periodic forays to Menlo Park.

The central station was nearly completed by June and awaited only the six huge dynamos. Sherburne Eaton visited the plant at the end of the month and came away "impressed with an accumulation of dirt, and with the indolence of the workmen," three-fourths of whom "were doing nothing but loafing."[8] The underground network was also largely finished, except for the installation of meters and the connection of main conductors to individual buildings. Edison redesigned the safety catches (fuses) in his system after discovering at the end of April that they were "worse than useless" and positively dangerous.[9] More vexing was the problem of regulating the dynamos to provide a constant voltage under varying loads. Edison gave this considerable thought at Menlo Park, from where he directed Charles Clarke's experiments at the Edison Machine Works. He also drafted several patent applications on the subject,[10] but did not settle on a method of regulating the First District until July.

With his lighting system so close to commercial operation, Edison confidently pursued a go-it-alone strategy concerning the work of other inventors. He continued to push Francis Upton to develop a high candlepower incandescent lamp that would rival arc lights for illuminating large outdoor areas. He entertained the idea of using arc lights in tandem with his incandescent bulbs because "we should encourage anything that will help us to sell current."[11] However, he declined to buy the arc patents of others and instead negotiated with an assistant, Otto Moses, to develop a suitable arc lamp. Edison declined to associate with a company promoting batteries for lighting (for fear of being "smirched" by their enterprise) but did begin his own experiments along these lines.[12] He also categorically refused any alliance with rival companies to challenge British inventor Joseph Swan's priority claims to the carbon incandescent lamp. In Great Britain, he took an important step to improve his own legal position by conditionally ratifying disclaimers of five fundamental patent specifications.[13]

Edison kept abreast of negotiations by Joshua Bailey and Theodore Puskas to form an investor's syndicate for his electric light in Italy. In May, he approved terms for organizing a new firm, Edison's Indian & Colonial Electric Company, to exploit his patents in British colonies. With so much money at stake in distant enterprises, he began subscribing to London financial newspapers to keep track of stock prices abroad.[14] Closer to home, Edison entered into a profit-sharing agreement (Doc. 2293) in June with both the superintendent and the treasurer of the Edison Machine Works. This was in anticipation of increased production, in part to fill a large order from the English Edison company that threatened to strain the plant's capacity and Edison's finances.

Though still putting in long work hours at Menlo Park, Edison also sought diversions unattainable amid the crush of business in New York. He inquired about purchasing one or more monkeys, likely as amusements for his three children, though he seems not to have carried this out.[15] In a conspicuous moment of attention to his wife, he invited his brother Pitt Edison, an experienced liveryman, to visit for the purpose of selecting horses for her.[16] In June, he began to seek out a steam yacht for sale or hire.[17]

1. Doc. 2263.
2. Doc. 2267.
3. See App. 5 and Doc. 2323 n. 1.
4. TAE to Charles Dean, 12 June 1882; TAE to C. W. Nieder, 22 June 1882; Lbk. 7:448B, 559A (*TAED* LB007448B, LB007559A; *TAEM* 80:610, 669).
5. See headnote, Doc. 2274.
6. Upton to TAE, 8 May 1882, DF (*TAED* D8230ZAK; *TAEM* 61:753).
7. Doc. 2267.
8. Eaton to TAE, 27 June 1882, DF (*TAED* D8226ZAG; *TAEM* 61:313).
9. Doc. 2265.
10. See headnote, Doc. 2242 and Docs. 2269 and 2272.
11. TAE to Eaton, 22 May 1882, Lbk. 12:363 (*TAED* LB012363; *TAEM* 81:697).
12. Doc. 2274; see also Doc. 2286.
13. See Doc. 2285 esp. n. 2.
14. TAE to Kingsbury & Co., 20 June 1882, Lbk. 7:525A (*TAED* LB007525A; *TAEM* 80:648).
15. Henry Seely to TAE, 19 May 1882, DF (*TAED* D8204ZBM; *TAEM* 60:143).
16. See Doc. 2257.
17. See Doc. 2311 n. 2.

–2251–

Memorandum: Electric
Railway Patents

[New York, c. April 1, 1882[1]]
Railroad

Dyer

Want to take out a new application in England and the US. If you do not think it sufficiently covered already[2]

Ask that draughtsman down stairs[3] for a picture of Mason Clutch = X is the mason clutch

R is resistance that are in the main circuit and permits of the slowing up of the train. This appears to be <u>essential</u> because in both mine & Siemens I dont remember where any means were shewn for slowing up the train. You might mention that the field magnet could be weakened or the commutator brushes moved around from the cab of the Locomotive by means of a rod with worm & worm wheel, but the method shewn is thought to be[a] preferable in practice.

Another thing is Double wound field magnet so that in ~~starting from a state of rest~~ cases where the locomotive is slowed down very considerably by heavy load the armature will not cut ~~all the~~ too much current away from the wire upon the feild If we had a single field & this was multiple arc'd across the armature It would have sufficient Current passing through it to produce the requisite field owing to the Counter EM force of the revolving armature but if this was slowed up very much the fall off Electromotive force in the armature would be so great that the field w~~o~~ire would get very little current hence the power of the machine would rapidly diminish   to prevent this I not only keep the fine wire on the field across & throw in a <u>coarse</u> wire coil right in the same circuit as the bobbin or[b] in other words make it a Dynamo for a while. this keeps the field up and when the load becomes lighter etc the lever G cuts the coarse wire out[b] of the main circuit[4]

d d is the fine wire regular field wire.[c]

Did Siemens take current off of more than one wheel. I think not. I remember I had another US application where this was shewn—get this & take[b] what is good out of it for the new English.[5]

I propose to put a ~~Large~~[d] pinion gear on the shaft where X is meshing into a large gear on the main driver— I wonder if we couldnt get a combination claim of belts, friction clutch or equivalent disconnecting mechanism and[b] gears.[6]

Have we in any our cases spoken off nickel plated ends of rails, nickel plated fish plates— also Japanned rails=[7]

AD, NjWOE, Lab., Cat. 1148 (*TAED* NM017:29; *TAEM* 44:389). [a]"to be" interlined above. [b]Obscured overwritten text. [c]Followed by dividing mark. [d]Interlined above.

1. On 4 April, Richard Dyer sent to London patent agent Thomas Handford the provisional specification for an omnibus British specification encompassing elements of this document. Handford filed it on 18 April. Brit. Pat. 1,862 (1882); Dyer to Handford, 4 Apr. 1882, Lbk. 12:26 (*TAED* LB012026; *TAEM* 81:512); Dyer memorandum, c. 1 May 1882, DF (*TAED* D8248ZAL; *TAEM* 63:457).

2. Edison filed a number of U.S. patent applications on electric railways during the spring and summer. Some were abandoned and their execution dates lost; they may have been among the cases withheld by Zenas Wilber. In addition to those identified below, they included Case 429 (Patent Application Casebook E-2537:212, PS [*TAED* PT021212; *TAEM* 45:760]) and U.S. Patents 273,490; 339,278; and 448,778; see also Doc. 2227.

3. The editors have not identified the draftsman.

4. Edison filed an application on 7 August for controlling the locomotive's speed by using a doubly-wound field magnet and by varying the resistance of the field coil circuit. It was later abandoned. Patent Application Casebook E-2537:208 (Case 428), PS (*TAED* PT021208; *TAEM* 45:759).

5. This design was implicit in Edison's U.S. Patent 263,132, filed in 1880. In the British specification completed in April (see note 1), Edison described means by which "two or more wheels . . . may be provided with brushes or springs to increase the extent of contact with the rails."

6. An application filed on 7 August covered mechanical means such as belts, gears, and friction clutches, for controlling the locomotive's speed. It was later abandoned. Patent Application Casebook E-2537:218 (Case 432), PS (*TAED* PT021218; *TAEM* 45:761).

7. Edison executed an application on 7 July for reducing resistance in the circuit by electroplating the ends of the rails with nickel, silver, "or other metal not easily oxidized by exposure to air and moisture." In August he filed an application, subsequently abandoned, for japanning the rails. About that time he sketched methods to japan spikes and rails and to nickel-plate fish plates and the ends of rails. U.S. Pat. 273,494; Patent Application Casebook E-2537:300 (Case 466), PS (*TAED* PT021300; *TAEM* 45:771); Cat. 1148, Lab. (*TAED* NM017:171–74; *TAEM* 44:531–34).

–2252–

Notebook Entry:
Facsimile Telegraphy

[New York,] Apl 2 1882

Inside roll of wet paper   Cover on patent in Kinny Transmitter[1]

Autographic try ~~douple~~ double paper   Starch between or other substance outside paper to be thin & glossy   this will do the biz for indentation[2]

Ask genl Eckert[3] for a standard table ~~& device~~   figure out a duplex for autogphic[a]

$205 per year each guarantee 300, if they use 10 $4000 per year if they dont use at all for control purpose[4]

T.A.E.

X, NjWOE, Lab., PN-82-04-01 (*TAED* NP016:21; *TAEM* 44:31). Document multiply dated. [a]Followed by dividing mark.

1. Patrick Kenny began collaborating with Edison on facsimile (autographic) telegraphy in the spring of 1878 and by that September had apparently produced a working instrument. Edison continued to pay for experiments at least through December 1882. See Docs. 1328 n. 3, 1388 n. 6, 1638, and 1757; Ledgers #4:157–58, #5:25, #8:13, all Accts. (*TAED* AB002:83–84, AB003:26, AB004:27; *TAEM* 87:292–93, 429; 88:29).

2. In July 1881, Edison and Kenny jointly filed a patent application for facsimile transmitting and receiving instruments (not issued until 1892). Sometime thereafter they also drafted a similar specification for Britain and continental countries. In this system an impression was created by writing on a sheet of paper. The paper was mounted on a wooden cylinder (revolved by a motor) underneath a carriage carrying two contacts. The contacts were arranged so that one of them passing into the indentations on the paper closed a circuit, thereby transmitting a current onto the line. The receiving instrument was similar but had a metal style mounted on a carriage that moved in unison with that of the transmitter. The style marked a piece of electrochemically sensitive paper on a metal cylinder as it received current from the line. Draft U.S. patent application, 19 July 1881; draft foreign application, n.d. [1882?]; both DF (*TAED* D8142ZBL1, D8248ZDR1; *TAEM* 59:349, 63:568); U.S. Pat. 479,184; "Télégraphe Autographique De M. Edison," *La Lumière Électrique,* 5 (1881): 418–21; "The Edison Exhibit at the Palais de l'Industrie," *Engineering* 32 (1881): 343.

3. Thomas Eckert became vice president and general manager of Western Union in 1881 when that company consolidated with the American Union Telegraph Co., of which he had been president. *ANB,* s.v. "Eckert, Thomas Thompson"; Docs. 389 n. 6, 1713 n. 4.

4. Nothing more is known of this proposal.

Figure from the foreign patent application completed by Edison and Patrick Kenny for their facsimile telegraph, showing the mechanism for synchronizing one instrument with another.

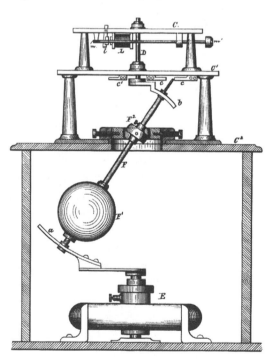

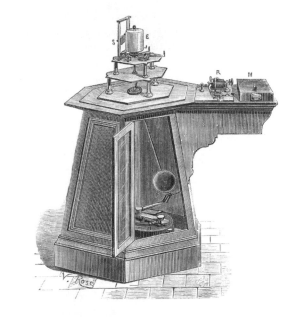

*Edison and Kenny's
facsimile telegraph
instrument exhibited in
Paris. The large pendulum
P regulated the motion of
the cylinder of paper **E**.*

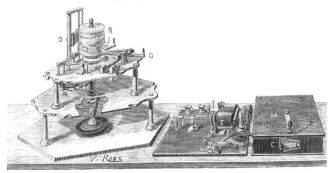

*Detail of the facsimile
telegraph, showing
commutators **B** and **G** that
changed the instrument
from receiver to transmitter
and vice versa.*

–2253–

To Alfred Reade[1]

[New York,] 4th Apl [188]2

Sir,

I have your favor of 10th March[2] and in reply beg to state that I think chewing tobacco acts as a good stimulant upon anyone engaged at laborious brain work. Smoking although pleasant is too violent in its action; and the same remark applies to alcoholic liquors

I am inclined to think that ~~for~~ it is better for Intellectual workers to perform their labors at night as after a very long experience at night work I find my [~~mind?~~][a] brain[b] is in my better condition at that time[c] especially for experimental work when so engaged I almost invariably chew tobacco as a stimulant   Yours truly

Thos A Edison   I[nsull]

L (letterpress copy), NjWOE, Lbk. 12:33 (*TAED* LB012033; *TAEM* 81:514). Written by Samuel Insull. [a]Canceled. [b]Interlined above. [c]"at that time" interlined above.

1. Alfred Reade (b. 1851) was a journalist in Manchester, England. He published several books about this time, including one on temperance. Kirk 1891, s.v. "Reade, Alfred Arthur."

2. Reade wrote to Edison about the proposition "that all great thinkers and popular authors find the use of stimulants a help to them in their intellectual efforts." His handwritten letter solicited Edison's "opinion as to the effects of tobacco and alcohol upon the mind and health." Edison's reply, apparently the only one to acknowledge chewing tobacco, was one of more than 120 published in Reade 1883 (45, 181). Reade to TAE, 10 Mar. 1881, DF (*TAED* D8204ZAF; *TAEM* 60:94).

-2254-

To Charles Batchelor

New York, 6th Apl *1882*[a]

My Dear Batchelor,

Your letter of 6th Mar came to hand while I was in Florida and the enquiry about engines had to await my return.[1]

I do not think you should use any more Porter engines The new Armington Sims Engine is better simpler & cheaper than Porter. I think we shall use Armington & Sims Engines altogether.

They will cost you $1800 each packed for Export f[ree] o[n] b[oard]. at Providence R.I.[2] Yours truly

Thos A Edison   I[nsull]

L (letterpress copy), NjWOE, Lbk. 12:43 (*TAED* LB012043; *TAEM* 81:523). Written by Samuel Insull; stamped Thomas A. Edison to Chas. Batchelor. [a]"*New York*," and "*188*" from handstamp.

1. Batchelor asked about prices for Porter-Allen engines for central station dynamos and how quickly they could be delivered in Paris and Hamburg. Batchelor to TAE, 6 Mar. 1882, DF (*TAED* D8238ZAF; *TAEM* 62:343).

2. Two days earlier, Armington & Sims had telegraphed Edison a price of $1,750 on engines for foreign shipment. Armington & Sims to Insull, 4 Apr. 1882, DF (*TAED* D8233Z; *TAEM* 61:1024).

-2255-

To Edward Johnson

New York, 6th Apl *1882*[a]

Personal

My Dear Johnson,

Referring to letter & Power of Attorney sent herewith I very strongly object to selling my future inventions & certainly think that in any event the sale should be limited to three years future inventions.[1] My reason for this is that I propose experimenting with a view to getting electricity <u>directly</u> from coal & I think there is great promise of my succeeding.[2] I should most strongly object to part with so valuable an invention to these Indian & Colonial Companies as if I ever succeed it will be of immense value. Have the agreement drawn so as not to include this but it should be done without making apparent what we want to do.

You had better say nothing at all about this to anyone else Yours very truly

T A Edison

Its a 3 year job— E[b]

LS (letterpress copy), NjWOE, Lbk. 12:40 (*TAED* LB012040; *TAEM* 81:520). Written by Samuel Insull. Stamped Thomas A. Edison to Edward H. Johnson. [a]"*New York*," and "*188*" from handstamp. [b]Postscript written and signed by Edison.

1. The power of attorney authorized Johnson to negotiate the sale of Edison's electric light and power patents in the British colonies in Australia and New Zealand. It applied to specified patents and "all improvements hereafter to be made." In a separate cover letter, Edison also instructed Johnson not to "make any use of this Power whatever unless you have distinct instructions so to do by cable or letter from Drexel Morgan & Co & myself." He approved terms for Australia and India in May. TAE power of attorney to Johnson, 5 Apr. 1882, DF (*TAED* D8239ZAO; *TAEM* 62:827); TAE to Johnson, 6 Apr. 1882, Lbk. 12:41 (*TAED* LB012041; *TAEM* 81:521); see Doc. 2280.

2. See Doc. 2230.

–2256–

From Joshua Bailey

[Milan,] April 7. 82

Edison N.Y.

Industrial and financial condition and customs country make absolutely impossible form companies giving us part of capital in fully paid up shares   can form syndicate for all Italy and sign contract Monday next embracing the bank Generals the banks of Milan Venice Genoa Turin Naples Rome and credit Lombards with numerous private bankers[1]   this combination constitutes Italy   close corporation for Edison with factory started immediately capital million and half with three millions more for small instalations and model stations for cities   we receive half profits factory and fifteen percentum cash on all machines tubing and patented material generally expect[2] lamps   prices there to be fixed in accord with us and profits divided equally with obligation take all material from our factories   we pay no commission to syndicate   will you account this contract in accord with French company   answer tomorrow[3]

Bailey   Hotel Continental   Milan

L (telegram, copy), NjWOE, LM 1:184C (*TAED* LM001184C; *TAEM* 83:964). Written by Charles Mott.

1. At the end of April, Bailey brought a committee of Italian bankers to London to examine the Holborn Viaduct central station. Bailey to TAE, 27 Apr. 1882, LM 1:200A (*TAED* LM001200A; *TAEM* 83:972).

2. Bailey probably intended "except."

3. Edison cabled Bailey on 11 April: "Fifteen percent on cost installation not enough   want continuing annual income additional." Puskas and Bailey apprised Edison a week later of a revised proposal which would grant him royalties for three years and, in lieu of additional royalties, 20% of the shares of the stock companies formed. In May they described the prospective market for electric lighting in Italy, pointing to high gas rates and the fact that Milan, with 330,000 inhabitants but only 30,000 gas lights, was nevertheless "the most generally lighted city in Italy." TAE to Bailey, 11 Apr. 1882, LM 1:188E (*TAED* LM001188E; *TAEM* 83:966); Puskas and Bailey to Sherburne Eaton, 18 Apr. and 15 May 1882, both DF (*TAED* D8238ZAV, D8238ZBF; *TAEM* 62:380, 404).

–2257–

Samuel Insull to Pitt Edison[1]

[New York,] 8th Apl [188]2

Friend Edison,

TA.E. would be glad if you could manage to come on here about 1st May. He wants to buy some horses & says he would like you to be here to pick them out for him.[2]

I will send you on a check on a/c of expenses should you require it   Yours truly

Saml Insull

ALS (letterpress copy), NjWOE, Lbk. 12:54 (*TAED* LB012054; *TAEM* 81:530).

1. Edison's older brother, William Pitt Edison, lived in Port Huron, Mich. A former liveryman, he had in recent years operated a horse-drawn omnibus service and was also involved with street railways there. Israel 1998, 6; Docs. 325 n. 5, 651, 1414 n. 2, and 1963 n. 2.

2. Edison had consulted Pitt about his purchase of ponies in 1878 (see Doc. 1360 n. 1). In late May, Pitt reported that he had "been over about half of Canada   I have got the single horse for Mrs TAE   tell her he is the best horse in the west can trot in 2.40 and a child can drive him cares nothing for cars or RR tracks." He was also looking at horse teams and "the most trouble I have is to get Horses that will stand the Steam Cars but think will have them this week so as to make a start by June 1st."

He requested $400 to buy horses, a car, and feed; the team was at Menlo Park by August. Pitt Edison to Insull, 24 and 31 May, 28 Aug. 1882, all DF (*TAED* D8214N, D8214O, D8214T; *TAEM* 60:713, 715, 718).

-2258-

From Edward Johnson

London, E.C. April 9 1882[a]

My dear Edison:—

Doubtless you will have noticed my correspondence has begun to slacken, and when one has lost a little ground it is difficult to regain it at a bound but I will endeavour to answer your various communications as well as those received from Insull, as briefly as possible, to convey to your mind the status of things in general here. I will not go into matters of money payments negotiations, organization and that sort of thing since I have just done that in a letter to Mr. Fabbri, which I have asked him to show to you.[1]

Yours of March 7th[2] you say you can probably arrange to come to England when you have the Central Station started in New York and providing the unexpected does not arise in the meantime. That conveys to my mind that I will see you only in America not that the Central Station will ~~ever~~ never[b] be started but that the unexpected will certainly[c] happen. What one must expect in estimating your movements is the unexpected. You must excuse the "bull" but I am near Ireland besides I have an Irish wife.[3]

Bar Armatures—. Hammer has had the Bar Armature in and had 250 B. Lamps on it working them satisfactorily.[4] I am of the opinion that this is the Armature for the small machines but since I have been unable to get Sir William Thomson to make a test of the machine now in his possession for some months, I have decided not to send this one to him for that purpose especially as the Company have so able an Electrician as Dr. John Hopkinson. Dr. Hopkinson and Dr. Fleming[5] together with Hammer will unitedly make a test of this armature during the coming week and if your figures are corroborated they will in conjunction with myself make a report to the Board recommending the adoption of these Armatures as the standard in which case I will have an order for at least 100 of them sent to you immediately in order that you may be justified in making the tools wherewith to produce them cheaply. I trust you will give us the figures if only approximately at which you can supply these Armatures for the following machines

B. machines
A. machines and
250 Light <u>A</u>. machines

I note all you say as to unequal electro motive force of machines connected in multiple arc. We have gone all through that and have taken care of it. Our Plant at Crystal Palace consistsing of 12 machines in simple[d] multiple arc works to perfection in fact as the Viaduct is occupying both Mr. Hammer and myself now we leave this little plant with its 1075 Lights to the tender mercies of Jim Holloway[6] and his assistants. It gives us no concern—we are afraid of nothing happening. We dont grow nervous or gray haired over the outfit and we know that when we send any one to see the Light at the appointed hour of starting they will see it and furthermore we as inspire everybody with absolute confidence that they will always[d] find the Edison Light running a little before that of any one else and continue to run a little after every body else have stopped and that between thiese times with by[b] no possibility will they ever see any sections of the Lights extinguished. This is a record impossible to any other exhibitor. In point of fact no other exhibitor has been able to sustain his Lights without failure during the exhibition. With some it is an even bet that at a given hour they will be lighted or extinguished.

Crystal Palace—Other Inventions— My recommendation in respect to this is that they be boxed and shipped to America immediately on the close of the exhibition. I have already had an application from some of your French people to have them sent to Vienna where there is to be an Electrical Exhibition. Electrical Exhibitions are going to be the order of the day for some years to come and if you consent to these things passing out of your the[b] hands of one of your immediate Assistants you can bet your life you would never see anything of them again. If I had to repeatt the Crystal Palace Exhibition I would make it Electrical Light[b] pure and simple therefore I strongly urge in fact for your sake I shall[d] insist that all these costly instruments that are of value to you and you alone shall be returned to America just as soon as the Palace Exhibition closes. I want you therefore to decline to send them to Vienna. There was a depreciation I suppose of fully 30% on the transfer from Paris to London by virtue of certain pieces and parts and certain entities[e] being lost. That depreciation would be greater when going into the hands of persons unfamiliar with[d] them than it was in going from the hands of Batchelor to my own. You can easily see therefore that by the time this exhibit would pass through two or three additional exhibitions it would be totally useless to you unless therefore you are prepared to contribute the large sum of money which this exhibit has cost you to the glory of a few individuals in whom you have no personal or direct interest you will comply with my suggestion.

I have overlooked your request for some Swan Lamps; it is an easy matter to get them and Hammer says that he will attend to it and get a dozen or so and ship them to you. I will also get for you the Motograph Telephones from the United Company. I can do this through Gouraud.

Insulls of 24th March—referring to the Edison Foreign Electric Light & Motive Power Company[7] this is a name only. Gouraud is in correspondence with Drexel, Morgan & Co on the subject of organizing Companies for Australia India and Norway, Sweden and Portugal. He considers that these countries are practically in his hands for disposal   His Australia negotiations on the basis of an outright sale fell through if indeed it ever attained the dignity of a negotiation. I have my own opinion on that. However I am of the opinion that we can make better terms through him than otherwise and that therefore the present correspondence in respect to the same is in order   As to the use of your name as well as of mine the cheek of the man is beyond expression. I have intimated to him that you did not care to have your name associated with any of his enterprises as a Director but I have positively demanded that my name be absolutely withdrawn from all his enterprises. He advises me that this will be done but does not do it. My last letter to him on the subject was to the effect that if it were not done I should write to his Companies explaining that my name was used without authority and request its immediate withdrawal. You should write him a letter to the effect that you will not permit your name to be used as a Director in any enterprise. If you care to send the letter to me I will deliver it[d] and see that it is enforced precisely as I mean to do in respect to my own immediately. The first intimation that I had that I was a Director in the Edison Foreign Light Coy. was in seeing it in the newspaper. The Duke's action in this matter is quite in harmony with his modus operandi in business matters generally. He uses anything that he thinks he may with impunity precisely as if he were dealing with his own personal property. In respect to the Domestic Company about which you ask you will find the explanation in my letter to Mr. Fabbri. You did quite right in declining to allow him to sell his 20 plants to the Domestic Coy.[8] He sought hard to inveigle me into a quasi permit to use these[d] plants in England but he did not succeed. He has thus far disposed of three or four of them—some to the Cape of Good Hope and some to Sweden

I do not see any immediate prospect of getting any money out of Gouraud. The United Telephone and the Edison Telephone have finally settled their affairs and the residue of our shares has become available. I obtained mine and sold them at 13¾ per share. I opened up correspondence with Insull by telegraphing with respect to yours and at present engaged[d] in trying to sell them. Of course I am doing it through Gouraud. He has now agreed that he will sell them for the next stock exchange ~~statement~~ settlement[b] which is about the 15th to the 18th of this month. Your money will be available probably between the 15th and the 20th.[9] You should get a good price— not less than 13 and possibly more. As to the wisdom of selling these shares there are two things to be said. 1st.— The general impression every where is that the shares will go very much higher but as against that is the fact that you and I and the rest[d] of us want the cash so I sold and think you would be wise in doing the same. As to that Oriental Telephone concern the money is now finally in hand but Gouraud tells me there is some formality yet to go through before it can be divided. You must judge as I do whether or no this is a fact or a mere pretext for holding onto the money yet a little while longer. My opinion strongly inclines to the latter supposition however you can rely on me doing all I can to hasten the dividend.[10]

I asked you in a letter a longtime since whether my recollection was right as to whether you had agreed to join Gouraud in allowing me 5 percent in this Oriental business. Gouraud as Insull knows did agree to it and he now says that of course its all right providing you agree. I am not able to say to him that you do except on general principles[d] for you have made no reply to my letter of enquiry on the subject. If you do not reply or if you do not care to [~~to?~~ -----][f] do it[g] I have nothing to say. In that event I will simply notify Gouraud to transmit to you one half of whatever he may have for division. In respect to the Electric Light I think there is but little doubt that he can organize a company for Australia,—another for India, and another for the three small countries, Norway, Sweden and Portugal on the basis telegraphed namely,[11]

Australia	£25 000 cash.
India	£25 000 "
and	£5 000 "

for the other three countries. That he will be able to get in addition to this cash the costs of the patents I have no doubt in respect to Australia and India though I have in respect to the other countries where they are very poor, no market for any invention in them and it may be difficult to obtain a company for £5000 cash however Gouraud and I are discussing the matter at present with Drexel's and I am doing what I can to have their wishes met by Gouraud. I do not expect however from the way such things generally run that even if the terms are agreed upon that Gouraud will be able to effect an organization for these several countries for some considerable time. I like therefore the stipulations contained in Fabbri's telegrams[12] to the effect that the terms are only agreed to on condition that an organization be effected within a limited time. I hardly think however it is worth while to limit him to the first of May as he certainly could not do much in so short a time but if thirty days was stipulated from the day ~~following~~ of the final consent as to terms I think we might reasonably expect that he would consummate the work within the time.

Crystal Palace— It is not yet definitely known when this Exhibition will close. It is paying the Crystal Palace company[d] very handsomely at the present time and I therefore fully expect that they will leave no stone unturned to keep it in operation as long as possible and that it will ultimately be broken only by the withdrawal of a number of the exhibitors. The appointment of Jurors to make the awards has been made in large part and they are now having their first sitting to arrange preliminaries. A young fellow of the name of Frank Sprague (I believe you have met him connected with the U.S. Navy has been detailed by that Department to attend the Electrical Exhibition has been appointed as one of the American Jurors on Electric Lighting.[13] He is a very competent young fellow and in my judgment[d] fully qualified to act in the capacity assigned him and furthermore he fully appreciates your work as you will see from an article which I will forward you either by this mail or the next, and which he has written for a Paris Technical Paper—I think La Lumiere Electrique.[14] At all events I think it is important that he should act as an American Juror that I am doing whatever I can to further that end. His leave of absence expired and he wrote home to have it extended. I engaged to use what influence I could command in Washington (through Painter)[15] to have it extended in case he himself failed, but I believe he has succeeded and that his leave has been extended to a time sufficient to allow him to act as a Juror. I shall have more to say on the subject of Mr. Sprague after the exhibition closes and he has performed the work assigned him. You will have observed of course long ere this that all the technical papers as well as the daily press are unanimous in praise of the perfection and completeness of your work. This is a revolution which I am proud to have had a hand in accomplishing and it is a revolution effected on such solid ground that a counter revolution is now impossible. You are therefore fully established before the British public as the one man who has brought the Electric Light within the reach of the people. We have had four opinions on the validity of your lamp patent number 5. Dr. Hopkinsons opinion was the last and we are now taking a fifth namely that of Sir Frederick Bramwell.[16] The four are of one mind that the patent is good. Our Directors have therefore taken the advantage of the opportunity presented by the appearance before the public of the British Company with a new Company for the West of England to notify [--]ᶠ the public that the Lane-Fox Lamp is an infringement.[17] This brought a reply from the British Company to which reply we respond by a general advertisement in all the papers as per enclosed[18] notifying the public and all parties concerned that we are advised that we have the exclusive right of the Incandescent Lamp under whatever name it may be made and that we propose to defend that right. This is bold ground but our Directors have voted that it shall be sustained to the extent of bringing suits and fighting them through the courts and that they will if necessary expend a sum of £30 000. for the purpose. As my baby would say, "how does that suit you, youᵇ old grand mother"? We have had a long interview with Bramwell and though it was his first examination of the patents we were very much gratified to find that he could not discover anything new in the way of obstacles or objections and that therefore his written report will in all probability confirm the opinions we have in hand.

Patents— I do not quite know whether you understand that patents now being taken out commencing at #43, are being taken out at your expense. That is to say that only those patents enumerated in the Schedule and assigned to the English Company have passed beyond you in respect to cost that you are liable for all fee's and costs of whatever character for all subsequent patents until such time as the Company may elected under the contract to take them over. This means simply (and it is for this purpose that I allude to it) that you must deal directly with Handford[19] in the matter of his payments   I have asked him to prepare a monthly statement of his account and forward it directly to Dyer who will audit it upon which you will remit the amount to him. This is only another way of doing just what you have been doing all along. Except that instead of paying an attorney in New York who hired another another attorney here you hire your own attorney here and pay him direct   There is but one opinion as to the wisdom of this course and it is summed up in the same words by every body, namely, now Edisons patents will be in accordance with patents Laws, and with common sense and will therefore be of some value. Handford will have nothing to do with the Company or the Company with him, he will deal only with you and with Dyer direct by personal communication.[20]

Viaduct— The long agony is well nigh over. Machine No 2. has been tried so often that we have become familiar with it and know just what it will do, namely, it will run 1000 lights without difficulty. Machine No 3 is new to us and although it comes with a big endorsement from you, Hood pronounces it to be nowt as good as the other one. The engine certainly makes a vast deal more noise and the coupling very little less. As for the commutator it sparks a damn sight more but you have explained about the commutator consequently we expected but I am disappointed in the quietness of the engine and of the coupling. However as Hood has made a vast improvement in engine No 2 since it originally came over I am expecting to see him similarly improving engine No 3.— At all events I am perfectly satisfied that I have two machines either one of which will do the work of sustaining 1000 lights   The night before last we ran No 3 satisfactorily and last night we tried some experiments with them both. I had telegraphed to Paris in accordance with a promise to Mr. Batchelor to let Mr. Batchelor know what I was going to do as he was coming over. Instead of comeing over he sent a lot of his Frenchmen who were present during the whole trial and who were exceedingly happy over the result   We rather astonished them by our triumphant yells and songs and beer drinking and smoking and smutty stories ala Menlo Park and Goerck Street. We first ran the two engines without a load together to see that[d] our steam pipes, exhausts &c were all ready satisfactory. They proved to be. We then ran No 3 with the load (all that was available last night many of the shops being shut up) namely 700 lamps. We then stopped it and ran No 2 with the same load. We then stopped it and starting them up very slowly attempted to connect them together but they would not have it—they kicked like hell. This instantly suggested that[d] the machines were opposed to each other. A tracing out of the main conductors however showed this not to be the case apparently but we concluded that they were oppositely polarized consequently we shifted the main conductors of one of the machines and then starting up quietly ~~connected~~ sought to connect them again and this time without a kick Hood took one machine I took the other Hammer took the commutators and we put on a little more steam—a little more and then the light began to appear—and then a little more and again and we had the whole thing in full bloom with two Dynamo's of 1000 lamps pure multiple arc for the first time in the history of electricity. Then and there went up some yells, some cheers for Edison and for all the boys when supper was immediately ordered. We ran this way for some time to our entire satisfaction. Then our Paris friends bid us good night leaving for home thoroughly satisfied that they might go to work and order any number of these machines without any fear that they would not assimilate properly   Tonight (Sunday) we propose to continue our experiments and this time to try and throw in and out the different machines as though one had given out. The switches we have provided are not exactly suited to the purpose so that we may not be able to do this in a satisfactory manner but Hammer has ordered new ones which will be provided in a day or two when we will[b] do it without difficulty The 32 candle street Lamps arrived and we obtained them late yesterday. We therefore slipped them into 7 lampposts just to see what the effect would be and all I have to say is that it was entirely satisfactory to me, and I think will make an impression exceedingly favourable to street lighting by Incandescence which is not universally believed to be out of the question. More of this in my next. My present opinion is that one of these 32 candle lamps placed in a suitable street lamp will be quite sufficient. I will have a special lamp post fitted up for illustrating it. The engines make a little more noise when they have a big load on than I thought would be the case but still not sufficient to constitute our work a nuisance by annoying our neighbors. We send you by this mail the plans of the following[21]

1   Installation of Crystal Palace
2   Circuits showing the lights at the Palace
3   Of the Holborn Viaduct showing the wire system—the arrangements of the street conductors—separate conductors for the street lamps and divided into different circuits so as to enable us to light and extinguish at pleasure to the right and to the left for illustration, also the arrangement of the street and house circuits showing the disposition that has been made of the lamps.

On the plan each street lamp is shown with two lamps whereas by putting two 32 candle lamps it will make it the equivalent of 4 A Lamps. Theise with some other additions will make the 1000 A Lamps which I am intending to supply We will send you a plan of the machinery installation in the basement of the Viaduct establishment probably by next mail.

We also send you photographic views of different portions of our exhibits, among others, the Dynamo installation and Chandeliers at Crystal Palace.[22]—and also detailed particulars of the Engines[h] & Dynamos with their countershafting at[b] Crystal Palace. Batchelor wanted a set of these made for himself & having obtained them for him we made tracings for you. You will see that this installation is capable of being divided into three distinct plans. It is proposed to either sell the things as an entirety or to sell it in sections of 1, 2, or 3. We also send you some prints of a small chandelier which I have in the exhibition room at the Viaduct & which is a model of the large chandelier at the Crystal Palace. If you would like to have some of these for distribution I can send you[d] any number. Say, White & I have embarked in a private enterprise by permission of the company, which is to have a lithographic picture of the Entertainment Court showing the entire interior with the arrangement of your various exhibits, the occasion being the visit of the Duke of Edinburgh & his party but the artists will take the usual liberties with such affairs & picture numerous other distinguished visitors as being present. Only those however who have actually visited us officially will be included. The idea is to make a picture which will be of value as a memento of the Crystal Palace Electrical[b] Exhibition and at the same time a picture of what is universally conceded to be the display par eccellence of the exhibition. And again to make a picture which will be of value simply as a picture from an artistic point of view— It will cost a very considerable sum to do this and White & I have determined to do this [on?][f] on[b] our own account relying upon the sale of copies to reimburse for our outlay. I would therefore like the various Electric Light Interests in America to subscribe for a certain number. It will be a very handsome thing & well worth framing & if a considerable number are taken can be sold for a dollar each. We are going to retail them here for two & a half dollars. Will you have Insull feel the pulse of the different concerns & see how many you can order. Strain yourself a little in the matter for I know you will be pleased with it.

We open up the Viaduct formally on Thursday next to the

people, from & after which time it will continue to run until it breaks down which let us hope will be at a very distant date. I have every reason to believe that no mishap can occur. You will have observed that the universal comment on our work is that it is thoroughly well done. We have of course had to deal with unskilled workmen & it has cost a great deal of time money & [energy?]ⁱ to get the work done in a satisfactory manner. But I think it is now in such shape that no mishap can occur. The Crystal Palace exhibit runs like clock work & I see no reason why the larger & more important one on the Viaduct should not do the same. At all events I am preparing to treat it as though it would. Dr. Hopkinson takes hold actively on Wednesday next, Dr Fleming tomorrow. Hammer as you will presently see has already has already at his finger ends so that I calculate by the end of the week that is^b after running Tuesday, Wednesday, Thursday, Friday & Saturday I may temporarily leave it without the fear haunting me that something may go wrong resulting in a failure during my absence. Of course this would not be pleasant even though I might be able to satisfy myself & others that it would not have been otherwise had I been present, but there is only one way by which I can convince others that the thing does not need the personal supervision of the principal & that is by running away from it. I therefore propose to go to Paris & spend two or three days with Batchelor at the end of this week, always providing of course that nothing goes wrong meantime.

Now as to the question of my finally leaving this work. You will have You will have^j learned from my letter to Mr. Fabbri what my views are as to my having accomplished the work for which I came here. I now wish to satisfy you as to the capabilities of my succcssors. In the first place, forseeing that you would not be happy if I left the work in the hands of men of little experience I determined long ago that the principal authority at all events should be one known to you as well as to general repute. I think you will admit that securing Dr. John Hopkinson F.R.S. that I have secured an electrician fully capable of understanding & appreciating the theory & practice of your system & distinguishing between its efficiency scientifically & economically & that of others. Hopkinson is not only an electrician but is an engineer & I was never more convinced of this than I was last Wednesday night when he spent the evening with me at the ~~v~~Viaduct & had several lengthy interviews & talks with ~~h~~Hood over the Porter engine. I knew by the manner in which Hood received him & talked to him that

Hood recognized him a practical engineer. Besides all this Sir Frederick Bramwell, Preece, Shelford Bidwell, & all parties conscerned pronounced him to be a combination of the theory & practice of Electrical Engineering. His exact position with the Company is this. While he does not undertake to give his entire time to the Company he does[d] undertake to be actively their electrician, to report on all their operations to the board once a month & to at any and all times hold himself ready to give them advice on any matter that may be submitted to him. And further to keep himself so thoroughly informed in respect to the work being done by the company as to take the initiative in any recommendation that his judgment might suggest and not wait to be called upon by the Company for the same. The consideration he is to receive is rather peculiar. He is to have a salary of not less than £500 nor more than £2,000. Precisely what it will be is to be determined in this way—he is to have 2% per cent of the profits which are available as dividends up to the maximum of £2,000 beyond which his 2% ceases.[k] The idea is to interest him directly in the success of the company. I have met him several times since I have been in England and my own opinion is that out of the United States where you hold supreme sway in electrical invention the Company will have a superior chief. Doctor Fleming is a man about whom I know less. That he possesses the knowledge both scientific and practical for the proper performance of his work I have satisfied myself. That he will display the requisite energy and interest in the business time only can tell. Those who know him say he will but of this I can offer no opinion myself. He is very much interested in it and in order to test him to a certain extent I asked him to write me a little treatise on the Edison system descriptive of its merits and so forth   He did so and I shall send you a copy of it.[23] I think you will admit that at all events he understands the system and has the power of setting it forth. The arrangement with him is of course for his entire time; he will receive £800 ~~pera~~ year.

And now we come to the important ~~thing~~ work[b] of putting all these various things into practical operation of making the installations and of directly supervising men and other wise personally conducting the practical work. In so far as the parent company will do this it will at least for the present be almost wholly in the hands of Hammer. You will remember perhaps that at Menlo Park he displayed a peculiar talent for doing his work in a methodical manner and considerable ingenuity and talent for devising original ways of doing it. This trait has been of greater value to me here than any other one thing that I could name with possibly the exception of Drexel Morgan & Cos money. You remember I brought Hammer with me for the purpose of making my measurements and attending to the regulating and photometric devices &c. and that I expected to have availed myself of Rose[24] for doing the practical work. Well, when I arrived I found that Rose had made a new contract with Gouraud and that he was not available and by the time that I discovered this I also discovered that the work here for a time was going to be comparatively slow and tedious and that I need not be in a hurry to send back to you for a proper assistant. In the process of doing the preliminary work I made the further discovery and one that I have never ceased to congratulate myself on, that in Mr. Hammer I had all the assistance I required. He not only understood his department but apparently had a thorough comprehension of all the departments. I therefore gradually entrusted everything to him and however the work which has been done here either at Crystal Palace or Holborn Viaduct may be adjudged to be good or bad I have this to say that Mr. Hammer shares with me equally at least in the result. His methodical ways have resulted in the[d] orderliness of the work and therefore in its completeness and therefore in its reliability which I never by[d] any possibility could have obtained from Rose. Beside and beyond all this is another and still more important fact namely that his energy is something prodigious; he works from early morning and that means early morning in the United States—not early morning in England until all hours of the night and the small hours in the morning when necessary without ever flagging. You can well imagine the amount of actual work that has had to be done in the putting down of over 1000 lamps at Crystal Palace (see your plans) and over 1000 lamps on the Viaduct and doing all this with men not at all familiar with this work or with any work in fact for they seem rather familiar with the ways and means of not doing ~~the work~~ a thing[l] than with those of doing it. The net result has been that he has acquired[d] a familiarity with every detail of the system whether the Central Station or Isolated so complete that if you were to ask me to nominate a man to go to some distant country and duplicate what has been done in England I should say of him as you said of me that there was only one man in my ma~~i~~nd for the work and that man was Hammer. Therefore[m] I think that I have been fully justified in paying him in accordance with what he has done & of recommending him for an important[d] official position to the

Company [-]^f at a still further increased compensation that you need have no doubt whatever [-]^f but^b that in his hands in conjunction with Drs Hopkinson & Fleming the work of installing the Edison Electric Light System in England will make more rapid progress and will be as well done as in any other part of the world not excepting the United States.

And now in regard to your humble servant, I have been made fabulous offers to remain in this country indefinitely. I have declined them. I have not been under pay from any source since the 1st March   I am remaining now only to see the Viaduct practically running for a month & to thoroughly & satisfactorily impart my views of the entire thing to Drs Hopkinson & Fleming. I think & say this honestly that if I leave here between the 15th May & the 1st June you will have to admit that I am not leaving before my work has been accomplished. I came here to do a specific thing. I have done that thing & a vast deal more. You will be wholly unable to find either in England or any other country a man who will even so much as criticise adversely a single^d thing I have done. Possibly I have not got as much cash out of this thing as I would have liked to have received. Possibly again I may have erred in throwing the thing into the hands of the present people rather than into the hands of Gouraud & his tribe, but the unanimous verdict here is that in respect to the first I have done wisely & that you will reap an immediate & larger return by virtue of the liberal character of your contract than you possibly could had you made it more onerous & received a larger sum down. And in respect to the second that it will not bear discussion. The character of the present management their standing in the community their energy and their fearlessness of action are all such as will command a position & respect that could not by any possibility have been accorded to any company which Gouraud might have organised. Besides this he never^n could have got such men as Hopkinson, Bramwell, Sir John Lubbock and Shelford Bidwell to have joined any of his enterprises. You have this to look to that not only have you these men interested with you in this company but you have their interests secured for your future enterprises and that means not only their personal interests but it means all that interest which they can command   Bramwell and Hopkinson are both intimate and personal friends of Siemens as well as of many other distinguished scientists and if ever there comes the tug of war in which you want to enlist a strong array of scientific talent on your side you have the pick of the country. Now I have said to you some time since (or rather to Insull) that when I was able to say that my work was done and that I proposed to return I would do so with your sanction. This I as confidently expect to receive promptly on the receipt of this letter as I would expect you to loan me a $\underline{V}^{25}$ if I was hard up. It is true I have not been able to add much to my banking account but what I have said to you in respect to your shares applies with equal force to mine. I consider therefore that ~~having~~ I have[o] done well in this respect   I have had another object in view all through this work and that was to enlist the interests directly and indirectly of men of such attainments and such character as would make it unnecessary for you to send me as a special envoy to this country with subsequent inventions. I therefore am returning to America with the fixed determination of never under any circumstances leaving it again except possibly on a flying trip. If any man were to say to me at any time in the future—"Go to England—Go to France—or go anywhere out of America for one year for us and here is half a million dollars placed to your account I would flatly decline the offer." Life is to short and riches are too little appreciated by me to tempt me to another prolonged absence from all that is enjoyable in it. Now my dear fellow say in reply to all this "It is all right, well done thou good and faithful servant enter thou into the joy of thy[d] Lord"   I am, my dear Edison, Very sincerely yours,

Edwd H Johnson

LS, DSI-MAH, WJH, Ser. 1, Box 20, Folder 2. Letterhead of Edison Electric Light System, Edward Johnson manager. [a]"London, E.C." and "188" preprinted. [b]Interlined above. [c]Repeated at end of one page and beginning of next. [d]Obscured overwritten text. [e]Written in left margin by Johnson. [f]Canceled. [g]"do it" interlined above by Johnson. [h]Text from "Engines" written in a different hand. [i]Illegible. [j]"You will have" repeated at end of one page and beginning of the next. [k]Text from "his 2% ceases" written in same hand as first portion of document. [l]"a thing" interlined above. [m]Text from "Therefore" written in same hand as second portion of document. [n]Text from "this he never" written in same hand as first portion of document. [o]"I have" interlined above by Johnson.

1. The letter to Egisto Fabbri has not been found.

2. Doc. 2239.

3. Johnson married the former Margaret Kenney of Philadelphia in 1873. *NCAB* 33:475–76.

4. See Docs. 2228 and 2239.

5. John Ambrose Fleming (1849–1945, knighted 1929) studied under Maxwell at Cambridge and, in 1881, became professor of physics and mathematics at the new University College at Nottingham. He left this position in 1882 at the request of Arnold White, his cousin, to join the Edison Electric Light Co. as a consulting electrician (after William Preece

declined the position). In 1885 he became a professor of electrical technology at University College, London, where he developed a "thermionic valve," the first vacuum tube, based on the flow of charged particles identified as the "Edison Effect." *DSB*, s.v. "Fleming, John Ambrose"; *Oxford DNB*, s.v. "Fleming, Sir (John) Ambrose"; Johnson to TAE, 19 Mar. 1882, Ser. 1, Box 20, Folder 2, WJH.

6. James Holloway (1839–1924), an English machinist, came to the United States in 1863 and began working for Edison at the Menlo Park laboratory machine shop in 1879. "Holloway, James," Pioneers Bio.; *TAEB* 5, App. 2.

7. Insull wrote about having found Edison and Johnson listed as directors of the Edison Foreign Electric Light & Motive Power Co., which he supposed was "one of G.E.G's creation[s] about which we have never heard of." He asked Johnson to determine if other companies included Edison among their directors because "it is not right at all that G.E.G. should be allowed to act in such a manner as if at any time any of his enterprises should collapse it would besmear Edison with mud." Edison wrote directly to George Gouraud about this company and subsequently asked to have his name removed from several other enterprises. Insull to Johnson, 24 Mar. 1882; TAE to Gouraud, 23 Mar. and 1 May 1882; Lbk. 11:486, 480; 12:204 (*TAED* LB011486, LB011480, LB012204; *TAEM* 81:462, 458, 607); Gouraud to TAE, 19 Apr. 1882, DF (*TAED* D8240J; *TAEM* 63:17).

8. See Doc. 2229.

9. After several cable messages to and from Edison, Johnson advised on 11 April that the shares had been sold. Edison received about $14,500 through Drexel, Morgan & Co. on 21 April. Johnson to TAE, 11 Apr. 1882, LM 1:189A (*TAED* LM001189A; *TAEM* 83:966); TAE to Gouraud, 21 Apr. 1882, Lbk. 12:139 (*TAED* LB012139; *TAEM* 81:577); Statements of Gouraud account, DF (*TAED* D8105ZZA [images 8–9]; *TAEM* 57:345).

10. Edison exchanged a number of cables in the next two weeks with Johnson and Gouraud about the proceeds from the Oriental Telephone Co. (see LM 1 [*TAED* LM001; *TAEM* 83:872]). Gouraud released Edison's money, amounting to £4,500 less portions to Johnson and Charles Batchelor, on 24 April. In a letter written the previous day Johnson explicated this complex transaction and the sale of United Telephone shares discussed above (Gouraud to TAE, 24 Apr. 1882; Johnson to TAE, 23 Apr. 1882; both DF [*TAED* D8256U, D8204ZAY; *TAEM* 63:872, 60:118]).

11. The cable has not been found.

12. Not found.

13. Frank Julian Sprague (1857–1934) studied engineering and electricity at the U.S. Naval Academy. He graduated in 1878 and apparently visited Edison at Menlo Park that summer. After two years at sea, he was assigned to the European Squadron and passed through London, where he secured the position of secretary to the awards jury of the Crystal Palace exhibition. Sprague joined Edison's central station planning department in 1883 and later the Construction Department, where he played an important role in setting up three-wire central station systems. He subsequently demonstrated the feasibility of electric streetcars. *ANB*, s.v. "Sprague, Frank Julian"; Passer 1952, 212–29.

14. Not found.

15. Uriah Painter was a Philadelphia reporter and lobbyist in Washington, where he enjoyed close relations with prominent congressional Republicans. Painter had previously been involved with Edison and Johnson in running the Edison Speaking Phonograph Co. See Docs. 672 n. 2, 1190, and 1583.

16. Sir Frederick Bramwell (1818–1903) began a distinguished engineering career in railroad and locomotive construction. About 1853 he began consulting on technical and related legal issues. He later became an authority on municipal engineering, especially water and sewer systems, and was knighted in 1881 (*Oxford DNB*, s.v. "Bramwell, Sir Frederick Joseph"). Bramwell's 12 June 1882 report on Edison's British Patent 4,576 of 1879 (Cat. 1321, Batchelor [*TAED* MBP019; *TAEM* 92:134]) is in DF (*TAED* D8239ZCF; *TAEM* 62:955).

17. Johnson likely referred to the British Electric Light Co.; the editors have not identified the subsidiary company or its relations with the Lane-Fox interests. An Edison company for Manchester was also formed about this time. See Doc. 2248 n. 4; memorandum of association, c. May 1882, DF (*TAED* D8239ZBZ; *TAEM* 62:939).

18. Not found.

19. Thomas J. Handford became Edison's British patent agent in March. No other biographical information has been found. Handford to Theodore Waterhouse, 3 Mar. 1882, DF (*TAED* D8248Y; *TAEM* 63:432).

20. Edison later explained to Brewer & Jensen, his former British agents, that on the advice of his English associates he had elected "to place the matter in the hands of one person directly under my supervision, instead of having it go through the hands of two parties as it did before at this point." TAE to Brewer & Jensen, 2 May 1882, DF (*TAED* D8248ZAN; *TAEM* 63:462).

The Electrician's *map of the Holborn installation; the colors have been reversed for clarity. Dashed lines appear to represent the conductors, and vertical slashes the street lights. Small dots represent the number of lamps within each establishment.*

21. The drawings and photographs described below have not been found. The *Electrician* published in its 22 April issue a map of the Holborn Viaduct installation showing the locations of conductors and the number of lamps in each establishment ("The Edison Light at Holborn," *Electrician* 8 [1882]: 368). A less detailed map drawn for the Edison Electric Light Co., Ltd., probably by William Hammer, is in Friedel and Israel 1986 (217).

22. A photograph of Edison's installation at the Crystal Palace is found in Friedel and Israel 1986 (216).

23. Not found.

24. In 1879 and 1880 Allen W. Rose was an assistant to Johnson, who was then chief engineer for the Edison Telephone Co. of London. Rose

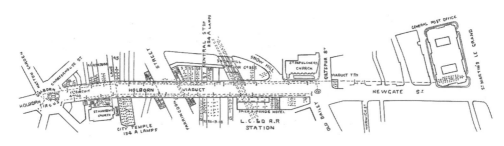

was evidently back in New York by the summer of 1880. He later returned to London in the employ of the Consolidated Telephone Construction & Maintenance Co., but left the company in March 1884 after a dispute with George Gouraud. Doc. 1808 n. 2; Bergmann's testimony, 33, *Edison v. Lane v. Gray et. al.*, Lit. (*TAED* W100DGA:17); Rose to Insull, 31 Mar. 1884, DF (*TAED* D8465U; *TAEM* 76:605).

25. Colloquial term for a five-dollar note. *OED*, s.v. "V," 4b.

-2259-

*Samuel Insull to
Charles Batchelor*

New York, 10th Apl *1882*[a]

My Dear Batchelor,

I have your favor of 27th March.[1] I certainly never told Dyer anything of the kind as I should have no reason whatever for doing so. He must have been mistaken that is all I can say.

I am sorry you should say "I know absolutely nothing of my own affairs over there since I left." I used to write you quite regularly but I seldom or ever got any reply and consequently left off doing so thinking you did not care to hear from me. However if you care for me to write you I shall be only too happy to do so. And in order to relieve myself of any blame (real or imaginary) I will proceed to give you a full account of your private affairs as I can off hand from memory & some little data before me.

Accounts. This is the one point that I feel I am to blame. I should have sent you forward a statement but you have never asked me for it. I promise you however it shall go within a week. Meantime I will tell you roughly how you stand.

The Balance to your credit when you left (including amounts still to be received by T.A.E. on sale European Bonds)[2] was

	$9997.04
Since then there has accrued to you as due on G[old]. & S[tock]. Telephone Royalty	850.
5% of a further $10 000 received on English Telephone	500
Sundry Bills you have paid	72.11
	$[11,419.15][b]
Against this we have disbursed	
Lamp Co Assessments	4000
Edison Machine Works	4750
Insurance (Carman)	20
Sundries	8.89
	8778.89
Balance in hand about	$640.26

Of course this is very rough as I tell you I will send you a statement next week. Then we have to pay the Lamp Co another $1000

on your assessments

the Machine Works [---]^b d[itt]o about   1700
& 50% on 7% shares new Issue Edison Electric Light Co
Stock   3650

                                                    $6350

Thus you will see with a balance in hand of about $650. we have to pay on your account about $[6350].^{b} Now as to how to do this. In a few days Edison will get about $44,000 on English Light. You will be entitled to about $4400. This will enable me to pay for your Parent Stock & Lamp Co assessment & then very shortly Australia India [Sweden?]^b & Portugal will doubtless be sold & that will enable us to pay up everything & remit you besides. As you probably know Gouraud made one arrangement as to Australia but has since backed out of it.

Now as to the future. The Machine Works will not if properly conducted require any more capital— & if they should your share cannot come over $1000 or $2000 in addition to the $1700 above referred to. They have been running with very little support from T.A.E. all this year. Their debts are not over $30 000. This is gross. [Outstanding?]^b bills [needing to be paid?]^b amounts to probably [$---- with?]^b liabilities^c My only fear is that Edison may order things in advance of requirements & if so that will put him in the same hole financially as he has just got out of as in such a case the Machine Works will [weigh?]^b on him heavily. However I hope that such a state of affairs will not occur again. The Machine Works should make in my opinion $50 000 at the least this year.^{3} ~~In my opinion it is~~ It is said to be^d a very expensively conducted place   I am told Dean pays high wages to poor workmen & that the shop is badly managed. I do not of course profess to be much of a judge of such things. They certainly turn out a good deal of work & the Central Station Dynamo work they are now engaged on seems to be very fine indeed. There are a great many fine tools laying idle which is owing in my opinion to the very reckless way in which Dean bought machinery. If I were a mechanic I could probably tell you more about such things.

As to the Lamp Co their bills payable over bills receiveable [amount]^b to about $10,000. The assessment just made will provide for payment of this. There has been paid on the Peters property up to the present time about^d $[11,000?]^b and there is still about $[12 000?]^{tb} to be paid. This is not included in the

$10 000 above referred to. If the orders come in, in the manner in which they have during the last few months I do not think the Lamp Co should require more than $40,000 or $50,000 & this should enable them to pay off the balance on the Peters property & increase their capacity. (Of course you know they are now moving to Newark) The question of the money they will require depends entirely on what demand there is for Lamps. Their expenditure [is abo?][b] at Menlo Park was about $1800 a week say $90 000 a year. Presuming they are the same at Newark (I am inclined to think that for some time they will be much higher than at Menlo) & that they sell about 2000 lamps a week they will require at least $40,000 to meet their obligations & current bills for the remainder of 82. However this is a bad view. It is to be expected that the Demand for lamps will greatly increase & should before long amount to 6000 a week regularly.

Isolated Co. As to your stock I told you that yours, your wifes, & Edisons was taken by the Machine Works & is being paid for by them in machines[5] So of course 10% of it is yours. This Company is doing a good business & should make money. Its stock is not dealt in. I do not think 20 or 30 shares have passed hands.

Parent Co. I also wrote you as to this As I have stated above your Stock will be taken up as soon as Edison has some money. The stock is quoted at 400 to 425 with quite a strong market.

Illuminating Co. There is still about 2 or 2½ miles of tubes to be laid. Kruesi will probably finish[e] in about ~~two~~ six weeks. The Dynamos will be finished in about two months. So in about three months you may begin to expect to hear something of the lighting up. Stock nothing at all doing

European no sales at all. Illuminating at 50 to 55. But I do not think much could be got at that. Ore Milling nothing at all doing. No bidders. No sellers.

Now do not think for a moment that I do not want to keep you posted or that I am too lazy to do so. It takes me from the time I get up till I go to bed to attend to my business & if you do not get what you want just write me & tell me so straight & I will try & give it attention. Only just remember that a letter in reply now & then is somewhat encouraging.

As far as the purely official part of your business is concerned I do not think you can complain of inattention. If you get somewhat ruffled at delays now & then remember that we at "65" are not omnipotent & that we have to depend on others to fill our orders & they often delay a good deal. All we have to get for you now is wire, drawings, special machinery, & Drawings of[f] various ~~sizes~~ Dynamos. The armature wire shipments commence this week. The special tools & machinery Drawings I am promised within a week. As to Dynamo drawings. I am now having the fourth set prepared of Central Station Dynamo Drawings   Each previous set have been discounted in consequence of alterations. There have been no drawings yet made for 250 light machines. Edison simply took three fields six cores & three keepers out of Z Dynamo stock made an armature three times length of Z armature. Of course he used a heavier shaft & larger bed plate than with Zs. As soon as the first test machine is made I will get Edison to give me a letter to you on the subject.[6] As to the 125 light machine nothing much has been done with it yet.[7]

Early next month Edison will go back to Menlo Park to live & to experiment   Dick Dyer myself & Hughes will be there with him. We will keep an office here for some time longer probably till the Fall. I expect to come into the City three or four days a week.

Hughes has the rail road running very well indeed. He has over [---][b] two miles of track finished now & runs over it with four or five tons of rail road material at 30 miles an hour with ease. We are having a Freight Locomotive built & 10 cars of four tons each for freight. In about a month everything will be running and the economy test be made. Yesterday I ordered from Brill of Philadelphia[8] a car about 20 ft long after the style of a Broadway Car only longer. Edison is going to put a motor underneath the car & by the use of a fly wheel to store power and a Masons Clutch[9] to throw the armature in and out of gear hopes to run with a small motor consume little power & overcome difficulties in connection with starting. I will send photo of car when it is finished. With the present Locomotive a Masons Clutch is used. They start the armature running & when they have got up good speed start throw the Clutch in and so start off. Edison considers this Clutch essential. It is covered by his first patent.[10] Yours very truly

Saml Insull

ALS (letterpress copy), NjWOE, Lbk. 12:62 (*TAED* LB012062; *TAEM* 81:534). Stamped Thomas A. Edison to Charles Batchelor. [a]"*New York*," and "*188*" from handstamp. [b]Illegible. [c]"This is gross . . . liabilities" written in left margin. [d]"It is said to be" interlined above. [e]Obscured overwritten text. [f]"Drawings of" interlined above.

1. The entirety of Batchelor's letter was the complaint that "Dyer informs me that you have ordered him to stop sending me reports as you are sending me them yourself— I do not like being cut off entirely in this

way. You never send me a report and I know absolutely nothing of my own affairs over there, since I left." He presumably referred to bookkeeper Philip Dyer. DF (*TAED* D8238ZAM; *TAEM* 62:357).

2. In December, Batchelor was credited with $1,500 from the sale of European bonds. These probably were debenture bonds of the Edison Electric Light Co. of Europe (Edison Private Ledger #2:28, 108, Accts. [*TAED* AB006:33, 98; *TAEM* 88:196, 261]; Edison Electric Light Co. of Europe, undated list of "Subscriptions to Debenture Bonds received," DF [*TAED* D8127ZAC; *TAEM* 58:194]). Batchelor's personal account book from this period is Cat. 1318, Batchelor (*TAED* MBA001; *TAEM* 93:855).

3. Charles Rocap had recently sent Batchelor a copy of a financial statement in which the Machine Works reported a profit for 1881 of $10,853.36. Rocap to Batchelor, 7 Mar. 1882, enclosing Machine Works statement, 1 Jan. 1882, Unbound Documents (1882), Batchelor (*TAED* MB064, MB064A; *TAEM* 92:422–23).

4. Edison made two payments, in May and November, totaling $11,660. Edison Private Ledger #2:4, Accts. (*TAED* AB006:16; *TAEM* 88:179).

5. See Doc. 2215.

6. The K dynamo (see Doc. 2269 and App. 3). In a letter to Batchelor the same day, Edison stated that "no drawings have ever been made for 250 light Dynamo that is except rough shop sketches. A standard machine is now being tested. As soon as this is done I will have a set of drawings made for you" (Lbk. 12:105 [*TAED* LB012105; *TAEM* 81:557]). The machine was not tested until mid-May (see Doc. 2279).

7. Francis Upton had written to Edison in January about a 150 light dynamo he was designing. Edison replied that he was designing a 125 light machine "with longer Armatures and one field magnet to do the business." This was designated the L dynamo; by the time the machine was tested in July it was rated for 150 lights. Upton to TAE, 4 Jan. 1882, DF (*TAED* D8220C; *TAEM* 60:752); Insull to Upton, 9 Jan. 1882, Lbk. 11:52 (*TAED* LB011052; *TAEM* 81:257); Sherburne Eaton to TAE, 10 June 1882; Charles Clarke to TAE, 15 July 1882; William Andrews to TAE, enclosing test results, 18 July 1882; all DF (*TAED* D8226U, D8227ZAC, D8235R; *TAEM* 61:254, 504; 62:51); see also Doc. 2322 and App. 3.

8. John G. Brill and his son George founded the J. G. Brill Co. (later J. G. Brill & Co.) in Philadelphia in 1868; it built railway cars, especially streetcars. The Brill company expected $850 cash upon delivery in early June and dunned Edison for payment throughout that month and July (Brill 2001; TAE to Brill & Co., 8 Apr. 1882, Lbk. 12:53 [*TAED* LB012053A; *TAEM* 81:529]; Brill & Co. to TAE, 6 and 10 Apr. 1882, Cat. 2174, Scraps. [*TAED* SB012BBP, SB012BAF; *TAEM* 89:477, 409]; Frederick Scheffler to TAE, 5 June 1882; Brill & Co. to TAE, 30 June, 21 July, and 27 July 1882; all DF *TAED* D8249U, D8249ZAI, D8249ZAM, D8249ZAN; *TAEM* 63:603, 613, 617, 618]). The motor that Hughes designed for the car in April was later built at the Edison Machine Works (Samuel Insull to Hughes, 27 Apr. 1882, Lbk. 12:177 [*TAED* LB012177; *TAEM* 81:596]).

9. See Doc. 2227 n. 7.

10. Edison's British patent 3,894 (1880), the provisional specification of which is Doc. 1987. Cat. 1321, Batchelor (*TAED* MBP029; *TAEM* 92:194).

-2260-

To Francis Upton

[New York,] 11th April [188]2

Dear Sir,

I think the best arrangement at Newark[1] would be to appoint Holzer as Superintendent of Manufacturing while Bradley should retain his present charge of the Fibre [~~dept~~Cutting?][a] and in addition have charge of all repairs & motive Power[2]

I am very anxious that everything should run smoothly and believe that Holzers interest with us is the best guarantee against their being any friction him and Bradley   Yours very truly

T A Edison

LS (letterpress copy), NjWOE, Lbk. 12:82A (*TAED* LB012082A; *TAEM* 81:547). Written by Samuel Insull. [a]Illegible.

1. On 7 April Upton wrote from the new lamp factory noting that he was "now at Harrison the greater portion of my time and everything is going on smoothly." Harrison was across the river from Newark and the Lamp Company used both Harrison and East Newark on its letterhead. Upton to TAE, 7 Apr. 1882, DF (*TAED* D8230ZAE; *TAEM* 61:744); TAE to Upton, 6 Apr. 1882, Lbk. 12:48 (*TAED* LB012048; *TAED* 81:525); see Doc. 2312.

2. In his 7 April letter (see note 1) Upton had suggested that "Holzer be made superintendant of the manufacturing of lamps and to take fibres from Bradley and turn lamps in for testing. Bradley to have charge of fibre room and of all repairs or improvements and of motive power. I will take the office and as connected with the office the testing and packing of lamps and the supply department. As now I shall reserve the right of directing the policy of the manufacture and the deciding whether repairs or changes in manufacture are needed." Upton considered Holzer too inexperienced to run the entire factory, but thought him "a most excellent man to get along with help to instruct them, cheer them up and keep them in order." When the Edison Electric Lamp Co. started production at the Harrison factory on 31 May, Upton was the treasurer, Holzer the superintendent, and Bradley the master mechanic. TAE to Edward Johnson, 1 June 1882, Lbk. 7:396 (*TAED* LB007396; *TAEM* 80:578); "New Factory of the Edison Electric Lamp Company," *Operator* 13 (1882): 321.

-2261-

From Sherburne Eaton

[New York,] April 12th. 1882.

Mr Edison,

May I beg of you to let me know when you receive important news from any part of the world of general interest.

Twice I have heard[a] important news down town on the street before I heard it in my own office

To day I learn in Drexel, Morgan & Co's that cables had been received recently from Mr Johnson stating that he was about to start his plant on the Holborn Viaduct. We telephoned from Mr Fabbri's[a] office but could not get connection. I presume no such cables[a] were received, else I would have been favored with a copy of them.[1] But my feeling at the time I heard of it was that even if such cables[a] had come I was by no means certain to know of it.

I would esteem it a great favor if you would let me know when important news of general interest is received.

S. B. Eaton   per Mc.G[owan].[2]

TL, NjWOE, DF (*TAED* D8224R; *TAEM* 61:24). Typed by Frank McGowan. [a]Mistyped.

1. Johnson had cabled Edison to this effect on 8 and 10 April (LM 1:186A, 187B [*TAED* LM001186A, LM001187B; *TAEM* 83:965]). After learning of a congratulatory cable on the opening of the Holborn station, Eaton asked Edison if he could see it (Johnson, John Hopkinson, and John Fleming to TAE, 12 Apr. 1882, LM 1:189B [*TAED* LM001189B; *TAEM* 83:966]; Eaton to TAE, 13 Apr. 1882, DF [*TAED* D8224S; *TAEM* 61:25]).

2. Frank McGowan (1849?–1890?) was Eaton's stenographer. He remained connected with the Edison lighting business and in 1887 embarked on an expedition to Brazil in search of fibers for lamp filaments. McGowan to Charles Curtis, 3 Mar. 1882, Cat. 2174, Scraps. (*TAED* SB012ACJ; *TAEM* 89:392); Jehl 1937–41, 935; Dyer and Martin 1910, 1:303; "Shot and Robbed," *Los Angeles Times*, 2 Mar. 1890, 8; "Tinged with Romance," *New York Times*, 2 Mar. 1890, 1.

–2262–

Draft Patent Application: Electrochemistry

[New York,] April 12 '82.[a]

Invention consists in effecting the economical decomposition of various substance electrolytically by submitting such substance to electrolysis when under high temperature[b] pressure.—[1c]

~~I have found that many substances~~

Heat tends to diminish the amount of electric energy required to effect the decomposition, and as This heat is obtained directly [~~separ?~~][d] from the combustion of fuel acting on the substance the process requires but a small amount of electric energy[b] [~~still altho much at~~ -----?][d] and this amount is still further diminished by the fact that nearly all compounds diminish the resistance by heat. [------][d]   The more particular object of this invention is to act on substances which do not conduct e at all at ordinary temperatures [~~but?~~][d] just conduct sufficiently at very high temperatures to permit of decomposition & as many of these substances are liquid at ordinary temperatures the receptable for effecting the decomposition must be such as to permit of high pressure[2e]

Claim ~~A cell for electrolytic decomposition~~[c]
~~The method~~[e]

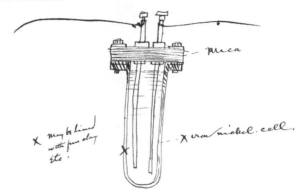

Carbon, or metallic Electrodes. Cell made to withstand several hundred lbs pressure to sqr inch[3]

Witness   R. N. Dyer.

ADf, NjWOE, Lab., Cat. 1148 (*TAED* NM017:106; *TAEM* 44:466). [a]Date written by Richard Dyer. [b]Obscured overwritten text. [c]Followed by dividing mark. [d]Canceled. [e]Followed by several blank lines.

1. It is not clear what prompted Edison's research into this subject. He executed a patent application on 19 May based on this draft pertaining particularly to "those substances which do not conduct electricity at all at ordinary temperatures or below the boiling-point. This is accomplished by subjecting the substances to the action of an electric current when under high temperature and pressure." The vessel described would be heated externally in a furnace so that "the substance to be decomposed is vaporized. The escape of the vapor being prevented, a high pressure is obtained within the vessel, and the heat of the substance can be increased accordingly. When the proper high heat and pressure have been obtained, an electric current is passed through the substance and the electrical decomposition takes place." The patent issued in 1892 as U.S. Patent 466,460.

2. In his application, Edison expanded this paragraph to explain that high pressure and the consequent lowering of resistance were essential for the economical operation of this process. When required by the Patent Office to provide an example, Edison added a paragraph explaining

the production of the metal aluminium from its chloride. The chloride of aluminium being a salt which volatilizes below the temperature at which it can be successfully decomposed electrolytically, the process permits of the increase in the temperature of the chloride to that point at which electrolytic decomposition readily takes place. The pressure exerted by the evolution of chlorine within the vessel under the influence of heat prevents the chloride from volatilizing, while it is maintained in a fused condition and at a much higher temperature than can be attained without pressure. The higher the temperature of the fused chloride, the more readily it is decomposed by the electric current. [Pat. App. 466,460]

Edison had briefly investigated aluminum smelting by electricity in February 1880; Otto Moses also did so in June of that year at Menlo Park (see Docs. 1902, esp. n. 5, and 1944).

3. Edison's sketch of the vessel is similar to that shown in the patent, where it was represented inside the furnace.

–2263–

Samuel Insull to
Edward Johnson

New York, 19th Apl *1882*[a]

My Dear Johnson,

By this same mail I send you two unmounted photographs one of the old locomotive & car and the other of our present locomotive & car.[1]

The later photograph was taken yesterday together with a number of other photographs of the road. But today there has been no sun & it has been raining all day & instead of having two complete sets of photographs one set mounted to be sent by express to you and the other set unmounted to be mailed to you I have only this solitary photograph to send you with the photo of the old road which was taken long ago.

Now there is another mail going tomorrow at 1:30 (S.S. [------][b] Cambria[c] via Plymouth) & I may have some more ready then to send you. On Saturdays steamer (S. S. City of Rome) I shall send you two sets if the weather favors us—one by mail unmounted—one by express mounted which I hope you will get in time.[2] Anyway you will see we are doing our best to get the photographs to you by the time required.

Now a few lines as to the road. Hughes has finished all but about 1000 ft of it & when this is done it will be 2¾ miles long—including three sidings & two turn tables. The grade is about 25 ft to the mile.[3] We find that we can attain any speed required. We have run it 35 miles an hour with a load of 39 piled on the car and engine & Hughes frequently takes a load of about six tons of material for use in construction at the same rate. The Locomotive weighs about four tons & I think Edison said it will develope 25 horse power. You will notice on the photograph that the Insulated Hub of the wheel comes out well & from this you can explain some of the connections. The [axle?][d] of two of the wheels forms one pole & the two other wheels and the other base forms the other pole   We are now having built a freight Locomotive and ten freight cars each one to carry four tons. As soon as these are ready the economy test will be made.

Edison is sick in bed with a terrible cold—too sick to attend to any kind of business—or else I would get him to give me a letter to you on this subject   He has been in bed for over a week & now Mrs E refuses to let anyone talk business to him so I have to run things as much on my own responsibility as when he was in Florida. Doctor says he will be all right again in a few days.[4]

You ask when will the First District be started. The Seventh Bulletin sent you by this mail will give you some information.[5] I think the station will be started inside of three months certainly not before

It has just struck me that it would be a good thing to send you photographs [used?][d] in the Siemens–Field–Edison Electric Railroad Interference. These may be useful also Clarke pamphlet.[6]

Cannot write more now or will miss mail   Yours as ever

Saml Insull

ALS (letterpress copy), NjWOE, Lbk. 12:95 (*TAED* LB012095; *TAEM* 81:552). Stamped Thomas A. Edison to Edward H. Johnson. [a]"*New York*," and "*188*" from handstamp. [b]Canceled. [c]Interlined above. [d]Illegible.

1. Edison's 1880 locomotive is illustrated in Doc. 1939 n. 1; one of the new 1882 locomotives is shown in Doc. 2237 n. 5. Johnson had requested them, perhaps as early as January or February, for a 3 May lecture at the Crystal Palace (Insull to Johnson, 20 Feb. 1882, LM 3:54 [*TAED* LM003054; *TAEM* 84:57]; Johnson to Edison, 11 Apr. 1882; Hopkinson, Fleming, and Johnson to TAE, 12 Apr. 1882; LM 1:189A, 189B [*TAED* LM001189A, LM001189B; *TAEM* 83:966]). Johnson's talk on Edison's light and motors was reported in London dailies ("Edison's Electric Light and Motors," 4 May 1882, *London Daily News;* reprinted in *Metropolitan* [London], 6 May; Cat. 1016:6a, 40d, Scraps. [*TAED* SM016006a, SM016040d; *TAEM* 24:82, 100]).

2. An unspecified number of photos went on 20 April aboard the *Cambria* and a complete set two days later on the *City of Rome*. TAE to Johnson, 24 Apr. 1882, LM 1:195B (*TAED* LM001195B; *TAEM* 83:969).

3. This was well within standard engineering practice and far less than the grade Edison had planned to construct in 1880. Doc. 1939 n. 3.

4. Edison apparently did not begin to resume normal activities until the end of the month. An interview published in the 1 May *St. Louis Post-Dispatch* noted that he had been "quite ill at the Everett House for a couple of weeks, but yesterday was well enough to sit up in an easy chair." "The Doom of Gas," *St. Louis Post-Dispatch*, 1 May 1882, Cat. 1327, Batchelor (*TAED* MBSB52192; *TAEM* 95:191).

5. This publication summarized the work of constructing the central station, laying conductors, and wiring buildings. It stated that "little now remains to be done, except to finish the laying of the underground conductors, before the First District will be entirely completed, and the lighting-up commenced." Edison Electric Light Co. Bulletin 7, 17 Apr. 1882, CR (*TAED* CB007; *TAEM* 96:692).

6. No copy of this pamphlet has been found. It was evidently com-

pleted by early March, when Insull planned to send a copy to a Philadelphia reporter. Insull marginalia on R. M. Hughes to TAE, 2 Mar. 1882, Cat. 2174, Scraps. (*TAED* SB012ACI; *TAEM* 89:391).

–2264–

To Edward Johnson

New York, 26th Apl *1882*[a]

My Dear Johnson

Your letter[1] in which you speak so emphatically as to your never leaving America when once you get back has scared me out of my idea of going to England. If you are so fearfully anxious to get back then I think it would be foolish for me to go over. In fact I would not give two cents to go over to England and if I can possibly get out of it by causing the unexpected to occur it will occur you may be sure.

My figures in the bar armature machine are the same as those of John Howells in his test at the Hoboken Institute[2] with this exception that I believe that all his tests were against the machine

I shall accept your recommendation to have all the Paris Exhibits other than those belonging to the Electric Light shipped to Menlo Park and you can take this as your authority to go ahead and do the same as soon as the proper time is arrived.

I am under the impression that I wrote you or else told you that you were to have 5% of the Oriental sale. If I did not I certainly intended you should have it & therefore cabled you & Gouraud to that effect.[3]

I fully understand about Hanford's Patent arrangement. I am to pay for Electric Light just as I would for entirely outside matters. Have Hanford send on his bills & I will remit for the amount.

I am glad you speak so well of Hammer. The [meagre?][b] treatment and work which the young man had to undergo at Menlo the absence of all silver spoon business and his intelligence and ambition accounts for it. I am glad that you are going to get him into a good position and hope you will get him as much salary as possible with a chance of increase.

I send you a little clipping from Sundays Times which will interest you as relating the collapse of the Bohm-Fox-Crosby electric lighting swindle[4]

With such men at the head of affairs and instructed as you say you have instructed them and with men we can send from here to aid in the construction of dynamos I do not see why you cannot come home. My thoughts do not flow freely today   I have been in bed two or three weeks and this is the second day

I have been able to get out to business & my thoughts are still very sticky so [we will?][b] save the taffy[5] till some other time.

I get great enjoyment from reading your letters so I think it would be just as well for you to stay where you are as when you get back here there will be no letters for me to read!!! Yours very truly

Thomas A Edison

LS (letterpress copy), NjWOE, Lbk. 12:170 (*TAED* LBo12170; *TAEM* 81:593). Written by Samuel Insull; stamped Thomas A. Edison to Edward H. Johnson. [a]"*New York*," and "*188*" from handstamp. [b]Illegible.

1. Doc. 2258.

2. John Howell prepared Howell 1881 as a student at the Stevens Institute in Hoboken, N.J.

3. The Oriental Telephone Co., Ltd., finally paid its patent vendors in the middle of April; Gouraud transferred Edison's portion to him on 24 April. That day Edison instructed Gouraud to "Pay Johnson five per centum proceeds Oriental then pay Batchelor one fifth of remainder," and he also cabled Johnson that he had done so. Gouraud to TAE, 15 and 24 Apr. 1882, both DF (*TAED* D8256P, D8256U; *TAEM* 63:866, 872); TAE to Gouraud, 24 Apr. 1882; TAE to Johnson, 24 Apr. 1882; Gouraud to TAE, 20 Apr. 1882; LM 1:195A, 195B, 192D (*TAED* LMoo1195A, LMoo1195B, LMoo1192D; *TAEM* 83:969, 969, 968).

4. The article described the financial collapse of the American Electric Lighting Co., of Massachusetts, allegedly resulting from the fraudulent transfer of the company's state charter from New York investors to a group in Boston ("Ruining a Boston Company," *New York Times*, 23 Apr. 1882, 9). Ludwig Böhm, a skilled glassblower who left Edison's employ on bad terms in October 1880, was the firm's electrician and superintendent of its incandescent light department (see Doc. 2000 esp. n. 3). Edison's acquaintance Edwin Fox served as secretary. Fox later testified that after an unspecified "misunderstanding" with Edison he had "joined with Elisha W. Andrews, president of the American District Telegraph Co., in the formation of a company that would be a rival to the Edison Electric Light Co." (Doc. 1668 n. 3; Fox's testimony, 12 July 1893, *Edison Electric Light Co. v. Electric Manufacturing Co.*, DF [*TAED* D9323AAF; *TAEM* 134:197]). George Crosby, a relative of Fox who had been an unpaid laborer at the Menlo Park laboratory, was also involved (Doc. 1926 n. 3; *TAEB* 5 App. 2). The company was not directly related to the American Electric Light Co., of New Britain, Conn., which marketed the arc lighting system of Elihu Thomson and Edwin Houston (Carlson 1991, chap. 5; Uriah Painter to TAE, 25 Dec. 1881, DF [*TAED* D8104ZFN; *TAEM* 57:331]). See also "A Queer Swindle: Collapse of the American Electric Light Co.," *New York Herald*, 23 Apr. 1882; and "Incandescent: American Electric Light Co. Stockholders," *Boston Globe*, 24 Apr. 1882; both Cat. 1327, items 2189 and 2186; Batchelor (*TAED* MBSB52189, MBSB52186; *TAEM* 95:190, 188).

5. American slang for flattery. *OED*, s.v. "taffy."

To Charles Batchelor

New York, 2nd May *1882*[a]

Dear Sir,

Referring to my cable of yesterday as to safety catch plugs & which was confirmed in my respects of yesterday[1] I beg to advise you that I have sent by this mail under separate cover two sample plugs—one marked "N.G." is as you have them at present; the one marked O.K. is how you must alter every one you now have, as if you do not your safety plug will be worse than useless & cause fire.[2]

I enclose you a list of the number of New Style of plugs sent you & where they were sent ~~you~~ to for your guidance   Yours truly

Thos A Edison   I[nsull]

L (letterpress copy), NjWOE, Lbk. 12:228 (*TAED* LB012228; *TAEM* 81:623). Written by Samuel Insull; stamped Thomas A. Edison to Chas. Batchelor. [a]"*New York*," and "*188*" from handstamp.

1. Edison cabled: "Just found new lamp tip when applied as safety catch plug, worse than useless as vapor lead crosses both poles   old style all right   new plug made right by connecting bottom button to screw by putting lead through hole bored in center plug   change all sent without fail   mailed samples." Edison referred to the base of the lamp as the "tip." He confirmed this message in a letter the same day (LM 1:202B [*TAED* LM001202B; *TAEM* 83:973]; Lbk. 12:196 [*TAED* LB012196; *TAEM* 81:602]; Doc. 2187 n. 30). Edison also sent this information by cable to Edward Johnson and by mail to William Hammer (TAE to Johnson, 1 May 1882, LM 1:202C [*TAED* LM001202C; *TAEM* 83:973]; TAE to Hammer, 2 May 1882, Lbk. 12:227 [*TAED* LB012227; *TAEM* 81:622]).

2. These safety catches were designed for use within buildings or rooms and not for use on feeders or at dynamo terminals. The new safety catches used the screw fittings recently designed for lamps. Dredge 1882–85, 614–15; Doc. 2187 n. 30.

The safety plug became the standard fuse for consumers.

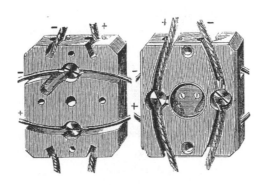

To Charles Batchelor

My Dear Batchelor

Taking out Patents. I do not think I have ever written you personally on the question of the taking out of patents on the continent of Europe

My reasons for taking the European business out of Serrells hands were 1st that I had engaged R. N. Dyer as Patent attorney exclusively in my interest and therefore no longer required the services of Serrell as all cases are prepared in my own office & 2nd that so many complaints had been made against Serrells Foreign Agents as to many glaring imperfections in Patents taken out that I thought it absolutely necessary that my Foreign patent business should go through other channels.

Both Puskas and Bailey have always stated that Brandon[1] is a very poor Patent attorney, that he is quite unacquainted with French Patent practice & that my patent interests in France were very much prejudiced by having my applications pass through his hands. Of course it is impossible for me to know whether this is so or not as I do not know whether he remodels the cases sent him before filing them, or whether he files an exact translation of what he receives from here. If the latter course is pursued he is in no way responsible. However this may be I think the whole business should be put into the hands of a first class Parisian attorney who should look after all new cases for the Continent of Europe & remodel the specifications sent from here so that they will comply with the local patent laws. He should also take in hand the paying of annuities & working of patents & in this connection your attorney should be placed in communication with Brewer & Jensen with the object of taking over this part of the business from them

I have heard Armengaud spoken of as a man extremely well suited for our business but of course I will leave the choice of a man to yourself as you are on the ground and far better able to judge than myself.

So far as payment for Patents is concerned of course your people in Paris pay for the Electric Light Patents for countries controlled by them. I particularly wish to pay myself for all Rail Road patents that Dyer may order to be taken out.

As to the payment made by Brandon on the Spanish Colonial Patents[2] I presume he must have received some instructions from Serrell. If so he should apply to Serrell for payment and not to you   I have never given Brandon any instructions except through you   Yours very truly

Thos A Edison

LS (letterpress copy), NjWOE, Lbk. 12:237 (*TAED* LB012237; *TAEM* 81:627). Written by Samuel Insull; stamped Thomas A. Edison to Chas. Batchelor. ᵃ"*New York*," and "*188*" from handstamp.

1. David Brandon (d. 1893), a native Englishman trained in engineering in France, worked in Paris as a patent attorney, primarily for English-speaking inventors. Doc. 1019 n. 2.

2. This concerned the extension of Spanish patents to the colonies, for which Batchelor had paid in the past. Batchelor to Serrell, 17 June 1882; Batchelor to Brandon, 17 June 1882; Cat. 1239:300, 303; Batchelor (*TAED* MBLB4300, MBLB4303; *TAEM* 93:679, 682).

–2267–

To Charles Clarke

[Menlo Park,] May 8 [188]2

Clarke

Please keep me posted by brief notes of the progress of things.[1] I am going to ~~d~~try great many things here that its difficult to do at Goerck. will send you results. Want you to be on hand and test that small bar armature machine yourself with various loads with EMF & margin on field so we can calculate the 100 light bar armature[2]

Want to get along with just as few bars & plates as possible my Impression is that the regular Z bar armature with say Considerably less than double the bars will give 100 lights (8½ per hp Lights) with margin to spare

AL (letterpress copy), NjWOE, Lbk. 12:250 (*TAED* LB012250; *TAEM* 81:637).

1. Edison also asked Charles Dean on this date to keep him informed about the new bar armature and test results on the 250 light dynamo under development. Lbk. 12:249 (*TAED* LB012249A; *TAEM* 81:636).

2. Edison was probably thinking of plotting the dynamo's characteristic curve, which represented armature voltage as a function of field excitation. See Doc. 2067 n. 2.

EDISON ELECTRIC LIGHT COMPANY BULLETINS   Doc. 2268

Sherburne Eaton, vice president of the Edison Electric Light Co., wrote a series of pamphlets about the activities of various Edison lighting interests. The company began irregular publication of these Bulletins in January 1882. The first four numbers were marked "Confidential, and for the use of the Company's Agents only." This admonition was dropped when the fifth one appeared in March; the company soon began sending

Bulletins to its shareholders.[1] It also encouraged agents "to communicate to the Vice President whatever practical points of general interest may be developed by their experience in installing or operating our light."[2] Eaton noted with some satisfaction that technical journals, especially in Great Britain, frequently excerpted them.[3] The Bulletins provide considerable detail about the commercial development of Edison lighting in New York, elsewhere in the U.S., and abroad. Sixteen were issued in the period covered by this volume, eight of them by the date of Doc. 2268.[4]

1. By late summer Eaton distributed the Bulletins to more than 1,700 individuals, including the stockholders of every Edison electric lighting company, customers and officers of related companies in Chicago and Lawrence, Mass., and others who had inquired about the light. Eaton to TAE, 27 Aug. 1882, DF (*TAED* D8226ZBM; *TAEM* 61:415).

2. Edison Electric Light Co. Bulletin 5:1, 17 Mar. 1882, CR (*TAED* CB005; *TAEM* 96:681).

3. Eaton to TAE, 6 June 1882, DF (*TAED* D8226R; *TAEM* 61:240).

4. Electric Light Companies—Domestic—Edison Electric Light Co. Bulletins (1882–1884), CR (*TAED* CB; *TAEM* 96:667).

-2268-

To Sherburne Eaton

[Menlo Park,] May 8 [188]2

Major

How would it do to mail all the back numbers of the bulletins[1] to the applicants for license that have been received and on file in statistical dept with a letter stating that the final experiment of a central station is fast approaching and that upon the completion & success of the experiment the Company will be ready to enter into negotiations.[2a]

We should keep the interest up among these people, and start the negotiations so that we shall have friends in every city looking out for us & themselves in awakening interest getting rights way etc. Yours

T A Edison

ALS (letterpress copy), NjWOE, Lbk. 12:248 (*TAED* LBo12248; *TAEM* 81:634). [a]Followed by dividing mark.

1. See headnote above.

2. Eaton replied that he had already started a list of recipients for this purpose. He also encouraged Edison to send other ideas as they occurred to him so that "I can always keep myself abreast of your views." Eaton to TAE, 9 May 1882, DF (*TAED* D8224W; *TAEM* 61:27).

-2269-

From Charles Clarke

New York, May 9 *1882*[a]

My dear Edison,

Today the 250 light machine was ready for a test. The arrangement for preventing the creeping and throwing of oil was perfect, would not now be in the least detrimental to silk hats or dresses, which is quite a point for exhibition purposes. The bearings became fearfully hot at speed (1250 per m.) but after the run we found journals slightly out of true. This would correct itself in a very short time and I think you are safe on that point, but I am pretty sure that at that speed and with that weight they will always be warm.[1]

The new oil arrangement admits of free use of the same, as it can be screened and used again.

The commutater worked much better and there seems nothing to improve there. The brushes, double width, behaved badly after running half an hour. They are of the standard thickness. It is difficult to adjust the brushes so that one will not take a good part of the load.

Dean is making them about one half thicker. Will try it again. Lamps were way above required E.M.F. at 1250 but then you must allow for drop in line in practice.

The bar machine is not quite ready or I should have heard from it.[2]

Shall test the fifth steam dynamo for Pearl St. tonight. Shall also determine the street safety catch, Kruesi having fitted up assorted sizes and a catch box which is now in place in testing room.

Dean is going to take his armature cage down to Pearl St. tomorrow and see what is to be done to get it in with the armature in it.

Bergmann's price for the field regulators for Central Station will be about $135 each—total $810. The general idea and details you know. Will it not be best to give Bergmann the order just as soon as I get his final bid, without delaying to submit it to you? ⟨Yes after model thoroughly tested⟩ We shall be in a great hurry for that, and will want it long before the feeder regulation is necessary.[3]

Soldan is busy upon C dynamo drawings for Batchelor.[4]

The Armington & Sims engine (auxilliary) is at Central Station. Four sole plates are in position with engines and lower fields.

Twenty-eight street safety catch boxes are in place. Kruesi is much delayed and the work has to be incompletely done for want of joints

Have urged Bergmann to hurry up his estimate for meters. Shall have something to say about Grower's results.[5]

There will be no great change in new meters, lugs for sealing with lead seal, longer flexible cord. Rubber tubes over the copper wires of zinc plates. That marine glue seems perfect. George gave me a sheet of G[erman]. Silver covered with it. Could not scratch or break by bending. A difficulty, is that it easily spreads over the zinc when applying. Should not Bergmann furnish with the meters short tubes with nuts, leading into meter thus?[6]

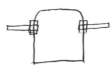

⟨Yes⟩ Believe he did in London meters. Will give you result of test of last experimental pressure indicator in a day or two. Should give order for twenty-three to Bergmann as soon as possible.

We ought to remodel the street boxes of all kinds for next district to save compound, weight of meatal, clumsiness.

So many lamps have failed in the testing room that the question of a fresh supply is serious. All the sockets are for the old style. Shall we not put in new style gradually as the old lamps fail?

Please read my notes when writing so that you will offer suggestions, as I want to forge right ahead and want to carry every thing out to your satisfaction. Ys. truly.

Chas. L. Clarke

ALS, NjWOE, DF (*TAED* D8227B; *TAEM* 61:430). Letterhead of Edison Electric Light Co. ᵃ"*New York*," and "*188*" preprinted.

1. This machine was designated the K model. William Andrews reported on 9 May that the dynamo sparked badly during that day's test. Charles Dean told Edison the next day that the machine ran hot and would require another pair of five-inch field magnet cores or two pairs of eight-inch cores. After an unsuccessful test on 12 May, Dean telegraphed Edison that he "must come to some decision at once regarding two hundred & fifty light cant go on Very necessary for you to come here today Will you." However, on 17 May Clarke reported that the machine worked well at 900 rpm and provided ample power; he added that the large brushes for the C dynamo did not spark on this machine and promised that Dean would design new brushes for the K. Andrews to TAE, 9 May 1882; Clarke to TAE, 12 and 17 May 1882; Dean to TAE, 10 and 13 May 1882; Andrews to TAE, 12 May 1882; all DF (*TAED* D8235H, D8227C, D8227E, D8233ZAQ, D8233ZAW, D8235J; *TAEM* 62:34; 61:438, 445, 1042, 1048; 62:37).

2. At least the first ten K dynamos, shipped in late May or early June

to London, used wire-wound armatures. Andrews tested a bar armature for the K machine during the rest of May but found its resistance too high. In early June he traced the problem to damaged insulation caused by the acid used with hard solder. He recommended using a softer solder and adopting rosin or paraffin flux, which would also create better connections and reduce resistance. TAE to Dean, 25 May 1882, Lbk. 12:432A (*TAED* LB012432A; *TAEM* 81:731); Rocap to TAE, 5 and 9 June 1882; Andrews to TAE, 12 and 27 May, 5 June 1882; all DF (*TAED* D8233ZBP, D8233ZBT, D8235J, D8235L, D8235M; *TAEM* 61:1071, 1075; 62:37, 40, 42).

3. Edison distinguished between two types of regulation in a central station district. One, generally referred to as field regulation, was to maintain a constant voltage without respect to load on the dynamo, usually by varying the current through the field magnets (headnote, Doc. 2242; see also U.S. Pat. 287,515). Edison intended what he called "feeder regulation" to maintain a uniform voltage throughout a central station's service area irrespective of the geographic distribution of the load, so that all lamps would provide the same intensity of illumination (see Doc. 2405). Feeder regulators were represented in a wiring diagram of the Pearl St. station but, according to a later reminiscence, were not installed (C. J. Leephart to James Bishop, 17 Dec. 1931, acc. no. 30.1507, accession file, EP&RI). It is unclear if they were used at Holborn Viaduct (see Doc. 2322). They were in place at the Milan central station which began operating in June 1883 but apparently could not be used because the operators could not measure voltage at the feeder ends (TAE to Société Électrique Edison, 7 May 1883, DF [*TAED* LB016263; *TAEM* 82:380]).

4. Gustav Soldan was in charge of the drafting room. On 22 May Clarke indicated that Soldan was occupied making drawings for the foreign companies and would soon complete those of the C machine. Clarke to Insull, 22 May 1882; DF (*TAED* D8227H; *TAEM* 61:460); Sprague to Hammer, 27 Dec. 1883, Sprague; Jehl 1937–41, 929–30.

5. George Grower (1862–1921) began working for the Edison Machine Works in early 1881. He specialized in electrochemistry and worked on the chemical deposition electric meter; he also helped test central station dynamos. He later took charge of the meter department of the Edison Electric Illuminating Co. "Grower, George," Pioneers Bio.

6. On 17 June Clarke wrote Edison that he wanted the meter parts made of brass but Bergmann preferred less expensive zinc. Clarke to TAE, 17 June 1882, DF (*TAED* D8227W; *TAEM* 61:484).

–2270–

From Edward Johnson

London, E.C. 9th. May, 1882.[1]

My Dear Edison,

I have yours of the 26th. April,[2] many communications from Insull of various dates, telegrams, etc., etc., etc. I cannot recall now all the subjects treated of but generally, have to remind you that you have been telegraphing and writing me for that which had already been accomplished in one way or the other invariably before the reception of your communications. You want work for the lamp[a] factory; you want orders for Goerck Street; Bergmann wants work; the lamp[a] factory could not be started till orders warranted it. My 32 & 52 candle lamp could not be completed because the lamp factory was shut down, and then—when the orders were sent, you did not appear to realize that you actually had them in hand.[3] The order for six Dynamos was delayed, as explained in my letters and original telegram for a little more consideration to be given to ~~the other parts~~ it,[b] and when the order was finally sent it was sent from the head office of the Co., 74, Coleman Street, which office has adopted "~~74~~ Sevenfour" as its cable code instead of "Fifty-seven"[4] which continues to be my private code address. Then you cabled me that you could not undertake so much work without some provision for money. Now all this I knew and, knowing it, provided for it.[5] At the very meeting at which these orders were discussed, I made a very considerable speech in which I took occasion to say that if orders were now given you for material they could be promptly filled but that if delayed until you had the New York station started your works would be so overwhelmed with home orders that practically we would be shut out.[6] On this line of argument I got the order made for a very considerable quantity of material and on this line of argument I finally succeeded in getting an order for six large Dynamos. So much for the order. I then called the attention of Mr. Bouverie to our former experience and to the general fact that you work all your establishments on such a close margin that you could not take an order of this kind in hand without advanced payment on account thereof. Mr. Bouverie promptly proposed and Sir John Lubbock seconded that a considerable sum be paid you in advance. It is all therefore arranged and was arranged before the receipt of your telegram that whatever money you require shall be paid to you through Drexel Morgan & Co., on account. Of course I quite understand your anxiety and I quite understand the object of your continual nagging in that you wish to assist me in carrying[c] your points. But I think you pushed it a little further than was necessary in your last cable. However, it is all now adjusted. There remains nothing to be done as between you, Drexel Morgan & Co., and the company, but for you to fill orders in hand and get your money therefor. The last payment of £1666 on account of Drexel Morgan & Co.'s disbursements under the contract has been made consequently the contract is now fulfilled in every detail. The moneys paid by Drexel Morgan & Co., since the closing of the contract will be returned to them just as fast as the accounts are received.[7] Pierrepoint Morgan has done me the honour of calling at the Viaduct to see me and a further honour of expressing himself as very well pleased [at?][d] with[b] all he has seen [--][d] in[e] England and in fact of all that has been done. Batchelor has been to see me and is delighted with everything, especially with my English company, its men and its lines. There is one thing certain you will get more money out of England alone than you will get out of all the remainder of this hemisphere. I do not like the contracts the French people are making—they appear to me to be drawn with one idea, and that is, endless competition for the purpose of ultimately euchreing somebody. On the whole therefore I can consider that the prospect in England is bright,[c] as compared with elsewhere.[c]

Colonel Goureaud has proved his inefficiency since he has been wholly unable thus far to negotiate even the two colonies India & Australia. I am doing what I can to urge upon him the importance of accomplishing something in this respect but make but little headway. He tells me that he is only awaiting the preparation of the papers by Mr. Waterhouse. I have asked Mr. Waterhouse to let him have them as soon as possible as I have given Goureaud to understand that as soon as the papers are in his hands I shall expect some demonstration of the progress to be made meantime there is nothing more to be said in regard to these colonies.[8]

Crystal Palace Exhibition closes on the 3rd. June. I have had the jury in tow and I am satisfied that the difference of recognition as between your work and that of others will be more marked than it was in Paris. The Viaduct continues to run satisfactorily, but we are in trouble with No. 3. machine: it is crossed with the base. A test made by [------][d] Hammer[b] & Francis[9] developed the fact that one of the bars of the armature is crossed with the iron discs. I immediately telegraphed to Paris for Coningham Cunningham[10b] to come by the first train, but the damned fool telegraphed back that Batchelor was away. I replied that I knew that, and was responsible for the order I gave him and asked him to come at once. I expect him to be here tonight. I then proposed to unwind the armature, remove the bar, reinsulate it, put it back and rewind the armature. This is the first real trouble we have had and I am[e] in hopes we shall be able to tide over the time necessary for remedying it without a breakdown of the other machines; for in case anything should happen to it while we are repairing this one the light on the

Viaduct will disappear and then you will hear such a howl as never went up from Babylon before. I quite realize what this would mean to me. All the good work that I have done and all the credit I have received therefor, would go as so much chaff before a strong wind. However, it is all in a lifetime and I am taking the chances. No doubt I could run this machine with one side crossed for a long time but I prefer not to do that sort of thing. You must in future, as no doubt you are fully aware, use even greater caution in the matter of insulation than you have heretofore. A slight thing of this kind with a machine so far away[a] is likely to result in serious damage to your interests. The machine costs too much money to be running constantly so near the verge of destruction. The insulation in the rods which lead from the brushes is very bad—it has totally given out of No. 3. machine and we shall have to put in a better arrangement. In all other respects we are running along nicely. I note what you say about the tit[f] having broken off of No. 2. by virtue of the sudden change of load, but I think you are wrong.[11] We had not done anything of that sort for a long time past. I think it is a gradual break—the result of vibration; however, that is a trifling matter and was soon remedied. Your telegram of to–day says that the new machines will be far superior to the old,[12] comes just at the right time but I can use it as a sort of salve to heal the injuries which our people will have received by the announcement of the trouble with No. 3.

This telephone trial is interfering somewhat with our progress in the matter of testing for cost, efficiency, and so on, since Dr. Hopkinson & myself are both locked up by it every day.[13] However, we hope it will be over [---][d] by[b] the end of this week; I have been on the witness–stand and have succeeded in making an impression although I came on as the last witness. I send you a copy of the Times so that you may see for yourself what it says.[14] I am satisfied we are going to win in this fight although it is a very much better fought battle than that at Edinburgh. Aldred,[15] Husbands[16] & crew are the real backers although they are not the people against whom the action is brought. I want to win this because of its bearing on the electric light patents. I find a very striking analogy between the lamp patent and the patents for this carbon button. I will point them out to you when I see you. But because of this analogy I am able as the trial goes along to make innumerable points with Webster[17] and our experts and thus educate them to a better understanding of the vital features of the lamp patent. My work in this respect is said by Waterhouse and others to be simply without price.

We have had another conference over the lamp patent at which Sir Fredk. Bramwell, Mr. Webster, Dr. Hopkinson, Shelford Bidwell, Theodore Waterhouse and myself were present. It is only one of many that have been held lately and it had the same result as our previous ones. The general verdict now is as follows;—the patent is not without its weaknesses—it is not a patent of which you can say—"on it we are sure to win"—it is not a patent which you shake in the faces of infringers, and cause them to quail and tremble—but nevertheless it is a patent which can be defended, and which ought to be defended. It is a patent which cannot be amended. There is no way by which it can be improved by process of disclaimer. There is in fact nothing in it which if distrained would render the remainder of the patent stronger. This fact is an important one since there is an enormous difference of value attached to a patent by the courts which comes to them in its original form rather than in a dress all tattered and torn. This being the general verdict and it having been reached now several times in succession it was finally presented to the board at their last meeting and Mr. Waterhouse was instructed to push the preliminaries forward with[a] the utmost dispatch, and prepare for battle all along the line. You must therefore look for some interesting proceedings before long and one of the greatest contests that ever was fought over an electrical invention. The Brush Co., Siemens, Swann, Lane-Fox Maxim, and in fact the whole array will be united against us. It is a money power that will be able to prolong the contest almost indefinitely; but on the other hand you have the support of the public. They believe that the thing belongs to you. You have proven that in its practical development you are a long way ahead of all of them. You will have the support of the best scientific and legal talent in England. You will have the moral and financial support of some of the best known and most highly [respectable?][d] names in English financial circles; and further than that your patent will be fought [until?][d] the bitter end at any cost [-------].[d] They have voted that £30,000 be set aside for the purpose as a starter and are fully prepared to [-----][d] take the[g] case [----][d] to[b] the House of Lords if necessary. Meantime they supplement this powerful assertion of their rights by holding out the olive branch of peace and it is not at all improbable that the most dangerous of all our antagonists (Swann) will be the first one to surrender and recognize our claims. Whatever may be done however, in this direction it is distinctly agreed all round, can only be done by the absolute recognition of our patents.

The present proposition is that a royalty be paid us [---]^d we granting a license to manufacture under our patent. It is too early, however, in this stage of the proceedings, to venture any opinion as to what [---]^d will^b be done.

Parliamentary Bill.[18] We are all before Parliament with several bills, asking for privileges for taking up streets—there are six of us [-- -----]^d in all, and the Government is before Parliament with a general bill asking that privileges be accorded. We have determined to unite in support of the Government bill and withdraw our several private bills. We have done this because we have learned that the entire gas interest is consolidated to fight us. A conference at Westminster last Saturday morning was called for the purpose of selecting someone to draft our case and present to Parliament a statement of the present stage of the art as contrasted with the status in 1879 (when Parliament last had the matter before them.) and then to follow up this statement by verbal evidence before the parliamentary committee in support thereof. After some little discussion they unanimously agreed that I was the man for the work—what do you think of this? In all England these six companies agreed that there is no man so capable of presenting and supporting their cause as the "young American," who at present represents you. Siemens had one objection to me and that was the fear that I would be too partisan. Several dissenting voices were raised, so my informant tells me—for I was not there, but it was finally accepted that I was quite capable of dealing with the question apart from Edison. I therefore yesterday, upon receiving a notice of my selection, dictated my preliminary^a statement and supplemented it with the assurance that^h I would^i come before a committee^c in support of every assertion contained therein. I have of course, as far as possible avoided any direct reference to your own work, but I have explained to my fir friends to-day that it is impossible to make a statement of the present state of the art without asserting that certain things are done which you, and you alone accomplish. Of course this will come out more clearly on testimony. I understand that giving testimony of this character before a parliamentary committee, is very like giving testimony in court. The gas companies will be represented by able counsel for the purpose of challenging every statement that is made and to break it down if possible. My ability to hold my own under cross-examination, as evidenced by my testimony in the telephone case but a day or two previous, is probably one of the chief reasons why I was selected for the work. At all events I am not much concerned about the matter; in point of fact I am rather eager for the fray. This is my one work which I feel it incumbent on me to do yet, before leaving England, and recognizing its importance, I have consented to remain in this country a month longer for the purpose of accomplishing it.[19]

Your consent to my return on the day fixed in my previous communication, you will observe has cost you nothing since I do not propose to avail myself of it; but nevertheless I am just as much obliged. Your remark about your not coming to England, because, forsooth, I am so anxious to leave—is a little strained.[20] It don't follow, because I find life in England very unsatisfactory, that a mere visit to that country would be unpleasant. On the contrary, I expect to visit England frequently myself and to derive a good deal of pleasure in doing so.

I have given orders already that all the exhibits at Crystal Palace, not strictly belonging to the electric light, shall be packed and shipped directly to you, so soon as the exhibition closes.

I have to thank you for your telegram authorizing Goureau to pay me 5 per cent of the Oriental.[21] It did not get me as much as I expected (as usual) for the reason that Goureaud exacted of me that I should pay 5 per cent of the £4,000 you and he subscribed to the Telephone Co., and 5 per cent of the £[----]^d [1000?]^j you and he subscribed to the automatic adjustment. This took £250 out of the £450 that I supposed I would receive; thus I received £200. ($1,000)   Batchelor was over here a few days since and received his share, which was to him a considerable windfall, and went away quite happy. He received from the sale of his Uniteds and from this Oriental together, £1600 and some odd pounds in cash. There is a matter however, which he has to adjust with you, and which he asked me to write to you about. It is this: Goureaud has charged you with your proportion of the £4,000 subscribed to the Supply Co., not deducting Batchelor's portion therefrom. The result is that Batch. gets the entire amount due to him while you have paid to Goureaud, on account of this Supply Co., a percentage on what he gets. You have therefore to receive from Batchelor back again, that which you have paid which of course will be 20 per cent of the amount that you actually paid on account of his £4,000. I will not go into the figures, as I understand from Goureaud that he has sent you a full statement of the whole affair.[22] Roughly, Batchelor owes you something in the neighbourhood of £400. I will leave the matter therefore for you to [-----]^d settle^b with him.

I have got Hammer the position of assistant engineer at

£450 a year—the salary to date from the formation of the Co., I think you will admit that that was doing pretty well. He has a splendid chance for advancing himself, and I think you may rely upon his doing it.

The article in the "Sunday Times" on the collapse of Bohm,[a] Fox & Co., is very satisfactory.

I am very sorry to hear you have been sick; in fact, you seem to have had a serious pull of it this time. Probably in the course of a few years, you will learn to take better care of yourself — if you don't, we shall have to appoint a committee to do it.[c] I will constitute myself one member of that committee, on my return to America.

I am now compelled to write a long and rapid account of affairs here to Major Eaton;[c] because for some reason or other, the information which is conveyed by my letters to you does not [---][d] reach him, and therefore the reference to our work in his bulletin is meagre, and consists principally of quotations from newspapers rather than from information from me direct.[23] I hope there is nothing amiss with the Major. Very truly yours,

Edwd H. Johnson

The trouble with No 3 proves to be in the Commutator. We have it all apart—one of the blocks has been dead grounded & a big hole has been burned in the Iron—the mica is eaten away   Cunningham is hard at work on it—& will probably have it OK again today. Its slow work as we have no conveniences for doing[b] it= Meantime we are running regularly & smoothly with No 2— Can you make a new commutator for No 3—on your new plan. I am afraid this cross is only a beginning & that we shall frequently have trouble of this kind—

We have found also a cross between 2 of the bars—they are too close together—

The armature is all right—& if we could only get at the comut[ato]r more easily the job would have been a simple one= I am keeping the misshap from the Public but our own people know all about it— Write me freely about the provisions against this sort of thing in the new machines   Yours   EHJ[k]

TLS, NjWOE, DF (*TAED* D8239ZBK; *TAEM* 62:878). Typed in upper case; cancellations and interlineations made in Johnson's hand. [a]Mistyped; corrected by hand. [b]Interlined above. [c]Mistyped. [d]Canceled. [e]Written by Johnson. [f]"the tit" written by Johnson. [g]"take the" interlined above. [h]Obscured overwritten text. [i]"that I would" written by Johnson. [j]Illegibly interlined above. [k]Postscript written and signed by Johnson.

1. This letter was stamped as received on 22 May.

2. Doc. 2264.

3. Postal confirmations of cable correspondence among Edison, Johnson, and the British company provide an overview of these and other matters. The company ordered a large amount of equipment, including one hundred 100-light dynamos with bar armatures at a price of about six dollars per light. TAE to Johnson, 21 Apr. 1882, Lbk. 12:132 (*TAED* LB012132; *TAEM* 81:572); Arnold White to TAE, 1 May 1882, DF (*TAED* D8239ZBC; *TAEM* 62:868).

4. When Edison received the unsigned order from "Sevenfour London" for six 1,000-light dynamos, he cabled Johnson to determine who had sent it. Edison Electric Light Co., Ltd., to TAE, 2 May 1882; TAE to Johnson, 2 May 1882; LM 1:203B, 203C (*TAED* LM001203B, LM001203C; *TAEM* 83:973).

5. Johnson had urged Edison in April to suggest terms for advance payments on large orders from the English company. He asked Edison to "so arrange it, that whatever is agreed upon in the start, may not have to be departed from, afterwards. It is much better to anticipate your needs than to have them unexpectedly cropping up in the form of demands." On 5 May, Edison cabled, "Official order not yet arrived. . . . arrange for me have payments weekly actual amount expended large dynamos during building   cannot finance such heavy work   can carry smaller work till completed but permanent arrangements must be made for payments here as work is finished." Johnson replied, "Order gone It provides for advance payments. credit me with some knowledge of your needs. assume them provided for and go ahead." The British company advanced $1,000 monthly to the Machine Works, an amount which bookkeeper Charles Rocap found inadequate by June. Johnson to TAE, 23 Apr. 1882 (p. 10); Rocap to TAE, 15 June 1882; both DF (*TAED* D8204ZAY, D8233ZBZ; *TAEM* 60:118, 61:1082); TAE to Johnson, 5 May 1882; Johnson to TAE, 6 May 1882; LM 1:206A, 206B (*TAED* LM001206A, LM001206B; *TAEM* 83:975).

6. In a letter that Johnson would not yet have received, Edison urged that the company "should send their orders right away. By the middle of next week the Dynamos for the Central Station here will be completed. . . . I am very much afraid that directly the station starts here the demand for Central Station machines will be very large. I should be sorry to see the English Company prejudiced owing to their not sending orders forward in good time." TAE to Johnson, 1 May 1882, Lbk. 12:198 (*TAED* LB012198; *TAEM* 81:604).

7. Under terms of the agreement providing for the establishment of the British company, Drexel, Morgan & Co. was both a major stockholder and the bank through which shareholders disbursed payments to Edison and others. TAE agreement with Drexel, Morgan & Co., Grosvenor Lowrey, Egisto Fabbri, and Edward Bouverie, 18 Feb. 1882, CR (*TAED* CF001AAE1; *TAEM* 97:295).

8. About a week later, Edison approved arrangements for a company to operate in India and Australia. TAE to Drexel, Morgan & Co., 17 May 1882, Lbk. 12:335 (*TAED* LB012335; *TAEM* 81:681); Drexel, Morgan & Co. to TAE, 18 May 1882, DF (*TAED* D8240L; *TAEM* 63:20).

9. Francis Jehl.

10. David Cunningham was a mechanic and machinist who began working at the Menlo Park laboratory sometime in the first half of 1879. In 1881 he assisted Charles Batchelor with Edison's installation at the International Electrical Exhibition in Paris. Cunningham then returned to the United States until Batchelor asked Edison to send him back to Paris in April 1882. He sailed later that month. Jehl 1937–41, 680, 682; *TAEB* 5 App. 2; TAE to Batchelor, 6 Apr. 1882; Insull to Cunningham, 8 Apr. 1882; Lbk. 12:39, 54A (*TAED* LB012039, LB012054A; *TAEM* 81:519, 530); Batchelor to TAE, 20 Apr. 1882, Lbk. 4:247, Batchelor (*TAED* MBLB4247; *TAEM* 93:649).

11. Johnson had reported the need to shut down dynamo No. 2 when the armature suddenly began sparking badly. It was found that "one of the little attachments which lead to the radial commutator rods was broken square off." Edison cabled his suspicion that "breaking disk tit probably due sudden load" with the field magnets energized. Johnson to TAE, 23 Apr. 1882 (p. 7), DF (*TAED* D8204ZAY; *TAEM* 60:118); TAE to Johnson, 7 May 1882, LM 1:206D (*TAED* LM001206D; *TAEM* 83:975).

12. Edison cabled: "New large dynamos far superior one you have more powerful Engine less spark removable commutator strips so can change two hours   carry fourteen hundred lights with ease and Economy." TAE to Johnson, 7 May 1882, LM 1:206D (*TAED* LM001206D; *TAEM* 83:975).

13. The United Telephone Co. was sued by the London and Globe Telephone and Maintenance Co., owner of the Hunnings carbon transmitter patent. The plaintiffs alleged that working transmitters could not be constructed according to Edison's July 1877 patent. In preparation for the trial, solicitor William Winterbotham asked Johnson to make instruments according to the specification. Edison sent materials and instructions for doing so. Winterbotham to Johnson, 8 Mar. 1882; Johnson to TAE, 8 Mar. 1882; both DF (*TAED* D8256H, D8256I; *TAEM* 63:856, 857); TAE to Johnson, 6 Apr. 1882, Lbk. 12:49 (*TAED* LB012049; *TAEM* 81:526).

The case concerned a notice published by the United Co. that no form of carbon transmitter could be manufactured, sold, or used in Britain except under license. The judge held that despite the vindication of Edison's patent in the recent Edinburgh decision (see Doc. 2231 n. 1) the United Co. could not make an unqualified claim to all carbon transmitters. Decision of Edward Fry, 22 June 1882, DF (*TAED* D8256ZAH; *TAEM* 63:887).

14. The *Times* referred to Johnson's testimony under cross-examination as "perhaps the clearest and most interesting of any" by a witness for the United Co. "High Court of Justice," *Times* (London), 3 May 1882, Cat. 1034:93, Scraps. (*TAED* SM034093a; *TAEM* 25:430).

15. Horace Eldred, formerly general manager of Western Union's telephone department, was an assignee (with José Husbands and two others) of the Hunnings telephone patent in Great Britain. Doc. 1823 n. 6.

16. José Husbands contracted with Edison in 1879 to form and operate a telephone company in Chile (see Doc. 1823). He left Chile in September 1881, apparently after a breach with Eldred, who was in London by that time. Husbands was in London by December (Husbands to TAE, 24 Sept. and 5 Dec. 1881, both DF [*TAED* D8147ZAA, D8104ZFE;

TAEM 59:736, 57:298]). About this time he severed relations with Edison over Edison's refusal to allow his name to be used in connection with the Anglo Pacific Electric Light Telephone & Power Co. (Johnson to TAE, 19 May 1882; Husbands to TAE, 20 May 1882; LM 1:214B, 213B [*TAED* LM001214B, LM001213B; *TAEM* 83:979, 978]; Husbands to TAE, 5 June 1882, DF [*TAED* D8254A; *TAEM* 63:766]).

17. Richard Webster, a member of the Queen's Counsel, had been engaged by Johnson to provide outside legal advice to the Edison Telephone Co. of London in 1879. Doc. 1833 n. 5.

18. Delegation of authority to lay underground conductors was one essential feature of a bill pending in Parliament, on which hearings began in April. A general act to regulate the supply of electricity for lighting and other uses, the Electric Lighting Act of 1882 became the first such regulatory effort when it took effect in August. Under its terms, companies that wished to lay lines below public streets were required to obtain a special act of Parliament, a license from the Board of Trade, or a provisional order from the Board. The provisional order, subject to Parliamentary approval, would specify the maximum rates and maximum voltage to be applied within the defined service area. Hughes 1983, 58–61 esp. n. 50; Bowers 1982, 155–58; Hannah 1979, 5–8; Johnson testimony, n.d. 1882, excerpted in Edison Electric Light Co. Bulletin 11:14–16, 27 June 1882, CR (*TAED* CB011118; *TAEM* 96:727); Gordon 1891 offers a usefully detailed polemical analysis of the history and terms of the legislation.

19. The testimony is in United Kingdom. Parliament. House of Commons 1882; see also Hughes 1962. Substantial excerpts from a press report of Johnson's testimony were published in the Edison Electric Light Co. Bulletin 11:14–16, 27 June 1882, CR (*TAED* CB011; *TAEM* 96:720). At the request of the English company's directors, Johnson postponed by several weeks his departure date given in Doc. 2258, and now planned to leave on 15 June (Johnson to TAE, 23 Apr. 1882 (p. 11), DF [*TAED* D8204ZAY; *TAEM* 60:118]).

20. See Doc. 2264.

21. See Doc. 2264 n. 3.

22. Gouraud to TAE, 3 May 1882, DF (*TAED* D8256X; *TAEM* 63:875). This probably refers to Edison's Foreign Telephone Supply and Maintenance Co. (see Doc. 1978 n. 1).

23. Samuel Insull quoted this sentence in a letter to Sherburne Eaton a few weeks later. He added that Johnson had blamed him, in a separate communication, for withholding information, but asserted that "whenever we have received letters of a general & not of a private character I have sent them to you for you[r] perusal." Edison asked George Gouraud soon after to "write me weekly a 'newsy' kind of a letter as to Electric Light . . . I want it to cull information for Bulletin." Insull to Eaton, 22 May 1882; TAE to Gouraud, 31 May 1882; Lbk. 12:367, 7:377A (*TAED* LB012367, LB007377A; *TAEM* 81:701, 80:570).

–2271–

Notebook Entry:
Miscellaneous

[Menlo Park,] May 10, 1882.

Experimenting on reducing iron sand to iron also mixing with coke, charcoal, etc & tar, Crude Pet[roleum]. Mould in brix and heating sufficiently to consolidate.[1a]

Thermo Experiments[2]

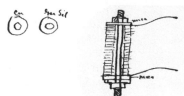

discs of Cu & Ger[man] Sil[ver] alternated. heating one End causes heat travel to other End and there is a gradual fall of heat until the ~~E~~other End is scarcely hotter than the atmosphere.

by using several thousands of these discs an exceedingly novel & Economical conversion of heat into Elect is obtained

I am plating copper sheet with German Silver also other sheets of Copper with nickel. Alternately[b] first Copper then Exceedingly thin coat nickel then Copper then nickel & so on until several hundreds thickness is obtained   the sheet is then cut out in circles holes drilled ~~in~~ centre & ~~all~~ several discs bolted together forming thermo pile.

Today we run the ~~R~~Electric Railway with 3 machines in series each working its own field independent of the other as shewn on next page

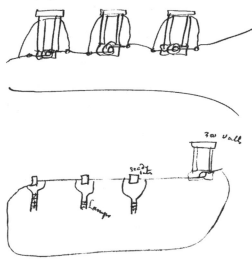

Reduce pressure by Second[ar]y batteries. No consumption[3]

John F. Ott   TAE

X, NjWOE, Lab., N-82-05-10:1 (*TAED* N198:1; *TAEM* 40:419). Document multiply dated. [a]Followed by dividing mark. [b]Obscured overwritten text.

1. Nothing more is known of these experiments, or even if Edison carried them out. His attempt to form iron ore bricks likely was a response to the difficulty of smelting the extremely fine particles yielded by his magnetic separation process; see Doc. 2393.

2. Edison had conducted extensive experiments on thermoelectric batteries in the spring of 1879 and returned briefly to the subject later that year (see Doc. 1724). There is no record of further experiments until October 1881, when John Ott "tested thermo battery made up of plates of Carbon and plated with a copper deposit." On 13 May 1882, Edison sketched two arrangements for "Thermo Experiments" using bars of copper and silicon. N-81-09-03:46, N-80-10-01:149, Lab. (*TAED* N235:24, N304:87; *TAEM* 41:533, 41:1137).

3. Figure labels are, from left: "Lamps," "secdy bats," and "300 volts." Edison executed on 22 May a related patent application for reducing relatively high voltage current; see Doc. 2276 n. 5.

–2272–

To Charles Clarke

[Menlo Park,] May 11 [188]2

Clarke:

Please send me result of George's test of the several meters in series.[1a]

In testing the cheese knife, look out that when you have thrown several Carbons out ~~that the~~ & get the EMF right that the Carbons dont gradually change their resistance necessitating another readjustment   if this is so it will keep the regulator man very busy—[2a]

I should put Campbell[3] or Soldan after he gets through with the C, 250 & 15 light drawings on remodeling the whole street box system [with an idea to economy ------ for the ---- station?][b] as you say—[---][b]

I think you will perhaps have to use mercury & brush rigging exactly similar to the C Dynamo on the 250 light. Do you not think it safe to go ahead on them as the spark biz is a matter of brushes because ~~if thi~~ the Commutator cannot be better & I understand the bobbin does not overheat, etc   please give attention to this Dynamo.[4] it is very important in view of the Borden contract for lighting large mills[5]   My Impression is that Andrews ought to have an assistant fully competent to test machines, so that Andrews can pay more attention to the new & first machines, hunting defects etc= I think you ought to see Maj Eaton about this=[6a]

Did you see the Jumbo run last night   Dean writes that

spark was nothing & that my conjecture [verified?][7b] what do you say to this[a]

I advise renewing test lamps at Goerck St with [new tips?][b] only as fast as they break be sure when you order of Upton to tell him when you [will -----?][b] & be very [preemptory?][b] in [requiring?][b] the same kind of lamps as before and to be <u>ab-solutely</u> sure that they are 8[$\frac{1}{2}$?][b] per hp & not 10 per hp. insist on this.[8a]

The Commutator speed of the 250 is less than $\frac{1}{3}$ more than the jumbo hence I dont see why the Current shouldnt be carried nicely if the <u>area</u> of <u>Contact</u> is sufficient I think the[c] mercury dodge should be worked up on these smaller ma-chines anyway if we do I assume we will get lots of [dodges without -----?][b]

I should not order the chese knife regs until I was thor-oughly satisfied by actual test. I[a]

I should order the field reg because I see no chance for mis-take there but please carefully consider it with a view of con-venience, <u>and the absolutely</u> [impossibility?][b] that connections in binding posts or elsewhere should ever get loose. This is of the greatest possible importance as you can imagine[9]

T A Edison

ALS, NjWOE, Lbk. 12:285 (*TAED* LB012285; *TAEM* 81:653). [a]Fol-lowed by dividing mark. [b]Illegible. [c]Interlined above.

1. Clarke gave Edison results of George Grower's tests the following day. Grower compared meters by "connecting their shunts in series and passing the current for about twenty lamps through them." Clarke warned that much care was required in weighing the plates, particularly with respect to temperature variations. Clarke to TAE, 12 May 1882, DF (*TAED* D8227C; *TAEM* 61:438).

2. The "cheese knife" was a feeder regulator for which Edison and Clarke jointly executed a patent application in October. It consisted of a variable number of high-resistance conductors interposed in the feeder circuit, in parallel with each other. As the knife switch contacted more of these conductors, the circuit's overall resistance decreased and propor-tionately more current would flow. U.S. Pat. 287,525.

3. Charles Campbell was a draftsman working under Gustav Soldan. Edison placed him in charge of the Engineering Department of the Edi-son Electric Light Co. in July 1883 but fired him in 1884. Jehl 1937–41, 930; Insull to Campbell, 17 July 1883, DF (*TAED* D8316AHL; *TAEM* 64:945).

4. On this date Edison also wrote to Charles Dean, "All you have to do is to experiment on the best kind and form of brushes to take the cur-rent off the commutator. I suggest that the experiment be tried with the present brushes with mercury on Afterwards put on two wide brushes like those on the large machine first <u>without</u> mercury and then <u>with</u> mer-

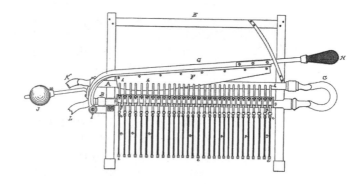

The "cheese knife" feeder regulator patented by Edison and Charles Clarke.

cury   These experiments should be hastened." TAE to Dean, 11 May 1882, Lbk. 12:277 (*TAED* LBo12277; *TAEM* 81:651).

5. Spencer Borden (1849–1921), agent and treasurer of the Fall River (Mass.) Bleachery, was the New England agent of the Edison Company for Isolated Lighting. By August 1882 he had sold $45,000 worth of dynamos, accounting for nearly half the company's sales. During the summer he resigned from the Bleachery to manage the recently-organized New England Department of the Edison Electric Light Co.; he played a major role in building a central station in Fall River. Phillips 1941, 2:113, 129, 188–90; Borden to TAE, 11 Jan. and 2 Feb. 1882, 24 Jan. 1883; Eaton to TAE, 8 Aug. 1882; all DF (*TAED* D8215C, D8221B, D8226ZAZ; *TAEM* 60:736, 853; 66:492; 61:362); Edison Electric Light Co. Bulletins 1, 4, and 13; 26 Jan., 24 Feb., and 28 Aug. 1882, CR (*TAED* CB001, CB004, CB013; *TAEM* 96:668, 676, 738).

6. Eaton consented to hiring an assistant so that Andrews could give his time to "more important matter[s]. He is a good man." Eaton to TAE, 8 June 1882, DF (*TAED* D8226S; *TAEM* 61:243).

7. Charles Rocap indicated on 10 May that Dean was about to try larger brushes on the machine, and Dean wrote Edison that the machine "worked splendid." It did not spark and the brushes showed only slight signs of wear. Rocap to TAE, 10 May 1882; Dean to TAE, 10 May 1882; both DF (*TAED* D8233ZAP, D8233ZAQ; *TAEM* 61:1041, 1042).

8. Francis Upton complained to Edison that Clarke had ordered 500 new-socket lamps with operating voltages of 97 and 98 volts. Upton thought that because the lamps would only be used to provide resistance for testing dynamos, it was preferable to use higher voltage lamps already on hand which were not suitable for use by customers. Edison demurred because "we have 1000 lamps of 97–8 volts & we want to put them all on the same line & any new lamps we have must be of the same volts as those now being used." Upton to TAE, 24 May 1882, DF (*TAED* D8230ZAR; *TAEM* 61:771); TAE to Upton, 25 May 1882, Lbk. 12:435 (*TAED* LBo12435; *TAEM* 81:734).

9. On the following day Clarke replied that he was "pushing Bergmann with the cheese knife model," though he did not expect to need it "until the consumption in 1st dist. is considerable." He added that the field regulators were ready to be produced and that he would "try to force it through in three weeks." Clarke to TAE, 12 May 1882, DF (*TAED* D8227C; *TAEM* 61:438).

-2273-

To Thomas B. A. David[1]

[Menlo Park,] 11th May [188]2

Dear Sir,

Your favor of the 5th[2] has followed me here where I am now permanently located.

We have one Locomotive here that can pull four cars containing thirty people each at the rate of 20 miles per hour. It takes about ~~16~~ 15[a] horse power. We are now building a single car which will have the motor underneath.[3] It will carry 30 passengers twelve miles per hour & consume about four horse power   This is the best thing for the purpose you name as it is better to have plenty of trains each[a] to carry a few people than to have few trains each carrying a great number of people

We have three miles of road built now & the whole thing works splendidly[4]   Yours truly

Thos A Edison   I[nsull]

L (letterpress copy), NjWOE, Lbk. 12:284 (*TAED* LB012284; *TAEM* 81:652). Written by Samuel Insull. [a]Interlined above.

1. Thomas B. A. David (1836–1918), formerly president of the Central District and Printing Telegraph Co. in Pittsburgh, was a general agent for the Edison light in Ohio. Doc. 923 n. 2; letterhead of David to TAE, 5 May 1882, DF (*TAED* D8249L; *TAEM* 63:596).

2. David wrote on behalf of William Willshire Riley, inventor of a monorail system, who proposed to build a short electric railroad at Columbus, Ohio. Riley sought information about the power requirements, cost, and likely delivery date for an electric locomotive. David described Riley as "quite taken with the idea of using your engine, and doubtless would be a very desirable party to introduce it." Edison's marginalia on the letter is the basis for his reply. David to TAE, 5 May 1882, DF (*TAED* D8249L; *TAEM* 63:596).

3. See Doc. 2259.

4. About a month later, Charles Hughes compiled for another party an itemized estimate of about $38,600 to build and equip a 3½ mile electric railroad. This included track, a 40 horsepower central station, and four motorized cars at $2,500 each (Hughes to Phillips Shaw, 9 June 1882, Lbk. 7:432 [*TAED* LB007432; *TAEM* 80:597]). A breakdown of construction costs and general estimate of operating expenses for Edison's Menlo Park railroad as of 25 February 1882 is in Cat. 2174, Scraps. (*TAED* SB012ACE; *TAEM* 89:380).

SHERBURNE EATON REPORTS   Docs. 2274, 2303, and 2318

Although Edison was again working at Menlo Park, Sherburne Eaton continued to keep him informed about the commercial progress of the electric light. In early May, Eaton began writing him frequently, sometimes more than once a day. Closing one of several messages on 9 May, he apologized "for giving you all these details. The fact is I miss your morning call, and not having you here to talk to I have to talk to the stenographer. By and by I shall get used to it and will not then trouble you with such a lot of little things."[1] Eaton funneled information assembled from weekly reports made by various Edison principals, including patent attorney Zenas Wilber.[2] Samuel Insull also passed on pertinent news gathered from Edison's own correspondence.[3] The weekly updates are not extant but Eaton's summaries are a rich source of information about the activities of the Edison enterprises. Edison did not always reply directly but he used Eaton's memoranda as the basis for decisions about business and some technical matters.

1. Eaton to TAE, 9 May 1882, DF (*TAED* D8226A; *TAEM* 61:147).
2. Eaton's reports to Edison are in Electric Light—Edison Electric Light Co.—Eaton, S. B.—Reports (D-82-26), DF (*TAED* D8226; *TAEM* 61:146); closely related correspondence is in Electric Light—Edison Electric Light Co.—General (D-82-24), DF (*TAED* D8224; *TAEM* 61:2).
3. See Doc. 2270 n. 23.

–2274–

To Sherburne Eaton

MENLO PARK, [N.J.][1a] 11th May [188]2

Major Eaton,

Your letter letter of 9th with extract from letter of John Moore & Co Rio de Janeiro has been received & I will take steps to prevent machines going into Brazil[2]

Returned herewith is telegram received from M. F. Moore Am glad to hear such good news[3]

I am glad to receive your daily chatty memoranda   It gives me the information I desire. Do not hesitate to call on me whenever you require information or require my opinion[4]

I have instructed Kruesi to experiment on a compound which will over come the difficulties referred to in Kendall & Cos letter of 1st April 1882[5]

I also return you herewith the Reports. You need not send these out here I can see them when I come in. Your memoranda above referred to supplies me with what I want in the way of the Departmental work of the various Companies[6]

I return you Hazards letters[7]

I will see what I have done in the way of accumulators & let you know. We should have nothing to do with the Faure Battery people. It will only turn out a Stock jobbing affair & we will get ourselves "smirched" if we go into it. Their agent sent a letter to Barker making a offer. He will show it you if you ask him & you might bring it before the Board but I would have nothing to do with it if I were them[8]

I have nothing to say as to Gramme Co organization[9]

Please call Mr Andrews attention to the work on Dynamos & ask him to more thoroughly inspect workmanship[10]

Wednesday Mornings Memo. Gas gives $98\frac{1}{2}$% heat[b] & $1\frac{1}{2}$ light whereas the Electric Light gives $6\frac{1}{2}$% of heat & $93\frac{1}{2}$% of light.[11]

I would not unite with anybody to go for Swan. I would let Swan light up whatever he desires but simply notify him that he is infringing

I would pay good attention to the underground conductor business

I enclose application from a man named Paine.[12] I do not know anything about him but thought it as well to send it to you.

Please send out here a man from Isolated Co to wire my house for 60 lights   He had better bring wire Switches Safety Catches &c &c

Thos A Edison   I[nsull]

L (letterpress copy), NjWOE, Lbk. 12:288 (*TAED* LB012288; *TAEM* 81:656). Written by Samuel Insull. [a]Place taken from Edison's laboratory handstamp. [b]Obscured overwritten text.

1. Insull occasionally used the handstamp acquired by Stockton Griffin in 1878. See Doc. 1292.

2. J. G. Moore & Co. was a large construction and civil engineering firm established by John Godfrey Moore, president of the Mutual Union Telegraph Co. Eaton enclosed an extract of the firm's 14 April letter from Rio de Janeiro stating that the importation of Edison equipment into Brazil would hinder their efforts to secure patent privileges. *NCAB* 5:247; Moore Obituary, *New York Times*, 24 June 1899, 7; "Not Mutual Union Property," ibid., 23 Jan. 1883, 8; Eaton to TAE, 9 May 1882, with enclosure, DF (*TAED* D8237K; *TAEM* 62:157).

3. Miller Moore's telegram has not been found. Eaton explained that after spending three nights aboard the new yacht of publisher James Gordon Bennett, Jr., Moore had reported that despite engine problems "the electric light works splendidly. One night they had fearful weather all night. The dynamo was run until four o'clock in the morning. Much of the time the magnets were describing an arc first at right angles on one side and then at right angles on the other. But they worked first rate" (Eaton to TAE, 10 May 1882, DF [*TAED* D8226B; *TAEM* 61:148]). The *Namouna*'s 120 B lamp installation was completed by early May (Edison Electric Light Co. Bulletin 9, 15 May 1882, CR [*TAED* CB009; *TAEM* 96:706]; see also Strouse 1999, 206). In 1880, Edison had made preliminary estimates for lighting a different Bennett yacht (see Doc. 1969 n. 3).

4. See headnote above.

5. Kendall & Co. was the agent of Fabbri & Chauncey in Valparaiso, Chile, and controlled the government's license for Edison's lighting system there. According to a typed extract of a 1 April letter, they suggested that the insulating compound should be altered, "as with the slightest exposure to the Sun it melts and runs out. . . . This is hardly likely to occur when once underground but during the voyage or landing, transport or process of laying, it may easily and will occur that they are exposed to the powerful heat of the Sun." Edison's directive to Kruesi has not been found. Kendall & Co. to TAE, 1 Apr. 1882, DF (*TAED* D8236E; *TAEM* 62:100).

6. Eaton enclosed in a letter on 4 May two of the weekly updates he received. He asked if Edison wished to see them in the future. DF (*TAED* D8224U; *TAEM* 61:27).

7. Rowland Hazard, a New York investor and banker, was president of the Gramme Electrical Co., the patent consortium. His letters have not been found but according to Eaton, he reported that "the Swan Company will light up the Madison Square Theatre or some other conspicuous building right away. He wants to know if we will unite with the United States [Electric Lighting] Company without prejudice to our controversy with them, to crush out the Swan Company. What do you say. Hazard has notified Mr Mallory & the Madison Square Theatre that the Gramme Company will go for the Swan light if they introduce it and go for Mallory too." Doc. 1595 n. 3; Eaton to TAE, 10 May 1882, DF (*TAED* D8226B; *TAEM* 61:148).

8. The letter to George Barker has not been found. Eaton subsequently reported that Barker believed the Faure patents were probably invalidated by the Plante battery. Eaton to TAE, 15 May 1882, DF (*TAED* D8226D; *TAEM* 61:158).

9. After having initially declined to join the Gramme Electrical Co., the Edison Electric Light Co. did so in March 1882. Its stated reasons were to present a united front against dishonest or extortionate lighting companies, to simplify settlement of patent litigation, and to coordinate other business of interest to the industry as a whole. "Electric Light Monopoly," *New York Times*, 27 Apr. 1882, Cat. 1327, Batchelor (*TAED* MBSB52190; *TAEM* 95:191); Edison Electric Light Co. Bulletin 9, 15 May 1882, CR (*TAED* CB009; *TAEM* 96:706).

10. Edison subsequently explained that he had "no specific complaint to make against Andrews. My only object, in requesting you to ask him to examine the work more closely, was merely precautionary." Edison may have been particularly concerned about loose binding wire on armatures. Charles Dean cautioned that Eaton would receive complaints about this, which he called a "a very difficult little defect to remedy. It has had a great deal of attention from us, as well as from Mr Edison." In a separate incident, Edison advised Dean to "have an inspector on those winders up stairs as I find nearly every armature is a little different somewhere." TAE to Eaton, 15 May 1882; TAE to Dean, 13 May 1882; Lbk. 12:320, 307 (*TAED* LB012320, LB012307; *TAEM* 81:673, 666); Dean to Eaton, 11 May 1882, DF (*TAED* D8233ZAS; *TAEM* 61:1044).

11. Eaton had inquired about Edison's interview in the *St. Louis Post-Dispatch*, in which "you said 'gas gives ninety eight and one half per cent more heat than light.' Do you mean that or do you mean that it gives

ninety eight and a half per cent as much heat as light?" Eaton to TAE, 10 May 1882, DF (*TAED* D8226B; *TAEM* 61:148); "The Doom of Gas," *St. Louis Post-Dispatch*, 1 May 1882, Cat. 1327, Batchelor (*TAED* MBSB52192; *TAEM* 95:191).

12. Probably Sidney Borden Paine (1856–1940), since October 1881 an assistant to Spencer Borden, the Edison company's New England agent. He designed and sold isolated lighting plants and, in September 1882, became an Edison agent in Boston. The item referred to has not been found. Eaton to TAE, 15 May 1882, DF (*TAED* D8226D; *TAEM* 61:158); "Paine, Sidney B.," Pioneers Bio.

–2275–

To Sherburne Eaton

MENLO PARK [N.J.]ᵃ 11th May [188]2

Major Eaton

Your letter 9th inst as to Mr Clarkes letter of 5th returned herewith[1]

I do not see that Mr Clarke need be at all afraid that an Arc Light Co if started would prejudice our chances of success if we do not take up that branch of the business & commence operations right away.

If an Arc Light Co is started in Lawrence & they light up all the streets & large buildings it would not make a difference (under the most favorable conditions to the Arc Coy) of more than 10% in our earnings when we start to light later on the private dwelling houses.

So far as the mere lighting is concerned there is nothing that can be done with the Arc light which we cannot do very much better & (when taken in connection with our General Distribution System where the expenses are spread over a wide area), certainly as cheap—to say nothing of the Safety and simplicity of our light as compared with the Arc Light

There can be no doubt as to our ability to light streets & large places as at this time we are now lighting over a quarter of a mile of t̶h̶e̶ one of the widest streets in London (Eng) under the most unfavorable conditions as the Stores on the Street (Holborn Viaduct) do no business after 6 p.m. & therefore display no light.

As I have often said the reason I have not given my attention to outside lighting is that it forms such a small factor in the Industry of Lighting a City & I am quite content that the Arc Companies should get all the business it is possible for them to obtain as I know that from the nature of their light I shall have at least 90% of the lighting to cater for. A Gas Co makes its profits out of the gas sold to consumers whose bills are less than $10 a month

We can make an incandescent light as powerful & certainly far more agreeable than the Arc Lights used. We are using in London 32 candle & 50 candle lamps & for weeks we had 100 candle lamps burning in our Lamp Factory here. At Geneva (Switzerland) the Coy that controls my light there had 3000 candle incandescent lamps burning for some time.[2]

When the Electric Light becomes an old story Corporations[3] will not pay the price they do for Arc Lights now. In New York City the Gas Contract calls for [-----][b] 4 ~~candle~~ feet[c] gas burners & the corporation only pay the gas Companies for this amount of Light. The Brush Lights on Fifth Avenue & Broadway give about 250 candles effective light.[4] If Gas burners of 250 candles were put up instead the Streets would be lighted just as effectively far more agreeably, and considerably cheaper as the cost of Lighting by gas becomes more economical as the candle power of the burner is increased

Then the Arc light is dangerous ˙ The wires erected in the Streets are but temporary—the chances of their coming in contact with other wires & thus causing fire are very great.

I feel firmly convinced that an Arc Coy could not live in Lawrence ~~of~~ very long after we have entered into competition with them— The Safety & completeness of our system—the agreeableness simplicity & superiority of our light & last but not least our ability to undersell them—will prove more than they can stand up against

We have no competitor except the Gas Co

Thos A Edison

LS (letterpress copy), NjWOE, Lbk. 12:292 (*TAED* LBo12292; *TAEM* 81:659). Written by Samuel Insull. [a]Place from Edison's laboratory handstamp; "N.J." not copied. [b]Canceled. [c]Interlined above.

1. Eaton enclosed a letter (not found) from F. E. Clarke, an agent of the Pemberton Mills in Lawrence, Mass. Eaton explained that he was "a very important man to us in that section of the country, and I feel some doubt about what to say to him"; he subsequently forwarded this document to Clarke. By 1883, Clarke had become president of the Edison Electric Illuminating Co. of Lawrence. Eaton to TAE, 9 and 16 May 1882; Clarke to TAE, 24 May 1883; all DF (*TAED* D8224X, D8226E, D8348A; *TAEM* 61:28, 162; 69:688).

2. Nothing is known of the 3,000 candlepower lamps. Edison made some experimental 100 candle lamps in February, which ran in the Menlo Park lamp factory. In late May, Edward Johnson advised that Swan reportedly had lamps of 60 to 100 candles, while Edison's 32 candle streetlights had "met with nothing but disaster" on Holborn Viaduct. Edison commented that while those were handmade, production versions of the 32, 50, 100, and 200 candle lamps would be as good as stan-

dard A lamps. Edison Electric Light Co. Bulletin 3, 24 Feb. 1882, CR (*TAED* CB003; *TAEM* 96:674); TAE to Johnson 15 Feb. 1882 (p. 14), Lbk. 11:276 (*TAED* LB011276; *TAEM* 81:351); Johnson to TAE, with TAE marginalia (pp. 5–6), 21 May 1882, DF (*TAED* D8239ZBU1; *TAEM* 62:909).

3. That is, municipalities.

4. At the end of 1880, the Brush Electric Co. installed 2,000 candle-power arc lights along Broadway from Madison Square, near Fifth Ave. and 25th St., to Union Square, at 14th St. In July 1881, the company began illuminating each square with 6,000 candlepower arc lamps suspended from 160 foot wooden masts. Friedel and Israel 1987, 192–93; "Rival Electric Lights," *New York Star,* 21 Dec. 1880, Cat. 1241, item 1559, Batchelor (*TAED* MBSB21559X; *TAEM* 94:623); "Electric Lighting in America," *Metropolitan* (London), 25 Dec. 1880, Cat. 1013:29, Scraps. (*TAED* SM013029a; *TAEM* 23:601); "New Lights in Broadway," *New York Times,* 20 Dec. 1880, 1; "The Electric Light in Union and Madison Squares," *Harper's Weekly* 26 (1882): 26.

–2276–

Notebook Entry:
Electric Light
and Power

[Menlo Park,] May 12 1882

Reducing E.M.F.[1]

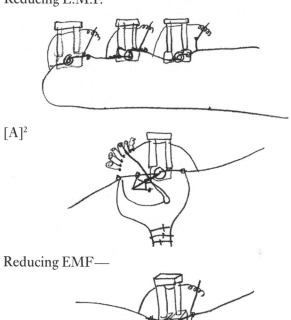

[A][2]

Reducing EMF—

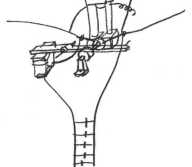

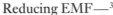

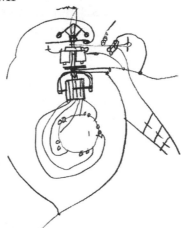

Tried today plate lead on pole plate Compressed peroxide lead other pole as secondary battery.[4] works elegantly. Zinc pole to peroxide.[a]

Try EMF reducing Experiment today[5]

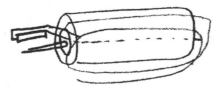

Worked Splendid[b]

Present Logan Hughes Dyer Insull Fred,[6] & Dynamo man[7a]

Improvement on Dynamo bobbins to reduce spark   double up the number of Commutators bringing the Extra ones from the other end through the bobbin ie[c] the lignum vitia & [alternate?][d] as front with a back connection with commutator blox

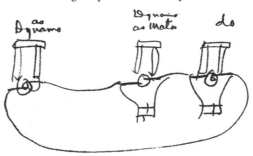

I am now trying to make Commutator bars of solid amalgam by plating ie[c] depositing the metal electrolytically in the presence of a mercury salt so it will be continuously amalgmted[e] as it is deposited. this will prevent spark & render the constant putting of Hg on the Commutator unnecessary

Make battery by using pole of peroxide Copper other pole ~~pero~~ Iodide [----][f] Copper.[a]

also try lead plate & peroxide Lead plate of ~~p~~moulded per-oxde only in Solution of Acetate Lead or other Soluable Salt of Lead.

Apparatus[e] used for determining the resistance of Contacts on surfaces at high velocities for Commutator of Dynamo machines[8]

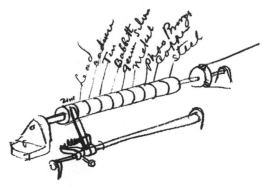

Reducing Emf[9]

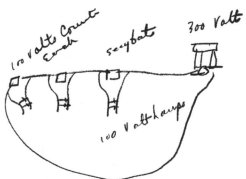

Regulating Field of force magnets in Isolated Dynamo ~~autl~~ automatically.[10g]

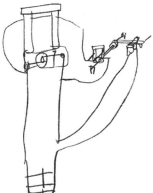

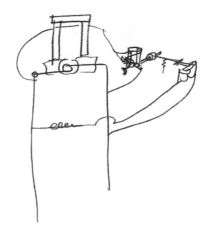

Reglating EM.F.[11]

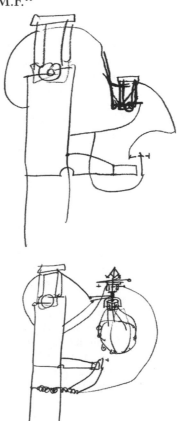

X, NjWOE, N-80-10-01:105, Lab. (*TAED* N304:62; *TAEM* 41:1113).
Document multiply signed and dated. [a]Followed by dividing mark.
[b]"Over" written as page turn. [c]Circled. [d]Illegible. [e]Obscured overwritten text. [f]Canceled. [g]"over page 125" written as page turn.

1. Several of the sketches in this document pertain to long-distance transmission of relatively high voltage currents and voltage reduction at the point of consumption. Edison executed a number of patent applications related to this between late May and early July; five resulted in patents. U.S. Pats. 265,786; 278,418; 439,390; 446,666; and 464,822; see Doc. 2295 n. 3.

The first sketch above is similar to one Edison made in March (see Doc. 2242). Both appear to be the basis for an application he executed on 22 May that became U.S. Patent 265,786. In this specification Edison described an electrical distribution system using high voltage from a distant source and several motor-generator machines. In each motor-generator, a single shaft carried armatures for a motor, run by the high-voltage line current, and a generator, which induced current at a lower voltage for local consumption.

2. This sketch showed a method for regulating the voltage of a dynamo by using a flyweight connected to a lever which switched resistances in and out of the field circuit, thus compensating for changes in engine speed. This sketch appears to have been part of the basis for an application Edison executed on 28 November 1882 that became U.S. Patent 278,413. In this specification, the flyweight arrangement was used to regulate a dynamo for changes in load and engine speed.

3. Edison represented a similar arrangement in an application he executed on 7 July that became U.S. Patent 265,783.

4. Other notebook entries from this time represent details of Edison's work on secondary batteries. On 13 May, he described a process for making "plates of spongy metallic Lead pressed up into plates" by precipitating lead chloride or lead acetate; this worked "bang up." He later testified that in late May he tried forming plates by pouring molten lead into water from a considerable height "so as to get enormous surface without the necessity of reducing an oxide to the metallic form electrically." He hoped that such a plate would "have less local action, less resistance per unit of peroxide of lead," and would therefore operate more efficiently than a Faure battery. In June and December he and Martin Force again investigated methods to make lead plates and other battery elements. N-80-10-01:131; N-82-05-15:1–7, 15–57, 77–79; both Lab. (*TAED* N304:75, N203:1–4, 8–29, 39–40; *TAEM* 41:1126, 40:473–76, 480–501, 511–12); Edison's testimony, 5, 14, *Edison v. Seymour*, Force.

Between May 1882 and January 1883, Edison executed seven patent applications on the construction of secondary batteries. Two issued as patents but Edison abandoned the others. The particular process described in this notebook entry, electroplating lead onto battery plates, was apparently the basis of an abandoned application, Case 420, filed on 26 June. U.S. Pats. 273,492 and 274,292; Cases 420, 452, 458, 480, 530; Patent Application Casebooks E-2537 and E-2538; Patent Application Drawings, Case Nos. 179–699; all PS (*TAED* PT021190; PT021266; PT021280; PT021328; PT022034; PT023:64, 77, 82, 85, 101; *TAEM* 45:755, 767, 770, 774, 791, 882, 895, 900, 903, 919).

5. This sketch formed the basis of an application Edison executed on 22 May that became U.S. Patent 379,772. The specification described the invention as providing "efficient and economical means for dividing an electric current of high electro-motive force or tension into a number

of currents of lower electromotive force or tension" by applying a "counter electro-motive force" at a number of points in the circuit. Electric motors could be used for producing this counter electromotive force but Edison preferred to use secondary batteries, which could also supply power to the load. See also Doc. 2295 n. 3.

6. Possibly Alfrid Swanson.

7. Possibly John H. Vail. See *TAEB* 5 App. 2.

8. See Docs. 2131 and 2134.

9. See Doc. 2295 n. 3.

10. Between May and December 1882, Edison filed at least four patents for regulating dynamo voltage by placing a motor in the field circuit to apply a counter electromotive force. U.S. Pats. 264,667 and 264,672; Cases 427, 523; Patent Application Casebooks E-2537 and E-2538; both PS (*TAED* PT021206, PT022018; *TAEM* 45:757, 788).

11. This sketch apparently formed the basis of applications that became U.S. Patents 264,667 and 264,672, both executed on 22 May. Both pertained to a motor in the field circuit and a relay to energize or de-energize the motor field coils.

–2277–

To Edison Lamp Co.

Menlo Park 13th May [188]2

Dear Sirs,

I think you should put in such tools as would enable you to make what we have been making here & also your moulds so as to relieve us of any more work for the Lamp Factory. I shall simply finish what we have on hand now

Let Bradley make out a list of tools that would be required for repairing Dynamos, Heaters, Annealers & Moulds & send the list to me and I will give you an estimate as to cost as I have facilities for obtaining tools at low rates[1]

Attend to this immediately as I want all the facilities I have here for Experimental work   Yours truly

Thos A Edison   I[nsull]

L (letterpress copy), NjWOE, Lbk. 12:312 (*TAED* LB012312; *TAEM* 81:669). Written by Samuel Insull.

1. Edison evidently sent a detailed inquiry (not found) on the factory's behalf to a New York machinery dealer a few days later. He then urged Francis Upton to order the tools specified there. Upton did so in early June, paying at least in part by consigning a steam engine for $1,000. Harry Livor to TAE, 20 May 1882; Livor to Samuel Insull, 29 May 1882; Upton to TAE, 6 June 1882; all DF (*TAED* D8204ZBO, D8204ZCE, D8230ZAX; *TAEM* 60:146, 162; 61:776); TAE to Upton, 24 May 1882, Lbk. 12:414 (*TAED* LB012414A; *TAEM* 81:723).

To Sherburne Eaton

Major Eaton,

Will send you [copy?]ᵃ about Faure Battery for Bulletin in a day or so.[1]

Your memo 15th inst. I return [Borden's]ᵃ letter.[2] He had better take [a]ᵃ patent out. If they stay in [the?]ᵃ office they will probably get [into?]ᵃ interference and he will probably have trouble whereas if he gets them out Sellon & Volkmann cannot go back of their first foreign patent[3]   I am working on Storage Batteries but not quite in the same line as Faure, S.&V., &c. Our company might advance the money to [take]ᵃ out the patents already allowed & make some arrangement, that if his patents should turn out to be essential to my work on Secondary Battery, or if we should want to acquire them for other reasons that we shall have the option to purchase them on some terms to be arranged.

I enclose letter from Wm Kline [----]ᵃ a smart man Supt of Lake [Shore and?]ᵃ Mich [Southern]ᵃ R.R. Telegraph over an immense territory & is considered by men along that line a great Electrician so he might be useful to us[4]

I enclose a sheet from Iron Age.[5] You may find something in it useful for your Bulletin

Your memo 16th.[6]

The ~~Deff~~ Defects Report is only meant for those engaged in manufacture & for the Coys Engineers   I do not think it should be sent round to the agents nor to Havana[7]

Referring to Bliss' order we have no 20 light Dynamo.[8] The "E." will run 17 As or 30 to 32 B lamps

Kruesi is now putting in the Service Pipes & if you drop a note to him he will cut in Winslow, Lanier & Co.[9] By proper representation I think we can get all the paraffine wire passed.

Kruesi has got a compound that will stand tropical heat.

We certainly do not want to have anything to do with Goebel.[10]

Thos A Edison   I[nsull]

L (letterpress copy), NjWOE, Lbk. 12:343 (*TAED* LB012343; *TAEM* 81:684). Written by Samuel Insull. ᵃFaint copy.

1. Eaton inquired again about this in June but there is no record that Edison supplied the information. Eaton to TAE, 19 June 1882, DF (*TAED* D8226ZAA; *TAEM* 61:288).

2. Eaton enclosed Spencer Borden's letter (not found) about storage battery experiments and related patent applications. Eaton asked Edison to provide comments from which he could write "a nice answer." He subsequently promised to negotiate terms for purchasing patents from

Borden and his two partners. Eaton to TAE, 15 and 24 May 1882, both DF (*TAED* D8226D, D8226H; *TAEM* 61:158, 175).

3. John Scudamore Sellon, an English metallurgist, was one of the founding partners of Johnson, Matthey & Co. He and Ernest Volckmar, a French inventor in London, independently improved the Faure battery by perforating the lead plates and inserting the active compound into the holes. Sellon also used an antimony-lead alloy, which increased the strength of the plates and their resistance to corrosion. Sellon and Volckmar obtained British patents in 1881 and combined their interests in March 1882 by forming the Electric Storage Battery Co. as a subsidiary of the Anglo-American Brush Co. McDonald 1960, 197–98; Schallenberg 1982, 61–65; Dredge 1882–85, cxcv, ccvii.

4. William Kline's 12 May letter has not been found. Edison replied to him at the Toledo, Ohio, telegraph office of the Lake Shore & Michigan Southern Railroad, which operated between Chicago and Buffalo. TAE to Kline, 18 May 1882, Lbk. 12:353 (*TAED* LB012353; *TAEM* 81:692); *Lake Shore & Michigan Southern Railway* 1900, 23–24.

5. This clipping has not been identified.

6. DF (*TAED* D8226E; *TAEM* 61:162).

7. Eaton had discussed the defect reports with José de Navarro who asked that they be sent to Havana, Cuba, to aid in the operation of plants being installed there. This discussion prompted Eaton to ask Edison "what do you think of my sending a copy round to most of our leading agents." On the defect reports see headnote, Doc. 2283.

8. George Bliss was a longtime Edison associate in Chicago previously involved in promoting and commercializing Edison's electric pen, phonograph, and telephone (Doc. 861 n. 1; *TAEB* 4, passim). Following a serious illness in early 1881, he wrote to Edison that he was "busted" and in need of work. He came to New York at Edison's invitation and became a general western agent for the Edison Electric Light Co. in December. By the beginning of 1882, he had established a Chicago office of the Edison Co. for Isolated Lighting and was inquiring about the company's plans for the northwest region generally (TAE to Bliss, 25 Oct. 1881; Bliss to TAE, 19 Apr., 14 Oct., 3 Nov., 9 Dec. 1881, and 3 Jan. 1882; all DF [*TAED* D8120ZBC, D8104ZAY, D8120ZBC, D8120ZBK, D8120ZBY, D8241A; *TAEM* 57:644, 109, 644, 653, 673; 63:84]; Andreas 1886, 598). The dynamo he ordered was for a Milwaukee saloon (Eaton to TAE, 16 May 1882, DF [*TAED* D8226E; *TAEM* 61:162]).

9. The banking house of Winslow, Lanier & Co. had pioneered a variety of investment services before the Civil War and made its reputation selling railroad securities. By this time the well-known firm participated actively in the management of companies for which it raised capital. Its offices were at 26 Nassau St., near the corner of Cedar St. Eaton reported that the bank wished "to be the first people lighted up in the down town district. I sent Greenfield to see about their fixtures and wiring. He reports that the old paraffine wire was used and that the insurance people will not pass it. It would be a pretty serious matter if we had to change all that wire in that district." When the Pearl Street station started their offices were lit with between 50 and 60 lamps. Carosso 1970, 12–13, 27–28; *Trow's* 1883, 1768; Eaton to TAE, 16 May 1882,

DF (*TAED* D8226E; *TAEM* 61:162). "Electricity Instead of Gas," *New York Times*, 5 Sept. 1882, Western Edison Light Co. Bulletin 1:9 (*TAED* CA005B; *TAEM* 96:309).

10. Henry Goebel, a watchmaker from Germany, had been working on improved vacuum pumps for the American Electric Light Co. in New York. He applied in January for a patent (issued in October) on an improved high resistance incandescent carbon lamp and exhibited the instrument in April. Goebel's lamp resembled Edison's except that the ends of the platinum lead-in wires were twisted into spirals and secured to the carbon by "a cement made of lamp-black or finely-ground plumbago." (Edwin Fox's testimony, *Edison Electric Light Co. v. Electric Manufacturing Co.*, DF [*TAED* D9323AAF; *TAEM* 134:197]; "A New Incandescent Light," *New York Times*, 30 Apr. 1882, 2; U.S. Pat. 266,358). Edison recommended that the company "pursue the same policy" respecting Goebel's claims as it had in similar cases because "if we were to buy up Goebel it would simply be an invitation to dozens of others to come forward with their fraudulent claims and make similar demands on our Treasury. That Goebels claims are altogether without foundation there cannot be the slightest doubt." Patent attorney Edward Dickerson and his son, Edward Jr., looked into the matter and recommended, according to Eaton, "that the only use in buying up Goebel is to prevent his being a witness against us. They say that on the witness stand he could treat his experiments as abandoned or not as he might think best and that that is about all we could buy in paying him money" (TAE to Eaton, both 15 May 1882, Lbk. 12:320, 323 [*TAED* LB012320, LB012323; *TAEM* 81:673–74]; Eaton to TAE, 16 May 1882, DF [*TAED* D8226E; *TAEM* 61:162]). A contemporary account claimed that Goebel had made an incandescent lamp almost thirty years earlier (Eaton to TAE, 19 June 1882, DF [*TAED* D8226ZAA; *TAEM* 61:288]; see also Passer 1953, 160).

–2279–

From Charles Clarke

New York, May 18 *1882*[a]

My dear Edison,

Your telegram at hand.[1] The 250 light machine is, in my opinion, all that could be desired. There was no cross in the armature. After the long run on the 16th the field was hardly warm although the binding wire on armature was hot. You have the notes of the test, which will show that the current was greatly in excess of what it will be in practice, the volts throughout test being very high. The armature wires themselves were not too warm. I think it safe for Dean to go ahead.[2] While I think of it, you should order that different kind of rails for the Z dynamo base be furnished with the machines. ⟨Write Dean ask him if he is using cast iron rails on 250 light⟩ They are now of wrought iron and too light. Much complaint comes from the Isolated Co. on account of them. I have, some time ago, called Dean's attention to it but nothing seems to have been done. The heavy cast-iron rail, first used, was best. It has additional advantage from a broad bearing face and will not bind in the grooves in sole-plate.[3]

Dean has no Z cores on hand with which he can try the experiment on the bar armature which you suggest. All things considered, we have thought it best to try it on the 4" cores which he says will be ready on Monday. ⟨ ~~I think~~ write ~~Dean ask if he cannot suggest a pattern for field peice for Z bar armature that will~~⟩[4]

Do you think that Stewart in S. America will want standard batteries, high resistance shunts, galvanometer, bridge &c with the rest of his plant?[5] It seems to me that with his photometer alone he can do all that is necessary for regulation, if he keeps on it a lamp with a clear globe. What do you say? ⟨OK   Think photometer ok if he is properly cautioned about clear globes⟩

What is to be done about Albert Swanson as a fireman at Pearl St. Station.[6]

I find that Stewart includes in his order for meters, all necessary chemicals, and apparatus, so that tells me just what to do.

I think that, as Central Stations are established of sufficient magnitude to require meters, a young man educated to the business should go with them as a <u>necessary part of the plant</u>. There are more <u>kinks</u> about the business than can be written but which are readily learned. How does this strike you. ⟨I think it absolutely necessary that young man should be sent who will understand & take charge of the meters.⟩

The sketch shows a method of wiring which I think will overcome the difficulty in respect to dispensing with safety-catches in sockets and fixtures.

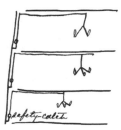

⟨Think this OK⟩   The old method would be

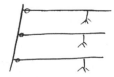

in which the safety-catches would have to be at the ceiling or on the floor above. The first case takes no more wire and would not cost much, if any more to install.

The complaints from the Isolated Co. on account of safety-catches in fixtures are loud. I think we should do away with them.[7] Yrs truly

Chas. L. Clarke

ALS, NjWOE, DF (*TAED* D8227F; *TAEM* 61:455). Letterhead of Edison Electric Light Co. [a]"*New York,*" and "*188*" preprinted.

1. On this date Edison telegraphed Clarke, "Do you think it safe for Dean to go ahead largely on two hundred and fifty light dynamos After they have found the cross cannot you make test with B bar armature with two Z cores and A lamps to ascertain the number of A lights this armature will carry without overheating." Lbk. 12:341A (*TAED* LB012341A; *TAEM* 81:682).

2. Clarke had described this test to Edison the previous day. Despite his confidence in the armature, in early June he reported that Andrews had tested a K dynamo with a resistance of only 1,500 ohms between armature and base, indicating poor insulation. He noted that some armatures "which are tested when green have been found faulty and became perfect after a few days. This may be a similar case but there is a possibility of its being a permanent cross which may increase. Andrews never passed a Z which measured less than 3000 ohms." Edison instructed the Machine Works on 5 June not to ship dynamos with a resistance between armature and base lower than 1,500 ohms. Clarke to TAE, 17 May and 3 June 1882; both DF (*TAED* D8227E, D8227O; *TAEM* 61:445, 470); Rocap to TAE, 5 June 1882, Lbk. 7:401A (*TAED* LB007401A; *TAEM* 80:581).

3. On 20 May John Vail, general superintendent of the Edison Co. for Isolated Lighting, informed Sherburne Eaton of customer complaints about the wrought iron skids on which isolated dynamos were mounted. Charles Dean promised to use cast iron rails. Vail to Eaton, 20 May 1882, in Edison Co. for Isolated Lighting, Defect Reports; Dean to TAE, 24 May 1882; both DF (*TAED* D8222 [image 13], D8233ZBG; *TAEM* 60:944, 61:1061); TAE to Dean, 23 May 1882, Lbk. 12:394 (*TAED* LB012394; *TAEM* 81:713); see Doc. 2203 n. 29.

4. Andrews tried the dynamo with four four-inch solid cores for the field magnets on 27 May, but Clarke was disappointed with the results and recommended trying six-inch cores made from steam pipes. He told Edison that the tests made by Andrews confirmed "your theory of cross section of metal being the important factor when the cores are of equal length." Andrews to TAE, 27 May 1882; Clarke to TAE, 29 May 1882; both DF (*TAED* D8235L, D8227J; *TAEM* 62:40, 61:462).

5. Willis Stewart was the first Edison employee to complete formal training on dynamo repair and installation at the Edison Machine Works, after which he went to Chile as the agent for Edison's light and power system in 1881. During the summer of 1882, he worked on a small central station in Santiago that opened in the fall with three Z dynamos running 150 lights. Edison Electric Light Co. Bulletins 1:1 and 20:37,

26 Jan. 1882 and 31 Oct. 1883, CR (*TAED* CB001:1, CB020:19; *TAEM* 96:668, 96:885); Jehl 1937–41, 963–64.

6. Alfrid Swanson remained at Menlo Park as night watchman and tender of the engines and dynamos for the electric railroad. He asked to be transferred to Pearl St. in February 1883, but remained at Menlo Park through at least the end of that year. See Doc. 2348; Swanson to TAE, 8 Feb. 1883; Swanson to John Randolph, 17 Dec. 1883; both DF (*TAED* D8313B, D8374ZAP; *TAEM* 64:555, 70:1210).

7. On the previous day, Clarke warned Edison of a "big bug in Bergmann's brass mounted sockets." The internal safety catch made a short circuit to the brass and "burns a good sized hole through the brass and uses up the wood work completely. I cannot say that a fire is at all likely but underwriters would doubtless think differently." Clarke suggested eliminating catches inside the lamp sockets but the New York Board of Fire Underwriters subsequently mandated safety catches in each line and also "at the immediate entrance to each fixture . . . on account of the safety catch being omitted from the newest improved sockets." Eaton then recommended putting catches only inside sockets because this would be cheaper. On 3 June Clarke asked Edison for a statement in writing confirming his verbal agreement to the elimination of safety catches in sockets and fixtures. Clarke to TAE, 17 May 1882 (pp. 16–17); Edison Co. for Isolated Lighting, Defect Reports, entry of 28 October 1882; Clarke to TAE, 3 and 8 June 1882; all DF (*TAED* D8227E [image 8], D8222 [images 56–57], D8227P, D8227R; *TAEM* 61:445; 60:987–88; 61:471, 478).

–2280–

From Drexel, Morgan & Co.

New York May 18 1882[a]

Dear Sir,

We received yesterday a cable from Col Gouraud worded as follows:[1]

One Company agreed for India, Australasia. Capital £200,000 A £100,000 B   These proportions necessitated by Stock Exchange rules. Profits equally between A and B after 6% cumulative to A. Vendors receive all B and £50,000 cash unconditionally. Purchase money half for India, half Australasia[2]   Good Board with Marquis Tweedale[3] Chairman. Will issue Co. Saturday if Waterhouse has agreement ready   Everybody objects limiting improvements clause to 5 years. No conditions exacted about new inventions but Company surely ought to have right to improvements at agreed price or[b] arbitration so long as working original patents   half capital on these terms guaranteed before issue. Success certain   Reply promptly.

and in accordance with your request we cabled him in reply last evening as follows:

"We approve proposition for Company India Australia. Edison consents to clause about improvements upon and during life present patents at agreed price or arbitration, provided no conditions exacted concerning or involving future inventions. which please confirm. Yours very truly

Drexel Morgan & Co.

L, NjWOE, DF (*TAED* D8240L; *TAEM* 63:20). Letterhead of Drexel, Morgan & Co. ᵃ"*New York*" and "*188*" preprinted. ᵇRepeated at end of one page and beginning of next.

1. An undated and unaddressed copy of this cable is in DF (*TAED* D8204ZKU; *TAEM* 60:458).

2. These terms were slightly altered under a 12 June agreement (not found) among Edison, Gouraud, and principal investors in Edison's Indian and Colonial Electric Co., Ltd. The company was capitalized at £250,000 by 40,000 A shares and 10,000 B shares at £5 each. Edison was to receive £25,000 in cash, 5,000 A stock shares, and 10,000 B shares. No dividend would accrue to the B shares until after payment of a 6% dividend on the A shares. The company held rights to Edison's patents for electric light, heat, and power (except for railways) in India, Ceylon, Australasia, and South Africa. Prospectus of Edison's Indian and Colonial Electric Co., Ltd., 12 June 1882, DF (*TAED* D8240X; *TAEM* 63:32).

3. William Montagu Hay, tenth Marquess of Tweeddale (1826–1911), formerly a Member of Parliament, was during his lifetime the chairman of a number of large firms, including several cable telegraph companies. Obituary, *Times* (London), 27 Nov. 1911, 11.

–2281–

Technical Note: Dynamo

Menlo Park, N.J., May 20 82ᵃ

Without increasing investment in armature   By con[cen]-tr[atio]n of field all lines of for[ce]

Patent[1]

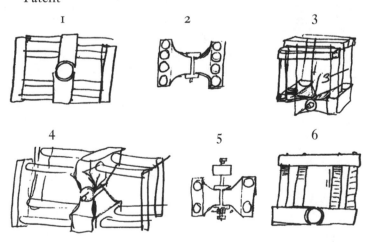

Concentration of field   Put in concentrated poles.

7   Patent[2]

  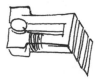

X, NjWOE, Scraps., Cat. 1148 (*TAED* NM017:122; *TAEM* 44:482). Letterhead of T. A. Edison; document multiply dated. [a]"Menlo Park, N. J.," preprinted.

1. Edison signed a patent application embodying these ideas on 3 June. He intended to make generators more efficient and less costly to build

> by contracting the field of force of the machines, and in this way increasing the lines of force or the strength of the lines of force per unit of surface of the armature. The armature is made to correspond in size with the active faces of the polar extensions, which nearly surround such armature. It will be seen that smaller armatures can be used, and that powerful machines may be built without the increase heretofore required in the size and cost of the armatures, and with but a small increase, comparatively, in the cost of the other parts of the machines, and also diminishing greatly the resistance of the armature. The contraction of the field of force is brought about by making the polar extensions smaller at their active opposing faces than at any other point, such polar extensions being made convergent in one or two directions. One or more pairs of electro-magnet cores are attached to one or to each side of the polar extensions. Two more more pairs of cores are preferably thus attached to the same polar extensions, and are placed either in a horizontal or in a vertical position. Each pair of cores is provided with its separate magnetic yolk or back piece, while the polar extensions, to which all the pairs of cores are attached, are made each of one piece magnetically. [U.S. Pat. 281,353]

Figure 2 represents a horizontal section of a machine with four pairs of magnet cores placed vertically as shown in figure 3, which has three pairs of magnets. A figure similar to this and another essentially the same as figure 3 were included in Edison's 3 June patent application.

Figure 4 shows four pairs of magnet cores arranged horizontally, as in the first patent drawing.

Figure 5 shows a section of this machine; the patent application included a corresponding drawing.

2. Edison signed another patent application on 3 June for a means of intensifying a dynamo's magnetic force without increasing its iron mass or electric consumption. This was attained by using flat field magnet cores of soft iron, each wound separately with wire, which would give "greater magnetic strength than the round cores wound separately, or cores made up of a number of flat plates covered with a common winding." The accompanying patent drawing closely resembled the magnet

sketch below except that it showed the convergent pole pieces for concentrating magnetic lines of force, which Edison stated would further improve efficiency. U.S. Pat. 287,523.

–2282–

To Edward Johnson

[Menlo Park,] 22nd May [188]2

EH.J.

I suppose you have some resistances in the field of the big dynamo when you have the load <u>full load on</u>[1]   <u>if so</u> try the experiment of setting both brushes one or more commutator blocks ahead of the [the?][a] centre in the direction of rotation. This will reduce the electromagnetic force and the <u>spark</u>. and you will reach a point where the volts will [jus?][a] be <u>just right</u> with all resistance cut out of the <u>field</u>. The spark will be less [-----][a] at this point with 105 volts than with the same volts [&?][a] when[b] brushes exactly opposite the centre of the slot in the field & a[b] resistance is[b] in field magnet. there is also a slight gain in economy ie[c] $^{38}/_{100}$ of a lamp per Indicated h.p.   ~~if the~~ when the load is light the least spark will be[d] at the point you have them now=

please write minutely how where & appearance & probable cause of cross in commuta[tor]=[e]

I wonder if use of too mu[ch][f] acid solution had anything to do with the cross.[2] ~~Yrs~~   Yrs

E

ALS, NjWOE, Lbk. 12:362 (TAED LB012362; TAEM 81:696). [a]Canceled. [b]Interlined above. [c]Circled. [d]Followed by "over" to indicate page turn. [e]Obscured overwritten text and copied off edge of page. [f]Copied off edge of page.

1. See Doc. 2292.
2. Shortly after this William Andrews found that acid used as solder flux damaged the armature's insulation. See Doc. 2269 n. 2.

ISOLATED LIGHTING COMPANY DEFECT REPORTS   Doc. 2283

As part of his effort to organize the information passing through his office, Sherburne Eaton began systematically collecting reports about all types of problems with isolated lighting installations in the spring of 1882. They were compiled in one or more Defect Books. Only weekly summaries (indicating corrective action, in some cases) from April to December 1882 have been found.[1] John Vail, superintendent for the Edison Company for Isolated Lighting, first prepared these for Eaton; Eaton began signing them himself in late October. Eaton sent them to Edison and apparently also to other principals (Francis Upton referred to reading one in November[2]). Eaton asked if he could comply with José Navarro's request to receive a copy each week; he also suggested sending copies to the company's agents. He hoped they would show "that defects do not amount to much if we handle them intelligently and have a system of weeding them out."[3] Edison discouraged this because the reports were "only meant for those engaged in manufacture & for the Coys Engineers."[4]

1. Electric Light—Edison Co. for Isolated Lighting—Defect Reports (D-82-22), DF (*TAED* D8222; *TAEM* 60:932).
2. Upton to TAE, 13 Nov. 1882, DF (*TAED* D8220ZAU; *TAEM* 60:820).
3. Eaton to TAE, 16 May 1882, DF (*TAED* D8226E; *TAEM* 61:162).
4. See Doc. 2278.

—2283—

To Francis Upton

[Menlo Park,] 24th May [188]2

Dear Sir,

I see in the extracts from Defect Book[1] for week ending May 20th 1882 a remark about the water going through the Plaster of Paris and condensing in the Inside Tube. I think this is all moonshine but I should try some experiment[s][a] by soaking these sockets in water and mix something with the Plaster of Paris—say white [wash?][b] that will make them impervious to water.[2] You should answer this Defect Book question Yours truly

Thos A Edison   I[nsull]

L (letterpress copy), NjWOE, Lbk. 12:415A (*TAED* LB012415A; *TAEM* 81:724). Written by Samuel Insull. [a]Copied off edge of page. [b]Illegible.

1. See headnote above.
2. The extract to which Edison referred is in DF (*TAED* D8222 [images 9–10]; *TAEM* 60:940–41). There is no evidence regarding the experiments he suggested.

[Menlo Park,] May 24 1882

Reducing emf[1a]

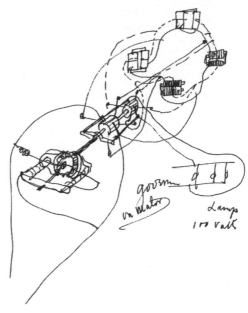

governor
on motor

Lamps
100 Volt

Line   500 volts   passes through efine wire coils on magnets. Lamp circuit comes from coarse wire on mags[2]

TAE   Saml Insull
                                                John. F. Ott

X, NjWOE, Lab., N-82-05-10:11 (*TAED* N198:6; *TAEM* 40:424). [a]Fol-
lowed by dividing mark.

 1. Text is, from left, "governor on motor" and "Lamps   100 volts."
 2. Edison made sketches on 10 and 12 May for reducing voltage by
electromagnetic induction (the operating principle of the modern transformer) but this document is apparently his earliest full representation
of the idea. He made a schematic wiring drawing and brief notes on 22
June (Unbound Notes and Drawings [1882], N-80-10-01:109, N-82-
05:13; Lab. [*TAED* NS82:8, N304:64, N198:7; *TAEM* 44:1030, 41:1115,
40:425]). On 7 July, he executed a patent application for this device. The
rationale was to use smaller conductors and reduce transmission losses,
as in his other voltage reduction arrangements. Edison explained in the
specification that he used magnetic cores "wound with two sets of wire
placed on different portions of the cores or coiled one upon the other on
the same portions of the cores. One set of coils is of high resistance, while
the other set is of lower resistance, and each set of coils is connected in a
closed circuit independent of the circuit of the other set." Each circuit
was connected to a stationary commutator having a bar for each coil. A
pair of brushes on a common motor-driven shaft revolved around each
commutator. The brushes for the high resistance commutator were connected to the high voltage outside line, those on the low resistance commutator to the consumption circuit. Rotation of the brushes would

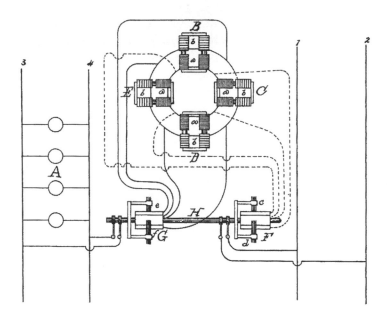

Edison's patent drawing for using electromagnetic induction to transform a relatively high voltage current on the main line to a lower voltage direct current in the consumption circuit.

cause the pairs of magnetic cores to have the connections of their coils reversed two at a time, thus causing a reversal of magnetic polarity. This change in magnetic polarity induces a current of lower tension in the coarse of low-resistance coils, the connections with which being advanced simultaneously with those of the fine-wire coils, the induced current is made a continuous current, or a current flowing in one direction. [U.S. Pat. 278,418]

–2285–

From Edward Johnson

London, E.C., May 27th 1882[a]

My dear Edison,

We are sending out to you tonight copies of 5 patents which we have prepared for Disclaimer.[1] In order that we may obtain our Disclaimers before what is known here as the Long Vacation, commences, and thus be prepared for prompt action, it has been considered desirable to make the applications immediately.

I have requested that you have a further opportunity to criticise before we take the final step but it is impossible to delay the filing long enough to permit this to be done. We find, however, that after having applied for Disclaimer we may withdraw and amend. If your criticisms come back promptly on receipt of these papers there will yet be time for this last course. I do not think there is anything that you will care to alter.[2] I send the papers to you on the chance that we may have overlooked some vital feature and that you will point it out while there is yet time. One thing is certain, the Telephone case has shown us that we must not be chary about disclaiming and that wherever there is considerable doubt about our ability to hold a claim it should be excised unless it happens to be something of very great importance. Had we disclaimed the Phonograph when we were on the Telephone case in the first instance we would not now have had our patent upset. Bear this fact in mind and the further fact that none of the conclusions in respect to disclaiming have been reached except through the most crucial examination by a number of Patent Experts.

I have particularly requested that such patents as may affect the Dynamo and motor be held back as long as possible as I do not feel the same confidence in my own ability to say what in them is vital and what is not. The Lamp Patent and those features which pertain to the General System are, however, pretty thoroughly thrashed out and I think we may proceed with them without fear of committing any grave mistake. Very truly yours,

Edwd H. Johnson

I have today taken my passage Tickets per S. S. Alaska[3] Sailing Saturday June 17th[4]   Will be in New York City at 4 P.M. on Sunday June 25th   Such is the present stage of Ocean traveling that you can predict with tolerable accuracy the <u>hour</u> of arrival at the end of a 3000 mile trip   My money goes on the Guion Boats Everytime[5]   EHJ[b]

LS, NjWOE, DF (*TAED* D8248ZAX; *TAEM* 63:474). Letterhead of Edison Electric Light Co., Ltd. a"London, E.C.," and "188" preprinted. bPostscript written and signed by Johnson.

1. John Henry Johnson sent copies the same day (John Johnson to TAE, 27 May 1882, DF [*TAED* D8239ZBW; *TAEM* 62:936]). The five affected British patents were 578 (1880); 1,385 (1880); 3,880 (1880); 3,964 (1880); all Cat. 1321, Batchelor (*TAED* MBP023, MBP026, MBP028, MBP030; *TAEM* 92:149, 166, 187, 212); and 768 (1881).

2. Edison replied that he approved most of the proposed changes but disagreed in some instances. He argued against excising the fifth claim from his British Patent 578 of 1880 (Cat. 1321, Batchelor [*TAED* MBP023; *TAEM* 92:149]) covering "an incandescent conductor formed of several separate conductors joined together." This would be a "mistake of grave importance since a lamp of this character is in the direction of future improvement in the art of lighting by electrical incandescence. I am working on lamps of this kind myself." Edison made several recommendations regarding the other patents, but conceded that "my views on this subject having been fully considered, I cannot do otherwise than submit to the judgment of the legal advisers of the Company." TAE to Arnold White, 15 June 1882, Lbk. 7:487 (*TAED* LB007487; *TAEM* 80:625).

Shelford Bidwell responded to a number of Edison's critiques but recommended that the disclaimer petitions not be altered; they were granted in early July. Bidwell memorandum of opinion, 30 June 1882; Waterhouse & Winterbotham to TAE, 1 and 8 July 1882; all DF (*TAED* D8239ZCY1, D8239ZCY, D8239ZDB; *TAEM* 62:1005, 1007, 1012).

3. The *Alaska*, one of the largest Atlantic steamers at the time, entered service at the end of 1881. It quickly established several speed records (somewhat over seven days) between Liverpool and New York. The *Alaska* was fitted with incandescent electric lamps manufactured by Siemens. Bonsor 1955, 238; Fox 2003, 288–89; "The Alaska's Great Record," *New York Times*, 17 Apr. 1882, 8; "Speed on the Ocean," ibid., 22 May 1882, 8.

4. Johnson sailed on the *Alaska* on Friday, 16 June. Edison Electric Light Co., Ltd., to TAE, 20 June 1882, LM 1:227A (*TAED* LM001227A; *TAEM* 83:985).

5. Stephen Guion, a native New Yorker and naturalized British subject, founded his own shipping line in 1866. The firm was capitalized in Britain and the U.S.; its vessels were built in Britain but carried American names. It quickly acquired a reputation for technical innovation and, in this period of sharpening trans-Atlantic competition, also for speed. It was formally named the Liverpool & Great Western S. S. Co. but commonly called the Guion Line. Fox 2003, 265—66, 276, 291; Hyde 1975, 58–79.

–2286–

To Sherburne Eaton

[Menlo Park,] [29th][1a] May [188]2

Major Eaton,

Enclosed please find letter from Professor Draper.[2] He wants 110 volt E machine & some "A" lamps the former to replace his present 55 volt machine. Will you please have this attended to.[3]

I also enclose letter of Dr. E. P. [Hussey?].[4a] I have written him that I have referred it to you.

I have also written to Wexel & DeGress (letter enclosed)[5] stating that if they communicate with you, you will put them into communication with the right people to negotiate with as to Mexican Light

Your memorandum of 25th inst   I think that it is absolutely essential that the company get a man of good address and business qualifications (whom they can easily afford to pay a large salary) to visit the different cities & work up the local companies. we are sadly behind hand in this respect as compared with other companies which are very active. Our station will soon be going & I think we should then form our companies all over the country with great rapidity & move at once before the things get to be an old tale[6]

In re Dr Moses. I think a quarter or a third is enough. He cannot have the use of my laboratory   When he gets a drawing which looks right & is free from patents it can be made by Bergmann & tested at Goerck St[7]

There will probably be danger in reissuing our patents in Canada as the surrendered patents would expire and probably affect the American Patents.[8]

In re Mott. we cannot buy thing when we do not know what they are. Let him give a diagram of what his things are with copies of patents already granted & copies of applications pending then we can tell something about it.[9]

It is a curious fact that Gordon[10] Barker & others do not see that there is another advantage in our lamps besides economy & that is its high resistance even for Isolated Lighting. For instance to lose only five % in wiring a building say the copper wire cost $100 with us whereas with Swan Maxim & Lane Fox they would have to spend $500 in copper[b] if they wanted to get the same percentage of loss & if they put in only the same amount of copper as we do they lose about 25% on their conductors. This is a point I have never heard spoken of. Just think it over a little. All these people are using wires about the same size as ours and they do not appreciate the enormous loss they have upon their conductors & it is just as well so far as our interests are concerned that they should not know.[11]

We wrote Painter. He says Barkers report is not on file yet but that he will get it when it is printed.[12]

How about forming a company for Philadelphia. I think it is about time.[13]

You might well ask "Where is Ananias"[14] when they talk about 10 per horse power each 20 candle [---][c] power with 1000 hour life for a Swan Lamp. These fellows are a gang of liars & will turn out just as the Maxim Crowd did.

Regarding your long interview with Stockley from what I know of him previous to his going into Electric Light business I look upon him as a man who deceives by excessive frankness.[15] I should not worry myself very much about the Brush Company & Mr Stockley using secondary Batteries & that sort of nonsense or Stockleys statement that Brush says he can do this & that, "compound multiple arc." &c &c.[16] If you ask Wilber to let you look at my applications for working incandescent lamps in multiple &[d] compound arc in connection with the Brush arc light & machine you will see they I have pretty well covered that matter up and the applications have been in quite a good while & no one can beat me on it [it?][c] in an interference as in 1878.[17] I worked arc and incandescent lights in conjunction in all kinds of ways.[18]

If I understand rightly they propose where they run arc lights with a great many on a circuit to put in Secondary Batteries storing up electricity in the day time and at night when the circuit is required for arc lights, using the Storage Battery to feed the Incandescent Lamps and this all for the sake of using smaller wires ~~up~~ suspended on poles.

This would be analagous to a new gas company starting with a one inch pipe suspended on poles, putting gasometers in each mans house & then carrying the gas under great pressure so that the customers can use gas⁵ at night from their own gasometer & not from the mains. As people who use light are only 25 ft apart I think you will see the absurdity of this when it takes only one length of tube to go from one house to the other, the first cost of which is infinitely less than to have secondary batteries as a substitute— In fact Mr Stockleys talk in this connection is just nonsense.[19]

Of course it can be done but there is no economy in it and it would be very unreliable   As I have said before I think we had better keep free from all arrangements with anybody <u>and save all our energies</u> for prosecuting with the utmost rapidity our own business at the same time getting ready to defend our rights before the court.

I think from reading your memo. that the Brush Co will buy the Swan Light and that they will go ahead with it. But that will do them more harm than good.

I cannot see how one can have any confidence in a man who states that his Electrical Adviser has said that Edison has covered up incandescent lighting yet also says they must have an Incandescent light & if not ours someone elses & consequently on their own words it is natural to surmise that they are knowingly going to the Swan people—Infringers of our rights.

To sum up Stockley has a good wife thinks her of great value to him and he thinks a second wife would be an advantage & that two would be better than one⁵ so he put up someone else to capture my wife for him & then protests he has nothing to do with it.

Above all things I would not have the Gramme Coy have anything to do with our litigation   They are perhaps handy for getting information[20]

As to your memo of 26th which came to hand this pm. I would not sell Isolated Plants to the American Coy. If they give you an order you could arrange to give them the usual com-

mission but I do not think it would be advisable to do anything
further.[21]

Thos A Edison   I[nsull]

L (letterpress copy), NjWOE, Lbk. 12:459 (*TAED* LBo12459; *TAEM*
81:748). Written by Samuel Insull. [a]Illegible. [b]"in copper" interlined
above. [c]Canceled. [d]"multiple &" interlined above. [e]Obscured overwritten text. [f]"& that two . . . than one" interlined above.

1. Date determined from contents of letter.

2. Edison first met Henry Draper (1837–1882), professor of chemistry and natural science at the University of the City of New York (later
New York University), in 1877. Draper had a 4 horsepower Otto gasoline
engine and three dynamos: one each by Gramme, Maxim, and Edison.
His enclosed letter has not been found but see note 3. See Doc. 967; *ANB*,
s.v. "Draper, Henry"; "Henry Draper," *Sci. Am.* 25 (1883): 89–96.

3. After the Edison dynamo with a "B" armature was installed in the
middle of May, Draper reported that he could not bring incandescent
lamps above two candlepower without running the engine at an unsafe
speed. Edison offered to have an armature wound specially that would
provide sufficient voltage at the engine's rated speed. On 22 May, Draper
wrote that the 55 volt machine was working well but he evidently complained again a few days later. Edison referred to that letter (not found)
in assuring him on 29 May that he had asked Eaton to replace it with a
110 volt machine. This was installed in the first week of June. Draper
received the entire plant free of charge. Draper to TAE, 14 and 22 May,
and 4 June 1882; Eaton to TAE, 3 July 1882; all DF (*TAED* D8220O,
D8220P, D8220Q, D8226ZAK1; *TAEM* 60:769, 771, 773; 61:324);
TAE to Draper, 18, 24, and 29 May 1882; TAE to Eaton, 7 July 1882;
Lbk. 12:347, 347, 443; 7:679 (*TAED* LBo12347, LBo12416, LBo12443A,
LBo07679; *TAEM* 81:688, 725, 738; 80:720).

4. E. P. Hussey was a homeopathic physician in Buffalo. The enclosure has not been found. King 1905, 1:50.

5. The partnership of Wexel & De Gress sold armaments in Central
and South America; it had an office in New York. The firm asked Edison about manufacturing electric light equipment in Mexico. Doc. 1394
n. 5; Wexel & De Gress to TAE, 5 May 1882, DF (*TAED* D8237M;
TAEM 62:160); TAE to Wexel & De Gress, 29 May 1882, Lbk. 12:443
(*TAED* LBo12443; *TAEM* 81:738).

6. After meeting with prospective organizers of lighting companies in
several cities, Eaton had expressed a wish for "some man of knowledge
about our system, of judgment, of good address, and a good talker, to follow up the seed which I sow in this office in such cases." Eaton to TAE,
25 May 1882, DF (*TAED* D8226I; *TAEM* 61:181).

7. Otto Moses had begun investigating the use in Europe of arc and
incandescent lights in the same system (see Doc. 2229 n. 2). After returning to the United States in late winter, he continued experiments
with Eaton's encouragement. He offered in May to develop the arc lighting business for Edison. Eaton found particulars of his proposal "objectionable" but recommended that Edison consider arc lighting generally
(William Andrews to Eaton, 17 Mar. 1882; Eaton to TAE, 20 May 1882;

both DF [*TAED* D8235F, D8226F; *TAEM* 62:31, 61:168]). Edison replied that

> when our Central Station Plant is running a great many parties will want to have arc lights in front of their buildings which they can have by connecting an arc light across our mains the electricity being recorded on the meters. Of course we should encourage anything that will help us to sell current   If Dr Moses can present drawings of an arc light which I think will work it might be well to authorize the making of one at the expense of the Company the arrangement being made for us to have the sole use of such a lamp in case of its being perfectly satisfactory. [TAE to Eaton, 22 May 1882, Lbk. 12:363 (*TAED* LB012363; *TAEM* 81:697)]

Eaton offered these general terms and the use of "our laboratory." Moses accepted but wanted half the proceeds from such a lamp, which Eaton refused (Eaton to TAE, 24 and 25 May 1882, both DF [*TAED* D8226H, D8226I; *TAEM* 61:175, 181]; TAE to Eaton, 25 May 1882, Lbk. 12:435A [*TAED* LB012435A; *TAEM* 81:734]). Moses ultimately filed seven U.S. patents on arc lighting, for which Edison paid the filing fees (see App. 5; Eaton to TAE, 10 Jan. 1883, DF [*TAED* D8370L; *TAEM* 70:925]).

8. According to the Patent Act of 1870, a U.S. patent would expire at the same time as a foreign patent taken out by the inventor. A bill relating to the creation of an Edison electric light company in Canada had recently failed in Parliament and Eaton had proposed the reissue of Edison's patents in Canada, apparently to provide a longer period for the introduction of equipment there. In order to keep Edison's Canadian patents in force, the Edison Electric Light Co. started a small lamp factory in Montreal with Edison's shop and lamp factory in Menlo Park supplying the necessary machinery. It was operating on a limited scale by the end of May 1882. Edison Electric Light Co. Bulletin 7:4, 17 Apr. 1882, CR (*TAED* CB007:3; *TAEM* 96:692); Samuel Insull to Francis Upton, 25 Oct. 1881; Eaton to TAE, 25 May, 2 June and 1 Aug. 1882; Lamp Co. to TAE, 27 May 1822; all DF (*TAED* D8131N, D8226I, D8226O, D8226ZAY, D8230ZAS; *TAEM* 58:400; 61:181, 227, 359, 773); see also Doc. 2303.

9. Eaton had forwarded to Edison a few days earlier Samuel Mott's statement "of all of his alleged inventions and patents. He wants to sell them to us and make some arrangement about them." He mentioned the matter again in his 25 May letter, noting that "as I read over his list of patents, I am in some doubt whether you invented the Edison light or he did?" Eaton to TAE, 22 and 25 May 1882, both DF (*TAED* D8226G, D8226I; *TAEM* 61:173, 181); TAE to Eaton, 24 May 1882, Lbk. 12:415 (*TAED* LB012415; *TAEM* 81:724); see also App. 5.

10. The basis for Edison's reference is not clear. British electrical engineer James Gordon (1852–1893) in 1880 completed a *Physical Treatise on Electricity and Magnetism*, which went through several editions in Britain and the U.S. He had started experimenting with incandescent electric lighting about the same time as Edison. After attending the 1881 Paris exhibition as a delegate he turned his attention to dynamo design

and in 1882 built what was reportedly the largest dynamo at that time. *Oxford DNB*, s.v. "Gordon, James Edward Henry."

11. In reply, Eaton expressed reluctance to state in the company Bulletins the advantages of high resistance lamps "because I think it would spur up the other people to increase the resistance of their lamp. As I understand it, they can make a high resistance lamp if they know how to and have the necessary skill as manufacturers. Is it not probable that they will gradually acquire that knowledge and skill? If so, I do not wish to hasten it." Edison responded in Doc. 2290. At the end of June, Charles Clarke completed a comprehensive economic analysis of central station operation that included the advantages of high resistance but less durable lamps. Eaton to TAE, 31 May 1882; Clarke memorandum, 30 June 1882; both DF (*TAED* D8226M, D8227ZAA1; *TAEM* 61:213, 493).

12. Edison requested Uriah Painter to get a copy of George Barker's report to the U.S. Mint on the use of the Maxim electric light at the mint in Philadelphia, which he said was "very favorable to us." Grosvenor Lowrey referred to having read it by 12 May but the report has not been found and may never have been filed in Washington. TAE to Painter, 25 May 1882; TAE to Eaton, 20 July 1882; both Lbk. 12:430A, 7:720 (*TAED* LB012430A, LB007720; *TAEM* 81:730, 80:731); Lowrey to TAE, 12 May 1882; Painter to TAE, 6 June 1882; both DF (*TAED* D8230ZAM, D8204ZCS; *TAEM* 60:181, 61:754).

13. The city of Philadelphia awarded street rights-of-way to the Maxim and Brush companies in June. Edison's backers there did not apply because, Eaton explained, "the Drexel people are not disposed to do anything until after our down town central station here is started." Eaton to TAE, 28 June 1882, DF (*TAED* D8226ZAH; *TAEM* 61:315).

14. A New Testament figure, synonymous with a liar. *OED*, s.v. "Ananias."

15. George Stockly (1843?–1906) was vice president and general manager of the Brush Electric Co. Eaton sent Edison a detailed report of a meeting with Stockly, occasioned by a report that the Brush company had purchased Swan's incandescent light patents. Stockly acknowledged that an affiliated Brush company not under its direct control had done so. Obituary, *New York Times*, 21 Apr. 1906, 13; Carlson 1991, 82; Eaton to TAE, 26 May 1882 (pp. 5–8), DF (*TAED* D8226J; *TAEM* 61:188).

16. According to Eaton's account (see note 15), Stockly claimed that Charles Brush had succeeded in operating several arc and incandescent lamps simultaneously on the same circuit. Stockly used the term "compound parallel" circuit (p. 11), which Eaton interpreted as multiple arc. Brush was reportedly also experimenting extensively with storage batteries for the purpose of improving the efficiency of electrical transmission. Eaton concluded that Stockly might consider

> bringing about a combination in some way between the Brush company and ourselves. We can possibly get along without an arc light. If we want one we can probably invent one ourselves. . . . But the Brush company are obliged to have an incandescent light. They feel that they have almost reached the outer boundaries of the area of arc lighting . . . It is for us to formally consider their necessities, and to decide whether to make an arrangement with the Brush Com-

pany . . . or to let them take up the Swan lamp and thus from that moment become hostile. [p. 17]

17. Edison may have been referring to the Telephone Interferences declared in 1878, which involved a number of his patent applications. See Docs. 1270 and 1358 n. 3.

18. Eaton replied that Zenas Wilber told him about four patent applications that Edison had executed in November dealing with arc and incandescent lamps in the same circuit. Two were rejected; the other two, pertaining to regulation of the movement of carbons in arc lamps, were allowed about this time. A fifth, similar to the latter two, issued in April 1884. On 12 June, Edison completed another application for operating incandescent and arc lamps in the same circuit; it issued in 1889. Eaton to TAE, 1 June 1882, DF (*TAED* D8226N; *TAEM* 61:217); Patent Application Casebook E-2537:74, 82; Patent Application Drawings, Cases 370 and 373; PS (*TAED* PT021074, PT021082, PT023:46–47; *TAEM* 45:740–41, 864–65); U.S. Pats. 263,138; 265,775; and 401,486.

19. Edison nevertheless completed several patent applications for using batteries to reduce voltage in the lamp circuit, so as to obtain the transmission economy of high voltage; see Doc. 2295 n. 3.

20. The Brush company, like the Edison Electric Light Co., belonged to the Gramme Electrical Co. consortium. In his 26 May letter (see note 15, pp. 1–4), Eaton raised the possibility of having the Gramme Co. take legal action against Brush. See also Eaton to TAE, 10 June 1882, DF (*TAED* D8226U; *TAEM* 61:254).

21. Eaton had written regarding a letter (not found) from the American Electric & Illuminating Co. He stated that Edison had not answered their query "namely, whether we will sell them complete plants for isolated lighting to be resold by them." Edison had sent the letter to Eaton with the remark that "I do not think we should sell lamps to these people or any other Company   I would however continue to sell small quantities to Professors & others." Eaton to TAE, 26 May 1882, DF (*TAED* D8224ZAD; *TAEM* 61:33); TAE to Eaton, 22 May 1882, Lbk. 12:363 (*TAED* LB012363; *TAEM* 81:697).

–2287–

To Uriah Painter

[Menlo Park,] 29th May [188]2

My Dear Painter,

Referring to the Bill, now before the Senate, for the protection of purchasers of patented articles buying from Infringers of the Patents,[1] if the act provides that the purchase shall have been made in good faith, and that no notice of such sale being fraudulent shall have been given,—I should see no great objection to its enactment if a clause was inserted giving the Inventor some remedy against the parties who wilfully pirated his invention by such sales. Such patent pirates should I think, (upon proof being given that they were fraudulently selling the Invention of another) be sent to the States Prison. France sets us a good example in this matter as Infringers there are, I believe, liable to be sent to Toulon to work in the Chain Gang for five years. I think if the proposed act passes in its present form it will simply be an Act for the Encouragement of Patent Pirates. I see not the slightest objection to protection being given to an innocent purchaser of a pirated patented article but I think it would be a great injustice to give that protection at the expense of the Inventor & Patentee.

Congress when attempting to give protection to the purchaser must not forget the interests of the Inventor, and should at the same time take radical measures to discourage the now too prevalent practice of patent pirating   Yours very truly

Thomas A Edison

LS (letterpress copy), NjWOE, Lbk. 12:448 (*TAED* LB012448; *TAEM* 81:740). Written by Samuel Insull.

1. Painter had sent Edison on 16 May a report (not found) of "the debate in the House yesterday over the passage of a bill to rape patentees. I wish you would write me a letter giving me your opinion on the Bill in such a manner that I can use it. I think it can be beat in the Senate Comm." In 1879, at Painter's urging, Edison had written against similar legislation. Painter to TAE, 16 May 1882, DF (*TAED* D8204ZBG; *TAEM* 60:136); Doc. 1684.

The bill in question, H.R. 6,018, was the result of broad efforts by the Patrons of Husbandry (the Grange) to reform patent laws. The Grange particularly wanted to indemnify purchasers of patented articles and attach all potential patent liability to their makers and vendors. (A similar proposal died in the Senate in 1880.) The Gramme Co. published the bill's major provision in a circular letter in June. Buck 1963 [1913], 119; Gramme Electrical Co. circular letter, 26 June 1882, DF (*TAED* D8224ZAN; *TAEM* 61:40).

Edison subsequently wrote Sherburne Eaton that he had sent "a strong letter" to Painter. Eaton informed him on 10 June that "the bill substantially repealing the patent law would be likely to pass after all." The Gramme Co. appointed a committee (including Eaton) to organize the opposition. The bill ultimately failed in the Senate. TAE to Eaton, 6 June 1882, Lbk. 7:406 (*TAED* LB007406; *TAEM* 80:586); Eaton to TAE, 10 and 15 June 1882, both DF (*TAED* D8226U, D8226Y; *TAEM* 61:254, 282); Buck 1963 [1913], 119.

–2288–

Samuel Insull to
Calvin Goddard

[Menlo Park,] 29th May [188]2

Dear Sir,

I have arranged with Mr Kruesi to put a service pipe into Drexel Morgan & Cos office from the Mains on the opposite side of Wall St. Would you kindly see that D. M. & Co. office is wired as I think Mr Fabbri and Mr Wright are very anxious

to have their office lighted up before any other consumer[1]
Yours truly

Saml Insull

ALS (letterpress copy), NjWOE, Lbk. 12:446A (*TAED* LB012446A; *TAEM* 81:739).

1. Samuel Insull had written John Kruesi a few days earlier that Egisto Fabbri felt "very strongly" about this. The Drexel, Morgan & Co. office on Wall St. (at Broad St., opposite Nassau St.) lay across the street from the Pearl St. station's distribution area. At Edison's direction, Insull asked Kruesi "to find out if a service could not be run across the road to their office. Please look into this at once as the office should be wired if the matter can be arranged." Kruesi promptly replied that it could be done. The firm of Winslow, Lanier & Co. also asked to be the first lighted in the district. Insull to Kruesi, 25 May 1882, Lbk. 12:432 (*TAED* LB012432; *TAEM* 81:731); Kruesi to Insull, 26 May 1882; Eaton to TAE, 16 May 1882; both DF (*TAED* D8236P, D8226E; *TAEM* 62:113, 61:162).

Jehl 1937–41 (1065) states that Edison, with several associates, joined Morgan in his office on the afternoon of 4 September to turn on the ceremonial first lamp. Jehl's account is apparently the basis for later elaborations of the story; see, for example, Friedel and Israel 1987, 222; Israel 1998, 206; Strouse 1999, 233; and Jonnes 2003, 85–85.

–2289–

*Edison Electric
Illuminating Co.
Memorandum:
Central Station
System*

[New York,] May 31st 1882

Cost of First District Per Statement (attached)[1]		295,302.35
Real Estate	36,375.66[2]	
Supplies	10,639.34	
Material	16,668.66	
a/c Oustanding	81,701.13	145,384.79
Canvassing	2,958.64	
Interest	1,720.20	
License &	109,760.66	
Light Apparatus and Expense		
65 Fifth Av	8,344.47	122,783.97
	$563,471.11	

Receipts

Instalments on Capital	544,960.	
Rents[3]	340.	
U H Painter a/c Sub[scriptio]n	360.	
J. G. Moore and Co[4]	.06	
Overdrawn a/c	4,812.48.	550,472.54
Bills unpaid	$12,998.57	

ADDENDUM[b]

[New York, May 31, 1882]

E[dison].E[lectric].I[lluminating]. Co of N.Y. Cost of Installation of First District May 31st 82

Street Conductors

Tubes	98,693 feet 18$^{69}/_{100}$ miles		118,019.60	
Boxes	5,993	Number	14,701.90	
Joints	8,153½ Number		7,013.65	
Wooden plugs	4,575	"	21.22	
Iron Plugs	4,700	"	181.75	
Iron Caps	313	"	154.35	
Card Board	4,894½ Pounds		502.06	
Compound	76,017	"	7,278.49	
Iron Bolts	27,140	Number	404.79	148,277.81
Opening Streets			5,095.64	
Repaving	53,544½ L[inear] feet	15¢	8031.68	
Payments to City	60,500 L feet	1¢	605.00	13,732.32

Sundries

Packages	176.33	
Tape	36.60	
Repairs Test App[aratu]s.	42.75	
Cartage	674.53	
Miscellaneous	2,859.73	3,789.94
		165,800.07

Station Structure

Foundations	1,767.22	
Iron structure & Erection	8,018.71	
Fire Proofing Floors	533.00	
" " Stacks	480.00	
Vaults	3,470.87	
Carpentry	1,130.14	
Hardware and Blacksmith	429.92	
Stairs and Platform	1,320.00	
Sundries	692.89	
Plumbing	156.28	
Sheet Iron Working and Piping	366.25	
Pay Roll and Petty Cash	5,327.95	
Setting Boilers	1,773.53	25,466.76

Section Apparatus

6 Dynamos	48,000.00	
Engines	6,056.76	
Sundries	1,068.48	
Boiler and Fittings	18,939.94	
Smoke Stacks	1,625.00	
Conveyors	3,069.56	
Shaftings	250.00	79,009.74
		270,276.57

General Expenses

General Expense	3,081.76	
Clecks and Collectors	1,175.00	
Stationery and Printing	1,021.01	
Experimental	82.23	
Insurance	408.90	
Taxes	864.60	6,633.50
Lamps[5]		7,500.00
Engineering Expenses		10,892.28
Estimated Amounts to Complete		295,302.35

D, NjWOE, DF (*TAED* D8223E; *TAEM* 60:1026). Commas supplied in large numbers for clarity. [a]"Light . . . Ave" enclosed by right brace. [b]Addendum is a D, written on paper ruled for double-entry bookkeeping.

1. See addendum below. Eaton summarized the expenses this way for Edison on 10 June: "The Pearl St. station has cost up to the first of June about $290,000 and it will be completed inside of $300,000   This includes only items strictly chargeable to the installation of the district. The underground plant, will cost a little over $180,000 of which about two thirds is for Kruesi's conductors, including both mains and feeders, and the other third is for the expense of putting in conductors, boxes &c. &c." DF (*TAED* D8226U; *TAEM* 61:254).

2. The Edison Electric Illuminating Co. reportedly purchased the Pearl St. buildings for about $30,000 (see Doc. 2145 n. 2). The figure cited here presumably includes legal fees or other charges.

3. The source of this income is unknown.

4. This was a New York contracting company established by the construction magnate and telegraph executive John Godfrey Moore (1847–1899), president of the Mutual Union Telegraph Co. The firm erected many of the Mutual Co.'s lines about this time. *NCAB* 5:247; "Not Mutual Union Property," *New York Times*, 23 Jan. 1883, 8; Obituary, ibid., 24 June 1899, 7.

5. The Edison Electric Light Co. paid 35 cents per lamp under its 1881 contract with the lamp company, and presumably transferred them to the Illuminating Co. at that rate. At Edison's urging, the contract was being revised at this time to provide for 40 cents per lamp but it was not ratified before 1883. See Doc. 2039 n. 1; Grosvenor Lowrey to TAE,

12 May 1882; TAE draft agreement with Edison Electric Light Co., undated 1882; both DF (*TAED* D8230ZAM, D8230ZAN; *TAEM* 61:754,
757); Samuel Insull to Lowrey, 5 Oct. 1882, Lbk. 14:217A (*TAED*
LB014217A; *TAEM* 81:912); see also Docs. 2343 n. 3 and 2387.

-2290-

To Sherburne Eaton

[Menlo Park,] 1st June [188]2

Major Eaton,

Your memorandum 31st May[1]

31A did not come to hand[2]   I return you herewith 31B[3] &
[31C?][4a] also newspaper cuttings.[5]

As most of our skill for the last two years has been devoted
not to making a lamp but in making a lamp of high resistance
It will be a long time before anyone with our present secrecy
will be able to make a high resistance lamp. Everybody else has
been at work coating their [---][b] carbons to cover defects and
this brings the resistance down but do not say anything about
this in the Bulletin.[6]

I am in favor of throwing open the patents in Brazil and
[opening?][a] up the business there right away. Never mind the
patents   Delay will injure us.[7]

I should like to have a copy of the Stager[8] contract here for
my information[9]

Thos A Edison

L (letterpress copy), NjWOE, Lbk. 7:384 (*TAED* LB007384; *TAEM*
80:573). Written by Samuel Insull. [a]Illegible. [b]Canceled.

1. DF (*TAED* D8226M; *TAEM* 61:213).
2. This was an unidentified letter which Eaton marked as 31A. The
designation apparently referred to the date of Eaton's letter and the sequence of enclosures; cf. Doc. 2318.
3. This enclosure has not been found. Eaton described it in his 31
May letter (see note 1) as "a circular issued by the Electro Dynamic Co.
of Phila. offering to sell double induction motors for all kinds of purposes." The Gramme Electrical Co. had sent it to him.
4. This enclosure has not been found. Eaton described it as a letter
from Mr. McCarty, an agent in Brazil.
5. Eaton referred in his letter (see note 1) to an article from the *San
Francisco Chronicle* and one from the *San Francisco Examiner.* He stated
that an interview by George Ladd, San Francisco agent for the Edison
Electric Light Co., "about the American Co. is first rate." He also referred to one by Thomas Rae as "slashing but good."
6. See Doc. 2286.
7. Eaton had reported (see note 1) that patents likely would not be issued in Brazil while the legislature considered revising the law, poten-
tially to require patentees to manufacture their articles in Brazil. Eaton
suggested that the company must "either go into business there without

patents, or we must send material and skilled labor and start factories, which is of course out of the question. If we do not do one of those two things we must keep out of Brazil for the present."

8. Anson Stager held high offices in the Western Union Telegraph Co. and was a founder and president of Western Electric. He helped organize the Western Edison Electric Light Co. and served as president until his death in 1885. Doc. 817 n. 13; *ANB*, s.v. "Stager, Anson"; Platt 1991, 37.

9. See Doc. 2299 nn. 1 and 5.

–2291–

*Draft Patent
Application:
Incandescent Lamp*

[Menlo Park,] ~~M~~June 2, 82

Claim: flexible[a] filiments of carbon formed of cellulose carbon obtained from cellulose or compounds of cellulose dissolved by proper solvents ~~to a~~ & made into sheets or membranes from which the ~~loop~~ carbon filiment may be punched or cut before or after [--][b] carbonization or made into solid masses & subjected to heat & pressure whereby it it forced through dies as a fine filiment [-][c]   pure[a] Cellulose may be disolved in ~~cupric ammonium~~ cuprammonic hydrate.[1]

By means of heat and pressure the viscous proxyline[2] may be formed with fine[d] filaments ~~bef~~ and then carbonized.

flow colodion over glass   allow dry ~~carb~~ put membrane between sheets of paper or metal and punch out in proper form & carbonize under strain & pressure— also esheet celluloid may cut in proper shape & carbonized

X, NjWOE, Scraps., Cat. 1148 (*TAED* NM017:131; *TAEM* 44:490). [a]Interlined above. [b]Canceled. [c]Illegible. [d]Obscured overwritten text.

1. Edison's draft is the first surviving evidence of his experiments with nitrated cellulose as a filament material. It is the basis for a patent application that he executed the next day and filed in August. It was rejected as having been anticipated by William Crookes's British Patent 2,612 (1881), covering treatment of cellulose by cupro-ammonia. It was amended and then placed in interference with a number of patents and pending applications, including those of Crookes, Joseph Swan, and Hiram Maxim. The interference was decided at least in part against Edison, but his application did ultimately issue in 1895. Pat. App. 543,985.

Edison's patent referred only to cellulose sheets or membranes and did not mention filaments made by extrusion. A related application, also executed on 3 June, covered making flexible filaments of an "oxidized drying-oil," such as "linseed, cotton-seed, poppy-seed, or nut oil." Edison described coating the liquid oil onto smooth plates, then drying or baking it into a "tough flexible sheet or membrane" from which flexible, high-resistance filaments could be cut or punched. Alternatively, oil in a "solid or semi-solid state may be forced out through dies under heat and pressure, or pressure alone, in the form of a long filament or a thin sheet, which is dried or baked" (U.S. Pat. 365,509; Cat. 1148, Lab. [*TAED*

NM017:136; *TAEM* 44:495]). It was evidently about this time that he completed a third application (Case 445), filed on 14 August, for techniques to prevent deformation of cellulose sheets. That application was rejected and abandoned. The claims and a single patent drawing, corresponding to a 5 June sketch, survive (Patent Application Casebook E-2537:250; Patent Application Drawings; PS [*TAED* PT021250, PT023:71; *TAEM* 45:762, 889]; Cat. 1148, Lab. [*TAED* NM017:135; *TAEM* 44:494]). On 12 June he experimented with cellulose compounds and oxidized drying oil as adhesives for attaching additional layers of carbon and possibly even lead-in wires to filaments (N-82-05-26:31, Lab. [*TAED* N204:15; *TAEM* 40:589]). See also Doc. 2306 and, for a related use of drying oils, Doc. 2307.

Edward Weston was experimenting at this time with filaments of cellulose dissolved and formed in ways similar to those described by Edison. He filed three patent applications in March which issued in September 1882 and were later transferred to the U.S. Electric Lighting Co. Joseph Swan began similar experiments in 1883. Filaments of squirted cellulose were used widely by the combined Edison and Swan company in England, and eventually by General Electric after the expiration of Edison's basic U.S. patents. Woodbury 1949, 119–27; U.S. Pats. 264,986; 264,987; 264,988; Swan 1946, 36–38; Howell and Schroeder 1927, 81–83; Israel 1998, 218, 317.

2. Edison presumably meant pyroxylin, a class of flammable nitrated cellulose compounds, often used as gun-cotton. *OED* s.v., "pyroxylin."

–2292–

From Edward Johnson

[London,] Sunday, April [June] 4, 1882.[1]

My Dear Edison,

Your autograph on adjustment of brushes at hand.[2] We have not had any resistance in the fields of the big dynamos at all; we have been regulating entirely by the steam. Ever since the first night or two, when the new machine gave us considerable trouble, we have been working the brushes at a different[a] angle from that which they had[a] when we received them, and this was done in order to enable us to adjust them to the very position that you now direct. If you will think backwards to one of my early letters, you will recollect the fact that I therein told you that the difficulty was owing to Hammer[a] having the brushes on too straight, and that we set them more obliquely, after which the trouble disappeared. The fact is, that we have had no trouble whatever with sparking, except that there was something wrong with the machine i.e., when the commutator got crossed and subsequently when the lug connections got loose. These two defects having been permanently remedied, we have now no trouble whatever and the machine is running beautifully. There is scarcely any sparking with 700 to 800 lamps, and we are using less and less coal all the time; the entire apparatus, in fact, gains in economy. You can draw a great many interesting facts out of Hood, by cross-questioning him, (I presume you know you will not get much from him unless you do pursue this course).[b] Notwithstanding the above facts I am none the less pleased to receive your letter and have framed it, gilt-edged, as an evidence that you do sometimes write autographically.

We are now discussing the question of erecting another boiler and another steam-dynamo on the Viaduct, for the purpose of securing additional reliability and of putting out another 1000 lights, in order to obtain higher economy.[3] Our shaft, which cost[a] nearly £600, our reserve dynamo, £2,000, the rent of the establishment, labour, and various other things now charged to the product on one machine, will then be chargeable to that of two, and will thus greatly increase our economical showing[c] besides which, we shall be able to make a much larger display and to secure greater reliability. It is not contemplated to make of this a permanent station, or to further increase the plant beyond this additional machine, but it is considered wise (at least, it is my proposition and is assented to by the board)[b] to make this addition, in order to [the?][d] effect an advantageous operation and showing in respect of this first installation. A larger installation will shortly be established, more nearly in the centre of the city proper; but my idea is, that it will never do in the general interest, to permit this Viaduct installation to be discontinued, and that, since it is is to be continued until a larger station can be put into operation and since that cannot be effected within a year—and since also, the great question of cost will be ~~returned~~ determined[e] by the work done here, I am of the opinion that the proposed addition is fully justified by the ends sought. As it is, we will even now be able to show the cost to be within the cost of gas, or to closely approximate thereto. By slightly increasing the economy, I am satisfied that I can make the Viaduct station show an actual profit—taking the price charged for gas—and after taking into account every item that can possibly be charged to the production of the light.

The 260 ohm lamps have been found, and will be tested by the Crystal Palace jury. I am also putting in one of your 250 light dynamos for testing. I hope, on these two later arrivals to be able to obtain a very favourable report. I like the new lamp very much—I have one in my 5–light chandelier at the present moment, and am hardly able to distingush it between it and the A lamps. Sir William Thompson, who has all along protested that Swan was in advance of you in respect to the lamp, as considered by itself, remarked to me a few days ago—upon my telling him that I had received a cask of lamps 260 ohm hot—10-candles "that is a very great stride indeed."

We commenced on the 1st. June, the accurate economy test on the Viaduct.[4] I leave here on the 17th., and will therefore not be able to bring with me the 30 days' record, but will do so in respect of the first 15 days, from which you can make your own calculations. On the 1st. July, we shall begin to charge for the light, by which time I sincerely hope that you will be able to furnish us with either 50, or 100, candle[f] street lamps. My private opinion is, that the 50-candle lamp will ultimately be adopted for street lighting, even though the 100 candle lamp should be forthcoming and prove satisfactory, as I do not believe the city authorities will care to pay for more than 50 candles, which is quite sufficient—seeing that the lamp posts are placed so close together, to give a sufficiently brilliant light for the purpose.

Last week we ~~found~~ finally concluded the negotiations for the S.E. and S.W. and the Western Districts of London. The Consolidated Telephone Construction & Maintenance Co., and Clark, Muirhead & Co., are the parties making the purchase, and they jointly pay us for these three districts, £75,000 and half profits. We have therefore brought two of our district organizations to a final conclusion, viz;—these districts and Lancashire.[5] Our cash revenue from these two organizations[a] is £125,000. Negotiations for other districts are in more or less forward state. Goureaud's colonial affair moves very slowly. I had a long session with the parties yesterday, and I think the small differences about which they have been quibbling for some time, are all now removed. The cash payment will be £25,000, and a like amount in fully paid A shares. It is expected, of course, that these shares will immediately go to a premium, in which event we will sell them, and thus be able to transmit to you the £50,000 originally contemplated[a] as the purchase money for India & Australia[6]

Crystal Palace Exhibition closed last night. We will immediately begin packing all the exhibits, other than those pertaining to electric light, and return them to you, addressed "Menlo Park" I have your communication about the consular certificates, and will attend to it, so that there will be no duty charged upon them.[7]

We have sent 5 of your patents into court, for disclaimer and hope to get the disclaimers allowed before the long vacation, and thus be prepared for the great contest which we propose to inaugurate without delay. The road[a] has been pretty well cleared and I am satisfied that all that it is possible to do has been done to make the contest a strong one. The entire theory of your system, both in general and in detail, is now understood, as it was not understood at first, and I think you may place entire confidence in the ability of the men in whose hands the matter is at present placed, to bring the thing to a successful issue. Meantime, negotiations of a more or less indirect character may be opened with Swan, but on one ground only; viz;—a complete recognition of the validity of your patent on the lamp. Very little, however, has been done in this direction as yet; in fact, nothing beyond the publication of a card by our solicitors, on the occasion of the launching of the new Swan Co.,[8] to the effect that the Swan lamp was an infringement of ours, and also occasional interviews which Mr. Waterhouse has with the solicitor to the Swan Co., with whom he is personally acquainted, in which both parties express a hope that matters may be so arranged that a legal contest will be avoided. You need not, however, fear anything like an amalgamation of the two interests—that will not be done.

I wrote you some time since about selling your Oriental Telephone shares; you need not trouble yourself about it, as I have been wholly unable to get a quotation for them at any price.

Waterhouse has written to Lowry, or Fabbri about the division of the B shares of the Electric Light Co., I hope it will be so arranged that I will be able to get some of mine before I leave here, as I desire to sell a few, though not many. I have been offered £10,000 for my 500 A shares, but by the advice of Sir John Lubbock and others, I have declined it. I want to keep them, and would prefer to sell the B shares. I am arranging before leaving, that White shall conduct any transactions in these shares for you and I, that we may hereafter desire. Very truly yours,

Edwd. H. Johnson

TLS, NjWOE, DF (*TAED* D8239ZBZ1; *TAEM* 62:822). Typed in upper case. [a]Mistyped. [b]Parentheses written in unknown hand. [c]"ing" written in unknown hand. [d]Canceled. [e]Interlined above in unknown hand. [f]Typed following "street"; correct placement indicated by arrow.

1. The letter was dated 4 April but clearly refers to events after that date; 4 June occurred on Sunday.
2. Doc. 2282.
3. According to a report printed in June 1883 for the Streets Committee of the London Commissioners of Sewers, a third C dynamo was

installed at 35 Snow Hill, near the Holborn Viaduct. The report summarized the first year of operation of the Holborn demonstration district. Haywood 1883, 8–9.

4. In July, John Fleming and John Hopkinson submitted a detailed report about the expenses and productivity of the Holborn demonstration district. They concluded that although the costs of electric and gas lighting were very similar at this time, expanding the Holborn station to gain economies of scale would produce a more favorable comparison with gas. Hopkinson and Fleming to Edison Electric Light Co., Ltd., July 1882, Ser. 1, Box 20 (manuscript copy in folder 1, printed pamphlet in folder 3), WJH.

5. The Lancashire organization was for the installation of isolated plants; presumably, the London districts were arranged for the same purpose. Arnold White had told Edison about these plans a month earlier, noting that half the profits were reserved for the parent British company. A similar agreement was reached for Croydon. Johnson had given some details about the formation of district companies in a 21 May letter. Johnson to TAE, 19 Mar. 1882, Ser. 1, Box 20, Folder 2, WJH; White to TAE, 5 May 1882; Johnson to TAE, 21 May 1882 (pp. 13–14); both DF (*TAED* D8239ZBD, D8239ZBU1; *TAEM* 62:870, 909).

6. Edison's Indian & Colonial Electric Co. prospectus, 12 June 1882, DF (*TAED* D8240X; *TAEM* 63:32).

7. Edison cabled instructions to have the exhibits certified in London as American-made, to avoid import duties in New York. TAE to Johnson, 2 June 1882, LM 1:220A (*TAED* LM001220A; *TAEM* 83:982).

8. The Swan United Electric Light Co., Ltd. was organized in mid-May with a broad charter for the production and distribution of electricity for lighting, power, and transportation, as well as for related manufacturing. The earlier firm, Swan's Electric Light Co., Ltd., had been organized in 1881 to manufacture lamps. *DBB*, s.v. "Swan, Sir Joseph Wilson"; Swan United Electric Light Co. memorandum of association, 19 Apr. 1882; Swan's Electric Light Co. memorandum and articles of association, both 3 Feb. 1881; all CR (*TAED* CF001AAG, CF001AAB, CF001AAC; *TAEM* 97:304, 227, 233).

-2293-

Agreement with Charles Dean and Charles Rocap

[New York,] June 7 1882

Copy of Contract[1a]

100	250 light
150	125 & 150
25	extra bar arma. 250
25	" " " 125 & 150
50	Z. mach.
15	Jumbos
25	30 light

All to be done in less than 4 months= 10 per cent of net profits to works is to be paid Dean & Rocap[2] of which Dean gets 7 pct & Rocap 3 pct. If not done within that time I shall make a reduction of the percentage.[3]

signed T. A. Edison

D (copy), NjWOE, DF (*TAED* D8233ZBQ; *TAEM* 61:1072). Written by Charles Rocap. [a]Followed by dividing mark.

1. The editors have not found the original of this contract.
2. Charles Rocap (c. 1856–1932) began working for the Edison Machine Works as a bookkeeper about August 1881 and at some point became secretary and treasurer (Obituary, *New York Times,* 4 Oct. 1932, 21). Edison fired him in August 1883 (see headnote, Doc. 2343 esp. n. 23).
3. Dean and Rocap expected to produce 15 to 20 K dynamos and 12 to 15 125-light dynamos a week. To meet this accelerated production schedule, the Machine Works added 100 men in early July and another 500 soon thereafter. In January 1883 the Machine Works credited Dean with $10,500 and Rocap with $4,500 as their share of the profits. Rocap to TAE, 15 and 28 June 1882; Dean to TAE, 19 July 1882; all DF (*TAED* D8233ZBZ, D8233ZCO, D8233ZCT; *TAEM* 61:1082, 1099, 1104); Account Statement, Edison Machine Works, 20 Jan. 1883, Miller (*TAED* HM830168C; *TAEM* 86:511); "Electric Light," *New York Tribune,* 14 Aug. 1882, 2.

–2294–

Notebook Entry:
Incandescent Lamp

[Menlo Park,] June 8 1882

Caveat[1a]
  Prevention E̶Carrying in Lamp

Condenser in high Vacuo
  Prevention Electrical Carrying—[2]

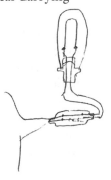

heated Condr
  Exprmts Elec Carrying[1]

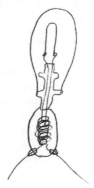

Wire Condr in or not in Vac
  Expts Electrical Carrying[3]

TAE

X, NjWOE, Lab., N-82-06-08:21 (*TAED* N197:11; *TAEM* 40:343).
Document multiply signed and dated. [a]Multiply underlined.

1. Edison applied this heading to each of the four figures below. The editors have found no evidence of caveats or patent applications incorporating these ideas. It is not clear how any of the arrangements represented here, each of them situated outside of the evacuated globe, would work.
2. Figure label is "Carbon."
3. Figure label is "metal."

—2295—

Notebook Entry:
Electric Light
and Power

[Menlo Park,] June 8[a] 1882

I propose to use 200 double[a] plate cells of lead only exceedingly close together probably like a dry pile with moistened wool between with SO_3—and charge these batteries across multiple arc in series & discharge in multiple arc ie[b] 4050[a] volts would give 50 or so[c] volts active on battery   it touches main line for an instant & charged in tension[1]   Commutating apparatus throws it in quantity the next instant & simultaneously throws lamps on   this is done several times a second.[2] I appartus is put in each house with meter which acts to discount when there is sufficient charge &but the instant it falls below a definite amount the rapid chgg & dischg amechanism works.[3] This can also be done by a macentered armature & mag across the lamp ckt

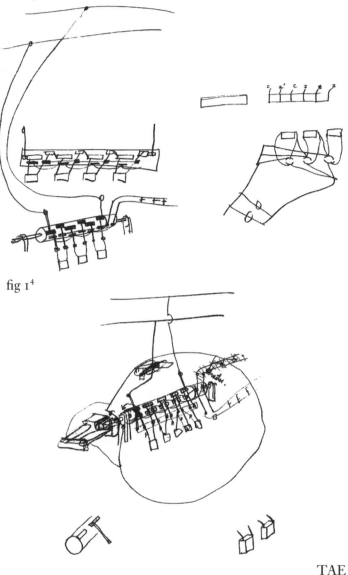

fig 1[4]

TAE

X, NjWOE, Lab., N-82-06-08:33 (*TAED* N197:17; *TAEM* 40:349). Document multiply signed and dated. [a]Obscured overwritten text. [b]Circled. [c]"or so" interlined above.

1. "Tension" refers to electrical potential or electromotive force, usually connoting relatively high voltage. Battery cells were said to be con-

nected either for intensity (in series) to produce high voltage or tension, or for quantity (in parallel) to maximize current.

2. This entry is preceded by several undated pages of rough sketches of commutators and related switching mechanisms. Those immediately follow unrelated drawings from 8 June and were probably made the same day. N-82-06-08:28–31, Lab. (*TAED* N197:15–16; *TAEM* 40:347–48).

3. Edison filed five patent applications in late June for closely related variations on arrangements of batteries to reduce voltage. Two were rejected and subsequently abandoned; only the claims and drawings remain. Case 450 pertained to batteries switched alternately between a series connection to a high voltage charging line and a parallel connection to the low voltage household circuit, so that one battery discharged while the other charged. In Case 451, this switching was carried out rapidly and continuously by a motor-driven commutator (Casebook E-2537:262, 264; Cases 450 and 451, Patent Application Drawings; PS [*TAED* PT021262, PT021264, PT023:74–75; *TAEM* 45:763, 765, 892–93]). Three similar applications executed on 19 June resulted in patents, though not until 1890 and 1891. One of these incorporated a balance consumption meter to switch pairs of batteries between the charging and consumption circuits (U.S. Pat. 439,390). Another pertained specifically to charging batteries in the circuit of an arc lamp and discharging them to incandescent lamps (U.S. Pat. 446,666). The third application described a battery (or condenser) used to reduce voltage to the consumption circuit. Its claims more broadly covered a high voltage distribution system with similar tension-reducing devices (U.S. Pat. 464,822).

4. Figure label is "meter." This drawing closely resembles the first figure in Case 451 (see note 3).

-2296-

To Lyman Abbott[1]

[Menlo Park,] 9th June [188]2

Dear Sir,

I have read the Article "How to succeed" which seems all right except that I am put down as the Author whereas the Article is the result of an interview with me.[2] I would prefer that it be stated that it is so as I make it a rule not to write anything for the Press   Yours truly

Thos A Edison   I[nsull]

L (letterpress copy), NjWOE, Lbk. 7:431A (*TAED* LB007431A; *TAEM* 80:596). Written by Samuel Insull.

1. Lyman Abbott (1835–1922) was a Progressive reformer, Congregational minister, and protege of Henry Ward Beecher. A prolific writer and editor, he succeeded Beecher as editor of *The Christian Union* (later *The Outlook*) in 1881. *ANB*, s.v. "Abbott, Lyman"; Brown 1953.

2. The June 1882 issue of *The Christian Union* contained the article attributed to Edison, "How to Succeed as an Inventor" (Edison 1882). The essay also appeared in Abbott 1882; a note below the table of contents stated that it was one of several given by dictation to *The Christian Union*. A copy is in Published Works and Other Writings (*TAED* PA012; *TAEM* 146:36).

Menlo Park, N.J., June 11 1882[a]

From Francis Upton

Dear Sir:

We have stopped today in order to resupport our shafting.

Mr. Bradley worked all day yesterday and hopes to finish to-morrow morning. We find that the building vibrates very badly from the jar as the walls are not built heavy enough for such quick running shafting as we use. We shall support the entire shafting on a false floor and brace the whole very firmly.

We shall put the donkey engine[1] for running the shafting to the mercury pump in next Sunday and take out the Volney Mason friction clutch,[2] and replace it with a plain pulley.

Holzer has brought the wages paid in the fibre room down to an extremely low figure, so that the girls will earn about half what they did. This will bring their weekly wages down to about $4 to $5 per week.

I spoke to you some time since about my loaning money to the Lamp Co. I enclose a statement of the amounts loaned.[3] I feel exceedingly anxious to have our credit A. No. 1 and have advanced money on that account. If you decide to make an assessment I will pay my share ~~from it~~ in, otherwise I feel obliged to charge interest on the amount loaned calling it a call loan.

I have tried in every way to make the money go as far as possible remembering that we shall have a very large business in the future, and it is good economy to make provisions for it. Yours Truly

Francis R. Upton.

ALS, NjWOE, DF (*TAED* D8230ZAY; *TAEM* 61:777). Letterhead of Edison Lamp Co. [a]"Menlo Park, N.J.," and "188" preprinted.

1. An auxiliary steam engine operating independently of the main engine. Knight 1881, s.v. "Donkey-engine. (Steam-engine)."

2. Volney W. Mason & Co. manufactured friction clutches, pulleys, and hoisting equipment in Providence, R.I. "Miscellaneous and Advertising," *Manufacturer and Builder* 5 (1873): 236.

3. Upton's enclosure has not been found. Philip Dyer's weekly account summaries show that Upton loaned $12,358.52 to the lamp company on or about 1 June, at the same time Edison advanced $6,000. The Edison Electric Light Co. had put up $7,500 in early May. Upton made a second, smaller advance by 5 June. The following week he put in about $7,000 more and, after a fourth installment by 19 June, was owed $22,845.75. He received 6% (later 10%) interest. Dyer to TAE, 8 May and 1, 5, 12, 19, and 26 June 1882; all DF (*TAED* D8231AAT, D8231AAX, D8231AAY, D8231AAZ, D8231ABA, D8231ABB; *TAEM* 61:893, 897, 902–905); Samuel Insull to Upton, 2 Mar. 1883, Lbk. 15:379 (*TAED* LB015379; *TAEM* 82:191).

To Sherburne Eaton

Major Eaton

The reason your memorandums have remained unanswered for some days past is that Mr Insulls time has been so much occupied that he has not been able to attend to correspondence.

Yours June 6th. I see no reason why we should buy Jno. D. Mullers patent 258795.[1]

I believe with you that it would not do for us to sell licenses to people with the idea of their dealing in such licenses in a wholesale manner   It would damage us considerably[2]

Yours June 8th   Insull will take up the European a/cs as soon as he gets a spare moment[3]

As to measuring the candlepower of lamps ours giving 16 candles in every direction [versus?]ª the candle power of Arc Light ([A-----])ª measured by the French method is quoted at four times this that is they take the candle power in one direction & multiply it by four   when there is a globe on an Arc Light it gives about 200 candles in all directions (the Brush people would call it 800). A gas Expert who tested the Brush Lights in the Park at Cleveland found the lights averaged 210 candles by our method of measuring.[4]

As to wiring Building here for Anderson[5] to "set on fire" we can use the old Lamp Factory. Let Clarke take it in hand & use good Isolated men for wiring.[6]

As to the material here belonging to Light Co I will have it sent wherever you direct & the rest can go with it to your store keeper[7]

Yours June 9th. Johnson made a mistake about prices[b] of [machines?][b] he quoted to Logan. [We have never?][b] sent a 200 Light machine to London & have only just commenced [sending?][b] 250 Light machines there & they are billed at $[1300?].ª Johnson made ~~them~~ the same mistake in quoting a price to Batchelor[8]

The notice you propose serving on Swan Maxim & American Cos has not come to hand; nor any of your other enclosures refd to in yours of June 9th.

I think we had better pass the German Contract & sign the Power of Attorney but nothing to be done except approved by Batchelor[9]

I would refer ~~to~~ the Mott papers to Wilber to report on.[10] Then the report can be submitted to me & then I will say whether we want to buy anything. I have not the slightest objection to Insull or Dyer holding patents on matters of detail in connection with Electric Lighting as long as they do [not?]ª attempt to cover up [broader?][b] things that I may want to [---].[b] It is much better to encourage our own people to work on details which I can never expect to touch than to leave them for others (outsiders) to work on as[c] after we have got well started there will be a great demand for these small things

Regarding Hannington I understand that Steringer[11] (engaged by Michell Vance & Co)[12] made some inventions in fixtures and asked Hannington to "put up" for Patent fees which I understand he did. Now do you not think it is a great deal better that, in the case of outsiders who invent these things, to have interested with them our own people who are affiliated to us in such a manner that we are sure to derive benefit as the inventions would be used in our interest. Furthermore I do not see how we can prevent, or what right we have to prevent our employees from investing their money in patent property. Rather I think we shou[ld][b] encourage it. I, myself, have been in the habit of encouraging the working up of [---][a] providing the claims are not [broad?][a] so as to cover principles I required Insull & Dyer have submitted [their?][a] various little things to me in every case. In one instance their claim involved a broad principle. I applied for the patent on this myself & left them to work up the detail for applying the principle so that we really have control of the broad principle & can thus regulate the royalty to be paid to the Patentees by those who use the device if necessary. The various devices will tend to further the demand for current which is certainly to our interest   Now Seeleys invention (Dyer & Insull are interested in it) for the Flat Iron which provides for the heating of the Iron by a carbon resistance will I think be of use & will consequently increase the demand on us for current. I think it is a very good thing to have our people work up these things as if they do not do it outsiders will as there are innumerable contrivances to which current can be applied & I certainly can never afford to go into these matters as they are decidedly outside my work. Motts case is different   He took advantage of his position to work up applications to cover principles (an insight of which he got by making my drawings) & so get round my invention. The case to which he refers when a patent was hurried to head off Mott is well known to me. Dyer & Insull told me of it & it was my application that was hurried forward—theirs (Is & Ds) being confined to the ~~use~~ detail[c] of this application of this principle which I wished to cover[d] to a specific object[13]

My time is altogether too valuable to allow of my going into the City to attend conferences on Dynamo orders. Let us get the model machine (125) first & then we can talk about what alterations shall be made— I think this whole thing when the time comes can be done by correspondence. Dean is too busy to attend Conferences as well as myself[14]

The engine I have here is simply borrowed from London Co until I can get another 8 × 10 which please hurry up[15]

Your enclosures with memo 10th June not to hand.

Do not make any contract with Moses until he has shown me Drawings which will lead me to reasonable expect something as the outcome of his work. So far he has nothing

Thos. A Edison   I[nsull].

L (letterpress copy), NjWOE, Lbk. 7:440 (*TAED* LB007440; *TAEM* 80:602). Written by Samuel Insull. ᵃIllegible. ᵇPaper damaged. ᶜInterlined above. ᵈ"which I wished to cover" interlined above.

1. Eaton asked if Edison wished to purchase or license John Muller's U.S. Patent 258,795, which he had seen listed in the Patent Office *Gazette.* Issued on 30 May, it pertained to a glass globe to soften the illumination of an electric light. Eaton to TAE, 6 June 1882, DF (*TAED* D8226R; *TAEM* 61:240).

2. Eaton reported in his 6 June letter (see note 1) that an associate of Spencer Borden (see Doc. 2303) sought to "buy from our parent company licenses for cities and towns with the intentions of reselling them at a profit," as the Brush company did. Eaton thought such an arrangement "would give us immediate cash and would also result in the formation of a good many local companies without delay, but ultimately there would be dissatisfaction. We ought to let the local companies get the licenses as cheaply as possible. . . . We may possibly allow a small commission in some cases, as we do with the Chicago company when they form subordinate companies."

3. The Edison Electric Light Co. of Europe was trying to reconcile its accounts with the Edison companies in Paris. Eaton stated that Samuel Insull had promised to send relevant information from Edison's personal accounts by late May but had not done so, despite several requests by William Meadowcroft. Eaton to TAE, 8 June 1882, DF (*TAED* D8226S; *TAEM* 61:243).

4. Eaton had asked for an explanation he could publish in the bulletin about measuring the luminous intensity of electric arc and incandescent lights. This paragraph is substantially the same as the reply Edison drafted on Eaton's letter. Two days later, Eaton intended to send Edison Charles Clarke's memo on the subject "which makes it look as though the less I said about the matter the better. . . . Please destroy Clarke's memo. after reading it." Eaton published instead a lengthy excerpt from William Preece's article on the difficulty of making reliable candlepower measurements of arc lights by standard photometric techniques. The Brush company had started a demonstration of twelve arc lamps in Cleveland's Monument Park on 29 April 1879. Eaton to TAE, 8 and 10 June, both DF (*TAED* D8226S, D8226U; *TAEM* 61:243, 254); Edison Electric Light Co. Bulletin 11:13, 27 June 1882, CR (*TAED* CB011; *TAEM* 96:720); Hammond 1941, 27–28.

5. William A. Anderson was chairman of the Committee on Police and the Origin of Fires of the New York Board of Fire Underwriters. Anderson to Eaton, 27 Apr. 1882, DF (*TAED* D8204ZAZ; *TAEM* 60:130).

6. Eaton had reported on a meeting with Anderson about the use of paraffin-covered wires in the First District, and the wiring of chandeliers and brackets. Anderson accepted paraffin except in cases of unusual hazard, but would not waive the Board's interior requirement of a half inch of wood or other insulation between two wires or wire and metal. He planned to visit Menlo Park to witness some tests. Edison promised to wire a building for Anderson's benefit, and Eaton reminded him to keep this pledge in light of a fire that had occurred in Anderson's presence and the overall skepticism of committee members. Edison had purchased the Lamp Co.'s Menlo Park factory building in early May for $500. Eaton to TAE, 1 and 8 June 1882; Upton to Insull, 27 Apr. 1882; all DF (*TAED* D8226N, D8226S, D8230ZAH; *TAEM* 61:217, 243, 749); TAE to Eaton, 6 June 1882; Insull to Upton, 1 May 1882; Lbk. 7:406, 12:188A (*TAED* LB007406, LB012188A; *TAEM* 80:586, 81:600).

7. Edison had promised to make a list of the Electric Light Co.'s property to be removed from Menlo Park. It has not been found. TAE to Eaton, 6 June 1882, Lbk. 7:406 (*TAED* LB007406; *TAEM* 80:586).

8. According to Eaton, Edward Johnson told James Logan that Edison had furnished a 200 light dynamo to the British company for $1,000. Eaton queried whether "the London Company pays precisely for dynamos what we pay?" Edison marked "Same price." Shortly after this, Edison quoted a "strictly confidential" discount on isolated plant dynamos to Mather & Platt, the large English engineering and manufacturing firm. Eaton to TAE, 9 June 1882, DF (*TAED* D8226T; *TAEM* 61:249); Insull to Charles Rocap, 15 June 1882, Lbk. 7:466A (*TAED* LB007466A; *TAEM* 80:615).

9. In two recent letters, Eaton discussed the pending contract between the Compagnie Continentale Edison and two German banks on behalf of Emil Rathenau. The Edison Electric Light Co. of Europe had already authorized Charles Batchelor and Joshua Bailey to execute it, and it was signed by the Paris parties on 15 July 1882. The agreement has not been found but Bailey presented a summary to the European Light Co. in August. It called for the establishment of a temporary syndicate to conduct isolated and central station lighting business through February 1883. Permanent companies were then to be formed as agents of the European Co., one for central station work throughout the German Empire, and another for isolated lighting and manufacturing in the Empire excepting Alsace and Lorraine. Eaton professed confusion over the complexity of the terms, some of which were embodied in a later agreement between the same parties. TAE to Bailey, 5 May 1882, LM 1:205C (*TAED* LM001205C; *TAEM* 83:974); Eaton to TAE, 29 May and 6 June 1882; Bailey and Theodore Puskas report, 25 Aug. 1882; Sulzbach Bros. et. al. agreement with Compagnie Continentale Edison et. al., 13 Mar. 1883; all DF (*TAED* D8226L, D8226T, D8228ZAC, D8337ZAL; *TAEM* 61:207, 249, 643; 67:662); see Doc. 2392.

10. This and the rest of the letter refers to Eaton's letter of 10 June. Eaton indicated that Samuel Mott declined to submit a second list of his electric lighting inventions (see Doc. 2286 n. 9) because Insull and

Richard Dyer, who had personal stakes in several patents (see note 13), had seen the first one and he did not trust them "to look over such a confidential paper as a statement of unpatented inventions." Eaton suggested that Mott either meet privately with Edison or allow Zenas Wilber to advise the company; he subsequently directed him to submit the list to Wilber. Eaton to TAE, 10 June 1882; Eaton to Mott, 14 June 1882; both DF (*TAED* D8226U, D8224ZAK1; *TAEM* 61:254, 38).

Mott apparently entered the employ of Ernst Biedermann, through whom Edison had been negotiating the sale of patents in Switzerland (see Docs. 1878 and 1962). When the two left for Geneva, apparently in connection with an electric railroad, Edison instructed Eaton to accept Mott's resignation and have no further contract with him because "he would have nothing at all but for fraud." He also disavowed Mott and Biedermann to Batchelor and other associates abroad. TAE to Eaton, 27 June 1882, Lbk. 7:592 (*TAED* LB007592; *TAEM* 80:686); TAE to Edison Electric Light Co., Ltd.; TAE to Charles Batchelor; TAE to Theodore Turrettini; all 27 June 1882, LM 1:230A, 230B, 230D (*TAED* LM001230A, LM001230B, LM001230D; *TAEM* 83:987).

11. Luther Stieringer was a gas engineer who began working for Edison about the fall of 1881. He devised and patented a number of lighting fixtures (see App. 5). Stieringer became a highly respected illuminating engineer and was responsible for artistic lighting exhibits at international expositions in Chicago (1893), Buffalo (1901), and St. Louis (1904). Edison later declared that "no account of the development of decorative lighting in America would be complete which did not embrace the brilliant and successful work done by Mr. Stieringer." Jehl 1937–41, 755–57; TAE to Thomas Martin, 9 May 1902, Lbk. 68:112 (*TAED* LB068112; *TAEM* 196:900); *Chicago Tribune,* 19 July 1903, 4.

12. Mitchell, Vance & Co. made gas fixtures in New York City. *Rand's* 1881, 225.

13. Eaton reported in his 10 June letter (see note 10) that Mott had complained of an occasion at Menlo Park in which "a patent was hurried in order to head off Mott. Mott says it was something which he was engaged on." He also discussed three patents in which several Edison associates had interests: U.S. Patent 259,054 for an electric iron, issued to Henry Seely and assigned in part to Dyer and Insull; U.S. Patent 259,115 for an electric toy, to Dyer and Seely and assigned in part to Insull; and U.S. Patent 259,235 for fixtures, to Luther Stieringer and assigned in part to Dyer and Charles Hanington. All issued on 6 June 1882. Eaton made no comment about Insull and Dyer "as they are not in the employ of this company. But I have something to say about Hanington. I disapprove of a policy of allowing employees of the company to be associated in taking out patents in that way." He also held that patents issued to Mott or others working closely with Edison belonged to the company. Eaton to TAE, 10 June 1882, DF (*TAED* D8226U; *TAEM* 61:254); see also App. 5.

14. Eaton had endorsed in his 10 June letter (see note 10) a request from Charles Clarke and Miller Moore that Edison confer in New York about dynamo orders and other matters.

15. Eaton had been trying to get a steam engine for Menlo Park and expressed irritation after learning that Edison already had an Armington & Sims engine there.

CHICAGO, June 14th, 1882.[a]

Friend Edison:

In addition to the directors of the Western Co.[1] considerable of the stock will be distributed among such men as Marshall Field,[2] Geo. M. Pullman[3] and others of that class.

A better body of men could not be brought together and if the Central[b] District in New York goes as well all expect any amount of money needed will be forth coming to press the business in Illinois, Wisconsin and Iowa.

Gen' Stager has been sick most of the time since the organization and is out today for the first time.

Arrangements are being made to push everything with vigor and some nice results will be reached ere long.

The Western Company are desirous of having the business in the non gas towns as well as in the gas towns.[4]

There is no doubt about one company being able to do all the work to better advantage than two can if a proper basis can be agreed upon.

I presume the matter will be properly negotiated in a short time.

Of course when I went with the Isolated Company I expected it would be a permanent thing[5] and that after the first of June they would pay me $8.000.00 and a percentage of profits or of stock in non gas towns where companies were started.

Then it did not seem to be contemplated to put more than a single city in one company.

I am not disposed to find fault for my great object is to advance the interests of the Edison electric light with a reasonable care for myself.

The General is very friendly and offers me $3.000.00 salary with the new company and will carry $5.000.00 worth of stock for me.

I think this stock will go to 200 with in a year which will be a favorable result for me.

I have had enough experience with the light to be satisfied that it will be a great success and I should not be afraid to take any territory with good backing to work up local organizations and get the business started.

If satisfactory to all concerned I prefer not to go away from here without considerable additional compensation.

My plants are going to stick here and I shall clean up a profit of the business already done.

Gen' Stager has made money for all his associates and friends in the telephone and other companies which he has handled and it pays for a man to stick to him.

It seems to be that Mr. Insull is very slow about sending along that cut[6] and it would be better to say it cannot be had than to allow the matter to drag along. Sincerely yours,

Geo. H. Bliss

TLS, NjWOE, DF (*TAED* D8241E; *TAEM* 63:88). Letterhead of Western Edison Co. Typed in upper case. [a]"CHICAGO," and "1882." preprinted. [b]Mistyped.

1. The Western Edison Light Co. was organized in late May to exploit Edison's electric lighting patents in Illinois, Iowa, and Wisconsin. The principal organizers were Anson Stager (president) and Bliss (general manager). The directors included Edison, Stager, John Crerar (associate of George Pullman), Samuel Merrill (former governor of Iowa), and Z. G. Simmons (president of Northwestern Telegraph Co.). The initial capitalization was $500,000; most of the stockholders were from Chicago, where the company had its headquarters. By the end of February 1883 it had installed 24 isolated plants with another 5 under contract. In its five years of corporate existence Western Edison installed 67 isolated plants in Chicago alone. It was superseded by the Chicago Edison Co. in July 1887. Edison Electric Light Co. Bulletin 10:5, 5 June 1882; "Edison's Illuminators," *New York Herald*, 5 Sept. 1882, reprinted in Western Edison Light Co. Bulletin, 12 Sept. 1882; CR (*TAED* CB010, CA005A; *TAEM* 96:714, 308); Western Edison Light Co. List of Stockholders, n.d. [1882]; Western Edison Light Co. Secretary's Annual Report, 19 Feb. 1883; both DF (*TAED* D8241ZAM, D8364A; *TAEM* 63:136, 70:525); Andreas 1886, 598; Hogan 1986, 11–19; Platt 1991, 33–37.

2. Marshall Field (1834–1906) operated one of the premier dry goods stores in the United States, recently reorganized as Marshall Field and Co. It was one of the first businesses in Chicago supplied with an Edison isolated system. Field acquired $10,000 of Western Edison stock. *ANB*, s.v. "Field, Marshall"; Edison Electric Light Co. Bulletin 10:5, 5 June 1882, CR (*TAED* CB010; *TAEM* 96:714); Western Edison Light Co. List of Stockholders, DF (*TAED* D8241ZAM; *TAEM* 63:136).

3. George Pullman (1831–1897) produced railroad sleeping cars at his firm, the Pullman Palace Car Company. He acquired $10,000 of Western Edison stock. *ANB*, s.v. "Pullman, George Mortimer"; Western Edison Light Co. List of Stockholders, DF (*TAED* D8241ZAM; *TAEM* 63:136).

4. Under its contract with the Edison Electric Light Co., the Edison Co. for Isolated Lighting had rights to install plants except within "the municipal limits of any town, city, village, or other territorial municipality wherein illuminating gas was or had, prior to" 1 January 1882, "been supplied for purposes of lighting to more than ten customers." The same contract granted the Isolated Co. rights to install isolated plants within gas limits until it was notified that Edison Electric had granted a license to another party. After protracted negotiations Western Edison received the exclusive agency for isolated plants in Illinois, Iowa, and Wisconsin in non-gas territories, with profits to be divided equally with the Isolated Co. Edison Electric Light Co. agreements with Edison Co. for Isolated Lighting, 26 Apr. 1881 and 1 Sept. 1884, Defendant's Exhibits, 4:2363–84, *Edison Electric Light Co. v. U.S. Electric Lighting Co.*,

Lit. (*TAED* QD012E2363, QD012E2371; *TAEM* 47:1004, 1008); Eaton to TAE, 15 Aug. 1882; Bliss to TAE, 18 Aug. 1882; Western Edison Light Co. Secretary's Annual Report, 19 Feb. 1883; all DF (*TAED* D8226ZBF, D8241M, D8364A; *TAEM* 61:392, 63:99, 70:525).

5. Bliss, who was the Edison Isolated Co.'s agent in Chicago, had learned in January of Stager's interest in acquiring rights for central stations there. Through the formation of Western Edison they combined the agencies for the central station and isolated businesses. Bliss to TAE, 3 Jan. and 12 May 1882; both DF (*TAED* D8241A, D8241D; *TAEM* 63:84, 87); Samuel Insull to Bliss, 10 Jan. 1882, Lbk. 11:59 (*TAED* LB011059; *TAEM* 81:259).

6. Having been asked by the editor of the *Railway Age* to write an article on Edison's electric railroad, Bliss requested Edison to send a copy of the woodcut used to illustrate an article in the 5 June 1880 *Scientific American*. Samuel Insull had written to Bliss on 18 May that he could not send an electrotype of it for about a week; he did not receive it from the engraver before 19 July. Bliss's article finally appeared in the 21 September 1882 issue of *Railway Age* with a different illustration than in *Scientific American*. Bliss to TAE, 14 Apr. 1881, Cat. 2174, Scraps. (*TAED* SB012BAH; *TAEM* 89:412); Bliss to TAE, 3 and 12 May 1882, both DF (*TAED* D8207U, D8241D; *TAEM* 60:539, 63:87); Insull to Bliss, 18 May and 24 July 1882, Lbk. 12:353, 7:750 (*TAED* LB012353A, LB007750; *TAEM* 81:692, 80:748); "Edison's Electrical Railway," *Sci. Am.*, 42 (1880): 354, Cat. 1241, item 1513, Batchelor (*TAED* MBSB21513X; *TAEM* 94:605); Bliss 1882.

–2300–

From Charles Clarke

New York, June 14 1882[a]

My dear Edison,

I have within a fortnight written Batch. two long letters in reference to wiring, safety-catches, connecting up dynamos &c, and shall continue the correspondence.[1] Please let me know the following.

In estimating on conductors for transmission of power in mountainous districts what do you consider to be the safe, and proper E.M.F. to use, and not so great as to make the armatures liable to cross? Yrs truly,

Chas. L. Clarke

⟨Increased emf dont make the armatures cross   its the short cktg causing <u>heavy load</u> that does the biz & this heavy load is terrible when there is a high emf. I think 1000 Volts for ordinary cases & never higher than 1500⟩[2]

ALS, NjWOE, DF (*TAED* D8227V; *TAEM* 61:483). Letterhead of Edison Electric Light Co. [a]"*New York*," and "*188*" preprinted.

1. Clarke's letters have not been found. About this time, Edison urged him to "open up correspondence for exchang[in]g Idea[s] with Batch." TAE to Clarke, 14 June 1882, DF (*TAED* LB007455; *TAEM* 80:613).

2. Clarke later inquired about the design of a plant to provide power several miles away. He pointed out that transmitting power below 1,100 volts over that distance would entail either unacceptable electrical losses or prohibitively large conductors. Edison replied that there was "no reason why you should not run up as high as 2000 volts." Clarke to TAE, 29 July 1882, DF (*TAED* D8227ZAO; *TAEM* 61:516); TAE to Clarke, 31 July 1882, Lbk. 7:810 (*TAED* LB007810; *TAEM* 80:774).

-2301-

From Charles Rocap

New York, June 16th 1882.[a]

Dear Sir,

Your favor of 15th inst. to hand. I should have said Mr Insulls' signed by you, as the letter is one of his kind.[1]

As to Mr Insulls' complaints about being unable to get information from the Works if he has reference to me, he has made representations to you which have, putting it mildly, no foundation. Mr Insull cannot say but that I have always treated him with becoming courtesy, and have always liberally furnished him with every information and assistance.

Regarding the statement of dynamos shipped abroad, I beg to enclose you herewith the original letter asking for such information, and request you to personally compare it with my statement, and if I have not complied in every sense with what was asked for, I do not understand the english language. All I have to say for Mr Insull is that either he does not know what he wants or how to ask for it.[2]

In future it is my desire to transact all business with yourself personally, and to avoid any occasion of ever again calling down on myself such disagreeable feelings as these letters have produced. I hope all your business and correspondence will be conducted in your name.[3] Yours faithfully,

Charles Rocap

ALS, NjWOE, DF (*TAED* D8233ZCB; *TAEM* 61:1085). Letterhead of Edison Machine Works. [a]"New York," and "188" preprinted.

1. The letter was written by Insull and signed by Edison. It stated that Insull "has been unable to get information from the Works which he is obliged to have for the proper conduct of my business." It referred specifically to Insull's 11 May request (see note 2). It claimed that the information was not received until 12 June and was "not at all what was asked for and is absolutely of no use." Edison stated that "whatever data of any kind Mr Insull may ask for is to be supplied to him as quickly as is possible. The non [compliance?] with his wishes in the particular case in question is likely to put me in a very false position with the Light Co." TAE to Rocap, 15 June 1882, Lbk. 7:469A (*TAED* LB007469A; *TAEM* 80:616).

2. Insull requested a "<u>complete</u> statement of all Dynamos shipped from the Works to foreign Ports since the start." Calvin Goddard had sought this information a week earlier. Insull also wanted copies of invoices for all items sent to Joshua Bailey and Theodore Puskas. On 25 May he again prompted Rocap, who reported in a 2 June letter that he was enclosing the desired statements and copies, but these have not been found. Insull to Rocap, 11 and 25 May 1882, Lbk. 12:273, 423A (*TAED* LB012273, LB012423A; *TAEM* 81:648, 728); Goddard to Insull, 4 May 1882; Rocap to TAE, 2 June 1882; Goddard to Insull, 13 June 1882; all DF (*TAED* D8224T, D8233ZBL, D8224ZAK; *TAEM* 61:26, 1066, 38).

3. Insull replied directly to this letter but made no reference to the allegations about himself. He acknowledged copies of the invoices sent on 2 June but said he did not receive the other information at that time. Insull to Rocap, 19 June 1882, Lbk. 7:512 (*TAED* LB007512; *TAEM* 80:639).

–2302–

To Charles Batchelor

MENLO PARK, N.J. June 19 1882[a]

Batch—

In putting up large installations & for that matter small installations look out for the drop of pressure. Our tables are made for a loss of 5 per cent. this is 5 volts   now a difference of 5 volts will make the lamp at the machine say 16 candles & the last lamp 6 candles or 25 or 30 candles at the machine and 16 on the last lamp— We have changed that here. We only allow 2 per cent loss from the 1st lamp to the last, and make the loss 5 8 or even 12 per cent on the wire leading from the machine, according to circumstances for instance[1]

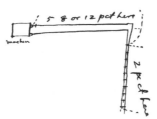

The Conductor leading from the machine having no Lamps on any difference dont effect while the circuit on which the lamps are having but 2 volts fall the difference in candle power is not[b] noticeable. Now in your large installations you must look sharp after this and I think in most places you will have to callculate for a loss of 2 per cent on the tubes carrying the lights and ten to 12 percent on your feeder thus[2]

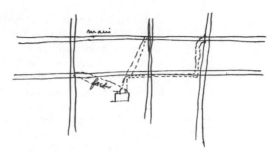

a loss of 5 per cent on your lamp condrs would be too much—
In square mile installations without a model is put up on a
board & the pressure & <u>position</u> of feeders acurately determined general distribution would be a failure as there would
in certain hours be a difference of 15 volts which[c] would bring
some of the lamps to 40 candles[3]

AL, NjWOE, Unbound Documents (1882), Batchelor (*TAED* MB069;
TAEM 92:430); a letterpress copy is in Lbk. 7:533 (*TAED* LB007533;
TAEM 80:655). Letterhead of Thomas A. Edison. [a]"MENLO PARK, N.J."
and "188" preprinted. [b]Followed by "over" as page turn. [c]Obscured
overwritten text.

 1. Text is "machine," "5 8 or 12 pct here," and "2 pc ct here."
 2. Text is "main" and "feeder." Placing feeder conductors at various
points in the main network for this purpose is reminiscent of a suggestion that Edison incorporated into an 1879 caveat (Doc. 1789) related to
several subsequent patents. He did not at that time give specific figures
for allowable losses on the lines. He filed a British specification in September 1880 that also embodied this principle. This, too, did not specify
losses but stated that "the greater portion, if not all, the drop occurs
in the feeding conductors, the pressure in the service or lamp circuits being practically uniform at all points." Brit. Pat. 3,880 (1880), Batchelor
(*TAED* MBP028; *TAEM* 92:187).
 3. See also Doc. 2327.

–2303–

From Sherburne Eaton

[New York,] Wednesday, June 21st. 1882. 5, P.M.
 <u>SEEING EDISON AT GOERCK ST</u>. I received word that you
were at Goerck St. this morning. I was about starting there to
see you, when Mr Hoffman, Borden's friend,[1] called and kept
me nearly two hours. I had nothing of especial importance to
see you about, merely to talk matters over generally.
 <u>PROPOSED COMPANY FOR MASS</u>. Mr Hoffman and Mr
Borden have now changed their policy regarding a proposed
company for large territory. They now propose to do just what
the Chicago company does, namely, exploit central stations
with their own capital unless good local companies can be or-

ganized. Hoffman wants to get up a Mass. company, and have it at once start a central station at Boston. He wishes to reserve the right to get up subordinate companies, just as the Chicago company, if he wishes to do so. Borden will take the general management of the business. I think the matter is worth earnest consideration.

INCONVENIENCE OF ISOLATED CONTRACT. I sometimes wish there were no Isolated Company. Our local company in Lawrence wishes to embrace two neighboring towns Andover and Methuen[a] in their Lawrence territory. These towns are practically[a] a part of Lawrence and in a few years will be united to the city of Lawrence. The factories in these towns are treated as Lawrence factories in all business details. But the parent company cannot sell them these two towns, and it is difficult to arrange[a] satisfactory terms for them because it involves another contract and different principle of compensation. The same question comes up again in South Carolina. The Charleston company will run the whole state, and they wish to make but one contract. The same question is now presented in Chicago. Stager's company[a] has secured one territory within gas limits, and he finds it awkward[a] in doing business.[2] He can secure the rest from the Isolated company, but he considers it a conflicting and unsatisfactory way of doing business. The same question was presented in San Francisco, and Ladd[3] is obliged to make two different contracts with two different sets of terms. This proposed Mass. company of Hoffman and Borden will bring the same question up again. Thus I sometimes wish there were no Isolated Company. Possibly the parent company may sometime before long absorb the Isolated Co., increase its own stock and buy up the Isolated Co. by giving the stockholders in the latter company some stock in an increased [---][b] capitalization of the parent company.

CANADIAN CO.[4] Swinyard[4] has gone back to Canada. I enclose a copy of the memo. which he takes back with him.[5] The most difficult matter presented there is what to do with Hearle.[6] We have decided that the best way [--][b] is for me to go to Montreal next Monday night and pay Hearle the compliment of coming there personally to see him. My purpose will be simply to induce him to accept his being superseded by Swinyard as pleasantly as possible. I shall expect him to meet me with the same smiling face with which the culprit greets his executioner.

POLICY IN CANADA. I do not propose to have an isolated co. formed in Canada. I find from experience that the division of territory between two different companies is very inconvenient in dealing with licensees. In Canada the parent company will itself do the isolated business, and I am not sure but that it had better do the mnfg. business. I am afraid that if a separate company is formed for mnfg., the profits will make the plant so expensive that it will be hard to compete with gas. If the parent company should also own the mnfg. business, they might be satisfied with less profit than an independent company would. But if that reason is true of Canada is it not also true of the United States? Am I right about this? Please give me your views.[7]

GODDARD GONE. Col. Goddard sailed for Europe at an early hour this morning.[8] Bon voyage.

BRUSH CO. Did you see the cable from London about the Brush Co. which was in the Tribune this morning?[9] The Evening Commercial last Thursday night had a long editorial on electric lighting, in which great praise was given to Mr Montgomery[10a] for making money for Brush company in England.

SUING SWAN. My present inclination is to give the general conduct of the suit against Swan[11] to the Dickersons,[12] but to secure Betts to assist them. They are not in many respects entirely satisfactory to me as counsel in such[a] a case, but on the whole I am not sure that we can do better than take them. Swan has even less than I supposed he had. Really I consider it almost a fraud to float an independent company on so little. Knowing that Mr ~~Swan~~ Hoffman[c] is well acquainted[a] with the proprietors of the Swan parent company, I spent nearly an hour this morning explaining to him our patents and showing him the Swan patents. He said that the people who were investing their money~~in~~ in the Swan company evidently had no idea of[a] what the facts were about the patents and that Swan had so little patented[a] in[a] this[a] country.

WATER SUPPLY AT PEARL ST. I believe Clarke has told you that Hornig will build a cistern in that little yard behind our station to hold a few hours supply of water so as to guard against the Croton supply being stopped. The Croton water is shut off sometimes to make repairs.

COAL FOR PEARL ST. We have arranged with Cavanagh[13] to furnish us coal at $4.85[a] a ton for the present at Pearl[a] St. Cavanagh is intimate with the officials who control the docks and he thus gets his coal barges at convenient spot and[a] also gets his ashes on the city dump. Our arrangement with him is only temporary, pending subsequent arrangement[a] on a larger scale.

LAWRENCE CO. The Lawrence company has organized. They wrote me to day that they have called in $30,000 for a[a] small central station right off.[14] Byllsby is to take charge of making estimates and doing the figuring in such cases. He has gone away to be married and will not be back here until the first of next week.

TO CLOSE EVENINGS. I have decided to close the office evenings during the fourth of July week. The ostensible reason is to give our employees[a] those evenings for themselves. The real reason is that we are going to put in a new engine down stairs which would prevent our lighting up at least two nights, so I thought I would make a week of it and treat it as a sort of vacation

VANDERBILT. I wish the Vanderbilt matter could be settled, especially for the reason that we need the money. I think I have asked you in a previous memo. if I could not make an appointment for you and me to call on Mr Vanderbilt in the matter. Please reply.[15]

ADDENDUM[d]

[New York,] June 21

Mr Edison,

Please excuse appearance of this memo. as typewriter is out of order. Respectfully,

F. Mc. Gowan.

TL, NjWOE, DF (*TAED* D8226ZAC; *TAEM* 61:296). [a]Mistyped. [b]Canceled. [c]Interlined above. [d]Addendum is a TL on separate sheet; date written by hand.

1. James H. Hoffman (1834?–1900), a Hessian-born inventor, manufacturer, and distributor of paper collars, whom Eaton elsewhere identified as Spencer Borden's "wealthy Israelite friend in New York." Hoffman had hoped to buy rights to Edison lighting franchises through Borden and resell them, a proposal which neither Eaton nor Edison favored (see Doc. 2298 esp. n. 2). Obituary, *New York Times*, 10 July 1900, 7; Eaton to TAE, 6 June 1882, DF (*TAED* D8226R; *TAEM* 61:240).

2. See Doc. 2299 n. 4.

3. George Ladd was president of the Pacific Bell Telephone Co. and the Gold and Stock Telegraph Co. of California. He had applied to Edison in 1881 to form a Pacific Coast lighting business and later joined the California franchise formed by Louis Glass and Frank McLaughlin. See Docs. 1393 n. 5 and 2176; Ladd to TAE, 29 Nov. 1881; Glass to Samuel Insull, 16 Feb. 1882; both DF (*TAED* D8120ZBT, D8220I; *TAEM* 57:662, 60:760).

4. Thomas Swinyard was vice president of the Dominion Telegraph Co. of Canada. William Preece to Swinyard, 6 Feb. 1882, in Edison Elec-

tric Light Co. Bulletin 9:13, 15 May 1882, CR (*TAED* CB009089; *TAEM* 96:712).

5. The memorandum prepared by Calvin Goddard outlined terms for the organization of an Edison lighting company to operate in Canada. It was to manufacture all needed items, including lamps and dynamos, and set up isolated plants; it was authorized to sublicense local illuminating companies for central stations. Goddard offered Swinyard a commission of 7.5% of the proposed company's $1,000,000 capital. Goddard memorandum, 17 June 1882; Goddard to Swinyard, 17 June 1882; both DF (*TAED* D8237T, D8237U; *TAEM* 62:166).

6. E. Hearle of Montreal was evidently trying to organize a Canadian Edison company, and later was among the original shareholders of Canadian General Electric. At this time he was also corresponding with Edison about luminous paint and ferrous sand deposits in Canada. See *TAED* and *TAEMG2*, s.v. "Hearle, E"; Kee 1985, 112.

7. Edison replied that although he thought it "a misfortune" for a patent holding company to conduct manufacturing, "I do not see how we can do otherwise in Canada where the wants are comparatively small." He thought it best for the Canadian parent company to build isolated plants. TAE to Eaton, 22 June 1882, Lbk. 7:554 (*TAED* LB007554; *TAEM* 80:668).

8. Goddard returned on 10 August "much improved in appearance," according to Eaton. Eaton to TAE, 10 Aug. 1882, DF (*TAED* D8226ZBC; *TAEM* 61:378).

9. The *Tribune* stated that directors of the Anglo-American Brush Light Co. planned to meet to divide profits from the sale of concessions to subsidiary companies. Edison replied that he had not seen this report. "Topic in London," *New York Tribune,* 21 June 1882, 1; TAE to Eaton, 22 June 1882, Lbk. 7:554 (*TAED* LB007554; *TAEM* 80:668).

10. Eaton referred to Thomas Montgomery; the editorial has not been found.

11. Eaton suggested taking legal action after learning that the Brush interests were vigorously promoting the Swan lamp in Chicago and Philadelphia. He reported that Zenas Wilber had examined Swan's patents but found no act of infringement as yet on which to bring suit. Edison recommended getting "ready to sue both Swan & Maxim & immediately Swan starts up go for him." Eaton to TAE, 12 and 15 June 1882, both DF (*TAED* D8226V, D8226Y; *TAEM* 61:264, 282); TAE to Eaton, 16 June 1882, Lbk. 7:482 (*TAED* LB007482; *TAEM* 80:620).

12. Distinguished patent expert Edward Dickerson had served as one of the main attorneys for Western Union in the Quadruplex Case, cross-examining Edison for several days. He and his son Edward Jr. operated a law practice in New York and they wanted either to have full control over the conduct of the Swan suit or no connection with it, in which case they would take other electric lighting clients. Eaton had asked Edison if he wished to retain the Dickersons on those terms. Edison consented, and suggested that Grosvenor Lowrey also participate in addition to Frederic Betts. Doc. 1011 n. 3; Letterhead of Dickerson & Dickerson to TAE, 23 Feb. 1882; Eaton to TAE, 13 June 1882; both DF (*TAED* D8126R1, D8226W; *TAEM* 58:26, 61:269); TAE to Eaton, 16 June 1882, Lbk. 7:482 (*TAED* LB007482; *TAEM* 80:620).

13. Unidentified.

14. Eaton saw the Lawrence plant as a prototype of the small central station. This was one of three types that he envisioned, the others being large plants (such as Pearl St.) and the village system, later built at Roselle, N.J. Eaton to TAE, 15 June 1882, DF (*TAED* D8226Y; *TAEM* 61:282).

15. William Vanderbilt evidently disputed a bill for his isolated lighting plant (see Doc. 2108) and wished to deal only with Edison. Vanderbilt was away from New York in July and the ultimate resolution of the matter has not been determined. Eaton to TAE, 15 June, 3 and 14 July 1882; all DF (*TAED* D8226Y, D8226ZAK1, D8226ZAN; *TAEM* 61:282, 324, 328); TAE to Eaton, 22 June 1882, Lbk. (*TAED* LB007554; *TAEM* 80:668).

−2304−

From Charles Dean

New York, June 24 1882[a]

Dear Sir,

We must insist on having the testing room without delay. I am terribly crowded for room now, and must have room for Andrews to test a dozen or more armatures at once.[1] I want plenty of room for my armatures and there is danger of having them injured in the Test Room now with so many in there. I shall have to crowd all armatures as soon as finished in that room, and they (the test force) will have to get out. Yours truly,

C L Dean   Supt

⟨Referring to yours of 24th June I would remind you that the Light Co pay us a rent for the Testing Room & have quite a large investment in fixtures &c there ~~which~~ and therefore we cannot ~~ask~~tell them to get out. They are entitled to proper treatment and all that can be done is to try & arrange for the Test Force to go somewhere else which I am trying to do— I doubt if we can have them moved for several[b] months yet⟩[2]

ALS, NjWOE, DF (*TAED* D8233ZCM; *TAEM* 61:1097). Letterhead of Edison Machine Works. [a]"New York," and "188" preprinted. [b]Obscured overwritten text.

1. On 13 June Dean telegraphed and wrote Edison to expect delays in shipping dynamos unless William Andrews devoted his whole time to testing dynamos instead of performing work for the Light Co. and the Ore Milling Co. The Machine Works completed 37 dynamos in June and 38 during the week of 5 August. Jehl 1937–41, 986; Israel 1998, 205; Dean to TAE, both 13 June 1882; Rocap to TAE, 19 July and 7 Aug. 1882; all DF (*TAED* D8233ZBW, D8233ZBY, D8234D, D8234D1; *TAEM* 61:1079, 1081; 62:9, 12).

2. No full reply to Dean has been found. Sherburne Eaton suggested freeing space by moving the armature plating operations from Goerck St. but Edison preferred to do nothing until business, particularly that for village plants, was "booming." The Testing Room remained at the

Edison Machine Works under supervision of the Edison Electric Light Co. until at least June 1883. TAE to Charles Edgar, 8 June 1883; Lbk. 17:99 (*TAED* LB017099; *TAEM* 82:577); Eaton to TAE, with TAE marginalia, 23 June 1882, DF (*TAED* D8226ZAD; *TAEM* 61:300).

–2305–

From Charles Batchelor

Ivry-sur-Seine, le June 26 1882[a]

My dear Edison,

Your numerous cables on details of machines to hand releiving our minds greatly; you cannot give us too much of this information— Your letter on factory also to hand and contents digested—[1]

I am sorry there should have been so much trouble about those payments on the large dynamos.[2] These people are very different to Americans or even Englishmen and it is necessary to work them in their own way— I find it a little difficult to get their full confidence until we have something to show them here. If I could have put up a couple of machines here like Johnson did in London there would have been no difficulty— We are getting up some good plants in Europe but we have not a single installation in France and shall not have until I make it here

As regards the advances made of $5 000 and $9 000 they wish it understood that these payments are to be applied on the first two large dynamos and only them. When you want an advance on the second 2 you must make a special demand for it and I should advise you to make that demand immediately in order that these two machines may be commenced and we shall not have to wait for them[3]

Of course all Z dynamos and 250 lights you collect for immediately they are finished— There will be no more trouble like this if we can get a really good installation up somewhere— I suppose you are worked to death just now— I would give a great deal to be at the starting of the New York station— My job here is no fool of a job what with lamps,[b] dynamos, chandeliers, and all the extras; I am just in up to my neck; then I have so much outside work of such a responsible nature and involving so much money, that I wear a hat about three sizes larger than when I left New York— Yours

"Batchelor"

ALS, NjWOE, DF (*TAED* D8238ZBS; *TAEM* 62:440). Letterhead of Société Industrielle et Commerciale Edison. [a]"Ivry-sur-Seine," preprinted. [b]Multiply underlined.

1. Neither Edison's cables nor his letter have been identified. Batchelor had been especially anxious for specifications of the 250-light dynamo, which he had wanted to start manufacturing in the spring; he urged Edison to "give me something for God's sake!!!" Batchelor to TAE, 19 Apr. 1882, DF (*TAED* D8238ZAW; *TAEM* 62:388).

2. The Machine Works rushed several H dynamos to completion in June in order to obtain cash until the Continental company advanced money on its Jumbos. Edison had requested the company to make a series of advances, but the directors refused. He protested that he had "no wish as a manufacturer to make these large machines & were my interests only that of a manufacturer I certainly should refuse to do so. There is scarcely any profit in doing so . . . [T]o tie up my own capital while building them is certainly out of the question." Edison's correspondence to and from Paris parties at this time contains transcriptions and explanations of cable messages on this subject. Charles Rocap to TAE, 17 May 1882; Batchelor to TAE, 29 May 1882; both DF (*TAED* D8233ZBA, D8238ZBJ; *TAEM* 61:1054, 62:420); TAE to Société Électrique Edison, 29 May, 1 and 29 June 1882; TAE to Bailey and Puskas, 20 June 1882; Lbk. 12:456; 7:391, 626A, 543 (*TAED* LB012456, LB007391, LB007626A, LB007543; *TAEM* 81:745, 80:574, 701, 661).

3. The Machine Works shipped two C dynamos to Genoa on 8 August, apparently destined for Milan. TAE to Société Électrique Edison, 8 Aug. 1882, LM 1:242D (*TAED* LM001242D; *TAEM* 83:993).

–2306–

Draft Patent Application: Incandescent Lamp

[Menlo Park,] June 27th 82

Minimum amount of water

flour paste or dough[a] well kneaded is rolled ~~out~~ or pressed out on polished surfaces in sheets   filiments are punched therefrom which are carbonized under strain & pressure.

Also the dough is forced by pressure ~~from~~ from a chamber through an orifice in the shape of a filiment   ~~extra pieces~~ it is cut in lengths & extra pieces put on the ends & then dried under strain or strain & pressure & then ~~bent in an~~ in any desired shape then carbonized under strain & pressure.

More or less or all[b] of the starch may be forced out & the glutin used. Describe glutin broadly[1]

ADf, NjWOE, Scraps., Cat. 1148 (*TAED* NM017:149; *TAEM* 44:508). a"or dough" interlined above. b"or all" interlined above.

1. This draft describes processes manifestly related to those in Doc. 2291. It was the basis for an application that Edison filed on 5 July covering "the use of gluten, or a mixture thereof with starch or other material. The gluten is obtained by removing the starch from the flour of cereals; but a part or all of the starch may be allowed to remain mixed with the gluten. A dough or paste is formed of the gluten or gluten and starch with the minimum amount of water." The patent issued in March 1883. U.S. Pat. 274,296.

*Draft Caveat: Electric
Light and Power*

Caveat.

The object of this caveat is to set forth various devices and inventions which I have perfected for use in my system of Electric Lighting, but which requires the inventions and arrangement for economical manufacture etc that they may be used commercially=

In Electric Lamps when very high candlepower and electromotive force is used the sealing in of the wires through the lamp should be very perfect   I use[a] a platina leading in wire wound in the form of a spiral and both of the wires are sealed[a] in tubes before final sealing to the inside part. See fig 1[2] ~~P~~Straight platina wires coated with Tin Lead or other soft metal may be used. Even platina may be dispensed with and nickel iron copper ~~and~~ silver cobalt, steel & other wires may be coated with tin lead or other soft metal & sealed in the glass, the expansion & contraction of the glass being allowable as the soft metal gives. In the case of platinum lead ~~gives~~ combines with both the glass and the platina & therefore makes a good seal. The platinum may be coated with an oxide of lead mixed with anhydrous oxide[a] of sodium and fused in the glass   the lead attacks the platina & the Lead oxide soda & glass combines at the high temperature to[b] form glass thus making a perfect seal other metals as heretofore mentioned may be used instead of platina.

finely divided platinum may be mixed with glass to such an extent as to cause the glass-platinum mixture to become of sufficient conductivity to be used as leading in wires in this case there is a very perfect seal— When platina alone or platinum-Iridium allow[3] is used for the leading in wires I prefer to put it through my process for freeing it of air patented in 1879—[4] I[a] have made lamps in which in addition to the glass seal a cement was poured into the inside part that was not in vacua; this cement is preferably a cement having no pores with vitrous fracture. Sealing wax & high boiling points tars resins are useful—[5]

To keep water from being absorbed by the carbons filiments after carbonization I dip them when they are taken out of the mould in a solution containing a disolved solid the menstrum having no water   parafine disolved in ~~Benzine~~ a hydrocarbon Liquid will cause the carbon when dipped to be covered with a skin of parafine; Beeswax, resins, gums Tar etc may be used all but the broadened Ends are so dipped.[6] afterwards the carbons are connected to the leading in wires & plated to them Electrolytically then before the carbons are sealed in the ~~insid~~ bulb of the lamp the parafine or other gum or substance is well disolved out by immersing the loop in a solvent of the substance on the carbon   The loop is then sealed in the globe and the top of the bulb has a long tube upon the end   on the end of this tube is a piece of rubber tubing. Small tubes closed at one end are filled with a drying agent such as sulphuric or phosphoric anhydride—pentachloride of phosphorus chloride calcium burned[a] sulphate[a] copper[c] or other drying agents and this tube is provided with a bevelled mouth into which the rubber of the tube on the lamp fits   The lamps are kept with these drying tubes on until ready for exhaustion on the pump when they are placed on a revolving heater and heated & then placed on the pumps   a very rapid way of driving out the moisture is to ~~heat the lamp by the revolving heater then~~ place the bulb in connection with an exhausting & forcing pump the bulb remaining on ice and exhausting the air & then forcing fresh air into the bulb which must pass over a long tube filled with phosphoric anhydride or other drying agents. When the moisture has ~~e~~been driven out of the lamp it is placed on the revolving heater and heated very hot and then placed on the pumps— In ~~a~~ two[b] side tubes connected to the exhaust tube of the Electric Lamp one tube may have a drying agent such as phosphoric anhydride while the other may have organic charcoal which has been made red hot & then allow to cool for an instant & plunged into dry chlorine gas. the charcoal absorbs chlorine. It[a] is then placed in the tube, the bulb being previously dried as mentioned   the lamp is exhausted, the charcoal tube sealed off at once or after a time while the phosphoric anhydride tube may remain on for a greater length[a] of time   of course in exhausting the lamp the filiment is at times brought to incandescence   I will mention that ~~the~~a single tube might be used containing the drying agent & the chlorinated charcoal, and it might be sealed off the lamp or kept in permanent connection therewith =[7]

a small ball of charcoal filled with phosphoric anhydride or other drying agent might be used within the lamp, the charcoal being impregnated with chlorine, pure anhydrous[b] alumina Silica in a finely divided state & pressed up into hollow[a] cubes balls or other shapes might ~~have~~ be filled with a drying agent. Even paper & many other organic fibres & [---][d] act as powerful drying agents & might be used.[8]

A good method of regulating the electromotive force of the lamp circuit automatically from the Dynamo or rather magneto electric machine, is to use seperate brushes on the commutator to work the field and so place these brushes that when the load increases by putting in lamp ~~the~~ advantage will be taken of the shifting of the neutral point on the commutator due to load to give a greater electromotive force across these two brushes employed to give the field magnet as current. When the two brushes are placed at right angles to the brushes that form the lamp circuit ~~it acts~~ [----][d] the field forms the bridge wire of a Wheatstone balance. ~~to ob o~~ this is when the load is very light hence to obtain sufficient emf to energize the few[b] lamps that are on the brushes must be set off[a] of the exact neutral point. Now if the load is increased by adding lamps the neutral point will shift & while this causes the emf of the mainline brushes to drop the emf ~~on~~ between[b] the field brushes increases hence the drop in emf on one is nearly compensated by[a] the rise in the other due to the action of the field magnet[9]

Another method of regulating is to use a small electric engine for rotating brushes around a commutator & have this commutator connected to a portion only of the wire around the field magnet this portion being more or less rapidly reversed by the rapidly revolving brushes the engine being controlled by a magnet ~~or therm~~ across multiple arc or by a thermostat playing between two points.[10] another device for regulating the emf of a Dynamic[a] machine consists in interpolating in the line a Resistance in the form of several small wire multipled or a sheet, and winding this around a cylinder of mercury with a long tube extending therefrom into which there a several platinum wires connected to resistance coils in the field of force magnets the rise & fall of the mercury column serving to cut in & out[a] more or less resistance as more or less lamps are put on[11]

this mercury controller might be surrounded with very fine wire & the wire placed in multiple arc or it might receive its heat from an Electric Lamp radiating upon it such lamp being placed in multiple arc across the line.

Another method of regulating the lamp circuit consists in using an extension from the polar field peice & between which extension a small induction bobbin is rotated by a belt from the dynamo   this belt is shifted by an electro magnet so it will either go slow[a] or fast the electromagnet being either in or across the line. Thermostatic devices may also control it = this little bobbin is connected to the field of force coils & serves to energize them.[12]

Another form of gives consists of ~~p~~the driving pulley either on the machine or from a countershaft or even[a] the engine being loose on the shaft and is connected to the shaft through the intermediary of a spring & lever which lever serves to throw resistance in and out of the ~~main~~ field of force magnets. If but ~~lit~~ few lamps are on the circuit the drag on the pulley is light the spring is scarcely depressed and all the resistance is in   if now the load is increased the spring will be dipressed and ~~more~~ resistance will be cut out of circuit[13]

Another device consists in making a small portion of the curved polar extension of the field magnet detached & moveable. have moons about 1/2 inch wide & 1/2 thick ~~are~~from part of the polar peices and are connected together at the neutral point opening—by brass. a lever is connected to the brass;—~~now~~ which leaves moving in the direction of rotation of the dynamo bobbin cuts out resistance   the tendency of the field or any part of it which is moveable is to move in the same direction as the rotating armature and this tendency increases with the load. Hence[a] this ring encircling the roating bobbin tends to rotate with the bobbin in proportion to the load hence it becomes a very acurate[a] device[a] in connection with resistance to regulating the lamp by the field magnet[14]

If plates of smooth lead between which are placed insulating material such as porous plates, and several pairs of these placed in a cell half of the plates being connected to one pole & the other half to the other pole and the whole of the plates kept pressed together by a wedge spring, gravitation or otherwise and a 20 percent solution of sulphoric acid be used and the liquid kept nearly to the boiling point & the battery while thus hot is connected to a powerful dynamo machine, thick coats of peroxide of lead are formed rapidly all homeogeneous & integral & in electrical con~~n~~tact with the electrode   this coating when~~ce~~ once formed remains integral   the great object in perfect storage[a] batteries always charging & discharging the elements while they are under pressure the more powerful the pressure is the better.[15]

In forming carbons for Electric Light arcs I use nonvolitile hydrocarbons such as the[a] thickened Rubber[a] like substance obtained from boiling the drying oils. The carbon finely powdered is mixed with the viscous drying oil either with or with asphaltum and well ground it in then dried and then put in a powerful press & forced out into sticks[16]   They are then baked & brought up to a white heat then when cool are placed in tubes with a volitile hydrocarbon or bituminous coil.[17] the tube is closed and the whole brought up to a white heat   the gases formed enter into the pores of the carbon & deposit dense carbon therein due to the enormous high pressure in the tube due to the gases themselves   Chlorides of carbon may be used The tubes may be made so strong that a liquid hydrocarbons may be used and be brought to a white heat. Extremely dense carbons are thus obtained which do not oxidize just above the arc like ordinary carbons hence will last ten or 15 times longer than ordinary carbons   besides they are better conductors of electricity

When several filiments are placed in series in one lamp sometimes one of the filiments has a higher resistance than the others   in this case a side tube being placed on the exhausting tube & containing chloride of carbon the particular filiment is brought up to incandescence for an instant to reduce its resistance by a deposit of carbon on it   it is then compared with the others & when all give the same light with the same Volts they are permanently connected in series and the lamp sealed off of the pump—[18] another method is to shunt the particular loop by a fine fibre filiment exterior to the lamp but of very high resistance & through which current passes insufficient to bring it to the oxidizing point   this resistance may be placed in the base of the lamp

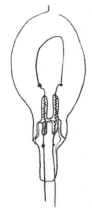

Edison's 18 June caveat sketch.

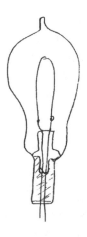

Edison's sketch for sealing the lamp with cement.

ADf, NjWOE, Lab., N-82-06-08:77 (*TAED* N197:39; *TAEM* 40:371). [a]Obscured overwritten text. [b]Interlined above. [c]"burned sulphate copper" interlined above. [d]Canceled.

1. Edison likely began preparing this draft soon after 18 June, when he dated several entries in this notebook describing related ideas.

2. Edison sketched this figure on a previous page on 18 June. At an undetermined time he marked it "fig 1   Caveat." N-82-06-08:41, Lab. (*TAED* N197:21; *TAEM* 40:353).

3. Edison probably meant "alloy."

4. Edison referred to the application he executed in April 1879 that issued in May 1880 (U.S. Pat. 227,229); the draft of it is Doc. 1695, which was derived from Doc. 1676. See also Docs. 1675, 1678, and 1796.

5. Edison sketched a lamp sealed with cement on 18 June. He also marked it "Caveat." N-82-06-08:43, Lab. (*TAED* N197:22; *TAEM* 40:354).

6. In a patent application he signed in October, Edison specified dipping carbonized filaments in a sugar solution or in camphor, shellac, or other readily carbonized substance, to prevent water absorption. U.S. Pat. 275,612.

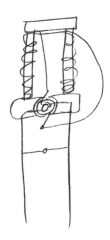

Edison's 18 June regulator arrangement using commutator brushes for the field circuit in a different position than those for the outside line.

7. Edison drew several sketches related to this method in the same book on 18 June and made related notes the next day. He filed a patent application covering this process (Case 453) on 26 June that he subsequently abandoned. N-82-06-08:45–47, Lab. (*TAED* N197:23–25; *TAEM* 40:355–57); Cat. 1148, Scraps. (*TAED* NM017:144; *TAEM* 44:501); Patent Application Casebook E-2537:268; Case 453, Patent Application Drawings; both PS (*TAED* PT021268, PT023:78; *TAEM* 45:768, 896).

8. Edison noted this idea, with a sketch, in this book on 18 June. N-82-06-08:62–63, Lab. (*TAED* N197:32; *TAEM* 40:364).

9. Edison made several sketches of this regulation scheme in this book. John Ott tested this type of regulator on 23 June. He found that "the nutral line changes proportial to load or nearly so. But the bridge as arranged is constant and does not change potentials proportial to load." On 12 September Edison executed an application for a patent on a regulator using an extra brush or pair of brushes to provide current to the field magnet. The brush or brushes could be moved around the commutator by a ratchet and pawl arrangement responding to changes in load on the main line. N-82-06-08:51, 56–59, 198–99; N-82-06-21:50; Lab. (*TAED* N197:26, 29–30, 71; N238:6; *TAEM* 40:358, 361–62, 403; 41:600); U.S. Pat. 273,487.

10. Edison sketched variations of this on 18 June. He described one as "using the mag[net] itself to make C[ounter] Emf." N-82-06-08:53, 55, Lab. (*TAED* N197:27–28; *TAEM* 40:359–60).

11. The editors have not found any contemporary evidence for this type of regulator.

12. Edison drew this variation on 18 June. N-82-06-08:61, Lab. (*TAED* N197:31; *TAEM* 40:363).

13. Edison executed a patent application for this arrangement on 13 November 1882. He pointed out the unusual nature of using entirely mechanical, rather than electrical, means to sense changes in the load and adjust resistance in the field circuit. U.S. Pat. 281,350.

14. Nothing more is known of this design.

Edison's 18 June sketch of a motor device to regulate dynamo field strength.

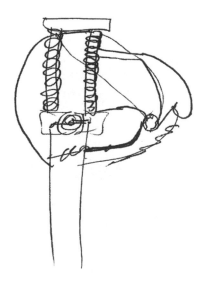

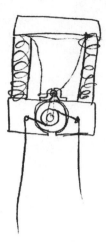

Edison's sketch of a regulator variation using a belt-driven induction bobbin to energize the field coils.

15. For contemporaneous battery experiments and related patent applications see Doc. 2276 esp. n. 4.

16. See Doc. 2291 n. 1.

17. Edison presumably meant "coal."

18. Edison filed a patent application on 9 November for testing and equalizing filament resistance by these means. It was placed into interference with Hiram Maxim and Edison subsequently abandoned it. Only the claims and drawings survive. Patent Application Casebook E-2537:390; Case 509, Patent Application Drawings; PS (*TAED* PT021390, PT023:93; *TAEM* 45:784, 901).

6

July–September 1882

Edison spent most of the summer at his Menlo Park home, but his movements away from there neatly frame the beginning and end of this quarter. In early July, he chartered a steam yacht and headed up the Hudson River, reaching Montreal via Lake Champlain, then traveling up the St. Lawrence River to Lake Ontario. Samuel Insull, his secretary, was with him and telegraphed the party's whereabouts to associates in New York but not, apparently, to Mary Edison, who may have been aboard with their children. Edison was away ten days. Then in late September, he and Mary prepared to return to New York, ending their Menlo Park interlude. He rented a house on Gramercy Park, about a dozen blocks from his Fifth Avenue office, for two years beginning 1 October. According to Insull, Edison declared he needed to be near the Pearl Street central station but "In the next breath he said he would never come near the city if it was not for the women constantly bothering him to do so." Privately, Insull thought that Edison "wants to come in just as much as the women do."[1]

Edison used the intervening weeks at Menlo Park much as he had since moving there in May, allowing his mind to play over problems and possible improvements in electric lighting. He gave particular attention to the residual atmosphere in lamps, and in early August undertook a substantial search of scientific literature in conjunction with experiments on various gases and chemical absorbents.[2] He also made improvements in storage batteries, but concluded that the battery generally was a "beautiful product of science but a most dismal failure commercially."[3] Francis Upton, having overseen the lamp factory's move to Harrison (near Newark), corresponded frequently about a variety of experiments, especially on a new 10-candlepower lamp suitable for village plants and the British market. In addition, Edison began planning for future central stations with the construction of a plant to demonstrate the "Village System" of distribution, which began in Roselle, New Jersey, in September[4] and his encouragement of Charles Clarke's efforts to design standardized urban central stations.

Edison again completed a remarkable number of patent applications in this period. The exact number is impossible to determine because, sometime in July, it became apparent that Zenas Wilber, who handled Edison's cases for the Edison Electric Light Company, had for some time failed to file papers at the Patent Office. Most, if not all, the of errant cases were submitted in late July and August, but Edison later recalled Wilber's malfeasance with uncharacteristic bitterness. He claimed that it had cost him a number of patents, and indeed the proportion of unsuccessful applications from this period is unusually high.[5]

During July, the elephantine C dynamos were installed and tested at the central station on Pearl Street. The underground network was completed, though not without interference from a company "playing havoc" by laying steam lines under the same downtown streets.[6] The First District system was tried discreetly and found to work. One can only imagine Edison's state of mind on 4 September when he gave the station a final inspection, then went to the office of J. P. Morgan to inaugurate it—and the commercial electric utility industry. It was almost four years to the day since his visit to a Connecticut factory had prompted him to begin intensive research into electric lighting, confident of quick success.[7] Despite its prominent location in the financial center of New York, the First District drew only modest public attention in its early days; one prominent technical journal noted the event in retrospect but declared that "the year 1882 was not characterized by the announcement of any great or revolutionizing discoveries or industrial processes."[8] Though the plant started smoothly, serious problems soon arose when Edison attempted to use two steam dynamos at the same time. Reacting violently against each other, the massive machines outraced their governors; suspended as they were on iron girders, they threatened catastrophic damage until Edison shut the throttles. It was several weeks until he devised a solution that allowed the station to run with more than one dynamo.[9]

Edison's lighting interests in Europe appeared to move

ahead. Italian investors formed a syndicate with 3,000,000 francs to start a small central station and factory at Milan.[10] Giuseppe Colombo, a principal planner of the Milan plant, placed a large equipment order and then sailed to New York to confer with Edison. Joshua Bailey, a tireless promoter with ambiguous allegiances, traveled with him and signed an agreement to work wholly for the Edison Electric Light Company of Europe. In Paris, Charles Batchelor succeeded in producing Edison lamps and dynamos for the first time outside of the United States. Edison thanked Batchelor for his reports from the factories and passed "them round to the people who can take advantage of them."[11] That was a select group of individuals working directly on his behalf; from others, even friendly associates like physics professor George Barker, Edison carefully withheld all but the most general information.[12]

In London, the Edison Electric Light Company, Ltd., still relied entirely on Edison's factories in the United States. Confident of success in New York, Edison showed little patience for complaints from Britain about the cost of equipment and creeping doubts about the economic feasibility of his system against cheap London illuminating gas. Edward Johnson, who had been Edison's personal representative in England, had returned to New York in June; without his mediation, Edison's correspondence with company secretary Arnold White became filled with disputes over equipment orders, payments, and the price of new lamps. The chilly formality of their letters accented the writers' growing mutual disdain.

In early July, Edison borrowed $37,000 from Drexel, Morgan & Company, pledging various stocks as security. Some of this money was likely needed to support expansion of production capacity at the Edison Machine Works, as well as continuing losses by the enlarged lamp factory. In early September, he formalized an agreement to buy (for about $38,000) an equal interest (with Edward Johnson and Sigmund Bergmann) in Bergmann & Company, which manufactured sockets and fixtures.[13]

1. Doc. 2343.
2. See Doc. 2324.
3. Doc. 2334.
4. See Doc. 2336 esp. n. 3.
5. See Doc. 2323 and Apps. 1.C.21 and 5.
6. Sherburne Eaton to TAE, 2 June 1882, DF (*TAED* D8226O; *TAEM* 61:227).
7. See Docs. 1423, 1433, and 1439.

8. See Doc. 2338; "1882," *Manufacturer and Builder* 15 (1883): 2.

9. See Doc. 2343.

10. Joshua Bailey and Theodore Puskas to TAE, 17 July 1882, DF (*TAED* D8238ZCC; *TAEM* 62:454).

11. TAE to Batchelor, 31 July 1882, Lbk. 7:804 (*TAED* LB007804; *TAEM* 80:769).

12. Cf. Docs. 2309 and 2337.

13. See Doc. 2343 n. 18.

–2308–

Memorandum: Electric Lighting Patents

[Menlo Park,] July 1, 1882

New patents—

feeder system for small town Installation.[1]

Mixing wood paper etc fibre with partially oxidized Linseed or other drying oils   drying & forming in sheets by Hydraulic pressure afterwards stamping or punching or cutting out filiments, before punching etc baking—[2a]

Carbon moulds made out of gas retort carbon by sawing etc.[3a]

depositing carbon on highly heated porcelain—polished[b] Lime plates, magnesia etc. or depositing on platina nickel, peeling off & punching filiments.

Heating filiments under enormous pressure in presence of a gas containing carbon—[a]

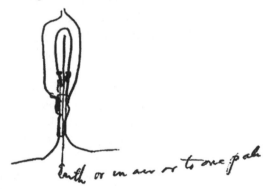

metallic wire to attract ~~Ca~~statically Charged Carbon Vapor[4]

Making insulating wire consisting of coating with cotton drawing through Japan Varnish baking recovering again thorough then baking & so on— cloth or paper may be used instead of ~~cloth~~ cotton braid or winding—

TAE

ADS, NjWOE, Lab., N-80-10-01:157 (*TAED* N304:91; *TAEM* 41:1138). [a]Followed by dividing mark. [b]Obscured overwritten text.

1. Edison did not complete such an application until November. This was for the basic patent on a three-wire distribution network for locales

in which it was "desired to employ electric currents of unusually high electro-motive force, so that the size of the conductors . . . may be diminished, thus economizing in metal." In contrast to the 330-volt, two-wire "village system" (later erected at Roselle, N.J.), the third wire permitted each lamp to operate independently at 110 volts. He made sketches of the system, some clearly related to this application, in late October. U.S. Pat. 274,290; Cat. 1148, Lab. (*TAED* NM017:87–89; *TAEM* 44:447–49).

2. See Doc. 2291.

3. Edison executed such a patent application on 7 July. The specification covered carbonizing molds made of "hard carbon—such as gas-retort carbon—which is powdered and mixed with tar" or other binding material. The raw molds were to be baked in the presence of a carbon vapor, which would form a "coating of deposited carbon [that] is exceedingly hard and compact, and does not readily absorb gases." Edison stated that these instruments were less expensive, more durable, and less able to absorb destructive gases than were the nickel forms in use (U.S. Pat. 334,853). He later described their manufacture and use in Doc. 2335.

4. This was presumably to abate the problem of carbon carrying by having a wire inside the globe to carry a charge independent of the filament circuit. See Docs. 2294 and 2313.

–2309–

To Francis Upton

[Menlo Park,] July 5—[188]2

Dear Sir

Referring to your telegram just received.[1] Wire Prof. Barker that you will be happy to have him personally visit the factory.

Please attend him yourself and let all descriptions be general, giving no detailed explanations, and admit <u>him only</u>. Yours Truly

If he should bring friends inquire who they are & if unobjectionable OK but if any object's say that the lamp Co have a rule to admit no one & you dare not disobay[2]   Edison[a]

L (letterpress copy), NjWOE, Lbk. 7:664a (*TAED* LB007664A; *TAEM* 80:714). Written by Samuel Insull. [a]Postscript written and signed by Edison.

1. Upton wired Edison that George Barker wished to visit the lamp factory on 7 July, and asked for instructions. Edison drafted a reply on the message form: "Telegraph you will be happy show him personally through factory." Upton to TAE, 5 July 1882, DF (*TAED* D8230ZBA; *TAEM* 61:780).

2. Edison advised similarly in December, when Barker asked to visit the factory with a friend. Edison consented, but told Upton to "dilute your explanations." In early 1883 Upton asked Edison to clarify whether visitors might also be admitted on Sherburne Eaton's authority. He also suggested discriminating on the basis of a prospective visitor's depth

of understanding: "I see no objection to taking men that are outside of the technical part of the business through. Men like Barker hurt us most." Upton to TAE, 28 Dec. 1882; Insull to TAE, c. 28 Dec. 1882; TAE to Insull, c. 28 Dec. 1882; Upton to TAE, 17 Jan. 1883; all DF (*TAED* D8230ZCT, D8230ZCT1, D8230ZCT2, D8332K; *TAEM* 61:849–50, 67:109).

-2310-

To Arnold White

Menlo Park N.J. July 5, 82

My Dear White

I have just seen your letter to Johnson regarding high prices of our machinery as compared with Swan & others.[1] We had and have that difficulty here but when put in the right light before the intending purchasers by figures we generally get the orders which give satisfaction and help future sales immensely.

Now if you will give me all the data so I can work understandingly I can bring you out all right. First send me Siemens price list of all machines used for incandescing lights, the number of lights each machine is guaranteed to run,—the candle power of each lamp (It is perhaps as well to know this point from actual observation)—the horse power they say is required— Also the same data from other makers   Also Swans and others' price lists and publications. Also a report of actual cost [--][a] what has been done for the money including power & boilers if put in as already in, etc. in some place where it has not been put in as an advertisement, I mean an honest sale for profit[2]

You know that our efforts have been to devise machines from which we are to sell Electricity or rather light and are highly economical and hence in the Isolated branch of the business we work to a slight temporary disadvantage.

You will notice that lately we have reduced the Dynamo cost per light down to [$5.40?][b] on the 250 machine and $6.00 per light on the 150.[3] Now my impression is that Swans installations have generally a 10 candle lamp (those on the Servia[4] were 6 candle lamps reduced to 4½ to 5 candles when all were on Siemen's alternating 200 light machine being used   if they are 10 candles then our 250 machine of 16 candles will give 400 of our 10 candle lamps

If an engine is to be provided but 35 horse power will be required. An Armington & Sim Engine for [$750?][b] fob will produce that power, thus we have 400, 10 candle lamps with engines for $2100. or [5.25?][b] per lamp including power.

The cost of copper to make the lamp [installation?][b] will be seven times less than Swan his lamps being, I believe 30 ohms and ours 250. I am curious to learn the particulars of the costs of his installations for profit and the opinion of an expert as to the extent, if any, of the profit. Machines can be made at almost any price, this is only a question of policy. Yours Very Truly

L (letterpress copy), NjWOE, Lbk. 7:665 (*TAED* LB007665; *TAEM* 80:715). Written by Charles Mott. Decimal points added to monetary values for clarity. [a]Canceled. [b]Illegible.

1. The editors have not found White's correspondence with Edward Johnson about the cost of Edison installations but several subsequent letters evidently passed between them. White summarized his side of the exchange in a 24 August letter to Edison:

> What the average public want is electric light and in the small installations they do not stay to examine very closely into the relative economy of maintenance of the competing systems. What they look at is first cost and however much we may all regret the folly & ignorance which are thus evinced the fact remains that we must either cater for this foolish ignorant public in their way or suffer others wiser in their generation to step into the breach. [DF (*TAED* D8239ZDU; *TAEM* 62:1037)]

2. See Doc. 2317.
3. White had written to Edison a week earlier for prices on dynamos and armatures. Edison replied on 24 July with a list of prices substantially lower than those offered on retail sales by the Edison Co. for Isolated Lighting, but consistent with the "strictly confidential" discount to another prospective British licensee. These figures are based on the price he quoted White for the K ($1,350) and L ($900) dynamos. White to TAE, 27 June 1882, DF (*TAED* D8239ZCV; *TAEM* 62:1002); TAE to White, 24 July 1882; Samuel Insull to Charles Rocap, 15 June 1882, Lbk. 7:760, 466A (*TAED* LB007760, LB007466A; *TAEM* 80:753, 615); see also App. 3.
4. The *Servia* entered service for the Cunard line in November 1881 as one of the largest and fastest transatlantic steamers. "The Servia," *Times* (London), 3 Mar. 1881, 11; "A New Cunarder," *New York Times*, 18 Mar. 1881, 8; "The Steam-Ship Servia," ibid., 9 Dec. 1881, 8.

–2311–

Telegrams: From/To
Samuel Insull

July 6 1882[a]
Rhine Cliff NY[1] 7:15 am

T A Edison,

Have seen boat   exactly what we require[2]   cost four fifty month including engineer   man on board cost you thirty dollars month   we shall have to get captain   in addition yacht carries two small boats has good accommodation goes twelve miles per hour easy. should advise taking it   if so telegraph owner John Aspinwall Barrytown N.Y.[3] immediately enable him get boat ready saturday   wire me fifth ave office and I will see about captain

Insull

Menlo Park Depot NJ 10:04 [a.m.]

Insull

have wired aspinwall for yacht and engineer month   ready saturday[4]

T A Edison

L (telegrams), NjWOE, both DF (*TAED* D8204ZEH, D8204ZEI; *TAEM* 60:223). Both messages on Western Union Telegraph Co. message forms. ªDate from document; form altered.

1. Rhinecliff, N.Y., is on the east bank of the Hudson River, about 80 miles from New York City.

2. Edison had evidently made at least two inquiries about buying or chartering a steam yacht. The Herreshoff Manufacturing Co. of Bristol, R.I., offered in early June to build one to his specifications for a price between $18,000 and $50,000. At the end of the month, Edison received a proposal through Sherburne Eaton to rent a vessel at Perth Amboy, N.J. Herreshoff Mfg. Co. to TAE, 1 June 1882; Samuel Holmes to Eaton, 28 June 1882; both DF (*TAED* D8204ZCK, D8204ZDY; *TAEM* 60:171, 211); Insull to Eaton, 28 June 1882, Lbk. 7:605A (*TAED* LB007605A; *TAEM* 80:698).

3. Aspinwall has not been otherwise identified. Barrytown also lies on the Hudson's east bank, about 8 miles upriver from Rhinecliff.

4. Edison and Insull separately telegraphed Aspinwall to await their arrival on the morning of Saturday, 8 July. The editors have not determined who (if anyone) joined them aboard the steam yacht *Arrow*. Edison also instructed Charles Dean to provide an engineer accustomed to marine engines. TAE to Aspinwall, 6 July 1882; Insull to Aspinwall, 7 July 1882; TAE to Dean, 7 July 1882; Lbk. 7:667B, 671B, 671C (*TAED* LB007667B, LB007671B, LB007671C; *TAEM* 80:717, 719).

The party reached Whitehall, N.Y. on 10 July by sailing up the Hudson River and Champlain Canal to the southern tip of Lake Champlain. Insull wired from there to Menlo Park that they were "just starting off for the unknown. we are having a splendid time." They reached Montreal on or about 14 July, where a local newspaper noted Edison's visit (Insull to Charles Hughes, 10 July 1882; TAE to Charles Mott, 14 July 1882 [*TAED* D8204ZEK, D8204ZEM; *TAEM* 60:226–27]; "Visiteurs," *La Patrie* (Montreal), 15 July 1882, [3]). They steamed up the St. Lawrence River into Lake Ontario, eventually reaching Oswego, N.Y., then passed through the Oswego and Erie Canals to Rome, from where Edison and Insull traveled by train to Menlo Park on 18 July. After the *Arrow* was returned on 8 August, Aspinwall claimed damage to the engine and stern from the lack of "a good competent engineer" (TAE to Mott, 14 and 17 July 1882; Insull to Mott, 18 July 1882; Aspinwall to Insull, 13 Aug. 1882; all DF [*TAED* D8204ZEO, D8204ZER, D8204ZET, D8204ZFN; *TAEM* 60:229–30, 232, 257]).

Dear Mr. Edison:

Regarding the formation of Lamp Co.

I have been to see Mr. Adams[2] and he has taken all the needed memos. He says that the law requires certificates issued for property to be stamped "Issued for Property only." Unless this is done the shares are assessible up to their par value and holders are responsible.

My idea is to issue a certain amount of stock to the presents partners in the concern to ~~buy out their~~ [all?][b] buy out their rights. Then after this to issue stock at par giving ~~presents~~ members of concern first choice.[3]

Regarding a different price for A and B lamps.

I think it would be advisable as establishing a precedant for other lamps.

We can make a B lamp ~~five~~ 3[c] cents cheaper than an A lamp and like to make them as they help out in the pump room to fill ~~out~~ up odd moments when there is not time to make A lamps. You could use this as a concession to acquire the 1/2 profit without change in price.[4]

Regarding making higher resistance and greater economy lamps.

I am having a set of $6 \times \dfrac{9^{1/4}}{1000}$ made running from 5" to 8" by 1/2" There will be 200 fibres in each ~~sett~~ set. I shall have a careful test made of them and report to you.[5] They will all be run with phosphorous tubes on them so as to lay drying for five days after coming off from pumps.

You know that these special lamps will entail an enormous expense on the Lamp Co. at the start until we can find a market for product in all covering all volts, and bring the breakage down.

600 hours life and 12 per H.P. is a big thing to do and we are the only ones that can touch it giving 320 ohms as a resistance.

My idea is, that anyway for next central station, we will make a lamp requiring 140 volts while giving 16 candles and of as high economy as it is commercial to make.

If we were not so much behind on our orders I should not have tried to run on the days near the fourth. As it is we have had hard work to make both ends meet, with the sleepy hands and picnics.[6]

The masons are making good headway with the foundation and promise to show a good deal of wall in a week's time. We have decided to put on a tin roof as the surest to get done in time.[7]

I try to restrain expenditures, but the money goes all the time. Every time I say yes, it cost on an average $10 and my noes bring discontent.

Holzer is going to buy one of Queen's lamps that you mentioned.[8] We can tell from the style when they were made. We do not see where the leak can be and suspect no one. We shall organize a breakage department and require all broken lamps to be returned to it. The[n] with our lamp accounts every lamp must be accounted for. Yours Truly

Francis R. Upton.

ALS, NjWOE, DF (*TAED* D8230ZBB; *TAEM* 61:781). Letterhead of Edison Lamp Co. [a]"*East Newark, N.J.,*" and "*188*" preprinted. [b]Interlined above and canceled. [c]Interlined above.

1. Although the lamp factory lay within the formal boundaries of Harrison, the company used both that city and adjacent East Newark on various forms of its letterhead.

2. Frederic Adams, an attorney in Newark, had assisted Edison in purchasing the Harrison property. *TAEM-G2*, s.v. "Adams, Frederic."

3. A draft of the Lamp Co.'s new contract with the Edison Electric Light Co. from this time specified that the Lamp Co. would be incorporated in New Jersey and supersede the individual interests of Edison and his partners. This did not occur until 1884. Draft agreement, n.d. 1882, DF (*TAED* D8230ZAN; *TAEM* 61:757).

4. Edison agreed to a three cent discount on B lamps but thought that five cents "would be too much of a reduction" (TAE to Upton, 7 July 1882, Lbk. [*TAED* LB007682A; *TAEM* 80:723]). Under the existing contract, the Edison Lamp Co. sold its lamps to the parent company for 35 cents apiece, without regard for the type of lamp. It paid to the parent company one-half of its profit in excess of 3 cents per lamp (see Doc. 2039 n. 1). Under the draft terms under consideration the price was 40 cents for "the standard lamp" with any profit greater than 5 cents on each one to be divided equally with the parent company.

5. Upton presumably meant to indicate raw fibers 0.006 × 0.00925 inches, varying in length from 5 to 8½ inches. The standard filament since February or March 1882 was 0.008 × 0.0135 inches (6 inches long for A lamps and 3 inches for B lamps); cf. Doc. 2085 n. 4 (Cat. 1302, Batchelor [*TAED* MBN008; *TAEM* 91:365]). Upton reported the results in Doc. 2319.

6. Factory notebooks refer to "day" and "night" production until 11 July, when records for three shifts first appear. Production evidently returned to day and night runs on 20 July. According to Philip Dyer's weekly reports around this time, the factory carried an inventory of more than 61,000 lamps but many of these likely required the wrong voltage. Cat. 1302 (orders 873–75, 912), Batchelor (*TAED* MBN008:48, 57; *TAEM* 91:413, 22); see e.g., Dyer to TAE, 10 July 1882, DF (*TAED* D8231ABD; *TAEM* 61:912).

7. This was probably the new building for the carbonizing department. The structure and furnaces cost about $5,000. Upton to Samuel Insull, 28 July 1882, DF (*TAED* D8230ZBH; *TAEM* 61:798).

8. Edison had apprised Upton that James W. Queen & Co., of Chicago, was making unauthorized sales of B lamps at 35 cents each. He warned that because "these lamps have certainly not gone through the channels of the Co. it looks as though they may have been surreptitiously removed from the factory, and I would advise you to take some steps to prevent any thing of this kind." TAE to Upton, 5 July 1882, Lbk. 667A (*TAED* LB007667A; *TAEM* 80:717).

–2313–

Notebook Entry:
Incandescent Lamp

[Menlo Park,] July 6 1882

Prevention Electrical Carrying—[1]

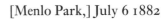

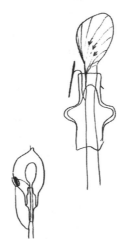

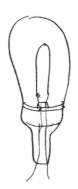

TAE

Edison's 5 July drawing showing platinum wires around the globe to reduce the static charge of the glass.

X, NjWOE, Lab., N-80-10-01:63 (*TAED* N304:38; *TAEM* 41:1097).

1. These drawings appear to represent means to reduce the difference in electrical potential—and static attraction—between the filament and lamp globe. Edison made three related sketches on 5 July. One appears to show a connection between two platinum wires placed circumferentially around the lower part of the bulb, one inside and the other outside. The others represent a wire or wires inserted through the glass near the top of the filament and connected to the lead-in wires, reminiscent of the lamp drawn first in Doc. 1898; see also Doc. 2346. N-80-10-01:71, N-82-05-26:23, Lab. (*TAED* N304:43, N204:11; *TAEM* 41:1101, 40:585).

Edison made two other drawings on 5 July of ways to create an electrostatic attraction of carbon particles away from the glass globe.

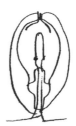

*From Charles
Batchelor*

Paris, le 17 July 1882.[a]

My dear Edison,

lamps[b]   I confirm my telegram to you last night for 5000 bulbs and enough glass to make all complete 5000 lamps—[1] I have struck a "bug" which I am sure is in the glass— I have not confined myself to one manufacturer but have got samples of three of the best makers here= The bug is this:—one of the wires in the inside part becomes black in the inside of glass— it is always the same wire as if it was an action like a plating action— I have tried for it in everything and I think it must be in the glass   I therefore ordered the 5000, if I find it is something else we can always use the glass. I have had a great ~~of~~ deal of trouble with the glass men here and when I have got them just where I want them I have struck this bug—they are making me different lots now—

⟨You will find this blackening of the platina wire on one pole in all lamps made by us that have run for some time= What is duties on these bulbs— the Corning folks make the best glass in the world perfectly reliable & we never have any trouble⟩[2]

Z Dynamos[c]   I have finished 6 Z dynamos and have 28 more almost finished   I have 100 more well underway— I have started 2 big dynamos and I want you to tell me which engine I must use   An order for them I will send officially—

⟨Armington & Sims 175 hp going up to 250. price is about 1850 @ 1875⟩[3]

Mercury tubes= leakage[d]   You know at Menlo Park we had considerable difficulty with leaky mercury tubes— I have struck a thing here that has just killed all that trouble— In all my tubing for 600 pumps there has never been the slightest leak. it is this:— I put no washer under the flange of mercury cock but screw it in to within a $\frac{1}{16}$ inch of shoulder and then I lap fine fish cord round it to fill up the gap— after this is fastened tight I soak it with shellac in alcohol and when this is hard & soaked in it makes a most perfect mercury joint—

⟨Send this to part Upton⟩[4e]

Socketing mould.[f] I have cast all my socketing moulds of type metal and they work very well   when the plaster comes out of mould they have the name on so EDISON on one side & Brevèté S.G.D.G[5] on other deep in plaster

Safety Catches=[g] I have made all my safety catches of plaster of paris and I think they are cheap   I believe you did this once but I dont remember why you did not use it— Was it too dear? I put a piece of glass tubing over the wire so as to have a big hole—

⟨Send this to Bergmann & EHJ⟩[6]

Screwing bolts=[h] I made a good strike here in the shape of a Brown screwing machine for bolts[7]   I have not cut a thread on the dynamos   the kepeeper bolts & steel head screws I cut in this machine at one cut— It takes my man just 4 minutes to finish a keeper bolt with a beautiful thread at one cut   I do everything on it from the pulley screw up to the keeper bolt— I make the belt tightening screw at one cut ~~of~~ without any turning

⟨We have the Large Buffalo bolt machine   1. $8 boy cuts all the bolts used in shop—⟩

Let me know your opinion on some of these things   Yours
                                                                                    Batchelor

Should like to have about 1 nights talk with you[8]   have not time to write much   B.

ALS, NjWOE, DF (*TAED* D8238ZBX; *TAEM* 62:447). Letterhead of Charles Batchelor. [a]"Paris, le" preprinted. [b]Written in left margin. [c]"Z Dynamos" written in left margin and multiply underlined. [d]"Mercury tubes= leakage" written in left margin; "leakage" multiply underlined. [e]Obscured overwritten text. [f]"Socketing moulds." written in left margin and multiply underlined. [g]"Safety Catches=" written in left margin. [h]"Screwing bolts=" written in left margin.

1. Batchelor's cable to Edison, for an immediate shipment of glass and bulbs, was received on 17 July. LM 1:235E (*TAED* LM001235E; *TAEM* 83:989).

2. This marginal note became the basis for part of Edison's reply to Batchelor on 31 July. He raised the possibility of shipping Corning glassware at prices that would be competitive in Paris. Lbk. 7:804 (*TAED* LB007804; *TAEM* 80:769).

3. In his 31 July reply (see note 2), Edison stated that he would notify Batchelor about the results of tests on the Armington & Sims engine he expected to receive in a week. He clarified that the price would be "about $1850 or $1875." Because of the inability of Armington & Sims to supply engines in time, he had been forced to use Porter-Allen engines on four dynamos that were nearly ready for shipment to Europe. Batchelor cabled an order for two engines a month later. Batchelor to TAE, 16 Aug. 1882, LM 1:244B (*TAED* LM001244B; *TAEM* 83:994).

4. Samuel Insull copied Batchelor's paragraphs about mercury tubes and the socketing tool (including drawings) into a letter he wrote to Upton on 10 August. Lbk. 7:881 (*TAED* LB007881; *TAEM* 80:804).

5. "Breveté Sans Garantie du Gouvernement," i.e., patented without government guarantee, a standard disclaimer in French patent notices. See, e.g., Doc. 2141 n. 3.

6. No correspondence with Bergmann or Edward Johnson on this subject has been found.

7. This presumably was a tool made by the Brown & Sharpe Manufacturing Co., a major machine tool manufacturer in Providence, R.I. since the Civil War. Rolt 1965, 171.

8. At the end of his reply to Batchelor (see note 2), Edison stated that he "like[d] to have your views on things & take advantage of them by sending them round to the people who can take advantage of them."

–2315–

To Joshua Bailey and Theodore Puskas

[Menlo Park,] 19th July [188]2

Dear Sir,

Your favor of 26th June[1] came to hand during my absence from home and hence the delay in replying thereto.

We receive the same complaints from England but I think if you look closely into the matter you will find that they are unfounded.[2] The Seimens machine is far from being economical—they are built without the slightest regard as to what it will cost to run them. I think when you complain of the prime cost of the machine you should at the same time give a statement of the comparative cost of running an Edison 60 Light Plant & a Seimens 60 Light Plant.

I am fully alive to the importance of bringing down the price of dynamos but as I am now selling them at next to cost there is not much chance of my doing so. I have however decided to quote all the small machines i.e 15 Light (E), 60 Light (Z), 150 Light (L), 250 Light (K), fob New York and this will make a very material reduction.[3] You will find I have billed the machines recently sent you on this basis.

You will also notice that the cost of the 150 & 250 Light machines is really much less proportionately than that of the 60 Light machine.

I will certainly reduce prices to you at the first opportunity but I think as you get deeper into the manufacturing business you will find my prices are not high   Yours truly

Thos A Edison   I[nsull]

L (letterpress copy), NjWOE, Lbk. 7:717 (*TAED* LB007717; *TAEM* 80:729). Written by Samuel Insull.

1. Puskas and Bailey had complained in their 26 June letter about the expense of the Z dynamo which, with shipping from New York and customs fees, proved to be the same as the entire cost of a 60 lamp Siemens plant. They asked Edison to reduce the price of the Z dynamo from $575 to about $400 and to give discounts on other types of dynamo machines. DF (*TAED* D8238ZBT; *TAEM* 62:442).

2. See Docs. 2200, 2211, and 2229 n. 4.

3. Edison quoted prices to Giuseppe Colombo in August that were the same as those charged to the Continental company. TAE to Colombo, 31 Aug. 1882, Lbk. 14:055 (*TAED* LB014055; *TAEM* 81:842); see also App. 3.

-2316-

From Charles Clarke

New York, July 19 *1882*[a]

My dear Edison:

In transmitting power in the country is in fact on lamp lines I do not believe in using a ground circuit. Do you? I wish to know your policy in this regard. ⟨No⟩

I am getting up a series of estimates for Central Sta. and am basing them upon 320 ohm lamp, 140 volts—12 per h.p. and C dynamo of 1400 of these lamps capacity. Also that the district is one half a mile square.[1] I think it best to keep within these limits rather than to run out farther and thus lose the advantage which the high resis. lamp give us. What do you say?

⟨I believe in a mile square for various reasons which will explain personally⟩

Have written you several letters since your absence. Please note contents and ans. the questions.[2]

Glad to hear of your safe return and improved health and weight. Yrs truly

Chas. L. Clarke

P.S. Shall I appoint any day for the person whom I have in view for Supt. to meet you with me and notify you of time appointed?[3] ⟨Yes⟩

ALS, NjWOE, DF (*TAED* D8227ZAG; *TAEM* 61:507). Letterhead of Edison Electric Light Co. [a]"*New York*," and "*188*" preprinted.

1. Clarke specified these design parameters in an earlier letter. He indicated that operating 1,400 lamps would require altering the C dynamo from 98 to 134 commutator bars. In August, the Edison Electric Light Co. prepared a comprehensive statement of the cost of constructing and operating a central station on similar terms, the major difference being a rating of 1,200 lamps on the dynamos. Clarke to TAE, 10 July 1882; Edison Electric Light Co. estimate, 3 Aug. 1882; both DF (*TAED* D8227ZAB, D8227ZAT; *TAEM* 61:504, 525).

2. Edison responded the next day, giving substantially the same an-

swers as in his marginal notes. A few weeks earlier, before Edison's boating trip, Clarke had pointed out to Samuel Insull that Edison was "considerably behindhand" in answering "many questions of importance to me." TAE to Clarke, 20 July 1882, Lbk. 7:724 (*TAED* LB007724; *TAEM* 80:734); Clarke to Insull, 14 June 1882, DF (*TAED* D8227U; *TAEM* 61:482).

3. Clarke had recently urged the importance of having a superintendent "at the start so that he can be with us and see the bugs and failures from the beginning. If we wait until the station is in good order, then he will not know what steps to take in case of trouble." Clarke put forward names of two men, a Mr. Lavery (of whom nothing else is known) and Joseph Casho, to whom Clarke offered the job on 9 August. Casho accepted and started on 1 September. The position included superintendence of the buildings and equipment, purchasing, and authority to hire and fire; it paid $2,400 a year. Clarke to TAE, 9 and 17 June, 31 July, and 15 Aug. 1882; all DF (*TAED* D8227S, D8227W, D8227ZAP, D8227ZBC; *TAEM* 61:479, 484, 519, 552); Clarke to Casho, 9 Aug. 1882, MiDbEI (*TAED* X001J7B) ; see Doc. 2403 n. 6.

–2317–

From Arnold White

London, E.C., 20th July, 1882[a]

My Dear Edison,

I have received your letter of July the 5th[1] on the subject of the cost of machinery and I note the points on which you require specific information. I am taking immediate steps to supply your wants in this respect and shall hope to communicate to you in the course of three or four days price lists of Siemens, Swan & Brush together with particulars of candle power, horse power & cost.[2]

You will be glad to hear that contrary to expectation the the Government have taken up the bill and on Saturday it was passed through committee at a sitting. This means that the bill is safe to pass the House of Commons, and the only question is as to whether the House of Lords will throw it out. Being a Government measure, I do not think there is any doubt but that that bill will become law this year.[3] It is quite absurd to watch the ignorance of parties nominally interested in electric lighting and their blindness to their own interests as contained in this bill. The Edison Company was the only one who protested against the term of years being restricted to 15 and we were accordingly met in the House of Commons by the unanswerable reply from the President of the Board of Trade[4] that although the Edison Company might be dissatisfied all the other companies were content. In order to overcome the bad effect produced by the hasty admissions of the Brush and Maxim Companies (of which latter I can hardly speak without contempt and disgust) I have prepared a[b] petition, copy of which is enclosed,[c] which is being circulated among the principal companies and which will be presented to the House on the third reading.[5] Sir John Lubbock will refer to this petition in his speech as evidence of the inaccuracy of the impression of the Government that all the electric light companies except the Edison Company are content with the period of 15 years.

I am happy to tell you that our prospects improve. We have got an order today to light the Waterloo Terminus of the London & South Western Railway, but as the official order has not come in please do not let Major Eaton publish this fact until I send it to him direct.[6] I shall endeavour to send you by this mail copy of Hopkinson's & Fleming's report on the Holborn Viaduct installation, which is a very able document, and the results of which are on the whole satisfactory.[7]

The financial troubles through which this country is now passing have had a great effect on electric lighting. The shares of the subsidiary companies of the Brush are not only at a discount but are wholly unsaleable.[8] Mr. Bouverie and Sir John Lubbock are glad that we have not found ourselves with a crowd of subsidiary share-holders clamorous and discontented, but that we have taken the more dignified and quiet course of solidifying our position before actually launching sub-companies. We shall have three more companies out by October.

Johnson is missed terribly here. He had a capacity for inspiring faith which exceeded that of any other man with whom I ever came in contact. Although my faith is equal to his I cannot pretend to more than a tenth of his power of imparting it to others. We are however, more than holding our own and we have most powerful influences with us.

Learning from some of the mistakes we made in the old telephone days, we have agreed to join an electric light bund composed of representatives of the principal companies.[9] The cardinal principle of this society is that we shall fight among ourselves as much as we like but agree together to defend ourselves against outsiders and that we unite against the gas companies and against small infringers. One of the effects of this arrangement is that the five disclaimers to the patents have been already passed by the attorney general without opposition from Swan or Lane Fox. You will remember, how many months were consumed before the disclaimers were allowed on the telephone patents.[10]

We are about to put in large installations at Westminster, at

Holborn[11] and other places, full details of which I will send you and Johnson from time to time. Yours very truly,

Arnold White

TLS, NjWOE, DF (*TAED* D8239ZDJ; *TAEM* 62:1020). Letterhead of Edison Electric Light Co., Ltd. a"London, E.C.," and "188" pre-printed. bMistyped. c"d" added by hand.

1. Doc. 2310.

2. White sent a price list from Crompton and Co., an electrical equipment manufacturing firm; the editors have found neither it nor the other information promised by White. White to TAE, 8 Aug. 1882, DF (*TAED* D8239ZDO; *TAEM* 62:1029).

3. The Electric Lighting Act, which passed on 18 August 1882, allowed individuals, companies, or local authorities to establish electrical supply systems. As originally drafted, the legislation permitted owners of a utility to enjoy a monopoly for seven years, after which local government authorities could make a compulsory purchase of the system based on the total value of its components, such as machinery and wires. This period was expanded before the law's enactment to fifteen and eventually, by the House of Lords, to twenty-one years (Gordon 1891, 357–62; Hughes 1983, 60). Millard 1987, chap. 4 compares the political and institutional conditions in which the 1882 Act was written with those in the U.S. and Germany.

4. Joseph Chamberlain (1836–1914), who had brought Birmingham's gas and water utilities under public ownership, headed the Board of Trade from 1880 to 1885. *Oxford DNB*, s.v. "Chamberlain, Joseph."

5. Joining the Edison Electric Light Co. in this petition to the House of Commons were the Anglo American Brush Electric Light Co., Siemens Brothers & Co., the British Electric Light Co., and the Electric Light and Power Generator Co. Edison Electric Light Co., Ltd. to Great Britain Parliament, 1882, DF (*TAED* D8239ZGK; *TAEM* 62:1186).

6. The order for Waterloo station has not been found; the installation was expanded in early 1883 from two Z dynamos to an L machine. Edison Electric Light Co. Bulletin 17:7, 6 Apr. 1883, CR (*TAED* CB017; *TAEM* 96:809).

7. See Doc. 2292 n. 4.

8. The Brush Co.'s practice of selling licenses to a large number of subsidiary companies fueled a speculative boom in electric lighting stocks in the early part of 1882, coinciding with a modest upturn in the British economy. What became known as the Brush bubble burst in May. Byatt 1979, 17–19; Hughes 1962, 29–30.

9. White likely recalled the uncoordinated responses of the Edison and Bell telephone companies to the British Post Office's general opposition to private telephone interests in 1879 and 1880. See Docs. 1870, 1919, and 1925.

10. The disclaimer of Edison's fundamental telephone patent was filed in December 1879 and allowed in February 1880, although disputes continued for some time. See Docs. 1870 and 1903.

11. White may have been referring to the 150-lamp isolated plant for the library and dining rooms at the House of Commons installed in April 1883. A larger station was completed about that time in the Holborn

Restaurant, an expansive establishment opened in 1874 at 218 High Holborn St., just west of the Viaduct. "Parliamentary Column," *London Daily News,* 11 May 1883, DF (*TAED* D8338ZAB1; *TAEM* 68:50); Edison Electric Light Co. Bulletin 17:27, 6 Apr. 1883, CR (*TAED* CB017; *TAEM* 96:809); *London Ency.,* s.vv. "Holborn Restaurant," "High Holborn."

—2318—

From Sherburne Eaton

[New York,] FRIDAY, JULY 21st. 1882.

EDISON'S RESIDENCE.[1] The point about wiring your house is that there is concealed work in it and that the Board of Underwriters have never yet made any rules for concealed work or even consented to pass it under any circumstances. The Mills Building[2] was an exception. In that case they made a special examination and passed that building but without establishing a precedent. ⟨My impression is that my house is wire well enough⟩[3]

THEY COPY US. The Scientific American of July 22nd. contains my item verbatim on The Edison Lamp Company published in the 11th. Bulletin.[4] They publish it as editorial without giving the Bulletin any credit for it. However, that does[a] not matter so long as they keep the truth before the public. The Bulletin is more and more a success every day. ⟨I Saw the article⟩

WILBER'S WORK. Here is a memo.[5] from Wilber regarding the specifications which Dyer has this day told me he has been sending in here for the last two or three months. Dyer will explain this memo. to you. From this memo. it seems that Wilber has done better than Dyer supposed he had[a] done. ⟨[p-----Have put to patent?]⟩[b]

BRUSH PATENT. Referring to your memo. of July 20th., which is herewith enclosed with a lot of other papers, please read Wilber's memo. lengthy memo. of this date pinned on at the end of the bunch of papers and please give me your comments.[6] The papers are marked 21B. ⟨The device Brush uses of compound multiple arc is worse than useless   its destructive to Lamps⟩

MY MEMO. ON SWAN LAMPS FOR BULLETIN.[7] The only correction in the proof which you made was with reference to the mention of the Canadian patent. I called Wilber's attention to the matter to day and asked him for the reference to the U. S. or English patent as you suggested. He sends me the enclosed memo. in reply.[8] It is marked 21–1 and need not be returned.

BOSTON INSTITUTE. Please look at 21C and then return it. Do we care to exhibit. I suppose not. ⟨Yes⟩[9]

BERGMANN'S[a] PRICES. Here is what Bergmann has been charging us on the following things net, that is the discount is taken off: Pipe and flange $1.13.[a] Holder nine cents. Shade eighteen cents. Here is what the same things cost us when we buy[a] them from other people: Three foot pipe fifteen cents. Flange eight cents. Shade thirteen cents. Holder fourteen cents   That is a saving of one dollars. I am surprised at Bergmann charging us such enormous prices on these things. I will call Johnson's attention to it.[10] By buying these things elsewhere[a] than at Bergmann's we saved over five hundred dollars on the King Phillip Mill[11] alone. ⟨Will See B   Must be some mistake⟩

ISOLATED CO'S STATEMENT. I hand you for your files a statement made from the balance sheet of the Isolated Company of June 30th. 1882.[12] It shows that on a gross completed installation of $132,340,84 there was a profit of only about thirty per cent or a little less than $40,000. But our profits now are larger and average more than fifty per cent. Consequently we estimate a profit of fifty per cent on the work in progress so that the total profit, on that basis up to June 30th., is $34,228,53. If the Isolated Company should wind up its business to day, and could realize on its material and accounts, it could pay back all its money and divide up about seven per cent on the half million of stock. The report which is for your files is marked 21−2

TL, NjWOE, DF (*TAED* D8226ZAS; *TAEM* 61:341). "21.", "21B.", "21−1.", "21C.", and "21−2." written at top of page. [a]Mistyped. [b]Canceled.

1. In his report to Edison of 14 July, Eaton bemoaned the "triumph of red tape" regarding his frustrated efforts to have the wiring at Edison's home approved (see Doc. 2274). He had asked John Vail, superintendent of the Edison Co. for Isolated Lighting, to accompany an inspector to the house. When Eaton raised the issue again on 22 July, Edison responded that "if the Insurance people do not like it they may cancel their policies. I do not think I need Insurance now I am able to dispense with kerosene." Vail's warning that the inspector would ask to move furniture and take up carpets and floors prompted Edison's declaration that "If an Insurance man wishes to inspect my House he must be contented with what he can see without moving anything otherwise he need not come at all." Eaton to TAE, 14 and 22 July 1882; Vail to TAE, 27 July 1882; all DF (*TAED* D8226ZAN, D8226ZAT, D8221ZAA; *TAEM* 61:328, 344; 60:881); TAE to Eaton, 25 July 1882; TAE to Vail, 28 July 1882; Lbk.7:765, 785A (*TAED* LB007765, LB007785A; *TAEM* 80:757, 80:762).

2. Designed by George Post for banker Darius Ogden Mills (1825–1910), later a director of the Edison Electric Light Co., the ten-storey Mills Building was being erected on Broad St. between Exchange Place

and Wall St. Following approval from "the Insurance people" in May, the Edison Electric Illuminating Co. had outfitted the building with 5,588 lamps, which the Bulletin of the parent company termed "the largest enterprise of the kind ever undertaken." *ANB*, s.v. "Mills, Darius Ogden"; "The Mills Building," *New York Times*, 3 Feb. 1882, 8; Eaton to TAE, 15 May 1882, DF (*TAED* D8226D; *TAEM* 61:158); Edison Electric Light Co. Bulletin 11:2, 27 June 1882, CR (*TAED* CB011; *TAEM* 96:720).

3. Edison's marginalia throughout this document formed the basis of his reply to Eaton on 23 July. Lbk. 7:744 (*TAED* LB007744; *TAEM* 80:747).

4. "New Factory of the Edison Electric Lamp Company," *Sci. Am.* 47 (1882): 55; Edison Electric Light Co. Bulletin 11:3–4, 27 June 1882, CR (*TAED* CB011; *TAEM* 96:720).

5. Zenas Wilber's memo has not been found but see Doc. 2323.

6. On 20 July Samuel Insull returned to Eaton a letter from Spencer Borden about a patent recently issued to Charles Brush. This was U.S. Patent 261,077 for "translating devices arranged in multiple-arc series" (see Doc. 2286 esp. n. 16). Insull requested that Wilber read the specification because "The device claimed is very old & was operated by Mr. Edison long time back when he was using Platinum lamps." Wilber's memo on the subject has not been found but he reportedly examined whether Brush's was truly a multiple arc system. Edison declined to comment further. Eaton to TAE, 18 and 27 July 1882, both DF (*TAED* D8226ZAP, D8226ZAW; *TAEM* 61:335, 61:355); Insull to Eaton, 20 July 1882; TAE to Eaton, 31 July 1882; Lbk.7:727, 812 (*TAED* LB007727, LB007812; *TAEM* 80:735, 776).

7. Eaton sent Edison his draft of an article critiquing the limitations of Joseph Swan's U.S. lamp patents on 17 July. Edison approved it for publication with minor revisions. It appeared in the 27 July 1882 Bulletin and, in an updated revised form, in the 18 December 1883 Bulletin. Eaton to TAE, 17 July 1882, DF (*TAED* D8224ZAW; *TAEM* 61:60); TAE to Eaton, 20 July 1882, Lbk. 7:720 (*TAED* LB007720; *TAEM* 80:731); Edison Electric Light Co. Bulletins 12:13–20, 21:54–62, 27 July 1882 and 18 Dec. 1883; CR (*TAED* CB012133, CB021507; *TAEM* 96:734, 920).

8. Not found.

9. This was the New England Manufacturers' and Mechanics' Institute Fair. Edison replied on 23 July (see note 3) that "it is an exhibition worthy of our attention & by exhibiting we would bring our Light before all New England manufacturers." Eaton described his plans for the exhibit in a later report and John Vail requested photographs and models for it. Managed by Spencer Borden, the Edison installation boasted four dynamos and nearly 700 lamps. The fair opened on 6 September with several different electric lighting systems displayed in the main hall. Eaton to TAE, 8 Aug. 1882; Vail to TAE, 16 Aug. 1882; both DF (*TAED* D8226ZAZ, D8221ZAF; *TAEM* 61:362, 60:889); Edison Electric Co. Bulletin 14:16–18, 14 Oct. 1882, CR (*TAED* CB014; *TAEM* 96:754); "The Boston Fair," *Manufacturer and Builder* 14 (1882): 257.

10. Eaton later relayed Edward Johnson's opinion that "Bergmann has no system whatever of making charges, and charges enormous profits on some things and no profit at all on others." He considered the Isolated Co. and Bergmann & Co. to be in a state of "friendly war." Eaton to Edi-

son, 22 and 27 July 1882, both DF (*TAED* D8226ZAT, D8226ZAW; *TAEM* 61:344, 355).

11. Variously the "King Philip Mill," under construction at Fall River, Mass. The plant had contracted in May for a 700 lamp isolated station to be completed in September. Edison Electric Light Co. Bulletin 10:5–6, 5 June 1882, CR (*TAED* CB010; *TAEM* 96:714); Eaton to TAE, 10 and 15 May 1882, both DF (*TAED* D8226B, D8226D; *TAEM* 61:148, 158).

12. Edison Co. for Isolated Lighting financial statement, 30 June 1882, DF (*TAED* D8221X; *TAEM* 60:878).

–2319–

From Francis Upton

East Newark, N.J., July 21, 1882.[a]

Dear Sir

Weighed fibres[1]

I have the record of the Volts on one set and can see no change from regulars. There were 200 in the order.

The resistance after carbonizing was the same as regular. We are going to run other sets through. ⟨What is the curves on regular now—⟩[2]

5"–6 × 9¼ fibres[3]

We tried several of these at 14 candles which is the incandescence that they must be run at to give 12 per H.P. of 16 candles each. They were a complete failure making no decent record.[4] I am so desirous of making the new 140 volt central station lamp a success that I hope 12 per H.P. Yet it is almost beyond hope at present as we have no record to encourage us of such a lamp lasting.

⟨Did you send some 32 candle London Volts to London & when⟩

A Village-Plant Lamps.

Mr. Moore[5] told me that he would want 3000 [-],[b] he said six candle, lamps in six weeks. We are running some of these 10 candle A lamps[c] each day. Holzer is having a thousand started under order number and an account kept in each department so that we shall be able to estimate closer on cost. None of the operations cost any more with a thin fibre than with a thick one, it is only the great shrinkage and the fact that the volts spread and there are a great many high lamps that makes them so expensive.[6]

If you will give us some directions we will pair them off as we test them. The lot of old ones tested at with[d] a maximum of 106 volts.

I think it would be as well to run these lamps at the first village plant at seven or eight candles as breakage is three times as apt to occur where three lamps are put in series. Any one of the three giving out extinguishes the rest. Then when a lamp breaks it may be any one of three [--]^b so that perhaps each will have to be tried before finding the troubles. Then 10 candles is the same incandescence as a 10 per H.P. lamp and this is considered risky for isolated. ⟨We can run the 3 10 candles at 8 candles—⟩

Regarding sockets.

Some time ago a standard size was to be given us to make the rim of our sockets in accordance with. We have heard nothing regarding the matter lately.

Nickle Forms

We are lucky to have some other promising substitute for them as we have had another piece of bad luck losing 200 forms by melting[7]

Old Furnace

Lawson reports that it is not working as it should and has no reason to assign for it, as everything has been put in thorough repair.

Pump Room

Holzer has now put the men on piece work 2¼ cts per A lamp 1¾ cts. per B lamp. We expect to raise this price ~~but~~ as soon as we have made better provision for taking care of broken lamps. We guarantee the men so much a day and as much more as they can make.

The new pump has arrived and we have ordered the fittings to go with it. The lamps seem to be about regular so far as life goes though we are not back to old standards.

Platina wire

We are now needing some more, having only a few weeks supply on hand. Johnson, Matthey quote us in 500 oz. lots the best price. It will take about ~~$2500~~ 3000^e—cash down. Shall I order? ⟨I shouldnt order only enough plat wire to get along with until we get in better shape⟩

Small Drill Press

Bradley wants one. Shall I order one.

Cost lamps

I know that the lamps cost now more than they did at one spell at the Park. Yet I feel very sure that we shall make them cheaper than ever in a short time. We shall have^e more complete facilities for carbonizing and more pumps ready in a few weeks.

K Machine

The Edison Machine Works charge us $1350 for machine $1.25 each for^f brushes. Is that right price. ⟨Yes   Same as to all others⟩

We[f] run the machine with wide brushes and trim them once a week when we stop on Sunday. The commutator is like a mirror.

Our engines ran all last week and so far this week without stopping. The bearing on the K machine runs warm still. We use the oil after straining over and over again and allow it to flow freely.

Money.

We need this badly   Yours Truly

Francis R. Upton.

We have an offer from Morris Tasker[8] to make two mercury pumps @ $400 each if order is placed immediately. ⟨I should place the order at once=⟩

⟨Please have C̶Lawson dip several f̶e̶l̶ Carbons in Kerosene for ¹/₂ hour then plate the ends if OK— take out a days run while the moulds are hot & put in Kerosene until ready for use then put in & plate & run through[g]   I think this will prevent Carbons from absorbing water & oxidizing & do no harm as the Kerosene will evaporate in the vacuum   E⟩

ALS, NjWOE, DF (*TAED* D8230ZBF; *TAEM* 61:787). Letterhead of Edison Lamp Co. a"East Newark, N.J.," and "188" preprinted. bCanceled. c"10 candle A lamps" interlined above. dInterlined above. eInserted in left margin. fObscured overwritten text. g"Followed by over" as page turn.

1. John Lawson's "extra weighed fibres" (probably made in forms like those described in Doc. 2335) were tested on 20 July but the results were not recorded. Lawson continued to work with weighted carbonizing forms in August. Cat. 1302, Batchelor (*TAED* MBN008:59; *TAEM* 91:424); Upton to TAE, 9 Aug. 1882, DF (*TAED* D8230ZBQ; *TAEM* 61:810).

2. Edison replied on 24 July, incorporating all the marginal notes on this letter. He pointed out that since the factory relocated to Harrison he no longer received copies of the curves indicating lamp life. Upton promised to have duplicate curve sheets made for him. He admitted that lamps being made at Harrison did not come up to the standards of those from the old Menlo Park factory, for no reason that he could discern. After Edison visited the factory at the end of the month, Upton addressed him as "Dear-Maker-of-Good-Luck   You were here Thursday last. There was no change made in Thursday's lamps or Thursday night lamps, yet they were the best we have made for several weeks. . . . Your luck is almost a proverb, otherwise it seems strange that your mere presence can change the run of lamps." TAE to Upton, 24 July 1882, Lbk. 7:752 (*TAED* LB007752; *TAEM* 80:750); Upton to TAE, 26 and 31 July 1882, both DF (*TAED* D8230ZBG, D8230ZBM; *TAEM* 61:794, 805).

3. See Doc. 2312.

4. Ten of these were tested at 14 candlepower on or about 8 July. Cat. 1302, Batchelor (*TAED* MBN008:44; *TAEM* 91:409).

5. Miller Moore.

6. That is, an unusually large number of lamps were unusable because the standard voltage would not produce their rated intensity. The 10 candlepower A lamps apparently had what Edison called a "6 by 8" filament, that is, 0.006 by 0.008 inch. In reply to Sherburne Eaton's inquiry about them, Edison stated that the factory was not prepared to make them in large quantities. These lamps cost sixty cents apiece to manufacture because nearly a third of the fine filaments failed. Eaton to TAE, 9 Aug. 1882, DF (*TAED* D8226ZBA; *TAEM* 61:368); TAE to Eaton, 11 Aug. 1882; TAE to Upton, 14 Aug. 1882; Lbk. 7:877, 899 (*TAED* LB007888, LB007899; *TAEM* 80:807, 816); see also Doc. 2321.

7. Upton may have been referring to the carbon forms discussed in Doc. 2308 n. 3.

8. Morris, Tasker & Co., Ltd. manufactured tubing and piping, mainly for locomotives and steam boilers. In October, Edison directed the Lamp Co. to send the firm drawings of the pump to be built. Ashmead 1884, 754; *Rand's* 1881, 76; TAE to Morris, Tasker & Co. 17 Oct. 1882, Lbk. 16:7 (*TAED* LB016007A; *TAEM* 82:249).

–2320–

*Notebook Entry:
Incandescent Lamp*

[Menlo Park,] July 21 1882

method exhausting lamps

finely divided

Metal such as Copper

Exhaust Lamp with common air pump then fill with chlorine then exhaust again then pinch off pump and allow finely divided metal to absorb the chlorine combining with it[1]

TAE

X, NjWOE, Lab., N-80-10-01:169 (*TAED* N304:97; *TAEM* 41:1144).

1. The procedure is broadly consistent with Edison's concern at this time for residual gases in lamp globes but there is no record that he carried out this experiment. He executed a patent application on 25 August that indicated a secondary benefit to this process in addition to removing oxygen from the globe. Filling the evacuated globe with hydrogen "or any other inert gas—such as chlorine gas," then heating the carbon

by electricity above its normal operating temperature and evacuating again, would produce a carbon that was "more compact" and more durable. Edison completed an application several weeks later for successively filling the globe with carbonic monoxide and evacuating it, to the exclusion of nearly all atmospheric gases. Also in the middle of September, he signed a third application for speeding the pumping process by introducing anhydrous compounds of chlorine, bromine, or iodine to combine with residual mercury vapor. U.S. Pats. 287,518; 297,581; 395,962.

–2321–

To Sherburne Eaton

[Menlo Park,] 23rd July *1882*[a]

Your memo July 19th.[1]

Ten Candle lamps. You must be careful about doing business in ten candle lamps. We are not fitted up yet to ~~do~~ make[b] them economically   They will cost not less than 50 cents the carbon being so hair like & our breakage so great.[2]

Bordens letter.[3] I am very glad to receive such letters & always read them with pleasure. The lamps will last more than 600 hours but in our estimate we do not put them right up to the limit. We make an average of 600 hours whereas we know they will last almost twice as long. There are many contingencies we wish to provide against. Bordens mistake about the four thousand A lights is owing to the fact that the South American figures were based on 10 candle lamps then of course the two Dynamos would give four thousand lamps but only of ten candles, which would not be equal to a 15 ft gas burner. In some places it requires 12 ft of gas to give ten candles in other places seven feet will give ten candles and in other places 7½ ft will give 16 candles according to the quality of the gas and dependant on whether the pressure in the mains is in proper relation to the slit in the burner. We count that 2,000 standard candles per hour are equal to a thousand feet of gas in any part of the country   In most places it is more than equal.

Safety catches in chandeliers & brackets   There is more danger with them in than out. [~~than?~~][c] I think we permit Osborne to dictate too much.[4] I am coming in every day this week & will go for him.

Italian Contract. I think we shall have to take the leap into the dark.[5]

Russell   I thought he was discharged long ago.[6]

Pearl Street District. Johnson & myself have taken this matter & I think we will start the station with as much dramatic effect as any of us can wish for.[7]

Clarkes Statement for Bulletin[8]   I return it herewith with pencil memos on it.

Baltimore estimate. I am communicating with Clarke about the estimate he prepared[9]

Complaints against Dean. To prevent any complaints on the part of the Isolated Co I will after the first 20 Ks have been delivered—deliver all everything from the Machine Works except "C" Dynamos f.o.b anywhere on Dock or Depot or Isolated Cos store[d] in New York City. So I will do all this work packing & carting at my own expense.[10] This means a considerable reduction in the cost of machines

Thos A Edison   I[nsull]

L (letterpress copy), NjWOE, Lbk. 7:741 (*TAED* LB007741; *TAEM* 80:744). Written by Samuel Insull; handstamp of Thomas A. Edison. [a]"*188*" from handstamp. [b]Interlined above. [c]Canceled. [d]"or Isolated Cos store" interlined above.

1. The memo to which Edison replied in fact comprises two reports written on successive days (19 and 20 July) and paginated continuously. The sequence of items in Edison's reply reflects that of Eaton's reports, in which the break between 19 and 20 July occurs between the discussion of safety catches and the Italian contract. Edison's marginalia on those documents is the basis for this reply. Eaton to TAE, 19 and 20 July 1882, both DF (*TAED* D8226ZAQ, D8226ZAR; *TAEM* 61:337).

2. Eaton had reported in his 19 July memorandum to Edison (see note 1) an order for "350 10 candle power lamps to light a silk mill at Winsted, Conn." Edison reiterated his caution about 10 candle lamps in an 11 August letter to Eaton, in which he suggested that 8 candlepower bulbs could run at 10 candles for 600 hours. Lbk. 7:888 (*TAED* LB007888; *TAEM* 80:807); see Doc. 2329.

3. Spencer Borden's letter has not been found. Eaton asked in his 19 July memorandum (see note 1) if he should continue sending such items to Edison.

4. Robert S. Osborne was an inspector for the New York Board of Fire Underwriters ("Miscellaneous City News. Edison's Electric Light," *New York Times*, 5 Sept. 1882, 8; *Trow's* 1884, 1292). Osborne had devised a safety catch for chandeliers that he offered to Edison Electric Light Co. for free. Eaton promised to have Sigmund Bergmann make one for him because "Of course we are only too glad to do whatever Osborne wants." Eaton also reported (see note 2) that the New York Board of Fire Underwriters "seem disposed to make us" include a fusible link within each fixture. This remained a requirement until at least November 1882, when the company protested to the Board (Eaton to TAE, 17 July and 4 Nov. 1882, both DF [*TAED* D8226ZAO, D8223L; *TAEM* 61:332, 60:1038]; regarding safety catches within fixtures see Docs. 2265 and 2279).

5. After protracted negotiations, Joshua Bailey had signed the contract creating the Comitato per le Applicazioni dell'Elettricita Sistema Edison on 10 July (see Doc. 2235 n. 5), with the intent to exchange rati-

fications on the 17th. The "leap into the dark," a phrase which Eaton used in his 20 July report (see note 1), refers to the ex post facto authorization of Bailey's signature on behalf of the Edison Electric Light Company of Europe despite confusion over the contract's specific terms and status. Eaton had a copy of the agreement and inferred that it had been signed, but commented that "Like most of the papers received from Puskas and Bailey it is unintelligible." Bailey to TAE, 15 July 1882; Bailey and Puskas to Eaton, 17 July 1882; both DF (*TAED* LM001236B, D8238ZCC; *TAEM* 83:990, 62:454).

6. Likely James A. Russell (b. 1837?), formerly a canvasser in the first central station district. Eaton stated on 20 July (see note 1) that Russell "had not done any work for a long time" but had been kept on the payroll. Russell had worked as a private detective and reportedly accompanied Edison through the tough Goerck St. neighborhood. Russell subsequently requested an appointment with Edison to plead his case "in the defence of my name and reputation." He was again in Edison's employ performing miscellaneous tasks from 1883 through about 1893, when he apparently relapsed into alcohol abuse and became ill. See Doc. 1995 and *TAEB* 5 App. 2; App. 1.B.67; Russell testimony, *Sawyer and Man v. Edison*, 191–92, Lit. (*TAED* QD006191; *TAEM* 46:248); Russell testimony, *Edison & Gilliland v. Phelps* (*TAED* W100DKD); Jehl 1937–41, 959, 975; Russell to TAE, 8 September 1882; Tate to Russell, 10 February 1893; both DF (*TAED* D8204ZGF, D9310ABS; *TAEM* 60:277, 134:57).

7. Eaton indicated (see note 1) that he had talked with Edward Johnson about arranging "the lighting of the Pearl Street District with some dramatic effect."

8. Charles Clarke's report, to which Eaton referred on 20 July (see note 1) as a "statement of coal consumption," has not been found.

9. Clarke's estimate has not been found. Eaton had informed Edison on 15 June of a possible order for a central station in Baltimore. The prospective investors there, however, balked at the Edison Electric Light Co.'s proposed terms because gas sold for less than $1. Edison advised against offering any concessions because "our people there could run the gas out & establish a monopoly within a few years even if they charged $1.25." The plan was declined. The company's simultaneous proposal for a central station in Cincinnati met similar local skepticism. Eaton to TAE, 15 June, 21 and 24 Aug. 1882; all DF (*TAED* D8226Y, D8226ZBI, D8226ZBL; *TAEM* 61:282, 399, 412); TAE to Eaton, 26 Aug. 1882, Lbk. 14:43 (*TAED* LB014043; *TAEM* 81:836).

10. Eaton enclosed with his 20 July report (see note 1) a copy of an unspecified complaint against Charles Dean (not found). Noting that he had no authority to judge the matter, Eaton turned the matter over to Edison. Edison's proposed resolution suggests that shipping charges may have been at issue, which would also be consistent with the stated reason for Dean's dismissal in 1883. See headnote, Doc. 2343.

Menlo Park, N.J., July 28 188±2.[a]

Hammer,

Did you recive a letter which I sent to E.H.J wherein I explained the great importance on the 250 Light machine of setting the brushes in the direction of rotation to the non-sparking point[1]   For fear that Johnson did not give it to you I will again[b] state that the[c] K or 250 Light machine will give [------][d] 130@135[e] volts when load is on and when the brushes are set a right angles to the slot or open part of the field but the sparking is very great   but If the arm is moved in the direction of rotation you will reach a point where there will be absolutely no sparking, and you will still have sufficient Volts hence in practice we find that the K machine gives ~~the least~~ less[b] trouble with the brushes than the Z[f]   This cannot be done on the 60 Light or Z for the reason that it does not give Volts enough to permit of setting the brushes in the direction of rotation to any great extent[g]

If it is attempted to set the Z brushes away up towards the top where there is no spark the Volts will fall as low as 85   But I have made the K and the new L or 150 light which you will soon receive so powerful & with such an excess of Volts that you can avail of and use the non-sparking point & still have sufficient Volts to bring the lamps up to proper candle power & also lose several per cent on the Conductors. if the brushes are set at the non sparking point with the full load & ½ are taken off it will spark slightly, as the non spkg[h] point is now[b] nearer to the old place— in one special Experiment on an L machine we ~~pla~~ adjusted the brushes in the direction of rotation so far that these were at right angles to the brushes position on the Z, & still we had enough volts to bring the lamps up=[2] You should "hammer" this dodge in all our men who put up plants there & impress on them the vital necessity of turning the brushes to the non-sparking point on the K & L. at all times.[i]

I have[c] just tested a duplicate of your No 3 for Milan Italy & tried [------- --][d] setting the brushes in advance ie[j] in the direction of rotation. I took out the adjusting devices & moved the brush rigging up as far as it would go until it struck the pillow block. this placed the brushes about 5 or 6 blocks beyond the regular position in the direction of rotation, and I found that the sparking was so greatly diminished that I kept adding lights until I put on 1630 all we had and the brushes carred the Current without trouble. we also found that there was a gain in economy[g] for the reason that in the old position 2[c] bar loops around the armature was short circuited all the time one brush touching two blocks of course short ckts one loop & the other does the same and this shortcircuiting is done in ~~the~~ a[b] strong part of the field [f----][d] & the powerful Current circulating in this circuit of[k] extremely low resistance produces powerful ~~Current~~ sparks on the brushes. If now the lat[ter][l] is moved in the direction of rotation [---][d] it approaches & shortckts loops which are in a[b] weaker field hence the sparking is diminished & the economy increased. this effect takes place of course on the K. & .L. on the K & K we get ½ a light more per indicated ~~p~~H.p. = The reason why we did not try it on the large machine before was the belief that the Volts or pressure would fall too low but by Connecting up the field in 3s we were enabled to put on the 1630 Lights at 20 candles; we use the set nut on the brush rigging to hold it in place of the old adjusting devices.

[---][d] The K & L machines I think are the 1st Dynamo machines specially[b] constructed to convert the required power ~~with~~ [~~two?~~][d] when their brushes are at the non sparking point which in these machines is the normal position.

If with any other machine such as Siemens etc it is attempted to set the brushes at the non sparking point the power of current conversion of the machine is reduced nearly ½= I am reconstructing the Z machine giving it more margin so that it will convert the proper power when the brushes are at the non–sparking point= As the meanest thing for the public to handle is sparking brushes this will eliminate that difficulty[g]

We have 3 different varities of Automatic regulators to go with Isolated plants working at Menlo going through the Evolution test & I think one of them will prove reliable   Did EH.J write you about running a feeder or feeders[m] out of your H[olborn] Viaduct station when you increase your lamps so as to keep the pressure on your distributing mains even & have all your drop of E.M.F in the feeder— Please show this letter to Dr Fleming— Yours

T A Edison

ALS, DSI-MAH, WJH, Series 1, Box 1. Letterpress copy in Lbk. 7:816 (*TAED* LB007816; *TAEM* 80:779). Letterhead of T. A. Edison. [a]"Menlo Park, N.J.," and "1881." preprinted. [b]Interlined above. [c]Obscured overwritten text. [d]Canceled. [e]"130@135" interlined above. [f]"than the Z" interlined above. [g]Followed by "over" to indicate page turn. [h]"non spkg" interlined above. [i]Followed by dividing mark. [j]Circled. [k]"circuit of" interlined above. [l]Illegible. [m]"or feeders" interlined above.

1. Doc. 2282, to which Johnson's reply is Doc. 2292.
2. On 11 July Andrews tested an L dynamo with a load of 150 lamps

and "brushes set well up on commutator." He found that the brushes performed well but were a little warm, as were the armature and main wires. However the journals were "quite hot" after one of the tests. Andrews test report, 11 July 1882, DF (*TAED* D8235S; *TAEM* 62:52).

-2323-

From Sherburne Eaton

New York July 31st. 188[2]ᵃ

Dear Sir:—

Immediately after your visit this afternoon, when you showed me Mr Dyer's report upon Major Wilber's report to me, dated July 24th., I sent for Major Wilber and had a talk with him. He denies in toto the charges sustained by Mr Dyer's report and by the accompanying exhibits obtainedᵇ from the various examiners in the Patent Office. Major Wilber pledges me his word,ᵇ franklyᵇ and boldly, that every statement he has made in his report is true, ordinary errors in clerical work excepted, and that he will satisfy me or any committee of our Board of Directors that he has been guilty of no dishonesty or irregularity, and that the charge that he has reported cases as filed when they have not been filed is utterly false. He leaves for Washington this evening to investigate the matter and promises to make a good report to me by the end of the week.[1]

I told Major Wilber that I thought a prima facie case had been made out against him, and that under the circumstances I could not continue him inᶜ his present responsible position until he had cleared himself. I told him that I should at once put Mr Dyer in charge of his office and asked him to consider himself suspended, until he shall have cleared himself. In reply to this he said he was entirely willing I should do so and that he could not see how I could do else.[2] He will accordingly report to Col. Dyer[3] in Washington on the taking of testimony in the railroad case of Edison vs. Siemens–Field, the testimony in which case will be taken in this city Thursday.[4] He will ask Col. Dyer to come to New York to take charge of taking the testimony. I had intended, in order to save the expense of taking outside counsel, to let Wilber take this testimony, a service for which he would not be paid outside of his regular salary. I will write M̶r̶Col.ᵈ Dyer about this by this evening's mail.

Regarding the alleged difficulty in obtaining copies of papers Wilber states that every paper that has ever been filed or been prepared for filing, is in his room properly arranged in boxes which Mr Dyer understands and that Mr Dyer is entirely at liberty to come there at any time and make any copies he may desire, on either your direction or mine. Wilber states that it will not be necessary for him to be here in order to enable Mr Dyer to get such papers as he will require to commence immediate work on these cases, but that Dyer will find them in the proper boxes in Wilber's room. He makes however the proper request that if Mr Dyer removes any papers whatever he will leave a receipt for them. I told him that Mr Dyer would of course do that in any event without being especially asked to do so.[5]

Regarding the patents already allowed (Wilber says there are 28), will you kindly request Mr Dyer to at once[b] take them out without any delay and to draw upon this Company for the necessary funds?[6] I told Major Wilber that I should ask Mr Dyer to do this. Will you kindly pass this letter over to him and ask him to attend to it. I will write him officially on the subject.

Wilber meets this charge in a way that puzzles me more than ever. He declares himself absolutely innocent, and he unhesitatingly makes a counter charge that the allegations sustained by the report and exhibits are false, both in general and in detail and that they were gotten up to injure him without foundation or truth. This puzzles me. However, we will get to the bottom of the thing in a few days, and we will then know just what the facts are. Very truly yours,

S. B. Eaton   Vice President.

TLS, NjWOE, DF (*TAED* D8248ZBT; *TAEM* 63:502). Letterhead of Edison Electric Light Co. [a]"New York" and "188" preprinted. [b]Mistyped. [c]Interlined above by hand. [d]Corrected by hand.

1. The reports and exhibits referred to in this paragraph have not been found. Since the dissolution of his partnership with George Dyer, Zenas Wilber had continued to work for the Edison Electric Light Co. (though not directly for Edison) with responsibility for filing patent cases drafted by Richard Dyer (Dyer to Wilber, 27 May 1882, Lbk. 12:438 [*TAED* LB012438; *TAEM* 81:737]). The editors have not determined which incidents cast doubt on the veracity of Wilber, who had been submitting weekly reports to Eaton. Concerns may have attended Richard Dyer's request in late May to return drafts of applications for Edison's files. At the end of the month Dyer instructed Wilber that Edison wanted to have "from day to day . . . notes of the filing of applications and of actions either by the Patent Office or yourself. He wishes to keep himself informed of the progress being made in his cases" (Dyer to Wilber, 24 and 30 May 1882, Lbk. 12:396A, 468C [*TAED* LB012396A, LB012468C; *TAEM* 81:715, 756]). Eaton, however, expressed no reservations for some time. He sent Dyer's bill to Edison on 3 July with the remark that "I naturally feel that we can hardly afford to pay additional sums like this to Dyer, for work which Wilber might possibly do." Edison noted next to this that he was "informed that W dont do much work while at Wash[ingto]n." He made this comment in a letter to Eaton a few

days later, also asking whether Wilber was not supposed to be engaged exclusively by the company: "I think if you enquire you will find that he is working for others notably the Postal Telegraph Coy." Near the end of July, still uncertain about which applications had actually been filed in Washington, Edison quietly asked the Patent Office to list for him the cases received there. Wilber reportedly prepared his own list for Eaton (Eaton to TAE, 3 and 25 July 1882, both DF [*TAED* D8226ZAK1, D8226ZAU; *TAEM* 61:324, 349]; TAE to Eaton, 7 July 1882, Lbk. 7:679 [*TAED* LB007679; *TAEM* 80:720]; Insull to Eaton, 24 July 1882, Lbk. 13:1 [*TAED* LB013001; *TAEM* 81:766]).

2. Wilber subsequently acknowledged withholding patent applications and misappropriating $1,300, which the company recovered. In early August, Edison and the company separately revoked their powers of attorney to him for interference cases pending at the Patent Office. Eaton to TAE, 8 Aug. 1882, DF (*TAED* D8226ZAZ; *TAEM* 61:362); TAE powers of attorney to Richard Dyer, both 8 Aug. 1882; Edison Electric Light Co. power of attorney to Dyer, both 10 Aug. 1882; *Edison Electric Light Co. v. U.S. Electric Lighting Co.*, Defendant's depositions and exhibits [Vol. IV], pp. 2218 and 2273, Lit. (*TAED* QD012E:74, 99; *TAEM* 47:934, 959).

3. Washington, D.C. attorney George W. Dyer (d. 1889), Zenas Wilber's former partner and the father of Edison associates Richard N., Philip, and Frank L. Dyer.

4. Presumably testimony on behalf of Siemens or Field as testimony by Edison and witnesses for him had been taken in November and December 1881. The testimony for Edison is in Lit. (*TAED* QD001; *TAEM* 46:5).

5. Dyer complained on 20 July that in response to requests for copies of applications Wilber had either not responded or sent only drafts. He asked permission to take Wilber's office copies or, if those did not exist, that Wilber obtain them from the Patent Office. Dyer to Wilber, 20 July 1882, Lbk. 7:732 (*TAED* LB007732; *TAEM* 80:738).

6. Specifications allowed by the Patent Office were not issued until payment of a fee. In June, Wilber reportedly was ready to take out twenty-three patents for Edison and one each for William Holzer and Samuel Mott. Only two patents had been issued to Edison so far in 1882, one in January and one in May. He received twenty patents on 22 August. Eaton to TAE, 6 June 1882, DF (*TAED* D8226R; *TAEM* 61:240).

More damaging to Edison was the fact that the Patent Office had received only a handful of applications from him since December 1881. Dyer filed more than fifty cases on 7 August, some executed as long ago as October and December 1881. Of these, nine were subsequently rejected and abandoned. At least one case was lost entirely, a dynamo which Edison had promised to patent for Henry Rowland (see Docs. 2021 n. 6 and 2391). Edison claimed years later to have lost 78 patents to Wilber's malfeasance, and recalled this as a singularly bitter experience (see App. 1.C.21–24); on Edison's patent activity generally at this time see App. 5).

Notebook Entry:
Incandescent Lamp

Lamp Experiments—[2]

Use cocoanut charcoal platinized by soaking 5 minutes in boiling solution Bichloride platinum then igniting in plat crucible or holding in flame spirit Lamp, put it in lamp hot & pump out quickly

Zinc Ethyl and[a] Sodium Ethyl mixed[b] instantly absorbs CO. turns black= Wanklyn says deposit not Carbon as it disolves in HCl. See Jnl Chem Soc p 13 Vol IV, No 19—[3c]

Karsten[4] (see Wm A Miller 107—)[5] say charcoal made from wet wood has but 14 ~~pet~~parts of the carbon while dry wood gave from 3.3 pts to 25 pts carbon but 100 parts wet only yielded 14 pts[c]

Charcoal charred by superheated Steam 536 fahr analysed

Carbon	71.42
Hydrogen	4.85
Ox & Nitrogen	22.91
Ash	0.82

Schonbein[6] found that ferric Salts reduced in cold by agitating their solutions with charcoal powder & mercuric are reduced to mercurous salts

Bertholet[7] has proved that by intensely igniting charcoal by the voltaic arc in a current of pure Hydrogen that <u>Acetelyne</u> C_2H_2 is formed[c]

Try lighting a lamp in pure dry Carbonic acid to see how long it stands   also exhausted in which Carbonic acid is the residual gas= ditto try these two Expmts with pure dry Carbonic Oxide CO.[c]

The residual gas of our lamps is Carbonic acid from the Lungs of the glass blowers[c]

Carbonic anhydride is not decomposed by heating with Sulphur or Chlorine & the Halogens ditto. but if mixed with hydrogen & submitted to a high temp water & CO produced=

Carbonic acid decomp by Sparks or the silent discharge making ~~CO_2~~ & CO & O.

A Solution of Cuprous Chl in HCl or a Cuprous Salt in ammonia gradually absorbs CO is agitated with it but if liquid boiled most CO comes off

According to Bottinger Deut. Chem. Ges. Ber 1877 x 1122[8] CO readily absorbed large quantity by well cooled dry Hydrocyanic acid.

CO unites with Potassium if latter heated to 176 Fahr   this is frequently employed in gas analysis—[9]

CO unites with equal vol Chlorine in sunlight forming Carbonic Oxydichloride, phosgene gas   CO Enters directly combn with Potassic hydrate when heated with it forms Formate.

Chloride[a] silver in powder exposed current dry ammonial gas rapidly absorbed— The Chloride increases $\frac{1}{3}$ in weight, does this at ordinary temperatures= by heating can be driven out & condensed by ice in another part of tube

Electric Spark decomposed ammonia gas 2 nitrogen 6 Hydrogen

Chlorine dec[omposes] ammonia at ordinary temp liberating N & forming under certain circumstances Chloride Nitrogen

Nitrogen combinesd with finely divided Boron at red heat also 1 pt anydrous Borax & 2 pts Chl Ammona at red heat gives nitride Boron, which is white powder feels like talc & not attacked by heating in either H or Cl.[c]

I think pentachloride phosphorous is even better than phos anhydride as when heated to ~~212~~ high temperature decomp to phos anhy & free Chlorine   See Miller 295—inorganic[10]

Mould up bottons ⊟ from our little mould of following some hard & some moderately hard

Phosphate Lime
Magnesia.
Alumina.
Silica
~~Hydrated~~ Ferric oxide
~~H~~ gypsum

a piece of freshly burned charcoal exposed to air condenses moisture rapidly so in 3 days increases in weight $\frac{1}{3}$

spongy palladium absorbs 686 volume of Hydrogen but exhibited no tendency to absorb Nitrogen or Oxygen

~~Gypsum~~

Graham[11]—see Miller physics p 120[12]—states that soft[a] rubber lets in O & scarcely any N. hence on pump residual after while be Entirely O. a square meter of surface of rubber on Sprengel let in at 20C 2.25 cub centmes per min of which 42 pc was O.[d]

Cooutchouc[13] like charcoal has power absorbing ~~charcoal~~ rapidly Ammonia, Nitrous Oxide, sulphurous anhydride   See Grahams Experiments in his Chemistry & Phil Trans 1866, p 399. also 1863 p 385[14]

The gases are frequently reduced in bulk as much as would be needed for liquification—

absorption by Caoutchoc equal times was CO_2 1   H 2.4—
O 5.3   air 11.8   CO 12   N 13.5   air passed through the membrane at[a] 4C[elsius] with only $^1/_{11}$ the velocity with which it passed through at 60C

Might try working sealing in plat wire by our process for freeing air as P takes 1.4 vols of H & retains it up to a red—

Copper takes up .3 to .6 its bulk of H.[c]

For Secondary Battery   Try red oxide of Hg   Lets go its O by slight heat—

In form of sponge platinum absorbs 1.48 its bulk while palladium sponge absorbs[a] 90 vols Hydrogen

The blackening of one of the platina wire ie[e] positive wire passing through the glass is due to fact that there is a constant spark here which decompg CO to C & free O. its probable that this spark also decomps some CO in vac & deposits on glass—[c]

The Hydrates of Sodium, Potastium & phosp anhydride retain their water at any temperature hence in our Experiments use anhydrous K & Na by burning the metal as well as the dried Hydrates[c]

Make an extraordinary long seal to get rid of that spark on p plat wire that decomposes the CO.

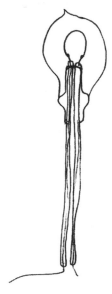

Meter[15]

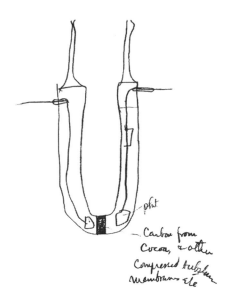

pht

— Carbon from Cocoa, & other Compressed tubular membrane Ele

Electrical dyalisis—transference by dif of EMf.
Bunsen[16] says syrupy phosphoric acid dont absorb gases
For preparing absorbent balls See p 53 Bunsens Gasometry
Fused Chloride Calcium[a] absorbs ammonia gas
Fused KO, absorbs[b] CO_2
Bunsen used balls of phosphorus Casedt under warm[b] water
absorption only occurs above 15 or 20 C if aethyl methyl elayl or simular Hydrocarbons be present phos must be heated nearly up to melting point before it absorbs O if at commencement exp[eri]m[en]t Phos ball whitens then its ok if no cloud then no O Phosphorus acid is formed[c]

Tension Vapor phos-ous acid which coats walls considerable if ngas not carefully dried (with ball potash Bunsen) think phos-ic anhydride best

Bunsen rather recmds Liebigs[17] plan of absorbng with Ball of paper pulp mould & dried at 100C & saturated with a concentrated solution of pyrogallate of Potash it absorbs slowly but completely particularly if Ball renewed, but this not essential with the small amount O in residual, everything must be well dried. The pyrogal of K is a thick syrupy solution.

to absorb CO_2 the potash ball must contain so much H_2O that its soft nuff to receive impression from the nail & must be moistened externally with H_2O before using

Sulpheretted H. absorbed by Coke ball cooled with sol Sul Cu Lactiate Ag— Tartar Emetic or other salt decompsble by Sulphtd H or chromate Hg[f] Sul Cu in form ball absorb it slowly

Dry Benoxide Manganeese or peroxide Lead decomp Sul H quickly & immediately but Bunsen says they are bad in gas analyses as they absorb Large quantities gas being porous bodies, thats ok for us

Phosphoric acid balls made by dipping plat wire in red hot cooling sol of phos-ic acid

~~Benoxide mang can be u~~

Ball pure caustic K moistened also absorbs Sulphtd H as well as CO_2 simultaneously

HCl gas absorbed by ball oxide Bismuth, oxide zinc which formed while moist & then ignited in flame.

Sul Magnesia or borax & especially Sul Soda answer extremely well   balls formed by dipped wire in melted salts melted in thcir own H_2O of Crystalzn specially good when small quantity HCl gas present   Everything dried with phos anhydrite

Bunsen[b]   Carbonic oxide CO can be seperated from Light carburetted H—Hydrogen, Nitrogen, carbonic acid, etc by means of a concentrated solution subchloride of Copper by saturating ball of paper pulp

Olefiant gas[18] (Elayl) absorbed coke ball soaked concentrated (but still liquid) solution of anhydrous sul acid in monohydrated Sul acid   The acid fumes which difuse through tube are absorbed by KO, fused

Ditetryl gas (Tetrylene) completely absorbed by fuming SO_3 acid   Aethyl[a] same process   End Bunsen[c]

Plat black absorbs acetylene

when the vapor of any volitile organic substance if submitted to the Electric Spark acetylene is produced. ~~it is~~ acetylene is formed in reactions at very high temperature   Berthelot An Chem & P 4[c] xxx 431[19]

Shuster absorbed O by heating electrically an iron wire in Vac— Shuster also used metallic Sodium 1874—in Vac tubes[20]

Favre[21] says denser charcoals dont absorb so much gas as porous

M P v Wilder, Deut Chem Ges Ber[22] says acetylene treated electric spark condensed to yellow oily liquid which after time turns brown & no solvent can be obtained

Wood charcoal placed in one end of tSealed tube & heated by boiling H_2O[g] & other in frezing liquid & saturated with dry chlorine gives liquified chlorine condensed but volitile liquids like Brome[23] hydrocynic a[cid] carbon ulphide Dioxide Sulphur Ether & alcohol retained with tenacity   nothing condensed

Phosphorous either in Vacuo or H CO_2 ArH or marsh[a] gas or N & when exposed to violet light it 1st volatilzes & then settles on[a] side glass in form[a] Brown red substance   its supposed that either O or water decomp does this but Gmelin[24] said precautions taken

phosphoric Oxide absorbs O & Cl either dry or moist phosous[h] acid or Chloride phos being formed— When dry this oxide yellow & [destitute?][i] or smell—[c]

Phosphorous shines from Oxidation in highest Vac—diminishes as Vacuum falls— The greater the rarification the less is it necessary to heat the P   at atmosphc press temperature must be 2oC

P. absorbs Chl nitrous ox vapor hyponitric acid, marsh gas olifant gas or vapour ether,[a] alcohol, rock oil, turpentine Eupion,[25] creosote & other volitile oils because it forms a compound with them

Chlorine Combines with H to form ~~an~~hydrated[a] Hydrochloric acid which is white—but How about H_2O in Lamps— guess white coating not this   Mercury absorbs the whole of the Chlorine

Cl combines with H by action Spark

Chl & CO, combine in bright light in few minutes forming phosgene gas &[b] ½ the Vol. its not decomp by the Elec spark when mixed with either O or H but when mixed with ½ vol O & equal vol H it explodes yielding HCl & CO_2 Potassium causes entire disappearance of the gas one portion[a] absbg the Cl the other the O of the CO   products ChlK KO & Carbon Phos sublime but produces no change in phosgene gas—

Suboxide lead absorbs O. made by heating glass tube in oil the tube containing Oxalate lead. Dr H[26] has bottle=

Acetylene is the only Hydrocarbon that can be prepared directly[j] from its free elements   This occurring when the Electric arc passes between Carbon poles in an atmosphere of Hydrogen= It is formed from nearly all organic Compounds at a red heat & occurs regularly in Coal gas   it can be obtained from the latter in considerable quantities by incomplete Combustion

Its a colorless gas—on passing acetylene over heated Potassium H is set free & voluminous colorless Compounds formed ie[e] CH CNa CNa=CNa which are violently decomposed by water.

If gaseous mixture acetylene passed over ammonical sol of Cuprous Chloride a red precipitate formed (Spose ball moistened would answer)   a strongly ammonical sol of argentic nitrate—white precip.

acetylene is absorbed by antimonic Chloride precip.

chlorine explodes when mixed with acetylene

acetylene is absorbed by cold Bromine   it unites with Iodine at 100C forming crystals which melt at 70C

platinized asbestos produces slow combustions by absorbing   try mineral wool also—

Pumice Stone & pipe clay are slower   they absorb the produced ie^c CO2

All seem to agree that CO is redily absorbed by Cuprous^a Chloride in HCl.

~~CO~~ CO & H acted upon by spark after a long time give an oily liquid which soon becomes amber color & on heating to 100C became Brown   it was slightly soluble in H_2O giving yellow opalescent Sol having odor of melacetone mixed with formic acid   it reduced <u>mercuric oxide</u>   (How about putting a reducing agent on walls of globe of lamp)

Cyanogen & H with Electric spark combine to form Hydrocyanic acid

Graham says CO rapidly absbd by Sol^k of Subchl Cu in HCl or ammonia, indeed by ammonical sols of cuprous salts in general   the sulphite it^k absorbs it as quickly as potash does CO_2

deposit on glass may be paracynogen—its black

CO_2 is decomposed to free O & CO by electric spark

Nitric Acid acts on wood charcoal to form a soluble in H_2O^l black stuff—^{c}

Hydrogen & nitrogen are very slightly absorbed by Cocoanut charcoal   Hunter$=^{27}$ power to absorb diminishes with heat

Chromic anhydride combines with acetylene— Wash bulb ie^e I mean coat bulb with it$=$

eCO may be oxidzd to CO_2 by chromic acid   introduce ball plaster paris soaked sol chromic acid   concentrate highly— try drying it thoroughly also higher temp increases effect

Anhydrous^a Chromic A combines directly with C according to Bertholet (dont believe it) in vac—

H is oxidized by chromic acid using plaster p[aris] ball [dry?]^i Carbon might answer   here is table of ball with Chromic A in H^{28}

original Vol	35.8
15 hours	29
19	27
26	23
65	7
86	00

when the Chromic A is diluted Oxdzn very slow

Ludwig[29] recomd balls of plaster Paris when absorption agents[a] are not fusible mixing with H_2O & moulding in bullet mould inside of which is oiled. The reason they coat with syrupy phos acid is prevent absorption gas[a] mechanically— (we want this)

Carbon monox uses ball gypsum steeped mixture 1 vol saturated Chromic A Sol & 2 vols H_2O the CO_2 must be absorbed by potash— The mould must be oiled as Chromic A attacks it[c]

1 vol cocnut charcoal only absorbs 4 of H & 15 of N = Hunter

Dry acetylene absorbed with Evolution heat by antimony pentachloride=

Acetylene is formed by action electric spark among hydrocarbons

Acetylene colorless gas combines with Cl explosively in diffused daylight result heavy oil—

Acetylene forms metallic derivates with fused potassium forms black powder=perhaps K if ash of bamboo works

Acetylene uniteds with copper form deva[s]tating compound   Acetylene in presence moisture rapidly attacks metallic Copper

Acetylene is absorbed by Copper becomes coated with black substance, (probably low Vac blue before [or?][i] dried air does it)

Napthalin absorbs acetylene

Best solvent glacial Acetic A & absolute alcohol   poorer Solvents Turpentine & Bisulphide Carbon ƀTetrachloride Carbon—[c]

Bertholett states acetylene only formed by arc not by induction spark (perhaps when loop high incand it would)   passage of Induction spark through CO gives acetylene=[c]

Cyanogen gas offer great resistance to spark

When Induction spark passes through marsh gas tarry black hydrocarbons are deposited on tube=

Acetylene is decomposed by Electric Spark & deposits Charcoal   Acetylene is formed in all organic carbonizations=

send for metallic potassium subchloride copper

X, NjWOE, Lab., N–81–03–09:65–81, 91–125, 129–145 (*TAED* N206:31–39, 44–61, 63–71; *TAEM* 40:682–90, 695–712, 714–22). [a]Obscured overwritten text. [b]Interlined above. [c]Followed by dividing mark. [d]"Continued to p. 91" written as page turn. [e]Circled. [f]"or chromate Hg" interlined below. [g]"& heated by boiling H_2O" interlined above. [h]"ous" interlined below. [i]Illegible. [j]Multiply underlined. [k]Preceded by "x" to indicate relationship between these parts of text. [l]"in H_2O" interlined above.

1. The first page of this notebook entry is dated 1 August; the next dated entry in the book is 17 August. Several pages of sketches, likely related to apparatus for experiments described here, intervene between this document and the 17 August entry. N–81–03–09:150–79, Lab. (*TAED* N206:73–80; *TAEM* 40:724–31).

2. John Howell had experimented in January with charcoal to absorb gases in lamp production (see Doc. 2212). During August, September, and October, Edison executed several patent applications related to the chemical composition of the gas in lamp globes or chemical methods for exhausting the globes. U.S. Pats. 274,293; 287,518; 297,581; 395,962; Patent Application Casebook E–2537:332 (Case 482); Patent Application Drawings (Cases 179–699); PS (*TAED* PT021332, PT023:86; *TAEM* 45:775, 1003).

3. James Wanklyn (1834–1906), British chemist, did pioneering research with organometallic compounds, particularly ethyl and methyl halides (*Oxford DNB*, s.v. "Wanklyn, James Alfred"). Edison cited Wanklyn 1866.

4. Karl Johann Bernhard Karsten (1782–1853), German chemist and inspector of mines, specialized in metallurgy and the chemistry of salts. Partington 1961–64, 3:603–604.

5. William Miller (1817–1870) was a professor at King's College and a Fellow of the Royal Society. He made fundamental contributions in astronomical spectroscopy, public health, and meteorology. His three-volume *Elements of Chemistry* was a standard text in Great Britain and the United States; Edison used the 6th edition. *Oxford DNB*, s.v. "Miller, William Allen."; Miller 1877, 1:106–109.

6. Christian Friedrich Schönbein (1799–1868) was a German chemist best known for his work on ozone and his co-discovery of guncotton. From 1835 to his death he was professor of physics and chemistry at the University of Basel, Switzerland. *DSB*, s.v. "Schönbein, Christian Friedrich"; Partington 1961–64, 4:190–96.

7. Pierre Eugène Marcellin Berthelot (1827–1907), French chemist and statesman, specialized in organic synthesis. He was professor at the École Supérieure de Pharmacie and the Collège de France. In 1862 he synthesized acetylene by setting up an electric arc between carbon rods in a current of hydrogen. Partington 1961–64, 4:465–77; *DSB*, s.v. "Berthelot, Pierre Eugène Marcellin."

8. C. Böttinger was a chemist at the Polytechnikum in Darmstadt, Germany. Böttinger 1877.

9. On 14 August Edison drew a sketch of an apparatus using potassium heated to this temperature to absorb carbon monoxide from lamp globes. Unbound Notes and Drawings (1879), Lab. (*TAED* NS82:22; *TAEM* 44:1044).

10. Miller 1877, 2:295.

11. Thomas Graham (1805–1869), Scottish chemist, did significant research on the diffusion of gases. His *Elements of Chemistry* was a standard text. Graham was professor at University College in London until 1855, when he became master of the mint. *Oxford DNB*, s.v. "Graham, Thomas."

12. Edison meant Miller 1877, 1:120.

13. Caoutchouc is a form of waterproof rubber. *OED*, s.v. "caoutchouc."

14. Graham 1863; and Graham 1866; see also Graham 1843, 74–75.

15. Figure labels are "plat" and "carbon from cocoa, & other compressed substances membranes etc."

16. Robert Bunsen (1811–1899), professor of chemistry at the University of Heidelberg, was renowned for his experimental skill in the field of chemical analysis; his pioneering spectroscopic research included the discovery of cesium and rubidium. Bunsen's book on gasometry was a standard reference. *DSB*, s.v. "Bunsen, Robert Wilhelm Eberhard"; Bunsen 1857.

17. Justus von Liebig (1803–1873) was a central figure in the development of organic chemistry, analytical chemistry, and the formation of chemical laboratories and agricultural research stations. He taught at the Universities of Giessen and Munich. *DSB*, s.v. "Liebig, Justus von."

18. Ethylene, C_2H_4. *OED*, s.v. "olefiant."

19. The editors have not located this reference in the *Annales de chimie et de physique*.

20. Arthur Schuster (formerly Franz Arthur Friedrich Schuster, 1851–1934) became an authority on spectrum analysis early in his career. As with other general attributions in this document to the findings of specific individuals, the editors have not identified the published source or sources. *Oxford DNB*, s.v. "Schuster, Sir Arthur."

21. Pierre Antoine Favre (1813–1880), French chemist and professor. Partington 1961–64, 4:610–11.

22. The editors have not located this reference in the *Berichte der Deutschen Chemischen Gessellschaft.*

23. A former name for bromine. *OED*, s.v. "brome."

24. Leopold Gmelin (1788–1853) spent his entire career as professor of chemistry and medicine at University of Heidelberg. His 13-volume chemistry handbook was a standard reference. Partington 1961–64, 4:180–82.

25. A volatile distillate of wood or tar; paraffin oil. *OED*, s.v. "eupione."

26. Alfred Haid.

27. John Hunter (1843–1872) experimented on the absorption of gases by charcoal and other porous solids. Partington 1961–64, 4:740.

28. It is unclear whether Edison obtained the values in this table from published works or by experiment.

29. Possibly German physiologist Carl F. W. Ludwig (1816–1895). *DSB*, s.v. "Ludwig, Carl Friedrich Wilhelm."

–2325–

To Charles Clarke

[Menlo Park,] 2nd August [188]2

Dear Sir;

Referring to Major Eatons memorandum to you of 31st July which I return you herewith I do not remember any instance in which a change has been made by you in my Electric Light system. without my knowledge.[a] On the other hand I have always found you very anxious to and painstaking in carrying out my wishes with reference to the system generall[y.][1b]

Major Eat[on']s[b] memorandum was certainly no[t][b] suggested by me as so far as my recollection serves me I have never complained of your taking the initiative without consulting me Yours truly

T A Edison

LS (letterpress copy), NjWOE, Lbk. 7:841 (*TAED* LBoo7841; *TAEM* 80:787). Written by Samuel Insull. ª"without my knowledge." interlined below. ᵇDocument torn.

1. The unsigned directive has not been found. It evidently contained instructions to record all changes to the electric light system so, as Clarke understood it, "the responsibility can be placed." It apparently also went to the Edison company in London, leading Clarke to believe that it originated not from Edison but "from some of the officers of the Company." After discussing it with Edison on the night of 31 July, he promised to take the order "in the proper, common sense spirit." He also asked Edison "to bear witness if you can, to the fact that I am painstaking with reference to the system generally and in detail, that I take particular attention to consult you on all points pertaining to your system and never take or have taken the initiative step without your assent." Clarke to TAE, 1 Aug. 1882, DF (*TAED* D8227ZAQ; *TAEM* 61:520).

–2326–

From George Hamilton

New York, Aug. 7th, 1882.ª

Dear Sir,

Referring to your favor of the 24th ulto.—¹ Please advise me what is size & weight of such of your machines as you would consider most suitable for local purposes—that is, for working sounders—and what you could supply them for.²

What is armature resistance of same? If not as low as we want could you supply others as low as .01 ohm and at what price? Truly Yours,

George A. Hamilton.³

⟨one of our Z machines could be wound to give say 3 volts. its resistance would be $1/10\,000$ of ohm. one ten thousandths we have two armatures nowᵇ that are used in our ~~or~~ regular Z machines one of .035 ohm resistance & giving 55 volts and one of .14 of ohm giving 110 volts. 500 lines of 100 ohms each could be worked from this machine making theᵇ external resist 2 ohms or 14 times higher ~~external~~ than theᵇ internal resistanceᵇ requiring ~~in~~ about 250,000 ft lbs of electricity or 9 horse power of which $1/14$ only is lost on the machine bobbin; $1/2$ hp for friction and 7000 ft lbs on field= Machine will ~~run~~ keep perfectlyᶜ cool—while in use   This weᵈ will guarantee ~~this~~= ~~get~~ price of regular Z as above \$1200 set up.= Current perfectly steady   We also make attachments to Commutators whereby you can run several circuits from the same bobbin each having different volts— E[dison])[4]

ALS, NjWOE, DF (*TAED* D8252M; *TAEM* 63:694). Letterhead of Executive Office, Western Union Telegraph Co. Miscellaneous calculations not transcribed. [a]"New York," and "188" preprinted. [b]Interlined above. [c]"keep perfectly" interlined above. [d]"while in use   This we" interlined above.

1. On 18 July Hamilton asked Francis Upton if Edison could build four replacement armatures, similar to those shown in Edison and Porter 1882, for Siemens dynamos. Edison replied on 24 July that it would cost too much "for us to change our sizes in order to build armatures for you" as it would to supply complete machines, particularly since he would offer a discount on the first one. Hamilton to Upton, 18 July 1882, DF (*TAED* D8233ZCS; *TAEM* 61:1103); Edison to Hamilton, 24 July 1882, Lbk. 7:758 (*TAED* LB007758; *TAEM* 80:752).

2. Although Edison and Martin Force sketched designs for using dynamos on Western Union telegraph lines in May 1884, the company elected to use dynamos designed by Stephen Field. The company installed fifteen dynamos in its New York headquarters by 1886 but did not completely replace batteries with dynamos until 1893. Edison and Force sketches, 23 and 24 May 1884, Unbound Notes and Drawings, 1884 (*TAED* NS84:10–13; *TAEM* 162:736–39); Israel 1992, 155; Reid 1886, 729–30; Western Union 1893, 5.

3. George Hamilton (b. 1843) was an electrician for Western Union from 1875 to 1889. He had promoted Edison's quadruplex in Britain in 1877 and, a few years earlier, had assisted Moses Farmer with dynamo experiments. Doc. 1015 n. 3; Taltavall 1893, 255; TAE to Preece, 3 Sept. 1877, WHP (*TAED* Z005AC).

4. Edison's marginalia was the basis of his reply of the next day. TAE to Hamilton, 8 Aug. 1882, Lbk. 7:866 (*TAED* LB007866; *TAEM* 80:795).

–2327–

To Charles Batchelor

Menlo Park, 10 August *1882*[a]

My Dear Batchelor,

I do not know whether I have called your attention to the importance of preventing a fall of pressure in your Isolated Installation as well as in the General Distribution business.[1] If with an Isolated plant the main wires are run out from the Dynamo all through the Building there will be as you know a gradual drop and there will be a difference between the first lamp and the last lamp of five or six volts which will bring the lamps up near the Dynamo to from twenty to twenty five candles provided a point is made of keeping these at the extreme distance from the Dynamo at sixteen candles. A difference of a few volts when our lamps are made to run at a given number of volts will make a terrible difference in the life of the lamps.

The proper way to do is to never have a loss of more than two and at most three per cent upon the wires and mains upon which the lamps are strung. This will cause a greater investment in copper but by not connecting the Dynamo to one end of the system but by running from the dynamo out to the <u>centre</u> of your system you can have a loss of what you please on these <u>feeder</u> wires say of five or ten or any per cent you may wish without affecting the loss on the mains carrying the lamps and thus you can bring the copper down so that it would be not more than the amount used in the present method and at the same time you can have the great advantage of only having a difference of two per cent between any two lamps on your circuit. In the installations with large steam Dynamos your calculations should be directed to taking advantage of this as it is of great importance not to have more than two or[b] at most three volts difference between the nearest and next distant lamps and this can only be done by the feeder plan as above detailed. Johnson's installation at Holborn Viaduct should have been put down on these lines. He should have run ~~out~~ two mains the whole length of the Viaduct and then run out in each direction a feeder cutting into the mains half way on each side.[2] This they will have to do when they increase their installation for if they attempt to load up the circuit they will find their volts will be irregular and if the lamps at the station are sixteen candle the lamps at the extremity of the circuit will not be more than seven or eight candles. It does not require very much more copper for the reason that you can calculate for a loss even as high as ten per cent on your <u>feeders</u> thus diminishing the copper invested in feeders an[d] calculate the loss on your mains at 2% The great drop on your feeders will not affect ~~the drop on your mains~~ the lamps at all as you can readily see. The feeders which they will have to run out on the Viaduct may be the same size as the mains the loss on the latter being but 2% while on the <u>feeders</u> it may be 8% or 10%. You will see how important this matters is if you think the matter over.

You know in our First District in New York we find by calculating that the copper tubes would cost $125,000 with a loss of 10% which means about ten volts fall. Now ten volts fall would cause the lamps to be all kinds of candle power in different parts of the district if the loss were upon the mains direct. Hence the necessity for adopting the plan of feeders so as to have the greater loss upon the <u>feeders</u> & the lesser loss upon the <u>mains</u> or in other words the <u>drop</u> in volts that is[c] the drop in candle power is upon the feeders & not upon the mains and so we maintain uniformity of candle power. The cost was increased but very little by so doing because we made the mains all alike, with a loss upon them of 2% and made practicable all our loss upon the feeders increasing the loss from 10% to 15% when all the lights are on so that when our District is in full play there will not be in any part of our District a difference of more than two volts between the lamps. Yours truly,

Thos A Edison

LS, NjWOE, Unbound Documents (1882), Batchelor (*TAED* MB075; *TAEM* 92:437); a letterpress copy is in Lbk. 7:871 (*TAED* LB007871; *TAEM* 80:797). Written by Samuel Insull; stamped Thomas A. Edison to Chas. Batchelor. [a]"*188*" from handstamp. [b]Repeated at end of one page and beginning of next. [c]Inserted later into line.

1. See Doc. 2302.
2. In their July report on the Holborn district, John Hopkinson and John Fleming remarked on the unusually long life of lamps at the extreme ends of the mains. They attributed this to a voltage drop up to 7% along the lines, which "seriously reduce[d] the Candle power" of lamps compared with those near the station. Hopkinson and Fleming to directors of Edison Electric Light Co., Ltd., July 1882, Ser. 1, Box 20, WJH.

–2328–

Technical Note:
Consumption Meter

[Menlo Park,] Aug 13 1882

Capilliary Meter[a]

X is platinum wire sealed in glass ~~so~~ the small jars are ¾ full of mercury[1]

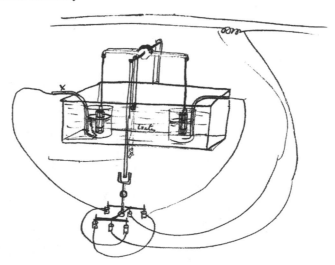

Have John Ott design & test on[2b]

fig 2

bundle of glass tubes
    Large square trough filled with water

                  TAE

X, NjWOE, Lab., Cat. 1148 (*TAED* NM017:159; *TAEM* 44:518). [a]Followed by dividing mark. [b]Followed by "over" written as page turn.

    1. Figure labels are "mercury" and "water."

    2. Details of this meter's design and operation are unclear. Edison apparently intended it to work by electro-capillarity, in which a voltage applied to a liquid changes its surface tension and, consequently, the height it will attain in a capillary tube (or group of tubes as shown here). Mercury responds particularly strongly to the electric polarization of its surface. In 1872, Gabriel Lippman applied this principle to a capillary electrometer, in which minute changes in voltage registered visually as fluctuations in the height of a tube of mercury. *Ency. Brit.*, s.v. "Capillary action"; *DSB*, s.v. "Lippmann, Gabriel Jonas."

    John Ott tested some form of mercury balance meter on 20 July. His drawing of this instrument does not clearly represent its operation, but

A mercury balance meter tested by John Ott on 20 July. Its unreliability, because of the variability in the resistance of mercury, may have prompted Edison to consider using capillary action as a measure of electric current.

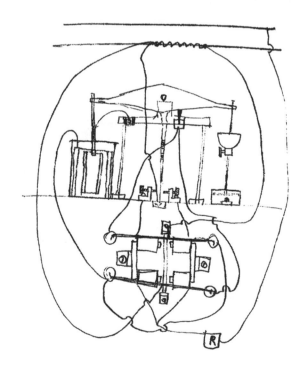

the meter shunt circuit apparently passed through the mercury. Ott found it unreliable because "the mercury changed its resistance while working, so much as to give one half the reading, after working half day." N-82-05-26:64–65, Lab. (*TAED* N204:32; *TAEM* 40:606).

–2329–

From Francis Upton

Harrison, N.J., Aug. 21, 1882[a]

Dear Sir:

We have just tried 10 candle A lamps at 32 candles, carbons from carbon forms, and the orders went all to pieces. It is a great disappointment. We shall try some more tomorrow.

Lawson is going to run some with special carbons.

Holzer and Saxelby[1] have been looking after the lamps in the pump room.

Regarding lamps for competing plants in England.

I wish you would take B lamp to meet Swan with. They are profitable to manufacture and give good life.[2]

If you want a special lamp you entail on us a great expense and a corresponding delay.

The B lamp is larger than Swans and has a higher resistance so gives you every chance in competing. The Lamp Co is under very heavy expense at present and new styles will more than take expected profits.[3] Yours Truly

Francis R. Upton.

ALS, NjWOE, DF (*TAED* D8230ZBU; *TAEM* 61:812). Letterhead of Edison Lamp Co. [a]"Harrison, N.J.," and "188" preprinted.

1. F. Saxelby had worked at the lamp factory since at least June 1881, when he apparently was in the pump department; he was still associated with the factory in 1887. Record of Lamps Sealed off and Broken, 11 June 1881, Unbound Notes and Drawings, MiDbEI (*TAED* X001G2AZ); TAE memorandum to Saxelby, 10 June 1887, Unbound Notes and Drawings (1887), Lab. (*TAED* NS87:36; *TAEM* 107:501).

2. For reasons of operating economy, the British company had ordered 10 candlepower A (110 volt) lamps. See Doc. 2339.

3. Upton reiterated this opinion two days later. He asked why Edison should not "use regular 8 × 17 B lamps, taking a range of 10 volts for various installations and rush the business? We can make money on B lamps @ 40 cts. as they go by the two tough places cheaply, carbonizing and the pump room, and then the fibres only cost ⅔ of A fibres." Upton to TAE, 23 Aug. 1882, DF (*TAED* D8230ZBW; *TAEM* 61:814).

*To Emma McHenry
Pond*[1]

[Menlo Park,] 22nd Aug [188]2

Madam,

In reply to yours of 17th I beg to state that the alternating machine of Gramme would kill instantly without suffering the very largest animals.[2] There is not the slightest difficulty in doing it.

The Electricians at the Torpedo Station at Newport[3] should be able to give you all the information you desire on this subject   Yours truly

Thos A Edison   I[nsull]

LS, NjWOE, Lbk. 14:10 (*TAED* LB014010A; *TAEM* 81:816).

1. Emma McHenry Pond (1857–1934), the wife of U.S. naval ensign (later rear admiral) Charles Fremont Pond, signed her correspondence with Edison as "Mrs. C. F. Pond." A painter and native of California, Pond traveled extensively; in later life she became actively involved with the Society for the Prevention of Cruelty to Animals. Hughes 1989, s.v. "Pond, Emma McHenry"; *WWW*-1, s.v. "Pond, Charles Fremont"; online catalog of Keith-McHenry-Pond family papers, Bancroft Library, University of California-Berkeley (http://sunsite5.berkeley.edu:8000, accessed May 2006).

2. Pond wrote to Edison from Newport, R.I., as a "Member of a 'Society for the Protection of Animals'" for information about electrocution as a humane method of slaughtering animals. Such organizations at this time were generally concerned with the painless destruction of injured, ill, or suffering animals. Both Pond and her sister, Mary McHenry Keith, were active later in life in the Society for the Prevention of Cruelty to Animals. Pond to TAE, 17 Aug. 1882, DF (*TAED* D8206E1; *TAEM* 60:491); McCrea 1910, chap. 3; Beers 1998, 138–39.

3. The Torpedo Station at the naval base on Goat Island, just off the fashionable summer resort city of Newport, R.I., was established in 1869 as an experiment station for the development of torpedoes, explosives, and electrical equipment. The noted electrical inventor Moses Farmer had been electrician at the station from 1872 until 1881. *WGD*, s.v. Newport (9); "Naval History in Rhode Island," Naval Undersea Warfare Center (http://www.nuwc.navy.mil/hq/history/0002.html, accessed July 2006); *ANB*, s.v. "Farmer, Moses Gerrish."

*To Frederick
Lawrence*[1]

Menlo Park N.J. Aug 24 1882.

Dear Sir,

I desire to know upon what terms annualy the Exchange will grant me the same facilities as are now enjoyed by the Gold & Stock Tel Co.[2] I desire to introduce a quotation instrument which shall furnish quotations much more rapidly than by the present system.[3] I propose to do this with my personal means.[4]

Awaiting an early reply[a] I am[5]   Yours Truly

Thomas A Edison

ALS, NNNYSE, Comm. Arr. (*TAED* X135AA). ªInterlined above.

1. Frederick (alternately Frederic) Lawrence (1834–1916), of the brokerage firm Lawrence & Smith at 30 Broad St., became president of the New York Stock Exchange earlier in 1882. Lawrence had attended Columbia College without graduating. Obituary, *New York Times*, 25 Dec. 1916, 9; Columbia University 1932, s.v. "Lawrence, Frederic"; *Rand's* 1881, 462.

2. An alliance of private companies (dominated by Gold & Stock) outside the Exchange's control provided stock quotations from the trading floor. The Exchange made various efforts to regulate these services until it succeeded in establishing direct supervision over them in 1885. Sobel 1977, 31–32.

3. Edison is apparently referring to the chemical stock quotation telegraph system for which he and Patrick Kenny filed a patent application in March 1884 and which issued a year later as U.S. Pat. 314,115. The design of the instruments and system drew on Edison's earlier work on Roman-letter automatic telegraphs as well as their joint work on facsimile telegraphs. See Docs. 184, 186, 349 (and headnote), 373, 394, 397, 482, 486, 566, and 2252.

4. Edison's letter was published in a newspaper the next day. A person associated with the *New York Graphic* sent him the clipping, with a request to see the instrument and an offer to help get it introduced. Edison replied that he did not "like to exhibit the apparatus until I get the requested permission from the Exchange." C. W. Goodsell to TAE, 25 Aug. 1882, DF (*TAED* D8252N; *TAEM* 63:696); TAE to Goodsell, 26 Aug. 1882, Lbk. 14:48 (*TAED* LB014048; *TAEM* 81:840).

5. Edison's request was referred to the Exchange's Governing Committee and then to its Committee on Arrangements. In October, he received the same privileges on the Exchange floor, on the same terms, as Gold & Stock. New York Stock Exchange to TAE, 31 Aug., 28 Sept., and 12 Oct. 1882; all DF (*TAED* D8252P, D8252Q, D8252R; *TAEM* 63:698–99).

–2332–

From Charles Clarke

New York, Aug. 25 1882ª

My dear Edison;

Will you please give to Mr. Claudius[1] the loss per cent. in feeders and also in mains, likewise resistance and candle-power of lamps,

(10 c.p. = 250 ohms.
16 c.p. = 140   ")

for Milan central station? I do not know just what you deem best under existing circumstances.

In my judgement, it will be best for you to inspect the map and decide, from the number and location of the lamps, the number of feeders and also their approx. location (the number of lamps not to exceed an average of more than 800 per feeder, although the one going to the theatre[2] may easily carry the full number required there, because it will be of considerable size on account of the distance) After the number of feeders is emperically decided upon, the size of mains can readily be determined.

They are now waiting only for this data and can then go ahead.[3] Yrs truly

Chas. L. Clarke

ALS, NjWOE, DF (*TAED* D8227ZBH; *TAEM* 61:558). Letterhead of Edison Electric Light Co. [a]"*New York,*" and "*188*" preprinted.

1. Hermann Claudius was responsible for building electrical models of central station lighting districts in order to calculate the dimensions of the feeders and mains (see Doc. 2028 n. 2). His calculations for projected stations in Milan, Italy; midtown Manhattan; Roselle, N.J.; Santiago, Chile; and Shamokin, Penn. are in N-82-08-28, Lab. (*TAED* N231; *TAEM* 41:450).

2. Giuseppe Colombo was in New York making arrangements for the 1,400 lamp Milan central station, which would serve the Teatro alla Scala. Colombo hoped to prove the practicality of electric theater lighting by "experimenting with not less than 1400 lamps on the stage of the largest theatre in the world." Colombo to Joshua Bailey, 12 Sept. 1882; Colombo to TAE, 10 Oct. 1882, DF (*TAED* D8238ZDA, D8238ZDT; *TAEM* 62:497, 540).

3. Before leaving New York, Colombo placed a detailed order through Joshua Bailey for the necessary equipment for a complete central station, with the proviso that Edison should personally adjust the amount of material based on his experience. Colombo to Bailey, 12 Sept. 1882, DF (*TAED* D8238ZDA; *TAEM* 62:497).

–2333–

Harry Olrick[1] to William Mather[2]

London, E.C. Augt 28 1882

Dear Sir,

In making a preliminary report to you,[3] as requested, of my enquiry into the Edison system of Electric Lighting it will be well, perhaps, to acquaint you first with my impression of Mr. Edison, as so much depends upon him for the success of incandescent electric lighting.[4]

Mr. Edison[a]   I was agreeably surprised to find him to be a man of great simplicity, & one whose earnestness and honesty could not be questioned by anybody after conversing with him for any length of time. Instead of being ready at the slightest inducement to put on the market any invention, original or suggested, as many people imagine, he makes a rule of carefully finding out whether his invention is "at least 10% better than anything else of a similar nature, previously known, & he has thrown away as useless many hundreds of ideas which, at first glance, were very promising, but did not come up to this standard.

He appears to be ready at all times to give an unbiassed opinion,[b] whether for or against himself, & he has, therefore, inspired me with the utmost confidence in his judgement & integrity. I thoroughly believe that he would not make a false statement or garble results relative to any experiment or enterprise with[b] which he was connected.

He has an unlimited capacity for work, providing he is interested, but loses interest when the great difficulties are surmounted, & ordinary intelligence can finish what he has commenced,—to this may be attributed the reason that the Central Station in New York has not long since been completed & running, for after completing the experiments with the Dynamos the size and location of the copper mains and feeders, and carefully formulating the plans for the guidance of the Company, he left them to finish the business.

Edison Electric Light Co[a]   The Edison Electric Light Co owns all Edison's American patents, & they provided the necessary money to pay for the costly experiments necessary before the system was completed.

Edison Illuminating Co.[a]   This Company licensed the Edison Illuminating Co. to put down plants for general distribution of light & power from Central stations, the first plant being now almost completed and another in contemplation.

Edison Isolated Company[a]   The Edison Company for Isolated Lighting has been licensed by the Edison Electric Co. for supplying plants outside the Gas Companies' limits, to mills & manufactories, but, under special permission they also supply to people who do not care to wait for the general distribution, which will necessarily be slow.

This Company has been eminently successful, and although they have not been organized more than 11 months, I was informed that they have earned 15% on their capital invested. Their work has been chiefly amongst Mills & Hotels. In all cases they have sold the plant outright, not having sufficient capital to instal on the basis of rental. They do not advertize for business at all, but employ travellers on salary & commission.

Central Station[a]   The first central station which is now nearly half finished is situated in the Southern part of New York, which is devoted to Offices and Warehouses, & is in consequence, a very unfavorable site for Lighting purposes, the locality being almost entirely deserted after 7 p.m. There is, however, a large amount of power required for driving elevators & small shops, & the Company hope eventually to supply these places with motors and cement.

The number of hours per annum when light is required does not exceed 300, whereas in the upper or residential part of the City the light would be required for about 1000 hours per annum, & it is in this district that the Illuminating Co. intend putting down the second plant.

The station consists of 2 buildings, ordinary warehouses, about 25 feet frontage each & about 75 feet deep, having 3 floors.

There are to be 6—1200 light dynamos in each house, but only one [-ince?]ᶜ house has been fitted at present, & that will not be running before about the commencement of next month.

In fitting this building they have in my opinion made the mistake of putting the 6 Dynamos, with their directly connected Engines upon the first floor, or rather upon a structure built independently of the walls of the building, instead of on the ground floor where the boilers have been placed.

This arrangement may give rise to serious difficulty from vibration; besides the heat from the boilers will increase the trouble of keeping the armatures cool. The 2nd floor is devoted, at present, to the apparatus for regulating the current and a high & low EMF Indicator.

The 3rd floor is to be used as a testing room   When the two houses are completed they will be capable of supplying current to about 14 400—16 candle power lamps, or the equivalent of a constant supply of 77 000 cubic feet of gas, estimating 3000 candle power for 1000 ft. of gas.[5]

This current will be supplied through a system of mains & feeders, that is, the mains are laid throughout the streets, & are supplied with electricity by the feeders which go from the dynamos to different parts of the system, the houses being connected to the mains only.

The system of laying these mains is shown on looking at the map which I sent you.[6]

It is necessarily very important that no leakage of current should take place, and therefore great care is taken that the conductors are perfectly insulated one from the other, & from earth. This is effected by putting the 2 conductors (½ round in section) into steampipes holding them in position by thin pasteboard, & then running in a special insulating material made of asphaltum, resin, & linseed oil; this compound which is the result of many experiments is so perfect, that each tube of some 18′ 0″ long shows a resistance of 300,000,000. ohms between either conductor & the pipe.

This station has cost up to the present time about £60 000,[d] but it has necessarily been more expensive than plants of similar size will cost in future.

The officials of the Company seem very confident of making a financial success of the undertaking. Mr. Edison has gone very carefully into the estimated cost of production, ~~interest~~ of the light & states that, including depreciation, interest, coal, wages, & maintenance, the cost will be 4s/6d[e] for every 3000 candle power, this being the estimated light given by the combustion of 1000 cubic feet of gas of good quality.

This price of 4s/6d[e] is based on using the light only 300 hours per annum, but as you are aware the cost of the light decreases at a very rapid ratio as the time of using increases, and therefore Mr. Edison estimates the cost for the residential portion of the City @ 3s/4d[e] per 3000 candle power.

These figures cannot be looked upon as a basis of calculation for England, since the coal costs 8/- per 2000 lbs., & labour & materials are very high.

The Engines used by Mr. Edison up to the present time have been the Porter–Allen high speed engine similar in all respects to those in use at the Holborn installation. These engines have, however, given some trouble, & Mr. Edison is trying the Armington & Sims engine. The first named engine is supposed to give an I[ndicated]HP with 4 lbs of coal, & 26 lbs. of water; the latter has not been tested yet.

Boilers.[a] Vertical fire tube boilers are used for small isolated plants, & Babcock & Wilcox boilers for the larger installations.

Laying Mains[a]   Laying mains in the street is a very simple operation, the only care required being in the soldering of conductors into the sockets, and of course connecting all negative poles with negative poles. The soldering is effected by the use of a hydrogen flame, this gas being made & contained in copper cylinders, easy of transportation. I possess drawings showing the different sections of tubes of different capacity.[7]

Meters.[a] Very careful experiments are still being made to perfect the meters, with every chance of success.

Transmission of power[a]   The experiments which have been carried out to determine the loss in transmitting electrical energy have proved conclusively that not more than 40% is lost, & thus the use of electricity is shown to be as economical as the use of compressed air on hydraulic power. The 40% of loss could be materially reduced with an increase of copper conductors, but, in Mr. Edison's opinion, the extra cost is not worth the increased economy.

I had every opportunity of seeing Mr. Edison's Electric

Railroad which seemed to worked perfectly in all respects. The current is carried ½ a mile from the Laboratory at Menlo's Park to the grounds where the 3 miles of railroad are laid, the rails then becoming the conductors of the current. The motor itself consists of a ordinary dynamo machine with the field laid horizontally instead of vertically in fact similar to the 1200 light machine. by a very simple arrangement the current is turned either on one side or the other from the commutator so as to reverse the motor when when desired. I had a practical proof that the current can be suddenly reversed without burning out the armature, as in running round a curve we came suddenly upon some freight cars on the track & it was necessary to stop & reverse the motor quickly. Experiments are being made to determine the sizes of motors for different purposes, for sewing machines, lathes, & elevators, but when I left these experiments had not been finished.

Manufacture of electric tubes.[a] The manufacture of the electric tubes or conductors is carried on by the Electric Tube Co. of New York at very considerable profit to themselves. This manufacture is of an extremely simple nature requiring only open boilers, air pumps & force pumps, besides the ordinary testing apparatus in the shape of a Wheatstone Bridge of high resistance & a Thompson Galvanometer.

Lamps[a] The lamps are made at a very large factory in Newark, which will shortly have a capacity of 100,000 lamps per day, but as you informed me that they would always make these lamps, I did not spend the time to go over this factory.

Dynamos.[a] The dynamo machines are made by the Edison Machine Works, Mr. Edison being the owner working under a license from the Edison Electric Light Co. There has been, according to Mr. Edison's private Secretary, some 300,000 dollars put into these works & up to the present time no profit of any description has been made by the manufacture & sale of the machines, Mr. Edison preferring to turn out a cheap & efficient machine without profit to himself, believing that his outside interest would be benefited to a greater extent than a commercial profit on the machines would give him. You will, therefore, perceive that in making these machines your only chance of profit will be the difference in the cost of materials and labour. I would say here that although the machine appears to be of very simple construction, the great care necessary in making the insulation has been a fruitful cause of considerable loss, even to the Edison Machine Works where a number of men are employed as foremen who have been working for many years for Mr. Edison.

Improvements.[a] I am not aware what official connection Mr. Edison has with the Edison Electric Light Co. but I know that he is working very hard at the present time to improve the lamps, the automatic regulation of the current, and in general, to increase the efficiency. His aim is to obtain a lamp having a much higher resistance than those previously made. At present a 16 candle lamp has a resistance of 140 Ohms, & Mr. Edison hopes to produce one with a resistance of 500 Ohms. He has already succeeded in making one of 280 Ohms resistance, which has stood the test of work remarkably well, & he is therefore encouraged to proceed with these experiments.

His success in this direction would mean a very good reduction in the cost of conductors, from the fact that the Electromotive Force & Candle power are not in direct proportion; that is to say the present 140 Ohms lamp absorbs .75 Webers of current, whereas the 280 Ohms lamp absorbs about .45 Webers of current.

Mr. Edisons Suggestions.[a] Mr. Edison impressed upon me to tell you that in his opinion, general distribution was the secret of success in Electric Lighting, & he strongly advises that you turn your attention to this in preference to attempting the introduction of isolated plants, pointing out that the current from a central station where economical engines can be used & where labour & supervision is concentrated, can be obtained at much cheaper rates than by small isolated plants.[8]

He strongly advises the sale of candle power or light, & not the sale of current, since improvements in lamps, &c., would then become a profit to the Co., in preference to the consumer. I pointed out to him that gas is very much cheaper in England than in America,[9] but still he is firmly impressed with the belief that the cheaper coal & labour here would allow the light to be supplied at the same rate as an equal candle power in gas.

Trusting that the above short sketch is sufficient for your present requirements, reserving details for my further, more exhaustive report, I remain, Dear Sir, Yours faithfully

L (copy), NjWOE, DF (*TAED* D8239ZGB; *TAEM* 62:1148). [a]Heading written in left margin. [b]Interlined above. [c]Canceled. [d]"£" interlined above. [e]"s" and "d" interlined above; format altered for clarity.

1. Harry Olrick (1851–1886) was a construction engineer at the London engineering firm of Lewis Olrick & Co. As part of the British company's preparations for domestic manufacture, Arnold White and Edward Johnson arranged for him to visit Edison's shops in New York. Obituary, *Iron* 28 (1886): 239 ; Arnold White to TAE, 16 May 1882, DF (*TAED* D8239ZBQ; *TAEM* 62:905).

2. William Mather (1838–1920), a mechanical engineer, was a senior

partner in the family textile equipment business of Mather & Platt, in Manchester. Mather 1926, 11–31; *Oxford DNB*, s.v. "Mather, Sir William."

3. At Samuel Insull's request, Olrick sent him this memorandum in November. See Doc. 2347; Olrick to Insull, 15 Nov. 1882, DF (*TAED* D8239ZGA; *TAEM* 62:1147).

4. Cf. Docs. 2092 and 2343.

5. See headnote, Doc. 2243.

6. Not found.

7. These drawings have not been found but presumably came from the Electric Tube Co. Charles Clarke also helped Olrick gather working drawings of dynamos from the Machine Works. The Edison company in London was anxious to have blueprints of the Z machine but Edison promised them that "With Olricks knowledge and Z dynamo as model you can build quicker than we can send drawings." Charles Clarke to Samuel Insull, 23 and 27 June 1882; Clarke to TAE, 19 July 1882; all DF (*TAED* D8239ZCS, D8239ZCU, D8239ZDH; *TAEM* 62:999, 1001, 1018); TAE to Edison Electric Light Co. Ltd., 17 Aug. 1882, LM 1:245A (*TAED* LM001245A; *TAEM* 83:994).

8. See Docs. 2374 and 2375.

9. Olrick later advised Edison that the price of illuminating gas in London was about 70 shillings. Olrick to TAE, 25 Sept. 1882, DF (*TAED* D8239ZEZ; *TAEM* 62:1091).

–2334–

From George Bliss

CHICAGO, 29th August, 1882.[a]

Dear Sir:

Please inform what is the maximum horse power you propose selling to any one party in the first central district in New York.

I am glad to be informed that you are going to soon give us a storage battery. Please have us kept posted in regard to particulars.

⟨I can give you storage battery better than Faure but if you desire to be a success dont touch a storage battery. its a beautiful product of science but a most dismal failure commercially—⟩[1]

When I was at the Goerck Street shops I had a long conversation with Prof. Moses in relation to the arc light and he was very sanguine that within a month the light would be in readiness to offer our patrons.[2]

He stated that it could be burned on the same circuit with the incandescent lights and that the maximum current would be only what is required[b] to burn 15 of the 8 candle lamps.

⟨Either he lied outrageously or you misunderstood him   he has nothing but an arc light which will burn on a B circuit; or 2 in series on a A circuit, it requires 1 hp each   it will ~~b~~soon be ready so will our 100 cp incandescent   the latter is preferable & more desirable.⟩[3]

Are you equally sanguine that the light can be so soon of-fered to the public?

J. V. Farwell and Co. are erecting an immense wholesale dry goods store in this city which promises to be the finest estab-lishment of the kind in the country.[4]

They are stock holders in the Western Company[b] and are going to use a considerable[b] number of the Edison lights but for large areas it is their intention to use the Fuller lamps.[5]

⟨We can give them arc or 100 cp incandsct⟩

If we were going to be in shape to furnish large incandescent or arc lights to be burned in connection with the Edison in-candescent system I could probably[b] shut other parties out of the field entirely. This is the reason why I am so anxious to get explicit information right away.

⟨you shall have them but it takes time to do all things⟩

I took an order for 310 A lamps to-day to be placed in two retail dry goods stores in this city.[6]

In order to get the contract we had to put in a forfeiture clause by which we will lose 100 dollars per day if the plant is not in operation by the time specified.

⟨it wont be on account of Dynamos or Lamps⟩

The only thing which will prevent us from getting the plant installed is getting the engine, but Armington and Simms have promised to ship it on the 15th prox.

⟨—Look out for Armington & Sims   their dont come to time. you better commence going for them now & keep it up⟩

I fear the engine question is going to be a very serious one in preventing us from filling our orders this fall.

⟨I am trying to arrange it so we can have large stock in hand.⟩

We are short four engines now and in one case, with the Academy[b] of Music in this city, where we agreed to have the plant in operation on a certain day on the strength of which the proprietor made promises to the public and we are having a good deal of trouble with it.[7]

I hope Armington and Simms will increase their capacity to meet the requirements of the business or that some provision[b] will be made to get out engines with the same promptness you now supply us with dynamos, as upon this will largely depend the rapidity with which business can be developed. Sincerely yours,

Geo. H. Bliss   Gen. Supt.

Please accept congratulations on the promising condition of matters in the first district.[c]

TLS, NjWOE, DF (*TAED* D8241Q; *TAEM* 63:105). Letterhead of Western Edison Light Co., George Bliss, general agent. Typed in upper case; "Pardridges Order. New Arc Light." typed in preprinted subject line. ᵃ"CHICAGO," and "1882." preprinted. ᵇMistyped. ᶜPostscript written by Bliss.

1. See Doc. 2389. Edison's marginalia throughout this letter formed the basis of his reply to Bliss on 1 September (Lbk. 14:85 [*TAED* LB014085; *TAEM* 81:857]).

2. See Doc. 2286.

3. Edison engaged in limited research of arc lighting in this period and filed a handful of U.S. patent applications (see Doc. 2286 n. 18; App. 5; also N-81-09-13.1, N-81-09-13.2, Lab. [*TAED* N329, 330; *TAEM* 42:646, 726]). He was first willing to quote prices for a production model in the spring of 1883. TAE to Ferro, 14 May 1883, Lbk. 16:319 (*TAED* LB016319; *TAEM* 82:415).

The lamp factory did not produce 100 candlepower lamps in quantity until the end of 1882; production problems persisted until at least the middle of 1883. TAE to Johnson, 1 June 1882, Lbk. 7:396; TAE to Edison Electric Light Co. Ltd., 12 March 1883, Lbk. 15:459 (*TAED* LB007396, LB015459; *TAEM* 80:578, 82:222); Upton to TAE, 13 June 1883, DF (*TAED* D8332ZBG1; *TAEM* 67:205).

4. John Villiers Farwell (1825–1908) owned one of the largest wholesale dry goods businesses in the U.S. His firm had a sales volume of about $20 million and a workforce of 600 in 1883, the same year that the company moved into a new eight-story, 400,000 square-foot building in downtown Chicago. Unlike many other dry-goods wholesalers, Farwell did not enter into the retail department-store business. The firm ceased operations in 1925. *DAB*, s.v. "Farwell, John Villiers"; *Ency. Chgo.*, s.v. "Farwell (John V.) & Co."

5. James B. Fuller (d. 1879) designed dynamos and arc lamps which the Fuller Electrical Co. sold after his death. Doc. 634 n. 2.; Crosby to TAE, 16 Feb. 1879; Fuller Electrical Co. to TAE, 21 June 1881; both DF (*TAED* D7903ZAR, D8138N; *TAEM* 49:102, 59:124).

6. The stores were Marshall Field & Co. and C.W. & E. Pardridge & Co. Edison Co. for Isolated Lighting, "List of Edison Isolated Plants," 1 Oct. 1885; PPC (*TAED* CA002E; *TAEM* 96:270).

7. In 1867 Florenz Ziegfeld, Sr. (father of the better-known Follies impresario) founded the Chicago Academy of Music, a musical college and concert venue. Its isolated lighting plant powered 125 A lamps. In June Bliss reported that there was "quite a mania here among owners and managers of theatres for the Edison light" and that he was preparing cost estimates for five theaters. By the end of 1885 three other Chicago theaters had installed Edison isolated plants. *ANB*, s.v. "Ziegfeld, Florenz"; *Ency. Chgo.*, s.v. "Classical Music"; Bliss to TAE, 21 June 1882, DF (*TAED* D8241F; *TAEM* 63:91); Western Edison Light Co. Bulletin, 12 Sept. 1882; Edison Co. for Isolated Lighting, "List of Edison Isolated Plants," 1 Oct. 1885; both PPC (*TAED* CA005, CA002E; *TAEM* 96:305, 270).

To Charles Batchelor

Batch=

Latest dodge Carbonizing—

We are using Carbon forms & weights for holding fibres instead of nickel and are getting even better Carbons=[1] I will send you a sample by mail of complete forms & weights so you can make a dies & punch=

They are pressed in Screw press little larger than ours= Use ground gas retort Carbon mixed with minimum quantity of Crude anthracene thinned slightly by Kerosene or[a] crude[b] ‡Coal tar. put a little more pressure[a] than we did on chalks=[2] then lay forms etc on flat iron plates & bake very gently up to say 400 or 500 Fahr so they dont blister & warp then put them in your Carbonizing Covers & bring up the same as you would fibres in nickel forms ie[c] just as slow then you are OK & will never have any more trouble about melting forms etc. We have run over 50 heats with them & they actually improve.[3] Another thing we do we use double covers thus[4d]

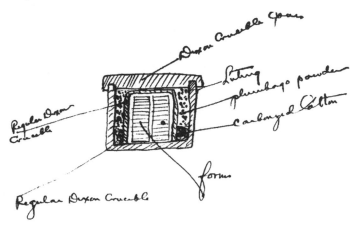

The Cotton should be carbonized first & only to a red heat it is then very sensitive to Oxygen   The object is to keep out oxygen.

with Carbon forms you can go up to as high a temperature as your fire brick will stand= we are using a special fire brick used by wharton to melt nickel[5]   you can doubtlessly get good ones there

I have ordered Upton to send you wkg drawings of our new furnace[6]   we have 4 going   they work splendid.

=We are running through several hundred Lamps the carbons of which were laid out straight in a bundle between Carbon blocks   also in bundles tied with thread & Carbonized afterwards the Carbon is bent as a loop & put in [Cla]mps[e] and plated. they seem to be as good as regular & make far nicer lamps as they follow the Contour of the bulb[7]

There is ~~now~~ no trouble putting them in bulb as the bulb is previously little bell mouthed & it acts as a squeezer

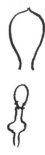

An improvement you might adopt in making your forms over ours is

overlapping= This will prevent circulation of oxygen

I forgot to say that we put two fibres in one mould. more anon

Edison

ALS, NjWOE, Batchelor, Unbound Documents (1882) (*TAED* MB076; *TAEM* 92:442). A letterpress copy is in Lbk. 13:2 (*TAED* LB013002; *TAEM* 81:767). [a]Obscured overwritten text. [b]Interlined below. [c]Circled. [d]Apparently followed by "over," canceled and partially torn, as page turn. [e]Paper torn.

1. See Doc. 2308 esp. n. 3. Upton sent one of the new moulds to Edison on 17 August (Upton to TAE, 17 Aug. 1882, DF [*TAED* D8230ZBT; *TAEM* 61:811]). In his patent covering these instruments, Edison specified that the "weights used within the mold to hold the filament under strain are preferably constructed with a core of nickel or other heavy element or compound fusible at high temperature only, which core is covered with the mixture of hard carbon and a binding carbonaceous or carbonizable material and baked, after which a coating of hard steel-like carbon may be deposited upon it or not, as desired" (U.S. Pat. 334,853).

2. Edison referred to the chalk cylinders for his electromotograph telephone receiver. He made hundreds at the Menlo Park laboratory complex for instruments sent to Britain and other foreign markets in 1879 and early 1880.

3. Workers at the Menlo Park laboratory complex made carbon forms and weights until late October. Because the weights broke easily, Francis Upton requested two sets for each form. Upton to TAE, 9 Aug. 1882; Thomas Logan to TAE, 9 and 15 Sept. 1882; Upton to Samuel Insull, 18 Oct. 1882; all DF (*TAED* D8230ZBQ, D8244P, D8244Q, D8230ZCG; *TAEM* 61:810; 63:197–98; 61:833).

4. The "Regular Dixon Crucible" marked in Edison's drawing refers to one made by the Joseph Dixon Crucible Co., which manufactured crucibles and other graphite-based products including, most famously, pencils at its Jersey City factory (*Oxford DNB*, s.v. "Dixon, Joseph"; Petroski 1990, 164–67). "Luting" is to cover with lute, a clay or cement, to protect from fire and provide an air-tight seal (*OED*, s.vv. "lute n.2, v.2").

Edison's drawing closely resembles one he made on 18 August and another for a patent application completed the following week (Cat. 1148, Scraps. [*TAED* NM017:163; *TAEM* 44:522]). The specification was for a carbonizing chamber from which oxygen could be excluded. It had

> a hollow space for containing the forms which hold the fibers to be carbonized, such space inclosed by walls of plumbago, and this in turn inclosed in outer walls of the same substance, the space between the inner and outer walls being filled partly with powdered plumbago tamped or driven firmly in and partly with a material which combines with oxygen. I may also place within the space which holds the forms a small quantity of plumbago saturated with anthracine or other hydrocarbon compound which decomposes at a red or even a lower heat. This compound decomposes and produces at atmosphere of hydrogen in the chamber. A small quantity of phosphoric anhydride may also be used to absorb the aqueous vapor, which it will retain at the highest temperatures. [U.S. Pat. 439,393]

The patent drawing and text clarified that "double covers" referred to an inverted crucible enclosed within a second crucible. This was first tried at the lamp factory in early August (Cat. 1302 [orders 956, 968], Batchelor [*TAED* MBN008:68, 70; *TAEM* 91:433, 435]).

5. Joseph Wharton (1826–1909) established the American Nickel Works in Camden, N.J. during the Civil War. He and a partner built it into the only producer of refined nickel in the United States, and he supplied nickel to Edison earlier in 1882. One of Wharton's earlier enterprises had been the manufacture of bricks in Philadelphia. He later became interested in Edison's experiments in iron ore milling. *ANB*, s.v. "Wharton, Joseph"; Yates 1987, 114–37, 54–67; TAE to Wharton, 31 Jan. and 19 May 1882, Lbk. 7:333, 12:359 (*TAED* LB007333, LB012359; *TAEM* 80:559, 81:694); *TAEM-G 3*, s.v. "Wharton, Joseph."

6. Upton sent the drawings before he acknowledged Edison's directive (not found) a few days later. Upton to TAE, 5 Sept. 1882, DF (*TAED* D8230ZCB; *TAEM* 61:826).

7. Edison executed a patent application on 9 October in which he stated several rationales for bending a straight filament to approximate the "pear-shaped" lamp globe. In addition to a pleasing symmetry, the filament was more equally near the glass at all points, resulting in more uniform heating of the globe and less breakage. It also required a simpler

carbonizing mold which could hold more than one filament at a time. U.S. Pat. 307,029.

–2336–

To Calvin Goddard

[Menlo Park,] 1st Sept [188]2

Dear Sir

Yours of 31st to hand.[1] I think that at some future date we might introduce the village plant system[2] in Roselle Rahway[3] and with that view suggested to Mr Taft some time back that the Council should not make a hard & fast contract with the Gas Coy.[4] Of course we can do nothing till the Roselle experiment is completed. Anyway I[a] would not propose to license the municipal authorities but would of course get up a small Company.

You had better write Mr Taft putting him off for the present. Yours truly

Thos A Edison   I[nsull]

L (letterpress copy), NjWOE, Lbk. 14:74 (*TAED* LB014074; *TAEM* 81:848). Written by Samuel Insull. [a]Repeated at bottom of one page and beginning of the next.

1. Goddard asked Edison to confirm that he and Miller Moore had chosen Roselle over Rahway for the site of the first village plant and that "we do not of course wish to do anything at Rahway." Goddard also advised against "grant[ing] any rights to municipalities as we might in that way prejudice the formation of local companies." Goddard to TAE, 31 Aug. 1882, DF (*TAED* D8224ZBC; *TAEM* 61:77).

2. Edison conceptualized this distribution arrangement in order to lower construction and capital costs in districts with relatively low population density (see Doc. 2343). The lines were placed overhead on poles rather than underground. To reduce the amount of copper, they carried the current at 330 volts. In homes, three lamps were grouped together in a series circuit, which allowed each lamp to operate at 110 volts. This meant that they could only be turned on or off together. The Roselle station was the only one constructed on this plan.

3. In consultation with Miller Moore, Edison selected Roselle, N.J. (a town of about 2,000 residents without a gas illuminating company) for a demonstration village plant because of its proximity to New York and because Moore and engineer Henry Byllesby lived there. One condition for obtaining the franchise was that the plant provide street lighting without charge. Work began in September 1882 under the direction of Moore and Wilson Howell. Although Edison expected it to be finished in the fall, it did not enter service until 19 January 1883. The Edison Co. for Isolated Lighting and the Edison Electric Light Co. shared the $25,000 installation cost. Three K dynamos (connected in series to obtain 330 volts) supplied 35 homes (500 lamps) and 150 street lamps. The plant operated for nearly ten years. Hicks 1979, 27–33, 75–79.; Jehl 1937–41, 1092–94; "Edison's 'Village Plant' System," *Electrical Review*

2 (12 Apr. 1883): 1–2; "Roselle To Be Illuminated by Electricity," *Elizabeth Daily Journal*, 7 July 1882, 3; Eaton to TAE, 24 June 1882, DF (*TAED* D8226ZAE; *TAEM* 61:308).

4. Edison meant John Tufts, a member of the Rahway Common Council. Tufts had urged Edison to set up a street-lighting plant there because of the gas company's high rates and poor service. Edison replied that although he could do nothing before completing the Roselle plant, Rahway should "make no contract with Gas Co. which cannot be easily revoked." Tufts to TAE, undated June 1882, DF (*TAED* D8224ZAP; *TAEM* 61:43); Insull to Tufts, 29 June 1882, Lbk. 7:620A (*TAED* LB007620A; *TAEM* 80:699).

–2337–

*Samuel Insull to
Miller Moore*

[Menlo Park,] 1st Sept [1882]

Dear Sir,

The man who will have charge of the Village plant at Roselle[1] came here today with an introduction from you to Mr Claudius. After consulting with Mr Ediso[n][a] I told the gentleman that Mr Claudius would no[t][a] be here today and he said he would call tomorrow. I at once telegraphed you[b] so as to preve[nt][a] his coming tomorrow.[2]

Mr Claudius' work here is kept secret even the employees of the Laboratory not being allowed in the room where it is conducted without a permit from Mr Dyer.[3] Furthermore your man would gain nothing of advantage to him in his work at Roselle as Claudius works out the data for the system & that is all that is required—how it is got at Mr Edisons prefers as few as possible should know.

Please instruct your Roselle man not to come here to see Mr Claudius' work & if at anytime you wish anyone else to come here for the purpose kindly communicate with Mr Edison before sending them   Yours truly

Saml Insull

ALS (letterpress copy), NjWOE, Lbk. 14:78 (*TAED* LB014078; *TAEM* 81:852). [a]Edge of original not copied. [b]Interlined above.

1. This may have been either William Oakley, who managed the Roselle plant, or Marcus Barnum, its chief engineer. Hicks 1979, 32.

2. Insull sent the telegram in Edison's name on this date. Lbk. 14:73 (*TAED* LB014073; *TAEM* 81:847).

3. Presumably Richard Dyer.

[New York, September 5, 1882]
EDISON'S ILLUMINATORS.[a]
THE FIRST DISTRICT BRILLIANT WITH THE INCANDESCENT LAMP—THE ISOLATED SYSTEM IS IN SUCCESSFUL OPERATION.[1]

In stores and business places throughout the lower quarter of the city there was a strange glow last night. The dim flicker of gas, often subdued and debilitated by grim and uncleanly globes, was supplanted by a steady glare, bright and mellow, which illuminated interiors and shone through windows fixed and unwavering. From the outer darkness these points of light looked like drops of flame suspended from the jets and ready to fall at every moment. Many scurrying by in the preoccupation of the moment failed to see them, but the attention of those who chanced to glance that way was at once arrested. It was the glowing incandescent lamps of Edison, used last evening for the first time in the practical illumination of the first of the districts into which the city had been divided. The lighting, which this time was less an experiment than the regular inauguration of the work, was eminently satisfactory. Albeit there had been doubters at home and abroad who showed a disposition to scoff at the work of the Wizard of Menlo Park and insinuate that the practical application of his invention would fall far short of what was expected of it, the test was fairly stood, and the luminous horseshoes did their work well. For a long time the company have been at work preparing for the lighting of the district. But there were obstacles to them which occasioned worrying delays. The insertion of meters had first to be attended to, and then came the inspection by the Board of Fire Underwriters. As there was but one competent expert encharged with this work it naturally lagged, and it is still being pressed forward in places where the lighting apparatus is not yet ready.[2] Then there were difficulties to be encountered in the laying of the wires and the establishment of connections. So, many people shook their heads at failure of the promised radiance and believed something was amiss. The company went on untiringly, however, and last night it was fairly demonstrated that the Edison light had a very fair degree of success.

THE LAMPS AGLOW.

It was early in the evening that the current was first transmitted over the wires and the carbon horseshoes became aglow.[3] The machinery worked well from the start and the marked difference of the electric and gas illuminators was apparent at once. Some had based their ideas of the Edison light upon the pole electric light made familiar to them on the thoroughfares. They were a trifle disappointed at first when they saw the soft, mellow radiance of the incandescent lamps devoid of the anticipated glare and brilliancy. But when they came to remember that the light is to be used indoors for purposes of business where the eye would suffer from the too trying glow—for the store, the counting house, the workshop and for domestic uses they came to appreciate how well the mellow yellowish light performed its functions. When the illumination was begun Mr. Edison stood in the workshop of the central office of the first district, at No. 257 Pearl street, in his shirtsleeves, superintending the work. Through the machinery the men flitted about busy as bees. Messengers came speeding in to say all was ready, and then the complicated apparatus was set going, and in a twinkling the area bounded by Spruce, Wall, Nassau, and Pearl streets,[4] where the incandescent lamps had been introduced, was in a glow. There had been scientists who claimed the lighting of such a space by such a method an impossibility. But the result proved the contrary, Edison was vindicated and his light triumphed. Over the lighted area were big buildings like the Drexel and little stores tucked away in dark corners, but the communication nowhere failed, and the practicability of the multiple arc method was attested. All the lights in this space were not started last evening. In some places only a few of the number in readiness for lighting were wanted, but about three thousand were aglow, and if everything goes well over five thousand illuminators will soon be in readiness for use.[5] Among the larger buildings in this section where the light was used are the Drexel Building, *Times* Building,[6] Polhemus',[7] Barnes',[8] Greene Sons',[9] Washburne & Moore's[10] and others.

ACTION OF THE LIGHT.

As the evening progressed the action of the light was curiously watched by those who had it close to them. But it rarely lessened its strength, and for the first night the illumination, except in very odd instances, was singularly powerful and even. The group in the company's office seemed perfectly satisfied, and expressed a full conviction that once it was fairly set in operation there would be no interruptions. Mr. Edison said that he was convinced such would be the case did not any unforeseen and unknown phenomena intervene.[11] Care would be taken to watch all influences that would offset the light, and doubtless new information tending to make it even more perfect would be gleaned. Altogether the experiment in district lighting was pronounced a success.[12]

The other method of introducing the incandescent lamp has also been well received. In some of the buildings down town, where an immense number of lights are used, the Edison Company for isolated lighting have put in their apparatus. They are able to put in plant in houses with a capacity of from fifteen lights upward. The HERALD building has the largest isolated apparatus for lighting in the city.[13] In it are 600 lights, of which 500 were used for illumination last night. To supply them are used a Babcock & Wilcox boiler and an Armington & Sims engine running with two of the 250 light dynamos, and with a capacity in boiler and engine to add three more dynamos. The isolated system of lighting has also been introduced in the American Bank Note Company Building, in Thurbers, E. S. Jaffray's, Everett's Hotel, Aitkinson's and Ams'.[14] If the light is made of thorough avail it is proposed, too, where great power is required, to introduce electric motors instead of steam.

PD, *New York Herald*, 5 Sept. 1882 [p. 6]. In Cat. 1016, Scraps. (*TAED* SM016006b; *TAEM* 24:82). ªFollowed by dividing mark.

1. This event was also reported in the *New York Times* ("Miscellaneous City News. Edison's Electric Light," 5 Sept. 1882, 8); the *New York Tribune* ("Electricity Instead of Gas," 5 Sept. 1882, 1); the *New York World* ("Edison's Incandescent Light," 5 Sept. 1882); Cat. 1018, Scraps. [*TAED* SM018029a; *TAEM* 24:248]); and the *New York Daily Graphic* ("The Electric Light," 5 Sept. 1882, 454); as well as the 15 September issue of the *Operator* ("Successful Inauguration of the Edison Electric Light System," 13 [1882]: 392). This document was reprinted with the *Times* and *Tribune* articles in the 12 September 1882 bulletin of the Western Edison Light Co. (Bulletin 1:6–11 [*TAED* CA005A, CA005B, CA005C; *TAEM* 96:308–10]).

These accounts generally expressed tempered enthusiasm (see Bazerman 1999, 232–33). Referring to electric lighting's uncertain cost relative to gas, the *New York Daily Graphic* concluded that it was "by no means certain that gas will be driven out of general use for lighting purposes even if this experiment with the electric light should meet all of Mr. Edison's predictions." In recognition of journalism's commercial imperatives, Eaton pointed out regarding the August *Tribune* article on various electric light enterprises that "the Tribune people are going to give the best notice to the company that pays the most money" for copies (Eaton to TAE, 22 July 1882, DF [*TAED* D8226ZAT; *TAEM* 61:344]; "Electric Light," *New York Tribune*, 14 Aug. 1882, 2).

2. The entire district was not yet illuminated because the Board of Fire Underwriters had not completed its inspections. According to Sherburne Eaton, Robert Osborne, the sole inspector, was "so pressed with work from various light companies that he will be unable to give our requirements exclusive attention." Osborne reportedly asked for a list of buildings that the company wished to light first, so that he could inspect them in time. "Miscellaneous City News. Edison's Electric Light," *New*

York Times, 5 Sept. 1882, 8; Eaton to TAE, 14 June 1882, DF (*TAED* D8226X; *TAEM* 61:277).

3. Other accounts state that the dynamos started at three o'clock. The lights in the *New York Times* building were in use by 5 p.m., though their effect was not fully appreciated for several hours. "Miscellaneous City News. Edison's Electric Light," *New York Times*, 5 Sept. 1882, 8; "Electricity Instead of Gas," *New York Tribune*, 5 Sept. 1882, 1.

4. The boundary of the First District extended several blocks southeast to the East River.

5. The Edison Electric Light Co. reported having 2,323 lamps installed by the middle of October. A newspaper account shortly before the station opened stated that the company expected to provide 7,916 A lamps and 6,395 B lamps, which would have exceeded its rated capacity (Edison Electric Light Co. Bulletin 14:1, 14 Oct. 1882, CR [*TAED* CB014; *TAEM* 96:754]; "Electric Light," *New York Daily Tribune*, 14 Aug. 1882, 2). Approximately 4,100 lamps of 8, 10, 16, 20, and 32 candle-power were connected by February 1883, but the number of each kind is unknown. The station had a record 2,214 of these in use at 5 p.m. on 30 January; the average number in use at any time was about 1,000. The Edison Electric Light Co. published a table in October 1883 showing the number of buildings wired for service and aggregate number of lamps in use at intervals up to that time. Its bulletins contain information about changes in the number of customers and lamps (Edison Electric Illuminating Co. report, 2 Feb. 1883, Miller [*TAED* HM830169A; *TAEM* 86:514]; Edison Electric Light Co. annual report, 23 Oct. 1883, CR [*TAED* CB020442; *TAEM* 96:887]; see also Charles Chinnock test report, 3 Nov. 1882, DF [*TAED* D8326V; *TAEM* 66:695]).

6. The offices of Drexel, Morgan & Co. at 23 Wall St. (see Doc. 2288) contained 100 lights; the *New York Times* building at 41 Park Row had 52. "A Successful Inauguration," *Operator*, 15 Sept. 1882, 392; Jones 1940, 183.

7. Probably the printer John Polhemus located at 102 Nassau Street. *Rand's New York City Business Directory* 1881, 383; *Trow's New York City Directory* 1883, 1310.

8. Probably the publisher A. S. Barnes & Co. located at 111 William Street. *Rand's New York City Business Directory* 1881, 79; *Trow's New York City Directory* 1883, 78.

9. Probably the printer S. W. Green's Son at 74 Beekman Street. *Rand's New York City Business Directory* 1881, 380.

10. This is the Worcester, Mass., wire manufacturer Washburne, & Moen Manufacturing Co., which had facilities located at 16 Cliff and 241 Pearl Streets. According to the *New York Times* (see note 1), which misidentified the firm as Washburn, Moen & Co., they had 50 lamps supplied by the Pearl Street station. *Trow's New York City Directory* 1883, 1707 and display ad.

11. For reference to Edison's whereabouts on this day see Doc. 2288 n.1. The *Operator* (see note 1) reported on Edison's mood: "Mr. Edison's countenance showed that he was greatly pleased. 'I have accomplished all that I promised,' he said. 'It was not without some fear that I started the machinery this evening. I half expected that some new phenomena would interfere with the working of the light.'"

12. In a draft reply to an inquiry at this time, Edison wrote that

"No one uses gas where our lights are in 1st dist." TAE marginalia on Bullard to TAE, 14 Sept. 1882, DF (*TAED* D8220ZAI; *TAEM* 60:799).

13. The Herald Building at 220 Broadway (near Ann St. and Park Row) was just outside the Pearl St. distribution district. The current for its lamps came through underground conductors from two K generators several blocks away. It began operating on 4 September. For unknown reasons, publisher James Gordon Bennett cabled from his home in Paris to stop the plant in late December. Edison offered to extend his central station lines to the building immediately; the *Herald* was connected to the network in the spring of 1883. Edison Electric Light Co. Bulletins 5:6, 14:10–11; 17 Mar. and 14 Oct. 1882; CR (*TAED* CB005, CB014; *TAEM* 96:681, 96:754); Sherburne Eaton to TAE, 23 Dec. 1882; TAE to Bennett, 23 Dec. 1882; both DF (*TAED* D8224ZDC, D8224ZDD; *TAEM* 61:136, 138); "The 'Herald' Building Lighted with Edison's Lamps," *New York World*, [5 Sept. 1882], Cat. 1018, Scraps. (*TAED* SM018029a; *TAEM* 24:248).

14. The American Bank Note Co., the leading engraver and printer of postage stamps, bonds, bank notes, and stock certificates, installed 125 lamps. They were headquartered at 142 Broadway from 1867 until 1882, when the firm moved to 78–86 Trinity Pl. It is unclear which building was lit at this time. Griffiths 1959, 45, 50; Edison Electric Light Co. Bulletins 2:11, 14:19, 7 Feb. and 14 Oct. 1882; CR (*TAED* CB002, CB014; *TAEM* 96:672, 754).

Headed by Horace W. Thurber (d. 1899) and Francis Beattie Thurber (1842–1907), the importer, wholesale grocer, food processor, and coffee roaster H. K. & F. W. Thurber Co. endorsed Edison's lights at their store on Reade Street where, after a trial run with a single Z dynamo, they quickly doubled to two, powering 120 lights. They doubled again by August and discontinued using gas lighting. By May 1883 there were 330 lamps in use at Thurber's New York establishment and 60 at their canning factory in Moorestown, N.J. *Rand's* 1881, 187, 237; *NCAB* 22:176; Obituary, *New York Times*, 22 July 1899, 7; Edison Electric Light Co. Bulletins 14:19, 8:8, 13:5; 14 Oct., 27 Apr., Aug. 1882, and 31 May 1883; CR (*TAED* CB014, CB008, CB013, CB018; *TAEM* 96:754, 698, 738, 827).

E. S. Jaffray & Co., a dry goods firm led by Edward Somerville Jaffray (1816–1892), installed 189 lamps at their main building located at 350 Broadway. Obituary, *New York Times*, 24 Apr. 1892, 5; *Rand's* 1881, 161; Edison Electric Light Co. Bulletins, 5 and 27 June 1882, 10:3, 11:8; CR (*TAED* CB010, CB011; *TAEM* 96:714, 720).

The Hotel Everett, at 84–90 Chatham St., installed lamps in a dining hall, parlors, a reading room, and an office. Samuel H. Everett (1836–1914) ordered a larger plant of two L dynamo as well as a station for a new property on New York's west side at about the same time. Edison Electric Light Co. Bulletin 5:2, 13:25; 17 Mar. and 28 Aug. 1882; CR (*TAED* CB005, CB013; *TAEM* 96:681, 738); Sherburne Eaton to TAE, 11 Aug. 1882, DF (*TAED* D8226ZBD; *TAEM* 61:380).

Aitkin, Sons & Co., a dry goods and importing house, lit their store at 873 Broadway with 120 lamps (Edison Electric Light Co. Bulletin 14:20, 14 Oct. 1882, CR [*TAED* CB014; *TAEM* 96:754]; "Firms and Companies in New York City," *New York Times*, 18 Sept. 1891, JS31). Max Ams Preserving Co. installed 63 lamps at their canning and packing business at 372 Greenwich St. (*Rand's* 1881, 103; Edison Electric Light Co. Bulletin 14:20, 14 Oct. 1882, CR [*TAED* CB014; *TAEM* 96:754]).

–2339–

To Arnold White[1]

[New York,] 12th September [188]2

Dear Sir,

I duly received your favor of 24th August.[2]

I would point out to you that my estimate for a 80 Light Plant was made to be compared with your estimate of net cost of same and I in no wise fell into the error of estimating the net cost of a plant "to the Company not even in London, but in New York, in the case of the Edison Co, with the gross price including profit charged by the Swan Co"

My estimate was made on the basis of packing and shipping expenses on[a] this side being paid by me—you being charged with the freight and insurance for which there is a margin in the case of the Dynamo of £2.16.8. My estimate is simply one of cost to you in London as compared with your estimate of cost to you there. From a paragraph in one of your letters to Johnson as to the preparation of this estimate I gain the impression that it is your idea that the estimate was not carefully prepared, which is certainly not justified as it was based on my figures worked out with the object of seeing whether I could not deliver Dynamos F.O.B in Port of London in case of large orders. It is my impression after reading your letter and again looking over my Estimate ~~must have~~ that your examination of the latter could not have been very great or else you would have discovered that I did compare "like with like" viz the net cost of an 80 Light Edison Plant delivered in London[3]

I did not mean to give you the impression that we can dispense with Electroliers—which I allowed for in my estimate.[4] What I meant you to understand was that however necessary we might the the Electroliers manufactured here it would be better policy to use those made in England in order to cheapen our plant. To sum up my figures I would state that where so poor a machine as the Burgon[5] is used it is possible for you to put in one of our plants at cost at a loss of about £2.0.0 where no engine and boiler is required, and where your customer requires the latter articles you can certainly make a small profit in consequence of the economy of my lamps and Dynamos as compared with those of my opponents so far as power is concerned. If a better machine than the Burgon is used (the Siemens) you can make a gross profit of at least 25%. I therefore consider your loud complaints in your letters to Mr. Johnson as to the ruinous cost of your machinery not only far fetched but scarcely in accordance with the facts of the case. We have had to meet here with the same competition and in the course of about eight months we have not only killed it but have earned profits amounting to a sum equal to our cash capital plus 7%. Mr. Olricks remarks to the effect that we have had no such competition as yourself are exactly the opposite to the facts of the case as we have had competition of the most severe character, which we have managed to kill and have made money in doing so and I am extremely disappointed at your complaints that you cannot follow in our footsteps.[6]

Referring to your letter of 18th August to Mr. Upton I would remark that the ten candle lamps sent you[b] were but samples and were made by hand and were necessarily very expensive.[7] In such cases it is our practice to charge our customer with one half and bear the balance of the cost ourselves. Of course when we make these lamps as regular articles of manufacture the cost will be very considerably reduced and we are now preparing tools with the view to achieving this object. If however you prefer it I will in the future refrain from sending you any new variety of lamp until they can be turned out at the bottom price which in every instance will be long after the first samples have been produced by hand. It is for you to decide whether you want this course pursued it being quite immaterial to me as I am glad to say my other companies both at home and abroad are only too anxious to get samples of anything new I may turn out, (with a view to improving our system) irrespective of immediate cost trusting in my ability and desire to cheapen the manufacture later on.

From your letter to Mr. Upton I gather that it is your opinion that I am compelled to supply your company with lamps of whatever character and candle power at 40 cents for "General Distribution" and 50 cents for "Isolated." If I am right in my conjecture I must say that I entirely dispute your assumption as I fail to see that the contract gives you any such right. At the time the contract was made the only lamps of regular manufacture were of sixteen and eight candle power for "General Distribution" and "Isolated" purposes. I am now working on a 32, 50, & 100 candle power lamps and it would be absurd to expect that these can be made at the same price just as it would be absurd for me to undertake to supply them at the same price as to the 8 & 16 candle lamps. I am also working on a higher resistance lamps with the object of reducing the investment in conductors. This lamp may prove expensive to manufacture and yet is economical for your company to purchase at say ~~three or four times~~ twice[c] the present cost of our regular lamps. It would be as impolitic for you as for to me to have such ~~an~~ one sided arrangement as to price of lamps as you assume exists as it could not but deter me from further investigation in the fear that I might devise a lamp expensive to manufacture but yet economical to use.

I must protest against the spirit which seems to underlie your letter to Mr. Upton and more especially your letters to Mr. Johnson. My object has always been to endeavor to cheapen the manufacture, increase the variety, and add to the completeness of everything in connection with my system and I must confess to a feeling of disappointment at finding, for the first time, that those associated with my ~~i~~enterprises of Electric Lighting in England do not seem (judging from your letters) to encourage my efforts in this direction

As to the extra L. armatures sent you I would point out to you that you accepted the "L" (150 light) machines sent you instead of 100 light machines and it was but natural that I should presume that you would require extra armatures for the same as I have not yet found that an armature will last for ever and that it is impossible to have an accident with them.[8] It is however a matter of little moment to me whether you keep the Armatures or not the question involved being not whether you or I were in error as to them but whether my credit with my Bankers should be impaired. Considering the nature of my authority from you to draw it would have been but mere business courtesy for you[d] to have met the Draft and cabled me to reimburse you the disputed amount in the first case instead of leaving it unpaid for several days. I do not know what may be the business usage under such circumstances in England but in this country this course is invariably pursued where relations such as those between us exist. I have already paid out an amount about equal to the sum in dispute (for which accounts go forward by this mail) and in consequence of your orders not to draw on your company again till I hear from you by letter I shall be compelled to pay, for goods shipped, from my personal funds as our various establishments have <u>my</u> personal order for the goods shipped and to be shipped to you.

I shall be glad to know if it is the intention of your Company to again refuse to honor my Draft on them in the manner pursued in this case as if so I shall be obliged to request that my bills against you be paid in New York <u>in cash</u> as I cannot afford to allow any doubt as to whether my draft on you will be honored as I sell them to my Bankers through whom I am constantly drawing on all my Foreign correspondents, nor can I expect my Bankers to negotiate them for me if such^c a doubt does exist.[9]

From your letters mainly to Mr. Johnson I gain the impression that you imagine that it is my desire to force on you the product of our factories here. I wish to disabuse your mind of any such idea and would add that I shall be equally as pleased as yourself when your Company can get their Lighting apparatus manufactured in England equally as good and cheaper than we can ship it from here. Yours truly

Thomas A Edison

LS (letterpress copy), NjWOE, Lbk. 14:102 (*TAED* LB014102; *TAEM* 81:866). Written by Samuel Insull; a typed copy is in DF (*TAED* D8239ZEF; *TAEM* 62:1062). ^aObscured overwritten text. ^b"sent you" interlined above. ^cInterlined above. ^d"for you" interlined above.

1. Samuel Insull addressed this letter only to "The Secretary   Edison Electric Light Co."

2. White discussed Edison's 11 August itemization of expenses for an 80-light isolated plant. Edison had written in response to correspondence between White and Edward Johnson. Those letters were sent, at Edison's direction, to Egisto Fabbri and have not been found. White to Edison, 24 Aug. 1882, DF (*TAED* D8239ZDU; *TAEM* 62:1037); TAE to White, 11 Aug. 1882; Insull to Fabbri, 4 Oct. 1882; Lbk. 7:892, 14:202 (*TAED* LB007892, LB014202; *TAEM* 80:811, 81:904).

3. White wrote Edison a formal response to this document, summarizing the company's official rebuttal of its contentions. White to TAE, 28 Sept. 1882, DF (*TAED* D8239ZFG; *TAEM* 62:1108).

4. "Electrolier" had recently come into use in Britain as a term for an electrified chandelier. It was adopted during the summer by Bergmann & Co. In his 11 August letter (see note 2), Edison encouraged White to purchase them from the same supplier as Joseph Swan if they were cheaper than those from Bergmann & Co. *OED*, s.v. "electrolier"; Edison Electric Light Co. Bulletin 12:11, 27 July 1882, CR (*TAED* CB012; *TAEM* 96:728).

5. Designed in 1875 by Swiss inventor Emile Bürgin as a single-lamp generator, this inexpensive machine was adapted by R. E. B. Crompton to supply several lamps. Dredge 1884, 1:218–23.

6. See Doc. 2333. In a 25 September letter to Edison, Harry Olrick explicated his comments to White. DF (*TAED* D8239ZEZ; *TAEM* 62:1091).

7. In July, Edison had declined to give White a price for the 10 candle-power lamps because they were still handmade experimental items. White explained to Francis Upton in his 18 August letter that the company had ordered them because they produced more light from the same power as standard lamps; he contended that 75 cents was an "excessively high" price that would "defeat their raison d'etre." He referred to the

agreement with Edison by which lamps were "not to exceed 50 cents each for isolated business & 40 cents each where the lamps are to be used in connection with central lighting stations." A few weeks later, the lamp factory quoted a provisional price of 60 cents for 10 candlepower and $2.50 for 100 candlepower, noting that few of the latter had yet been made. Insull later brought the matter to the attention of Egisto Fabbri, whom he asked to negotiate a new clause in the British contract governing the price for improved lamps. White to Upton, 18 Aug. 1882; Edison Lamp Co. to Insull, 5 Oct. 1882; both DF (*TAED* D8239ZDS, D8230ZCF; *TAEM* 62:1035, 61:832); TAE to White, 24 July 1882; Insull to Fabbri, 4 Oct. 1882, Lbk. 7:760, 14:202 (*TAED* LB007760, LB014202; *TAEM* 80:753, 81:904).

8. The company had refused to honor Edison's draft to pay for twelve L armatures, billed at £679, that it had not specifically ordered. (White also protested at the same time about charges for extra sockets from Bergmann & Co.) By 9 September the company reversed its position and agreed to honor the draft, provided that Edison would reimburse the contested amounts (White to TAE, 5 Sept. 1882 enclosing White to Edward Johnson, 5 Sept. 1882; White to TAE, 9 Sept. 1882; all DF [*TAED* D8239ZEA, D8239ZEB, D8239ZED; *TAEM* 62:1049–50, 1055]).

Edison later explained his understanding that:

Your original order was for 25 spare bar armatures. After this order was given you were notified that it was commercially impracticable to make bar armature machines at the present time in consequence of the heavy cost of the special tools requisite for the work and that I had decided to make the smaller machines up to 250 lights with wire wound armatures. I received your cable assent to this alteration. You also ordered 100 one hundred light dynamos. At the time your order was given I had never built a 100 light dynamo and in experimenting on the model machine I found that it would be cheaper for my customers if I made it a 150 light machine. I did this after consulting with the officials of my American companies who informed me that a 150 light machine would be fully as useful if not more so to them in their business. Had I made a 100 light machine it would have cost you considerably more than $6 per light. . . . In refusing to honor my draft, I consider that you acted in a manner not at all justifiable and in afterwards reconsidering the course you had taken and accepting the same you simply extricated yourself from a position which, had you pursued the proper course, you ought never to have occupied. [TAE to White, 26 Sept. 1882, DF (*TAED* D8239ZFA; *TAEM* 62:1095); see also Docs. 2258 and 2270 n. 3.]

9. Citing his unfriendly relations with the British company, Edison declined to resume the draft privileges he had enjoyed. He instead instructed White to "make the necessary arrangements to have my bills paid and accepted here in New York on shipment of the goods." TAE to White, 26 Sept. 1882, DF (*TAED* D8239ZFA; *TAEM* 62:1095).

*To the Société
Électrique Edison*

[New York,] 13th September [188]2

Dear Sirs,

I duly received your favor of 24th ~~inst~~August[1a]

You are somewhat mistaken in your ideas as to Mr Dyers duties in connection with the cases sent you. He prepares them from notes & rough sketches made by myself and his charges are so arranged as to be proportionately born by the various companies taking out my patents in different countries. The preparation of these cases often involves a great deal of study & research and I cannot well see how he can reduce his charge to you. The American Company pays the same as yourself and the English Company double the amount as Mr Dyer has to give their cases more attention. My Patent affairs have often been a cause of great loss to me in the past in consequence of of the care bestowed on them having been of a very indifferent character.[2] Since Mr Dyer has had charge of my cases the services rendered have been of a very different character and the charges made by him are very reasonable as compared with those of other Patent Attornies   I would add that whenever you may require assistance of any kind Mr Dyer will always be most willing to give it to you.

Trusting this explanation will prove satisfactory   I remain, Dear Sirs   Yours very truly

Thos A Edison   I[nsull].

L (letterpress copy), NjWOE, Lbk. 14:114 (*TAED* LB014114; *TAEM* 81:878). Written by Samuel Insull. [a]Obscured overwritten text.

1. Not found.
2. See Doc. 2323.

*Alden & Sterne to
Samuel Insull*

New York, Sept 21st 1882[a]

Dear Sir

Mr Pryor[1] called at our office this afternoon and will meet Mr Edison at Mr Pine's[2] office 110 Broadway at 12 oclock AM on Saturday (Sept 23d) next to sign leases— He prefers to make the lease for 1 year & 7 mo's from Oct 1st 1882 and we have told him that that will suit you as well and the leases will be made out for that term—[3] He seems pleased to have Mr Edison for a tenant & will no doubt make him very comfortable— Yours truly

Alden & Sterne[4]

L, NjWOE, DF (*TAED* D8204ZGO; *TAEM* 60:292). [a]"New York," and "188" preprinted.

1. This was likely James Williamson Pryor (1858–1924), a graduate of both Columbia College and Law School, and probably the brother of Caroline Pryor (1858?–1934). Together they leased the house at 25 Gramercy Park to the Edison family. Pryor was evidently associated at this time with patent attorneys Betts, Atterbury and Betts, a firm that provided some services to the Edison Electric Light Co. of Europe. He later served as secretary of the Edison Electric Illuminating Co. Classified advertisement 5, *New York Times*, 17 Sept. 1882, 10; Obituary, ibid., 11 Apr. 1924, 21; U.S. Bureau of the Census 1970, roll T9-480, p. 370.1000, image 704 [Mount Desert, Hancock, Maine]; ibid. 1967, roll M653-813, p. 886, image 432 [New York, Ward 7, District 10]; Columbia University 1932, s.v. "Pryor, James Williamson"; Pryor to Samuel Insull, 9 Nov. 1882, DF (*TAED* D8204ZIP; *TAEM* 60:379); Tate to Pryor, 28 Jan. 1893, DF (*TAED* D9310ABL; *TAEM* 134:49).

2. Attorney John B. Pine had offices at 110 Broadway. Classified advertisement 5, *New York Times*, 17 Sept. 1882, 10; *Trow's* 1883, 1305.

3. Edison signed the lease on 23 September for two years from 1 October. The $400 monthly rental fee included furniture and fixtures. Edison frequently paid late, prompting Pryor in early September 1883 to request two months' rent in advance. Edison instructed Insull to pay only the current month with the remark that he "would get out of the dam'd hole if could." Soon afterward he asked Pryor to terminate the lease a year early because of Mary Edison's need to "give up housekeeping in accordance with the Doctors instructions." Pryor permitted him to sublet the house, subject to approval of the new tenant. TAE agreement with Caroline and James W. Pryor, 23 Sept. 1882; Pryor property inventory, [1882]; Pryor to Insull with TAE marginalia, 6 Sept. 1883; Pryor to TAE, 24 Sept. 1883; all DF (*TAED* D8204ZGT, D8204ZKY, D8303ZFU, D8303ZGD; *TAEM* 60:297, 465; 64:271, 280); TAE to Pryor, 20 Sept. 1883, Lbk. 17:255 (*TAED* LB017255; *TAEM* 82:654).

4. William H. Alden, Jr. and Morris E. Sterne were real estate brokers who had previously arranged the purchase of the Edison Electric Light Co's. headquarters at 65 Fifth Ave. On 16 September, they wrote Mary Edison that they could show her the Gramercy Park house. At that time they also sent a list of nineteen other available properties, both furnished and unfurnished. These were located between 20th and 53rd Sts. in New York. Alden & Sterne to TAE, 31 Jan. 1881; Alden & Sterne to Mary Edison, 16 Sept. 1882; both DF (*TAED* D8126H, D8204ZGL; *TAEM* 58:15, 60:289).

–2342–

Frank Sprague to Edward Johnson

London, Sept. 21st, 1882.

My Dear Johnson:—

Returning from Southampton,[1] I find no letter from you, so suppose you are so busy with the Central Station that you have not the time to write. I cannot well help writing you about the ways the[a] Company's affairs seem to look here. I do not go to the office as much as formerly, but I cannot but notice the fact everything seems dead at 57,[2] and were I[a] a Shareholder there would soon be a row. I do not need to apologize to you for any criticism I may[b] make. The Company's business here of course does not affect me pecuniarily or otherwise, but I have that faith in Edison's system, that desire to see it properly extended, that it hurts me to see golden opportunities thrown away, splendid chances lost, through the supineness and selfishness of those on whom the immediate success of the Company depends—through the neglect of duty on the part of those[a] to whom the work has been intrusted. I had hoped, and you and Edison had also, after the energetic and successful launching of the Company last winter, the establishment of the 57 Station, and the good work at the Palace,[3] would come an active and prosperous business, and so there should. But what are the facts? I go to 57, and I see in your old office Hammer, occasionally despondent, working away at some estimate, Scott[4] debating the cost of some dozen or less petty fixtures, and Glover[5] perhaps penning some 30s man's dismissal in order that another may be found for 28. One other is sometimes present—Fleming—but he is at heart thoroughly given to insulite, and spends half his time away from the office attending to it, half the remainder at 74[6] consulting with White about it.[c] He has developed no business capacity whatever, and White instead of staying in the position you delegated him to, that of Secretary, is practically Manager, while old Bouverie is more interested in debentures and Egyptian Bonds than Electric Light. But the fault lies with Fleming and White. The latter is more[a] active in defeating any good tender by his damned legal cuteness than he is in advancing the interests of the Edison Electric Light Company. The Company is doing nothing at all. No Company ever started under more favorable circumstances, none ever had better opportunities than this one, but it seems as if everything is being thrown away. Witness the "Alaska" which they might have lighted;[7] the Holborn restaurant, the neglect to light which will be one of the worst possible blunders; it is one of the most finely fitted buildings in London. Gordon[8] is not exacting, and Verity[9] has worked hard for it. The Westminster affair[10] is very apt to fall through. Another false move is the going into the Aquarium Exhibition,[11] for which they are to be paid, but at which they are simply the bait to tempt, or the lever to move other companies. With proper management they might have two or three hundred linemen at work, and more orders that they could fill. Thousands of people want the Light, but are met by a dilly-dallying policy which disgusts and maddens them. What the Company wants is an active, competent, manager, who can put his foot on

White and Fleming to keep them in their place, and if they have more interest in insulite than electric lighting then they should find their business there, for the interests of the Company are so great, and the time too pressing to be trifled with.[b] Now these are facts. They are not the reflections of any one man, but I am not deaf nor[a] blind, and can see for myself what is going on, and I must say that no more suicidal policy can be adopted than that at present pursued by the Company, nor can any greater mistake be made than trusting to favorable patents issues[a] rather than a sound business basis. So much about the Company. I know it is not pleasant news, but Quixotic as it may seem to others, I think you will believe me when I say I have your own and Edison's[a] interests at heart, not merely for my own proposed connection, but because I know as well as any other man the value and extent of Edison's work, and also because of our own personal friendship.

The Swan Company is pushing forward, and in this connection let me say that Edison must not abate the work on the lamp, Swan is making continual advance, and has recently commenced making lamps by a new process, which are the finest lamps I ever saw. They are not yet public, but Crompton,[12] who is now one of the Swan Directors showed me one, and he says I will be able to get some soon; I will bring over one or two, if ever I get them. This new lamp is claimed to have 100 ohm resistance <u>hot</u>, and it can easily be increased, and to require but <u>.6</u> of an ampere, and 36 volt-amperes, for 20 candle power.[d] I think perhaps this is a little high,[13] but it is a wonderfully fine lamp, and I think can be made more cheaply than Edison's, and has fully as much life. He is making other lamps of <u>40 candle</u> and has recently made lamps of <u>300 ordinary</u> candle power. My position here has given me splendid opportunities to know the weak and strong points about the lamps of different makers, and I wish with all my heart that I was in a lamp factory for three months.

Crompton informed me privately that the Company were turning out <u>15,000</u> lamps a week, and I am inclined to believe him.

My work is keeping me here longer than I expected. Some of the experiments after all are hard work, are worthless, among others that of the 250 lights, and I shall not allow such ones to become public.[14]

I have not heard from Mrs. J. but trust you are all well. With love to all, and in hopes of an early reply, Your sincere friend,

F. J. Sprague.

TL, NjWOE, DF (*TAED* D8239ZEV; *TAEM* 62:1082). ᵃMistyped. ᵇInterlined above. ᶜ"half the remainder . . . about it" interlined above. ᵈShorthand notation in margin not transcribed.

1. The British Association for the Advancement of Science met this year at Southampton from 24–30 August. Sprague presented a paper on the "Demands of a System of Electrical Distribution," subsequently published in two parts as Sprague 1882a and 1882b. *Report of the British Association for the Advancement of Science* 1882, 448–49.

2. 57 Holborn Viaduct, location of the engineering department of the Edison Electric Light Co., Ltd.

3. The Crystal Palace electrical exhibition.

4. Claud Scott, an otherwise unidentified company employee who corresponded with New York on matters of orders and shipping. Scott to TAE, 12 Jan. 1883 and 15 Feb. 1884, both DF (*TAED* D8338D, D8437D; *TAEM* 68:11, 74:258).

5. Not identified.

6. The company's executive offices at 74 Coleman St., about a mile east of Holborn Viaduct.

7. While the steamship S. S. *Alaska* was under construction in 1881, Charles Batchelor and Grosvenor Lowrey cabled Edison that it was wired for lighting and urged him to make arrangements to provide the lamps. The ship's lighting plant ultimately came from a Siemens firm. Batchelor and Lowrey to TAE, 31 Oct. 1881, LM 1:83C (*TAED* LM001083C; *TAEM* 83:913); Fox 2003, 289.

8. Possibly electrical engineer James Gordon; his role in the restaurant installation has not been determined.

9. George Verity (b. 1831?) and his young nephew John B. Verity (1864?–1905) of the lighting fixtures company B. Verity & Sons had fabricated the chandelier for Edison's Crystal Palace exhibit (see Doc. 2226 n. 4). John later recalled that "my Uncle and I helped Mr. Johnson to get up the Grand Show"; by early 1884 he was a personal friend of both Johnson and Samuel Insull and was proposing various installations in London in collaboration with the Edison company. John Verity to TAE, 3 Jan. 1889, DF (*TAED* D8942AAA; *TAEM* 126:669) and, e.g., John Verity to Insull, 9 Feb. 1884, Insull to John Verity, 1 Mar. 1884; both DF (*TAED* D8437C, D8416ARW; *TAEM* 74:254, 72:15); see also George H. Verity to TAE, 29 Dec. 1905, DF (TAED D0506ACR; *TAEM* 189:1009).

10. Probably the plans referred to in Doc. 2317.

11. This was an extensive electric lighting exhibition which opened in March 1883 at the Aquarium Hall, predominantly an entertainment venue, on Tothill St. near Westminster Abbey. It featured arc and incandescent lighting displays from a large number of companies; the Edison installation included 240 lamps and two Z dynamos. "Electric Lighting Notes," *Engineering* 35 (1883): 104; "The Aquarium Electric Light Exhibition," ibid., 35 (1883): 254; Beauchamp 1997, 166–67.

12. Rookes E. B. Crompton (1845–1940) was a leading British electrical engineer. He received a gold medal for his electric lighting plant at the 1881 Paris Electrical Exhibition. *Oxford DNB*, s.v. "Crompton, Rookes Evelyn Bell."

13. White had recently complained that Swan's exaggeration of his lamps' candlepower misled the public and made it impossible to com-

pare the two systems fairly. White to TAE, 24 Aug. 1882, DF (*TAED* D8239ZDU; *TAEM* 62:1037).

14. After the exhibition closed, members of the jury conducted a long series of experiments on the engines, dynamos, and lamps, overseen by Professor W. G. Adams of the Wheatstone Laboratory, King's College, London. In November, Adams wrote a letter to help justify Sprague's extended absence from the Navy, confirming that he had an essential role in the tests required for the jury's final report. Sprague submitted a long report of the jury's experiments the following March; it was published by the Navy as Sprague 1883. Sprague to Adams, 14 Nov. 1882; Adams to Sprague, 18 Nov. 1882; both Sprague (*TAED* X120CAD, X120CAE).

EDISON'S MANUFACTURING OPERATIONS
Docs. 2343 and 2368

Edison had largely completed his rapid transformation from full-time inventor to major manufacturer by the summer of 1882. Though not permanent—it was just one episode in a long career—the change anticipated the exponential growth of the electrical industry and its future dominance in American manufacturing. He assumed this role because his financial backers in the Edison Electric Light Company preferred to hold and license his patents rather than invest directly in the factories needed to make equipment for his electric light and power system.[1] Drawing on all the financial successes of his inventive career, Edison became a major partner in four capital-intensive enterprises that made nearly every item for his system, from heavy equipment like dynamos, to the small items used by individual consumers, such as fuses, switches, and lamps. He controlled two of these concerns, the Edison Lamp Company and the Edison Machine Works. He had a large stake in the others, the Electric Tube Company and Bergmann & Company. Together these shops employed more than a thousand workers representing a wide range of skills in a variety of processes and materials, from cast metals to glass to slender bamboo. The glass lamp globes and the steam engines to drive dynamos were the only significant components of Edison's light and power system not under his direct control. The Corning Glass Works met his increasing demand but neither of the two engine builders whose designs he trusted could keep pace. The availability of engines became a major bottleneck in building both central generating stations and small isolated lighting plants, and he obtained a license to manufacture the preferred

Armington & Sims engine himself.[2] Edison and his principal associates—Edward Johnson, Charles Batchelor, Francis Upton, and John Kruesi—had provided start-up capital in early 1881. They kept supplying additional funds for operation and expansion of the Machine Works and, especially, the Lamp Company. The prices these companies charged other Edison interests for their goods sometimes became points of contention.[3]

Edison delegated to Samuel Insull, nominally his secretary, the task of integrating the production of his shops with respect to demand for their goods in the United States and abroad. Three factories were located in different sections of New York City and the other across the river in Harrison, New Jersey. Because he was either at his office uptown or at Menlo Park, Insull conducted most business with them by correspondence although he frequently traveled to the Machine Works to confer with Edison.[4] It was a big assignment in addition to his other duties, and he once complained privately that while Edison "has millions at stake in working so hard I have nothing."[5] He relished the authority, however, and expected of the superintendents the same acquiescence they gave Edison. His oversight often involved coordinating the production and shipments of goods. The shops sold their products to U.S. companies operating under license from the Edison Electric Light Company, chiefly the Edison Electric Illuminating Company of New York and the Edison Company for Isolated Lighting, and to Edison concerns in Great Britain, France, and elsewhere. Insull kept track of their accounts, particularly those of the overseas companies. These companies generally expected to receive shipments promptly and make payments slowly. Because the large investments in material and labor resulted in cash shortages, he often made advances on Edison's behalf to meet payroll and keep production going at the Machine Works and the Lamp Company. (Edison obtained large amounts of cash by liquidating most of his telephone interests in Britain. He also borrowed about $37,000 from Drexel, Morgan & Company in July using electric light stocks as collateral.)[6] He coordinated foreign orders, which were routinely divided among separate ships and even different ports, though he passed request for discounts directly to Edison. He also was charged with providing drawings of special machine tools and dynamos to Charles Batchelor, who was starting a factory outside Paris for the Electric Light Company's licensee in continental Europe. He was frustrated by numerous design changes in the course of dynamo production and, especially, by the perceived un-responsiveness of the Machine Works. Though freely admitting his lack of technical expertise, Insull formed his own opinions about the management and workmanship at the shops, frequently to the detriment of the reputation of the Machine Works.[7]

EDISON LAMP COMPANY

Edison organized the Edison Lamp Works during the summer of 1880 with Francis Upton and Charles Batchelor. The factory began operating by the end of that year under Batchelor's supervision in a building convenient to the Menlo Park laboratory. Upton relieved Batchelor at the beginning of 1881, about which time its name changed to the Edison Lamp Company and it became a formal partnership including longtime Edison associate Edward Johnson. By that summer the factory had over a hundred workers and could turn out 1,000 lamps each day and night, although it rarely did so because there was as yet no commercial need for so many. However, it could not turn a profit nor come close to breaking even. Several circumstances contributed to this problem. One was the extensive ongoing experiments to improve lamp efficiency, lifetime, and manufacturing processes.[8] Another was the partnership's contract to sell lamps to the Edison Electric Light Company, its largest customer, at a price that did not cover the costs of relatively small-scale production. Exacerbating this situation was the fact that many lamps, otherwise satisfactory, could not provide their rated intensity within a specified narrow voltage range; these contributed to the factory's unsold inventory. At bottom, though, lamp fabrication was irreducibly labor-intensive at this time. This led Upton and Edison to search almost immediately for cheaper labor. On the first of April 1882 the factory ceased production at Menlo Park and began moving to Harrison, near the much larger Newark labor market. The renovated plant there could accommodate daily production of 1,200 lamps immediately and, ultimately, up to 40,000 per day. At the time of the move the factory employed somewhat over 100 men; about 150 hands worked in Harrison.[9] When production resumed at the end of May, Upton and his new superintendent of manufacturing, William Holzer, found that female workers could substitute for men. They reduced labor costs in some operations by half. To their dismay, however, they found rival manufacturers luring away skilled glassblowers, forcing them to raise wages at least until new hands (including women) could be trained.[10] Edison also guarded against visitors who might give

*The Edison Lamp Works
in Harrison, New Jersey.*

sensitive manufacturing information to the public or competitors.[11] By late August 1882, Upton had enough confidence in his manufacturing routines to order tools to get the factory "ready for any demand." He did this by inside contracting with his own employees in the belief that "idle help is far more expensive than idle tools." The factory's secretary reported its output and financial condition directly to Edison at weekly and monthly intervals. While Edison was at Menlo Park in the summer of 1882, Upton went there every Friday morning to consult with him.[12]

Operating expenses and the move to Harrison gave the Lamp Company a large appetite for cash, which the partners supplied through frequent assessments on their shares. Edison and Upton also made personal loans totaling almost $30,000 in the middle of 1882. The opening of the Pearl Street station in September 1882 and the prospect of a more favorable contract with the parent company gave Upton reason for optimism in the fall, but Insull reported that the factory continued "to absorb money right along," nearly $200,000 of it by this time.[13] In September 1882, Upton predicted the factory would not earn a consistent profit for at least a year and possibly two.[14] The Lamp Company was incorporated in 1884. Its output eventually grew so large and production costs relatively so small that Edison, according to one account, gleefully declared a dividend every Saturday night.[15] It merged with two other Edison firms in 1889 into the new Edison General Electric Company.

EDISON MACHINE WORKS
At the end of February 1881 Edison organized a company to produce his dynamos. It was originally established as a part-

nership with Edison providing 90% of the capital and Charles Batchelor 10%. In early March 1881 Edison leased the plant of the Aetna Iron Company, a firm owned by the shipbuilder John Roach, located on a 200-foot frontage at 104 Goerck Street in lower Manhattan near the East River.[16] It came with some equipment but over the next six months Edison spent about $18,000 to refurbish the building and about $125,000 to equip it with machinery and tools.[17] He built it into a major manufacturing establishment employing hundreds of workmen under the imperious supervision of Charles Dean, a machinist with several years of experience working for Edison. During the summer of 1882 he leased a small plot at the rear of the Works and constructed a four story storage building so as to clear all available "room for manufacturing so that I can boom the business all over the world."[18]

The Machine Works shipped its first dynamo in September 1881. By the end of May 1882 it had finished nine of the big C machines (including the prototype for Paris), nearly 300 Z dynamos, the workhorse the of isolated lighting business, and a handful of others. During the summer of 1882 bookkeeper Charles Rocap reported that he and Dean expected to produce 15 to 20 K dynamos and 12 to 15 125-light dynamos each week. In mid-July the Works had under construction 15 Jumbo, 50 K, and 150 L dynamos, to be sold for a total of $352,500.[19] Payroll records no longer exist but about 100 new men started in July after problems with the supply of materials were solved; another 500 reportedly were added later in the summer, bringing the total employment to 800 men. In November Rocap valued the capital equipment of the Machine Works at about $185,000 and its inventory between $180,000 to $230,000.[20]

Alongside its manufacturing operations, the Machine Works was also a major site for design and testing. Charles Clarke, the Edison Electric Light Company's chief engineer, designed new models of isolated plant dynamos there. Gustav Soldan supervised the Machine Works drafting room, which made drawings of dynamos and related equipment. George Grower designed and developed a new model of consumption meter. Francis Jehl and William Andrews ran the Testing Room, a department under the control of the Edison Electric Light Co. which tested finished dynamos, wire conductivity, insulation, and consumption meters. Workers also learned in the Testing Room how to wire buildings and to set up and repair dynamos.[21]

In the summer of 1882 Insull grew worried about the lack of

Edison Machine Works,
at 104 Goerck Street,
New York.

oversight of the shop's accounts. Early in 1883 he persuaded Edison to give him full control of its finances, over objections from Goerck St. Having forced the issue, he reported that Edison "supported me in a bully fashion and I came out on top of the heap."[22] Subsequent investigations revealed that Dean had been taking kickbacks from suppliers. Edison was reluctant to believe the accusation but Insull persuaded him to fire Dean and Rocap in August 1883. Gustav Soldan took over the manufacturing operations and Insull managed their financial affairs from 65 Fifth Avenue.[23] The Machine Works was incorporated in 1884 with Edison as president, Batchelor as treasurer and general manager, Kruesi as assistant manager, and Insull as secretary. During a strike a year later, it moved production to Schenectady, N.Y.[24] It was consolidated into the Edison General Electric Company in 1889.

ELECTRIC TUBE COMPANY
The Electric Tube Company manufactured insulated underground electrical conductors. These were the least visible elements of the Edison system but perhaps its most symbolically significant because of Edison's desire to emulate gas distribution and his public denunciations of hazardous overhead wires. Its first large orders were for the First District in New York. Edison and John Kruesi, a longtime associate and skilled machinist to whom Edison entrusted the manufacturing, personally supervised installation of the tubes in the trenches emanating from the Pearl Street station. The company fabricated conductors for other plants like those in Milan, London, and Paris.

Unlike Edison's other manufacturing concerns, the Electric Tube Company was incorporated at its outset. This may explain the relative lack of correspondence with either Edison or Insull, and the consequent dearth of information about it. Edison, Kruesi, and Charles Batchelor each had a one-fifth interest; the other two-fifths were owned by partners in Drexel, Morgan & Company. Edison seems to have given full operational responsibility to Kruesi though retaining disbursement authority himself.[25] The company was capitalized at $25,000, nearly all of it invested in equipment and drawings. It paid out $24,000 monthly for materials and wages but was earning "a very considerable profit" in 1882, according to a newspaper account. Up to 100 men worked at its shop at 65 Washington Street, on the west side of lower Manhattan.[26] Kruesi began considering a new site in Brooklyn in late 1882, hoping to double his capacity of one-half mile of tube per week.[27] The Electric Tube Company moved to Brooklyn in April 1884 and was absorbed by the Edison Machine Works around 31 December 1885.[28]

BERGMANN & COMPANY

Edward Johnson, a longtime Edison associated, entered into a silent partnership in 1879 with Sigmund Bergmann, a former Edison employee who had started his own machine shop. S. Bergmann & Company manufactured a variety of items for Edison, including early sockets and fixtures. In April 1881 Edison joined the partnership, which became known as Bergmann & Company, though his participation was not formalized until September 1882.[29] The firm was the sole manufacturer of sockets, fixtures, fuses, switches, instruments, and related material for the Edison lighting companies. Bergmann

The new Bergmann & Co. factory on Avenue B in New York, where Edison set up a laboratory on the top floor late in 1882.

took out a number of patents for these items. Insull described him as "sharp as chain lightning" and Edison reportedly admired his shrewd (and sometimes deceptive) business practices.[30] These were exemplified by the secretive manner in which Bergmann at this time acquired, through a third party, a larger building on Avenue B from a rival lighting firm. Bergmann had employed about 50 men at his old shop on Wooster Street; he employed about 300 in the new plant, which was enlarged in late 1882 or early 1883.[31] The firm was absorbed into the new Edison General Electric Company in 1889.

1. For Samuel Insull's overview of Edison's manufacturing in the spring of 1881 see Doc. 2092. An exception to the Edison Electric Light Co.'s detachment from manufacturing occurred in Canada (see Doc. 2286 n. 8).

2. See Doc. 2343. Armington & Sims struggled to expand their capacity to meet Edison's anticipated demand, and Edison seems to have had at least a small role in helping them. Edison enjoyed a close relationship with them and favored their engine but would not publicly endorse it for fear of alienating builders of the Porter-Allen engine, the only other candidate to power his large direct-connected steam dynamo. Armington & Sims to TAE, 1 Oct. 1881, DF (*TAED* D8129ZBO; *TAEM* 58:283); TAE to Armington & Sims, 4 Aug. 1882, Lbk. 7:852 (*TAED* LB007852; *TAEM* 80:791).

3. See, e.g., Docs. 2310 and 2318 n. 10.

4. Jehl 1937–41, 986–87.

5. Insull to Edward Johnson, 17 Jan. 1882, LM 3:20 (*TAED* LM003020; *TAEM* 84:21).

6. TAE promissory notes to Drexel, Morgan & Co., 1 and 6 July 1882, Miller (*TAED* HM820164C, HM820164D; *TAEM* 86:477, 480).

7. See, for example, Doc. 2259.

8. See headnote, Doc. 2177. Edison estimated in early 1883 that starting the lamp factory had cost $250,000, "most of which has been sunk in experimental work." It was only about that time that the factory began to break even. TAE to Société d'Appareillage Électrique, 6 Mar. 1883, Lbk. 15:414 (*TAED* LB015414; *TAEM* 82:203).

9. The move and facilities at Harrison are briefly described in Edison Electric Light Co. Bulletin 11:3–4, 27 June 1882, CR (*TAED* CB011; *TAEM* 96:720); Harrison employment from *Edisonia* 1904, 141.

10. Upton to TAE, 11 June and 26 July 1882, both DF (*TAED* D8230ZAY, D8230ZBG; *TAEM* 61:777, 794).

11. See Docs. 2160, 2309, and 2312 esp. n. 8.

12. Upton included a list of the tools and equipment he expected for each department in a 23 August letter to Edison (DF [*TAED* D8230ZBW; *TAEM* 61:814]). Philip Dyer's reports are in Electric Light—Edison Lamp Co.—Accounts (D-82-31), DF (*TAED* D8231; *TAEM* 61:854). On the weekly meetings see Upton to TAE, 17 Aug. 1882, DF (*TAED* D8230ZBT; *TAEM* 61:811).

13. Doc. 2343; Insull to Grosvenor Lowrey, 14 Nov. 1882, Lbk. 14:421 (*TAED* LB014421; *TAEM* 81:1006).

14. Upton to Batchelor, 9 Sept. 1882, Unbound Documents, Batchelor (*TAED* MB077; *TAEM* 92:445). Upton enclosed an itemized summary of the factory's production expenses and income from 1 January 1881 to 1 July 1882.

15. Jehl 1937–41, 816; see also App. 1.B.54.

16. John Roach (1813–1887) was a prominent shipbuilder who played a leading role in the U.S. Navy's transition to iron vessels after the Civil War. He acquired the Aetna works in 1852. In the late 1860s he obtained several small marine-engine plants in New York and consolidated his operations at the Morgan Iron Works on the East River, developing what one naval historian has called "the finest marine-engine works" in the U.S. In 1871 he moved his shipbuilding operations to Chester, Pa. (*DAB*, s.v. "Roach, John"; Swann 1965, 14–26). Before New York City built housing projects over it in the 1940s, Goerck St. ran north and south from Grand St. to East 3rd St. between Mangin and Lewis Sts., two blocks from the East River (Spewack 1995, 81); see also Docs. 2055 and 2060 and Jehl 1937–41, 957–1029.

17. Rocap to TAE, 18 Nov. 1881, DF (*TAED* D8129ZCQ1; *TAEM* 58:317). For details of the building renovations and equipment purchases see Electric Light—Edison Machine Works (D-81-29), DF (*TAED* D8129; *TAEM* 58:203).

18. TAE to Eaton, 27 June 1882, Lbk. 7:592 (*TAED* LB007592; *TAEM* 80:686).

19. See Doc. 2293. Rocap to TAE, 15 June and 19 July 1882; Edison Machine Works list of completed dynamos, 10 June 1882; all DF (*TAED* D8233ZBZ, D8234D, D8234B1; *TAEM* 61:1082, 62:9, 7). According to Doc. 2343, the Machine Works used dynamos to pay, in part, for substantial stock holdings in the Edison Co. for Isolated Lighting.

20. TAE to Dean, 5 June 1882, Lbk. 7:403A (*TAED* LB007403A; *TAEM* 80:584); Rocap to TAE, 28 June 1882; Rocap to Insull, 14 Nov. 1882; both DF (*TAED* D8233ZCO, D8233ZEJ; *TAEM* 61:1099, 1147); "Electric Light," *New York Tribune*, 14 Aug. 1882, 2. Weekly reports of dynamo production and monthly balance sheets are in Electric Light—Edison Machine Works—Accounts (D-82-34), DF (*TAED* D8234; *TAEM* 62:2); for a general description of the workflow in the plant see TAE memorandum, undated 1883, DF (*TAED* D8334ZBS; *TAEM* 67:434).

21. For general accounts of the Testing Room see Electric Light—Edison Machine Works—Testing Department (D-82-35) and Electric Light—Edison Electric Light Co—Testing Department (D-83-30); both DF (*TAED* D8235, D8330; *TAEM* 62:25, 66:934) and Jehl 1937–41, 959–66.

22. See Doc. 2400; Insull to Edward Johnson, 6 Mar. 1883, LM 3:109 (*TAED* LM003109; *TAEM* 84:92).

23. Insull also suspected Dean of taking kickback from lucrative inside contracts with his employees. He believed that Rocap had in some way facilitated Dean's malfeasance, which he estimated cost the Machine Works about $50,000. Both Edison and Insull wrote recommendation letters for Rocap in 1884. Insull to Batchelor, 5 Nov. 1883; Insull to E. Myers & Co., 18 March 1884; TAE to Charles Warner, 12 May 1884; all DF (*TAED* D8316BEG, D8416AXZ, D8416BON; *TAEM* 65:393; 72:189, 683); Insull to Batchelor, 21 Aug. 1883, LM 3:173 (*TAED* LM003173; *TAEM* 84:147).

24. Jehl 1937–41, 1000, 1009–1010.

25. See Doc. 2058 and, e.g., Kruesi to TAE, 14 Dec. 1881, DF (*TAED* D8130ZAD; *TAEM* 58:372).

26. The *New York Tribune* ("Electric Light," 14 Aug. 1882, p. 2) reported its employment at 50 men in August, near the completion of the First District.

27. Electric Tube Co. memorandum, 14 Nov. 1882, DF (*TAED* D8236ZAP; *TAEM* 62:137); Doc. 2333.

28. Kruesi to TAE, 27 June 1882; Electric Tube Co. to TAE, 22 Apr. 1884; both DF (*TAED* D8236Z, D8433O; *TAEM* 62:124, 73:730).

29. See Docs. 2091 and 2343.

30. Insull to Johnson, 17 Jan. 1882, LM 3:20 (*TAED* LM003020; *TAEM* 84:21); see *TAEB* 5 App. 1.F.5, 7–8.

31. See Doc. 2343. The building was at 292–298 Ave. B at 17th St., an intersection that no longer exists. The Edison Electric Light Co. described extensively the building and its equipment in the 28 August 1882 Bulletin. Bulletins 13:23–24 and 16:25–26, 28 Aug. 1882 and 2 Feb. 1883; both CR (*TAED* CB013, CB016; *TAEM* 96:738, 789).

–2343–

Samuel Insull to
Charles Batchelor

New York, September 28th. 1882.

My Dear Batchelor:—

I have been going to write to you for about six weeks past, but some how or another circumstances have occurred to prevent me inflicting on you an epistle of any considerable length. About two weeks ago I got half way through a long letter to you, was called away from it, and was never able to finish it. You must excuse my negligence on the plea of the very great press of business that we have had here for some considerable time past. I will try now to wipe out the whole score.[1]

LAMP FACTORY. I think it as well to deal with the most expensive subject first. The Lamp Factory still continues to absorb money right along. They are turning out at the present time from about 800 to 1000 a day and still lose on everything that they sell. Only two or three weeks ago Edison paid an assessment to them on your account of $3,750.[2] They have, of course, been at considerable expense in moving from Menlo Park and the place has been fixed up on a scale not to meet the present demands but rather to be in a position to deal with future requirements when the rush for lamps comes which must be in a very short time. Upton claims that as soon as he turns out 1500 lamps a day, he can do it without loss; 2000 at a slight return and 2500 at a snug little profit always providing that we can get a re-arrangement of the lamp contract allowing us to charge 40 cents for our lamps and giving a profit to the Lamp Factory of 5 cents per lamp before dividing any profits that may be left after that equally with the Lamp Factory and the Light Company. This contract business I took up in the early part of the summer before Mr Lowrey went away on his summer vacation, but since he has been back, although I have [a--]ᵃ made half a dozen appointments with him he has never kept them, always promising to take up the question the very next week.[3] If we get this increased price on lamps it would appear that everything would be solid at the lamp factory. In addition to this we are going to claim from the Light Co. an amount equal to about $50,000 for purely experimental work. We propose to debit this sum to the Light Co's account and wipe it out by means of crediting them with any future profits on manufacture that they may be entitled to after the Lamp Co. has received 5 cents per lamp as its regular profit. It would appear that at the present time the Lamp Factory absorbs about from 5000 to $7500 a month in addition to the amount of their sales. Of course, as against this they have their [-----]ᵃ stock of lamps but for some time past their stock instead of increasing has considerably diminished and at present is not more than 50 or 60,000 lamps.

MACHINE WORKS. We have turned out a very great deal of work at Goerck St. this last few months. Whether it has been turned out cheaper and good your experience at Ivry will very soon inform us. All that I can say is that instead of losing money here now as heretofore there is some slight amount being made. Edison has taken out of the business $38 000ᵇ and you have been credited in account on my books with $3800ᵇ being 10 per cent of same. The Machine Works carries about 550 shares of Isolated Co. stock and they pay their assessments by turning in machines as the Isolated Co. requires them. About $35,000 (that is on the basis of the selling price of the machines) has at present been paid into this stock. We have got a large number of K, L, and Z machines on hand and no less than 7 central station machines. Our present work will be all cleared up in the course of a week or two and we shall shut down. We shall then probably have on hand about 40 Z machines, from 60 to 70 L machines and about 30 K machines and 5 or 6 central station machines. You will see that this is a very heavy stock for us to carry representing as it does from 130 to 140,000 worth of stuff. When this is all sold and our books balanced there should be a further amount of profit to be divided amounting to probably $40,000ᵇ Now, if you add these various amounts together including sum invested in Isolated Stockᶜ you will see that the Machine Works is not in quite so bad a condition as it was some months ago and there is hope that there can be money made there. Mr Dean seems to give more attention to his business, has most decidedly improved, and is not now at loggerheads with everybody as was his wont when he first started in owing to an impression that he used to have that every~~thing~~ one[d] in this world must bow down on their bended knees and worship him.

CENTRAL STATION ENGINES. Our great trouble just now is with reference to the engines for our central station dynamos. At Goerck St. we have 6 machines building for account of the Illuminating Co. and the same number which the Works is carrying as stock and up till a week ago we had not an engine to go with them and Edison was quite in a quandary as to what engine he would use and where he could get it. The Porter-Allen engine did not govern at all as he expected, those at the central station having given very considerable trouble.[4] The Armington & Sims Co. have been months turning out a model engine for us, but they have at last got one to Goerck St. The tests on this will be made within a few days. I went down to Providence and made an arrangement by which the Machine Works is to build their own central station engines on a royalty of $200 an engine, the engine to be used being $14^{1}/_{2}{}^{e} \times 13$ cylinder, such as the new model we now have in Goerck St. Whether Edison will avail himself of this license or will use Porter-Allen engines is just at the moment a matter of doubt. He wrote to you the other day stating that he should have to discard the Armington engine and go back to the Porter engine entirely in consequence of it being necessary to connect the governors of all engines in a central station with a special coupling arrangement so as to lock them together and make them run at the same speed irrespective of whether the load is thrown from one to another. At first he thought that he could not apply this device to an Armington engine.[5] I have the impression now that within the last 2 or 3 days I have heard him state that he can apply it, but I will advise you of this later. Sims claims that two of their engines will run together, that you can throw the loads ~~together~~ from one to the other with perfect ease and that they will regulate absolutely. They are now rushing their second engine which will probably be at Goerck St. within 10 days and then they will test two of them together and see if they obtain the perfect results which can be got with two Porter engines unless this coupling device is used. Edison says he is confident that they will not be able to do so. My electrical knowledge is too slight to ~~give you any evidence as to my ability to~~

describe the causes of this difficulty to you, but I will try to get Edison to dictate a letter to you on the subject within the next few days. All I can do is to state the facts and leave it to Edison to give you the theory. Up to 2 or 3 days ago it was quite impossible to run more than one C dynamo in the central station. The night before last Edison had this coupling arrangement for locking engines together tried and it worked with perfect success. He could vary the load just as much as he liked and yet not get a difference in the ~~engine~~ speed of the engine of more than 5 or 6 revolutions a minute. Paton,[6] the French engineer, is now making the drawings of this device and [-----][a] and they will be sent forward to you as quickly as possible. I think it would be very difficult for us to state precisely as to whether the Porter-Allen or Armington & Sims engine will be used until after the test has been made at Goerck St. and the two Armington & Sims engines working together. Now this bug has been eradicated it would appear that the central station machines will work perfectly. We have been running with one machine at Pearl St. now for about 4 weeks and it has given excellent results. There is no reason why as soon as this coupling device has been fixed up permanently we should not connect up 4 or 5000 more lights. At present we have consumers using our light constantly from Drexel, Morgan & Co's to the Times Building and from there down to the East River. Cowles, of the Ansonia Co.[7] told me the other day that he would not on any account have the light taken out even if it were to cost twice the price of gas. Edison told me two or three days ago that the light is costing the Illuminating Co. at the present time six dollars and a half a thousand feet but you must remember that we are only running 1000 lights; that the staff at present at Pearl St. could as easily run 6000; that the depreciation on the whole 16 miles of conductors and the general expenses is chargeable to one sixth of what will be the capacity of the present half of the station when it is running in full bloom. So Edison is very confident that everything will be all right so far as the economy goes. Johnson is assisting Edison in cutting in consumers and immediately Edison gives him the word to connect 4 or 5000 more he will put a large force of men at work and then we shall have probably the Edison boom that we have been so long hoping for.

PARENT STOCK. The parent stock has not sold at very high prices as yet in consequence of our lighting up. There were some few sales reported at $625 but to day it is passing hands at about $600[f] As, however, within the last two months there has been at least 150 shares of stock thrown on the market, I consider this price extremely good and shows public confidence in the enterprise. About 4 or 5 months ago about 80 shares of stock were offered with the object of putting the market down and there was a drop of about 2 to $300 in the selling price of the stock. I think as soon as 4 or 5000 lights are running we shall see a very considerable rise in Edison's stock, although as I have just stated it is passing hands at about $600 there is practically none offering, everybody holding for higher prices. If I do not hear from you again on the subject I shall look upon your order to sell about 30 shares at $850 as holding good.[8]

ILLUMINATING STOCK. There is nothing whatever doing in this stock. There is a considerable quantity of it offered at par but there are no buyers and sellers will not let it go at a lower figure.

ISOLATED CO. STOCK. This is the favorite Edison stock at the present time. The Isolated Co. have just called up their last assessment and it is now fully paid stock and is selling at from 135 to $140, that is from 235 to 240 a share. The Isolated Co. are doing an extremely good business. They do not put in any plants except on the basis of a gross profit of about 50 per cent. Their last great card has been an installation in the New York Herald[f] Building at a cost[f] to Mr James Gordon Bennett[9] of about 16 or $18,000. The whole ~~floor~~ Building[d] is lit up from the ground to the garret with about six hundred 16 candle lights. The Herald[f] people are awfully pleased with it. The engine and dynamos (Ks) are situated in Ann St. the current being taken from there to the Herald Building. Kruesis tubes[g] Mr Bennett has been in the habit of paying a gas bill of about $20,000 a year for the Herald Building. Most careful estimates show that his electric light will not cost [----][a] him this sum by 7 or $8,000. The Herald people are keeping an exact account of the cost of running to them, and I expect at the end of a year they will come out in their columns with a very strong endorsement of our isolated plants.[f] This Isolated Co. two or three days ago closed a contract for lighting the Pilgrim, new Fall River boat.[10] The contract is for $14,000 and the profit on the job will be about 7 or $8,000 gross. They have got more orders than they can fill at the present time and cannot undertake to install any plants under about three months. Edison has great hopes in the future of this stock. He bought some himself at $110 premium. You will be entitled to over 50 shares of it from the Machine Works and as soon as they have paid up all their assessments the stock will be sent to you and the account wiped off to profit and loss. They have about $20,000 more to pay before this can be done. The Isolated Co. could declare quite a substantial dividend on their first years work. In case you may not have received them, I send you by this mail price lists of the Isolated Co. and a list of the places where their plants are at present installed.[11] At Roselle, on the New Jersey Central Road, we are at present installing what is known as a village plant. I do not know whether Edison or myself has written to you on this subject. The idea is to use pole lines as conductors with a current of high[d] electro motive force running three 10 candle A lamps in series. This will be a miniature central station plant. It can be put up very cheaply and as soon as it is running and the tests have been made, I will let you know the result. Edison expects very great things of it and says it can be put in every village with very little capital and his figures show that it will return a profit of over 30 or 40 per cent on the cash investment. Moore, the General Manager of the Isolated Co., says that so far as he can see from actual experience with isolated plants Edison's figures are more than justified.

EUROPEAN CO.   PUSKAS AND BAILEY. You have received Bailey's report to the Directors of the European Co.[12] He has made quite a good impression here since he has been over and seems to have wiped out entirely that prejudice which has heretofore existed against him. The Puskas & Bailey contract for representation of the European Co. in Paris has been closed, their 5 per cent of the parts of founder has been given them by contract and there seems to be a general feeling of confidence in the outcome of the European Co. which certainly did not exist here before Bailey[f] arrived.[13] Prof. Columbo[14] has created an extremely good impression on Edison and Edison paid him the compliment of telling him that he must have mistaken his nationality,[f] that he could not by any possibility be an Italian and he thought that if he looked up his ancestry it would prove that he was a Down East Yankee. Edison has very great hopes of the Milan station. He thinks that it cannot fail to come out all right if Colombo gives his personal attention to it and we are using every effort to rush through the necessary supplies to enable Colombo to start his station on the 1st. of January.[15] There is a movement here which has hardly assumed definite shape yet to put the European Co's stock on a strong basis by syndicating the shares of the large holders. If this comes to anything I will let you know. At present the stock is selling at about $65, that is the last sale took place at that figure. There is not, however, any amount of it offered at a price that will average below $75 and I doubt if a larger buyer[f] came into the market if much could be picked up under 80.

BERGMANN & CO. Some two or three months ago Bergmann heard that the United States Co's[16] factory at the Cor. of 17th. St. and Ave. A. was for sale. It is about 100 × 1000 and has 6 stories.[17] The reason that the United States Co.[f] wanted to sell it was, they said, that they had better facilities in the Weston Co's shops at Newark, the Weston Co. being controlled by the capitalists of the United States Co. The fact is the United States Co. is somewhere about on their last legs and they were obliged to sell. Bergmann set a brother Dutchman to buy the property ostensibly for a cigar factory and he obtained it for $77,000. With the property goes a magnificent 150 H.P. Corliss engine, boiler capacity for this and another engine of the same size, and all the main shafting, benching, and a lot of sundries which altogether with the property cost the United States Co. about $170,000. You may judge of the chagrin of the U.S.[f] Co. when they discovered who had purchased the property as they had stated that they would not on any account sell it to a rival electric light Co. They offered Bergmann $5,000 to let them off the bargain as notwithstanding that they were so terribly hard up they could not well afford to let the Edison interests obtain possession of their property. Of course, Bergmann & Johnson were deaf to any such entreaties as they had already got their contract. Johnson came to Edison and asked him to go into partnership with him and the firm of Bergmann & Co. at the present time is composed of Johnson,[f] Edison and Bergmann, each holding one third interest.[18] Of Edison's one third you own 10 per cent. For this one third interest including interest in good will of the business stock on hand, machinery, the new property, in fact everything, Edison is to pay about $39,000[h] so that your 10 per cent of that will amount to $about $3900.[i] I have already debited to your account the sum of $3500[b] Edison having paid on account $35,000.[b] There is going to be a good deal of money made in this business. It is an extremely good thing and I consider the firm a very strong one and you will no doubt endorse my views. When the partnership contract is made I am going to propose to Edison that a paper is drawn up in which he assigns to you your 10 per cent interest of his one third. At the same time I shall get a paper drawn showing your 10 per cent interest in the Machine Works. I remember your telling me before you went away that you did not care for any contracts with Edison. But these cases are different from any others. You invest actual money in the business and you ought to have something to show your legal rights in case of anything happening to Mr Edison. I think you will endorse my views on this subject which I consider are as much in Edison's interests as your own.[19]

MENLO PARK. You will notice that this letter is dated from 65 Fifth Avenue. We have again moved in here from Menlo Park. Edison has taken a house at 25 Grammercy Park for 2 years. He instructed me to move my office in here and to move his library just as soon as I have a room fitted up at his house at Grammercy Park to receive it. He and his family move in this week. He has taken the top floor of Bergmann & Co's new premises as an experimental shop and will probably move his experimental force there immediately Bergmann obtain possession which will [b]e in about a month. Do you not think this looks very much as if he will never go back to Menlo Park again? In informing me of his decision on this matter he told me that it was in consequence of the necessity of his being close to the central station. In the next breath he said he would never come near the city if it was not for the women constantly bothering him to do so. Johnson and myself are of the opinion that it is six of one and half a dozen of the other and that he wants to come in just as much as the women do.

YOUR ACCOUNTS WITH EDISON. I have been wanting for a month or so past to render you complete accounts for the year since you left here but about five days a week I have been obliged to be here in New York while my office and papers were at Menlo Park. Immediately I get things straightened out here I will strike a balance and send you a copy of the account. I could not have set about rendering it earlier as there has been outstanding matters in connection with English Light and the Indian Colonial Light which require settlement before rendering your accounts. Your assessments are all paid up to[f] the Lamp Factory, your assessment for 10 per cent for Edison's one third interest in Bergmann & Co's is paid and all that is outstanding is about $1600 payable to the Machine Works. I cannot tell you off hand which way the balance is but I imagine that it is considerably in your favor. Please address all communications for Edison and myself to 65 Fifth Avenue for the future[20] and believe me, Very sincerely yours,

TL (carbon copy), NjWOE, DF (*TAED* D8243I; *TAEM* 63:167). Handwritten emendations by Samuel Insull. [a]Canceled. [b]Handwritten in space provided. [c]"including sum invested in Isolated Stock" interlined above by hand. [d]Interlined above by hand. [e]"1/2" handwritten in space provided. ᶠMistyped. ᵍ"Kruesis tubes" interlined above by hand. ʰ"about $39,000" handwritten in space provided. ⁱ"$about $3900" handwritten in space provided.

1. See headnote above.

2. Samuel Insull sent the lamp factory a check for $5,000 on 1 September to cover Edison's and Batchelor's assessments. Edison's transactions with the lamp factory at this time are recorded in Ledger #8:39, 205, Accts. (*TAED* AB004:40, 86; *TAEM* 88:41, 88).

3. On manufacturing costs cf. App. 1.B.54. Discussion of a revised contract had been underway since at least May (see Doc. 2289 n. 5), when Grosvenor Lowrey sent a draft for Sherburne Eaton's review. Late in June, he told Edison he wished to settle the remaining questions shortly. Insull wrote Lowrey in early October that Edison had been "pressing me to get a settlement of the Lamp Contract and he seems to feel very sore that the matter should have run on so long. I think that it is but due to him that it should be closed up right away as the longer it remains open the more seriously will be his loss." Lowrey to Insull, 29 May 1882; Lowrey to TAE, 27 June 1882; both DF (*TAED* D8224ZAF, D8224ZAO; *TAEM* 61:35, 61:42); Insull to Lowrey, 5 Oct. 1882, Lbk. 14:217A (*TAED* LB014217A; *TAEM* 81:912).

4. The Pearl St. plant began operating with one Jumbo dynamo. Edison decided to try putting a second machine on the line soon afterward. Quoted in a November *Operàtor* article, he recalled that

> The result of the first experiment was such as to astonish the engineers and every one who witnessed it and came near proving serious. When the current from the second dynamo was turned on, first one engine and then the other would go like lightning, and first one and then the other was converted into a motor. One of the engineers, witnessing the effect, shut off steam, and still his engine was running just as rapidly as before. He came running towards me as white as a sheet, exclaiming: "My God, Mr. Edison, what is the matter? I have shut off the steam, and yet my engine is running at lightning speed and I can't stop it!" . . . The next thing that happened was the melting in a second of six or eight pounds of copper, which was thrown off in a sort of vapor, filling the room and nearly blinding every one present. ["The Edison System. How the Inventor Overcame an Unforeseen Difficulty," *The Operator* 13 (1882): 528]

Each machine had apparently alternated between generating current and being driven by the other as a motor. Charles Clarke subsequently attributed the problem to an oscillation of the iron floor beams "acting in the same direction as the action of gravitation, which was one of the two controlling forces in the operation of the Porter-Allen governor" (Clarke 1904, 47, 49; see also App. 1.B11–12).

The date of the experiment is uncertain; one source gives it as 8 July (Josephson 1992 [1959]: 262–64; cf. Clarke 1904, 47). Edison began sketching corrective devices on 8 September, roughly the time the Armington & Sims engine arrived (PN-82-09-04, Lab. [*TAED* NP017:2, 4, 6; *TAEM* 44:67, 69, 71]; Armington & Sims to Insull, 31 Aug. 1882, DF [*TAED* D8233ZDT; *TAEM* 61:1131]). Clarke recalled that

Dash-pots were applied to the governors with no practical success. Finally, Mr. Edison overcame the difficulty for the time being, and the machines were operated in multiple, by means of a complicated system of pivoted rods, levers and shafting . . . by which the governors of all the engines at any time in operation were mechanically connected together so that the tendency of any governor to control its engine was communicated to the governors of all the other engines. The shafting was made light, and yet with comparatively great torsional rigidity, which was essential to prevent lost motion in the action of one governor upon another, by constructing it of tubing with a central rod, the tube and rod being first pinned together at one end, then twisted in opposite directions at the other end, and there pinned together in the twisted positions. [Clarke 1904, 47]

Edison filed six patent applications as a result of this experience. One (U.S. Pat. 271,614) covered the rigid shafting described above; the others concerned means for regulating the engine throttle or cut-off (U.S. Pats. 271,615; 271,616; 273,491; 273,493; and 365,465). Another application, filed in March 1883, for Edison's form of central station generally, encompassed means for economically regulating two or more direct-driven steam dynamos by cut-off and various governors (U.S. Pat. 281,351).

5. Batchelor had urged Edison to send him specifications for the Armington & Sims engine. No reply has been found but Edison noted on the letter that "owing to the fact that we cannot work 2 engines together without coupling the governor we shall have to use the Porter Engine" (Batchelor to TAE, 6 Sept. 1882, DF [*TAED* D8238ZCW; *TAEM* 62:488]). Armington & Sims used an inertial governor mounted on the engine shaft at the flywheel. Generally called a shaft or spring governor, this type of mechanism was increasingly popular because of its greater sensitivity than the older flyweight governor on a separate shaft, such as that used on the Porter-Allen engines. The cause of the difficulty that Edison anticipated in linking multiple governors is not clear. Hunter 1985, 473–80.

Armington & Sims governor with weights (1) mounted on the engine shaft inside the flywheel. This mechanism was used on a large A&S engine and illustrated in an 1884 trade catalog.

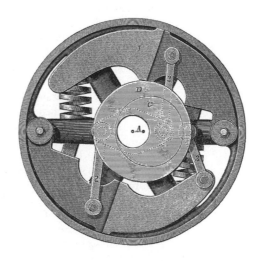

6. Identified elsewhere as Patin, he and a Mr. Picard worked in Paris to draw plans for the underground conductors at Milan. Bailey to TAE, 2 Sept. 1882, DF (*TAED* D8238ZCS1; *TAEM* 62:484).

7. Alfred Cowles (1845–1916), a longtime friend of Edison, was secretary of the Ansonia Brass & Copper Co. The firm had shops in New York and was a major supplier of the Edison Machine Works and Electric Tube Co. *NCAB* 18:66; TAE to Cowles, 13 Dec. 1916, Lbk. 115:350 (*TAED* LB115350); Ansonia Brass & Copper Co. to TAE, 21 Jan. 1882, DF (*TAED* D8235C; *TAEM* 62:29).

8. Batchelor had asked Insull in June to sell thirty shares at no less than $850 because "my being here cost an enormous amount and I have to spend three times what I get here from these people— So I should like to get a little money which I can put at interest." Batchelor to Insull, 19 June 1882, Cat. 1239, Batchelor (*TAED* MBLB4304; *TAEM* 93:683).

9. James Gordon Bennett (1841–1918), the flamboyant publisher of the *New York Herald* since 1867, resided in Paris. *ANB*, s.v. "Bennett, James Gordon, Jr."

10. The *Pilgrim*, a 3,500-ton sidewheel steamer, was the largest ship built in 1882 by the shipyard of John Roach & Son. Equipped with 910 Edison lights, it began making overnight runs on the historic Fall River Line between that city and New York in 1883 (Fairburn 1945–55, 2:1484; Dunbaugh 1992, 104–105, 210, 273–75; Edison Electric Light Co. Bulletin 17:24–26, 6 Apr. 1883, CR [*TAED* CB017; *TAEM* 96:809]). There was some disagreement whether Miller Moore, as head of the Isolated Co., or Spencer Borden, whose relatives had a financial stake in the ship, should negotiate the contract. Borden had been working towards a deal over the summer and Edison thought he should continue, despite having claim to a 10% commission on the dynamos (Eaton to TAE, 9 Aug. 1882, DF [*TAED* D8226ZBA; *TAEM* 61:368]).

11. Edison Co. for Isolated Lighting brochure, 1 Sept. 1882, PPC (*TAED* CA002A; *TAEM* 96:103).

12. There are two extant reports by Joshua Bailey and Theodore Puskas to directors of the Edison Electric Light Co. of Europe on 25 August. One, written by hand, provides a narrative of their efforts to organize the Edison companies in Paris. The other, a printed document, outlines terms of the Paris contracts and arrangements in Germany, Italy, Austria, Hungary, Russia, Alsace, Lorraine, Belgium, Holland, and France. Eaton helped Bailey redact the latter document for distribution to recipients of the Edison Electric Light Co. bulletins. Bailey and Puskas reports, both 25 Aug. 1882; Eaton to TAE, 27 Aug. 1882; all DF (*TAED* D8228ZAD, D8228ZAC, D8226ZBM; *TAEM* 61:658, 643, 415).

13. Joshua Bailey and Giuseppe Colombo left Milan for New York on 2 August. Eaton scheduled a long conference with Bailey, Edison, and James Banker to discuss European matters on 16 August. Under terms of a contract signed on 26 August, Bailey agreed to work entirely for the European Co., while Theodore Puskas was engaged on a part-time basis. They were to receive collectively five percent of the founders shares held by the European firm in the Paris Edison companies. Edward Acheson to TAE, 1 Aug. 1882; Eaton to TAE, 15 Aug. 1882; Bailey and Puskas agreement with Edison Electric Light Co. of Europe; all DF (*TAED* D8238ZCA, D8226ZBF, D8228ZAE; *TAEM* 62:452; 61:392, 668).

14. Giuseppe Colombo (1836–1921) was an engineering educator,

entrepreneur, and statesman. He held the chair of mechanics and industrial engineering at the Regio Istituto Tecnico Superiore, Milan's engineering school, from 1865 to 1911, and published an important engineering textbook, *Manuale dell'Ingegnere*, in 1877. After seeing Edison's electrical lighting system at the Paris exhibition in 1881, he then secured the backing of several banks to form a syndicate to exploit Edison's lighting patents in Italy. This group, the Comitato per le Applicazioni dell' Elettricita Sistema Edison, set up the Milan central station in June 1883; it was reorganized in December 1883 as the Societa Generale Italiana di Elettricita Sistema Edison. Guagnini 1987, 291–93; Pavese 1987, 391–96; *DBI*, s.v. "Colombo, Giuseppe."

15. The plant did not enter service until 28 June 1883. It initially used four Jumbo dynamos, powered by three Porter-Allen and one Armington & Sims engine. In August 1883 two more dynamos with Armington & Sims engines were added. The station eventually operated a total of ten Jumbos powering some 5,500 lamps; it remained in service until February 1900. Colombo to Bailey, 12 Sept. 1882; Lieb to TAE, 1 July 1883; both DF (*TAED* D8238ZDA, D8337ZCO; *TAEM* 62:497, 67:789); Edison Electric Light Co. Bulletin 19:25, 15 Aug. 1883, CR (*TAED* CB019; *TAEM* 96:847); "The Edison Central Station at Milan," *Engineering*, 31 Aug. 1883, Cat. 1018, Scraps. (*TAED* SM018094a; *TAEM* 24:281); Pavese 1987, 391–96; Clarke 1904, 53.

16. United States Electric Lighting Co.

17. Contemporary photographs show a nearly square building, probably 100 × 100 feet. It was on Ave. B (not Ave. A); see headnote above.

18. This agreement had been reached in principle in April 1881 (see Doc. 2091 esp. n. 6). It established the firm of Bergmann & Co. on 4 September, the day the Pearl St. station began operating commercially. Edison purchased a one-third share in the business for $38,290 cash (TAE agreement with Johnson and Bergmann, 2 Sept. 1882, Miller [*TAED* HM820165; *TAEM* 86:489]). A 2 September 1882 receipt to Edison for equipment and fixtures, as well as statements of the company's financial condition in August and September 1883, were enclosed with Bergmann & Co. to Insull, 16 Nov. 1883, Miller (*TAED* HM830200; *TAEM* 86:682).

19. Charles Batchelor's share presumably derived from his interest in all of Edison's electric lighting enterprises. Batchelor recorded his stake in Edison's share of Bergmann & Co. sometime before the end of the year but no formal declaration of his right has been found. Cat. 1318:142, Batchelor (*TAED* MBA001:57; *TAEM* 93:855).

20. Batchelor's 10 October reply is Doc. 2351.

With his electric light system working successfully in New York's financial district, Edison moved again with his wife Mary and their family from New Jersey to New York City, ostensibly to be near the generating plant. Edison rented a house in the fashionable Gramercy Park area on 1 October. The two eldest Edison children, Marion and Thomas, Jr., with Mary's youngest sister, Eugenie Stilwell, enrolled at a private school just a short walk from there. With a two-year lease, Edison seemed resolved to stay in New York.

Edison quickly dismissed all but a few employees from the laboratory at Menlo Park, where he had done most of the research and development for his electric lighting system. He had already arranged to rent the top floor in the large factory building recently taken over by Bergmann & Company, of which he was a partner. For $2,500 a year, Edison made this his new laboratory; according to his secretary Samuel Insull, he could usually be found there each day.[1] Edison had been accustomed to a large staff of assistants at Menlo Park; though it is not clear how much help he now had, this was a productive period for him.

Edison tackled a variety of questions, not all of them related to electric lighting. Adapting techniques from the manufacture of lamp filaments, he devised a method for plating decorative natural materials like cloth or wicker with gold or other metals, after first reducing them to carbon. He applied for patent protection on "a new material" that could substitute for hard rubber, produced by treating vegetable fibers with hydrofluoric acid.[2] He also entertained the possibility of devising a faster and more efficient process for separating cream from milk. This would have found a ready market in the dairy industry, but Edison seems not to have given it much time.[3]

The incandescent lamp remained the most complex and delicate part of Edison's lighting system, and the problems of manufacturing lamps and extending their life continued to occupy his mind. He again sought to prevent the carbon carrying that left filaments attenuated and vulnerable and also blackened the glass, thus reducing the effective life of the lamp. He explored the use of cellulose filaments, chemical reactions to fill lamp globes with inert gases, and ways to reduce the moisture inside.[4]

Although Edison predicted that the future of electric lighting lay in central station service, he recognized the short-term importance of isolated lighting plants, particularly for cultivating the large manufacturing capabilities needed to construct central station districts. Theater owners were particularly enthusiastic about isolated stations, and George Bliss, an Edison agent in Chicago, had reported a "mania" among stage managers there.[5] In addition to its advantages in safety and comfort, electricity offered opportunities for colored lights and independent control of banks on different areas of the stage. Without the hiss of gas jets, patrons seemingly enjoyed better acoustics as well.[6] Edison companies had a number of theater projects completed or under consideration by this time. Edison himself went to Boston in November and again in December, apparently to plan the installation at the Bijou Theater that was in operation by the end of the year.[7]

Edison devoted some time to problems particular to isolated plants. He continued to experiment with storage batteries, useful in such installations, then abandoned the subject in December. He had more success with voltage regulation, which was especially important for the satisfactory operation of lamps in isolated systems. He announced in late November, after months of work, that he had "a really reliable Automatic Regulator."[8] Gas engines, though otherwise attractive, were hardly suited to drive small dynamos because of their pulsating rotation. He devised a regulator adapted to these machines, completing it in time to make a "great success," he claimed, of Professor Henry Draper's soiree for members of the National Academy of Sciences in November.[9]

The Academy's meeting in New York presented an opportunity to open the Pearl Street station for the distinguished attendees' inspection.[10] Having solved the difficulty of running two or more of the Jumbo steam dynamos simultaneously,

Edison was apparently satisfied with the plant's operation. He declared privately to noted chemist Benjamin Silliman, Jr., that Pearl Street had proven "that we can compete with gas at the same price and it is further shown that if necessary the public will gladly pay 50% more" for the same intensity of light from his lamps than by gas.[11]

Edison confidently staked the financial success of his electric light system on central station business. In response to the rapidly worsening condition of the Edison company in London, he emphatically laid out a business strategy based on central station service, urging the company to delegate its isolated plant operations to a separate entity. He also expressed his philosophy that electric utilities should sell light (as gas companies did), rather than electric current, so that future efficiency gains would benefit the company rather than the consumer.[12] (Edison impressed these views on company secretary Arnold White, who visited New York in November and December in an effort to mend transatlantic relations.) Another measure of the strength of Edison's conviction was his categorical insistence that his companies retain all central station rights when negotiating with the Siemens interests for a patent-sharing arrangement in Germany.

While the British company floundered, business looked much brighter on the Continent. Orders came in for dynamos and other equipment for a small Milan central station, although these orders, like those earlier from London, became a major source of contention over prices and the authority to make quantitative and qualitative changes in the equipment. The difficulties of an overseas supply chain did not threaten business in France, where Charles Batchelor was making lamps and dynamos "equally as good as American" ones and beginning to plan for a Paris central station.[13] Edison's agents also continued to install isolated plants throughout Europe. Encouraged by these prospects, financier Egisto Fabbri exercised an option sometime during the fall or winter to buy a large block of shares, on behalf of Drexel, Morgan & Company, in the Edison Electric Light Company of Europe.[14]

1. Bergmann & Co. to TAE, 18 Dec. 1882, DF (*TAED* D8201Y; *TAEM* 60:31); Insull to James Kelly, 29 Nov. 1882, Lbk. 14:481A (*TAED* LB014481A; *TAEM* 81:1028).

2. U.S. Pat. 543,986.

3. See Doc. 2361.

4. See Docs. 2346, 2349, and 2370; U.S. Pats. 274,293 and 274,295.

5. Bliss to TAE, 21 June 1882, DF (*TAED* D8241F; *TAEM* 63:91).

6. Edison Electric Light Co. Bulletin 5:6, 17 Mar. 1882, CR (*TAED* CB005; *TAEM* 96:681).

7. Samuel Insull to John Taylor, 9 Nov. 1882; Insull to Seth Low, 13 Dec. 1882; Lbk. 14:407A, 15:35 (*TAED* LB014407A, LB015035; *TAEM* 81:1003, 82:45); Sherburne Eaton to TAE, 16 Dec. 1882, DF (*TAED* D8221ZBE; *TAEM* 60:922); Edison Electric Light Co. Bulletin 15:8–9, 20 Dec. 1882, CR (*TAED* CB015; *TAEM* 96:766).

8. Doc. 2371.

9. See Doc. 2359, esp. n. 2.

10. See Doc. 2367.

11. TAE to Benjamin Silliman, Jr., 20 Oct. 1882, Lbk. 14:309 (*TAED* LB014309; *TAEM* 81:951).

12. See Docs. 2357, 2374, and 2375.

13. Doc. 2366.

14. See 2356.

–2344–

From Camille de Janon

[New York,] Oct. 2, 1882.

Thomas A. Edison Esq.
        To Mademoiselle C. de Janon[1] Dr[2]
For first half year's tuition of

Miss Eugenie L. Stilwell[3]	$100
Miss Marion E. Edison[4] One Year	100
Master Thomas A. Jr.[5] One Year	100
Fuel for the season for two–	12
	$312

Received Payment

                        C. de Janon

ALS, NjWOE, DF (*TAED* D8214W2; *TAEM* 60:723).

1. Camille de Janon (d. 1890) succeeded her partner Henrietta Haines in 1879 as director of the school at 10 Gramercy Park, a short walk from the Edisons' rented home. She wrote this receipt on the reverse of a printed brochure describing instruction and fees of her "English and French School for Young Ladies and Children." Obituary, *New York Times*, 28 May 1890, 5; "Testimonial from the Teachers of New York City to Miss Haines," *The New York Evangelist* 50 no. 27 (3 July 1879): 5; Mademoiselle de Janon's school circular, Sept. 1882, DF (*TAED* D8214W1; *TAEM* 60:722).

2. Book-keeping abbreviation for debtor, used as a heading for the left-hand or debit column of an account. *OED*, s.vv. "dr," "debtor."

3. Eugenia Stilwell was Mary Stilwell Edison's youngest sister. The half-year fee corresponds to that listed in the brochure (see note 1) for the "2d and 3d Classes" for children twelve years or older, consisting of an unspecified "Elementary Course" plus "Tuition in Latin, French,

Drawing, and all the studies comprised in an Extended Literary and Scientific English Education." Edison continued to pay for her schooling after Mary's death in 1884. Israel 1998, 233.

4. Thomas and Mary Edison's oldest child, born in February 1873. The fee is that for the "Preparatory Class," which the brochure did not describe further. Prior to attending the English and French School, she attended the Misses Graham school for Young Ladies at 63 Fifth Ave. Graham School to TAE, 24 Oct. 1881, DF (*TAED* D8114E1; *TAEM* 57:539).

5. The Edisons' second child, born in January 1876. Thomas would have been in the "Boys' Class," on which the brochure did not elaborate.

–2345–

To Spencer Borden

N.Y., Oct. 4. 1882

Dear Sir

Please accept thanks for letters of S. B. Paine referred to me and which are returned herewith.[1]

I think that upon investigation you will find that the Maxim people use an extra coil either around their field magnet or around their exciter this extra coil being in the line and as lamps are added it strengthens the field so as to compensate for the drop of Electro-motive force in the armature. [---][a] They are compelled on account of the great drop within the armature due to load and small mass of iron, to use a separate exciter otherwise a few lamps would make a great difference, while the drop in our armature due to load is comparatively small, but too large where there is great variation in consumption.

The trouble with the Maxim device is that it does not regulate for speed but only for more or less lamps, and a variation of speed causes a greater variation in candle power than it would if the device was not used. Another defect is the use of an Exciter which, if it should break down, would stop one or all the main line dynamos.[2]

Where you get an order with regular speed but variable load we can send you a hollow bobbin to slip over your fields which will regulate for load but not for speed—of course like Maxim's device there will be a loss of energy on this device.[3]

About thirty different kinds of Automatic regulators have been made and tested at Menlo Park, some that regulate for speed only, some for lamps only, and some regulate lamps and make their candle power constant, independent of the number or speed.[4] Out of all these I have selected one which I think can be made satisfactory in the hands of the public; so far it has worked perfectly. It will be just the thing for irregular power. I will be able to supply them in about six or eight weeks—they go inside the regular resistance box and cause no perceptible loss of energy.[5] Please investigate the Maxim device and report if my surmises are correct as to <u>not</u> regulating for both speed and lamps. Yours Very Truly

Thos A Edison   I[nsull]

L (letterpress copy), NjWOE, Lbk. 14:214 (*TAED* LB014214; *TAEM* 81:909). [a]Canceled.

1. Not found.

2. In response to a changing load, the Maxim excitation dynamo varied the amount of current supplied to the field coils of the main dynamo by adjusting the position of its commutator brushes. "Recent Developments in Electric Lighting," *Sci. Am.* 43 (1880): 255–56; U.S. Pats. 228,543; 255,310; 255,311; 269,805.

3. It is unclear which specific regulator scheme Edison had in mind, but he may have been referring to a design which used two coils on the field magnet core: a higher resistance coil wound with fine wire and a lower resistance coil wound with thicker wire placed over the other. The two coils were wound in opposite directions, thus making the field current responsive to changes in electrical load. Edison worked on this design intermittently from January to October 1882 and it formed elements of three U.S. patents executed in February and May. Cat. 1148; entry of 5 April 1882, PN-82-04-01; N-82-05-26:49, 56, 62; all Lab. (*TAED* NM017:34, 61, 167, 42; NP016:31; N204:24, 28, 31; *TAEM* 44:394, 421, 526, 402, 41; 40:598, 602, 605); U.S. Pats. 264,668; 264,671; 264,662.

4. For a summary of Edison's work on voltage regulation see headnote, Doc. 2242.

5. The specific design of this regulator is unclear. Edison placed the first orders for production models by the end of November. TAE to Société Électrique Edison, 21 Nov. 1882, Lbk. 14:463 (*TAED* LB014463; *TAEM* 81:1021).

–2346–

*Memorandum:
Incandescent Lamp
Patents*

⟨493⟩[2a]
  ⟨Fig 1⟩

[New York, c. October 9, 1882[1]]

Run conductor in spiral directly on glass

⟨Fig 2.⟩

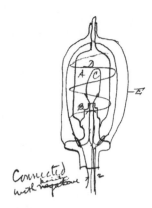

Connected with negative ↑₂

Wires or large carbon or other conductor[b] in vacuum—charge retained—

C̶a̶ Carbon positive—Earth negative or zero— Hence glass
so   Current makes glass positive—charges it & carrying is prevented

Dyer:

Shew this in double globe application[3]

⟨Fig. 3⟩

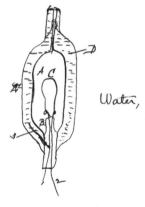

Water,

also mention that the rapidity of the discharge of the static Current may be diminished by even filling the globe with a transparent heavy oil or substance like Canada Balsam, t̶h̶e̶ ̶w̶ olive oil the whole of which would be charged by the immersed in it=

Glass is brought to same potential as carbon.

⟨Fig. 4.⟩

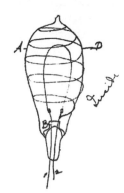

Connect with negative wire or sheet held by its spring   Or if platinum it may be [burned?]ᶜ to the glass

Hydrochloric acid gas residual Resists electrical carrying

Exhausted by hand-pump & hg gas allowed to flow in at at-mospheric pressure[4]

Patent[5d]

tube filled with aᵉ filiment of Carbon & then packed around it very finely powdered Zirconia, Magnesia Alumina Lime or other oxide or even metalic boron & then exhausting heating the glass by extraneous source heat while exhausting   then when exhausted the Carbon is gradually brot to incandescence & then brought up to such a temperature as to melt the oxideᵉ in proximity to the Carbon which coats the Carbon with an oxide.— the filiment is then taken out & put in clear globe ex-hausted & brought to incandescence & sealed from the pump = object is to cover filiment with wire conductor & prevent Elec-trical Carrying of Carbon—the oxide being carried instead

Carbon does not decompose oxide except in presence of aqueous vapor.

The Coating may also be obtained by soaking the filiment in an acetate of the oxide & then immersing the filiment in a hydrocarbon Liquid & bring it up so as to decompose & set free the acetic acid Liberating lime upon the Carbon—

⟨Case 508⟩[6a] Patent

Loops put together & put through X—afterwards spread apart by a tool put through .a.

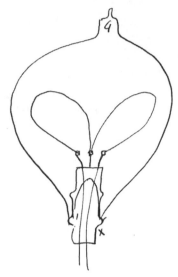

⟨⟨Case No. 509⟩⟩[7]

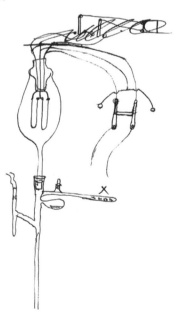

two Carbon flix[8] filiments in series; br̶Vacuum obtained. Each heated by Current, and brought up to incandescence seperately. after the air worked out of each Carbon & vacuum

obtained both Carbons are connected in series and brought up
~~so the bright~~[f] ~~one shall be at 16 Candles or other proper~~ [stan-
~~dard?~~][g] to incandescence. If one is brighter than the other, It
may be reduced until its brilliancy is the same as the other, by
disconnecting the two[e] & and bringing the brightest one up to
incandescence and heating X which contains Cyanide of mercury   Cyanogen is set free which being decomposed deposits
carbon upon the filiment and reduces its resistance; this is
done momentarily then the two Carbons are connected in series by a switch & watched to see if both are equal in illuminating power; if the one that was the brightest is still too bright
more cyanogen is set free and deposited & this goes on until
both filiments are equal in illuminating power when connected
in series; afterwards the two filiments are brought up to a
greater incandescence than they are afterward to be burned a
high vacuum obtained and the Lamp is sealed off—
Patent 1[9d]

⟨Fig 1⟩

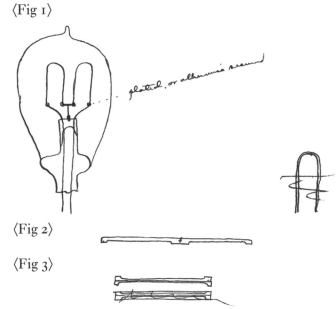

⟨Fig 2⟩

⟨Fig 3⟩

Patent Lamp consisting of two or more seperate filiments of
~~C~~flexible carbon cut from the same material and Carbonized
together so as to ~~e~~Insure same quality as to resistance & economy to permit the two or more filiments to be worked in
series.—

Patent 2[10]

gas set free passes out   X not filling—

Filiment is brought up to dull red in the hydrocarbon oil Bisulphide[e] Carbon or other Liquid[e] whose gas is not decomposable at a low red heat. This permits of ascertaining promptly if a Carbon filiment is perfect before putting in lamp the filiment is not allowed to reach a temperature sufficient to depcompose in[h] the slightest the gas or deposit anything upon the Carbon; if a weak or rather bright or duller spot is seen the filiment is not used in a lamp= If it is desired to change the resistance of the filiment rendering it of lower resistance, it is taken out of soaked in sugar or ~~mat~~ carbonizable material not soluable in the menstrum used and then immersed in the bath and brought up to red heat by the current the sugar or other substance being decomposed and Carbonized within the pores of the Carbon   at the same time it is seen that the Carbonization is even for if ~~e~~uneven spot will be seen & the filiment is not used—

Patent 3[11]

⟨Fig. 1⟩

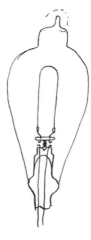

〈Fig. 2〉

[Men]tion[i] 2 or more filiments may be put in this way wked in M[ultiple] arc or series

〈Fig 3〉

〈Fig 4〉

〈Fig. 5〉

〈Fig 6〉

Patent.[12]
⟨Fig 1⟩

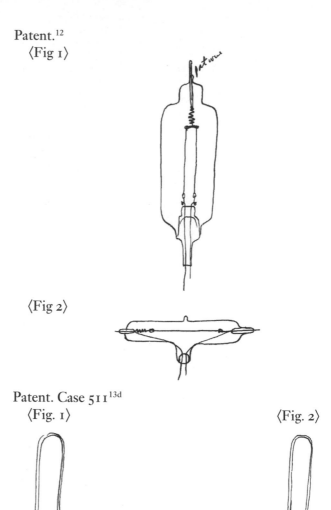

⟨Fig 2⟩

Patent. Case 511[13d]
⟨Fig. 1⟩   ⟨Fig. 2⟩

Patent=

~~Treating parchmentizing~~ gelatinizing[h] vegitable fibre by Hydrofluoric acid.[14]

The material may be thread paper bambo or other vegitable fibre in the form of a flexible filiment ready for Carbonization or sheets & sticks from[e] which the filiments may be cut or the process may be carried to such an extent as to completely gelatinize the vegitable materials the gelatine like mass being pressed free as possible of acid & then pressed in sheets by heavy pressure from which the filiments may be cut or punched. No washing of the material is required as Hydrofluoric acid gradually decomposes

[Witness:] H. W. Seeley[15]

AD, NjWOE, Lab., Cat. 1148 (*TAED* NM017:80, 82, 81, 78, 74, 69, 71–72, 76–77, 70, 73, 79, 75; *TAEM* 44:440–41, 444, 438, 434, 439, 431–32, 436–37, 430, 433, 439, 435). Document multiply signed and dated; marginalia probably written by Henry Seely. ᵃPreceded and followed by dividing marks. ᵇ"or large carbon or other conductor" interlined above. ᶜIllegible. ᵈFollowed by dividing mark ᵉObscured overwritten text. ᶠ"t̶h̶e̶ ̶b̶r̶i̶g̶h̶t̶" interlined above. ᵍCanceled. ʰInterlined above. ⁱObscured by ink blot.

1. Henry Seely signed and dated these loose papers when he received them from Edison on 9 October. The editors have arranged the pages by subject and patent application case numbers.

2. Edison executed a patent application (Case 493) on 12 October that embodied similar drawings and this general idea for preventing carbon carrying by neutralizing the static attraction of the glass globe for carbon particles (U.S. Pat. 273,486). Where a wire or other conductor was laid outside the globe, a second evacuated globe was placed around it "to prevent the discharge of the electricity." In figure 1, one end of the conducting wire is free; in figure 2, both ends are connected to the lead-in wires. The case applied specifically to wires in direct contact with the glass globe, which Edison stipulated was more a efficient construction than that shown in another application completed two days earlier (U.S. Pat. 268,206).

3. The essence of the following drawing and description were incorporated into Edison's Case 493 (see note 2).

4. Edison completed a patent application on 14 October for reducing carbon carrying by leaving a residual atmosphere of hydrochloric acid in lamps. This was accomplished by

> first exhausting the air from the globe to as great an extent as this can be done with an ordinary air-pump, and then allowing the hydrochloric-acid gas . . . to flow into the globe to replace such air. I then re-exhaust the globe and repeat the operation of refilling and re-exhausting several times until the small residue which remains consists almost entirely of hydrochloric-acid gas. The final exhaustion should be done by means of a Sprengel pump, so that as little gas as possible will remain in the globe. [U.S. Pat. 274,293]

5. Edison incorporated the drawing and substance of the description below into a patent application executed on 19 October. The patent pertained to preventing carbon carrying by "covering the flexible carbon filament . . . with a coating of insulating material not decomposable by carbon, and fusible at the highest temperatures only." (The section concerning the acetate of oxide was withdrawn during the examination process.) He recorded this idea in a brief note on 4 October; it is reminiscent of his research in 1878 and 1879 on coatings for wires in incandescent lamps. Pat. App. 492,150; Unbound Notes and Drawings (1882), Lab. (*TAED* NS82:25; *TAEM* 44:1047); see *TAEB* 4 chaps. 5–8, and 5 chaps. 1–4.

6. The drawing below is like one accompanying a patent application that Edison executed on 26 October. The application covered a lamp whose "light-giving body shall be of a broad, flat shape similar to that of a gas-flame." Two ordinary filaments were connected in series to produce

"a lamp of very high resistance, and of an ornamental and desirable construction." The unusually large nipple at the top of the globe accommodated a tool for spreading apart the carbons during manufacture. Edison completed an application for a related design patent in February 1883. U.S. Pat. 273,485 and Design 13,940.

7. This sketch and description formed the basis of a patent application that Edison filed on 9 November. It became the subject of an interference proceeding with Hiram Maxim and was later abandoned. Patent Application Casebook E-2537:390 (Case 509), PS (*TAED* PT021390; *TAEM* 45:784).

8. That is, like hair, though it is possible Edison intended "flex[ible]." *OED*, s.v. "flix."

9. On or about 4 October, Edison made notes for a patent application for testing and adjusting the resistance of two or more filaments connected in series in a lamp. He proposed to place them in an exhausted vessel, heat them with a current, and if they appeared uneven, then "pass cyanogen or HCarbon gas or Chl Carbon prefbly cyanogen as it leaves no white & equalize them then connect together in one socket so as to get 20 candles from both & have 500 ohms." He filed the application (Case 510) on 9 November with drawings similar to figures 1, 2, and 3 below. The application did not result in a patent, but the last of its three claims covered processes for forming and testing carbons made under the same conditions to obtain the same resistance. Unbound Notes and Drawings (1882), Lab. (*TAED* NS82:28; *TAEM* 44:1050); Patent Application Casebook E-2537:392, PS (*TAED* PT021392; *TAEM* 45:786).

10. The following figure and description were incorporated into a patent application that Edison executed on 19 October. U.S. Pat. 411,016.

11. Figures 1–5 closely resemble sketches that Edison made on 4 October. Figures 1–6 formed the basis of drawings for a patent application that he executed on 26 October covering methods of replacing broken carbons while "using the glass and metallic portions of the lamp over again with the new filament." Hooks and springs secured the ends of the detachable filaments to the lead-in wires. Unbound Notes and Drawings (1882), Lab. (*TAED* NS82:25–26, 28; *TAEM* 44:1047–48, 1049); U.S. Pat. 317,632.

12. In a companion application executed and filed with the one described in note 11, Edison claimed a lamp with a tension spring designed to prevent the filament from bending or breaking from thermal expansion. The patent included a single drawing, closely resembling figure 1 below.

13. Figures 1 and 2 formed the basis of a patent application Edison executed on 3 November for a method by which a carbon filament "may be removably attached to the leading-in wires." Edison placed small holes in the filament ends, then electroplated them "in such manner that the insides of the holes will be plated. By this means the strength of the ends is increased, the metal covering preventing the carbon from splitting when the hooks or other connecting devices attached to the ends of the leading-in wires are placed in the holes." U.S. Pat. 287,520.

14. "Parchmentizing" was a relatively new term for a process using acid to change a material's texture to resemble that of parchment (*OED*, s.v. "parchmentize"); Edison used it in a brief note of things to try on or about 4 October (Unbound Notes and Drawings [1882], Lab. [*TAED*

NS82:27; *TAEM* 44:1049]). He filed three patent applications related to it on 20 October. All were rejected; two were reinstated and eventually issued as patents in 1895. One of these patents pertained generally to a process using hydrofluoric acid "quite distinct from that of parchmentizing or vulcanizing vegetable fiber by the use of sulphuric acid or chloride of zinc, the resulting products being entirely different." The other patent was for making lamp filaments by this method, either by treating individual fibers or cutting them from sheets (Pat. Apps. 543,986 [Case 498] and 543,987 [Case 499]; see also Doc. 2291). The unsuccessful application, pertaining specifically to parchmentizing cellulose by sulphuric acid, claimed both the process and the resulting lamp filament (Patent Application Casebook E-2537:360 [Case 496]; PS [*TAED* PT021360; *TAEM* 45:782]). Edison gave no evidence of pursuing this subject further until Thomas Conant made a brief series of experiments in January 1883 (N-82-11-14:140, Lab. [*TAED* N143:68; *TAEM* 38:740]). Other inventors, notably Joseph Swan, were experimenting with parchmentizing techniques about the same time. Swan 1882, 357.

15. Henry W. Seely was one of Edison's patent attorneys from 1881 to about 1897. For most of this time he was a partner with Richard Dyer in the firm Dyer and Seely. Seely memorandum, 4 June 1881, DF (*TAED* D8142ZAT; *TAEM* 59:328); TAE to Dyer & Seely, 29 Oct. 1884, Lbk. 19:316 (*TAED* LB019316; *TAEM* 82:929).

–2347–

Samuel Insull to
Harry Olrick

[New York,] October 10 [1882]

Dear Olrick:—

Your letter to Mr. Edison[1] came to hand yesterday morning, and he asked me if I would write you a few lines in reply thereto. I do not remember exactly what it was that Arnold White said, but I believe it was something with reference to our business here and how very favorable are our chances of doing work as compared with those of the Light Company in England, and he made several statements and quoted you as his authority which were hardly in accordance with the exact facts.[2] Anyway whatever he said is of little importance, and Edison was extremely pleased to get your letter. We have been having quite some trouble with the London Company as to our machinery &c., and it seems to be their impression that we want to force things on them and are roping in piles of money at their expense. I know you can disabuse them of any such idea as this and if they think we are treating them in such a way we have told them that the best thing they can do is to get their machinery elsewhere. We like customers, but we do not like them of that kind. You might send us that dose of taffy[3] which you say you wrote on your return giving an account of your electrical inquiries here.[4] We sometimes like to see ourselves as others see us.

The central station has been started all right and we have now been running for upwards of a month and everything works first class. The conductors are just bang up and the dynamo machines are just as good. We run one of the Porter-Allen engines for 84 hours at full speed and full load on.

I am sorry that things are not going on so well in England but I suppose it will come out all right eventually.

We got from Armington & Sims two or three days ago their large engine 14 by 13. It is just a daisy. When you first look at it it looks like one side of a locomotive and works like a charm. It will greatly interest us, if, when Mather & Platt have been making machines some little time so as to give them a fare idea of prices, you can send us their figures.[5] We should very much like to compare English and American manufacture and we want to see where one is cheaper than the other and vice versa. Clarke has just gone off for 3 weeks holiday, played out. Edison looks bang up and is extremely delighted with the starting of his station. Everything promises here a very great success. The Isolated Co's business is something tremendous. They are simply coining money and the central station business looks as if it is going to be a great success also. You might write either to me or Edison now and then giving us an idea of things electrical in England from your point of view. I am sure Edison would appreciate it and so would I very much. Very truly yours,

TL (letterpress copy), NjWOE, Lbk. 14:262 (*TAED* LB014262; *TAEM* 81:930).

1. Olrick to TAE, 25 Sept. 1882, DF (*TAED* D8239ZEZ; *TAEM* 62:1091).

2. In his letter (see note 1) Olrick explained that after Arnold White had read Doc. 2339 to him that morning he decided to write immediately "to perhaps correct a misapprehension, if any exists, as to what I am credited with in your letter to him." White's correspondence attributing to Olrick the statements to which Insull referred has not been found; it may have been addressed to Edward Johnson. TAE to White, 11 Aug. 1882, Lbk. 7:892 (*TAED* LB007892; *TAEM* 80:811); White to TAE, 24 Aug. 1882, DF (*TAED* D8239ZDU; *TAEM* 62:1037).

3. American slang word for flattery. *OED*, s.v. "taffy."

4. Probably Doc. 2333, which Olrick sent in November.

5. Insull probably had in mind dynamos, not steam engines. Olrick took with him from New York drawings for Mather & Platt, and Edison had an understanding with that firm to build dynamos. At some time in the fall the arrangement changed to the construction of armatures only. Mather & Platt stated in November that although they still planned to do so, they would not organize their shop "for making these armatures until the demand is more lively." Olrick to TAE, 12 June 1882; Mather &

Platt to Charles Batchelor, 25 Oct. 1882; Mather & Platt to TAE, 2 Nov. 1882; all DF (*TAED* D8239ZCG, D8238ZDZ, D8239ZFT; *TAEM* 62:985, 548, 1137); TAE to White, 27 Sept. 1882, Lbk. 14:183 (*TAED* LB014183; *TAEM* 81:899).

-2348-

To Thomas Logan

[New York,] 12th Oct [188]2

Dear Sir,

Please discharge immediately all the men under your charge except Alfred Swanson (whom I want to attend the fires run the engines & the Dynamos for Rail Road) the two men engaged on Carbon mould for Lamp Factory[1] and the night watchman who must be paid watchmans wages & not Engineers wages as he has no Engineers work to do now.[2] As soon as the present order for moulds is finished the men working on them are to be discharged   Yours truly

Thos. A. Edison   I[nsull]

L (letterpress copy), NjWOE, Lbk.14:274 (*TAED* LB014274; *TAEM* 81:937). Written by Samuel Insull.

1. See Doc. 2335 n. 3.
2. Cf. Doc. 2121. Edison wrote to Charles Hughes the same day asking him to reduce expenses and cut by half the number of workers on the electric railroad, excepting Alfrid Swanson. Samuel Insull instructed John Randolph at the Menlo Park laboratory office to ensure that this document and the letter to Hughes were promptly received. He expected the staff changes to be made in about a week. Francis Upton suggested on 18 October that Edison decrease the Menlo Park workforce, noting that "There is a decided tendency to picnic showing itself" there. Swanson was discharged in late November and his duties taken by Cornelius Van Cleve. TAE to Hughes, 12 Oct. 1882; Insull to Van Cleve, 22 Nov. 1882; Insull to Logan, 22 Nov. 1882; Lbk. 14:273, 467, 467A (*TAED* LB014273, LB014467, LB014467A; *TAEM* 81:936, 1024); Insull to Randolph, 12 Oct. 1882; Upton to Insull, 18 Oct. 1882; both DF (*TAED* D8244U, D8230ZCG; *TAEM* 63:205, 61:833).

-2349-

Memorandum: Incandescent Lamp Patent

[New York, c. October 13, 1882[1]]

Patent on guarded or insulated clamps or terminals=[2]

The Metallic portion of the leading in wire & the clamp and ~~nearly~~ all of the ~~brand~~ broadened end of the Carbon filiment is covered with several Coats of japanned varnish well baked ~~so that it is nearly partially~~. This protection ~~pres~~ rest[r]ains to a considerable extent the Electrical Carrying from the Carbon Loop by diminishing the area of Contact ~~between~~ of[a] the Electrodes with the residual air thus increasing the resistance of the discharging space.

Claim Insulating the metallic[a] terminals within[b] the globe of an incandescent Electric Lamp—

Insulating the terminals [--][c] of the filiment of Carbon within the Vacuum chamber

~~the use of japan varnish or other varnish having a drying oil as a basic ingredient~~

Mention Collodion may be used & other compounds such as ~~gluten~~ glue which can be applied in a plastic state.

[Witness:] H. W. Seely

AD, NjWOE, Lab., Cat. 1148 (*TAED* NM017:44; *TAEM* 44:404). [a]Interlined above. [b]Obscured overwritten text. [c]Canceled.

1. Henry Seely signed and dated the document on this date when he received it from Edison.

2. This memorandum apparently served as the basis of a patent application that Edison filed on 20 October; it was subsequently abandoned. The application contained four claims, including an incandescent lamp with "metallic portions within such lamp coated with insulating material" and a lamp with "the leading-in wires and the ends of the carbon filament coated with Japan varnish." Patent Application Casebook E-2537:378 (Case 504); Patent Application Drawings (Case Nos. 179–699); PS (*TAED* PT021378, PT023:92; *TAEM* 45:783, 110).

-2350-

From Archibald Stuart

Cincinnati, Oct 14 1882[a]

Dear Sir

You will perhaps remember I met you two or three times at the Park in company with Mr Shaw,[1] at different times when I was East investigating your light. we are now figuring with the folks at 5th avenue in hope of having our company here in Cin organized & under way at no distant day.[2] during the progress of negotiations, and at sundry times when in New York they have spoken to me of your "multiple arc or village plant" system of Lighting which as I understand is one of your latest improvements, but in my talk with Mr Goddard and Mr Lowry, both of whom mentioned it to me, I was unable to get a very clear idea of it. as I understand it the current is conducted on

wires elevated on poles, & is intended for small villages of from 1000 to 5000 inhabitants and can be applied for both indoor & outdoor lighting and that the burners inside used for domestic purposes, can be operated in the same manner as your central station system. please inform me if I am right in this idea, & if not be kind enough to explain what its is & how it works & about the cost of investment for 1600 light plant of that kind, of 16 candle power.[3] I take the liberty to address you on this subject as I wish to have as clear an understanding of just what it is as possible, hence thought it better to write you [there?][b]
New York   Respectfully

A Stuart[4]

PS   I was out to see the first of your plants in Cin[cinna]ti last evening at the Mill Creek Distilling Co[5]   it works finely
A.S.[c]

LS, NjWOE, DF (*TAED* D8220ZAN; *TAEM* 60:807). Letterhead of Planters Leaf Tobacco Commission Warehouse, Worthington & Co. [a]"Cincinnati," and "188" preprinted. [b]Illegible. [c]Postscript written and signed by Stuart.

1. Phillips Shaw (c. 1848–1937), a Williamsport, Penn. manufacturer and financier, was Edison's agent in Pennsylvania. He helped organize several Edison electric light and power companies there, including the first three-wire central station at Sunbury. He remained associated with Edison light and power interests until 1897. "Shaw, P. B.," Pioneers Bio.; "Two Former Aides of Edison Are Dead," *New York Times*, 2 Feb. 1937, 23; Beck 1995, 55–56.

2. On 10 August 1882 Stuart entered into an agreement with the Edison Electric Light Company to set up an illuminating company in Cincinnati. From 1883 to 1885 Stuart served as Secretary of the Ohio Edison Electric Installation Co., a firm which set up lighting plants in several Ohio towns, including Piqua, Middletown, Tiffin, and Circleville. Stuart agreement with Edison Electric Light Co. and Theodore and Cook, 10 Aug. 1882; Electric Light—TAE Construction Dept.—Stations—Ohio—General (D-83-53); Electric Light—TAE Construction Dept.—Stations—Ohio—General (D-84-47); Electric Light—Edison Electric Illuminating Cos.—General (D-85-23); all DF (*TAED* D8224ZAZ, D8353, D8447, D8523; *TAEM* 61:63, 69:946, 75:82, 77:598).

3. Edison's reply is Doc. 2353.

4. Stuart wrote on the letterhead of his father-in-law, Henry Worthington, the owner of a large tobacco farm and wholesale business in the Cincinnati area. Stuart was later associated with the Thomson-Houston Electrical Co. *History of Cincinnati and Hamilton County*, 521–22.

5. In August 1882 this distillery ordered a Z dynamo and sixty A lamps for its re-distilling room. Because of the alcoholic vapor, the lamps were enclosed in glass globes of water to prevent an explosion. Edison Electric Light Co., Bulletin 13:23, 28 Aug. 1882, CR (*TAED* CB013; *TAEM* 96:738).

Charles Batchelor to
Samuel Insull

Ivry-sur-Seine, le 16 Oct 1882[1a]

My dear Insul,

Many thanks for your letter—[2] It eases my mind a little to see that my assessments over there are being met— From your letter I[b] should think the prospects good for both Lamp Co and Goerck St Works— Am glad I am in a little (indirectly) at Bergman and Co; of course it will make money!— Anything of that kind that EHJ. is in is bound to make money and I'll take my chance every time[c]

We are getting along here nicely now; only <u>glass</u> bothering us at factory[3] and those engines for C Dynamos,[4] but I tell you <u>I have had a good time on it!!</u>

I am getting some tests on my lamps for life now and shall be able to give you the results shortly[c]

I think there is now every likelihood of a Central station at Paris—[5] We are working like the devil for it and before long we shall have the right for laying the wires in the streets for two kilometers square in the heart of pParis— I have selected the place and figured out numerable plants from 10 000 lights up to 40 000 on it, and I dont hesitate to say that at this part I can get a 7 hours average burning for each light— I dont suppose you can do it any where else in the world than here[c]

I expect hourly now to have word that our engines are finished at Brünn when I shall leave to start the theatre there[6c]

I have ordered the Galvanometers and will send them on as soon as possible;[7] also the Berthoud Borcel Cable Co shall have my immediate attention as regards Edison's request.[8c]

Armington has been here and from his talk I see he wants us to pay a royalty for making his engines— He said he had seen Bailey in New York and I therefore told him that we would not do anything about it until he arrived— I telegraphed Edison ~~him~~ and of course you know you advised 8 per cent on his selling price there— We will settle the matter as soon as Bailey comes and he returns from Berlin[9c]

<u>Seubel's account.</u> When I go to Brünn I shall call at Munich and see Seubel and get this settled[10c]

Do you think it at all likely that Fabri will take up the option on the shares of European that he has from Edison and what is the term of option?[11] Yours truly

Chas Batchelor

ALS, NjWOE, DF (*TAED* D8238ZDR; *TAEM* 62:536). Letterhead of Société Industrielle et Commerciale Edison. [a]"Ivry-sur-Seine, le" preprinted. [b]Obscured overwritten text. [c]Followed by dividing mark.

1. This letter was stamped with date received (26 October), date answered (30 October), and file number (33).

2. Doc. 2343.

3. See Doc. 2314.

4. Unsure which type of steam engine to expect, Batchelor wrote to Edison a week later that he was "considerably in the dark as regards engines. My patterns are all made for the C Dynamo with the exception of that part of the base where the engine goes. I have got all my castings and all my forgings for two machines. You will see therefore that I am exceedingly anxious to know exactly what engines you are going to send me." Edison replied that he likely would select Armington & Sims but would cable definitely as soon as possible (see Doc. 2366). Batchelor to TAE, 9 Oct. 1882, DF (*TAED* D8238ZDQ; *TAEM* 62:532); TAE to Batchelor, 25 Oct. 1883, Lbk. 14:331 (*TAED* LB014331; *TAEM* 81:962).

5. The Compagnie Continental Edison, which absorbed the Société Électrique Edison in early 1884, planned to build a central station in Paris but abandoned its plan later that year. Edison Electric Light Co. of Europe, Ltd. report to stockholders, 7 Mar. 1884; Bailey to Edison Electric Light Co. of Europe, 28 Nov. 1884; both DF (*TAED* D8428H, D8428ZAI1; *TAEM* 73:298, 337).

6. The newly constructed municipal theater at Brünn, Austria (now Brno in the Czech Republic) was designed for electric lighting exclusively and had no facilities for gas illumination. The plant consisted of four K dynamos located in a separate machine house. Underground conductors similar to those installed in New York's first District connected the dynamos to 1,600 lamps inside the theater, of which only 1,000 were expected to be lit at any given time. The plant began operation in mid-November 1882. Batchelor intended it to serve as a model to demonstrate the Edison system for theater lighting, and it generated a great deal of publicity. Francis Jehl, who oversaw the plant's installation, also started a small lamp factory in this city in early 1883 in order to preserve Edison's lamp patents in the Austrian empire. Jehl 1882b; Jehl to TAE, 29 Oct. and 30 Nov. 1882; Jehl to Insull, 22 Mar. 1883; all DF (*TAED* D8238ZEC, D8238ZEQ, D8337ZAS; *TAEM* 62:552, 577; 67:686); Bulletins 12:12 and 16:12, Edison Electric Light Co., 27 July 1882 and 2 Feb. 1883, CR (*TAED* CB012, CB016; *TAEM* 96:728, 789); Batchelor to Jehl, 29 Nov. 1882; Batchelor to TAE, 15 Dec. 1882; Cat. 1239: 378, 405; Batchelor (*TAED* MBLB4378, MBLB4405; *TAEM* 93:734, 756).

7. Batchelor was referring to an improved galvanometer devised by Arsène d'Arsonval and Marcel Deprez in 1880. Batchelor first saw this galvanometer at the Paris Electrical Exhibition in the fall of 1881 and began using one for lamp and dynamo testing in August 1882 at the behest of Francis Jehl. Batchelor sent one to Edison in early December. Batchelor to TAE, 22 Aug. and 2 Dec. 1882; Jehl to TAE, 1 Sept. 1882; all DF (*TAED* D8238ZCI, D8238ZEV, D8238ZCS; *TAEM* 62:463, 584, 479); Batchelor to Clarke, 26 Aug. 1882, Cat. 1239:330, Batchelor (*TAED* MBLB4330; *TAEM* 93:700); Atkinson 1893, 809–17; *DSB*, s.vv. "Arsonval, Arsène d'" and "Deprez, Marcel."

8. Insull had asked Batchelor to try to arrange for Edison to acquire U.S. manufacturing rights for the Berthoud Borel insulated under-

ground cable (29 Sept. 1882, Lbk. 14:187 [*TAED* LB014187; *TAEM* 81:901]). Having evidently declined to participate in a new company to manufacture it, Edison had had tried to obtain a U.S. license in June through an associate in Switzerland. He also tried during the summer to interest the Ansonia Brass and Copper Co. in the project (Frederic Esler to TAE, 11 Jan. 1882; Daniel Lord to TAE, 13 July 1882; both DF [*TAED* D8236A, D8236ZAC; *TAEM* 62:91, 125]).

9. Armington & Sims licensed the Edison Machine Works to build their steam engines for central station use for a royalty of $200 each. Armington met with Batchelor in Paris on 7 October to discuss a similar licensing arrangement, and Batchelor and Edison exchanged telegrams that day. Edison directed Batchelor to offer a royalty of 8% on the selling price. Although no agreement was reached before mid-1883, the company allowed Batchelor to build their engines on the same terms as the Machine Works. Batchelor to TAE, 9 Oct. 1882; Pardon Armington to Edison, 30 Jan. 1883; all DF (*TAED* D8238ZDQ, D8322J; *TAEM* 62:532, 66:188); Batchelor to Armington & Sims, 2 Dec. 1882, Cat. 1239:389; Batchelor to Bailey, 3 July 1883, Cat. 1331:67; Batchelor (*TAED* MBLB4389, MBLB3067; *TAEM* 93:744, 353).

10. Before leaving for Europe in the summer of 1881 Seubel instructed Edison to send his wife in Canton, Ohio $30 a month out of his salary. Mrs. Seubel often received these payments late. Philip Seubel to TAE, 21 June 1881; Batchelor to TAE, 5 Oct. 1881; Lina Seubel to TAE, 23 Feb. 1882; all DF (*TAED* D8135O, D8135ZCF, D8243222; *TAEM* 58:900, 1087, 63:148); Batchelor to Seubel, 22 Nov. 1882, Cat. 1239:364; Batchelor (*TAED* MBLB4364; *TAEM* 93:724).

11. See Doc. 2356.

–2352–

From George Bliss

CHICAGO, Octr 17 1882.[a]

Dear Sir:

This will introduce Mr. S. L. Smith[1] who is interested in some copper property adjoining and on the vien of the celebrated Calumet & Hecla mine at Lake Superior.[2]

I am told the mine has been well opened and is in about the same condition of the Calumet shortly before it became so largely remunerative.

Mr. Smith's associates are putting in 500 Edison lights at Appleton Wis and expect to put in some thousand in that vicinity as they control an immense water power extending several miles along the Fox river.[3]

They are impressed with the need for copper conductors and think they have a source from which the raw material can be obtained to great advantage[4]   Sin. Yrs.

Geo. H. Bliss

I have known Mr Smith for several years as a reliable gentleman of who has had great business experience.

ALS, NjWOE, DF (*TAED* D8245W; *TAEM* 63:256). Letterhead of Western Edison Light Co.; "Introducing Mr. S. L. Smith" written in preprinted memo line. ᵃ"CHICAGO" "188" and "." preprinted.

1. Bliss likely referred to Augustus Ledyard Smith (b. 1833), an organizer of the Edison Electric Illuminating Co. of Appleton. Smith was president of the First National Bank of Appleton and a principal in the Appleton Iron Co. He previously had been associated with the Fox and Wisconsin Improvement Co., which worked to improve navigation on the Fox and Wisconsin rivers. Aikens and Proctor 1897, 174–76.

2. Smith's mine was located between the Calumet and Hecla and Osceola mines on Lake Superior. Bliss to Insull, 14 Nov. 1882, DF (*TAED* D8241ZAG; *TAEM* 63:126); Benedict 1952.

3. In the summer of 1882 Augustus Smith, his brother Henry Daniel Smith, H. J. Rogers, and Charles Beveridge organized an illuminating company in Appleton. Some accounts give the first date of operation as 15 August, nearly three weeks before the startup of New York's Pearl St. station. However, the K dynamo was not tested until 27 September and did not work properly until Western Edison engineers insulated the wires. It began working successfully on 30 September, powering 250 lamps in the Appleton Paper and Pulp Co. mill (owned by Rogers), the Vulcan paper mill, and the Rogers residence. The dynamo was driven by water power, which caused wide fluctuations in dynamo voltage until a gearing arrangement was devised to reduce variation in armature speed. A second 250-dynamo was installed in November to light the homes of the Smith brothers. Because it was the first hydroelectric power station in the United States, a group of three engineering societies declared it a National Historic Engineering Landmark in 1977. Edison Electric Light Co. Bulletins 13:15 and 15:19, 28 Aug. and 20 Dec. 1882; both CR (*TAED* CB013, CB015; *TAEM* 96:738, 766); A. C. Landstadt address to

Appleton's 250-light station which entered service on 30 September.

Wisconsin Electrical Assoc., 24 Mar. 1922, EP&RI (*TAED* X001FA); Aikens and Proctor 1897, 123–24, 174–76; "Did Not Work," *Appleton Crescent*, 30 Sept. 1882; "Electric Light," ibid., 7 Oct. 1882; "The Electric Light," *Appleton Post*, 5 Oct. 1882; IEEE Milestones, Vulcan St. Plant, http://www.ieee.org/web/aboutus/history_center/vulcan.html.

4. On 24 November Smith urged Insull to "impress upon Mr E's mind the importance of early advising Mr. Stager & Mr Bliss at Chicago of his views in regard to our copper project & the general advantage of a secured uniform supply from Lake Superior." He promised that the mine would yield "a very large profit" as well. Smith to Insull, 24 Nov. 1882, DF (*TAED* D8245X; *TAEM* 63:257).

–2353–

To Archibald Stuart

[New York,] 20th Oct [188]2

Dear Sir,

Referring to your letter of 14th inst[1] your understanding of our Village plant system is correct. The intention is to supply a Central Station System for Village use at a cost which will enable the Investor to make money even when the demand for light is comparatively slight.

We are putting the plant into operation at Roselle and I expect it will be running in about ~~two or~~ three or four weeks

Yours truly

Thos A Edison   I[nsull]

L (letterpress copy), NjWOE, Lbk. 14:308A (*TAED* LB014308A; *TAEM* 81:950). Written by Samuel Insull.

1. Doc. 2350.

–2354–

To Charles Batchelor

[New York,] 25th Oct [188]2

My Dear Batch,

Yours of 9th inst came to hand this a.m.[1]

I can scarcely enlighten you yet as to engines. I have coupled the engines at the Station (Porters) with the Device of which you have heard successfully, and this week I am going to test two Armington & Sims Engines with a load on, & see how they act under the same conditions as Porters Engines.[2] Then I shall be able to cable[a] as to which engine we will use. I certainly do not think it will be the Porter. Immediately I decide I will send you the necessary drawings.

I think the best thing your people can do is to pay Seimens a royalty in Germany. The English Co pay him a royalty of 6% on the selling price but if your people make any such arrangement you must see that it is only for the life of the Armature Patent. Seimens will try to make the license for all his patents but do not allow this or you will have to pay royalty for a considerably longer period than if the license is confined to the Armature patent.[3]

I will have the information you ask for got out & sent you as early as possible   Yours truly

Thos A Edison   I[nsull]

L, NjWOE, Lbk. 14:331 (*TAED* LB014331; *TAEM* 81:962). Written by Samuel Insull. [a]Obscured overwritten text.

1. Batchelor to TAE, 9 Oct. 1882, DF (*TAED* D8238ZDQ; *TAEM* 62:532); see Doc. 2351 n. 4.

2. Edison told a reporter about this time that after the near–disaster occasioned by placing more than one of the Pearl St. dynamos on the line (see Doc. 2343), he had been forced to

> invent some plan by which all the engines could be made to run at the same speed, so as to make them practically one engine. And that is just what has occasioned the delay. As many of our customers had dispensed entirely with gas and were relying upon us for light, it was necessary for us to furnish them, and hence we were forced to manufacture some electricity. I at once applied myself to the task of perfecting an apparatus whereby all the automatic regulators of the several engines could be connected, and this plan was soon perfected. But it has taken a month to construct and perfect the apparatus. It is now completed and in perfect order, and it works to perfection. We are turning on lights every day and I can safely assert that all obstacles have been overcome, and that the success of the district system has been placed beyond doubt or question. ["The Edison System. How the Inventor Overcame an Unforeseen Difficulty," *Operator,* 13 (1882): 528]

3. Batchelor advised in his letter (see note 1) that Siemens was planning to sue Edison for infringing dynamo patents. He also thought that Edison also had grounds to sue Siemens for infringing lamp patents. Batchelor recommended paying Siemens "a small royalty" on dynamos to avoid litigation. The Edison interests entered into a comprehensive cross-licensing agreement for continental Europe with Siemens & Halske in March 1883. Joshua Bailey to TAE, 10 Nov. 1882, DF (*TAED* D8238ZEE; *TAEM* 67:644); see Doc. 2392.

[New York, c. October 27, 1882[1]]

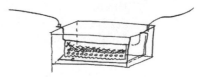

Dyer=
New patent.[2]
I take cloth with various figures, Lace, baskets[a] of wicker-
work or any article of manufactured made of Carbonizable
material, Carbonize the same under strain pressure or both to
prevent distortion, & make them into articles of carbon still
preserving their form. I then connect them to the poles of a battery or other source of Elec in an electrolylic bath & deposit
Copper, Gold or other metals over their entire surface of the required thickness. They are then taken from the bath & straightened if distorted and the same is a new article of manufacture
by a new process useful for ornamental & other purposes
Claim. NA new article of manufacture Electroplated Carbonized good.
2nd   process [----][b]
3rd   Carbonizing woven or plated goods or a articles from
manufactured from Carbonizable materials & Electroplating
the same after Carbonization
Etc   Etc
T A Edison   [Witness:] H. W. Seely

ADS, NjWOE, Cat. 1148, Lab. (*TAED* NM017:35; *TAEM* 44:395).
[a]Obscured overwritten text. [b]Canceled.

1. Henry Seely signed and dated the document on this date when he
received it from Edison.
2. Neither a patent nor an abandoned patent application has been
found. However, Edison had previously worked on similar ideas for
manufacturing miscellaneous products like this earlier and did so again
later in his career. Doc. 829 n. 10 and 11; Israel 1998, 266–70.

New York, October 30th. 1882.

My Dear Batchelor:—
Your favor of the 16th. inst. came duly to hand.[1] With reference to Fabbri's option on European stock, I do not know
whether he will take it or not. It will greatly depend I think on
the market price of the stock. The terms of the option are as
follows; Edison is to give Fabbri notice that he considers the
first station in successful operation and Fabbri is to have the

right to take the European stock at par within three months of that date. Now, Edison has not yet given Fabbri that notice.[2] He has been waiting to get two or three dynamos running together. This he has successfully done experimentally, and we shall have them running next week with the lamps on and as soon as[a] they have been going 3 or 4 days I propose suggesting to Edison the necessity of at once giving Fabbri the necessary notice. It is my personal impression that Fabbri will exercise this option and take the stock. I have not got any definite grounds for this idea except that Winslow, Lanier & Co. who are interested with Mr. Fabbri in the syndicate which took the first block of stock busied themselves considerably about European Light Co. affairs when Bailey was here.[3] I rather fancy that Drexel, Morgan & Co. will endeavor to get the control of the European Co. when they see that the first district is going to be a success. Bailey has some ideas as to this matter as they had frequent conversations with him on the subject. You might sound him about it but do not tell him that I have given you this information as I promised him I would say nothing whatever about it. However, as Edison is my original source of information and not Bailey I have a perfect right to tell you what I know about it as long as I know that Edison is willing that I should.

Bergmann & Co. are in full occupation of their new establishment and it is running splendidly. It is undoubtedly one of the most elegant factories in New York. They started moving two weeks ago yesterday (Saturday) and on the Thursday following had all their men at work again in the new place. They are preparing to double up their capacity in most departments and if your people propose ordering much from them I think you will be served far more promptly than heretofore. They are going to make money undoubtedly and I think your 10 per cent interest on Edison's one third [----][b] share will bring you in dollars at least during the first year of the partnership.

Goerck street is pretty hard pressed just now. We have about 160 to 180,000 dollars worth of stuff on hand and if you can send us some orders for L,[a] Z, or C dynamos you will be doing us a very great favor.[4] The Lamp Factory is going along pretty[a] well. They are not making money on their lamps but I do not think they are losing any now. I have got Mr. Lowrey to agree to a clause in the new contract providing that the Lamp Company shall have 40 cents for every lamp and we are to settle the details of the contract one day early next week.[5] When it is all settled, signed and delivered I shall send you a copy for your information.

Menlo Park is looking terribly desolate. Everything except the Brown engine and boiler and dynamo machines necessary to run the railroad is being brought in. Library and office affairs are all cleared out and the whole contents of the laboratory from floor to attic are being sent in to[a] Bergmann & Co.[6] Bergmann & Co. are to buy the machinery from the machine shop at Menlo Park and are going to do Edison's experimenting for him, his laboratory being on the top floor of their building.[7] I think this is about the last of Menlo Park. I am sorry for you owners of real estate there. Of course, the railroad experiments will still be continued out at Menlo. But this cannot last more than six months or at most a year. Did I write and tell you that we have got out a model and are now completing a locomotive for England which will be the size of an ordinary Penn. steam locomotive?[8] It will be built on exactly the same plans as an English locomotive and to look at the model that is what you would think it is.

Johnson has been elected Vice President of the Light Company Eaton being President.[9] Of course, Eaton's functions will be about the same as they have been heretofore. It gives Johnson an official standing and I suppose he will take about as much interest in the business as he has been doing since he returned home. This is looking after things in general in Edison's behalf, criticising work and suggesting improvements whenever necessary. We all think this a very good move. Johnson went in as the nominee of Mr.[a] Edison and Drexel, Morgan & Co.

The Isolated Company still continues to thrive. I think that on the first of December they will declare a dividend of 10 per cent.[10] They are getting orders every day and their plants give entire satisfaction. The first district is still running with one dynamo. Edison has got that rigging for the coupling device all fixed up and all last week the boys were practising throwing in and out machines.[11] He promises to put on over a thousand more lights tomorrow or Tuesday and another thousand will follow very soon after. From the figures I do not think it will be more than a month before instead of the first station being a drag on the resources of the Illuminating Co. it will be a source of income to them. Of course not very much at first but they will come all right after. It will be a great thing to be able to make some money three months after the first station is started.

ENGLISH AFFAIRS. English Light is getting into a delightful pickle. We have had some very sharp correspondence indeed with the English Co.[12] In fact about the sharpest business correspondence that I ever had anything to do with. It has ended by Mr. Arnold White starting for America with the object of trying to smooth things over. The trouble is though that A. White is not capable of running an electric light company. I told Johnson this at the time of his appointment.[13] Johnson admitted the truth of it and said that A. White was to be Secretary and not business manager. Unfortunately he is also the latter and still more unfortunately old Mr. Bouverie thinks White is capable of filling the position. But from private information Johnson gets from London[14] it would appear to us that the whole business is being bulled. Do you hear anything on this subject?

SWITZERLAND. We hear that the Societe Electricque has appointed agents in Switzerland and proposes to do business there. We complained of this to Bailey when he was over here, and stated that [Mr.?][c] Edison had made an agreement long before the French Companies were formed giving the right to use his name in Switzerland.[15] The action of the French companies reflects discredit upon Edison as it gives Lourrettine[16] the impression that we are instigating this movement from here. Now this is not so and Edison feels very much displeased about the matter. Bailey promised to put it right. Would you mind seeing that he does?

With kind regards to Mrs Batchelor, believe me, Very sincerely yours,

TL (carbon copy), NjWOE, Scraps., Cat. 2174 (*TAED* SB012BBL; *TAEM* 89:467). Underlining and marginalia added by W. J. Hammer in 1898 not transcribed. [a]Mistyped. [b]Canceled. [c]Illegible.

1. Doc. 2351. In the present letter, Insull referred to subjects discussed in Doc. 2343; see also headnote, Doc. 2343.

2. This agreement, reached in July 1881, gave Egisto Fabbri the right to buy 1,690 shares of the Edison Electric Light Co. of Europe, "at any time at his option between now and within three months from the date when the first Electric Light Station in New York City . . . shall have been in successful operation," with the provision for written notice added to prevent any ambiguity about when the three-month period would begin. The editors have not found such a notice to Fabbri by Edison (Samuel Insull memorandum, 6 July 1881; John Tomlinson memorandum, 6 July 1881; Fabbri to TAE, 7 July 1881; TAE to Fabri, 8 July 1881; TAE agreement with Fabbri, 8 July 1881; all Miller [*TAED* HM810149, HM810149A, HM810150, HM810151, HM810152; *TAEM* 86:404, 406, 407, 408 , 409]). Batchelor was to sell 10% of the shares to Fabbri, and Edison the remainder. Fabbri evidently had the 1,690 shares in his possession at the end of March 1883, when he transferred them to Drexel, Morgan & Co. (Insull to Batchelor, 6 Feb. 1883, Lbk. 15:264 [*TAED* LB015264; *TAEM* 82:134]; Edison Electric Light

Co. of Europe stock transfers, 31 Mar. 1883, DF [*TAED* D8331I; *TAEM* 67:18]).

3. See Doc. 2343 n. 13.

4. The Machine Works sold only 3 Z and 5 L machines in the month ending 7 November, leaving in stock several dozen of each kind. Edison Machine Works inventory list, 7 Nov. 1882, DF (*TAED* D8234M; *TAEM* 62:23).

5. Insull corresponded with Grosvenor Lowrey about completing a contract for the lamp factory but no formal agreement was reached at this time. Insull to Grosvenor Lowrey, 14 Nov. and 21 Dec. 1882, Lbk. 14:421, 15:58 (*TAED* LB014421, LB015058A; *TAEM* 81:1006, 82:54); see Doc. 2395.

6. John Ott packed and shipped the library and laboratory contents in the second half of October. Ott to TAE, 17, 23, and 26 Oct. 1882, DF (*TAED* D8244ZAA, D8244ZAD, D8244ZAG; *TAEM* 63:209, 211, 212); TAE to Ott, 18 Oct. 1882, Lbk. 14:302 (*TAED* LB014302A; *TAEM* 81:947).

7. In December, Bergmann & Co. submitted a request for $2,500 in annual rent for the 6th floor of their building, prompting Insull and Sherburne Eaton to demand a fuller accounting of the basis for the charge; Insull also wrote to Eaton a few months later about timely payment of the weekly laboratory bills. Bergmann & Co. to TAE, 18 Dec. 1882; Eaton to TAE, 19 Jan. 1883; both DF (*TAED* D8201Y, D8369A; *TAEM* 60:31, 70:868); Insull to Bergmann, 28 Dec. 1882; Insull to Eaton, 23 Feb. and 10 Mar. 1883; Lbk. 15:86, 342, 445 (*TAED* LB015086, LB015342, LB015445; *TAEM* 82:67, 178, 219).

8. This project was designed by L. Porsh, a draftsman who had worked previously for Edward Johnson in England. The parts were fabricated by the Cooke Locomotive & Machine Co. in Paterson, N.J. Edison submitted bills for the work to his financiers in September 1882 and February 1883, but around this time the project was suspended and Porsh left Edison's service. Cooke Locomotive & Machine Co. to Porsh, 29 July 1882; Cooke Locomotive & Machine Co. to Edison Electric Light Co., 4 Aug. 1882; TAE and Porsh specifications, n.d. [1883?]; Porsh to Edison, 29 Jan. 1883 and 7 May 1891; all DF (*TAED* D8249ZAP, D8249ZAQ, D8372N, D8372A, D9111AAG; *TAEM* 63:620, 621; 70:1116, 1090; 130:969); TAE to Drexel, Morgan & Co., 13 Sept. 1882 (copy), Cat. 2174, Scraps. (*TAED* SB012BBC; *TAEM* 89:445); TAE to Porsh, 1 Feb. 1883; TAE to Egisto Fabbri, 15 Feb. 1883, Lbk. 15:220, 324 (*TAED* LB015220, LB015324; *TAEM* 82:112, 171).

9. "Annual Meeting of Stockholders of Light Company," 24 Oct. 1882, Edison Electric Light Co. Bulletin 15:37, CR (*TAED* CB015235; *TAEM* 96:784).

10. Company secretary Calvin Goddard announced this to shareholders in a printed circular, and Fabbri requested that it also be advertised in New York newspapers. Edison Co. for Isolated Lighting report, 6 Dec. 1882; Eaton to Edison, 8 Dec. 1882; both DF (*TAED* D8221ZBA, D8221ZBB; *TAEM* 60:918, 919).

11. See Doc. 2354.

12. See Doc. 2339.

13. See Doc. 2221.

14. See, e.g., Doc. 2342.

15. See Docs. 1962 and 1985 n. 5. Ernst Biedermann and Théodore Turrettini had twice cabled Edison that the Paris company was encroaching on their Swiss rights from Edison (30 Sept. and 26 Oct. 1882, LM 1:252C, 259A [*TAED* LM001252C, LM001259A; *TAEM* 83:998, 1001]). On the same date as this letter, Edison instructed the Société Électrique: "you must suspend operations Switzerland   Bailey explain"; he also informed Turrettini that "Bailey promised withdraw Paris company." However, at the end of December, Turrettini complained that agents of the French company were still active in Switzerland and that the British company had been selling lamps there as well. He requested a formal statement, which Edison provided in February 1883, affirming the exclusive Swiss rights of the newly-organized Société d'Appareillage Électrique of Geneva (TAE to Société Électrique Edison, 30 Oct. 1882; TAE to Turrettini, 30 Oct. 1882; LM 1:261A, 261B [*TAED* LM001261A, LM001261B; *TAEM* 83:1002]; Turrettini to TAE, 25 Dec. 1882; TAE agreement with Société d'Appareillage Électrique, 13 Feb. 1883; both DF [*TAED* D8249ZBN, D8337S; *TAEM* 63:648, 67:619]).

16. Typist's error for "Turrettini." Théodore Turrettini (1845–1916) was a Swiss electrical engineer and a director of the Société Genevoise pour la construction d'instruments de physique et de mécanique, with which Ernst Biedermann had made arrangements to manufacture Edison lighting equipment for sale in Switzerland. Turrettini had previously visited Edison at Menlo Park in November 1880, and was apparently involved with an installation of Edison lamps at the Rothschild home at Pregny, near Geneva. In September 1882, he made plans to build electric railroads around Geneva, for which Edison had just sent him drawings and instructions. See Doc. 1962; Turrettini to TAE, 6 May 1882, DF (*TAED* D8238ZBB; *TAEM* 62:394); Turrettini to TAE, 1 Sept. 1882, LM 1:248B (*TAED* LM001248B; *TAEM* 83:996); TAE to Turrettini, 27 Oct. 1882, Lbk. 14:345 (*TAED* LB014345; *TAEM* 81:966); Favre 1923, 43–44, 78.

–2357–

Draft Memorandum to Edison Electric Light Co., Ltd.

[New York, October 1882?[1]]
Board of Trade Provisional Orders— Dr. Hopkinsons Report Thereon.[2]

I have carefully read the Requirements of the Board of Trade & Dr. Hopkinsons Report relating thereto, and am forced to the conclusion, after perusal of the latter, that Dr. Hopkinson fails entirely to appreciate the nature of the business upon which your Company is about to enter. It must be remembered that your competition ~~is~~will not be with rival Electric Light Companies—it will be with the Local Gas Companies. You propose to enter the lists ~~with~~ against[a] them in the business of public illumination. You will have to cater for the public and in order to get their custom you must be prepared to do it ~~on a basis~~ in ~~such a manner as to enable the~~ your ~~Customers to~~ a manner as near as possible like that at present adopted by the Gas Companies. You must charge your customer for exactly what he uses, the basis of that charge must be elastic enough to allow ~~only~~ of his paying only ~~at~~ for[a] the time he uses it. Furthermore your charges must be uniform whether a man uses one or a hundred lights. You must be prepared to supply light at all times whenever your customers choose to take advantage of your ~~light~~ system[a] and in order to do this you must never take the pressure off your mains for an instant.[3]

I consider[b] Dr Hopkinsons proposed basis of charge entirely erroneous. I fail entirely to see the necessity of[b] making two different classes of charges.[4] ~~What your Customers require is light I think~~ The proper ~~cost~~ course[a] to adopt is[b] to estimate what your current will cost you add to that the maximum amount of profit you ~~pr~~ require and then undertake to supply a given number of units of light (~~candles~~)—(standard candles)—~~for a~~ and [-][c] (not a given number of units of electricity). for a given sum[d] with ~~gas the charge is for a given number of~~ The Gas Companies takes units of gas (feet) as the basis of charge irrespective of the amount of light that gas gives. Theoret~~a~~ically of course a given amount of gas will give a certain amount of light but in practise it never does so owing partly to uneven pressure ~~often~~ sometimes[a] in the mains sometimes in the house gas pipes, ~~deterioration of burners~~[e] always to the deteoration in the gas burners. Now our basis of charge must in this respect differ from that of the Gas Companies. Dr Hopkinson is entirely wrong[a] in stating that units of ~~ligh~~ electricity should be the basis of charge. To adopt this course would greatly militate against the future ~~of~~ advancement of your Companies. Units of light (candle power) must be the basis of charge & not units of electricity. If in the future improvements are made in the economy of our lamps the public would be the only ones ~~to ha~~ who would have the advantage of it if Dr Hopkinsons proposal were carried out; whereas if units of light are taken as the basis of charge you will reap the advantage of ~~a~~increased economy

As to Dr Hopkinsons proposal "that "each consumer should be required to declare beforehand the maximum current he would wish to draw & that the current should be cut off immediately that amount is exceeded"[5] of course has no point if you ~~adm~~ refuse to admit the soundness of his views as to proposed basis of charge.[6]

~~I think~~ It should be provided that we should be allowed to use any pressure on our mains[f] up to 200 volts we undertaking never to have a variation of more than seven volts. The limits of our pressure should not be from 103 to 110 volts as at some future time we may find it to our advantage to ~~run~~ work our system at a higher ~~or a lower~~ pressure   I think however the limit of 200 volts will answer all practical purposes.[7]

There is not the slightest necessity to demand a certain specified[a] time during[b] which pressure can be taken off our mains for the purpose of Electrical testing. My Central Station here has now been running continuously since September 2nd without stopping for one instant and all electrical tests are made that are required while running[g] & this is the way your local Illuminating Companies must run as they must at all times be ready to supply light—with not even an intermission of ~~two minutes~~ an instant[h] per day.

I do not think you should require more than an ordinary notice from a would-be-consumer that he requires your light ~~pro~~ nor should you require of him an undertaking that he will use the Electricity for a certain time.[8] ~~All   What~~

What you have to do is select a district of say a mile square canvass it thoroughly so as to arrive at the amount of gas used within that District & then you can estimate the maximum amount of light you could under the best circumstances get customers. Then choose your central station ~~figure~~ determine[i] on the size ~~requ~~ of your mains & feeders & when you have installed your plant supply those with light who are prepared to pay the expense ~~of bein~~ (cost of house wiring & connecting with mains) of being cut in, supply light on a meter taking units of light as the one and only one basis of charge & when your customer is dissatisfied take the light out— In short perform the functions of a Gas Company—the only difference being that your basis of charge is units of light (candle[b] power) while that of the Gas Company is units of gas (feet).[9] This is the only practical & satisfactory way to do the business—to persue any other course will involve you in hopeless trouble with the public & be an absolute failure[j]

Dr Hopkinsons remarks that the prime cost of our machinery is probably higher than that of our competitors are erroneous.[10] He surely must have forgotten all that Mr. Johnson told him as to my system being especially devised for Central Station lighting. You have no competitor in this Dept of your business. Your investment in copper will be but half of that of any rival E. L. Co. your investment in boilers & Dynamos will be far less than that of any others when you cons[id]er[k] the economical results to be obtained from them. If you have a com-

petitor why have not other Companies made such an installation as you have on Holborn Viaduct which is Central Station Lighting on but a very small & imperfect scale[l] scale. The only competitor you have to fear is the Gas Company. Leave all other Electric Light Cos out of the question.[m]

On reading Dr Hopkinsons report I am forced to the conclusion that your company has no one in its service who understand the first rudiments of Cental Station lighting and if you wish to make a success of this Dept you cannot too early select[b] an engineer (not necessarily an electrician)[n] of ener[g]y & ability & with[a] no prejudisces & send him over here to get a thorough knowledge of that Department of Electric Lighting to which my System particularly applies viz Central Station Lighting[11]

Df, NjWOE, DF (*TAED* D8239ZFS1; *TAEM* 62:1174). Written by Samuel Insull. [a]Interlined above. [b]Obscured overwritten text. [c]Canceled. [d]"~~for a~~ given sum" partly enclosed by looping line. [e]"~~deterioration of burners~~" interlined above by Edison. [f]"on our mains" interlined above. [g]"while running" interlined above by Edison. [h]"an instant" interlined above by Edison. [i]Interlined above by Edison. [j]"& failure" interlined by Edison. [k]Obscured by ink blot. [l]"& imperfect scale" interlined above by Edison. [m]"Leave question." interlined by Edison. [n]"(not electrician)" interlined above.

1. Arnold White sent a copy of John Hopkinson's report to Edison on 22 September, with a request that he reply to the British company's directors. Edison likely drafted this response sometime in October. No copy of a final, mailed version has been found, suggesting that he may have discussed these matters during White's visit in November and December (cf. Docs. 2374 and 2375). White to TAE, 22 Sept. 1882; Hopkinson report (copy), 9 Sept. 1882; both DF (*TAED* D8239ZEW, D8239ZEX; *TAEM* 62:1086, 1087).

2. The Electric Lighting Act, which took effect in August, was accompanied by a set of "Rules made by the Board of Trade with respect to applications for Licenses and Provisional Orders, &c." These specified the bureaucratic processes for companies to obtain approval to provide electricity in the district of a local authority. Although most of the rules covered procedural and technical questions, one required a statement of the proposed "Conditions of supply, including price, nature and amount of supply, obligation to supply, &c." This was the only item addressed in the copy of Hopkinson's report sent to Edison. Hopkinson noted (see note 1) that the company "should begin to consider what conditions we would wish to propose and what we should be prepared to accept. Although it is probable that further knowledge may alter even my own views it may not be altogether useless to note down conditions which appear most appropriate to day. They may serve as a basis for discussion." His report was thus more a discussion of central station business than an analysis of the Board's rules per se. He had expressed similar ideas in a June memorandum following his testimony before the

House of Commons Select Committee on the Electric Lighting bill. Board of Trade Rules, August 1882; Hopkinson to Edison Electric Light Co., Ltd., 7 June 1882; both DF (*TAED* D8239ZDZ1, D8239ZCE; *TAEM* 62:1045, 950); see Bowers 1990, 92–94.

3. Hopkinson proposed (see note 1) that the company promise to supply current "at all hours of the day and night with the exception of certain short specified times" for testing the mains.

4. Hopkinson argued (see note 1) that

> as it will in general be much more costly to supply a consumer who uses a large current of electricity for a short time than a consumer who uses a smaller quantity for a longer time it is essential that we should be empowered to divide the charge to the consumer into two parts, one part being a fixed rental proportional to the maximum current he is able to draw from the mains and based upon the dead expense involved in providing machinery and other plant to meet such possible demand, a second charge proportional to the quantity of electricity actually used by the consumer as registered by a Meter . . . and based upon the current expense in Coal, wages and oil involved in producing that electricity.

5. Edison directly quoted from the report only the passage "each consumer . . . to draw." Hopkinson proposed to install at each customer's premises "a fusible plug which would break down and cut off his supply if that maximum were materially exceeded."

6. Edison did not respond to Hopkinson's suggestion that reduced charges might be negotiated for highly predictable consumption, such as street lighting, and for usage when demand was low, such as motive power during the day.

7. Hopkinson recommended that the company promise in its application to provide current between 103 and 110 volts. He had also raised this issue in his June memorandum (see note 2).

8. Hopkinson stated that "The order will no doubt require us to supply any Consumer who may wish it within a specified distance from our main conductors but I think we should be protected by the requirements of an ample notice from such Consumer and an undertaking that he will continue to use the electricity for a reasonable time."

9. See headnote n. 10, Doc. 2163.

10. Hopkinson argued that the Edison company could nevertheless negotiate prices high enough to ensure "an ample margin of profit" because it could "offer the most stringent guarantees of efficiency."

11. This idea appears at the beginning of Doc. 2374, in which Edison reiterated his essential contention that electric light should be sold on the same terms as gas.

–2358–

From George Bliss

Chicago, Nov. 1 1882[a]

Friend Edison:

I hear you intend making a 750 Volt dynamo and presume it is for use in Connection with the Village plant system where 3 ten candle lamps are to be burned in series.

Such a current must be dangerous to life and would be a departure from the corner stone of your system viz safety to every mann, woman & child who use it.

I sincerely hope you wont do it.[1] Yrs truly

Geo. H. Bliss

ALS, NjWOE, DF (*TAED* D8241ZAA; *TAEM* 63:120). Letterhead of the Edison Co. for Isolated Lighting, Geo. H. Bliss, Agent. a"Chicago," and "188" preprinted.

1. Edison's reply is Doc. 2364.

–2359–

Memorandum to Henry Seely

[New York, c. November 1, 1882[1]]

Seely,

Write patent on method of render [---]a equalizing the Electromotive force of Dynamo machines driven by Gas Engines—[2]

Gas Engines are worked by explosions & The Speed is consequently irregular   The Engine accelerating at each explosion causes the light to increase in intensity   To obviate this I place upon the reciprocating rotatingb portion of the Engine an eccentric or Cam, which cam may be set in any position   a strap over the Cam carries a rod or arm which serves to close and openc an electric circuit.

The cam is so set that at the moment of explosion the circuit is closed & a number of incandescent lamps are thrown across multiple arc in a circuit whereon there are other Lights. The effect of adding these several lights or wire resistance equal thereto is to check the rise in Electromotive force due to acceleration due to explosion & thus render the lights constant notwithstanding the Engine runs unequally.

If anb Otto Gas Engine[3] is used the gas and air should always have the same properd proportion   as the amount of gas is diminished the relative proportion of the air should be diminished   in this manner an explosion can be made to take place at every 2dd revolution and the compensation due to throwing on lamps or resistance actse perfectly—

TheInstead of throwing resistance across multiple arc at the moment of explosion the circuit controller may throw resistance in the circuit & thus accomplish the same purpose

Wh In casesf Where explosions do not ocurr at every 2dd revolution but irregularly;e a cylind springd governor may be used the movement of the sleeve of which throws resistance in or out iof circuit or across multiple arc.

It is obvious that motion may be derived from any part of the gas Engine or attachments connected therewith to control an electric contact device whereby compensation for difference of speed may be attained=

~~Even an Electromagnet~~ The electric compensating device may be placed on any mechanism controlled or operated by the gas engine for instance on the shaft of the dynamo itself—

~~Even an Electromagnet with armature~~

Claim very broadly a circuit controlling device on a gas engine or mechanism controlled thereby for compensation for intermitting due to explosions

This is entirely novel & Prof Draper reports that it works beautifully   it is very valuable in England so prepare an English ~~French~~ & continental set=[4]

Get Schleisser & shums Catalogue[5] or cut from Clarkes Books[6]   I think he has a description of the Otto Engine with drawing in some of his books & on the rotating shafts that lifts the inlet cock put an eccentric

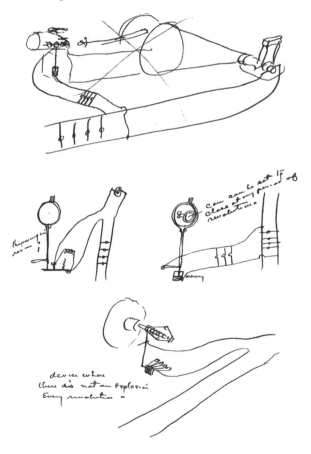

device where there ~~an~~is not an explosion every revolution=
~~If R~~

[Witness:] H. W. Seeley

AD, NjWOE, Cat. 1148 (*TAED* NM017:20; *TAEM* 44:380). Letterhead of Bergmann & Co. ªCanceled. ᵇObscured overwritten text. ᶜ"and open" interlined above. ᵈInterlined above, possibly by Edison. ᵉInterlined above. ᶠ"In cases" interlined above. ᵍ"but irregularly" interlined above, possibly by Edison.

1. Henry Seely signed and dated the document on this date when he received it from Edison.

2. Edison later claimed to have made this regulator for Henry Draper's dynamo, driven by a gas engine, to eliminate the flicker from the Edison lamps he planned to show to members of the National Academy of Sciences later in November. In early 1883 Edison heard that *New York Tribune* editor Whitelaw Reid credited Draper with having solved the problem. Edison denied this, claiming that his own compensating device had made Draper's event a "great success the lights being perfectly steady." Edison executed a patent application on 20 December for a variable resistance mechanism actuated by the engine valve shaft. Docs. 2286 nn. 2–3 and 2367; Sherburne Eaton to TAE, 5 June 1882; Pardon Armington to TAE, 5 Feb. 1883; both DF (*TAED* D8226Q, D8322L; *TAEM* 61:235, 66:190); TAE to Armington, 6 Feb. 1883, Lbk. 15:245 (*TAED* LB015245; *TAEM* 82:122); U.S. Pat. 276,232.

3. German engineer Nikolaus Otto (1832–1891) brought out a four-cycle internal combustion gas engine in 1876, apparently independently of Alphonse Beau de Rochas, who had patented a similar design in 1862. Otto's engine operated much more economically than others, and within a few years more than 35,000 were in use. Singer, Holmyard, Hall, and Williams 1958, 157–59; *WI*, s.v. "Nikolaus August Otto."

4. This was the subject of Edison's British Patent 1,019 (1883), filed on 24 February.

5. Schleicher, Schumm & Co. of Philadelphia, later known as the Otto Gas Engine Co., manufactured Otto gasoline engines under license. Mull 1969.

6. Not found.

–2360–

*Samuel Insull to
Seth Low*[1]

[New York,] 2nd Nov [188]2

Sir

I enclose you herewith permit to visit our Central Station in Pearl St in case you should wish to enquire into the nature of the work for the conduct of which in the near future there will be a great demand for Electrical Engineers

I will send you a copy of Mr Edisons letter to the Trustees of Columbia College[2] as soon as he has signed it   Respectfully
Saml Insull   Private Secretary

ALS (letterpress copy), NjWOE, Lbk. 14:377 (*TAED* LB014377A; *TAEM* 81:986).

1. Seth Low (1850–1916) was mayor of Brooklyn (1881–1885) and later of New York City (1901–1903). He became a trustee of Columbia College in 1881 and, in late 1882 and 1883, headed a committee of the Board of Trustees that explored Edison's offer to help establish a department of electrical engineering. Low and Edison were unable to schedule a mutually satisfactory meeting date until late January 1883. *ANB*, s.v. "Low, Seth"; *Ency. NYC*, s.v. "Low, Seth"; Low to TAE, 5 Dec. 1882, DF (*TAED* D8204ZJH; *TAEM* 60:407); TAE to Low, 25 Jan. 1883, Lbk. 15:199 (*TAED* LB015199; *TAEM* 82:106).

2. Edison's letter has not been found, but Columbia's President Frederick A. P. Barnard received it on 6 November. This was apparently the first direct correspondence between Edison and Barnard on this subject, though George Barker had acted as an intermediary. While affirming that he was "deeply interested" in the project, Barnard cautioned that trustees of the heavily-indebted College would not take on additional financial burdens. Barnard to TAE, 7 Nov. 1882, DF (*TAED* D8204ZIN; *TAEM* 60:374).

–2361–

Samuel Insull to David Burrell[1]

[New York,] 3rd Nov [188]2

Dear Sir,

In reply to your favor of 1st inst Mr. Edison has directed me to state that he will try some experiments on the Separation of cream from milk as soon as he can spare the time to do so.[2]

He has a great many suggestions made to him but few of them are of a practical character or refer to experiments the success of which would be of no advantage to anyone. It therefore affords him great pleasure to have his attention directed to a subject the solution of which would be advantageous to many and lucrative to himself.

Mr. Edison has asked me to remind him of the matter at some time when he has the time [to][a] try the experiment. If any result comes of it I will make a point of writing you.[3] Yours truly

Saml Insull   Private Secretary

ALS (letterpress copy), NjWOE, Lbk. 14:388 (*TAED* LB014388; *TAEM* 81:992). [a]Copied off edge of page.

1. David H. Burrell (1841–1919), son of a prominent cheese producer, was a principal in the dairy equipment supply business of Burrell & Whitman in Little Falls, N.Y. Credited with importing the first centrifugal cream separators to the U.S., Burrell made a number of other significant innovations and inventions in commercial dairying. *NCAB* 19:126.

2. Burrell asserted in his 1 November letter that nothing "is of greater importance or will prove more lucrative than some means for the instantaneous separation of Cream from milk." He asked Edison to undertake experiments with electric currents because "there is a difference in the specific gravity of Cream and Milk, that electricity, by its action upon the milk, possibly would cause the heavier parts . . . to settle to the bottom." He pointed out that even centrifuges could not operate on the scale he wished, and in any case consumed too much power for individual farmers to operate. Conventional practice was to chill the milk, often in water-cooled pans, until the cream rose. Burrell to TAE, 1 Nov. 1882, DF (*TAED* D8204ZIF; *TAEM* 60:359); McMurry 1995, 133–71; Willard 1877, 484–88.

3. Insull wrote to Burrell a few days later that Edison would like to meet him in New York; it is uncertain if this meeting took place. Burrell wrote several times in the next few months, asking with some urgency in April for news of any progress. Edison noted on that letter that "My man who was working on it been sick." Although Burrell corresponded sporadically until 1889, Edison apparently did not take up the subject in earnest. Insull to Burrell, 8 Nov. 1882, Lbk. 14:404A (*TAED* LB014404A; *TAEM* 81:1001); Burrell to TAE, 27 Apr. 1883, DF (*TAED* D8303ZCV; *TAEM* 64:166); see *TAED* and *TAEM-G2*, s.v. "Burrell, David H."

–2362–

*Samuel Insull to
Henry Draper*

[New York,] 3rd Nov [188]2

My Dear Sir,
    Mr Edison would be glad to know if you are a member of the Committee of Arrangements for the proposed dinner at Delmonico's to Mr Herbert Spencer[1]   Yours very truly
                                                    Saml Insull

ALS (letterpress copy), NjWOE, Lbk. 14:386 (*TAED* LB014386A; *TAEM* 81:991).

1. British philosopher and sociologist Herbert Spencer (1820–1903) championed individualism and opposed state interference in society and the economy. His application of Darwin's biological theories to contemporary social conditions made him a favorite of the American business classes. Spencer arrived in New York on 21 August for a three-month visit to the U.S. and Canada. *Oxford DNB*, s.v. "Spencer, Herbert"; Kennedy 1978; Russett 1976, 89–96; Beckert 1993, 212–13; Spencer 1904, 2:457–81.

Draper replied the next day that he was a member of the committee arranging the banquet to be held on 9 November. Although Edison admired Spencer, news accounts listing prominent participants did not mention him among the approximately 200 guests. Draper to TAE, 4 Nov. 1882; George Iles to TAE, 23 April 1883; both DF (*TAED* D8204ZIK, D8303ZCP; *TAEM* 60:367, 64:155); Doc. 1315, nn. 6, 7, 9; "Philosophy at Dinner," *New York Times*, 10 Nov. 1882, 5; Youmans 1973 [1883], 22–24.

[New York, c. November 6, 1882[1]]

Patent.[2]

I take several long thin fibres such as Ramie twist them together to form a thread[a] and secure their ends by a plastic compound of carbon & sugar or Hydrocarbon & carbonize the filiment under strain ~~or~~ pressure or both.—

This produces a filiment consisting of a great number of ~~seperate~~ individual but <u>continuous</u> fibres and gives a very elastic & even filiment

Claim   A flex[b] filiment for Carbonization formed of a number of continuous fibres.

2nd   A flexible[b] filiment of Carbon formed of a number of continuous fibres.

3rd   Fastening the Ends of such[c] an aggregated filiment by a plastic Carbonizable Compound.

4= A flexible filament of Carbon for giving light by incandescence[d] formed of a number of continuous fibres placed in a ~~glass~~ chamber made entirely of glass from which the air has been exhausted etc.

[Witness:] H. W. Seely

ADDENDUM

[New York, c. November 7, 1882[3]]

Seely

I gave you a patent to write where I used several thin continuous fibres= add—

The fibres may be laid straight and Carbonized and afterwards ~~the proper number are taken any~~ they[b] may be twisted together tightly by means of the ~~pl~~[b] Carbon Ends and while so twisted are clamped in the clamps upon the leading in wires of the lamp and may then be electroplated thereto while the fibres are under torsion. If the fibres have been twisted previous to carbonization they must be against twisted & placed under torsion while being sealed into the Lamp.

AD, NjWOE, Lab., Cat. 1148 (*TAED* NM017:15; *TAEM* 44:375). [a]"to form a thread" interlined above. [b]Interlined above. [c]Obscured overwritten text. [d]"for giving light by incandescence" interlined above.

1. Henry Seely signed and dated the document on this date when he received it from Edison.

2. This memorandum (including the addendum) served as the basis of a patent application that Edison executed on 13 November. The rationale was "to produce flexible carbon filaments . . . which shall be of high resistance and of even resistance throughout their length, and shall have great flexibility and toughness, so that they will not be liable to be

fractured by the expansion and contraction . . . which take place from the lighting and extinguishing of the lamp." U.S. Pat. 274,294; see also Doc. 2398.

3. The approximation of this date is based on contents of the addendum.

–2364–

To George Bliss

[New York,] 8th Nov [188]2

Dear Sir,

Referring to your favor of 1st inst[1] the machine I am making is a <u>300</u> volt machine to run three lamps in series having a total resistance of 750 ohms.[2] You can take right hold of the terminals without the slightest danger[3]   Yours truly

Thos A Edison   I[nsull]

L (letterpress copy), NjWOE, Lbk. 14:398 (*TAED* LB014398; *TAEM* 81:997). Written by Samuel Insull.

1. Doc. 2358.

2. Edison and Charles Clarke did preliminary design work to modify a K dynamo armature to generate a 300-volt current but such a machine was apparently not built. Instead, Edison generated high voltages by connecting 110-volt machines in series. This arrangement was first used in the Roselle, N.J., plant which began operating in January 1883 with three dynamos. A drawback to this design was that one switch controlled three lamps. Edison then devised a 220-volt three-wire distribution system that both reduced the amount of copper used in the distribution wires and permitted customers to control individual lamps. The first three-wire system entered service at Sunbury, Penn., in July 1883. TAE to Clarke, n.d. [1882] 1882, Hodgdon (*TAED* B017AC); Josephson 1992 (1959), 231–32; Jehl 1937–41, 1089–1120; Schellen 1884, 356.

3. Bliss replied that he was "a little suspicious of a 300 volt current but if you say it is safe shall rest content." Bliss to TAE, 11 Nov. 1882, DF (*TAED* D8241ZAF; *TAEM* 63:125).

–2365–

To George Bliss

[New York,] 8th Nov [188]2

Dear Sir,

Your favor of 1st as to Litigation came duly to hand.[1]

I do not at all agree with you as to the advisability of immediately bringing suit against Infringers. To do so would require me to give my personal attention to the matter & take me off other far more important work, besides involving us in a great deal of expense & giving our opponents a notoriety which it is hardly desirable they should gain at our expense.

Two years [ago?][a] the Directors of the Light Co wanted to

bring suit against the Maxim people. I objected giving the above reasons.[2] Today the Maxim people are worse off than they ever have been before   they have lost at least half a million dollars, their fine Factory is now used for our work & every lamp they put out is an expense to them whereas every one we put out is a profit to us. Or in other words we are doing a good paying business & they are doing none that is of practical commercial value.

You can show economy & give guarantees which they can never give & on this our business should be [-----][a]   When they [----- ---][a] affect our business then we shall have reason to sue them but so long as their work is conducive to their own ruin I see no reason for attacking th[em?][3b]   Yours Sincerely
T A Edison.

They are far more active in New England, but we plough right ahead, & have all we can do. Edison.[c]

LS (letterpress copy), NjWOE, Lbk. 14:399 (*TAED* LB014399A; *TAEM* 81:998). Written by Samuel Insull. [a]Faint copy. [b]Copied off edge of page. [c]Postscript written by Edison.

1. Bliss warned that "spurious" lighting companies would drain capital away from legitimate patent-users unless they faced aggressive litigation. He regretted the Edison company's participation in "that infernal" Gramme Electrical Co., which he said made its members' interests appear weak because it could not defend them (cf. Doc. 2286). Bliss to TAE, 1 Nov. 1882, DF (*TAED* D8241ZAB; *TAEM* 63:121).

2. Edison elaborated this point in reply to Sherburne Eaton's inquiry around this time concerning the advisability of suing the Maxim or Swan interests in the U.S. He stated that Maxim's business had lost $500,000 in incandescent lighting, and Swan's even more. He argued against "the necessity of our entering on costly & lengthy lawsuits against Infringers [whose] rivalry has in no wise affected our business. We should only be expending money and time which our business most certainly requires and give a prominence and publicity to Infringers who are not at present doing sufficient business to cause us the slightest alarm." TAE to Eaton, 2 Nov. 1882, Lbk. 14:383 (*TAED* LB014383; *TAEM* 81:989); cf. Doc. 2178.

3. Bliss acknowledged in reply that business was booming, but he warned that "If you don't want to be forced into consolidation and lose a large amount of the credit and profit in the Electric light field as was the case with the telephone put your foot on the enemy now." Bliss to TAE, 11 Nov. 1882, DF (*TAED* D8241ZAF; *TAEM* 63:125).

From Charles
Batchelor

Brüenn Austria Nov. 10 1882.

My dear Edison,

I acknowledge your cable of 3rd as follows:— "Armington engines work together perfectly   will send drawings base"[1] This eases my mind considerably— In the defect report for Oct 21st 1882 I notice that you are having trouble with your binding wires on armature— Every armature that you send me I always instruct my men to solder all binding wires all round before putting in and I never have any trouble with any of them   I think it is very important to have Dean do this on all—[2] I worked all night the first night I came here to do this although I had to take all armatures out & put them back again, since being here I have carefully watched these K's learning the man that has to mind them myself to set brushes and with three in circuit and 800 lamps on I have failed to see a spark at commutator yet. I consider this machine as perfection for 250 lamps— I have made some now and got them working in Paris they are equally as good as American— Everything works splendid at the theatre here and on Sunday we are to have a grand rehearsal and open monday— I have got Munich, Vienna, & Berlin theatre directors here and shall give estimates for lighting the principal opera houses in those Cities   As I have now something definite to base my calculations on better than before we shall be able to draw a revenue from these.— I note what you say on royalty to Siemen's   Yours

Batchelor

ALS, NjWOE, DF (*TAED* D8238ZED; *TAEM* 62:554).

1. This is the full text of Edison's cable on 3 November. LM 1:262C (*TAED* LM001262C; *TAEM* 83:1003).

2. Charles Batchelor had reported this problem with Z dynamos over a year earlier and had recommended that the binding wires be soldered more securely during manufacture (see Doc. 2179). In September 1882 Insull noted the same problem with the binding wires of C dynamos, attributing their breakage to the strong magnetic fields which attracted them and caused them to break at the point they were soldered. The 21 October defect report to which Batchelor referred described a loose binding wire which caused a short circuit between the field magnet and armature of a Z dynamo, causing it to burn out. Insull to Batchelor, 15 Sept. 1882, Lbk. 14:144A (*TAED* LB014144A; *TAEM* 81:890); Defect Report, 21 Oct. 1882, DF (*TAED* D8222 [image 49]; *TAEM* 60:932).

[New York, c. November 13, 1882[2]]

Sir.

Please say to the members of the Academy that if they desire to see Central Electric Lighting Station of the first district 255 and 257 Pearl Street they will be welcomed there at any time. I will myself be at that station on Thursday afternoon from 4 until 6 pm. to explain it in operation.[3] Respy

Thos A Edison

ALS, CtY, Marsh (*TAED* X121DA).

1. Edison addressed this letter to "President National Academy Sciences." Othniel Charles Marsh (1831–1899), a preeminent paleontologist and professor at Yale University, was the Academy's acting president since May 1882. He was elected president in April 1883, a position he held for twelve years. In 1878, Marsh had given Edison an official welcome to the meeting of the American Association for the Advancement of Science. *ANB*, s.v. "Marsh, Othniel Charles"; Cochrane 1978, 139–40; see Doc. 1406.

2. Edison evidently wrote this shortly before or during the National Academy of Sciences meeting in New York from Tuesday, 14 November to Friday, 17 November. "Explorations in Science," *New York Times*, 15 Nov. 1882, 3; "Recent Researches in Science," ibid., 18 Nov. 1882, 2.

3. Edison also attended a dinner hosted by Henry Draper for Academy members that featured an elaborate display of Edison lights (see Doc. 2359 n. 2). Draper had made a similar exhibition of Maxim lights to Academy members in 1880. "Prof. Draper's Guests," *New York Times*, 16 Nov. 1882, 2; Doc. 2022 nn. 3–4.

–2368–

*Samuel Insull to
Sherburne Eaton*

[New York,] 15th Nov [188]2

Dear Sir

By direction of Mr Edison I give you below the figures you ask for as to manufacturing in your memo of 8th inst.[1]

Electric Tube Co.

Investment in Plant	$23,000	
Completed Stock & Material	16,000	
		$39,000

Average monthly bills and pay rolls $24,000

Edison Machine Works

Investment in Plant	$198,000	
" Finished Stock	150,000	
" Unfinished "	35,000	
		$383,000

Average monthly bills and pay rolls $100,000

Edison Lamp Company

Investment Real Estate Buildings &
    Fixtures $54,665
      "   Machinery & Tools   79,412
      "   Finished & Unfinished Stock 25,000
      "   Sunk in Experiments   95,334

                                                  $214,411

Average monthly bills and pay rolls $15,000.00

Bergmann & Co

Investment in Real^a Estate $47,500
      "   Plant   55,000
      "   Finished & Unfinished Stock 53,000

                                                $155,500

In addition to the amount of cash invested by the Lamp Co in real estate their property cost a further sum of $30,000 for which amount they gave a mortgage. Bergmann & Cos property cost $40,000 beyond their cash investment & they gave a mortage for this amount   Yours truly

Saml Insull

ALS (letterpress copy), NjWOE, Lbk. 14:432 (*TAED* LB014432; *TAEM* 81:1009). Commas added to large numbers for clarity. ^{a}Obscured overwritten text.

1. Eaton asked Edison for this information in order to answer questions that he anticipated at the next meeting of the Board of Directors of the Edison Electric Light Co. Eaton to TAE, 8 Nov. 1882, DF (*TAED* D8224ZCE; *TAEM* 61:117); see also headnote, Doc. 2343.

–2369–[New York,] 17th Nov [188]2

To Giuseppe Colombo

Dear Sir,

Your letter of 10th October would have been answered before but for the fact that Mr Clarke has been away recuperating his health after the extreme exertion consequent on the experiments at the Central Station[1]

Mr Clarke says that Mr Paton[2] has kept you fully posted as to the work at the Central Station. Everything there is now running perfectly and we throw machines in and out without any trouble whatever.

The supplies for your work at Milan were pushed forward as rapidly as possible. There was great delay in the first shipments but this was owing to the bad steamer service to the

Mediterranean. Almost all your tubes have now been shipped. The work at Bergmann & Cos is almost finished. The regulation of the field apparatus will go tomorrow.[3]

I enclose you herewith memorandum with curves & calculations from Mr Clarke which will supply you with the information you ask for[4]

I also enclose you the contract for the Boilers. If you have not received them yet please communicate with Babcocks & Wilcox 107 Hope Street Glasgow Scotland.[5]

I think we will let the Dynamo matter stand just as it is.[6] The Société Electrique Edison have already taken two more large Dynamos from me. The two machines have attached to them the Armington & Sims Engines which are vastly superior in every way to the Porter-Allen Engines. The former do not require any coupling arrangements to be attached to the Regulator & govern absolutely perfect under the most varying conditions   The economy of running them will be considerably greater than that of Porter Allen— In fact the Armington Engine is <u>the</u> one thing requisite to make our Central Station complete

We have not had one single stoppage at our Central Station since we started Sept 4th. We have now 142 consumers connected with 3300 lights and we are cutting in about 100 lights per day.

Mr Lieb[7] leaves here for Milan on the Alaska on Tuesday next 21st inst. The terms I have arranged with him are as follows:— He is to receive his travelling expenses to & from Milan and while there the amount of his living expenses (not to exceed $15.00 per week) and a salary at the rate of $1500.00 per year. If after you have got through with him and he is required anywhere else some fresh arrangement as to remuneration must be made with. It is of course understood that I only <u>lend</u> Mr Lieb & if he should be required for other services after he has got through at Milan my permission must be obtained Yours truly

Thos A Edison   I[nsull]

L (letterpress copy), NjWOE, Lbk. 14:442 (*TAED* LB014442; *TAEM* 81:1015). Written by Samuel Insull.

1. Anxious to place the Milan central station into service in early January, Colombo asked Edison to ship the necessary equipment "with the greatest possible speed." He also posed several questions to Edison and Clarke about the operation of the Pearl St. station, the use of carbon rods for regulating a large group of lamps, and the proper means to cool dynamos. Colombo to TAE, 10 Oct. 1882, DF (*TAED* D8238ZDT; *TAEM* 62:540).

2. See Doc. 2343 n. 6.

3. In August 1882 Charles Clarke urged Edison that if the Milan "plant is to be gotten ready in Dec. Bergmann & Co. should immediately start on his regulating apparatus, which should include, large circuit breakers, field regulating resistances, switch board &c." While he was in New York in September, Colombo placed orders for the station's equipment, including underground tubes. Colombo and Joshua Bailey both insisted on receiving the same price that Edison himself paid for similar equipment. At the end of October Bailey cabled Edison to ask him to expedite the shipment of the conductors. Edison promised to send tubes on 5 and 16 November, and to ship the remaining equipment in weekly installments. Materials were still going to Milan in the spring of 1883, and the station did not open until the end of June. Clarke to TAE, 22 Aug. 1882; Colombo to Bailey, 12, 13 and 14 Sept. 1882; Bailey to TAE, 15 Sept. 1882; Kruesi to Insull, 27 Mar. 1883; Lieb to TAE, 5 Apr. 1883; all DF (*TAED* D8227ZBF, D8238ZDA, D8238ZDC, D8238ZDI, D8238ZDH, D8335J, D8337ZBB; *TAEM* 61:554; 62:497, 507, 515B17; 67:448, 696); Bailey to TAE, 29 Oct. 1882; TAE to Société Électrique Edison, 30 Oct. and 7 Nov. 1882; LM 1:260B, 261A, 263A (*TAED* LM001260B, LM001261A, LM001263A; *TAEM* 83:1002–1003).

4. The editors have not found the enclosures.

5. On 14 September Babcock & Wilcox sent Edison a contract and detailed specifications for four boilers rated at a total of 656 horsepower at a price of £2,098 for the Milan central station, the same discounted price that Edison paid. Babcock & Wilcox to TAE, 14 Sept. 1882; Colombo to Bailey, 12, 13 and 14 Sept. 1882; all DF (*TAED* D8238ZDE, D8238ZDB, D8238ZDC, D8238ZDH; *TAEM* 62:505–509, 515).

6. On 12 September Colombo confirmed in writing his verbal understanding with Edison that "the first of the new Dynamos of 1400 lamps now in construction shall be reserved for Milan and shipped as soon as possible." However, the Société Électrique Edison sent an older dynamo from Antwerp instead of the new model. Colombo restated his preference but agreed to let Edison decide the matter. Colombo to Bailey, 12 Sept. 1882; Colombo to TAE, 10 Oct. 1882; both DF (*TAED* D8238ZDA, D8238ZDT; *TAEM* 62:497, 540).

7. John William Lieb (1860–1929) graduated from Stevens Institute of Technology in 1880 and briefly worked as a draftsman for the Brush Electric Co. of Cleveland. In early January 1881 he entered the employ of the Edison Electric Light Co. as a draftsman, but soon transferred to the Edison Machine Works. He was placed in charge of the Pearl St. central station when it opened. Lieb's understanding of the station and his fluent French so impressed Giuseppe Colombo that he asked to have Lieb assigned to oversee installation of the central station in Milan. Lieb arrived there on 8 December. He was promoted to chief engineer of the Italian Edison Company in late 1883; he returned to the U.S. in 1894 as an official of the Edison Electric Illuminating Co. of New York and later became vice president of the successor New York Edison Co. *ANB*, s.v. "Lieb, John William"; Baldwin 2001, 138–41; Obituary, *New York Times*, 2 Nov. 1929, 13; Colombo to Bailey, 12 Sept. 1882; Lieb to TAE, 13 Jan. 1883; both DF (*TAED* D8238ZDA, D8337C; *TAEM* 62:497, 67:596).

[New York, c. November 20, 1882[1]]
exhaust with hand pump at a. The~~mn~~ heat wire, which absorbs O ~~oxi~~by oxidation leaving N at pressure needed.[2]

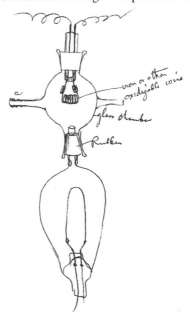

Cyanogen introduced by separate tube after exhaustion.[a]
Carbon instead of iron. Cyanogen decomposed and carbon[b]
deposited, nitrogen left.[c]

If carbon used with air carbonic oxide would be formed.
n[o].g[ood].[c]

~~Originally filled with chlorine or bromine, which~~[d]

[Witness:] H. W. Seely

X, NjWOE, Lab., Cat. 1148 (*TAED* NM017:8; *TAEM* 44:368). [a]"after exhaustion." interlined above. [b]Interlined above. [c]Paragraph written vertically in left margin. [d]Paragraph written vertically in right margin.

1. Henry Seely signed and dated this loose page when he received it from Edison on this date.
2. Edison executed a patent application on 28 November for two related processes of introducing nitrogen into lamp globes to prevent carbon carrying. In one, a "gas containing nitrogen and capable of being decomposed under certain conditions" was admitted to a partially-evacuated globe. When the gas was decomposed, oxygen and other undesirable constituents would be absorbed by the heated iron spiral, leaving an atmosphere of nitrogen at the proper pressure. The other process was specifically for cyanogen gas. In this case, a carbon filament was substituted for the iron spiral; when the cyanogen decomposed, its carbon would be deposited on the filament, leaving nitrogen in the globe. U.S. Pat. 438,298; see also Doc. 2346 n. 7.

[New York,] 21st Nov [188]2

Dear Sir,

Referring to your favor of 19th inst[1] I had two men working on Accumulators for three months and I have given a great deal of time to the matter and the conclusion I came to was that Accumulators are (in the present state of the art) of no value commercially. If I should at any time strike anything in this direction I will notify you. My inclination however is to discourage the use of Accumulators as I think they will rather damage us than do us good.[2]

With reference to Automatic Regulators I have spent a great deal of my time for months past in experimenting with a view to getting a good regulator and I have at last got what I consider a really reliable Automatic Regulator & one which I think will repay you for the waiting.[3] I have now ordered those you requested me to sometime back   As these will be the first made (except of course the model) there will be some delay in turning them out but I will see that the work is pushed forward as fast as possible   Yours truly

Thos A Edison   I[nsull].

L (letterpress copy), NjWOE, Lbk. 14:463 (*TAED* LB014463; *TAEM* 81:1021). Written by Samuel Insull.

1. Rau asked Edison for information on accumulators and automatic regulators. He believed that storage batteries "would prove most valuable for the introduction of our light in Europe. In our cities, and owing to the mode of construction of houses, it is impossible to put any machines in many places where the electric light would be accepted quite willingly; in other places . . . we are also prevented doing so by the police-laws on boilers, which are never allowed to be put under any inhabited apartments. The invention of a practical accumulator would open us a new and immense field of action." Rau to TAE, 19 Oct. 1882, DF (*TAED* D8238ZDX; *TAEM* 62:546).

2. See Doc. 2389.

3. See Doc. 2345 n. 5.

[New York,] November 27th. 1882.

Mr. Edison,

The talk about the installations of the Isolated Company[1a] leads me to suggest the following plan for your criticism.

FIRST. Let us officially establish the main features of the usual house and factory installation of the Isolated Company. For that purpose I will call a meeting of the Directors of the Isolated Company, and will invite Greenfield[2a] ~~to be pre~~ and

Clarke to be present. Among the Directors are yourself and Johnson. At that meeting let ~~the~~ a[b] standard installation be fixed.[3] ⟨OK⟩

SECOND. Let some plan be arranged at the meeting for such an emergency as a departure from the standard installation, where the peculiar circumstances of the proposed plant require it. Perhaps that can best be left to the discretion of the officers of the Isolated Company. ⟨Yes adopt a flexible rule—⟩

THIRD. Let the Board of Directors of the Isolated Company pass a resolution to the effect that no change in the details of the standard installation determined upon shall be made, nor shall the same be altered, without the written approval of the President (or any other officer you may think best) of the Isolated Company. Let the resolution further provide that when any change in any of the appliances of an installation is made, as for instance by Bergmann & Co., the same shall be submitted to the President of the Isolated Company, who shall at once confer with the engineering officers of the Isolated Company, and with such others as he may think best, and after doing so, shall notify Bergmann & Co., or such other officers as may have brought the proposed improvement to his notice, what the decision of the Isolated Company is. If this decision is not satisfactory to Bergmann & Co., let a Board meeting then be called, to which Clarke, Greenfield &c. shall come.

Does this plan meet your approval? If not will you kindly suggest another.

S. B. Eaton   per Mc.G[owan].

⟨Perfectly Satisfactory   Edison⟩

TL, NjWOE, DF (*TAED* D8221ZAX; *TAEM* 60:913). [a]Mistyped. [b]Interlined above by hand.

1. The first annual meeting of the Edison Co. for Isolated Lighting had been held on 21 November; the directors' report to shareholders stated that the company had installed 137 plants in its first year. Edison Electric Light Co. Bulletin 15:43–46, 20 Dec. 1882, CR (*TAED* CB015241; *TAEM* 96:787).

2. Edwin Greenfield was involved in wiring buildings in the first District. TAE to Eaton, 20 June 1882, Lbk. 7:527 (*TAED* LB007527; *TAEM* 80:650); see also Doc. 2125, n. 7.

3. A "conference as to what shall be considered a standard installation" was arranged for 7 December, with further meetings scheduled for later that month "to decide as far as possible upon the type of moulding to be used, sizes and types of cut outs." The committee was chaired by Charles Clarke, with William Meadowcroft as a secretary. A group called the "General Committee on Standards" continued to meet in January 1883; Samuel Insull later sent a complete set of its minutes (not found)

to Charles Batchelor and Edward Johnson. An undated 1883 document lists the materials required for a generic 100-lamp installation, totaling $4,544.22. Eaton to TAE, 5 Dec. 1882; Clarke to TAE, 14 Dec. 1882; Meadowcroft to TAE, 14 Dec. 1882, 5 Jan. 1883; Edison Co. for Isolated Lighting cost estimate, 1883; all DF (*TAED* D8224ZCV, D8224ZCZ, D8224ZDA, D8327B, D8325ZDI; *TAEM* 61:130, 134, 135, 66:704, 653); Insull to Johnson, 1 Apr. 1883, LM 3:115 (*TAED* LM003115; *TAEM* 84:98).

–2373–

*Samuel Insull to
William McCrory*[1]

[New York,] Nov. 29 [188]2

Dear Sir

Your favor of 21st inst with cutting from the "Minneapolis Tribune" came duly to hand.[2]

Mr. Edison makes it a rule not to answer adverse comments and misstatements of facts, but has suggested that I should write you pointing out the errors into which Dr. Perrie[3] has fallen

1st   Dr. Perrie says the Edison system of Electric Rail Road lacks economy but fails to give his authority for such an assertion. As a matter of fact tests are being daily made under all possible conditions to prove the exact economy of the Edison system over steam locomotion and the results so far have been most favorable.

2nd   Dr. Perrie says that a third rail is necessary. This is not so. At Menlo park we have a track two and a half miles in length and there are only two rails which are used as the conductors.

3rd   Dr. Perrie says that if a horse or man gets his foot upon the rails the shock is instant death. Our experimental track at Menlo Park crosses a country road very much used and during last Summer I frequently driven a very restive horse over the tracks whilst the Rail Road was being operated without the slightest inconvenience. Our men are frequently working on the track whilst the rails are charged with the electric current and of necessity have to touch the rails, but in-as-much as our current is of comparatively low pressure it is impossible for any such accidents as Dr. Perrie depicts to occur.

4th   Dr. Perrie says that "when you produce electricity by steam and then use the current to drive machinery you lose a great deal of power." But he omits to state that with a locomotive steam engine the power obtained is far more expensive than with a Stationary Steam Engine and in as much as Mr. Edison uses the latter he has a large margin to work on to make up for the loss occasioned by the conversion of power into electricity and back into power again.

Mr. Edison's motor is not "lying on an old heap of scrap iron—discarded" but is being used daily for experimental purposes.

Dr. Perrie's remarks show his total ignorance of Mr. Edison's Electric Rail Road and I would suggest that he should enquire into what Mr. Edison has done before making assertions such as those referred to.

You may make what use of this letter you please[4]   Yours Truly

Samuel Insull   Private Secretary

LS (letterpress copy), NjWOE, Lbk. 14:482 (*TAED* LB014482; *TAEM* 81:1029). Written by Charles Mott.

1. William McCrory (1839–1893) was the president and owner of the Minneapolis, Lyndale & Minnetonka Railway (known as the Lyndale Railway Co. from its opening in May 1879 until 1881). The company operated a narrow gauge line about 20 miles between downtown Minneapolis and Lake Minnetonka. In July, after the Minneapolis council approved a prohibition of steam power on the line, an associate of McCrory contacted Edison about the feasibility of electric traction. McCrory himself subsequently inquired for more information on several occasions. W. C. Baker, Jr. to TAE, 5 July 1882; McCrory to TAE, 14 Oct. 1882; both DF (*TAED* D8249ZAJ, D8249ZBA; *TAEM* 63:614, 632); McCrory to TAE, 31 Aug. and 5 Oct. 1882, Cat. 2174, Scraps. (*TAED* SB012BBA, SB012BBG; *TAEM* 89:440, 455); Olson 1976, 67–75; Olson 1990, 29–34.

2. McCrory forwarded the *Tribune* clipping, dated 21 November 1882, to Edison as an enclosure to his letter of that date. The article stated that no electrician had devised a practical alternative to steam locomotion. McCrory found this "annoying to me because I have told our council you were rapidly progressing with your electric motor and would have it done soon." He appealed to Edison to "have the statement contradicted—it will settle our people & give confidence to our friends in an early completion of the new power." McCrory to TAE, 21 Nov. 1882, DF (*TAED* D8249ZBK; *TAEM* 63:645).

3. The article included an interview with "Dr. Perrie of this city" who was reportedly involved with developing an electric railway motor, but no such name appears in a Minneapolis directory. This may have been a mistaken reference to Cyrus Perrine, a Minneapolis resident and principal in Laraway, Perrine & Co., a large plow manufacturer. Jarchow 1943, 297–98; Davison 1881, 443.

4. In response to subsequent inquiries, Edison referred McCrory to the Electric Railway Co. of the United States, which took control of Edison's and Stephen Field's electric railroad patents in 1883. The line was electrified experimentally with Charles Van Depoele's traction-motor system in late 1885, by which time McCrory had sold his controlling interest. McCrory to TAE, 30 Jan. 1883, Cat. 2174, Scraps. (*TAED* SB012BBY; *TAEM* 89:491); TAE to McCrory, 21 May 1883, Lbk. 16:389 (*TAED* LB016389A; *TAEM* 82:449); Prosser 1966, 97–98; Olson 1976, 67–75; Olson 1990, 29–34.

–2374–

Draft Memorandum to Edison Electric Light Co., Ltd.

[New York, November 1882?[1]]

Parent Co—

Cut down your expenses as far as possible. Dispense with the services of all Electricians or if contracted with on salary [----][a] dont use them[b]   Engage a good mechanical Engineer who has had practical Experience & a fair education, ~~send h~~ one having some executive ability, not older than Olrick. Send him ~~at on~~ to America, to learn the business practically & scientifically, he to be recalled only when needed.

permit me to learn a man here to be sent hereafter to England capable thereafter to determine the net works conductors of the general system, as well as small installations, salary say £20 per month= These 2 men with what assistants they can obtain [-------][a] superintend any installation in England 1 or 10 miles square=[c]

~~Now~~ arrange with the manufacturers in England or[d] America or both[e] to manufacture for your licensees everything needed[f] so that should you sell a certain area & your licensee desires to establish a ½ or[g] mile area, your Engineer can work out the whole thing & from his data   ~~your~~ The[h] manufacturers you have arranged with[i] can bid for the whole work. ~~therefore the cost of installation can be known beforehand to you licensees~~, If desirable

Bergmann & Co will bid against any one[j] to deliver the whole of the regulating apparatus & devices at the Central Station, & every kind of supplies for wiring.

The Electric tube Co will bid for the underground tubes for a square mile with all the services, & lay the same & deliver the whole over in working order, (except digging & paving)

The Goerck St Works will bid for ~~Dynamos & Eng~~ Steam Dynamos, set them in position & guarantee them

The Lamp Co will furnish of course the Lamps=

or the ~~Lamp~~Tube Co will learn an agent of any manufacturer you desire to make your tubes they & so will Goerck St if it is so desired.[k]

My impression is strong that but one Company should be formed for the whole of London—but I suppose that is now impossible[2l]   let them take over the Holborn installation, & arrange to Light ~~a~~ ½[h] square mile in London. ~~After the plant~~ Their capital should be small at first with power to increase as they expand after our plan here[m]   ~~Only the~~ The first outlay will be the[n] real estate [n--][a] then the plant can be gradually increased until the whole of the square mile is lighted   Your investment will be gradual & only made if warranted. But you must prepare your plans just as if you were going to make the whole installation otherwise, it will be a sink hole for money.[3] The small stations you are now installing will ~~only be a~~ [with with?][a] have[h] a hopeless future.[o]

The parent Co must do no business it must act as the agent of its licencies, doing all by its Engineer dept & arrangements with manufacturers for its license that is necessary to make a complete installation & [----][a] men for continuing its running any attempt to permit the licenses to do what they please will result in bankruptcy of the licenses & no future[h] profit to the parent[d] Co= I dont believe in the policy of taking ~~a large~~ initial sum in money make it small take more shares[p]

You have 50 000 invested in large Dynamos at Goerck St. We will try and get rid of some of them—[4]

You have a lot of Isolated machinery Will will try and get rid of some of this but I think you better not ~~permit this~~ accept this offer[q] if it is at all possible to form an isolated Co.[5c]

You sell Lamps to any one who wants them= I dont believe in this, the policy is bad, its a small fry business. it will hurt ~~your~~ the reputation you will acquire in the future, if the Isolated policy is adopted. people judge ~~of~~ by the lamps if the engine is unsteady or the Dynamo bad the criticism falls on the Lamps.

You have licensed the British Co to use their lamps.[6] ~~they~~ I dont believe in this. in 5 or 6 years without you get it back in the meantime[r] you will be immensely hampered.

I believe in having nothing to do directly or indirectly (whenever such a thing is possible) with any Electric Light Co in England— To make no trades alliances of any kind or character, ~~to~~ To sustain the patents whenever you have time, to do this leisurely & with great previous preparation. ~~That the~~ [term?][a] ~~of the~~ [pat?][a] That if ~~we are beaten in one point~~ you fail[s] to stop an infringer, never mind, when the proper time comes try him on another point= If we are proved infringers on any point, I can probably take care of that myself— We are getting new patents These patents are well drawn they will probably stand the test. it is future patents that will secure the system that replaces gas with great profit

The patents I am now taking are more valuable than those already taken. Those already taken were to secure if possible, the essence[d] of the thing. Those I am now taking are ~~those which covered it secured it to~~[h] commercially

[We?][a] ~~do~~Never stop an infringer who is daily losing money by so infringing give[d] him ~~rope~~ plenty of Rope,[d] & [---d -----].[7t]

Appoint & pay some one here connected with our[d] Co to keep you posted daily in every advance scientifically & commercially to give you all the experience, data as they come up—to gather information for you[u]

Conduct[d] the business of selling light exactly as the business of gas lighting is carried. Their system is admirable. <u>Sell light</u>, or power[v] <u>only</u>, sell it on a meter. never sell ~~current~~ electricity.[8] Keep the pressure <u>always</u> on the mains—[w] It has been stated that my estimate, already made shews that we cannot compete with gas; very well I will make a new estimate. The cheapening process has been going on since that estimate was made.[9] before you get legislative permission to open the streets it will be cheaper ~~yet~~ still. This process of cheapening will continue to go on until there will not be a place ~~in England~~ where gas is so cheap that the Electric Light cannot compete[d] with it at a profit if it gets $\frac{1}{2}$ the consumption.[10]

If owing to peculiar conditions you want something different or something changed give me minutely every detail, The why & wherefore and I will probably be able to[x] give you what you want. I have nursed the baby so far & I believe I can continue to do so without any extraneous aid especially from those who said the baby would never be born & when born would never live, & now that it lives wants to change the manner of nursing. if I should fail in any particular it will then be time to call in other inventors

ADf, NjWOE, DF (*TAED* D8239ZGB1; *TAEM* 68:214). [a]Canceled. [b]"of if . . . use them" interlined. [c]Followed by dividing mark. [d]Obscured overwritten text. [e]"or both" interlined above. [f]"everything needed" interlined above. [g]"$\frac{1}{2}$ or" interlined above. [h]Interlined above. [i]"you have arranged with" interlined above. [j]"against any one" interlined above. [k]"or the . . . so desired" added later at bottom of page. [l]"but I . . . impossible" interlined above. [m]"Their capital . . . plan here" interlined. [n]"The first outlay will be the" interlined above. [o]"The small . . . hopeless future." interlined below. [p]"I dont believe . . . more shares" interlined below. [q]~~"permit this~~ accept this offer" interlined above. [r]"in the meantime" interlined above. [s]"you fail" interlined above. [t]Canceled and followed by dividing mark. [u]Followed by several canceled words and dividing mark. [v]"or power" interlined above. [w]"Keep . . . mains—" written in right margin. [x]"probably be able to" interlined above.

1. Edison likely drafted this outline for reorganizing the British lighting business during Arnold White's visit to New York from late October to the middle of December. White came to discuss the British company's disputes with Edison and its state of affairs generally. Much of Edison's manuscript was subsequently typed as a comparatively clean draft, likely for a memorandum that White carried back to London (not found). White reported shortly after his return that the company's situation had

worsened during his absence but that he hoped "to announce the practical acceptance of Edison's programme nearly en bloc." Edward Bouverie to TAE, 15 Oct. 1882; White to TAE, 17 Oct. 1882; White to Samuel Insull, 13 Dec. 1882 and 1 Jan. 1883; TAE draft to Edison Electric Light Co., Ltd., n.d. [1882]; all DF (*TAED* D8239ZFP, D8239ZFQ, D8239ZGC, D8338A, D8239ZGB2; *TAEM* 62:1128, 1132, 1157, 68:3; 62:1160).

2. The Electric Lighting Act of 1882 delegated regulatory authority over electric lighting to local jurisdictions. As a distinct political entity, the City of London, encompassing a relatively small area, could act independently of other jurisdictions in the metropolis.

3. About this time the company submitted a map to the City of London for a proposed central station district that was never built. Edison Electric Light Co., Ltd., Map of City and District of London, 29 Nov. 1882, City of London Sessions, Clerk of the Peace, Plans CLA/047/LC/04 (UkLMA).

4. This apparently represents the six C dynamos ordered in May (see Doc. 2270) at about $8,000 apiece (see App. 3). They were still counted in the inventory of the Machine Works in October; see note 5. Edison Electric Light Co., Ltd., to TAE, 24 Apr. 1883, DF (*TAED* D8338U; *TAEM* 68:39); TAE to Edison Electric Light Co., Ltd., 7 May 1883, Lbk. (*TAED* LBo16268; *TAEM* 82:385).

5. See Docs. 2270 and 2375. Edison had not sold any of the British company's isolated or central station dynamos by May 1883 because of his own large inventory; he promised to continue trying to do so. Edison Electric Light Co. to TAE, 24 Apr. 1883, DF (*TAED* D8338U; *TAEM* 68:39); TAE to Edward Bouverie, 4 Dec. 1882; TAE to Edison Electric Light Co., Ltd., 7 May 1883; Lbk. 14:497, 16:268 (*TAED* LBo14497, LBo16268; *TAEM* 81:1039, 82:385).

6. The British Electric Light Co. manufactured Edison lamps under license. See Doc. 2248 n. 4.

7. Cf. Doc. 2365.

8. A rationale for this statement is given in Doc. 2333; Edison outlined the argument more fully in Doc. 2357.

9. The editors have not found Edison's previous estimate. It likely would have been similar to those made for the Edison Electric Light Co. in January for isolated plants and in August for large central stations (Cost estimates, 5 Jan. 1882 and 3 Aug. 1882; both DF [*TAED* D8221A, D8227ZAT; *TAEM* 60:845, 61:525]). Edison's July price list of dynamos for the company is discussed in Docs. 2310 and 2339.

10. Edison wrote an undated page of notes, possibly about this time, about the effect that a fifty percent drop in consumption would have on the price of London gas. He conjectured that companies presently selling gas for 75 cents (per thousand cubic feet) would have to charge $1.36 to cover their fixed costs and additional leakage. Draft memorandum, n.d. [1882], DF (*TAED* D8239ZGI1; *TAEM* 68:223); for an analysis of the economics of gas production and sale in Britain at this time see Shiman 1993 and Matthews 1986.

-2375-

Draft Memorandum to Edison Electric Light Co., Ltd.

[New York, November 1882?[1]]

~~Put company in as good shape as possible~~.

Organize an Isolated Company with good Mechanical Engineer such as Olrick at head of it. Let this new Co operate over whole England with restrictions as in U States. Turn over to Co all isolated machinery & supplies. Conduct business as in US; no advertising. Make the best installations possible & see that whatever you do make continue to give satisfaction. [~~ie?~~][a] adopt a mobile policy[b]   Get services two or more good commercial travellers preferably those who have been employed in machinery houses; instruct these men minutely how to present the subject arguments to use & methods of figuring Engage them on Salary at first keep ~~seperate~~ accounts of parent Co & Isolated seperate. ~~go slow &~~ [e------- ---][c]   ~~Dont be in a hurry to make money. when you mak   after receiving an order execute it quickly   pay great attention to your supply department. do not get out of supply of small[d] supplies.~~ [-------][c] ~~order ahead as the business warrants~~.

Say nothing against your competitors but explain the reasons why our prices are ~~higher, but~~ at the moment higher. if you lose the order, very well, try another. ~~sell~~ do all business by personal solicitation where ever possible. Whatever ~~is done~~ orders you do get execute quickly and well. Dont be in a hurry to make money   ~~expect to make none   at the end of two years you will have a fair business   in a few years a large business~~ Now let then let[e] the Isolated Co take care of itself. dont[f] interfere with the head of your Isolated dept in details ~~except~~ of business— dont hamper him to much[g] with ~~the degree~~ Red tape of a board of directors, ~~which business which they know nothing generally~~ [---------][c] ~~do the wrong way~~

ADf (fragment), NjWOE, DF (*TAED* D8239ZGI; *TAEM* 62:1171). [a]Circled and canceled. [b]"adopt a mobile policy" interlined below. [c]Canceled. [d]Obscured overwritten text. [e]"then let" interlined below. [f]Text from "dont" to end of document interlined. [g]"to much" interlined above.

1. The contents of this document suggest that Edison drafted it at about the same time as Doc. 2374.

-2376-

From Charles Batchelor

Ivry-sur-Seine, le Dec 1st 1882[a]

No 381[1b]

My dear Edison,

I do not know whether you have noticed that some of our lamps (I speak of those from America for I have now remedied

it in mine) are actually short circuited in the plaster socket—
I could not believe at first that such a thing could happen until I came across no less than three in the lamps at the Brünn
theatre— I put them in my pocket for investigation as I could
not understand how they could come out of your testing room
marked 110 Volts with a dead cross in the socket. I cut them
carefully to pieces after making sure of their resistance and I
found No 1 crossed at X—
Fig 1

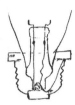

No 2 was touching as shown though when I had cut the plaster away sufficiently to see inside they did not look as if they
quite touched— No 3 touched as marked—
Fig 2.

I would also ask to have better varnishing on the lamps that
come to Europe—the moist atmosphere goes into the plaster— I notice that you put no varnish on the bottom part of
the plaster but it is very essential here— Since finding these
crossed lamps I have carefully investigated those lamps which
suddenly explode (a phenomenon I have never explained satisfactorily to myself) and in many cases I believe it is due to the
wires being very close together, and aA little extra current
being sent through sometime expands the wires in the shank,
and as they are held tight in the seal at one end, and the plaster at the other, they move sideways and touch—
Dampness in the Plaster will account for high volt lamps in
some cases— My volts used to range up to 112 but now I keep
all my lamps at about 90 Fahr for 2 days before varnishing and
testing and now I very seldom get one above 106 or 107—
Yours truly

Chas Batchelor

⟨Explosion is due I think to arcing   We have fixed the wires
in inside part by using insulated[c] wire= will tell Upton about
varnishing[2]

There must have been some change in your carbonzation to
bring volts down as Upton cant see how a drop of 2 volts could

take place in the wet plaster as the ft lbs would be sufficient to eat the wire off in few days—[3]

I have just got a new Compound that I think will be good for sockets of Lamps instead of plaster paris have not tried it= am trying it to stick mica together which it does splendidly= We roughen the mica pieces with sand paper & punch few holes in peices then with these we build up anything we want by putting thinly over mica the stuff= Bergmann is going to make mould try if we cannot mould the Socket displacing wood— its Common Oxide Magnesia mixed with saturated solution Chloride of Magnesia. There[c] may be several times its own bulk of inert stuff mixed with it such as finely powdered sand emery etc= It sets in 12 hours harder than marble—[4] I send sample—

boy[5] says sample made with 10 cubic Cent of Chloride Magnesia Solution to 10 grammes of oxide magnesium— You can experiment on proper amount of Oxide & Chloride= E⟩[6]

ALS, NjWOE, DF (*TAED* D8238ZER; *TAEM* 62:579). Letterhead of Société Industrielle et Commerciale Edison. [a]"Ivry-sur-Seine, le" pre-printed. [b]Multiply underlined. [c]Obscured overwritten text.

1. Batchelor had recently started numbering his outgoing letters.

2. Francis Upton reported to Samuel Insull on 29 December that "The lamps now are varnished all over the exposed part of the plaster in the socket." DF (*TAED* D8230ZCU; *TAEM* 61:851).

3. That is, if the increased voltage needed to run Batchelor's lamps at their rated light output had been due to current leakage between the two wires through damp plaster.

4. Laboratory employee George Gibbs noted these properties of the magnesium cement on 2 December. He and Thomas Conant began experimenting about that time with various cements to secure wires through layers of mica in the construction of commutator brushes (N-82-11-14:67, 77–93, Lab. [*TAED* N143:33, 36–44; *TAEM* 38:705, 708–16]). Conant first recorded a cement lamp socket on 19 December, describing it as "good." He made related experiments into the second week of January (N-82-11-14:95–139, Lab. [*TAED* N143:45–67; *TAEM* 38:717–39]).

5. Unidentified; probably neither Conant nor Gibbs, who was 21 and a recent mechanical engineering graduate from Stevens Institute of Technology. Conant's age is unknown, but Edison assigned him to enter the village plant business the following summer. *ANB*, s.v. "Gibbs, George"; TAE to Charles Chinnock, 6 July 1883, DF (*TAED* D8316AEZ; *TAEM* 64:884).

6. Edison's marginalia formed the basis of his 28 December reply to Batchelor, who promised to experiment with the sample and formula. TAE to Batchelor, 28 Dec. 1882, Lbk. 15:83 (*TAED* LB015083; *TAEM* 82:65); Batchelor to TAE, 23 Jan. 1883, DF (*TAED* D8337H; *TAEM* 67:605).

Draft Patent
Application:
Primary Battery

[New York, c. December 4, 1882[1]]

Patent ⟨Case 531⟩[2a]

New process for obtaining electrical suitable for large industrial operations=

I form electrodes by moulding an oxide of Lead around a lead core and Reduce the oxide of Lead by a reducing agent & heat, to the metallic state to form ~~the~~ one of the electrodes.

The other electrode is subjected to a chemical process whereby the oxide of Lead is raised to the higher oxide. When current is required the two different electrodes are ~~lowered~~ placed in a cell containing H_2O & acidulated with SO_4—after ~~the~~ a great portion of the peroxide of lead[b] has been reduced to Red Lead[3] & the metallic lead of the other pole oxidzed.

The plates are taken from the cell & the one which is oxidzed is dried & reduced to metallic Lead again by heat & a reducing agent while the other is again peroxidzed—

The peoxidation of one of[c] the electrodes is obtained by acting on it with chlorine in an acidulated solution, the Lower oxide being raised by the action of the Chlorine to peroxide of lead[d]

Peroxide made in this way is not inert Being recently precipitated & moist[e]

[Witness:] H. W. Seely

ADf, NjWOE, Lab., Cat. 1148 (*TAED* NM017:1; *TAEM* 44:361). [a]Marginalia by Henry Seely. [b]Followed by "over" to indicate page turn. [c]"one of" interlined above. [d]Followed by dividing mark. [e]Final paragraph written later or with a different pen.

1. Henry Seely signed and dated the document on this date when he received it from Edison.

2. He filed this application as Case 531 in January 1883; it was rejected and later abandoned. Among its eight claims were those for "a primary battery having one electrode capable of oxidation and the other electrode capable of reduction" (both electrodes formed of the same metal), and "the process of forming battery electrodes consisting in chemically reducing a metallic oxide to form one electrode, and chemically raising an oxide of the same metal to form the other electrode." Patent Application Casebook E-2538:36–39, PS (*TAED* PT022036; *TAEM* 45:792–93).

Martin Force made experiments on 2 and 4 December and George Gibbs on 21 and 22 December related to the fabrication of lead electrodes as described in this document. N-82-05-15:77–79, N-82-12-04:9–13; Lab. (*TAED* N203:39–40, N145:6–8; *TAEM* 40:511–12, 38:815–17).

3. Red lead is a higher oxide of lead (Pb_3O_4). Vinal 1955, 21.

[New York, c. December 4, 1882[1]]

Patent.[2] Improvements in plante accumulator[3]

I ~~prepare~~ make[a] the plate by moulding an oxide of Lead preferably Litherate[4] ~~around~~ which is to be mixed with water or gum water around a sheet of Lead having a roughened surface & having projections[b] & containing many perforation. ~~A~~The whole is made solid by great pressure in a powerful press. The plate or plates are then placed in a chamber or tube and a reducing gas such as Hydrogen ill[uminating] gas or other reducing agent[c] passed through the chamber; heat is externally applied and the whole of the oxide is reduced to metallic Lead which being very porous and integral ~~j~~gives a great[---][d] capacity of storage[e] & economy per lb of Lead than ~~by~~ [-----][d]   Care should be taken that the temperature should not be allowed to reach as point where the lead becomes liquid but just sufficient to produce perfect reduction to the metallic state. If greater porosity is required ~~ana~~ Light[a] Earthy oxide may be mixed & moulded with the oxide of lead the lead being reduced the earthy oxide disolves[f] out as a soluable salt by acid=[f]

A Battery made in this manner should not be charged to a point where the whole of the Lead[f] of the plate is converted electrically into the oxide— a[f] central webb of lead is not essential but it is more practical [-----][g] breaking

Claim: The formation of electrodes for secondary batteries consisting in moulding by pressure or other means[h] ~~ana~~ Salt of Lead into[g] the form desired & then chemically reducing the Lead to the metallic state for purpose etc—

Central webb to make good contact, Integral etc etc

Musn't overcharge: ½ leaving Integral[a] web of lead

[Witness:] H. W. Seely

ADf, NjWOE, Lab., Cat. 1148 (*TAED* NM017:3; *TAEM* 44:363). [a]Interlined above. [b]"& having projections" interlined above. [c]"ill[uminating] . . . agent" interlined above. [d]Canceled. [e]"of storage" interlined above. [f]Obscured overwritten text. [g]Illegible and interlined above. [h]"by pressure or other means" interlined above.

1. Henry Seely signed and dated the document on this date when he received it from Edison.

2. This and the preceding document marked the end of Edison's investigation of storage batteries for several years because of his skepticism about their technical and economic practicability (see also Doc. 2389 and TAE memorandum, undated 1882, DF [*TAED* D8220ZBI; *TAEM* 60:841]). The editors have found neither a patent application nor a specification pertaining to this process for making battery electrodes. Edison did file one other application (Case 530) related to secondary bat-

teries in January; it was rejected and ultimately abandoned. Patent Application Casebook E–2538:34, PS (*TAED* PT022034; *TAEM* 45:791).

3. Gaston Planté (1834–1889) was a French chemist who demonstrated the first lead-acid storage battery in 1860. His key contribution was the design and construction of lead electrodes. Planté 1887; Schallenberg 1982, 24–33.

4. Edison presumably meant "litharge" or "lithargite," a reddish protoxide of lead (PbO) formed by exposing melted lead to an air current. *OED*, s.v. "litharge."

–2379–

To Joshua Bailey

[New York,] 11th Dec [188]2

My Dear Sir,

Your favor of 10th ult came duly to hand and also your cable on the same matter sent subsequently.[1]

You will have learned my views on the subject of a fusion with Seimens from my cable sent in answer to yours.[2] I think that it is very important indeed that any business arrangement made with him should be so worded as not to include the Central Station business. The[a] Seimens' (judging from the utterances of several members of the family) do not think the Central Station business of any commercial importance.[3] I therefore think you would not have much difficulty in excluding Central Station business from such an arrangement as you might make with them. If this cannot be done in an open manner you should direct your endeavours towards wording any agreement in such a manner as to secure to ourselves the Central Station   My reason for desiring this is that I wish to preserve the monopoly of the Central Station business. The Isolated business may for the moment appear very attractive but the main[b] business upon which the success of our companies will depend will be that of General Distribution. Seimens will undoubtedly find this out eventually; in which case it would be very dangerous for us to have given him an agreement which would enable him to enter into competition with us in Central Station Lighting.

I place the greatest importance on this matter and would beg of you to exercise the greatest possible care in any deal your Directors may deem it wise to make with Seimens.

As to a fusion with Seimens here our Directors would of course consider any proposition which Seimens might make but so far as I can see at present he has no grounds on which to make one. If he had a solid Patent here I thi which would enable to close out everybody else I think our Company would pay him a royalty & vigorously prosecute all infringers if they could get an exclusive license. These remarks apply both to the Dynamo & the Rail Road System.

As to your cable from London about a fusion with Swan in England if such is the case I have no knowledge whatever of it and Mr. White the Secty of the English Co. who is here just now says there is no truth whatever in it. Whatever information you have on this subject I shall be glad to receive from you.[4]

I shall be obliged if you will kindly show this letter and a copy of yours of 10th ult to Mr. Batchelor   Yours truly

Thos. A. Edison   I[nsull].

L (letterpress copy), NjWOE, Lbk. 15:29 (*TAED* LB015029; *TAEM* 82:41). Written by Samuel Insull; "Per S. S. Abyssinia" written above dateline to indicate ship carrying letter. [a]Interlined above. [b]Obscured overwritten text.

1. Bailey met with Werner Siemens on 10 November hoping to come to a general agreement between the Edison and Siemens interests in continental Europe, particularly for France, Germany, and Russia. Bailey proposed a cross-licensing arrangement whereby Siemens would pay a royalty to manufacture lamps, while Edison would do likewise for Siemens's dynamo patents. Bailey expected that such an agreement would result in their joint domination of the European market. They also briefly discussed the American patent interferences between Edison and Siemens on the dynamo and electric railroad; Bailey promised to consult with Edison so as to reach an understanding. Bailey concluded by urging Edison to reply as soon as possible, "as it is quite important that an understanding be concluded for Germany the soonest possible." Bailey to TAE, 10 Nov. 1882, DF (*TAED* D8238ZEE; *TAEM* 62:555).

2. On 19 November Bailey cabled Edison that he had negotiated an agreement whereby Siemens would pay Edison a ten percent royalty on incandescent lamps sold in Germany. He also expected to reach agreement on the outstanding patent disputes between them in France and the United States, and told Edison that he had closed a contract to set up an Edison company in Germany. Edison cabled back on 22 November that he accepted the terms for Europe but would not include the United States in this agreement. On 3 December Bailey cabled from London that "Berlin parties accepted payment royalties." Bailey to TAE, 19 Nov. 1882; TAE to Bailey, 22 Nov. 1882; Bailey to TAE, 3 Dec. 1882; LM 1:265B, 266B, 268E (*TAED* LM001265B, LM001266B, LM001268E; *TAEM* 83:1004–1006).

The Edison interests later entered into an agreement with Siemens & Halske whereby Siemens & Halske agreed not to solicit or install lighting plants and the Edison interests agreed to purchase all equipment (except incandescent lamps) for lighting plants from Siemens & Halske. Edison Electric Light Co. of Europe agreement with Siemens & Halske, 13 Mar. 1883, DF (*TAED* D8337ZAK; *TAEM* 67:644); Insull to TAE, 23 Feb. 1883, Lbk. 15:355 (*TAED* LB015355; *TAEM* 82:182); Hughes 1983, 66–78; Feldenkirchen 1994, 107–10; Siemens 1957, 93.

3. Siemens preferred to build equipment for, but not to establish or

invest in, companies to operate central stations. Feldenkirchen 1994, 114–15.

4. In his cable of 3 December (see note 2) Bailey reported that Joseph Swan's recognition in England and the threat of lawsuits "demoral-ize[d]" Edison's European business. Edison replied that he was "un-aware Swan recognized anywhere. Rivals circulate reports   White sec-retary London Company here says fighting suits vigorously." TAE to Bailey, 4 Dec. 1882, LM 1:269A (*TAED* LM001269A; *TAEM* 83:1006).

–2380–

Draft to Charles Speirs[1]

[New York, December 12, 1882[2]]

Don Quixote—
Gil Blas.[3]
Bulwers Novels.[4]
Victor Hugos Novels
Dickens Novels.
Hawthornes Novels
Longfellows poems in full.
Abbotts Life of Frederick the Great—[5]
Hans Andersons tales for Children[6]
M̶George Eliots novels.
Coopers I̶n̶Novels.
Carlyles French Revolution
    "   Life Frederick the Great[7]
Macauleys Essays.[8]
Phillips (Irish orators) Speeches.[9]

Speirs, can you fill this bill   want good solid binding only nothing fancy— How Long will it take— good ordinary size type=[10]

Edison

ADfS, NjWOE, DF (*TAED* D8211K1; *TAEM* 60:622).

1. Charles Speirs was an agent of D. Van Nostrand & Co. publishers and booksellers in New York, with whom Edison corresponded often, primarily for scientific publications.

2. On this day, Samuel Insull revised this draft into a letter to Speirs from Edison. Lbk. 15:34A (*TAED* LB015034A; *TAEM* 82:44).

3. *The Adventures of Gil Blas of Santillane,* a novel by Alain-René LeSage (1668–1747), written between 1715 and 1735 and published in English in 1749. *Col. Ency.,* s.v. "Le Sage, Alain René."

4. Edward Bulwer Lytton (1803–1873), English novelist, essayist, and politician. *Oxford DNB,* s.v. "Lytton, Edward George Earle Lytton Bulwer."

5. Edison probably referred to *History of Frederick the Great,* com-pleted in 1871 by John Stevens Cabot Abbott (1805–1877), a college

classmate of Hawthorne and Longfellow. It is also possible that he intended *The Youthful Days of Frederick the Great,* a melodrama adapted from the French in 1817 by British actor William Abbott (1790–1843). *DAB,* s.v. "Abbott, John Stevens Cabot"; *Oxford DNB,* s.v. "Abbot, William."

6. English translations of *Tales for Children* by Hans Christian Andersen (1805–1875) were published in London in 1861 and 1869. *Col. Ency.,* s.v. "Andersen, Hans Christian."

7. British biographer and historian Thomas Carlyle (1795–1881) completed the last volume of *The French Revolution* in 1837. He finished *Frederick the Great* in 1855 after thirteen years of work. *Oxford DNB,* s.v. "Carlyle, Thomas."

8. British historian and politician Thomas Babington Macaulay (1800–1859) published a collected volume of his essays, *Critical and Historical Essays,* in 1843. *Oxford DNB,* s.v. "Macaulay, Thomas Babington."

9. Charles Phillips (c. 1786–1859) was an Irish-born barrister reputed for extravagant oration. He published *The Speeches of Charles Phillips* in 1817. *Oxford DNB,* s.v. "Phillips, Charles."

10. Speirs replied the same day that he could send the books on one day's notice. Speirs to Insull, 12 Dec. 1882 (*TAED* D8211L; *TAEM* 60:624).

–2381–

To Charles Batchelor

[New York,] 13th Dec [188]2

Dear Batch:

You will notice in the London Electrician for November 25th an article by Francis Jehl on "Electric Lighting at Brunn. It is a copy of the article sent by Francis to Van Nostrands & contains all the passages which I struck out & about which I wrote you[1]

Francis gives away to the public everything that has cost us time and money to find out & others can go ahead & use our formula by just reading the article.

There is not particular objection to Francis writing such articles but there is an objection to his giving us away. I question whether we want other people to know anything about our methods   Yours truly

Thos A Edison   I[nsull].

L (letterpress copy), NjWOE, Lbk. 15:36 (*TAED* LB015036; *TAEM* 82:46). Written by Samuel Insull.

1. This article (Jehl 1882b) described the installation in detail, including specifics of the construction and operation of the K dynamos. On 15 November Edison wrote Batchelor that Jehl "must not write to public Journals the why & the wherefore of making our machines as we do. It is sufficient to give sizes & weights but not explanation of theories. I do not mind Francis airing his knowledge but it must not be at our expense. I cut out of his article that part relating to the advantage of the K

machine." Jehl apologized but pointed out that Edison had revealed similar details of the central station dynamo in Edison and Porter 1882. Jehl later recalled that he "got quite a scolding from Edison." TAE to Batchelor, 15 Nov. 1882, Lbk. 14:423 (*TAED* LB014423; *TAEM* 81:1007); Jehl to TAE, 30 Nov. 1882, DF (*TAED* D8238ZEQ; *TAEM* 62:577); Jehl 1937–41, 985.

–2382–

From Josiah Reiff

[New York, December 18, 1882[1]]

Dear Edison

Mrs. Seyfert[2] has won for time— Will know more in a day or two what future prospects[3]

JCR[4]

ALS, NjWOE, DF (*TAED* D8252ZAA; *TAEM* 63:708).

1. On this date Edison's attorney Garrett Vroom telegraphed Josiah Reiff at his New York office that this case had just been decided. Vroom to Reiff, 18 Dec. 1882, DF (*TAED* D8252Z; *TAEM* 63:707).

2. Lucy Seyfert was the wife of William Seyfert, an investor in the Automatic Telegraph Co., who had conveyed to her an unredeemed 1874 promissory note from Edison. Edison repudiated the debt when she tried to collect; she filed suit against him in November 1880. See Docs. 516 and 2014.

3. After a postponement from April, *Seyfert v. Edison* went to trial in early December. According to Edison's attorney, the outcome hinged largely on whether Edison's $300 check to Mrs. Seyfert in 1875 was, as she alleged, a partial payment on the note or, as he claimed, a personal loan. The trial ended on 18 December with a directed verdict for the plaintiff. Edison appealed the judgment of $5,065.84 against him but the case continued to haunt him for years. Vroom to TAE, 13 Apr., 5 and 18 Dec. 1882; all DF (*TAED* D8252G, D8252W, D8252ZAB; *TAEM* 63:686, 704, 709); TAE to Reiff, 8 Dec. 1882; Lbk. 15:22A (*TAED* LB015022A; *TAEM* 82:38); Israel 1998, 233.

4. Railroad financier and telegraph entrepreneur Josiah Reiff was a longtime Edison associate who had provided most of the funds for Edison's work on automatic telegraphy. It was through his connection with the complex finances of companies formed to exploit the resulting patents that he had knowledge of the contested note held by Lucy Seyfert and he had been involved in negotiations seeking a settlement of the suit in 1880. See Docs. 141 n. 7, 452 n. 3, 676, 1713, and 2014, esp. n. 2.

–2383–

From Sherburne Eaton

[New York,] Dec. 21st. 1882.

Mr. Edison,

Mr Tracy[1] has made his first report giving the result of his personal call upon subscribers in the First District. He told them that they would have to pay for their light on and after date.[2] He finds almost unanimous satisfaction with the light and thinks that all the subscribers will become customers. On the whole his report of his experience thus far is very encouraging.

S. B. Eaton   per Mc.G[owan].

TL, NjWOE, (*TAED* D8223ZAH; *TAEM* 60:1076).

1. Probably H. L. Tracy, who apparently worked in the office at 65 Fifth Ave. at this time. Tracy to Insull, 14 June 1882, DF (*TAED* D8249W; *TAEM* 63:604).

2. The Edison Electric Illuminating Co. of New York initially provided power free of charge, owing in part to Edison's expectation of continued modifications and possible service outages. By December, with the system's stability and utility sufficiently established, the company began to charge customers for the current. Billing began in late January 1883 but was not done regularly for several months. Jones 1940, 209–11; Dyer & Martin 1929 [1910], 1:409; Jehl 1937–41, 1086–87; Eaton to TAE, 1 Feb. 1883; S. Allin to TAE, 6 Feb. 1883; both DF (*TAED* D8326B, D8326C; *TAEM* 66:655, 666); see Doc. 2409 and App. 1.B.29.

–2384–

Notebook Entry:
Incandescent Lamp

[New York,] Dec 21 1882

Tips of Carbon filiments Carbonized[1]

> Lampblack & sugar dont hold in filiment. ~~porous~~ hard[a] Carbon[b]
>
> Lampblack & Rosin good   Carbon holds well—not as well as Lamp Black and Tar[c]
>
> Lampblack Syrup & Anthracine dont stick to Carbon at all—porous Carbon
>
> Lampblack & Pitch— porous   no good—[2d]
>
> Lamp black and Tar quite good   adhears to filement.
>
> Lamp Black Gun copal Pulp Paper & Burnt Sy[ru]p Nothing left
>
> Pitch and Tar   Verry good
>
> Lamp Black & Anticene[3e]   Nothing left
>
> Plumbago & Pitch   Nothing left
>
> Plumbago & Syrup   Nothing left
>
> Plumbago & burnt sugar   No good
>
> Plumbago syrup[e] & Anticine   fair
>
> Plumbago & Anticine   No good
>
> Plumbago & Syrup   No good
>
> Plumbago, Tar & Pitch   fair

TAE

J. F. Ott
M.N.F[orce].

X, NjWOE, Lab., N-82-12-21:3 (*TAED* N150:2; *TAEM* 39:82). Document multiply signed and dated. [a]Interlined above. [b]Followed by dividing mark. [c]"not . . . Tar" written by John Ott and followed by dividing mark. [d]Remainder of document written by John Ott. [e]Obscured overwritten text.

1. Edison investigated the following substances for securing the ends of multiple twisted carbon fibers as an aid in manufacturing (see Doc. 2363). John Ott, who was working on the twisted filaments, made a "device for holding carbon and inside parts while admitting them to the flame for soldering" in mid-January. He explored other procedures for holding the parts together and, in February, completed a machine for twisting a fine copper wire to hold the filament ends until they could be electroplated to the lead-in wire. Edison executed another patent application on 17 February for temporarily securing filament ends with a "viscous carbonizable gum." N-82-12-13:9–17, 25–29, Lab. (*TAED* N150:5–9, 13–15; *TAEM* 39:84–88, 92–94); U.S. Pat. 476,528.

2. John Ott continued the list of results from this point.

3. Ott presumably meant "anthracine."

–2385–

From Pitt Edison

Port Huron Dec 27 1882

Dear Bro

I received leter from Insul saying he was agoing to Se[e] Bogart[1] I owe him 2000 but it is not due for two years and a half and as long as I pay the intrest he will let it run and by the time it is due I think the place will[a] bring at least $5000. for Port Huron is building up verry fast   I have got 180 acres of land up black river 6 miles which I only owe $750.00 due in yearly payments for the next four years and is worth at least $3000.00 I also have some personal property say $1000.00 besides the Rail Road stock which I consider yours as much as mine   also some property in Gratiot[2] worth ab[out] $1200.00 making in all property worth at least at this time $15 000.00 and I owe less than $4000 and $2750.00 of that is running on long term so you see what is troubling me at present is about $1200.00 or $1300.00 dollars to put me in easey shape   Insul says that you are hard pushed for money at present & I do not like to ask you to do to much for me under the circumstances and if you cannot of course I must sell something for what it will bring for thay are pressing me verry hard here every day and I must rais mony or thay will make me trouble and costs so let me know by return mail for I cannot put them off any longer.[3]

WPE

All well

ALS, NjWOE, DF (*TAED* D8214ZAD; *TAEM* 60:730). [a]Obscured overwritten text.

1. Adam Bogart was a lithographer at 27 Park Pl. in New York, whom Pitt explained held "a mortgage on my Home which I have been unable to pay the interest and he has Paid my Taxes for 1881 and unless I do something he will foreclose." He asked Edison to visit Bogart to request a reprieve. Insull promised to do so but cautioned that Edison was "very short of money." *Trow's* 1881, 140; *Rand's* 1879, 296; Pitt Edison to TAE, both 18 Dec. 1882, DF (*TAED* D8214ZAB, D8214ZAC; *TAEM* 60:728, 729); Insull to Pitt Edison, 21 Dec. 1882, Lbk. 15:059 (*TAED* LB015059; *TAEM* 82:55).

2. Fort Gratiot, Mich., a few miles north of Port Huron.

3. William Pitt Edison had made several requests to Edison for at least $1,000 to meet immediate obligations so that he would not have to sell either his home or stock in the street railway that he operated. In his 17 November 1882 reply Samuel Insull had noted that Edison was "very much pressed for money" but would try to send some in early December. Several months earlier, Pitt had understood Edison to promise capital for a prospective horse-selling venture, one of several occasions when he looked to his brother for financial help. Insull made several attempts to contact Bogart before succeeding about the middle of January 1883. In February, after having received the note and receipt for taxes (made in the name of Pitt's wife, Ellen J. Holihan Edison), Bogart sent payment instructions to Edison. Pitt Edison to TAE, 14 and n.d. Nov. 1882; Pitt Edison to Insull, 21 Sept. 1882; all DF (*TAED* D8214X, D8214Z, D8214W; *TAEM* 60:724, 726, 720); Insull to Pitt Edison, 17 Nov. 1882, Lbk. 14:440 (*TAED* LB014440; *TAEM* 81:1014); Insull to Pitt Edison, 18 Jan. 1883; Insull to Bogart, 18 Jan. 1883; Lbk. 15:172A, 173 (*TAED* LB015172A, LB015173; *TAEM* 82:95–96); Bogart to Insull, 22 Jan. and 14 Feb. 1883, both DF (*TAED* D8314C, D8314D; *TAEM* 64:683, 686).

—2386—

J. Pierpont Morgan to
Sherburne Eaton

[New York,] Decr 27th [188]2

Dear Sir,

Fully three weeks ago, I requested your Company to have my plant arranged in such a manner as to cause the minimum annoyance to my neighbors.[1]

Since that time, so far as I can discover, absolutely nothing has been done: in any event the annoyance, whatever it is, is far greater than it was then.

I must frankly say that I consider the whole thing an outrage to me, as well as the neighbors—& I am unwilling to stand it any longer.

Please let the matter have immediate attention. Yours truly

J. Pierpont Morgan

LS (letterpress copy), NNPM, Morgan (*TAED* X123BAH).

1. In September 1881 John Pierpont Morgan arranged for the Edison Electric Illuminating Co. to install a lighting plant in his house at 219

Madison Ave. The company charged only for the cost of the equipment and wiring plus 10%. The plant consisted of a Z dynamo and a K machine (among the first of this model placed in service), and 250 A lamps and 50 B lamps (increased to 269 and 116, respectively, by the end of the year). It operated only while an engineer was on duty between 3 and 11 p.m. Morgan was reportedly "delighted" when it began working on 7 June but was infuriated at the end of that month when Miller Moore took visitors through the still-unfinished house without permission. Morgan to Calvin Goddard, 12 Sept. 1881, Morgan; Johnson to TAE, 14 Dec. 1882; Eaton to TAE, 8 and 29 June 1882; all DF (*TAED* D8201X, D8226S, D8226ZAI; *TAEM* 60:30; 61:243, 317); Edison Co. for Isolated Lighting Brochure, pp. 6 and 10, 1 Sept. 1882, PPC (*TAED* CA002A; *TAEM* 96:103); Edison Electric Light Co. Bulletin 15:5–6, 20 Dec. 1882, CR (*TAED* CB015; *TAEM* 96:768–69).

After a neighbor, James M. Brown, complained about the noise and smoke, Morgan requested Edward Johnson and Miller Moore to place the machinery on india rubber supports to dampen vibration, to line the whole house with felt to minimize noise, and to reroute the smoke into another chimney further from his neighbors. Furthermore, in 1883 faulty wiring caused a fire in the library. Morgan's son-in-law Herbert Satterlee later recalled that the plant gave Morgan "a great deal of trouble" and suffered "frequent short circuits and many breakdowns." Satterlee concluded that it "was not in favor, either in the family or with the neighbors." Morgan to Brown, 1, 8, and 28 Dec. 1882; all Morgan (*TAED* X123BAG, X123BAH, X123BAJI); Satterlee 1939, 207–16; Strouse 1999, 230–35; Jonnes 2003, 3–11.

January–March 1883

The new year began with Edison working, as usual, on multiple fronts and in multiple guises. Although Sherburne Eaton was obliged to report in January that the previous year's attempt to develop an iron ore separation business at Rhode Island had met with a series of misfortunes, better news that month was the opening of the first, long-awaited, "village plant," using overhead wires, in Roselle, New Jersey.[1] Meanwhile, with the Pearl Street central station now well-established and running normally, attention turned to recruiting new customers in the First District, from the Stock Exchange to Fulton Market to federal government offices. At the same time, however, it was also found necessary to give a substantial discount to existing users who complained that the new service was costing more than gas.[2] Edison kept busy during the winter by drafting a detailed essay denouncing the Brush storage battery system of electric lighting, pursuing an unrealized plan for an electrical engineering department at Columbia University, and drafting a defense of electric lighting against the charge that it caused fires.[3]

Edison traveled in February to Boston (where an interview featuring more of his withering comments on the storage battery system had just been published), with Henry Villard, Charles Clarke, and Sherburne Eaton to establish an illuminating company there.[4] Edison then commenced a vacation in the South; Samuel Insull expected him to be away until mid-March, but he returned instead on 27 February, "rather unexpectedly."[5] He also began to plot an intrigue with patent attorney Richard Dyer against certain Patent Office personnel around this time, with a view to obtaining more favorable outcomes in interference cases. Certainly, Edison's patenting activity continued unabated during this period; seventeen applications he executed in these months eventually issued, covering lamps, dynamos, and distribution systems (particularly voltage regulation). Most urgent, however, and perhaps more worrying than he wished to admit, was the news from Edward Johnson in England that John Hopkinson had anticipated Edison's three-wire distribution system in a British patent. Hopkinson was also working on significant improvements to the Edison dynamo, which spurred Edison's own efforts on dynamos.[6]

This unwelcome surprise prompted Charles Batchelor to press ahead with planned European patent filings. This caused Edison to worry about the implications for his U.S. patents, while Batchelor complained about the lack of a coherent policy governing foreign patents. Across Europe, progress on some fronts—the establishment of an isolated plant company in Germany; plans to open a lamp factory in Brünn, Austria[7]—was more than offset by continuing obstacles to success in England and Italy, where competing financial and business interests among Edison and his many partners led to misunderstandings, conflicts, and hard feelings. As Edison told Joshua Bailey in a billing dispute involving the Milan station, "I am but a simple Stockholder in the European Coy & were I to give in to such a request as you make I would eventually be a Bankrupt Manufacturer." Edison's own thinking was succinctly expressed in his advice to Johnson to leave England and return to the United States at once, for it was "best to concentrate our efforts on an American certainty rather than an English possibility." Articulating his independence (and his frustration with British capital), he declared that "if the business is to be made a success it must be by our personal efforts and not by depending upon the officials of our Companies."[8]

On the domestic front, Edison had been implicated in a struggle between the interests of the parent Edison Electric Light Company, represented by directors Sherburne Eaton and Grosvenor Lowrey, and the manufacturing companies, particularly the Edison Lamp Company, over the question of how much of the net sales profit each party should realize. Francis Upton, who managed the lamp factory, wrote to Edison frequently about the unfavorable terms upon which lamps were sold to the parent company, complaining that "we are making asses of ourselves to give a soulless corporation ten cents on every lamp out of our own pockets." Delicately described by the Dun credit reporting agency as being "subject to the ups and downs that distinguish all the Companies based on the patents of Thomas A. Edison," the lamp factory was now in danger in being taken over by the parent company, just as it was finally becoming profitable.[9] When negotiations for a new lamp contract with Eaton and Lowrey fizzled, Edison, Upton, and Johnson issued a joint statement renouncing their efforts to obtain better terms for the manufacturers from the Electric Light Company.[10] The Edison Machine Works, by now a relatively stable enterprise, had been run with a free hand by superintendent Charles Dean and treasurer Charles Rocap. Early in the new year, Edison vested authority for the shop's finances in Samuel Insull. The idea probably originated with Insull, who bragged that after some contention, Edison "supported me in a bully fashion & I came out on top of the heap."[11]

In March, Edison turned his attention to the new project he had in mind with Johnson and Insull—the establishment of a Construction Department to build village plant central stations.[12] This was consistent with Edison's idea that he ought to refocus his attention from technical innovation to managing the business details of his enterprises, in order to see them prosper as he felt they deserved. And thus, by the summer of 1883, Edison would be claiming repeatedly in the press that he would be "simply a business man for a year. I am now a regular contractor for electric light plants, and I am going to take a long vacation in the matter of inventions. I won't go near a laboratory."[13]

1. See Doc. 2393; Edison Electric Light Co. Bulletin 17:5, 6 Apr. 1883, CR (*TAED* CB017; *TAEM* 96:809).

2. Sherburne Eaton to TAE, 2 Mar. 1883, DF (*TAED* D8327S; *TAEM* 66:732); Doc. 2409 n. 3.

3. Docs. 2389 and 2408; TAE, untitled essay, 29 Mar. 1883, DF (*TAED* D8320F; *TAEM* 66:23).

4. See Doc. 2403 n. 3.

5. Samuel Insull to Thomas Mendenhall, 27 Feb. 1883; TAE to Edward Johnson, 28 Feb. 1883, Lbk. 15:363A, 368 (*TAED* LB015363A, LB015368; *TAEM* 82:186, 187).

6. See Docs. 2402, 2407, 2414, and 2416.

7. See Doc. 2392; Francis Jehl to Insull, 22 Mar. 1883, DF (*TAED* D8337ZAS; *TAEM* 67:686).

8. Docs. 2399 and 2407.

9. Docs. 2387 and 2394.

10. See Doc. 2395.

11. See Docs. 2400 and 2409.

12. Doc. 2417.

13. "Mr. Edison to Be a Business Man for a Year," *American Gas Light Journal*, 3 Sept. 1883, Cat. 1016, Scraps. (*TAED* SM016047b; *TAEM* 24:103).

–2387–

From Francis Upton

EAST NEWARK, Jan. 4, 1883[a]

Dear Mr. Edison:

The Lamp Co. are about 25,000 lamps behind its orders. As soon as the Isolated Co. place their order for 50.000 lamps we can see daylight, and if we can have 40 cts. for lamps and the money due us we shall be able to run one more row in the pump room without danger of incurring increased liabilities.

It is exceedingly important that we have the five cents soon as they are now putting in very large orders for the U.S.[1]

The contract that Insull showed me I acknowledge to be poor as protecting the Lamp Co. from the Light Co. breaking the contract.[2]

From all I can learn regarding Maj. Eaton and Mr. Lowrey they do not intend to give us any contract but one that they can drive through in the future.

Taking this view I assent to the present version, in order that there may be a settlement.

If it were my private matter, I would notify the Light Co. that until we had a contract to protect us that we should charge 50 cts. for lamps, which is, as we are now manufacturing, only a fair price, giving us a reasonable profit.[3] As it is now, we are giving them about ten cents on each lamp, with every disposition shown on the part of those directing the policy of the company to take all the profits when any are to seem.

For example: The contract to give us only fair rights has now been hanging for a year and a half and the prospects are that it will hang indefinately, for Maj. Eaton will require even more time than Mr. Lowrey.—[4] The leading directors of the Edison Co. say that they wish to draw all companies into one grand company. That means that they will take the Lamp Factory and all its rights. Now that we have risked $200.000 in this place and pointed out the way that it can be made profitable they wish to reap the profits.[5]

I am not naturally suspicious but in this case I think that Maj. Eaton and Mr. Lowrey are leading us a wild goose chase and that they never intend to tie the Light Co's hands to any one place to get lamps.

I am looking to the future in wanting the contract made binding for some day it will be exceedingly profitable.

For the present we are making asses of ourselves to give a soulless corporation ten cents on every lamp out of our own pockets with no guarantee for the future, not even a permit to use the apparatus we have for this special ~~use~~ purpose. Yours Truly

Francis R. Upton.

ALS, NjWOE, DF (*TAED* D8332E; *TAEM* 67:97). Letterhead of Edison Lamp Company. ᵃ"EAST NEWARK," and "188" preprinted.

1. The lamp company's statement for January showed new domestic orders of 50,784 lamps, more than twice those of the preceding month. Edison Lamp Co. to TAE, 2 Jan. and 1 Feb. 1883, DF (*TAED* D8332B, D8332T; *TAEM* 67:92, 150).

Upton referred in a previous letter to "the five cent rule" and "the five cents advance," but his meaning is not clear. He may have been alluding to the guaranteed five-cent profit per lamp specified in drafts of the new contract, one from 1882 and the other from 1883 but retroactive to 1 September 1882. Upton to TAE, 13 Dec. 1882; TAE draft agreements with Edison Electric Light Co. and Edison Lamp Co., 1882 and [1883]; all DF (*TAED* D8230ZCQ, D8230ZAN, D8332ZDV; *TAEM* 61:845, 757; 67:280).

2. The proposed contract (see note 1) contained provisions by which the Light Co. could circumvent or terminate the agreement, or make purchases from other manufacturers. Upton also worried that the longer the contract was unresolved "the harder it will be to fix. Then now that our business is growing profitable their greed will be excited and we shall be at their mercy." Upton to TAE, 13 Dec. 1882, DF (*TAED* D8230ZCQ; *TAEM* 61:845).

3. The average lamp manufacturing cost over the previous six months exceeded 44 cents. This included experimental expenses of about $5,000 as well as fixed costs. Upton to TAE, 15 Jan. 1883, DF (*TAED* D8332J; *TAEM* 67:107).

4. Insull sent the most recent version of the contract to Sherburne Eaton on 16 January, who quickly produced 22 pages of critical comments. Insull to Eaton, 16 Jan. 1883, Lbk. 15:152A (*TAED* LB015152A; *TAEM* 82:84); Eaton memorandum, with TAE marginalia, 19 Jan. 1883, DF (*TAED* D8332M; *TAEM* 67:116).

5. Eaton asserted in his 19 January memorandum (see note 4, pp. [18a]–19) that "the Light Co. ought really to do all its manufacturing itself, including the manufacture of lamps." He urged strongly that it retain the right to purchase the lamp factory's business for twice its appraised value. A recent inventory put the value of the Edison Lamp Co.'s property, stock, and outlays "Sunk in Experiments" at $209,411.70. Edison Lamp Co. memorandum, 14 Nov. 1882, DF (*TAED* D8231ABR; *TAEM* 61:946).

-2388-

To Francis Upton

[New York,] January 5 1883—

Upton

Please have Lawson put ½ doz Carbon forms with Bamboos in ready[a] for Carbonztn in dish & cover with Boiled Linseed Oil & then heat oil[b] very gradually up to say 500 deg Fahr— when most of the Carbonztn will take place. Then you can either wash the forms in turpentine after taking out of dish & then put in regular run & Run up to high heat as regular or you can wash with turpentine & take out loop & put in new forms— The theory is that the heat is <u>applied</u> perfectly <u>even</u>[c] & thus you prevent internal stress, & also prevent oxidation at the point where the Carbon is the most sensitive[1]   Let me know how you come out with it

T A Edison

⟨Filled Jan 12/83   J.W.L[awson].⟩[d]

ALS, NjWOE, DF (*TAED* D8333A; *TAEM* 67:300). [a]Obscured overwritten text. [b]Interlined above. [c]Multiply underlined. [d]Marginalia written by John Lawson.

1. John Ott recorded the next day that bamboo carbonized in linseed oil at 600 degrees "worked well"; he also tried manila fibers. Ten days later he soaked forms in linseed oil, boiled them, then carbonized fibers in them. Ott did not give results and Edison apparently did not pursue this line of experiment. However, Edison executed a patent application in February for a sealed carbonizing form that used different means to the same ends: keeping the strain "even and constant upon every part of the filament" and sealing out oxygen. N-82-12-21:9, Lab. (*TAED* N150:5, 8; *TAEM* 39:84, 88); U.S. Pat. 287,522.

-2389-

Essay: Storage Battery

[New York, c. January 5, 1883[1]]

STORAGE BATTERIES.

Statements are made to the subsidiary Brush companies that they have a large investment laying idle 10 hours daily and that with the storage batteries and incandescent lamps they would be enabled to utilize their plant all day long and thus earn a great deal of money. This appears to be a self evident proposition, but the fallacy lies in the fact that the batteries required to store the product of the station would call for an increased investment twice as great as already invested in the station—which increased investment has an enormous depreciation 25%. The negative plates of all secondary batteries are soon eaten away and have to be renewed frequently. The batteries rapidly diminish their capacity for storage and have to be made up fresh.[2]

Only about two thirds of the current in a battery can be taken out as the current otherwise would be too weak to bring up the lamps. Hence there must be one third greater investment in batteries on this account. In addition there must be extra batteries which have to be thrown in from time to time to keep up the candle power of the lamps. The batteries are liable to all kinds of disorders there is creeping across the space between the electrodes of metallic lead produced by electrolytic depositions; there is local action in the battery which causes the stored up energy to be rapidly dissipated; the sulphuric acid gradually combines with the oxide of lead which covers the active material of the battery increasing its resistance—the exhaustion of the sulphuric acid stops the action of the battery.

The experiments on the best form of secondary by Tresca of the Conservotiore des Arts et Metres[3] shows that 190 lbs. of material is required to commercially store one horse power for one hour—[4] The cost of the completed battery with its mechanism—the labor—the power required for forming—will bring its cost up so that it cannot be sold with any degree of profit for less than 35 cents per lb., making say $6.50 per light.

With this data take a Cotton mill requiring 250 lights, where the longest time of lighting in any one day is two hours, and with 10 lamps of 16 candles per horse power of current.

Battery for 250 lights at $6.50 per light, per hour for two hours amounts to $3250. About 50 per cent was the return obtained by Tresca, from the dynamo but it would not be more than 45 per cent from the indicated horse power of the engine, but say 50 per cent. 250 lights require 25 horse power per hour to work the lamps, 50 per cent loss allowed for between the engine and the lamps. 50 indicated horse power per hour is required. Hence for two hours 100 horse power is requisite.[5] If now a small dynamo is used to charge for 8 hours, it will require 12 horse power; price of dynamo $1200., making total investment $4450. 100 horse power for one hour (or 12 for 8 hours) consumes 400 lbs. coal. This for 300 days would require 60 short tons at $4. $240.

Coal . . . . $240.
Depreciation and interest on small dynamo . . . . 120.
Depreciation on batteries (25 per cent) . . . . 812.
Interest on batteries . . . . 195.
Hence expenses . . . . 1367.
To work direct would require one 250 light
    machine at 3000 . . . . 3000.
35 horse power for one hour or 70 for two hours;

4 lbs. coal, 280 lbs. 300 days, 42 short tons at $4. Coal . . . 168.

Interest and depreciation on dynamo . . . . 300.

Investment . . . . 3000.

Running expenses . . . . 468.

If the hours of lighting are increased to 3 hours or any more hours, the investment by the direct system does not increase, only the coal.

While by the storage, the investment both in dynamos and batteries increase directly to the hours of lighting. If instead of two hours lighting we require four hours lighting, the figures would stand thus:

BATTERY SYSTEM.

Investment in dynamo . . . . $2400.

Investment in batteries . . . . 6500.

Coal . . . . 480.

Depreciation and interest small dynamo . . . . 240.

Depreciation of batteries . . . . 1624.

Interest on batteries . . . . 390.

Hence investment . . . . 8900.

Running expenses . . . . 2734.

Direct system dynamo . . . . $3000.

Coal . . . . 336.

Interest and depreciation—Extra depreciation for
    extra two hours . . . . 350.

Hence investment . . . . $3000.

Running expenses . . . . 686.

With six hours lighting the dynamo required to charge the storage batteries in eight hours would be as large as the dynamo used for lighting direct, hence the charging dynamo could light the works direct without the batteries.

Even supposing that Mr. Brush had obtained a theoretically perfect battery, the figures would be enormously against him. The fact of the matter is that the "shorter the period of lighting required from storage batteries the greater their economy, and this reaches a maximum when they are not used at all."[6]

TD, NjWOE, DF (*TAED* D8320ZAQ; *TAEM* 66:106).

1. This is likely the paper on "defects in the storage battery system" that Edison was writing in early January and sent to Henry Howard, former governor of Rhode Island and president of Armington & Sims, for his talk at a 15 January dinner of railroad officials. After receiving the paper Howard asked Edison for permission to publish "the substance of it in the Prov[idence] Journal striking out all reference to Brush and not

mentioning your name at all." Edison apparently did not approve of the publication but included some of this material in an extensive interview that appeared in the *Boston Herald* on 28 January. Miller Moore to TAE, 5 Jan. 1883; Howard to TAE, 8 and 11 Jan. 1883; all DF (*TAED* D8303C, D8303F, D8303K; *TAEM* 64:19, 21, 27); "Mr. Edison on Storage Batteries," excerpted from the 28 Jan. *Boston Herald* in the 2 Feb. 1883 Edison Electric Light Co. Bulletin 16:31–36, CR (*TAED* CB016275; *TAEM* 96:804).

Edison's essay may have been prompted by efforts of the Brush-Swan Electric Co. to publicize their storage battery system of electric lighting during the last half of December 1882. Sherburne Eaton reacted by writing letters to the *New York Times* and the *New York Journal* in which he accused the Brush system of bringing the "death-current" of high-voltage arc light systems into homes. Brush-Swan Electric Co. to TAE, 18 Dec. 1882, DF (*TAED* D8220ZAZ; *TAEM* 60:826); Eaton to the Editor, 25 Dec. 1882, *New York Times,*27 Dec. 1882, Cat. 1327, item 2229; Eaton to the Editor, 25 Dec. 1882, *New York World,* 27 Dec. 1882, and "Edison and Storage Batteries" *New York Tribune,* 23 Dec. 1882; Cat. 1343:1–2; all Batchelor (*TAED* MBSB52229, MBJ002002, MBJ002001; *TAEM* 95:219, 237).

This document is substantially the same as Edison's handwritten draft in DF (*TAED* D8220ZBH; *TAEM* 60:835); see also "Defects of Storage batteries employing Lead Electrodes & oxides" in DF (*TAED* D8220ZBI; *TAEM* 60:841).

2. In his draft on "Defects of Storage batteries employing Lead Electrodes & oxides" (see note 1) Edison detailed eight technical problems plaguing storage batteries:

> 1st— Finely divided lead decomposes water.
> 2nd   A sulphate of Lead is formed taking the acid from the solution
> 3rd   The sulphate of Lead depositing upon the peroxide renders it partially inert
> 4th   The peroxide formed gradually becomes detached from electrical contact with the electrode.
> 5   Lead eats off at the surface of the solution.
> 6th= Resistance of the battery increases as the battery is discharged—making a constantly diminishing economy.
> 7th   gradually diminishing storage capacity after certain number of times charging to to gradual inertness of the peroxide of Lead.
> 8th. gradual coating of surface of electrode with inert peroxide not reducable=

3. Henri Edouard Tresca (1814–1885) was a French mathematician and engineer who taught at the Conservatoire National des Arts et Métiers in Paris. In early January 1882 he was president of a commission that conducted tests there of the Fauré storage battery. The commission presented its conclusions to the Académie des Sciences in March but earlier reports appeared in *Engineering* and *La Nature. ABF,* s.v. "Tresca, Henri Edouard"; "The Faure Accumulating Battery," *Engineering,* 20 Jan. 1882, Cat. 1033:124, Scraps. (*TAED* SM033124a; *TAEM* 25:372); M.C. Faure, "Capacité d'Emmagasinement et Rendement des Accumulateurs," *La Nature,* 10 (April 1882), 303.

4. A Fauré battery weighing 75 pounds could supply 1 horsepower at an efficiency rate of 80%. Vinal 1955, 4.

5. Edison presumably meant the cost to produce 50 horsepower per hour for two hours as the horsepower to run the lamps for each hour would remain the same.

6. The source of this quotation has not been identified.

–2390–

From New York Department of Taxes and Assessments

New York, January 8th, 1883.

Hours for correction of Assessment 10 A.M. to 2 P.M.

Please bring this notice with you.[a]

R. Book. Line 4081[b]   Page 259[b]

Mr. Thomas A Edison[c]

No. 65 5th Ave[d]

DEPARTMENT OF TAXES AND ASSESSMENTS,

Staats Zeitung Building, Tryon Row.

You are hereby notified that your Personal Estate[1] for 1883 is assessed at $10,000[b] exclusive of Bank stock, and that the same, if erroneous, must be corrected before the Commissioners, on or before the 30th day of April next, or it will be confirmed at that amount; from which there will be no appeal. By order of the Board,

JAMES C. REED, Secretary.[2]

PD, NjWOE, DF (*TAED* D8303E; *TAEM* 64:20). [a]Line preceded by small printed icon of right-pointing fist or index. [b]Number handwritten in space provided. [c]"Thomas A Edison" handwritten in space provided. [d]"65 5th Ave" handwritten in space provided. [e]Number handwritten in space provided.

1. Edison's "personal estate" may have included equipment at the Edison Machine Works, stock holdings, and personal assets. New York City's rate of taxation varied according to the city's appropriations and the total valuation of property, personal and real; it was set at 2.25% for 1882. *Ency. NYC,* s.v. "taxes"; "The Mayor's Message," *New York Times,* 2 Jan. 1883, 3.

2. Col. James C. Reed (1838–1897), a Union Army veteran, was private secretary to Chester Arthur when he was appointed Secretary of the Board of Tax and Assessment Commissioners in February 1882. "City and Suburban News," *New York Times,* 11 Feb. 1882, 8; Obituary, ibid., 24 Apr. 1897, 7.

–2391–

To Henry Rowland[1]

NEW YORK   Jany 10 1883[a]

Dear Sir

Referring to yours of Dec. 10th

All the papers relating to your case were in the hands of Maj Wilber and I had reason to believe that the application had long since been filed. Since the receipt of your letter I have looked into the matter and find that the Major still has the papers in his possession but as he is absent from the City a great part of the time I have been unable to get them but have the promise that he will send them to me when he returns to the City. Other papers are now in course of preparation by Mr. Dyer who will call upon you with them in the course of a few days in case we do not get the original ones from Major Wilber who is not now in the employ or service of myself or our Company[2]   Yours Very Truly

Thos A. Edison   I[nsull]

Please excuse long delay in replying to your letter   It is owing to the fact that I have been endeavoring to get possession of the papers made out by Mr Wilbur

L, MdBJ, HAR (*TAED* X100AC1). Letterhead of Thomas A. Edison. A letterpress copy of this letter, with a line of text missing, is in Lbk. 15:130 (*TAED* LB015130; *TAEM* 82:79). ᵃ"NEW YORK" and "188" preprinted.

1. Henry Rowland (1848–1901), an experimental physicist, became the first professor of physics at the Johns Hopkins University in 1875. He held this position until his death and was instrumental in training the next generation of American physicists. During his career he studied magnetic and electromagnetic phenomena, thermodynamics, and spectroscopy. *ANB*, s.v. "Rowland, Henry Augustus"; Doc. 1910 n. 2.

2. While a student in 1868, Rowland had designed a dynamo with an armature similar to that later adopted by Siemens & Halske. In July 1880 Edison offered to defray his expenses for filing and defending a patent on his prior design in order to prevent Siemens from obtaining a U.S. patent. Rowland's case was apparently among the many which Wilber failed to file with the Patent Office. See Docs. 1951, 1964, 2021, and 2323.

On 1 December 1882 Edison wrote Rowland that "in order to legally establish and maintain your right as the discoverer of the Siemens-pattern of dynamo electric machine, it may be necessary to bring the matter forward within a short time. The application papers, contract &c. do not seem to be in my possession. If you have them, will you kindly forward them to me . . . Siemens is prosecuting his case in the Patent Office vigorously, and may obtain a patent with broad claims if not stopped by an interference with an application filed in your name." Rowland replied on 10 December that he had sent "all papers relating to the magneto-electric machine to Mr. Wilbur long ago and even forgot to reserve one of the copies of the contract, which I would like sent back." Dyer visited Rowland on or about 13 December and subsequently filed a new application for him. It was rejected as unoriginal, and Rowland evidently did not complete an affidavit required to have it examined again. TAE to Rowland, 1 Dec. 1882, Lbk. 14:491 (*TAED* LB014491; *TAEM* 81:1035); Rowland to TAE, 10 Dec. 1882; Dyer to Insull, 8 Jan. 1883; both DF (*TAED* D8248ZDJ, D8370J; *TAEM* 63:559, 70:922); TAE to Rowland, 13 Dec. 1882, HAR (*TAED* X100AD); Doc. 2021 n. 6.

Charles Batchelor to
Sherburne Eaton

[Paris,] Jan 16 1883

My dear Major,

I have just returned from a week in Berlin with Mr. Porges and Bailey both of whom I have left there. The result of this visit will be a contract with parties there for the Isolated business   The contract will be made with the German National Bank[1] and the house of Jacob Landau of Berlin[2] and Sulzbach of Frankfort[3]   The terms were not fully agreed on when I left but there is I think no doubt of its satisfaction. The terms are <u>roughly</u> these= Cash payment 250,000 marks 17½ percent shares of jouissance[4] and a small royalty on lamps and dynamos—[5]

Whilst in Berlin I took the opportunity to send for Mr. Knoop of the firm of Thode and Knoop of Dresden[6] so as to go over our German patents with him; and found much to my astonishment that Mr. Knoop was a man who knows absolutely nothing of electricity. I found that whenever Edison put in an application at the Patent office Siemens immediately filed as objection to it and without asking my advice at any time Knoop has replied to these objections and simply taken from the office whatever they gave densely ignorant of the fact that in some cases the essential principle has been cut out Siemens has now seven or eight objections in against our pending patents   I gave him some points on the most prominent & which have to be replied to quick and instructed him [never?][a] to reply without consultation with us.

Bailey is now making arrangements to have our patents in Germany arranged and put in good shape before application is made; by an employé of the Patent office in his own time; ~~aft~~ also he is going to engage the services of a first class expert—

The best part of our visit will be the contract with Siemens which I believe will go through—he is paying us a "prime d' inventeur"[7] (for all incandescent light plants that he manufactures or sells) per absorbed horse power—[8]

In our contract with the German Company it became necessary to define what was an isolated plant and what was a central station and I gave the following:—

Central Stations are establishments for manufacturing, measuring and delivering electric current for the production of light or power[b] in a given district to any or all inhabitants in such district who may desire it. Any installation made for supplying light or power to a single individual or several individuals and not comprising the general public is an isolated plant—

A number of times I have been going to write you regarding

our President M. Porges. Quite a number of times he has spoken to me about the want of confidence the Edison Light Co had in him   He seems to feel very sore that although he has founded the 3 Edison Companies he is never allowed to have any intercourse between yourself and Edison without using Puskas and Bailey as mouthpiece— He is certainly one of the most disagreeable men it was ever my lot to be cast with but if you could send him bulletins, items of news and above all a letter on the Central Station it would please him greatly   Respectfully Yours

Chas Batchelor

ALS (letterpress copy), NjWOE, Batchelor, Cat. 1239:431 (*TAED* MBLB4429; *TAEM* 93:771). "<u>414</u>" written at top to indicate sequential letter number. [a]Illegible. [b]"or power" interlined above.

1. The Nationalbank für Deutschland was founded in Berlin in 1881, with collaboration from Jacob Landau. Emden 1938, 210–11.

2. Originally from Breslau, Jacob Freiherr von Landau (1822–1882) founded the prominent bank of Berlin that bore his name. He backed Edison's three French companies in 1881 and provided capital for Emil Rathenau in 1882. Emden 1938, 210–11; *NDB* 13:481–82; Mosse 1987, 138, 244–45.

3. Gebrüder Sulzbach, formed in 1851 by Siegmund Sulzbach (b. 1813), achieved prominence as an issuing house under Rudolph Sulzbach (Siegmund's brother, 1827–1904). Rudolph participated actively in the establishment of major firms such as the Allgemeine Elektrizitäts Gesellschaft (AEG) and the Deutsche Bank. Emden 1938, 398.

4. That is, dividend paying shares.

5. The Deutsche Edison Gesellschaft für angewandte Elektrizität (DEG, or German Edison Co. for Applied Electricity) was incorporated in March 1883 with a capitalization of 5,000,000 marks. The DEG simultaneously signed a contract with Siemens & Halske concerning sublicenses and their relative fields of endeavor. German Edison Company for Applied Electricity articles of association, 13 Mar. 1883; Compagnie Continentale Edison Co. agreement with Société Électrique Edison, Sulzbach Brothers, Jacob Landau, and National Bank of Germany, 13 Mar. 1883; both DF (*TAED* D8337ZAO, D8337ZAL; *TAEM* 67:670, 662); Feldenkirchen 1995, 15–16, 489 nn. 8 and 9; Hughes 1983, 67–68.

6. This firm was Edison's patent agent in Dresden since at least 1878. In April 1882 Batchelor ceased being an intermediary and requested that the firm communicate directly with the Compagnie Continentale on patent matters. See Doc. 1625; Batchelor to Edison Electric Light Co. of Europe, 13 Apr. 1882, Cat. 1239:230, Batchelor (*TAED* MBLB4230; *TAEM* 93:637).

7. A French term meaning an inventor's royalty.

8. A March 1883 contract gave Siemens & Halske rights to manufacture under Edison's electric light patents in the German Empire. For lamps, Siemens & Halske was to pay a royalty of one-third the cost price per lamp in the U.S. On complete incandescent lighting systems installed by Siemens & Halske, Edison was to receive a royalty of 25 marks

per horsepower for the first 50 horsepower and 32 marks per each additional horsepower. TAE agreement with Siemens & Halske, Edison Electric Light Co. of Europe, Compagnie Continentale Edison, Gebrüder Sulzbach, Jacob Landau, and Nationalbank für Deutschland, 13 Mar. 1883, DF (*TAED* D8337ZAK; *TAEM* 67:644).

—2393—

Sherburne Eaton
Report to Edison Ore
Milling Co.

[New York, January 16, 1883[1]]

To the Stockholders of the Edison Ore Milling Co. Limited:

At your last annual meeting a report was submitted to the stockholders of the progress made with the separation of iron ore from the sea shore sand at Quonocontaug, Rhode Island, and also with the general business of the Company.[2]

It appeared from such report that Mr. Edison was carrying on experiments regarding the milling of gold and silver bearing ores, also that many inquiries were being made for our magnetic ore separators. Some of these experiments on gold and silver bearing ores have been made at Menlo Park under Mr. Edison's direction, but owing to pressure of business, he has not yet been able to accomplish much in this direction. He has, however, made several inventions as to the separation of magnetic ore and of free particles of gold, for which patents are applied.[3] We have also received a great number of inquiries as to the capacity, &c, of the magnetic ore separators belonging to this Company, to which the fullest replies have been given, but we have not as yet succeeded in disposing of any machine.[4] So that except for our ore separating works at Quonocontaug, R.I., mentioned below, there is very little business to report for the past year.

It also appeared in the last annual report that separating works had been established on the sea beach at Quonocontaug, Rhode Island, and that a quantity of iron had already been separated and sold. At that time we had just secured our first customer for this ore, namely, The Poughkeepsie Iron and Steel Company.[5] We had already sold them a cargo of iron, which had given such satisfaction that we received from them a further order for 200 tons, pending the shipment of which a proposal was made to us to enter into a contract with them to supply them with all our product at the rate of about 200 tons per month. About the time we were ready to ship 150 tons of the order above spoken of, in fact just as the vessel was about to start for the beach, we received orders from the Poughkeepsie Company not to ship this order. On inquiring into the cause of their rescinding this order, we found that they had burnt out their furnaces, were in financial difficulties, and had closed their works at Poughkeepsie. Therefore, negotiations with them were at an end. We thus met with a great drawback. This was the only customer we had, and this one had been obtained after a great deal of trouble and some expense.

The reason of this lack of demand for this ore was and is that the ore is in such small particles and requires so great a heat to melt it that it is not possible to work it in an ordinary furnace, and can only be smelted in a furnace of special construction.

The ore which should have been shipped to the Poughkeepsie Company was thus thrown on our hands and we were compelled to seek another market for our product. Pending our attempts in this matter, we had to keep on a part of our working force at the beach, so as to be ready to start up at such time as a demand could be created. We also sent a great number of samples ofto iron and steel manufacturers all over the country, not only to produce an immediate market, but also with the intention of creating a very large demand for this ore, in order that there might also be a demand for our separators in different parts of the country. All this, of course, made it necessary to spend money for our pay roll at the beach and for hauling the samples to the nearest railway station, about 6 miles away.

After some time we received an order from Messrs. Shimer & Co., Philadelphia,[6] for 30 tons of ore, as a sample.[7] They at the same time asked us to enter into a contract with them giving them the option to take 700 tons more within a specified time (30 days) at $9 per ton. Inasmuch as we were informed that these were good parties, and as we desired to get a contract for a definite quantity of ore, we accepted such contract and shipped the sample 30 tons. Then came another period of waiting and endeavoring to sell more ore. At the expiration of 30 days Messrs. Shimer & Co requested 30 tons more for a further trial before ordering a large quantity. This also was shipped and we again waited. We were informed, however, about the expiration of the time that they could not use this ore, and we were thus thrown again on our resources to find a market.

With this end in view a great number of people were seen and samples sent out, and after some time we obtained an order to ship 50 tons of ore to the American Swedes Iron Company at Rockaway, N.J.[8] Preparations were made to ship this cargo, and a vessel went down to the beach for the iron, but parted her cable, went ashore on the rocks and was lost. Another vessel was obtained and the iron shipped. This last vessel, however, only got us far as Stonington, when it was discovered that she was leaking badly and could not go any further. Then came another transfer of the cargo to still another vessel, by which the iron was at length delivered. All this, of course, meant delay and expense.

In the meantime, as before, our payroll and some other minor expenses had to be met. Then after the ore was received at Rockaway, the American Swedes Company said that it was not equal to sample and that they could not take it. This involved the necessity of sending for our Superintendent from Rhode Island[9] to meet them and demonstrate that we had shipped the ore according to our contract and sample. He came on here and went to Rockaway and after some trouble succeeded in bringing them to accept the ore.

The American Swedes Iron Company smelted this ore in their furnaces, which are specially designed for ores of this character, and were so well pleased with the results that they ordered another cargo of 50 tons a week if this shipment was good. The ore was delivered to them and they again objected to it as not being in accordance with their sample. This sample was one which they claimed had been given by us to them and was entirely clean. It was so clean that our Superintendent and others gave it as their opinion that it had been cleaned with a hand magnet. Our cargo of 50 tons was not as clean as the sample, but was as we agreed to furnish it. The upshot of the matter was, after another visit of our Superintendent to them, that we came to the conclusion to make a reduction on the price of the ore rather than go to the expense of moving it away and running it through the separator again. This decision was also based on the fact that these people seemed disposed to give a great deal of trouble and did not pay promptly, besides, on the last visit of our Superintendent to their works, he saw that their mills were stopped and they were doing no work. This was not during the recent strike in the iron trade, but was soon after.[10]

Our experience with the American Swedes Iron Company, not only with these two cargoes of ore, but for some time previously, warranted us in coming to this conclusion that we should not be justified in attempting to do further business with them, as they have been uncertain and not to be depended upon.

Inasmuch, therefore, as we had no other customers for ore, and it being uncertain when a demand therefor would exist, it has been thought best to close our works at Quonocontaug, bring back the separator to New York and so conclude the business of separating ore ourselves. This has been done, and our expenses at Rhode Island are now at an end.[11]

In commencing this undertaking in August, 1881, we were assured by people in the iron business that there was an unlimited market for all the ore we could produce, and it was thought that we could realize the money invested, together with a profit. Our separator had not at that time been practically tested. It is true that Mr. Edison had made a great number and variety of tests at Menlo Park, and the machine had also been worked on the beach at Quogue, L.I. for a short time,[12] but none of these tests had demonstrated what work the separator was capable of doing when it came to run week in and week out on a commercial basis. It was thought both necessary and desirable that an experiment should be made to furnish us with such data, and, therefore, the operations at Rhode Island were commenced. At the start and until after the first cargo of iron had been shipped, there did not appear to be much difficulty in obtaining a market for our ore, and we expected to make a profit of about $4 per ton, but our market failed as above set forth and sales became spasmodic, thus putting us to expense without being able to realize any immediate returns.

Again, we have been placed under nearly every possible disadvantage in our operations at Rhode Island; we were 6 miles away from the nearest railway station, and 10 miles away from the nearest large town. Hauling was expensive and not always to be obtained, we had no harbor or inlet in which to load vessels, our beach being right on the ocean and without shelter of any kind. It has for this reason been difficult to get a vessel to take a cargo from the beach. In fact there have been an innumerable host of petty details, annoyances and drawbacks connected with the operations at Quonocontaug which, apart from pecuniary matters, have been met and overcome with no small degree of patience and energy.

Notwithstanding all these disadvantages, however, we have made a practical demonstration of what can be done with the Edison Ore Separator, not only in regard to how much ore can be separated in a day and at what expense, but also in regard to the best method of setting up the necessary machinery to be worked in connection with the separator. All these conclusions have been arrived at, however, by means of numerous experiments, the cost of which, is included in the plant and running expenses at Quonocontaug.

It should be said, that if we had had a regular market for our separated ore we could have made a profit of about $3.50 or $4 per ton, even after taking into consideration all the disadvantages under which we were placed at Quonocontaug. It was upon this basis that the plant was kept in existence, and the expenses continued for so long a time, for the reason that if we could, at any time before closing, have found a regular market for all our product, we could have realized all the money we have spent and possibly made a profit besides. The difficulty is not with the ore, as that is of the best quality, but there is no furnace in existence which will successfully and continuously smelt this ore, both on account of its fineness and of its being exceedingly hard and tough.

We submit herewith a summarised statement of the money spent on these experiments at Rhode Island and of the money received from sales of ore, &c.[13] This statement shows the expenditures to have been as follows:

Plant	$2,672.39	
Running Expenses	4,144.00	
Miscellaneous	508.17	$7,324.56

And the receipts for ore sold

to amount to	1,513.12	
Account due	250.00	
Sale of building	129.00	
Sale of engine	600.00	2,492.12
Balance		$4,832.44

Showing the cost to the Company of the experiments at Rhode Island to amount to $4,832.44.

It is not without a feeling of regret that your Board has felt obliged to cease operations at Quonocontaug without having at least realized enough from the sale of ore to pay back the money invested. Such is the fact, however, and after the most strenuous efforts which have been made to interest possible purchasers of ore, and after the diligent enquiries which have been made to find a suitable furnace in which to smelt the ore, both without success, we have deemed it most expedient for the best interests of the stockholders to close out the business and save any further expense.

We believe, however, that during the present year there will be a number of separators required for large deposits of magnetic iron in Canada and elsewhere. Mr. Edison also believes that during the present year he will be able to devote some time to the continuation of his experiments for the profitable milling and treatment of low-grade silver and gold ores, in which, if he is successful, there will be a future of vast profit for the Company.

D, NjWOE, CR (*TAED* CG001AAI3; *TAEM* 97:420). Written by William Meadowcroft. A typed copy is in DF (*TAED* D8368C; *TAEM* 70:802).

1. This is the date of the stockholders' meeting at which Eaton presented the report.

2. Eaton report to Edison Ore Milling Co., 17 Jan. 1882, CR (*TAED* CG001AAI2; *TAEM* 97:416).

3. Edison had at least two pending patent applications for ore separators. He filed Cases 339 and 340 in August 1882 (both were likely completed a full year earlier). Each was rejected, but the latter one resulted in a patent in 1889 (Patent Application Casebook E-2537:9–10 [*TAED* PT021008, PT021010; *TAEM* 45:733–34]; U.S. Pat. 400,317). In response to a stockholder's inquiry in February 1882, Edison stated that he was continuing his experiments "with every promise of substantial results." In answer to a similar question in July, however, he said that he had dropped the subject to give all his time to electric lighting. In 1884, Eaton announced to stockholders that "it is not probable that any further considerable experiments will be made" in the near future. By that time the company owed Edison about $1,650 for equipment and experiments. Edison did not return seriously to the subject until 1887. TAE marginalia on George Fitton to TAE, 6 Feb. 1882; TAE marginalia on W. O. Arnold to TAE, 31 July 1882; both DF (*TAED* D8246C, D8246P; *TAEM* 63:272, 289); Eaton Report to Edison Ore Milling Co., 15 Jan. 1884, CR (*TAED* CG001AAI4; *TAEM* 97:426); Israel 1998, 341–62.

4. Although the company received several inquiries about the capacity and operating cost of the separator, it was unable to sell or lease any machines by January 1884. Eaton report to Edison Ore Milling Co., 15 Jan. 1884, CR (*TAED* CG001AAI4; *TAEM* 97:426).

5. See Doc. 2246.

6. This firm was likely the same as, or related to, Messrs. Shimer & Co. of Milton, Pa. (near Philadelphia), a noted maker of cutting tools. "The Shimer Cutter Heads for 'Ship-Lap,'" *Manufacturer and Builder* 15 (1883): 250.

7. In April 1882 Shimer & Co. bought a 25-ton sample of ore for $9 a ton, with the option to buy another 700 tons if it proved satisfactory. Shimer & Co. agreements with Frank Froment and Edison Ore Milling Co., both 28 Apr. 1882; DF (*TAED* D8246L, D8246M; *TAEM* 63:282, 283).

8. This company began operations in 1868 and attempted to make iron bars directly from ore without the intervening step of pig iron. It also used black-sand ore from eastern Long Island and magnetic ore separators made by a Rockaway, N.J. foundry. By 1882 the firm was in financial trouble. Halsey 1882, 357; Conley to Eaton, 12 June 1882, DF (*TAED* D8247A1; *TAEM* 63:341).

9. M. R. Conley.

10. The Amalgamated Association of Iron and Steel Workers called a strike in the summer of 1882. Brody 1960, 50–51.

11. At the end of December the company instructed Conley's assistant, John Beebe, to close the works and ship the separators and Fuller dynamo to the Edison Machine Works. Meadowcroft to Beebe, 29 Dec. 1882, 6 Jan. 1883; Eaton to Charles Dean, 9 Jan. 1883; Eaton to Robert Cutting, Jr., 15 Jan. 1883; LM 5:258, 261, 267, 271 (*TAED* LM005258, LM005261, LM005267, LM005271; *TAEM* 84:288–89, 291, 293).

12. See Doc. 2093.

13. For accounts and equipment inventory at the Quoncontaug site see Mining—Edison Ore Milling Co.—Accounts (D-82-47); Treasurer's Report, Jan. 1883; Conley to Meadowcroft, 25 Dec. 1882; all DF (*TAED* D8247, D8368D1, D8246T; *TAEM* 63:301, 70:813, 63:296).

-2394-

*R. G. Dun & Co.
Credit Report, Edison
Lamp Co.*

[Newark?][1] Jany 19/83

W.D.W.[2] This concern seems to be subject to the ups and downs that distinguish all the Companies based on the patents of Thomas A. Edison   sometimes there is plenty of money & again they are very short but parties selling say their is so much money committed to his projects that they see no special risk in crediting in reason   at the same time the interests involved are so large & diversified there are so many elements of speculation in the whole business of Electric lighting & especially in the system for which this concern supplies Lamps that without some explicit assurances from those interested no general recommendation would be of value as a basis of credit   so far they have paid & are in credit with those who are in a position to know as much about them as any one outside 7693. 5284. 400[3]

D (abstract), New Jersey, Vol. 36, p. 118, R. G. Dun & Co. Collection, Baker Library, Harvard Business School.

1. See Doc. 2097 n. 2.

2. Presumably the initials of the agent making the report. The book containing the agents' names has been lost.

3. The editors have not identified the meaning of these numbers.

-2395-

*And Edward Johnson
and Francis Upton
Draft to Sherburne
Eaton*[1]

N York Feby 1/83

Sir:

Since all our efforts ~~at~~ to obtain a more satisfactory definition of our relations as manufacturers with our other interests as purveyors of Light & power seem to breed nothing but hostility and suspicion on the part of those gentlemen who represent the latter, they apparently being wholly unable to see anything in our actions but an ulterior design upon them we beg leave hereby to withdraw from any & all further controversy or discussion in respect thereto—[2]

In doing this we beg leave to call your attention to a few facts

1st. Ere embarking in ~~these~~ our various manufacturing enterprizes we sought in a most determined way to induce our friends in the Parent Co—as well as others, to join us in the undertakings   In this we signally failed— We thereupon determined upon taking all the Risks & carrying the burden ourselves to the best of our ability—more with an object to make the EEL a success than to make money by mfg[a]   This is especially true in respect to the business of the Goerck st and Bergmann & Co's Factories—

2nd. These factories having[b] necessarily attained a stage of activity prior to the Light Company proper—the impression has grown that in our Factories we have the larger proportion of the profitable[c] business of the System— This is perhaps but natural, but is none the less erroneous— The practical effect is however to place us in a false position— ~~We desire simply to secure to ourselves in the future the benifits of the~~[c] Labor & Risks[d] ~~of the past. We do not think the E L Co should arbitrarily take these from us and pass them over without consideration to the very people who declined to join us in the earlier stages of our work~~ —[3e] If the Board or any Committee appointed by it—would give sufficient time & attention to this subject to thoroughly master its various bearings we are of the opinion that they would agree with us that in our propositions we are actually sacrificing something rather than grasping for more— We simply desire to make <u>safe</u> the capital we have invested & were ready to make some present sacrifices to that end—[f] ~~And~~ one thing is certain that[g] in the future should the absence of a contract with the various factories work an injury to the Electric Light Co, They can depend upon us at any future time[h] to make just and equitable arrangements ~~with it & be firm & honest under all circumstances~~[i]

3rd. As parties interested in the success of the system as a whole we are most anxious that no deterioration of the Character of the work shall occur and that no beniefit of our experience shall become available to other than the Edison System— This can in our judgment be accomplished only by keeping the manufacture—at least for some time to come—in present hands ie in the hands of parties vitally interested in the Edison System—

The expression given vent by one of the Directors of our Company to the effect that "they should like to feel that Mr

Edisons interests[c] were identical with theirs & not antagonistic" does not only him, but the rest of us great injustice—beside unless the Edison System becomes an absolute & unqualified success the business of manufacturing for that system ~~would~~will not be worth an effort to maintain it— That we hold to ~~an~~the opposite view is evidenced by the fact that we have ourselves invested more money in our various Factories than has as yet been put into the business of developing the system in other directions   We have not invested this money for today nor yet for tomorrow—but for the future—

As we believe in the future & have confidence in our own ability to maintain our lead in the manufacturing business & as we do not wish to foster the suspicions which are already too apparent we repeat, we are content to allow matters remain as they are— Very Truly Yours

Thomas A Edison   Edwd H Johnson
F. R. Upton   ~~& oth~~

Df, NjWOE, DF (*TAED* D8327K; *TAEM* 66:717). Written by Edward Johnson; an alternate typed version of the same date is in DF (*TAED* D8327J; *TAEM* 66:714). [a]"more with . . . by mfg" interlined above by Edison. [b]Interlined above. [c]Obscured overwritten text. [d]"~~& Risks~~" interlined above. [e]Remainder of paragraph written on separate page; placement indicated by notation (partially torn off) and the typed version. [f]Remainder of paragraph written by Edison. [g]"one thing . . . that" interlined above. [h]"at any future time" interlined above. [i]Bottom of page evidently cut off at top of following line.

1. Edward Johnson wrote (and signed) this draft, except as noted, probably in close consultation with Edison and Upton.

2. Eaton had recently given Edison two long memoranda on the Edison Electric Light Co.'s relations with the Edison Lamp Co. and Bergmann & Co., prepared for its 29 January board meeting. He expressed concern that the proposed contractual terms would favor the interests of the manufacturing shops over those of the parent company. Eaton did not include these proposals in his list of "unfinished business" for an executive committee meeting on 2 February. Eaton memoranda, 19 and 24 Jan. 1883; Eaton to TAE, 27 Jan. and 2 Feb. 1883, with enclosure; all DF (*TAED* D8332M, D8324C, D8327H, D8327M; *TAEM* 67:116; 66:395, 710, 723); see also Israel 1998, 226.

3. The directors of the Edison Electric Light Co. had recommended in October that the stockholders consider taking over the manufacturing shops, though it is not clear that any specific proposal to do so had been made. According to the board's report, the growth of electric lighting had shown "that the business of manufacturing apparatus is one of great importance as well as of great profit." Board of directors report to annual meeting, 24 Oct. 1882, DF (*TAED* D8224ZBJ; *TAEM* 61:86).

To Arnold White[1]

Dear Sir

I am in receipt of your letters of 13th & 16th January.[2]

I am much obliged for your Report of Defects which I will see have careful attention.[3] Many of the Defects you mention have already been rectified. The Switches that Mr Johnson will take with him to England overcome I think all the difficulties you complain of.

As to the trouble with the Terminals of the Lamp I am inclined to think the trouble must be with your Acorn Sockets (English make) as in process of manufacture, each lamp, which goes out of our Factory has to be tested by a guage six times[a] and is lighted in so doing. If the terminal was not of exactly the right size the lamp would not light and hence the trouble you complain of would be detected in our Factory.[4] Will you kindly look into this matter again and advise me further.

As to the life of the lamps of course a great deal depends on whether they are run at their proper electro motive force. Most of the trouble here we find is caused by running them above their normal candlepower. Your customers however should take the average life during the year and not on one particular lot of lamps as sometimes for reasons entirely unknown to us the life of our lamps will run down and again for reasons equally unknown we are able to turn out lamps of extraordinarily long life. We have noticed this at our Lamp Factory several times within the past year.[5] For some time past the experience here[b] has been just the opposite to yours viz that the life of lamps has been unusually good, the reason being I presume that you are using lamps made at a different date to those now being used here.[6]

With reference to the Armington & Sims Engines the A&S Co are now preparing a catalogue & as soon as it is ready we will send you one.[7] Finding that for the present we can get along with only three A.&S. engines in our Central Station (using ⟨three Porter engines in conjunction therewith) I have arranged so that you can get such Armington engines as you may desire for your Holborn Viaduct Station, hence my cable to you of 30th ult confirmed elsewhere.[8] I now find that it is necessary for you to have a new bed plate for these engines and I accordingly cabled you to this affect this a.m.[9] At our Central Station here the new Engine bedplates have been sent to the Station without any bolt holes we having found it far more convenient to bore the holes by hand after the arrival of the base at the Station in order to insure their being absolutely in the right place.

Your small engines are being pushed forward quickly. The first one although only ordered in Providence about four days ago is now on the dock in New York and will be shipped on S.S. leaving for London the day after tomorrow.[10] The other three I shall make great efforts to ship within ten or fifteen days.

In your Storekeepers order 345 of 16th January 12 E.M.F. Indicators are asked for.[11] As both myself & Mr Johnson think this must be an error I have only ordered two.

Please consult Mr Johnson on the various matters referred to in this letter as he can enter into far greater detail than is possible by letter. Yours truly

Thos. A. Edison   I[nsull].

L (letterpress copy), NjWOE, Lbk. 15:223 (*TAED* LB015223; *TAEM* 82:113). Written by Samuel Insull; "Per S. S. Alaska" written above dateline to indicate ship carrying letter. a"six times" interlined above. bInterlined above.

1. Like Doc. 2339, this letter was addressed to "The Secretary   Edison Electric Light Co."

2. White sent two letters. One enclosed John Fleming's report on repairs to one of the Holborn dynamos and his inquiry about substituting Armington & Sims engines for the Porter-Allen machines at the station. White to TAE, both 13 Jan. 1883; Fleming memorandum, c. 13 Jan. 1883; all DF (*TAED* D8338E, D8338H, D8338F; *TAEM* 68:12, 18, 14).

3. Enclosed in one of the letters, this three-page report identified problems with Z dynamos, switches, and lamps. Edison Electric Light Co., Ltd., report of defects, Jan. 1883 (*TAED* D8338H1; *TAEM* 68:19).

4. Variations in the size of the lamp base reportedly could prevent a good connection with the socket. White promised in reply to send examples to prove this point. White to TAE, 20 Feb. 1883, DF (*TAED* D8338N; *TAEM* 68:29).

5. See Doc. 2319 n. 2.

6. A customer letter cited in the defect report (see note 3) claimed that lamps of recent manufacture had broken under "no extra pressure or current of electricity."

7. White indicated in one of his 13 January letters to Edison that the company could not order Armington & Sims engines by number without a price list. DF (*TAED* D8338E; *TAEM* 68:12).

8. Edison cabled: "Can ship Central Station Armington Saturday answer"; the company replied for him to go ahead. TAE to Edison Electric Light Co., Ltd., 30 Jan. 1883; Edison Electric Light Co. to TAE, 1 Feb. 1883; LM 1:284C, 284D (*TAED* LM001284C, LM001285D; *TAEM* 83:1014).

9. Edison cabled "Armington engine requires new bed plate   shall we send it" and received an affirmative reply. TAE to Edison Electric Light Co., Ltd., 1 Feb. 1883; Edison Electric Light Co., Ltd., to TAE, 2 Feb. 1883; LM 1:285E, 286C (*TAED* LM001285E, LM001286C; *TAEM* 83:1014, 1015).

10. The English company had cabled on 25 January for three Armington engines to drive K dynamos, such as were used in the successful

Waterloo station installation. Samuel Insull determined from Armington & Sims that these could be shipped within a few days, and the order was confirmed by cable from London. Insull to Armington & Sims, 25 Jan. 1883; TAE to Edison Electric Light Co., Ltd., 6 Feb. 1883; Lbk. 15:201, 259 (*TAED* LB015201, LB015259; *TAEM* 82:107, 129); Armington & Sims to Insull, 27 Jan. 1883, DF (*TAED* D8322H; *TAEM* 66:186); see cables between TAE and Edison Electric Light Co., Ltd., 25–29 Jan. 1883, LM 1:282D–284A (*TAED* LM001282D–LM001284A; *TAEM* 83:1013–14).

11. This order has not been found.

–2397–

Francis Upton to Edward Johnson

EAST NEWARK, Feb. 1, 1883.[a]

Dear Sir,

A few facts regarding lamps may be of use to you.[1]

<u>Complaints of</u> arcing.

When we first started making lamps at East Newark we found it impossible to prevent arcing. We spent a large amount of money experimenting and burned up many lamps testing them at high candle power. As a result we now make lamps that show very little of this tendency.

<u>Life of lamps</u>

The lamps made during the summer were not up to the standard as we had green hands in every department.[2] England probably had more than her share of these as we filled the order received last fall mostly with lamps made at Newark.[3] We did the very best we could at the time. The pains taken in making the lamps may be realized when we say that the lamps cost us over sixty cents at that time, as we took every precaution that could be suggested.

England also has lamps of low volts. This fact will also tend to make their lamps run high candle power and show short life. We could not send them anything else at the time.

<u>Range of volts</u>

We are now in position to give them higher volts. Most of their Isolated lamps were below 95 V. we can now give them the same range as the Isolated Co.

The lamps we sent them for Central Station use were 106–7 volts.

As lamps are now made the low volts indicate a poorer economy and longer life, while the higher volt indicate a better economy and shorter life.

You understand that we cannot make lamps of any one volt, the best we can do is to make a lamp that will show more of any one volt than of any other.

We will do the best we can to give every customer just the volt he wishes, yet in order that we may not accumulate too large a stock of unused volts, we are compelled to insist on the right to ship the nearest volts we can to an order.

Special lamps

We have now ordered from Japan Bamboo of extra length. When this arrives we feel that we can give 50 and 100 C.P. lamps of long life that will be of the volts wanted at the Holborn Viaduct Station. I hope to make a 150 C.P. lamp of about 50 volts that can be used to replace arc lamps. With this lamp a considerable Isolated business can be done using machines already installed.[4]

Small lamps we can make in any quantity desired and at a lower figure than for regular lamps, if they are ordered in quantities sufficient to guarantee the making of the special tools requisite.

Price of lamps

B lamps can be made cheaper than A lamps. If Mr Edison agrees I would like to have you carry a proposition to England to make them for 42 cts (forty two cents) in lots of 25,000 (twenty five thousand)[5]

We shall make a very good B. lamp using a new dimension carbon that we have found gives excellent results. We have now a 16 candle B and a 12 candle B lamp that we can recommend.

You spoke of an arrangement with Siemens. I suggest. Give him for use with his machines the 12 candle B lamp. This will give a most excellent life and about the proper quantity of light. By running over 12 candles there will be margin enough in life and the economy can be increased. If Mr Edison agrees, I would offer them these lamps in lots of 50 000 (fifty thousand) at a time at 42 cts. (forty two cents.). This is less than the lamps have cost us during the past six months, yet in large orders I hope to make a small profit on them.

I should like to run the volts of the 12 candle B lamps at about 55, which I think would about suit the Simen's machine.

Life of lamps.

I wish you would explain what is meant by 600 hours average life and specially call attention to the fact that some lamps will go at once, even if the lamps are exceedingly good.

The record of nearly every Isolated plant in this country has shown that the lamps are as good as the guarantee made by the Isolated Co. and there are numerous instances of 1000 to 1500 hours average life. Yours truly

Francis R. Upton.

L (copy), NjWOE, DF (*TAED* D8332W1; *TAEM* 67:155). Letterhead of Edison Lamp Co. a"EAST NEWARK," and "188" preprinted.

1. Upton sent this copy of the letter to Samuel Insull in response to an admonition to "send good lamps" to England. Upton to Insull, 7 Feb. 1883, DF (*TAED* D8332W; *TAEM* 67:153); Insull to Upton, 6 Feb. 1883, Lbk. 15:267 (*TAED* LB015267; *TAEM* 82:137).

2. See Doc. 2319 n. 2.

3. The lamp factory's summary of its foreign accounts between July and November 1882 reflected three large London orders in late summer. Edison Lamp Co. to TAE, undated 1882, DF (*TAED* D8231ABT; *TAEM* 61:953).

4. The English company ordered three hundred 100 candlepower lamps shortly thereafter. The special bamboo did not arrive from Japan before mid-March (Edison Electric Light Co., Ltd. to TAE, 13 Feb. 1883, DF [*TAED* D8338L; *TAEM* 68:27]; Insull to Edison Lamp Co., 28 Feb. 1883; TAE to Edison Electric Light Co., Ltd., 28 Feb. and 12 Mar. 1883; Lbk. 15:368A, 370, 459 [*TAED* LB015368A, LB015370, LB015459; *TAEM* 82:187, 189, 222]). The design of these lamps is uncertain. Edison had filed a patent application in November 1882 for a lamp of very high resistance made with coils or loops of a "considerable length of very fine fiber." Upton subsequently planned to "make a spiral 100 candle power lamp and treat it so as to give 150 candles and try it in place of calcium lights" (U.S. Pat. 353,783; Upton to TAE, 3 May 1883, DF [*TAED* D8332ZAQ; *TAEM* 67:188]).

5. Upton had previously proposed to sell B lamps at forty cents to England, noting that "prospects there are of vital interest to me." Upton to TAE, 23 Aug. 1882, DF (*TAED* D8230ZBW; *TAEM* 61:814).

–2398–

Notebook Entry:
Incandescent Lamp

[New York,] ~~Jan~~February 3 1883.

Twisted Continuous fibre Lamp or rather Compound filiment lamp—

about 30 made & Exhausted so far.[1] Those having greatest number fibres best in fact 8 fibre filiments stand while the clamps melt from arcing= filiment OK= Ends Copper plated & put in platinum clamp. The resistance is such that one[a] gave 10[a] Candle while a 95 volt Lamp only came to yellow   most of them stand with 185 volts.

We find that the principle of making a filiment for the Lamp Compound so that every seperate fibre is seperately expansible as correct and that this kind of a filiment can be made even & cheap

The fibres (Manila) are twisted by a machine then the fibre double & allowed to twist on itself—hence it does not open in carbonizing— The Carbons of[a] this kind carbzd at lamp factory come out remarkably even—

owing to small mass & the high incandescence which these

~~fibres~~ filiments of carbon are intended to be worked at commercially the blackening & wasting of the filiment is now the only remaining difficulty   I have made a great number of experiments in low vacuum in connection with the compound filiments   We exhaust the Lamp to 27 inches[2] & then seal off & absorb oxygen by heating iron spiral wire coated with finely divided iron   The oxygen oxidzing the finely divided iron, but we have not succeeded in getting rid of all the oxygen[3] We are now completing a device for aspirating or moving the air in the lamp & spiral iron Wire tube so as to get up a circulation & permit all the air to have access ~~of~~ to the iron wire=[b]

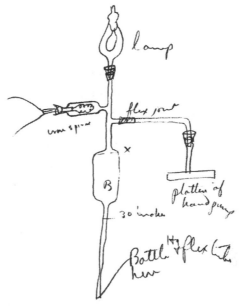

Hg up to X then Exhaust. Then move Hg in B up & down while spiral heated   Then seal off Lamp when O absorbed[4]

Edison   M N Force

                                                                    J.F.O[tt].

X, NjWOE, N-82-12-04:27, Lab. (*TAED* N145:12; *TAEM* 38:821). Document multiply signed and dated. [a]Obscured overwritten text. [b]"over" written as page turn.

1. John Ott had continued to work on the multiple twisted carbon fibers described by Edison in Doc. 2363; he sent 54 of them to the lamp factory on 24 January. Edison completed a patent application on 17 February for a process to coat the individual fibers with a "viscous carbonizable gum . . . which welds them all closely together into an integral mass." He executed another application the same day for a mold to carbonize the especially thin fibers for this type of filament. N-82-12-21:13–25, Lab. (*TAED* N150:7–13; *TAEM* 39:86–92); U.S. Pats. 446,669 and 287,522.

2. That is, inches of barometric mercury.

3. See Doc. 2370.

4. At Edison's direction, John Ott had a glassblower make this instrument the same day. N-82-12-21:21, Lab. (*TAED* N150:11; *TAEM* 39:91).

-2399-

To Joshua Bailey

[New York,] 8th Feby [188]3

My Dear Mr. Bailey,

Your letter of 20th January came to hand yesterday.[1]

With reference to the Milan Engines I presume you will have seen my letters to the Company on the question of Porter & Armington engines.[2]

You all seem to have lost sight of the fact that at the time the "C" Dynamos were sent to Milan we had no other Engine except the Porter. Again and again we tried to get other engines and could not. It was not till December that we succeeded in getting any Armington Engines for our Central Station & from the time we started our Station (4th Sept) up till December we used Porter Engines which ran together all right.

The Illuminating Co did not take the Porter Engines out of the Central Station because they would not work but because I told them I had at last got a better Engine and they therefore made the change—not at my expense however but at their own recognizing that it would be a great injustice to expect me to pay for the engines simply because I recommended their use. If the Italian people had an engine which it was quite impossible to work they would have reason to complain but such is not the case. They have engines which we have been using here for months & which they refuse to put in operation simply asserting that (which is entirely the opposite to our experience) they will not work, without even going through the form of trying them.

You say that it will cost all of us very dear if I do not do just what the Italians want. If such is the case why do not the Compagnie Continentale Edison bear the loss. If they think it of such great importance to give in to the unjust demands of the Italian Coy why do not they bear the expense instead of asking me to. I am but a simple Stockholder in the European Coy & were I to give in to such a request as you make I would eventually be a Bankrupt Manufacturer.

I have given instructions to have two Armington Engines sent to Milan but I wish it clearly understood that I have done this not because the Porter Engine will not work but simply because the Armington Engines will work better. This act must not be taken as any admission on my part that the claim of the Italian Company is just but simply as an earnest of my desire to see the Italian Station a success in every way. I shall look to the Compagnie Continentale Edison for reimbursement

The Italian Company have no right morally or legally to call upon me to make the change & if it were not for the fact that I have succeeded in getting the Armington engine we should never have heard of this controversy. They would have gone ahead and used the Porter engine & they would have been quite satisfied with them.

I have just turned the K (250) machine into a 325 Light machine.[3] Dont you think the Italian Company have a perfect right to demand that I should take back all the old K machines & give those in their place that will carry 75 more lights. I think that if they have a right to demand Armington engines they have a right to demand new K machines. But simply because they make a demand & threaten that if we do not give [---][a] in it will cost us very dear is no reason why we should give way. There ought any way to be a semblance of justice in their demand. If there was I should be only too happy to meet them and I think you will bear me out in the assertion that I as a rule I get the reputation of being just in my business dealings.

As to the question of my not[b] having paid Porter the reason I have not done so is that we have claims against them[b] & so have[c] the Light Co in other matters and consequently under advice I withheld part of the payment for some engines of which the Milan engines are part. I am responsible to the Porter Coy & will have to pay them as soon as other matters have been settled & I most indignantly repudiate the assertion that I have charged the Italian Coy with engines that I have rejected myself. Such a remark comes with bad grace considering that the bargain made at the time the goods were ordered has not been adhered to by the Italian Co as at no time since I commenced shipping have their bills been met promptly on presentation of invoice & B/L[4] as agreed. They have owed me for two months upwards of $6700 & I have cabled to the Compagnie Continentale again and again for the money without even getting a reply.[5]

The Italian Company must not expect to get everything to run perfectly in a new business without any changes or extra expense. I am spending money right along & your companies get the benefit of this expenditure of money & also of my time & I get no compensation for it & I think the least they can do is to bear the expense of taking advantage of such improvements as I may make. It surprises me somewhat that you should take the same view of this matter as the Italian Coy & I cannot but think that immediately you look carefully into the question you will alter your mind.

I should like Mr Batchelor to see this letter & also the correspondence between myself and the Paris Company on the subject.[6]

Major Eaton sent me a cable from you the other day as to these Armington engines in which you asked for a reply. I do not know what to say as I had answered all cables sent me from Paris on the subject. I would suggest that all cables as to supplies be sent to <u>me</u> as it is very confusing to have two people cabling on the same subject. You may rely on everything being answered just as quickly as we can get the data on which to reply, & you should not get impatient and telegraph to someone else to find out why I do not reply to you the very moment your first cable arrives.[7] Very truly yours

Thomas A Edison

LS (letterpress copy), NjWOE, Lbk. 15:274 (*TAED* LB015274; *TAEM* 82:141). Written by Samuel Insull; "Britannia" and "Queenstown" written above dateline to indicate ship carrying letter and port of arrival. [a]Canceled. [b]Interlined above. [c]Obscured overwritten text.

1. Bailey recommended that Edison agree to Giuseppe Colombo's request to replace "without charge" two of the four Porter-Allen steam engines for the Milan central station with Armington & Sims engines. Bailey also urged Edison to respond to the Italian company's complaint that he had not paid half the money owed on the Porter-Allen engines (which Colombo believed had been rejected in the U.S.) while still pressing the Italian company for full reimbursement. Bailey feared that "it will cost you and all of us very dear if the matter is not treated in such a way as to leave them satisfied.— There is also this: that it leaves a very nasty responsibility on Colombo, who, wholly on his confidence in and enthusiasm for you went ahead and made these orders." Bailey to TAE, 20 Jan. 1883, DF (*TAED* D8337F; *TAEM* 67:602).

The four-sided controversy over the engines among Edison, the Southwark Foundry, Colombo and the Edison interests in Milan, and Bailey and the French parent company began in late November. Charles Porter wrote Colombo that he could not ship new governors until Edison settled his account with the Southwark Foundry. Edison protested to Colombo that "There is no truth whatever in Mr Porters statements. The engines we have received from the Southwark Foundry and Machine Co. are not in accordance with contract and as soon as we can get other Engines we are going to take the Porter engines out of our central Station here." Edison refused the Italian company's request to reimburse the entire cost, cabling instead that the Porter engines could be used alongside the new Armington & Sims machines. Porter to Colombo,

28 Nov. 1882; Compagnie Continentale Edison to TAE, 6 Jan. 1883; both DF (*TAED* D8238ZEP, D8337B; *TAEM* 62:576, 67:593); TAE to Colombo, 6 Dec. 1882, Lbk. 15:7 (*TAED* LB015007; *TAEM* 82:30); TAE to Comitato Applicazioni dell'Elettricita, 12 and 15 Jan. 1883; LM 1:277E, 278D (*TAED* LM001277E, LM001278D; *TAEM* 83:1010–11).

2. Edison's letters to the Compagnie Continentale Edison on 16 January and the Société Électrique Edison on 2 February covered substantially the same points as this document. Lbk. 15:163, 233 (*TAED* LB015163, LB015233; *TAEM* 82:90, 118).

3. In July 1882 Edison noticed that the 35-horsepower K dynamo could operate one extra lamp per every two horsepower. In November William Andrews reported test results showing that the K machine run at 1100 rpm could power 350 lamps. In May 1883 Edison ran 530 lamps on it for six hours; by June this more powerful version had been designated the H dynamo and rated for 450 lamps. TAE to William Hammer, 28 July 1882, Lbk. 7:816 (*TAED* LB007816; *TAEM* 80:779); Andrews test report, 20 Nov. 1882; TAE to J. F. Tafe, 22 June 1883; both DF (*TAED* D8235Y3, D8316ABW; *TAEM* 62:66, 64:792); Batchelor to Edward Johnson, 14 May 1883, LM 1:311A (*TAED* LM001311A; *TAEM* 83:1027); see also Doc. 2416.

4. That is, bill of lading.

5. Edison cabled for payment six times in November, once in December, and three times in January. See cable correspondence with the Société Électrique Edison in LM 1 (*TAED* LM001; *TAEM* 83:872). The Italian company continued to dispute this amount, claiming both that Edison had shipped more equipment than they had ordered and that the disagreement over the engines was not settled. At the end of February Samuel Insull directed Bergmann & Co. to delay shipment of regulators to Milan because he had had "so much trouble about getting money on Bills for goods shipped there." At the end of March Edison complained to Batchelor that he had received only $3,325 of the $9,883.77 owed him. Compagnie Continentale Edison to TAE, 6 Jan. 1883; Comitato Applicazioni dell'Elettricita to Compagnie Continentale Edison, 21 Mar. 1883; both DF (*TAED* D8337B, D8337ZAY; *TAEM* 67:593, 692); TAE to Société Électrique Edison, 2 Feb. 1883; Insull to Bergmann & Co., 23 Feb. 1883; TAE to Batchelor, 29 Mar. 1883; Lbk. 15:233, 346; 16:40 (*TAED* LB015233, LB015346, LB016040; *TAEM* 82:118, 180, 274).

6. Edison wrote Batchelor on this date asking him to "try and straighten this matter out for me." Batchelor agreed in reply that Edison "cannot be expected to suffer a loss whenever you find a new and better thing." Batchelor also forwarded a translation of a letter from the Italian company suggesting "that they are disposed to settle the thing amicably." The dispute dragged into the summer; its resolution is not certain, but in September Edison was arranging the shipment of a new Jumbo to Milan. TAE to Batchelor, 8 Feb. 1883, Lbk. 15:282 (*TAED* LB015282; *TAEM* 82:149); Batchelor to TAE, 22 Feb. 1883; Comitato Applicazioni dell'Elettricita to Compagnie Continentale Edison, 15 Feb. 1883; both DF (*TAED* D8337Z, D8337W; *TAEM* 67:630, 623); TAE to Société Électrique Edison, 6 Sept. 1883, LM 2:13E (*TAED* LM002013E; *TAEM* 83:1039).

7. Neither Bailey's cable nor Eaton's transmittal of it has been found.

Edison had recently directed the Société Électrique Edison to "send all cables referring to supplies to me & not to Major S. B. Eaton. . . . This error has occurred a number of times before." TAE to Société Électrique Edison, 28 Dec. 1882, Lbk. 15:82 (*TAED* LBo15082; *TAEM* 82:64).

–2400–

To Charles Rocap

[New York,] February 10th. [1883]

Dear Sir:—

Inasmuch as my business is so large that it is quite impossible for me to give close attention to the details of it and as I consider it better for my general interests to have all my financial matters under one charge, I desire for the future that all checks of the Edison Machine Works shall be "vouchered" for by Mr. Insull before presentation to me for signature, and that you shall consult him on all questions connected with the finances of the Works.[1]

You will please, therefore, make arrangements with Mr. Insull to be at the Works such days as will be necessary in order to enable him to have a proper insight into such matters.[2] Yours truly,

TL, NjWOE, DF (*TAED* D8334F; *TAEM* 67:360).

1. See Doc. 2301.
2. Rocap replied three days later that he would give Insull all the assistance he required, but would not "surrender my independence or become subservient to Mr Insull. The moment this is done my position becomes worthless." Rocap to TAE, 13 Feb. 1883, DF (*TAED* D8334G; *TAEM* 67:361); see headnote, Doc. 2343.

–2401–

From Cyrus Brackett

Princeton, Feb. 13, *1883*.[a]

Dear sir:

I have just had a proposition to act as an expert and advisor in the matter of storage batteries for the application of electricity to lighting. I have refused to have anything to say or do about it till I hear from you in regard to the question. Do you consider that there is any claim or binding obligation on me as respects that question?[1] As I remember, the question of such batteries had not come up in relation to the Edison system of lighting and if so, it would seem that there would not be any impropriety in my rendering any service I may be able to, especially if you do not, and do not intend to make use of such batteries in your system.[b]

Please let me hear how you understand the matter. Yours truly.

C. F. Brackett.[2]

⟨No objections accepting retainer on storage batteries against everyone but Edison—⟩[3]

TLS, NjWOE, DF (*TAED* D8320B; *TAEM* 66:4). Letterhead of College of New Jersey, Department of Physics. [a]*"Princeton,"* and *"188"* preprinted. [b]Mistyped.

1. In March 1880, Brackett and his Princeton colleague Charles Young had served as independent experts to measure the efficiency of Edison's lamps and dynamos (see Docs. 1914 and 1916). They were later placed on retainer to the Edison Electric Light Co., but Sherburne Eaton and Samuel Insull were unable to locate the details of this arrangement in the summer of 1882. Eaton to TAE, 23 and 29 June 1882, DF (*TAED* D8226ZAD, D8226ZAI; *TAEM* 61:300, 317); TAE to Eaton, 3 July 1882, Lbk. 7:648 (*TAED* LB007648; *TAEM* 80:712).

2. Cyrus Fogg Brackett (1833–1915) was the Joseph Henry Professor of Physics at Princeton, where he had been a mentor to Francis Upton. In 1889, he began teaching some of the earliest courses in electrical engineering; he also consulted on the development of storage batteries. Doc. 1914 n. 2; Magie 1915; Condit 1952, 12.

3. A docket note indicates that a reply was sent on 14 February, but it has not been found.

–2402–

From Richard Dyer

WASHINGTON, D.C. Feby 14th, 1883[a]

My dear Mr Edison,

Hearing from Seely that you intend to leave N.Y. for the South on the 19th. inst., I have had several talks with Father in regard to the change which you said some time since you would try to bring about in the Examiner of Interferences of the Patent Office.

The following ~~are~~is a mixture of his and my own[b] views upon the matter.

The Commr[1] could make the change himself without the matter going further, but Church[2] might appeal to the Sec'y,[3] and hence it would be better to attack the Sec'y, and bring such influence to bear upon him that he would call Marble in and settle the question, without causing you to take the regular course, which would be to file a written request which the Sec'y would refer to Marble and upon which in the course of time Marble would make a report, smoothing the thing over and making your efforts amount to nothing.

The change can be brought about only by the best efforts, and if you intend to try your hand at it, you must come here prepared to spend several days, and bring with you letters to Senators and Representatives which ~~would~~ will influence them to go with you and see the Sec'y personally. The best course would seem to be to bring letters to Lapham[4] and ~~Warner~~Miller,[5] to the republican senator from N.J. Genl Sewell,[6] to the representative from the district in which Menlo Park is situated,[7] and to Cox,[8] Belmont,[9] Flower[10] and Hewett.[11] Get letters of the right kind to as many senators[c] and representatives as possible. You will have to interest these men in your story and take their advice as to how to proceed. The Secy is interested at present more especially in the appropriations, and nothing will command his respect so much as a strong eCongressional backing. You have a good foundation for the request in the inconsistencies of the decisions in your several cases, but, since any man is liable to err in judgment, a better case might be made out on the delay on the big telephone interferences.[12] They were argued 16 months ago[13] and Church hasn't decided them yet. He has not decided any other cases either, but has turned them over to incompetent clerks. Now it is absolutely absurd and unjust that an inferior tribunal whose decision will be appealed from by all parties, should make such delay. It shows the incompetence of Church.

His assistant Brown[14] should also be removed. He is a very light-waisted man, and would hamper a new Exr for a long time.

I am convinced that you can accomplish nothing by simply making the request yourself of the Sec'y, and I think also that you would weaken your cause by calling in the services of Mr Painter, if you have thoughts of so doing, since the Secy is an old Congressman and probably knows ~~him~~ his character[d] well.

It would be a remarkably good thing to bring about the change you propose, and would I think well repay you for the time you would[c] have to spend in accomplishing your object. If you succeeded, in addition to the benefit from the change, the moral ~~influence~~ effect[b] in the Patent Office of a showing of influence on your part and a disposition to make use of it would I think be good. I shall be in N.Y. on Monday next.[15] Yours very truly

Richd. N. Dyer.

ALS, NjWOE, DF (*TAED* D8370U; *TAEM* 70:940). Letterhead of George Dyer. [a]"WASHINGTON, D.C." and "188" preprinted. [b]Interlined above. [c]Obscured overwritten text. [d]"his character" interlined above.

1. Edgar M. Marble (1838–1908), an Interior Department attorney, was appointed Commissioner of Patents in June 1880. He held the post until October 1883; he continued to hold Interior Department positions for part of that time. *NCAB* 13:310–11.

2. J. B. Church was appointed as a second assistant examiner in the Patent Office in 1872 and examiner in 1877; he became Examiner of Interferences in 1879. He left the Patent Office by the end of 1883. In later years he practiced patent law with his brother Melville E. Church in the firm Church & Church, which included Edison companies among its clients. "Renewed Activity at the Patent Office," *Sci. Am.* 27 (1872): 168; "Notes from the Capital," *New York Times*, 26 June 1877, 1; *Congr. Dir.* 46.2 (Dec. 1879): 110; ibid. 48.1 (Dec. 1883): 135; *TAED* and *TAEM-G4*, s.vv. "Church & Church," "Church, Melville E."; Melville Church Obituary, *New York Times*, 12 Oct. 1935, 17.

3. Henry Teller (1830–1914), served in the Senate from Colorado between 1876 and 1882, when he was appointed Secretary of the Interior. That office encompassed responsibility for the Patent Office. *ANB*, s.v. "Teller, Henry Moore."

4. Elbridge Lapham (1814–1890), a Republican, resigned from the U.S. House in 1881 after being elected to one of New York's vacant Senate seats. *BDUSC*, s.v. "Lapham, Elbridge Gerry."

5. Warner Miller (1838–1918), a Republican, left the U.S. House in 1881 after winning election to fill New York's other vacant Senate seat. *DAB*, s.v. "Miller, Warner"; *BDUSC*, s.v. "Miller, Warner."

6. William Sewell (1835–1901) was breveted to the rank of Major General during the Civil War. A Republican, he entered the Senate from New Jersey in 1881. *BDUSC*, s.v. "Sewell, William Joyce"; *Ency. NJ*, "Sewell, William Joyce."

7. John Kean (1852–1914), a Republican, had not yet taken his seat in the House of Representatives from New Jersey's third district, which included Middlesex county. In the recent election he defeated Democrat Miles Ross, the incumbent since 1875. *Congr. Dir.* 47.2 (Feb. 1883): 49, ibid., 48.1 (Dec. 1883): 55; *Ency. NJ*, s.v. "Kean, John"; *BDUSC*, s.vv. "Kean, John," "Ross, Miles."

8. Longtime representative Samuel Cox (1824–1889), a Democrat, held a House seat from New York City at this time. *ANB*, s.v. "Cox, Samuel Sullivan."

9. Perry Belmont (1851–1947), a Democrat, had represented the New York counties of Suffolk, Queens, and Richmond (Staten Island) since 1881. Prior to that, he had briefly practiced law with Porter, Lowrey & Stone, the firm of Grosvenor Lowrey. *ANB*, s.v. "Belmont, Perry"; *Congr. Dir.* 47.2 (Feb. 1883): 51.

10. Roswell Flower (1835–1899), a Democrat, was nearing the end of his first term in a House seat from New York City. Having declined renomination, he was replaced in March by Orlando B. Potter. *Congr. Dir.* 47.2 (Feb. 1883): 53; ibid., 48.1 (Dec. 1883): 59; *ANB*, s.v. "Flower, Roswell Pettibone."

11. Iron manufacturer Abram Hewitt (1822–1903), a Democrat, was elected to the House from New York City in 1874 and had served continuously with the exception of one term; he was later mayor of New York. *ANB*, s.v. "Hewitt, Abram Stevens."

12. The Patent Office had declared two related groups of interference cases in March 1878 and August 1879 to sort out the competing priority claims of several telephone inventors (see Docs. 1270 and 1792). Some cases were decided in June 1881, but Church did not rule on eleven remaining questions until July 1883. His decision (reportedly 700 pages) was then appealed to the Examiners-in-Chief, who considered the matter until October 1884 and reversed his decision on the question of priority for the telephone receiver. Lemuel Serrell to TAE, 28 June 1881, DF (*TAED* D8142ZBF; *TAEM* 59:341); "Decision Regarding the Telephone Inventions," *New York Times*, 22 July 1883, 5; "The Telephone Interference Case Decided," *Sci. Am.* 44 (1883): 64; Decisions of the Examiners-in-Chief, Miscellaneous Interferences, TI 5 (*TAED* TI5 [images 39–63]; *TAEM* 11:891–915).

13. On the preparation of Edison's arguments in October 1881 see Serrell to TAE, 23 Sept. and 10 Oct. 1881; Serrell memorandum, n.d. [1881]; all DF (*TAED* D8142ZCF, D8142ZCM, D8142ZCZ; *TAEM* 59:382, 392, 417).

14. Probably Frank T. Brown, who was appointed third assistant examiner in 1877 and promoted to second assistant the following year. He eventually reached the position of Examiner in Hydraulics and then in Electricity, where he heard additional arguments about telephone priority in 1887 ("Notes from the Capital," *New York Times*, 26 June 1877, 1; ibid., 29 Aug. 1878, 1; "At the Nation's Capital," ibid., 25 May 1887, 1; "Affairs at the Capital," ibid., 15 July 1887, 1; Congr. Dir. 48.2 [Feb. 1885]: 140; ibid., 50.1 [Apr. 1888]: 174). George Dyer had recently written to Edison about appealing a decision in "which the Examiner of Interferences (or rather Mr Brown his clerk) awarded priority of invention" to Elisha Gray for a magneto telephone (Dyer to TAE, 25 Jan. 1883, DF [*TAED* D8370R; *TAEM* 70:936]).

15. Samuel Insull replied on 15 February that Edison would do nothing before his return from the South. Lbk. 15:325 (*TAED* LB015325; *TAEM* 82:172).

–2403–

Samuel Insull to Edward Johnson

New York, 17th Feby *1883*[a]

My Dear Ed,

Meadowcroft came up to me this am. with a London Company check for £5.0.0 & asked me how he could get it cashed. This shows that White pays Meadowcroft to send them information—or rather the London Co pays him. I do not know whether you object to this but thought I would let you know.

In a letter from the London Co which arrived this morning the following paragraph occurs:—

"We had heard that the Armingtons were not giving satisfaction in the Central Station & that you were reverting to Porter Allens"

Now this is the third or fourth time such reports have got to London   Where do they hear such things.[1]

On Thursday at 5 pm. I stopped an Armington Engine at the Station which had been running day & night without stopping for seven days and two hours with a load varying from 50 to 170 Horse power.

Edison, Villiard[2b] Eaton & Clarke are in Boston working up an Illuminating Coy with every promise I understand of good success.[3] Things in Boston look very well; that article of Edisons in the Boston Herald knocked the bottom out of the Storage business.[4] Bergmann is complaining that orders are growing slack. We are giving him very few Foreign Orders & the Isolated Co are falling off some, in their orders to B & Co—

Isolated Coy's business is somewhat slack. They are getting very few orders in just now although they have lots of Agents out—

The Station is running well although the collections are not coming in very well.[5] But for this I blame the collector more than anything else. Up to Feby 10th

Bills rendered were	$3201.37
" Collected	1291.07
leaving uncollected	$1910.30

I am arranging to have regular reports of Bills presented & collected & you shall have copies of them. Edison has suggested that I send you copy of Casho's Daily[b] report.[6] I will see that you get this also.

Estevez[7] would not take the Tenant we offered him for 56 W 12th St. because he is a Cuban & Estevez being a Spaniard does not like Cubans. Then we offered him an unmarried Lady & he will not take her on principle   I presume he is afraid the House might be used for "improper purposes" if he did. By the bye I heard the other day that at one time 56 W. 12th used to be the abode of one of the Leaders of the Demi-monde & was quite an El Dorado for the Boys.[8]

I have sent some bills to the London Co today. We want a cable transfer directly they get them. Yours Sincerely

Saml Insull

I guess if the House is not let I shall be able to strike some bagrgain with Estevez so that you will not have to pay much rent after 1st March— Any way I shall try to & a letter from him this morning makes me think I shall be successful   I[nsull].

ALS (letterpress copy), NjWOE, LM 3:104 (*TAED* LM003104; *TAEM* 84:89). Stamped letterhead of Thomas A. Edison to Edward H. Johnson; "Per SS Elbe" written below handstamp to indicate ship carrying letter. ᵃ"*New York*," and "*188*" preprinted. ᵇInterlined above.

1. White told Edison that two employees had informed him "that you had discarded the Armington engine in your central station and reverted to the Porter-Allen; because of the excessive wear of the machine." White accordingly decided to postpone shipment of the latest Armington & Sims engines to London. After an exchange of cables, Edison wrote White that these reports were "in error in saying we had dispensed with Armington engines. On the contrary experience only strengthens my belief in them. Johnson will tell you what good work they are doing for us here." White to TAE, 6 Feb. 1883, DF (*TAED* D8338J; *TAEM* 68:25); TAE to White, 28 Feb. 1883, Lbk. 15:369 (*TAED* LB015369; *TAEM* 82:188).

2. Henry Villard.

3. Edison and the others left for Boston on 15 February and returned two or three days later. The Boston company was organized at the end of March. Insull to Dyer, 15 Feb. 1883; Insull to Armington, 29 Mar. 1883; Lbk. 15:325, 16:55 (*TAED* LB015325, LB016055; *TAEM* 82:172, 286); Eaton to TAE, 23 Feb. 1883, DF (*TAED* D8327Q; *TAEM* 66:729).

4. Edison disparaged the storage battery at considerable length in an interview in the 28 January *Boston Herald*, reportedly deriding it as "a catch-penny, a sensation, [and] a mechanism for swindling." Edison Electric Light Co. Bulletin 16:31–36, 2 Feb. 1883, CR (*TAED* CB016275; *TAEM* 96:804).

5. See Docs. 2383 and 2409.

6. Joseph Casho (1840–1924) served as superintendent of the Pearl St. Central Station from 1 September 1882 until the spring of 1883 when Charles Chinnock replaced him. Before entering Edison's employ he worked at a Chester, Pa. machine shop which built slow-speed Corliss engines. Francis Jehl later recalled that Casho left Pearl St. because he was uncomfortable working with high-speed steam engines. During his tenure, Casho sent Edison daily reports on the number of lamps in use, running times of the dynamos, and consumption of coal, water, and lubricating oil. "Casho, Joseph," Pioneers Bio.; Jehl 1937–41; 950, 1057; Eaton to TAE, 8 Dec. 1883, enclosing Casho daily report, 5 Dec. 1882, DF (*TAED* D8223S, D8223T; *TAEM* 60:1062–63).

7. Unidentified.

8. This was apparently Johnson's former residence rented to him by Estevez. While Johnson was in England, Insull oversaw the move of his household to a new location. Johnson to TAE, 27 Jan. 1883, DF (*TAED* D8303Y; *TAEM* 64:56); see Doc. 2409.

–2404–

To Charles Batchelor

[New York,] 3rd Mar [188]3

My Dear Batchelor,

Will you please let me know what your arrangements with the French Companies are. Have you a time contract with them? If so when does it expire? Can you arrange to close up with them before its expiration & come home.

My impression is that you have a three years contract with them from some time in 1881 or early in 1882.[1] Yours truly

Thos A Edison   I[nsull].

L (letterpress copy), NjWOE, Lbk. 15:386 (*TAED* LB015386; *TAEM* 82:193). "Werra" and "Southampton" written above dateline to indicate ship carrying letter and port of arrival.

1. Batchelor replied on 16 March that he had no formal contract with the French companies, "but it is generally understood that it would take me about 3 years to get things underweigh. Everything goes well as long as I continually boost it but if I should leave here before we have a Central Station working in Paris I really believe it would go to the devil." Batchelor left Paris on 19 April to visit New York, returning on 1 June. He left Paris permanently in May 1884, without having set up a central station, to take over management of the Edison Machine Works. Batchelor to TAE, 16 Mar. 1883, DF (*TAED* D8337ZAP; *TAEM* 67:683); Batchelor journal, 19 Apr. and 1 June 1884; Cat. 1343:56, 77; Batchelor (*TAED* MBJ002:31, 37; *TAEM* 90:266, 272); Batchelor to TAE, 30 Apr. 1884, LM 2:59C (*TAED* LM002059C; *TAEM* 83:1061); Batchelor entry, 16 May 1884, Cat. 1306, Batchelor (*TAED* MBN011:30; *TAEM* 91:632); Israel 1998, 215.

–2405–

Draft Patent Application: Electrical Distribution System

[New York,] Mch 4 1883

The object of the invention is to automatically ~~bal~~ maintain a balance or equilibrium in the number of Lamps between two ~~se~~circuits ~~co~~

The invention consists in the Same[1]

A. & B.[2] are double electromagnets one magnet being placed across one circuit & the other across the other circuit. when there is the same emf across the terminals of each circuit the lever of the electromagnet is ~~att~~ is constantly attracted to that side which it had been previously attracted by reason of a disimilarity of emf—while in this position the lever & contacts put its multiple arc circuit of Lamps across say the No 1 circuit. now if No 1 circuit is weakened ie[a] the emf falls due to heavy loading on that side, the magnet weakens across that circuit and as the No 2 is raised in emf the lever is attracted to the other side & this throws the circuit of lamps across No 2 raising the emf of No 1 & lowering that of no 2 & by having say an automatic balancing device in every 10th house the moment any tendency of overloading one circuit occurs some of the automatic devices will throw Lamps from the weak to the strong side until equlibrum is attained= with these devices the centre wire ~~might~~ need not pass to the station but it is preferable & by its use the number of automatic devices can be greatly lessened. where there are a great number of consumers the automatic device is of course unnecessary but is valuable[b] where there are but[b] few Consumers & occasions[b] arise where the circuits are very much out of balance hence there is a large ~~o~~fall of pressure in the Conductors & the lights show great disimularity on the two circuits & this is corrected[b] by an automatic balancing device or Compensator.

A clockwork to shift the circuits of lamps from one ckt to the other could be used, which is controlled by a diffrential magnet   The expansion of wires deflection of galvanometer expansion of air mechanism controlled by magnetism, or electrolytic deposition could evidently be used to effect the Shifting of the Lamp circuits=

Seely make some broad claims on this

E[dison]

ADfS, NjWOE, Cat. 1149 (*TAED* NM018:6; *TAEM* 44:540). [a]Circled. [b]Obscured overwritten text.

1. This was one of several methods Edison devised to maintain a uniform voltage on central station district feeder lines regardless of the geographic distribution of the load (see Doc. 2269 n. 3). This draft patent application formed the basis of U.S. Patent 283,983 which Edison executed on 5 April.

Edison received at least six other patents on methods to regulate the voltage on the feeders. U.S. Patent 266,793 (executed on 25 October 1881) and U.S. Patent 287,525 (executed by Edison and Charles Clarke on 4 October 1882) both pertained to regulating by a variable resistance in series with each feeder. U.S. Patents 274,290 and 283,984 (executed on 20 November 1882 and 5 March 1883) applied to a 220 volt three-wire distribution system with two 110 volt feeder circuits sharing a common return wire to distribute the load equally. U.S. Patents 287,517 and 287,515 (executed on 14 March and 25 June 1883) described regulation by switching individual feeders in or out of service. Edison drew a notebook sketch of this method on 5 June 1883. Cat. 1149, Scraps. (*TAED* NM018:37; *TAEM* 44:566).

2. The editors have not found a sketch accompanying this draft.

–2406–

To Charles Batchelor

[New York,] March 5 [188]3

Dear Sir,

I beg to confirm cables as follows

To you March 3. "Impossible spare man"   Edison

To you March 3   If ordered large quantities forty cents lowest   recent improved life enhance cost   poor policy start factory as in year or so very much cheaper here. Edison[1]

Bailey has cabled Eaton to endeavour to persuade me to send Upton to Europe for three months. I suppose he wants him to start the German Lamp Factory   I suppose Bailey thinks our Lamp Factory business can take care of itself in this country or else he would not make such a request.

I cannot quote lower than 40 cents for lamps. For the past Six months our lamps have cost us 48 cents & are now costing us just about 40 cents. Of course we shall get them down cheaper in course of time[2] I think your people make a great mistake in starting so many factories You cannot expect them to have that skill in lamp manufacture that our experience has given us here; in as much as our own people get away off here & make bad lamps sometimes & for reasons that we know nothing about, su surely the same trouble will occur in Germany.

Furthermore if the lamp is to be made cheap it should be made in large quantities & in order to do this the less the number of factories the better Yours truly

Thos. A. Edison   I[nsull]

L (letterpress copy), NjWOE, Lbk. 15:406 (*TAED* LB015406; *TAEM* 82:197). Written by Samuel Insull; "Alaska" and "Queenstown" written above dateline to indicate ship carrying letter and port of arrival.

1. These transcriptions are the full text of messages cabled to Batchelor in Berlin on 3 March; this letter was mailed to him at Ivry-sur-Seine (LM 1:296A, 296B [*TAED* LM001296A, LM001296B; *TAEM* 83:1020]). Batchelor had cabled from Berlin on 2 March: "Best man you have wanted first May start lamp factory. whom will you send." On 3 March, he inquired, again from Berlin, if "Lamp Price Be Lowered in quantities for Germany instead making factory." Edison's marginal notes on the latter message are the basis of this reply (LM 12:95B [*TAED* LM001295B; *TAEM* 83:1019]; DF [*TAED* D8337ZAG; *TAEM* 67:638]).

2. See also headnote, Doc. 2343. The syndicate in Switzerland had requested a reduction in lamp prices, to which Edison made a similar reply at this time. Société d'Appareillage Électrique to TAE 14 Feb. 1883, DF (*TAED* D8337V; *TAEM* 67:621); TAE to Société d'Appareillage Électrique, 6 Mar. 1883, Lbk. 15:414 (*TAED* LB015414; *TAEM* 82:203).

-2407-

To Edward Johnson

[New York,] 5th March [188]3

My Dear Johnson,

Your letter of 16th Feby came duly to hand.[1]

From what you state I very much dispair of any great good being done by your visit to England. From your letter and from letters from White to Lowrey[2] I judge that our London Friends think they can get on better without us If so the best thing is to let them do so and to give our attention to the "Western Empire" alone. I think you had better do the best you can to get through by the end of this month and then come right back here as there is plenty to do and few to do it.

Just now we are doing all we can to rush the Village business.

There is immediate money and plenty of it in that business and I propose getting up some big Syndicates in order to boom the business, and by this means provide work for our shops

I am more than ever convinced however that if the business is to be made a success it must be by our personal efforts and not by depending upon the officials of[a] our Companies. For that reason I think you should hurry back as we want your assistance. We <u>know</u> there is plenty of money to be made in our business in this Country and that here we can get things done just as we say, and I therefore think it is best to concentrate our efforts on an <u>American certainty</u> rather than an <u>English possibility</u>

Besides the light there is the matter that you were working on before you went away[3] and the Stock Printers, both of which will be ready shortly.

It was unfortunate that Hopkinson should have got ahead of me on the new Conductors.[4] We are however "solid" so far as this country is concerned. My patent will be out in a few days. I can go away behind Hopkinsons first English date so fthere is no fear of his beating me on interference if one should at any time be declared[5]   Yours very truly

Edison

PS   [1 wire has middle?][b] between the other two[6]

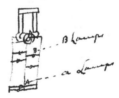

This is the arrangement I had at Menlo & worked [-]t[b] 2 weeks in 1881[7]   Nov 6 started it = also old caveat in Pat Ofs Dec 1881[8]   E

LS (letterpress copy), NjWOE, Lbk. 13:12 (*TAED* LB013012; *TAEM* 81:777). Written by Samuel Insull; postscript written by Edison on separate sheet and copied onto bottom of second letterpress page. "Alaska" and "Queenstown" written above dateline to indicate ship carrying letter and port of arrival. [a]Interlined above. [b]Illegible.

1. Johnson would have mailed this letter (not found) from England, where he arrived on or about 11 February. Johnson to TAE, 11 Feb. 1883, LM 1:289B (*TAED* LM001289B; *TAEM* 83:1016).

2. Not found.

3. Johnson filed a patent application on 24 January covering a thermoelectric telephone transmitter, for which he had asked Edison to order 10 ounces of tellurium. The editors have not determined the outcome of this application. Just before leaving for Britain, he executed a separate application for a transmitter with a compound high-resistance elec-

trode; this issued in June 1885 as U.S. Patent 319,415. Johnson to TAE, 17 Jan. 1883; Johnson application (copy), 24 Jan. 1883; both DF (*TAED* D8303N1, D8370Q; *TAEM* 64:34, 70:934).

4. John Hopkinson had filed a provisional British specification in July 1882 on a copper-saving "three-wire" electrical distribution system very similar to the one Edison had been developing (Brit. Pat. 3,576 [1882]; Hopkinson 1901, xlvii; Greig 1970, 32–34; Hughes 1983, 84; Israel 1998, 219; see also Doc. 2414). Richard Dyer was "very much astonished" by the patent and warned Edison that it seemed "to anticipate everything of importance in the case we have forwarded [to Britain and Europe]. . . . If the foreign cases are to be dropped we can have our U.S. patent issued immediately before we meet with trouble in the Patent Office." Johnson had evidently discussed the Hopkinson system with Frank Sprague, who sent him a long analysis of its features (Dyer to TAE, 26 Feb. 1883; Sprague to Johnson, 2 Mar. 1883; both DF [*TAED* D8370ZAC, D8320C; *TAEM* 70:953, 66:5]). Hopkinson was also making experimental improvements to the Edison Z dynamo; Johnson's report on this to Sherburne Eaton has not been found, but it drew a dismissive response from Edison (TAE to Johnson, 6 Mar. 1883, Lbk. 15:423 [*TAED* LB015423; *TAEM* 82:209]).

5. Edison was probably referring to U.S. Patent 274,290, issued on 20 March 1883; its figure 5 is similar to the drawing in the postscript below. Edison signed the application on 20 November 1882 but Richard Dyer witnessed other drawings for it on 28 October. Cat. 1148, Lab. (*TAED* NM017:87–92; *TAEM* 44:447–452).

6. Figure labels are "B Lamps" and "A Lamps."

7. Dyer indicated that this placement of the third wire on a commutator of a single dynamo, rather than between multiple dynamos, was the only configuration not already covered by Hopkinson's British specification. He suggested it as a basis upon which English and Continental filings might proceed. Insull reported to Johnson Edison's belief that "the original idea is embodied in his experiments on running 'A' & 'B' Lamps from the same machine in multiple arc" and that "Upton says this was done away back at Menlo." Dyer to TAE, 26 Feb. 1883, DF (*TAED* D8370ZAC; *TAEM* 70:953); Insull to Johnson, 1 Apr. 1883, LM 3:115 (*TAED* LM003115; *TAEM* 84:98).

8. Not found.

–2408–

*Samuel Insull to
Seth Low*

[New York,] 5th Mar [188]3

Dear Sir

I enclose rough estimate of apparatus necessary for starting a Dept of Elect Engrg at Columbia amounting to $14 469.[1] In this Mr Edison has not allowed for furniture fittings & Statione[ry],[a] but only[b] for the special apparatus

Mr Edison thinks the best thing to do, if there is not room in the College Bldgs, would be to take a private house near by & fit it up for the purpose[2]   Yours truly

Saml Insull   Private Secy

ALS (letterpress copy), NjWOE, Lbk. 15:401 (*TAED* LB015401A; *TAEM* 82:196). ªCopied off page. ᵇInterlined above.

1. The itemized estimate for equipment, including boiler, steam engine, three dynamos, machine tools, and laboratory instruments, is given in an undated and unsigned memorandum. DF (*TAED* D8303ZBK1; *TAEM* 64:106).

2. Columbia's old 49th Street campus lacked space to expand; Finch 1954 (68–69) cited this as a reason why the college did not start its electrical engineering program until 1889; see also McMahon 1984, 44. Edison promised to help first the Stevens Institute and then the Cooper Union to establish such a program if Columbia did not do so. TAE to Abram Hewitt, 12 Apr. 1883; TAE to Andrew White, 23 Apr. 1883; Lbk. 16:131, 173A (*TAED* LB016131, LB016173A; *TAEM* 82:317, 341).

–2409–

Samuel Insull to
Edward Johnson

[New York,] 6th March [188]3

My Dear Johnson,

Your letter of 21st Feby came to hand yesterday & I have sent you by this mail a letter from Edison giving you the authorization you ask for.[1] I have written Drexel Morgan & Co giving them copy of the paragraph in your letter referring to this matter & also a copy of the letter Edison has sent you & have asked them to do likewise.

I have requested Mr Wright to forward you the Oriental Shares directª but have not yet got his answer yet.[2]

I send you under separate cover some memorandum from Eaton which will give you some information about matters in general; tear them up when you have read them.

A committee meeting was held last night & it was decided to knock off 25% of the bills in the 1st District w[h]ere people are dissatisfied.[3]

I think we shall arrive at some decision as to the wiring organization this week. We have got to in fact in order to carry out our idea of a Construction Syndicate for erecting Village Plants.

Last night I brought a man to Edison & we had three hours session. I want him to form a Coy of $500,000 to work Village Plants. We propose get one started & when that runs successfully start or to put in about 25 or 30 more having an option on these latter from the Light Co, said option to be exercised on the first one running profitably & successfully. I think my friend will form the Company   If he does not we will get some one else. We hope to make this business boom.[4] More of this by Saturdays mail

The Illuminating Co do not propose to give their future to

Mr Morgan. If his scheme is carried it will only refer to the one District where his Station is started.[5]

Isolated Co are doing very badly just now. It is their dull Season & they have not had an order in two weeks. However this cannot last very long

Since you have been away I have spent for you about $1570 & have received from you $1200 & from Bergmann $250 in all $1450 so we are square on that amount advanced to Joe.[6] I will send my accounts forward next mail.

Estevez has not let the House yet. He refuses almost every tenant sent him & when I complain, as I do pretty lively, he says he will take anyone whom I will personally endorse & this of course I cannot do. He says the house will possibly be let on 15th & if so he will return me half the rent paid on the 1st but I doubt it. He seems as if he is going to make you pay your full rent to 1st May & if he is so inclined I do so[b] not see what can be done although it is monstrous.

The new house is not ready yet nor do I know when it will be as I have had nothing to do with that business at all. I hope however to get things moved up there this week. Mrs Harrison[7] is in Phila & I ~~as~~have asked her to remain there till the moving is over. She is in no condition to be around empty houses, & I can look after the moving. I have already had all the China Glass Pictures ornaments & breakables moved up to the new house & securely locked up. <u>Nothing whatever</u> was broken in transit[8]

I do not know what your relations are with Benton[9] but it is damnably annoying ~~to find that~~ when I go to attend to a thing or give instructions in connection with your affairs to be told Mr Benton has instructed us to do so & so. I understood you wanted me to look after your affairs in your absence same as I did on a previous occasion & I went ahead on that understanding & I do not liked to be asked who "told me to give any instructions as Mr Benton has been here from Mr Johnson." Of course I have said nothing to Benton as I do not know what instructions you may have given him, nor is there any necessity for[a] you to say anything to him as the thing will regulate itself   But I thought I would mention it more especially as the fact of Bentons living at your house during Mrs. Harrisons absence in Philadelphia gave people the impression that he was the man to go to & that I was interfering with matters which did not concern me.

I have no idea what it is costing to fix up your house as I did not make the Contracts so I do not know at all what I shall be called upon to pay during the next few months on your a/c. But I propose to try & find out.

I do not want you to look upon this as complaining that looking after your affairs is a burden. You know I am only too glad to attend to anything you may ask me.

It is just over two years since I came here & I am the worst broke man you ever saw. Matters have not improved with me any since you left & I do not think they will as I am kept so close at my business that it is quite impossible to look around outside to make money   I have not got the time to. Sometimes I think I am just on the verge of kicking then I remember that you told me to hold my tongue & I do so. But this cannot go on much longer. It worries me too much & I cannot stand that and the business to.[10]

I am <u>Financially</u> in charge of the Machine Works at last. IEdison got very uneasy because he said neither he nor I knew anything of their affairs. I reminded him that I warned him such would be the case if he did not support me. He offered to do anything to set it right & he did so by correspondence & when Rocap kicked Edison supported me in a bully fashion & I came out top of the heap. It is a great relief to me to have examined everything which I have done most thoroughly. Edison has got to pay those fellows (Dean & Rocap) that share of the profits of the Machine Works I told you about.[11]

Foll I will write more in a few days   Yours as ever

Saml Insull

ALS (letterpress copy), NjWOE, LM 3:109 (*TAED* LM003109; *TAEM* 84:92). "Alaska" and "Queenstown" written above dateline to indicate ship carrying letter and port of arrival. aObscured overwritten text. bInterlined above.

1. The editors have not found Johnson's letter. Edison authorized him on 5 March "to speak on my behalf on any matters that may come up for consideration" because Johnson's "views in relation to my English light interests are so fully in accord with my own." TAE to Johnson, 5 Mar. 1883, Lbk. 15:421 (*TAED* LB015421; *TAEM* 82:207).

2. Insull quoted to Drexel, Morgan & Co. Johnson's request to "Please get from Mr Fabbri or Mr Wright my 1500 shares of Oriental Telephone Co Stock. I forgot to bring them— They are of but little if any use to them as security for my loans but I may be able to realize on them here and thus help pay off my loans." Drexel, Morgan & Co. promised to send the shares to J. S. Morgan & Co. in London. Insull to Drexel, Morgan & Co., 6 Mar. 1883, Lbk. 15:419 (*TAED* LB015419; *TAEM* 82:206); Drexel, Morgan & Co. to Insull, 8 Mar. 1883, DF (*TAED* D8374J; *TAEM* 70:1167).

3. Some First District customers expressed dissatisfaction because their initial electric lighting bills (see Doc. 2383) were higher than their

prior gas bills. The company had 367 paying customers in March, resulting in income and profit of $5,294 and $386, respectively. Joseph Casho to Charles Clarke, 2 Apr. 1883; Frank Hastings to Charles Chinnock, 11 and 17 July 1883; all DF (*TAED* D8326F, D8326K, D8326L; *TAEM* 66:666, 678–79).

4. Insull and his friend W. A. Graves of Albany, N.Y., arranged a meeting on 5 March with Edison and several other Albany investors. They reportedly discussed formation of a central station company for that city. Graves to Insull, 4 Mar. 1883, DF (*TAED* D8303ZBK; *TAEM* 64:104); Insull to Graves, 6 Mar. 1883, Lbk. 15:422A (*TAED* LB015422A; *TAEM* 82:208); Insull to Johnson, 11 Apr. 1883, LM 3:135 (*TAED* LM003135; *TAEM* 84:118).

5. The editors have found no information about this plan.

6. Possibly Insull's older brother Joseph (1858–1941), for whom Samuel arranged a job at the Edison Machine Works in 1887. McDonald 1962, 6; Obituary, *New York Times*, 13 Oct. 1941, 17, 35; "Insull, Joseph," Pioneers Bio.

7. Unidentified.

8. A month later Insull told Johnson that work was proceeding well on his house and that the attendant expenses were manageable. City directories gave Johnson's residence addresses for the years ending 1 May 1883 and 1884 as 269 E. 10th St. and 139 E. 36th St. respectively. Insull to Johnson, 11 Apr. 1883, LM 3:135 (*TAED* LM003135; *TAEM* 84:118); *Trow's* 1883, 824; ibid., 1884, 844.

9. Charles Abner Benton (1847–1939), a friend of Edward Johnson's, began working for Edison in 1880 and remained active in Edison's electric lighting companies for the rest of the decade. Johnson apparently asked Benton to look after his affairs while he was in England. "Benton, Charles Abner," Pioneers Bio.; Jehl 1937–41, 950–51; Insull to Johnson, 11 Apr. 1883, LM 3:135 (*TAED* LM003135; *TAEM* 84:118); TAE to Johnson, 23 Apr. 1883, Lbk. 13:15 (*TAED* LB013015; *TAEM* 81:780).

10. Insull discussed his straitened finances and reached an agreement with Edison about 1 April; see Doc. 2417.

11. See Doc. 2293.

–2410–

From Francis Upton

EAST NEWARK, March 8, 1883[a]

Dear Mr. Edison:

We[b] are now in full running order so that all departments are good for 12,000–15,000 lamps a week according to the style of lamps run.[1]

Our orders on hand and prospective orders will keep us busy until the latter part of April. Then we will have to slack down in work unless we can bring in new orders.

I spoke to Insull yesterday about offering lamps at 40 cts. for May and June if ordered in lots of 50,000. We could make money on them as we are now in fine trim.

Everything is running in a regular manner with hands in each department that are trained for the work required of them.[2]

Holzer has a fine set of glass blowers coming on, we have taken on Germans and married men as far as possible in this department so as to have reliable help. We are very anxious to keep busy until July[b] 1, then we propose to shut down until our fall orders come in, probably for six weeks.[3]

There is no question in my mind that next fall we shall have all the business it is possible for us to do.

With the new village plants, a new station in N.Y., the new styles of dynamo machines, the new catalogue of Bergmann,[4] and the renewals that must come from lamps now burning there is no question about demand. Then abroad, England is bound to do some business and the Continent will take a regular amount.[5]

20,000 lamps a week is a large number, but everything points to it next fall. To take good care of this business we must have our trained hands. For example in putting in carbons. Glass blowers that have been with us six weeks can ~~put~~ seal[c] in 30 carbons in the globe a day with a loss of 20%. One of our old hands will put in 240 carbons in ten hours with a loss of less than 4%.

You know the advantages of running full, it brings down the expenses per lamp very materially.

I have been always very careful to study my orders before launching out, yet now everything points to an immense business next fall, and we need some orders to carry us through to July.[d]

Resistance lamps.[6d]

We have on hand

Res	10 c.p.	800
Low	"	2775
Old	"	668
High	B	5950
Old	"	6150
Res	"	2300
Old	A	2325
High	"	4575
Res	"	6800
Old	16 c.p. B	875
"	32 "	920
		33,938

These lamps are now entirely dead stock. If we can sell them at any price it would be so much profit to the Lamp Co.

If you could write to your agents abroad they might find uses for them.

The worst part of these is that we are making more constantly.

The W.U. could use a number if proper arrangements could be made with them.[7]

Bergmann might study out a theatre box that would employ the 16 c.p. B. Yours Truly

Francis R. Upton.

ALS, NjWOE, DF (*TAED* D8332ZAC; *TAEM* 67:169). Letterhead of Edison Lamp Co. a"EAST NEWARK," and "188" preprinted. bObscured overwritten text. cInterlined above. dFollowed by dividing mark.

1. The factory's monthly statements showed that it produced 64,216 lamps in March, a 50 percent increase over February and nearly twice as many as in January. Upton later explained that during March he had "tried to make a sample month of what we can do if we had orders enough." Edison Lamp Co. to TAE, 1 Feb., 1 Mar., and 2 Apr. 1883; Upton to TAE, 6 Apr. 1883; all DF (*TAED* D8332T, D8332Z, D8332ZAG, D8332ZAI; *TAEM* 67:150, 165, 174, 176).

2. Upton told Edison on 30 April that "In March when running perfectly full . . . we had on 324 hands, quite a number of them learners," but that this number had since declined to 127. DF (*TAED* D8332ZAL; *TAEM* 67:182).

3. Upton notified Edison on 4 May that he intended to close the factory for July and August "On account of the hot weather," but Edison disapproved of this plan unless it was "absolutely necessary." DF (*TAED* D8332ZAR; *TAEM* 67:189); TAE to Upton, 7 May 1883, Lbk. 16:239 (*TAED* LB016239; *TAEM* 82:367).

4. The 82-page fourth edition of the Bergmann & Co. catalogue of lighting appliances was published in April or May. Edison Electric Light Co. Bulletin 18:5–6, 31 May 1883, CR (*TAED* CB018; *TAEM* 96:827); Bergmann & Co. catalog, [1883], PPC (*TAED* CA002C; *TAEM* 96:185).

5. Six weeks later, Upton admitted he was "exceedingly disappointed in the demand for A lamps from Hamburg, Antwerp &c, and from the English colonies and other points. It is below anything my statistics show it should be if they are making constant advances in the same ratio as in the past." Upton to TAE, 18 Apr. 1883, DF (*TAED* D8332ZAJ; *TAEM* 67:178).

6. See Doc. 2085 n. 9.

7. See Doc. 2201.

Notebook Entry:
Electric Light
and Power

perm magnet   prevent Carrying[1a]

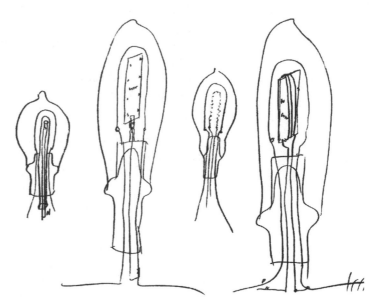

Lamps
    passed thread through tragacanth & die—[2a]
    Made Lamp with P terminals large surface   ditto N—
ditto both N .&. P   dont see much change regular[a]
    Have ascertained that blackening of one of the platina wires
in inside part[3] is due to Electrolysis & conductivity of the
glass—& that Lime glass has less & Potash glass still less than
our regular Lead glass   am going make some bulbs of these
glases[4a]
    Have ascertained that thickened Ends on Carbon can be only
done with perfection by the use of a Carbohydrate Cement=
No Hydrocarbon will answer as well—[5a]
    Am trying lot experiments get simple indicating meter
Electrolytically with Hg & Hg amalgm of Zinc= It is apparently difficult=[6a]
    Reading Scientific Amn I see there is a great need of process
of compressing Bran—. I propose due this in Hydraulic press
full Bran from which the air is exhausted ie[b] the chamber
containing the Bran— this will allow it to pack—[7a]
TAE

                           M[artin]. N. F[orce].
                           J[ohn]. F. O[tt].

X, NjWOE, Lab., N-82-05-26:65 (*TAED* N204:32; *TAEM* 40:606). Document multiply signed and dated. [a]Followed by dividing mark. [b]Circled.

1. The editors have found no evidence that Edison pursued the idea of placing a magnet in the globe, but cf. Doc. 1944 n. 8. In other efforts to attract carbon particles away from the glass, John Ott recorded a number of experimental lamps about this time with platinum or iron at various locations in the globe. In April, Edison proposed to draft a caveat covering "glass or mica between limbs of carbon," reminiscent of experiments made in late 1880. Cat. 1149, N-82-12-21:53–63; Lab. (*TAED* NM018:12, N150:27–31; *TAEM* 44:546, 39:106–110); Doc. 1944 n. 6; see also Doc. 2346.

2. Three days earlier Edison, John Ott, and Martin Force put fibers in grooved pieces of the natural gum tragacanth, placed them in carbonizing molds, and had them carbonized at the lamp factory. Thomas Conant made a few related experiments; he tried carbonizing tragacanth in a zinc chloride solution on 17 March, but found that "the carbon and gum were simply eaten." On 3 April Ott sketched a machine to draw manila fiber through tragacanth. N-82-12-21:43, 81–83; N-82-11-14:177, 191; Lab. (*TAED* N150:22, 41–42; N143:86, 93; *TAEM* 39:102, 120–21; 38:758, 765).

3. The short piece of tubing molded to the globe, through which the lead-in wires were secured. See Doc. 1887 n. 3.

4. Edison was aware of this phenomenon in July 1882, when Charles Batchelor wrote him about it (see Doc. 2314). He used leaded glass because that made with lime or other alkali was too brittle. Upton to TAE, 9 Jan. 1882, DF (*TAED* D8230C; *TAEM* 61:721).

5. See Docs. 2363 and 2384.

6. Conant's notes from this time are in N-82-11-14:155–213, Lab. (*TAED* N143:75–104; *TAEM* 38:747–76). At the end of March he tried more experiments "to determine whether or no a plate of lead amalgamated will either reduce mercury from or be oxidized by a solution of a salt of mercury, the aim being to obtain some metal which may be amalgamated. . . . The plate in question to be used as a carrier of mercury in a meter in which the quantity of current shall be measured by the quantity of mercury removed from the one electrode or deposited on the other." N-82-11-14:213–15, Lab. (*TAED* N143:104; *TAEM* 38:776).

7. The 10 March issue contained a description of a thousand-dollar prize offered by the Millers' National Association for a machine to compress bran, a valuable animal feed, to make it easier to transport. The guidelines stipulated that the machine should compress one ton per hour into packages smaller than 15 inches square. Edison apparently did not pursue the matter. "Compressed Bran—One Thousand Dollars Reward for a New Invention," *Sci. Am.* 48 (1883): 144.

[New York,] Mch 8 1883

Meter[1]

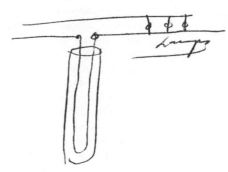

amalgamated Zinc wire in Sul Zinc deposits on one Side takes from other until it breaks   Each strip breaks at certain ~~L~~No Lamp hours   an Electromagnet auto replaces a second cell in ckt & destruction Each cells mean certain Bill for Light[2]

TAE   J.F.O[tt].

X, NjWOE, Lab., N-82-05-26:77 (*TAED* N204:38; *TAEM* 40:613).

1. Figure label is "Lamps."
2. This design for an easily-read meter that would register discrete units is very similar to John Ott's 6 March description and drawing of a safety plug operating on the same principle. N-82-12-21:47, Lab. (*TAED* N150:24; *TAEM* 39:103).

–2413–

To George Ballou[1]

[New York,] 13th Mar [188]3

Dear Sir,

Are you inclined to go into the business of lighting small towns by Electricity about which I spoke to you whilst going to Boston recently.

If you are disposed to look further into the matter I shall be glad to make an appointment to meet you here any evening to talk it over[2]   Yours truly

Thos. A. Edison   I[nsull].

L, NjWOE, Lbk. 15:465A (*TAED* LB015465A; *TAEM* 82:223).

1. George William Ballou (1847–1929) was a leading investment banker of New York and Boston whose main areas of activity were the railroad and telegraph industries. In 1880 he installed former President Grant as president of a New Mexico copper and gold mine and in 1882 he led an investors' syndicate which unsuccessfully opposed Jay Gould's merging of the Mutual Union Telegraph Company into the Western Union company. "Gen. Grant's New Business," *New York Times*, 25 July 1880, 1; "Jay Gould Aggrieved," ibid., 2 Nov. 1882, 8; "George W. Bal-

lou, Banker, Dies at 82," *New York Times*, 1 May 1929, 25; Grodinsky 1957, 453.

2. Ballou replied on the same day that he was "disposed to look into the matter of lighting small cities and towns by electricity" and promised to discuss the matter with Edison after an unidentified associate returned from Chicago. The editors have found no information about a subsequent meeting. Ballou to TAE, 13 Mar. 1883, DF (*TAED* D8320D; *TAEM* 66:20); see also Doc. 2417.

–2414–

From Charles Batchelor

Ivry-sur-Seine, le 13 Mch 1883[a]

455.[b]

My dear Edison,

In answer to my cable to you from Brussels, "(On Johnsons advice am filing distribution") I have received at London yesterday yours as follows:— "follow dyers directions concerning distribution" and have answered it today with:— "Fifty nine[1] filed in France   can stop other countries   if do englishman will be ahead of us"[2]

The fact of the matter stands thus; as soon as Johnson informed me of the fact that Hopkinson had anticipated you in this matter of distribution[3] I immediately decided to file this patent, and it was well I did; as Hopkinson I learn now was in Paris all last week, and I suppose that would be his business here; now he cannot file his before us in the other countries as ours have been sent some days ago; some of them are allowed by this time and if I take steps to stop them he will certainly be ahead of ~~this~~ us;[c] I shall however act in accordance with your wishes if I receive a telegram in answer to mine of today requesting me to still act in accordance with Dyer's letter[4]   Yours

"Batch"

ALS, NjWOE, DF (*TAED* D8370ZAG; *TAEM* 70:958); letterpress copy in Batchelor, Cat. 1239:478 (*TAED* MBLB4478; *TAEM* 93:806). Letterhead of Société Industrielle et Commerciale Edison. [a]"Ivry-sur-Seine, le" preprinted. [b]Multiply underlined. [c]Interlined above.

1. French patent application case number 59.

2. Batchelor's transcriptions are the full text of these cable messages. Batchelor to TAE, 8 and 13 Mar. 1883; TAE to Batchelor, 8 Mar. 1883; LM 1:296D, 297E, 297A (*TAED* LM001296D, LM001297E, LM001297A; *TAEM* 83:1020).

3. Edward Johnson knew of John Hopkinson's British three-wire distribution patent by 16 February (see Doc. 2407) but his 19 February letter to Batchelor (not found) apparently did not provide any details. Batchelor immediately instructed Joshua Bailey to report "what the

improvements are that Hopkinson has made. . . . Johnson thinks that these (whatever they are) are of great importance." Batchelor to Bailey, 20 Feb. 1883, Cat. 1239:470 (*TAED* MBLB4470; *TAEM* 93:797).

4. Richard Dyer's letter to Batchelor has not been found; he had advised Edison to abandon the foreign applications in order to expedite issuance of his U.S. patent. Edison cabled Batchelor on 14 March to "file fiftynine if patent for fifteen years and full fee paid   otherwise jeopardize American Patent" (Dyer to TAE, 26 Feb. 1883, DF [*TAED* D8370ZAC; *TAEM* 70:953]; TAE to Batchelor, 14 Mar. 1883, LM 1:298A [*TAED* LM001298A; *TAED* 83:1021]). Edison charged Batchelor in late March with having "busted" his three-wire system patent in the U.S. by filing European applications. Batchelor strongly objected, claiming also that Edison and the Edison Electric Light Co., which had its own concerns about foreign specifications shortening the term of U.S. patents, had ignored his efforts to establish a coordinated policy for filing patent cases (TAE to Batchelor, 26 and 27 Mar. 1883, Lbk. 16:21A, 26 [*TAED* LB016021A, LB016026; *TAEM* 82:260, 263]; Batchelor to TAE, 30 Mar. 1883; Sherburne Eaton to TAE, 4 and 7 Dec. 1882; all DF [*TAED* D8370ZAO, D8224ZCU, D8224ZCX; *TAEM* 70:968; 61:129, 132]).

–2415–

Notebook Entry:
Electric Light
and Power

[New York,] Mch 27 1883

Mch 24 [Mem?]—[1a]

Ask Clarke why cant deposit [first?][a] when reading is taken from other plate= Ask Seely abt. Auto compensater, also clamps—

Try heating at 300 or thereabouts[b] fahr bamboo for 6 hours in unboiled Linseed—Oil Merbane, Aniline oil. Sulphuric acid—Linseed containing Rosin= also ammonia Sol of tragacanth= also glycerin, also Parafine, Coal tar, petroleum residue Regular turpentine also Venice turpentine.

Send to Holzer for 1000 splints to prepare this way= ask clarke if 100 meter[2] is found show on 30 lights if correction cannot be made=

Have ~~Bergm~~ our draughtsman[c] design new brushes!=

Try flour paste on 8 ply fibres= get pectic acid

Try hard boiled linseed on fibres= make Cupric ammonia

Ascertain if Zinc immersed in Sul Zinc with lot hard rubber shavings oxidize—

Thermostat for Central Station[3]

Machines for clamp

Boil water in Meter bottle or rather put Zinc plate in tube
with Sul Zinc   boil & seal by fusion—

investigate that white stuff inside globes where tip is sealed
in—[4]

Tellerium battery—for meter to make Hydrogen pressure
polish new plat clamp   less carrying—

One carbon within the other in glass   2nd carbon to be
heated by 2ndy—& use as a plate to work relay

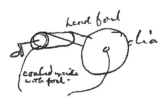

use induction coil see if get sound— try pad metal contain
cylinder & amalgamated surface

Dechanel[5] make Faraday[6] & Arago[7c] Ampere[8] continuous
rotation for meter[9]

Try with Shellac, Cup ammonia paper, paste hard caou-
etchin[10] Linseed & other things to stick fibres together & then
carbonize in closed potash glass tubes to see if stick[d]

Write Upton to Cut some fibres with paper on each side to
prevent [Crushing?]—[a]

Try some dilute Hydrofluoric on inside of blackened globe
get some small potash glass tubes   Carbize fiber in Linseed &
in sat sol chloride Zinc—

To make peroxide Lead 4 parts Crystallized acetate Lead
3 parts Carb Soda   pass chlorine
TAE   M[artin]N.F[orce]
                                             J[ohn]FO[tt]

X, NjWOE, Lab., PN-82-04-01 (*TAED* NP016:36; *TAEM* 44:46). Document multiply signed and dated. [a]Illegible. [b]Interlined above. [c]Obscured overwritten text. [d]Followed by dividing mark.

1. Text is "Zinc," "Hg," "glass," and "prevent oxidation differential."
2. Presumably a 100-light meter.
3. Text is "glass," [illegible], and "metal."
4. Text is "gold leaf."
5. This probably refers to Augustin Privat Deschanel, whose *Elementary Treatise on Natural Philosophy* appeared in several editions translated and edited with extensive additions by J.D. Everett.
6. Michael Faraday (1791–1867), English experimental researcher in chemistry and electromagnetism. *DSB*, s.v. "Faraday, Michael."
7. François Arago (1786–1853), French physical researcher in optics and electricity. *DSB*, s.v. "Arago, Dominique François Jean."
8. André-Marie Ampère (1775–1836), French chemist and physicist, pioneered the modern study of electrodynamics. *DSB*, s.v. "Ampère, André-Marie."
9. In early April, Edison completed a patent application for a consumption meter that may be related to this suggestion. The meter was operated by a "peculiar electro-dynamic motor, which is a non-commutator or uni-polar machine of such character that I apply to it the term 'mono–electro-dynamic motor,' its inductive or rotating part being a straight or one-part conductor, the current passing through it in one direction only" (U.S. Pat. 370,123). Such a motor is reminiscent of experimental non-commutating direct-current devices made by Faraday, Arago, and Ampère in their explorations of induction phenomena. For Edison's earlier efforts to develop a unipolar generator see Docs. 1641 n. 1 and 1683, esp. n. 1.
10. Caoutchin is a transparent liquid hydrocarbon distilled from caoutchouc and gutta percha. *OED*, s.v. "caoutchin."

–2416–

*Samuel Insull to
Charles Batchelor*

[New York,] 29th Mar [188]3

My Dear Batchelor,

Edison is making some radical changes in Dynamos now. What the exact outcome will be I cannot say at present. I have asked him several times to give me notes on which to base letters to you & Johnson on the subject. He has promised to do so as soon as his experiments are ~~thr~~ a little further along   I write this in case you may be going ahead with Dynamo building & thinking that what Edison is doing might possibly affect your action I write you this. He [----][a] has[b] already made the K machine a 350 light machine but expects to get even more lights than this out of it.[1] In haste   Yours Sincerely

Saml Insull

ALS (letterpress copy), NjWOE, Lbk. 16:41 (*TAED* LB016041; *TAEM* 82:275). "Gellert" and "Plymouth" written above dateline to indicate ship carrying letter and port of arrival. [a]Canceled. [b]Interlined above.

1. A few days later Insull wrote Edward Johnson that Edison was increasing the efficiency and capacity of his dynamos because he feared that John Hopkinson was doing the same and would try "to bulldoze" the Edison interests in Britain "into buying his Dynamo improvements." (Hopkinson's later work with the British engineering firm of Mather & Platt resulted in a general design known as the Edison-Hopkinson dynamo.) Insull asserted that "Edison has done as much as Hopkinson" to refine his dynamos, mainly by adding more windings to the field coils and improving the commutator brushes. The new brush design eliminated sparking and the need to apply mercury to the commutator. The alterations enlarged the capacity of the Z dynamo to 120 A lamps, the K to 350 lamps, and the C to about 1,500 lamps. Insull to Johnson, 3 Apr. 1883, LM 3:120 (*TAED* LM003120; *TAEM* 84:103); TAE to Rocap, 5 June 1883, DF (*TAED* D8334U; *TAEM* 67:382); Andrews 1924; Bowers 1990; Hopkinson and Hopkinson, 1886; cf. App. 3.

-2417-

Memorandum:
Village Plants

[New York,] Mch 29 1883

Village Plant Biz—[1]

~~Net   50~~

Office Expenses out.

Of the nett, where no sacrifices made:

100 cents Edison gets	60.
EH.J.	20 +
Insull	20.

Batch to be taken in when he returns, ~~add~~,

Add 50 per week for ~~Dist~~[a] Insull before any division—but charged to plant at .75.[2]

AD, NjWOE, Miller (*TAED* HM830172B; *TAEM* 86:542). Miscellaneous calculations and doodles not transcribed. [a]Interlined above.

1. The editors have not found a more formal version of this memorandum. It apparently served as the basis in May 1883 for the formation of the Thomas A. Edison Construction Department for the erection of direct-current electric power stations in towns and cities throughout the United States. In exchange for a license to use Edison's patents, local illuminating companies paid 20% of their capital stock and 5% in cash and agreed to buy Edison equipment at 12% over cost. Israel 1998, 220–25.

Edison wished to take charge of the village plant business because, as Samuel Insull explained to Johnson in April, "the Edison Electric Light business is not run well & so much impressed is Edison by this fact that he has practically left his Laboratory & now makes my Office his Head-quarters & is attending to purely business matters. . . . [H]e could see that this Village Plant business would go to ruin unless he came up & attended to matters. He says he has been told again & again by that they could get nothing satisfactory from the Executive Officers of the Light Co. So Edison decided to 'drop science & pursue business.'" Insull to Johnson, 3 Apr. 1883, LM 3:120 (*TAED* LM003120; *TAEM* 84:103).

2. Insull assumed day-to-day management of the village plant and Construction Department business and received an annual salary of $2,400 and a 20% share of profits beyond this amount. Johnson was to receive 20% and Edison 60% of the profits after paying Insull's salary. Insull to Johnson, 3 Apr. 1883, LM 3:120 (*TAED* LM003120; *TAEM* 84:103).

Appendix 1

Edison's Autobiographical Notes

From 1907 to 1909 Edison wrote a series of autobiographical notes to assist Thomas C. Martin and Frank L. Dyer in their preparation of his authorized biography.[1] Edison first produced Document D, including notes on queries posed by Martin, probably about October 1907.[2] This was followed by the recollections in books A and G, made in September and October of 1908. This material was incorporated into the initial chapters of the biography, which were complete by February of 1909; Martin then requested additional personal reminiscences from Edison in order to flesh out the remaining chapters.[3] William Meadowcroft, who was coordinating the project, acknowledged in May 1909 that the continuing lack of Edison's additional material was a "very serious affair," and the next month Edison produced the notes in books E and F.[4] Some of these formed the basis for oral interviews with Martin, the typed transcripts of which became documents B and C; together, these four documents served as the basis for anecdotes related in later chapters of the published biography.

Six of the documents contain sections related to events of the period of Volume Six; those sections are published here.[5] Edison sometimes referred in the same paragraph to the periods covered by more than a single volume; these paragraphs will be reprinted as appropriate. Each document has been designated by a letter and each paragraph has been sequentially numbered.

1. Dyer & Martin 1910. The designations A through F were assigned to these documents in Volume One, which also contains a general editorial discussion of them. See *TAEB* 1 App. 1; document G was discovered later.

2. An Edison notebook entry from this time reads "Martins book take Lab note bk 4, 1 Extra . . . answer Martins immediate notes." PN-07-09-15, Lab. (*TAED* NP077; *TAEM* 178:343).

3. Martin to TAE, 23 Feb. 1909, Meadowcroft (*TAED* MM001BAP; *TAEM* 226:660).

4. Meadowcroft to Martin, 24 May 1909, Meadowcroft (*TAED* MM001BAQ; *TAEM* 226:662).

5. The autobiographical document designated A does not refer to the period of this volume. The sections from A published in Volumes One and Four were drawn from a typed version of Edison's notes prepared by William Meadowcroft. However, a copy of Edison's original manuscript, in a notebook labeled "Book No. 1 September 1, 1908 Mr. Edison's notes re. Biography," was published in Part IV of the microfilm edition (*TAED* MM002; *TAEM* 226:787).

B. FIRST BATCH (JUNE 1909)

The following is from a typescript that Edison revised. At the top of the first page is a handwritten note by William Meadowcroft: "First Batch   Notes dictated by Mr Edison to T. C. Martin   June, 1909.— Pencil indicates Mr. Edison's revision." Twenty-five of its eighty-one paragraphs pertain to the period of this volume; items 3, 37, and 54 were previously published in Volume Five.

FRANK THOMSON AND ELECTRIC RAILWAYS.

[3] One day Frank Thomson, the president of the Pennsylvania Railroad came out to see the electric light and the electric railway in operation. The latter was about a mile long. He rode on it. At that time I was getting out plans to make an electric locomotive of 500 horsepower with 6 foot drivers with the idea of showing the railroad people that they could dispense with their steam locomotives. Mr. Thomson made the objection that it was impracticable, and that it would be impossible to supplant steam. His great experience and standing threw a wet blanket on my hopes. But I thought he might perhaps be mistaken, as there had been many such instances on record. I continued to work on the plans and about three years later I started to build the locomotive at the works at Goerck street, and had it about $^{1}/_{4}{}^{a}$ finished when I was switched off on some other work. One of the reasons why I felt the electric railway to be eminently practical was that Henry Villard, then president of the Northern Pacific Railroad, said that one of the greatest things that could be done would be to build right angle feeders into the wheat fields of Dakota and bring in the

wheat to the main lines, as the farmers now had to draw it from 40 to 80 miles. There was a point where it would not pay to raise it at all, and large areas of the country were thus of no value. I conceived the idea of building a very light railroad of narrow gauge, and had got all the data as to the winds on the plains, and found that it would be possible with very large windmills to supply enough power to drive these wheat trains.

THE LARGE DYNAMO FOR PARIS EXPOSITION.

[9] I built a very large dynamo with the engine directly connected which I intended for the Paris Exposition of 1881. It was one or two sizes larger than those I had previously built. I only had a very short period to get it ready and put it on a steamer to reach the Exposition in time. After the machine was completed, we found that the voltage was too low. I had to devise a way of raising the voltage without changing the machine, which I did by adding extra magnets. After this was done, we tested the machine and the crank shaft of the engine broke[b] and flew clear across the shop. By working night and day a new crankshaft was put in, and we only had three days left from that time to get it on board the steamer; and had also to run a test. So we made arrangements with the district Tammany leader and through him with the police to clear the street— one of the crosstown streets—and line it with policemen, as we proposed to make a quick passage of it and didn't know how much time it would take. About 4 hours before the steamer had to get it, the machine was shut down after the test and a schedule was made out previously[a] of what each man had to do. Sixty men were put on top of the dynamo to get it ready and each man had written orders as to what he was to perform. We got it all taken apart and put on trucks and started off. They drove the horses with a fire bell in front of them to the French Pier, the policemen lining the streets. Fifty men were ready to help the stevedores get it on the steamer and we were one hour ahead of time.

[10] In those days the Tammany Hall people were very convenient. The shop I had was in Goerck street, formerly owned by John Roach and was known as the Aetna or[c] Architectural Iron Works. The street was lined with rather old buildings and poor tenements. We had not much frontage. As our business increased enormously, our quarters became too small, so we saw the district Tammany leader and asked him if we could not store castings and other things on the sidewalk. He gave us permission,—told us to go ahead and he would see it was all

right. The only thing he required for this was that when a man was sent with a note from him asking us to give him a job he was to be put on. We had a head laborer foreman—"Big Jim"—a very powerful Irishman, who could lift above half a ton. When one of these Tammany aspirants appeared he was told to go right to work at $1.50 a day. The next day he was told off to lift a certain piece and if the man could not lift it he was discharged. That made the Tammany man all safe. Jim would pick the piece up easily. The other man could not, and so we let him out. Finally the Tammany leader called a halt, as we were running big engine lathes out on the sidewalk, and he was afraid we were carrying it a little too far. The lathes were worked right out in the street and belted through the windows of the shop.

GOVERNING THE PEARL STREET ENGINES.

[11] When I started the Pearl Street Station, the largest at that time in the world I met with a very serious difficulty. I had three Porter-Allen 200 hp. engines direct connected to the dynamos. These engines had gravity governors. When we started up the first engine and the current passed out over the network of wires it worked successfully. We lit lamps in all parts of the area covered by the station, and ran for many hours. Finally, after experimenting two or three days I thought I would put two engines together. Then something happened that had never happened before. The moment we threw in the second engine the first engine slowed[b] way down and the second engine jumped up to speed almost in an instant, and then went to two or three times its speed, until we thought the ~~belt would come off~~ building would collapse.[d] Then the other engine would speed up, and they would see-saw, from 50 revolutions a minute to 800 revolutions a minute. Nothing of steel or iron could stand it. The commutator brushes burned and red-hot globules of copper flowed down on the floor and began to burn the wood. Smoke poured all over. The building was apparently going to come down, and everybody made for the stairs. Finally I yelled to shut down, and two of the men left jumped in and closed the throttles.

[12] Now here was a problem: What caused it? Nothing was known at that time of one dynamo running another as[a] a motor, and exchanging with each other according to the speed. So I sat up all that night to figure out what the trouble was and found it. It was necessary to connect those three governors together so that one could not get away from the others. Next morning I put on all the men I could crowd at the works and

made a long shaft—70 feet long—and hangers, and all, and brought it down to the works. I put it up against the wall at the side of the engines and thus connected all the governors together. I then started the first engine but had the same trouble. I then found that this was due to the fact that the two arms at the two extremes of the shaft would not move together because the torsion of the long shaft allowed one arm to move 18 inches more than the other. It took me several days to find out how to obviate that. I went back to the shop and got some hydraulic pipe and some steel shaft that would go inside the pipe. I put the shaft in the pipe and pinned the two together at one end. I then twisted the pipe in one direction and the shaft in the opposite direction, until I could not twist them any more. Then I pinned it again. There was not a ¼-inch difference between the ends,[e] and the engines were thus all controlled together. This allowed me to start the station. Some time after that we got the Armington & Sims Company to build three engines which had centrifugal governors which controlled the engines independently of each other, and the shaft was removed. ~~The building looked like hell that Sunday~~. After this exciting test[f] I felt something wrong at the stomach. We went to a saloon across the way to get something to drink. Johnson poured me out a little glass full of liquor. I said: "Am I to drink the whole of that?" "Yes," he said. It fixed me all right and had no effect. Johnson took a big drink with me.

SLEEPING ON PILES OF PIPE.

[13] When we put down the tubes in the lower part of New York, in the streets, we kept a big stock of them in the cellar of the station at Pearl street. As I was on all the time, I would take a nap of an hour or so in the day time—any time, and I used to sleep on those tubes in the cellar. I had two Germans who were testing there, and both of them died of diphtheria, caught in the cellar, which was cold and damp. It never affected me.

LIGHTING VANDERBILT'S HOUSE.

[14] While at 65 Fifth Avenue, I got to know Christian Herter, the largest decorator in the United States. He was a highly intellectual man and I loved to talk to him. He was always railing against the rich people for whom he did work, for their poor taste. One day Mr. W. H. Vanderbilt came in to "65," saw the light and decided that he would have his new house lighted with it. This was one of the big "box houses" on upper Fifth Avenue. He put the whole matter in the hands of his son-

in-law, Mr. Twombly, who was then in charge of the telephone department of the Western Union. Twombly closed the contract with us for a plant. Mr. Herter was doing the decoration and it was extraordinarily fine. After a while we got the boilers and engines and wires all done and the lights in position, before the house was quite finished, and thought we would have an exhibit of the light. About 8 o'clock in the evening we lit it up and it was very good. Mr. Vanderbilt, his wife and some of his daughters came in and were there a few minutes when a fire occurred. The large picture gallery was lined with silk cloth interwoven with fine metallic tinsel. In some manner two wires had got crossed with this tinsel, which became red-hot and the whole wall was soon[b] afire. I knew what was the matter and ordered them to run down and shut off. It had not burst into flame, and died out immediately. Mrs. Vanderbilt became hysterical and wanted to know where it came from. We told her we had the plant in the cellar, and when she learned we had a boiler there, she said she would not occupy the house; she would not live over a boiler. We had to take the whole installation out. The houses went afterwards on to the New York Edison system.

REMINISCENSES OF REMENYI

[15] Years ago one of the great violinists was Remenyi: After his performances were out, he used to come down to "65" and talk economics, philosophy, moral science and everything else. He was highly educated and had a great mental capacity. He would talk with me but I never asked him to bring his violin. One night he came in with his violin about 12 o'clock. I had a ~~library~~ laboratory[a] at the top of the house and Remenyi came up there. He was in a genial humor, and played the violin for me for about two hours—$2,000[b] worth.[a] The front doors were closed, and he walked up and down the room as he played. After that every time he came to New York he used to call at 65 late at night with his violin. If we were not there, he would come down to the slums at Goerck street, and would play for an hour or two and talk philosophy. I would talk for the benefit of the music.

A PLANT FOR BENNETT.

[16] One night at 65 James Gordon Bennett came in. We were very anxious to get into a printing establishment. I had caused a printer's composing case to be set up with the idea that if we could get editors and publishers in to see it we could

show them the advantages of the electric light. So ultimately Mr. Bennett came and after seeing the whole operation of everything, he ordered Mr. Howland, general manager of the Herald, to light the newspaper offices up at once by electricity.

ELECTRICITY ON THE RAMPAGE.

[17] One afternoon after our Pearl street station started, a policeman rushed in and told us to send an electrician at once up to the corner of Ann and Nassau Streets—some trouble. Another man and I went up. We found an immense crowd of men and boys there and in the adjoining streets—a perfect jam. There was a leak in one of our junction boxes and on account of the cellars extending under the street the top soil had become insulated, and by means of this leak powerful currents were passing through this thin layer of moist earth. When a horse went to pass over it he would get a very severe shock. When I arrived I saw coming along the street a ragman with a dilapidated old horse, and one of the boys immediately told him to go over on the other side of the road—which was the place where the current leaked. When the ragman heard this he took that side at once. The moment the horse struck the electrified soil he stood right straight up in the air, and then reared again, and the crowd yelled, the policeman yelled and the horse started to run away. This continued until the blockade got so serious the policeman had to clear out the crowd, and we were notified to shut the current off. We got a gang of men, cut the current off for several junction boxes and fixed the leak. One man who had seen it came to me next day and wanted me to put apparatus in for him at a place where they sold horses. He said he could make a fortune with it, because he could get old nags in there and make them act like thoroughbreds.

A STEAM HEATING EPISODE.

[18] While I was digging the trenches and putting in the tubes in the several miles of street in the first district, the New York Steam Heating Company were also digging trenches and putting in steam heating pipes. Mr. C. E. Emery, then the chief engineer and I would meet quite frequently at all hours of the night, I looking after my tubes and he after his pipes. At the same time that Emery was putting down his pipes, another concern started in opposition to the New York Steam Heating Company and were also working nights putting down their pipes in Maiden Lane. I used to talk to Emery about the suc-

cess of his scheme. I thought he had a harder proposition than[b] I had, and he thought that mine was harder than his; but one thing we agreed on and that was that the other steam heating engineer hadn't any chance at all and that his company would fail. If he, Emery was right, the other fellow was wrong. Emery used mineral wool to surround his pipes, which was of a fibrous nature and was stuffed in boxes to prevent the loss of heat and pressure; whereas his competitor was laying his pipes in square boxes filled with lamp black. Before Emery had finished all his pipes and was working in the street one night, he heard a terrible rush of steam. It seems that his competitor had put on steam pressure to test out his pipe. There was a leak in the pipe, the steam got into the lamp black, and blew up, throwing about three tons of lampblack all over the place and covering the fronts of several stores in Maiden Lane. When the people came down next morning everything was lampblack—and the company ~~was~~ busted.

THE BOX OF CIGARS.

[19] When at "65" I used to have in my desk a box of cigars, I would go to the box four or five times to get a cigar, but after it got circulated about the building everybody would come to get my cigars so that the box would only last about a day and a half. I was telling a gentleman one day that I could not keep a cigar. Even if I locked them up in my desk they would break it open. He suggested that he had a friend over on Eighth Avenue who made a superior grade of cigars and who would show them a trick. He said he would have some of them made up with hair and old paper, and I could put them in without a word and see the result. I thought no more about the matter. He came in two or three months after and said: "How did that cigar business work?" I didn't remember anything about it. On coming to investigate it appeared that the box of cigars had been delivered and had been put in my desk, and I had smoked them all. I was too busy on other things to notice.

A LAMP <u>Patent</u>[g] INTERFERENCE.

[20] Soon after I had got out the incandescent light, I had an interference in the Patent Office with a man named Walter K. Freeman, of Kenosha, Wis. He filed a patent and entered into a conspiracy to swear back of the date of my invention, so as to deprive me of it. Detectives were put on the case and we found that he was a "faker," and we took means to break the thing up.

Eugene Lewis had this in hand for me. Some years afterwards this same man attempted to defraud the Parke Davis Company, chemists, and was sent to State prison. A short time after that a Jew syndicate took up a man named Goebel and tried to do the same thing, but again our detective work was too much for them. This was along the same line as that of Drawbaugh, to deprive Bell of his telephone. Whenever an invention of large prospective value comes out these cases always occur. The lamp patent was sustained in the New York Federal Court. I thought that was final and would end the matter but another Federal judge out in St. Louis did not sustain it. The result is I have never enjoyed any benefits from my lamp patents although I fought for many years.

THE MAN HIGHER UP.

[26] When I was laying tubes in the streets of New York, the office received notice from the Commissioner of Public Works to appear at his office at a certain hour. I went up there with a gentleman[b] to see the Commissioner, H. O. Thompson. On arrival he said to me: "You are putting down these tubes. The Department of Public Works requires that you should have five inspectors to look after this work, and that their salary shall be $5 per day payable at the end of each week. Good morning." I went out very much crestfallen, thinking I would be delayed and hampered in the work which I was anxious to finish and was doing night and day. We watched patiently for those inspectors to appear. The only appearance they made was to draw their pay Saturday afternoon.

PRIMITIVE LIGHTING BUSINESS METHODS.

[28] When we started the station at Pearl street in September, 1882, we were not very commercial, we put many customers on but did not make out many bills. We were more interested in the technical condition of the station than in the commercial part. We had meters in which there were two bottles of liquid. To prevent these electrolytes from freezing we had on each meter a strip of metal, when it got very cold the metal would contract and close a circuit, and throw a lamp into circuit inside the meter. The heat from this lamp would prevent the liquid from freezing, so that the meter could go on doing its duty. The first cold day after starting the station, people began to come in from their offices, especially down in Front street and Water street saying that the meter was on fire. We

received numerous telephone messages about it. Some had poured water on it and others said "send a man right up to put it out".

[29] After the station had been running several months and was technically a success, we began to look after the financial part. We started to collect some bills but we found that our books were kept badly and that the person in charge who was no business man had neglected that part of it. In fact he did not know anything about the station anyway. So I got the directors to permit me to hire a man to run the station. This was Mr. Chinnock, who was then superintendent of the Metropolitan Telephone Company of New York. I knew Chinnock to be [-]^h square and of good business ability and induced him to leave his job. I made him a personal guarantee that if he would take hold of the station and put it on a commercial basis and paid 5 per cent on $600,000, I would give him $10,000 out of my own pocket. He took hold, performed the feat, and I paid him the $10,000. I might remark in this connection that years afterwards I applied to the Edison Electric Light Company asking them if they would not like to pay me this money, as it was spent when I was very hard up and made the company a success, and was the foundation of their present prosperity. They said they "were sorry," that is "Wall Street sorry," and refused to do it. This shows what a nice, genial, generous lot of people they have over in Wall street.

NEW YORK REAL ESTATE.

[34] While planning for my first New York station—Pearl Street—of course,—I had no real estate, and from lack of experience had very little knowledge of its cost in New York, so I had assumed a rather large, liberal amount of it to plan my station on. It occurred to me one day that before I went too far with my plans I had better find out what real estate was worth. In my original plan I had 200 by 200 feet. I thought that by going down on a slum street near the water front I would get some pretty cheap property. So I picked out the worst, dilapidated, deserted street there was, and found I could only get two buildings each 25 foot front, one 100 feet deep and the other 25 feet deep. I thought about $10,000 each would cover it, but when I got the price I found that they wanted $75,000 for one and $80,000 for the other. Then I was compelled to change my plans and go upward in[----]^h in the air where real estate was cheap. I cleaned out the building entirely to the walls, and built my station of structural iron work, running it up high.

TESLA IN FRANCE.

[35] I sold the electric light patents in France to a syndicate which ~~built~~ rented[a] works outside Paris at Ivry on the Seine and started to make dynamos and other electric light apparatus. I sent Mr. Batchelor, my principal assistant, over to start the shop. To meet the French conditions we had to redesign the different dynamos and other apparatus, and Batchelor from his training worked night and day. Among those who drifted into the Ivry shop was one Nikola Tesla. He was a tall, lanky man, and was much interested in what Batchelor was doing, and was of great assistance to him, especially as he did not seem to want any more sleep than Batchelor did. After Batchelor had started up and everything was running in good shape, he said to Tesla one day: "We have worked hard, and now I am going to take you down to the Cafe Mignon—a recherché[i] restaurant in Paris—and blow you off to a fine dinner."[j] Whether Tesla had a tape worm is not now known, but they did go to the Cafe Mignon. Batchelor ordered a steak Chateaubriand—a thick steak of large dimensions, broiled between two other generous steaks—and he ordered all the other things which follow.[b] Batchelor, very nervous over his responsibilities, had little appetite, but Tesla got away with most of the steak and the bulk of the trimmings. After he had finished Batchelor said: "Now Tesla, this is on me; if you want anything more, say so." Whereupon Tesla, with the modesty peculiar to him said: "Well Batchelor, if you don't mind, and insist on it, I should like to have another Chateaubriand steak."

SELLING PATENTS IN EUROPE.

[37] I endeavored to sell my lighting patents in different countries of Europe and made a contract with a couple of men. On account of their poor business capacity and lack of practicality, they conveyed the patents all right to different corporations but in such a way and with such confused wording of the contracts that I never got a cent. One of the companies started was the German Edison, now the great Allegemeine Elektricitaets Gesellschaft. The English Company I never got anything for because a lawyer had originally advised Messrs. Drexel, Morgan & Co. as to the signing of a certain document and said it was all right for me to sign. I signed and I never got a cent, because there was a clause in it which prevented me from ever getting anything. One of my associates was Theodore Puskas, who was undoubtedly the first man to suggest the use of the telephone in a central exchange. He made the sug-

gestion to me when the telephone was still on exhibition, and was very enthusiastic over the subject. He started a telephone exchange in Buda Pest.

UNPROFITABLE LAMP MANUFACTURE.

[54] When we first started the electric light, it was soon seen that we had to have a factory for manufacturing lamps. As the Edison Light Company did not seem disposed to go into manufacturing, with what money I could raise from my other inventions and royalties, and some assistance, we started a small lamp factory at Menlo Park. The lamps at that time were costing about $1.25 each to make so I said to the company "If you will give me a contract during the life of the patents I will make all the lamps required by the company and deliver them for 40 cents." The company jumped at the chance of this offer and a contract was drawn up. We then bought at a receiver's sale at Harrison, N.J. a very large brick factory which had been used for an oil cloth works. We got it at a great bargain and only paid a small sum down, and the balance on mortgage. We moved the lamp works from Menlo Park to Harrison. The first year the lamps cost us about $1.10. We sold them for 40 cents, but there were only about 20,000 or 30,000 of them. The next year they cost us about 70 cents and we sold them for 40. There were a good many and we lost more the second year than the first. The third year I had succeeded in getting up machinery and in changing the processes until it got down so that they cost us somewhere around 50 cents. I still sold them for 40 cents and lost more money that year than any other because the sales were increasing rapidly. The fourth year I got it down to 37 cents and I made all the money up in one year that I had lost previously. I finally got it down to 22 cents and sold them for 40 cents and they were made by the million. Whereupon the Wall street people thought it was a very lucrative business, so they concluded they would like to have it and bought me out.

VISITORS TO 65.

[63] I have spoken of Remenyi's visits. Henry E. Dixey, then at the height of his popularity, would come in in those days, after theatre hours, and would entertain us with stories— 1882-3-4. Another visitor who used to give us a great deal of amusement and pleasure was Capt. Shaw, the head of the London Fire Brigade. He was good company. He would go out among the fire laddies and have a great time. One time Robert Lincoln and Anson Stager, of the Western Union, interested

in the electric light, came on to make some arrangement with Major Eaton, president of the Edison Electric Light Company. They came to 65 in the afternoon and Lincoln commenced telling stories—like his father. They told stories all the afternoon, and that night they left for Chicago. When they got to Cleveland, it dawned upon them that they hadn't done any business, so they had to come back on the next train to New York and transact it. They were interested in the Chicago Edison Company, now one of the largest of the systems in the world. I once got telling a man stories at the Harrison lamp factory, in the yard as he was leaving. It was winter and he was all in furs. I had nothing on to protect me against the cold. I told him one story after the other—six of them. Then I got pleurisy and had to be shipped to Florida for cure.

EPISODE WITH A GAS ENGINE.

[66] At an early period at 65 we decided to light it up with the Edison system, and put a gas engine in the cellar, using city gas. One day it was not going very well and I went down to the man in charge and got exploring around. Finally I opened the pedestal,—a storehouse for tools, etc. We had an open lamp, and when he opened the pedestal it blew the doors off and blew out the windows and knocked me down and the other man.

AN ALL NIGHT RESTAURANT.

[67] When we went to Goerck street, the neighborhood[b] was so tough I had to have a special detective take me down through the slu[m]s[b] there at night—Jim Russell. He knew all the folks there and they knew him, and I was not molested. We used to go out at night to a little low place, an all-night house— 8 feet wide and 22 feet long—where we got a lunch at 2 or 3 o'-clock in the morning. It was the toughest kind of restaurant ever seen. For the clam chowder they used the same four clams during the whole season, and the average number of flies per pie was seven. This was by actual count.

THE POODLE AND THE BELT.

[68] In testing[k] dynamos at Goerck street, we had a long flat belt running parallel with the floor, about 4 inches above it, and traveling 4,000 feet a minute. One day one of the directors brought in three or four ladies to the works to see the new electric light system. One of the ladies had a little poodle led by a string. The belt was running so smoothly and evenly the poodle did not notice the difference between it and the floor, and got

into the belt[b] before you could do anything. The dog was whirled around forty or fifty times and a little flat piece of leather came out—and the ladies fainted.

HANDSOME DIVIDENDS.

[77] When we formed the first lamp works at Harrison we divided the interests into one hundred parts or shares at $100 par. One of the boys was hard up after a time and sold two shares to Bob Cutting. Up to that time we had never paid anything, but we got around to the point where the[b] board declared a dividend every Saturday night. We had never declared a dividend when Cutting bought his shares, and after getting his dividends for three weeks in succession he called up on the telephone and wanted to know what kind of a concern this was that paid a dividend weekly. The works sold for $1,085,000.

TD (transcript), NjWOE, Meadowcroft (*TAED* MM003; *TAEM* 226:863). [a]Interlined above in pencil. [b]Mistyped. [c]"Aetna or" interlined above in pencil. [d]"building would collapse" interlined above in pencil. [e]"s" interlined above in pencil. [f]"After this exciting test" interlined above in pencil. [g]Interlined above in pencil and multiply underlined. [h]Illegible. [i]Accent added in pencil. [j]Quotation mark added in pencil. [k]Mistyped and corrected by hand.

C. SECOND BATCH (JUNE 1909)

The following is from a typescript that includes Edison's revisions. At the top of the first page is a handwritten note by William Meadowcroft: "Second Batch   Mr Edison's notes dictated Mr Martin   June 1909   Pencil indicates revision by Mr Edison." Six of its twenty-four sections pertain to the period of this volume; items 21–24 were previously published in Volume Five.

IRON SANDS OF LONG ISLAND.

[3] "Some years ago I heard one day that down at Quogue, Long Island, there were immense deposits of black magnetic sand. This would be very valuable if the iron could be separated from the sand. So I went down to Quogue with one of my assistants and saw there for miles large beds of black sand on the beach in layers from one to six inches thick,—hundreds of thousands of tons. My first[a] thought was that it would be a very easy matter to concentrate this, and I found I could sell the stuff at a good price. I put up a small plant, but just as I got it

started, a tremendous storm came up; and every bit of that black sand went out to sea. During the 28 years that have intervened it has never come back."

THE VANISHED LID.

[12] When we first put the Pearl Street Station in operation in New York, we had cast iron ~~p~~junction[b] boxes at the intersections of all the streets. One night or about 2 o'clock in the morning a policeman came in and said that something had exploded at the corner of William and Nassau. I happened to be at the station, and went out to see what it was. I found that the cover of the manhole, weighing about 200 pounds, had entirely disappeared, but everything inside was intact. It had even[a] stripped some of the threads of the bolts, and we never could find it. I concluded that it was either leakage of gas into the manhole, or else the acid used in pickling the casting had given off hydrogen and ~~aid~~ air[c] had leaked in, making an explosive mixture. As this was a pretty serious problem and as we had a good many of them it worried me very much for fear that it would be repeated and the company might have to pay a lot of damages, especially in districts like that around William and Nassau where there are a good many people about; and if an explosion took place in the daytime it might lift a few of them up. However, I got around the difficulty by putting a little bottle of chloroform in each box, corked up, with a slight hole in it. The chloroform being volatile and very heavy settled in the box and displaced all the air. I have never yet heard of an explosion in a manhole where this chloroform had been used. Carbon tetrachloride now made electrically at Niagra Falls is very cheap & would be ideal for this purpose[d]

A DISHONEST PATENT SOLICITOR.

[21] Around 1881–2 I had several solicitors attending to different classes of work. One of these did me a most serious injury. It was during the time that I was developing my electric lighting system, and I was working and thinking very hard to cover all the numerous parts in order that it would be complete in every detail. I filed a great many applications for patents at that time, but there[a] were 78 of the inventions I made in that period that were entirely lost to me and my company[e] by reason of the dishonesty of this patent solicitor. Specifications had been drawn, and I had signed and sworn to the applications for patents for these 78 inventions, and naturally I supposed they had been filed in the regular way.

[22] ~~As time passed, I was looking for some action of the Patent Office, as usual, but none came~~. He reported fictitious actions by the Patent Office and in many cases reported that patents had been allowed & obtained the final fees.[f] I thought it very strange but had no suspicions until I began to see my inventions recorded in the Patent Office Gazette as being patented by others. Of course, I ordered an investigation and found that the patent solicitor had drawn from the Company the fees for ~~filing~~ all these applications, but had never filed them. All the papers had disappeared. ~~However, and what he had evidently done was to sell them to others who had signed new applications and proceeded to take out patents themselves on my inventions~~. I afterwards found that he had been previously mixed up with a somewhat similar crooked job in connection with telephone patents.

[23] I am free to confess that the loss of these 78 inventions has left a sore spot in me that has never healed. They were important, useful and valuable and represented a whole lot of tremendous work and mental effort, and I had had a feeling of pride in having overcome through them a great many serious obstacles.

[24] It is of no practical use to mention the man's name. I believe he is dead, but he may have left a family. The occurrence is a matter of the old[b] Company's records.

TD (transcript), NjWOE, Meadowcroft (*TAED* MM004; *TAEM* 226:926). [a]Mistyped. [b]Mistyped and corrected by hand. [c]Interlined above in pencil. [d]"Carbon . . . purpose" added in pencil by William Meadowcroft. [e]"entirely lost . . . company" interlined above by typewriter. [f]"He reported . . . fees." interlined above in pencil by Meadowcroft.

D. "BOOK NO. 2" (1907)

The copy of the internal pages of this notebook held at the Edison National Historic Site is in a folder labeled "Book No. 2," but none of the pages are so labeled and no copy of the cover is available. It has been referred to as Book No. 2 since Volume One, and for consistency is still designated so here, but see also the introduction to document G.

The notebook contains a mix of narratives, questions, and notes in Edison's hand. The first two pages are a memo by Meadowcroft, dated 9 January 1920, recounting the preparation and use made of this material between 1907 and 1920. The next sixty-six pages alternately present narrative passages and brief references to various anecdotes. The next nine-page section is labeled "Martin's Questions." The remaining twenty-one pages contain only notes. Some pages have been initialed at the bottom "M.M.E.," by Edison's wife Mina Miller Edison, while several others bear check marks at the top of the page. Altogether fifty-eight items pertain to the period of this volume; items 128, 133, 140, 257, 289, 300, 301, 343, and 390 were previously published in Volume Five.

[103] Remenyi—

[108] Walking down to Lab Ave B & 17 school hours saluted by children thinking I was a priest—[a]

[111] 65 5th ave ofs hours 24— Remenyi   H N Dixie   Duke of Sunderland— Bull Run Russell— Insull—

[112] W H Vanderbilt— Mrs Vanderbilt ordering Engines out   Tinsel on Fire— Insull—ran everything   talk shop at [mothers?][b] funeral[a]

[121] Horse shocked Cor Naussau & Ann[a]

[122] Man hole blew up   never found Cover— got scared— Chloraform—[a]

[123] Hughado[c] Thompson Comr public wks—Inspectors never turned up—[a]

[124] Goerck St, Tammany Hall District manager—running Lathes in street—Castings etc for storage—[a]

[125] Dean—great executive ability—[a]

[126] Jumbo—race across NY—police arrangements—barrell beer—6 horses—just in time[a]

[127] Sims & broken Eng[a]

[128] Porter & 1st High speed Eng shook Hill—[a]

[133] Wilbur & 100 patents lost—[a]

[134] Explosion gas Engine 65 5th Ave cellar   stunned—[a]

[136] 1st starting pearl st station   Engines run away[a]

[137] A & S run 24 hours day   365 days never stopped—[a]

[138] Melted Cobble stone at feeder box—[a]

[140] Started 1st Lamp Wks—loss of money—finally making money[a]

[148] Jumbo across town to Paris aided by police

[149] Running Lathes in Goerck st—

[152] A & S Engine Crank burst   no sleep for s[---][b] 3 days— spit blood— Dean

[160] ground on big feeder 1st station melted Cobble stones[c]

[166] W H Vanderbilt house lighting—tinsel on fire etc—

[257] Villard, Frank Thomson on RR— V̶i̶l̶ Villard & wheat RR   Thomson said main lines never be run—

[267] 1st Jumbo Paris Elec Exh— breaking Crank shaft Eng— Volts too low raised by extra magnets racing across NY 6 horses   all[e] police cleared streets—bbl Beer

[268] Running Lathes outside Goerck St. Tammany Dist Leader— big Jim & Tammany Leader—

[270] Bldg Elec Loco   6 ft drivers

[271] [Sors?][f] starting Pearl St stn   torsion shaft—

[274] Christian Herter—Vanderbilts house   Vanderbilt coming into 65 5th ave   Twombly in WU & telephone   Tinsel on fire WHV house Mrs V ordering out of cellar Eng & B—

[275] Remenyi—

[276] Jas G Bennet came into 65[e] 5th ave said go around the world ordered Howland put it in Herald immediately

[277] Elec leaking   ann & nassau   old Horse—

[278] Emery & Steam heat & myself[e] laying[e] tubes— Lampblack Opstn—

[279] Smoked my own cigars—hair horn & cotton—

[281] Jim _____ detective for Fisk & Gould queer methods—

[282] Walter[e] Freeman & attempt steal Lamp—discomfiture—his fraud on Parke D & Co yrs after & prison

[283] Gobels attempt, Jews syndicate—

[289] Copper mine Menlo—[d]

[294] Laying tubes, Hugdo O Thompson & inspectors—[d]

[299] Great cost buldings 1st station surprised—[d]

[300] Elec RR Menlo, financing Sprague & E H.

[301] Insull.

[303] Johnson & Holbern Viaduct[d]

[304] Batch starting Ivry shop   Tesla—[d]

[306] Puskas & Bailey—failure   German Edison Co now allegmein geshelshaft— English Co never got anything[d]

[319] Lab over Bergman, poker[d]

[324] Tube shop in Washn st   Pierpont Morgan only investment made repaid 10 times over[d]

[343] How got gas bill every man below Central park— motors—insurance maps[a]

[344] Duke Sutherland, Bull Run Russell—Capt Shaw.

[353] 65 5th ave ofs hours 24 daily[a]

[354] Explosion gas Engine in cellar hurt me—[a]

[356] Big Jim—Dist Tam Leader   Lathes run on sidewalk— Dean— Tesla here, Sims Engine—broke shaft. All night house lunch 1 clam for season in chowder— [---------- -- ---][f] 6 flys for each pie— Bad neighborhood,

theives— Accompanied Jim Russell   late night, Spit blood, Deans boy—[a]

[358] fun on Cor Ann & Nassau leaky pavement, horse.[a]

[362] Invented the multiple tubes & gave plans to the Tammany Boss in 1881,= They adopted it[a]

[365] Incandescent Lights at Niblos & Iolanthe Bijou theatre Boston[a]

[369] Lamp Contract, [B]ob Cutting on telephone wanted know what kind of a Co it was to pay div every week[a]

[376] Moved Lab from Menlo to Ave B & 17 then to Lamp factory

[390] Wilbur & loss of 60 patents

AD (photocopy), NjWOE, Meadowcroft (*TAED* MM005; *TAEM* 226:944). [a]Paragraph overwritten with a large "X". [b]Illegible. [c]Paragraph overwritten with a large "X" and followed by dividing mark. [d]Paragraph overwritten with a checkmark. [e]Obscured overwritten text. [f]Canceled.

E. NOTES (JUNE 1909)

Taken from a notebook that has five pages in Edison's hand, these "Notes" are numbered consecutively from 1 to 33. They evidently formed the basis for part of Edison's interview with T. C. Martin, as recorded in document B. Thirteen items pertain to the period of this volume.

[14] 14   Jumbo Paris Expstn. breaking Crank shaft   Houston St Lunch 4 clams used all summer   guess wrong on Volts. Extra mag + spark jumped new insulation. 6 Horse [--]ed[a] clear across town 60 men police. bbl beer made it 1 hour to spare[b]

[15] 15   Running Lathes & storing material on side walk— Tammany dist Leader, Big Jim & taking on men[b]

[16] 16   Bldg Electric Loco 6 ft drivers[b]

[17] 17   Starting Pearl st station. twist on shaft[c]

[20] 20.   Christian Herter, Vanderbilt house fire, tinsel[d] V coming into 65 5th ave— Twombly took chge instaltion, T in WU.[b]

[21] 21   Remenyi[b]

[22] 22   Jas G Bennett came 5th ave heard he was coming set up & lighted printers case. ordered Howland equip Herald at once[b]

[23] 23   Electric Leak Cor Ann & Nassau[b]

[24] 24   Emery & I nights   Emery on steam heat. lamp black competition.[b]

[25] 25   Smoked my Cotton & hair cigars[b]

[26] 26 = Jim ___ Detective for G in my employ   relate queer methods

[27] 27   Walter Freeman attempt steal lamp   yrs after defrauded Parke D & Co sent prison

[28] 28   Goebels attemp[t] Jew syndicate   detective work broke up conspiracy

AD, NjWOE, Lab. N-09-06-27 (*TAED* MM006; *TAEM* 226:1034). [a]Text obscured. [b]Paragraph overwritten with a checkmark. [c]Paragraph overwritten with an "X". [d]Written in margin.

F. NOTES (JUNE 1909)

This notebook includes sixteen pages in an unlabeled section in Edison's hand relating to the Dyer and Martin biography. These pages are preceded by a memo to Edison from William Meadowcroft dated "June 28/09" stating that these notes had been copied, and by Edison's comments on the omission from the draft biography of several of the incidents he had related in documents A and G. There is a typed version of the notes in the William H. Meadowcroft Collection at the Edison National Historic Site. Four of the twenty-four sections pertain to the period of this volume; items 6–8 were previously published in Volume Five.

[6] Soon after the B.[a] business [---][b] had grown so large that ~~he~~ E H Johnson & myself went in as partners & Bergmann Rented an immense factory building at the Cor of Ave B & 17th St covering ¼ of a block & 6 stories high. Here was made all the small things used on the Electric Lighting system such as sockets, chandiliers switches meters, etc   In[c] addition stock Tickers, Telephones, telephone switchboards, Typewriters, (The Hammond type writer was perfected & made here)— over 1~~0~~500 men were finally employed—

[7] ~~B was~~ This shop was very successful both scientifically & financially   Bergmann was a man of great Executive ability and carried economy of manufacture to the limit. Among all the men I had associated with me ~~Bergmann~~ he[a] had the ~~most~~ commercial instinct, [- e--gy][b] most highly developed—

[8] Soon after this shop was started I sent a man named Stewart down to Santiago Chili to put up[c] a Central Station for Electric Lighting   Stewart after finishing the station returned to NYork with glowing accounts of the Country & an order from Madame Cousino the richest woman in Chili for a complete plant with chandeliers for her palace in the suburbs of Santiago. Stewart gave the order to Bergmann, & the price was to be for the chandiliers alone $7,000. Stewart having no place to go generally managed to stay around Bergmanns place recounting the emmense wealth of Madame Cousino, and Bergmann kept raising the price of the outfit until[c] Stewart realized that these glowing accounts of wealth was running into money when he kept away & the chandliers went billed for 17 000, Cash on bill of Lading, as Bergmann said he wasnt sure Stewarts mind wasnt affected[d] & he wanted to be safe.

[13] When I gave up the Laboratory at Menlo Park I moved the apparatus to the top floor of the Bergmann shop at Ave B & 17th st. here I carried on the various Experiments required to further perfect the lighting system. I also devised a system of telegraphy useful for Railways whereby each RR wire was made into two independent circuits no matter how many offices were connected to the line   This was called the phonoplex system & was adopted by over 20 Railways [--][b]   It had the merit that the added devices would work in all kinds of weather where the regular Morse system failed & would even work when the wires were blown down

AD, NjWOE, Lab., N-09-06-28 (*TAED* MM007; *TAEM* 226:1040). [a]Interlined above. [b]Canceled. [c]Obscured overwritten text. [d]"wasnt affected" interlined above and followed by a line of erased, illegible text.

G. "MR. EDISON'S NOTES" (OCTOBER 1908)

These reminiscences come from a notebook labeled "Book No. 2, Mr. Edison's notes re. Biography   October 1908." The sections from G published in Volumes Two through Five were drawn from a typed version prepared by William Meadowcroft. However, a copy of Edison's original manuscript was located and published in Part IV of the microfilm edition, and is the basis for the transcription below. Only one paragraph pertains to the period covered in this volume; a version of this paragraph was also published in Volume Five, based on the typed copy.

[32] The[a] men I sent over were used to establish Telephone Exchanges all over the Continent & some of them became wealthy. It was among this crowd in London that Bernard Shaw was employed before he became famous— The [--------][b] the Chalk Telephone was finally[c] discarded in favor of the Bell receiver it being more simple & cheaper   Extensive litigation with new comers[d] followed   My[a] Carbon Transmitter patent was sustained & preserved the monopoly of the telephone in England for many years   Bells patent was not sustained by the Courts. Sir Richard Webster now the Lord[e] Chief Justice of England was my counsel & sustained all of my patents in England for many years. Webster had a marvellous Capacity for understanding things scientific & his address before the Courts was lucidity itself. ~~My~~ His ~~mind~~ [--- ---][b] Brain is highly organized. My experience of the legal fraternity is that scientific subjects are distasteful & it is rare in this Country for a judge to interpret the statements of the experts correctly & inventors scarcely ever get a decision in their favor. In England the judges seem to be different they are not fooled by the experts but tear their testimony all to pieces & decide the case correctly   Why this difference between English and American judges I cannot explain   It seems to me that scientific disputes should be decided by Some Court containing at least one or two scientific men— If justice is ever to be given an inventor. Inventors acting as judges would not be very apt to decide a [p--][b] complicated law point, and per contra, it is hard to see how a lawyer can decide a complicated scientific point rightly. Some inventors complain of our patent system & the patent office. I consider both are good & that the trouble is in the Federal Courts. There should be a Court of patent appeal with at last two scientific men thereon who could not be blind to the sophistry of paid experts. Men whose inventions would have created wealth to the country of millions have been ~~made dry~~ ruined & prevented from making any money whereby they could continue their career as creators of wealth for the general good just because the Experts befuddled the judge by their misleading statements.

AD (photocopy), NjWOE, Meadowcroft (*TAED* MM008; *TAEM* 227:1). [a]Obscured overwritten text. [b]Canceled. [c]Interlined above. [d]"with new comers" interlined above. [e]Written in margin.

Appendix 2

Isolated Lighting Plant Installations, May 1883

The editors compiled these charts from a published list of all 330 Edison isolated plants in operation as of May 1883. These included 64,615 lamps, a total that is more than an order of magnitude larger than the lamps in the New York First District.[1] The figures indicate the different uses to which the Edison light was applied, for both practical and promotional purposes, since the installation of the first commercial plant in February 1881.[2] They also show the relative geographic distribution of Edison plants, both within the U.S. and around the world.[3]

Comparing the number of plants on each line with the total number of lamps installed in those plants yields the average ratio of lamps to plants, thus giving a rough gauge of the relative intensity of electric lighting use in various regions and situations. For example, it appears that industrial installations in the United States were considerably larger than their foreign counterparts, while institutional, hotel, and restaurant installations were larger abroad, and theatrical lighting was intensive everywhere. The dominance of large-scale plants in Massachusetts is obvious, while the systems installed in Edison's home state of New Jersey and in the Midwest were apparently more modest in size. Similarly, there is a pronounced gap between the ambitious plants installed in the United Kingdom and those elsewhere in Europe and overseas, which although numerous were not built on any great scale.

ISOLATED PLANT INSTALLATIONS BY TYPE

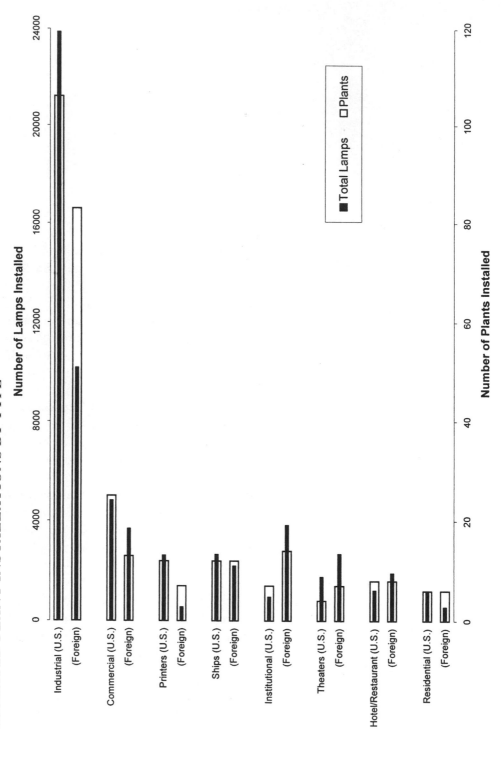

U.S. ISOLATED PLANTS BY LOCATION

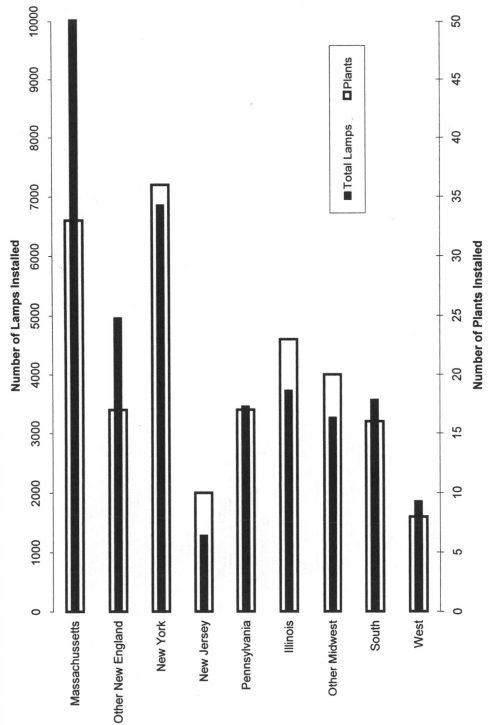

Number of Lamps Installed

Number of Plants Installed

- Total Lamps
- □ Plants

Massachussetts
Other New England
New York
New Jersey
Pennsylvania
Illinois
Other Midwest
South
West

FOREIGN ISOLATED PLANTS BY LOCATION

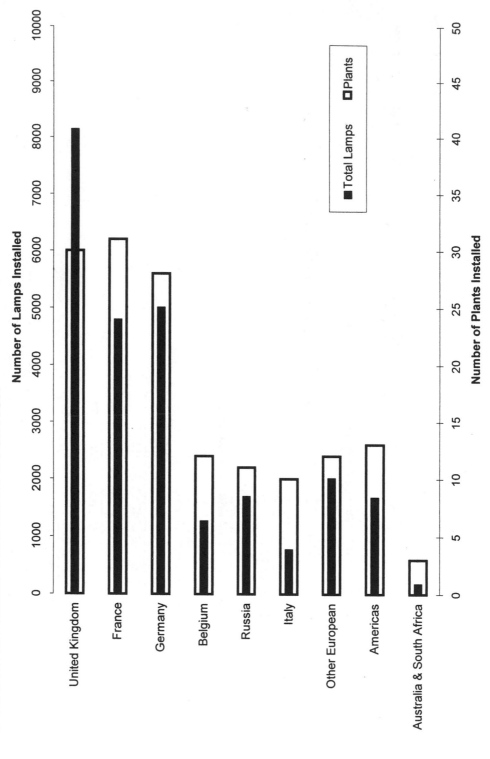

1. The computed totals vary slightly from those given in the original, Edison Electric Light Co. Bulletin 18:30–38, 31 May 1883, CR (*TAED* CB018:16–20; *TAEM* 96:842–46). These data do not include central stations.

2. The categories used for the various types of installations are based closely on those used in the published list, with slight modifications. Among the facilities included in the broader categories are:

Industrial: mills, factories, mines, agricultural estates, ship yards and dock yards;
Commercial: stores, offices, docks, and multiple-use facilities;
Institutional: government facilities, schools, rail stations, post offices, museums;
Hotel/Restaurant: also includes social clubs.

The first plant to open was that of Hinds, Ketcham & Co.; see Doc. 2053.

3. The regional geographical groupings include:

Other New England: Connecticut, Maine, New Hampshire, Rhode Island, Vermont
Other Midwest: Iowa, Michigan, Minnesota, Missouri, Ohio, Wisconsin
South: Arkansas, District of Columbia, Georgia, Kentucky, Louisiana, Maryland, North Carolina, Virginia
West: California, Oregon
Other European: Austrian Empire, Finland, Netherlands, Spain
Americas: Canada, Chile, Cuba.

Appendix 3

Specifications of Dynamos Produced at the Edison Machine Works, April 1881–March 1883

This table provides basic specifications for the five dynamo models manufactured for isolated plant and central station installations during the period of this volume.[1] Both physical and electrical characteristics of the commercially-available models are given, along with their published prices. The isolated plant models were available in two electrical configurations, designed to run either 110 volt A lamps or 55 volt B lamps.

Model	HP[2]	RPM	Field Coils	Ht. × Base (in.)	Wt. (lbs.)	Volts	Amps.[3]	Lamps	Dynamo/ Installation Price ($)[4]
E[5]	4	2,200	2	35 × 25 × 17.5	700	110	11	15 A[6]	500 / 561.25
						55	22	30 B	530 / 642.50
Z[7]	10	1,200	2	72 × 45 × 39	3,000	110	44	60 A	1,200 / 1,595
						55	88	120 B	1,320 / 1,860
L	19	900	4	78 × 60 × 39	6,000	110	92	150 A	2,000 /2,637.50
						55	184	300 B	2,300 / 3,425
K	35	900	6	78 × 70 × 39	8,250	110	183	250 A	3,000 / 4,062.50
						55	366	500 B	3,500 / 5,375
C	125	350	12[8]	78 × 168 × 105[9]	60,336[10]	110	900	1,200 A[11]	8,000[12] / not applicable

1. Although an Edison Electric Light Co. Bulletin (Edison Electric Light Co. Bulletin 10:3, 5 June 1882, CR [*TAED* CB010:2; *TAEM* 96:715]) mentioned a model R, rated at 500 A lamps, the editors have

found no evidence that this or any other model than those listed here were produced or sold in this period. The Z, L, and K models served as the basis for improved models rated at 25, 50, 100, 200, 300, and 400 A lamps that were developed during the summer and introduced by the end of 1883. Commonly referred to as Edison-Hopkinson machines, these dynamos featured shorter and thicker field coils, which reduced the reluctance of their magnetic circuits and increased their efficiency. Doc. 2416; Rocap to TAE, 6 July 1883, DF (*TAED* D8334W; *TAEM* 67:384); Andrews 1924.

Principal sources used to compile this table are Catalog of Edison Co. for Isolated Lighting, 1 Sept. 1882, PPC (*TAED* CA002A; *TAEM* 96:103); Edison Electric Light Co. Bulletins, 1882–84, CR (*TAED* CB; *TAEM* 96:667); Charles Batchelor notebook, 1882–83, Cat. 1311, Batchelor (*TAED* MBN010; *TAEM* 91:575); William Hammer Notebook 8, 1882, Ser. 1, Box 13, Folder 2, WJH; Schellen 1884, 348–57; Andrews 1924; Dredge 1882–85; and Clarke 1904. For electrical characteristics such as armature and field coil resistances and field coil connections see headnotes for Docs. 2122, 2126, and 2238; also William Hammer Notebook 8, 1882, Ser. 1, Box 13, Folder 2, WJH.

2. Horsepower figures are from the Edison Co. for Isolated Lighting's catalog and apparently give the rated horsepower of the engine and not the horsepower applied at the armature shaft. William Hammer gave different values, possibly because he used figures for engines readily obtainable in England or because he gave the horsepower at the armature shaft. His values were: E, 2.5 horsepower; Z, 8; L, 18; K, 30; and C, 150. One horsepower equals 0.746 kilowatts. Electrical power in watts is the product of voltage in volts and current in amperes. Catalog of Edison Co. for Isolated Lighting, 1 Sept. 1882, PPC (*TAED* CA002A; *TAEM* 96:103); William Hammer Notebook 8, 1882, Ser. 1, Box 13, Folder 2, WJH.

3. Current output for the 110 volt dynamos are from William Hammer Notebook 8, 1882, Ser. 1, Box 13, Folder 2, WJH. Current output for the 55 volt dynamos are estimates obtained by doubling the 110-volt current.

4. These were retail prices charged end-user customers and were more than double the prices for internal dynamo sales directly from the Edison Machine Works to other Edison companies, such as the Edison Electric Light Co., Ltd., in London. Retail prices included the dynamo, resistance for regulating voltage, six commutator brushes, lamps and sockets, but not the engine, counter-shaft, belt connections, foundations, or installation. Installed prices also included lamp fixtures, conductor wires, switches, safety catches, and miscellaneous accessories. Catalog of Edison Co. for Isolated Lighting, 1 Sept. 1882, CR (*TAED* CA002A; *TAEM* 96:103); TAE to Arnold White, 24 July 1882, Lbk. 7:760 (*TAED* LB007760; *TAEM* 80:753); TAE to Rocap, 5 June 1883, DF (*TAED* D8334U; *TAEM* 67:382); see Doc. 2310.

5. The E dynamo was an experimental model not widely sold. Edison regarded them "as unsatisfactory to myself as to the users of them, but the large demand for that size machine necessitated sending them out." TAE to Bliss, 29 Nov. 1882, Lbk. 14:481 (*TAED* LB014481; *TAEM* 81:1028).

6. William Hammer rated the E dynamo at 17 A or 34 B lamps. William Hammer Notebook 8, 1882, Ser. 1, Box 13, Folder 2, WJH.

7. The model Z was a refinement of the 60-lamp model A machines used in demonstration installations at Menlo Park in 1880 and early 1881. Two model A dynamos were shipped to South America in early 1881 to power 75-lamp plants. In an 11 July 1881 notebook entry Charles Clarke referred to the Z as an "upright" version of the A. Jehl 1937–41, 855–63, 870, 884; Docs. 2048 and 2144; Rocap to Eaton, 5 May 1882, DF (*TAED* D8233ZAO; *TAEM* 61:1040); N-81-07-11:1, Lab. (*TAED* N220:1; *TAEM* 41:3).

8. The first C dynamo, used at the 1881 Paris Electrical Exhibition, had 8 field coils. Clarke 1904, 37–45.

9. The baseplate dimensions included both dynamo and direct-coupled steam engine. Edison Machine Works Test Report, undated 1882, DF (*TAED* D8235ZAG; *TAEM* 62:86).

10. This weight included the direct-connected Armington & Sims steam engine, baseplate, and nonmagnetic zinc base. The weight of the dynamo alone was 42,822 pounds. Edison Electric Light Co. Bulletin 10:7–8, 5 June 1882 (*TAED* CB010:4–5; *TAEM* 96:717–18).

11. This was the nominal rating for 16 candlepower A lamps. However, Edison advised that the dynamo could be "forced" to power 1,400 lamps if run at a higher speed. This apparently was after modifications to the armature and commutator in the summer of 1882. TAE to Arnold White, 27 July 1882, Lbk. 7:781 (*TAED* LB007781; *TAEM* 80:759); see headnote, Doc. 2238 and Doc. 2316 n. 1.

12. As a central station dynamo the C was not offered for sale by the Isolated Co.; its price is that given (free on board in New York) by the Edison Electric Light Co. in an August 1882 memorandum. It did not include the price of the engine, given variously as $2,000 or $2,500. Rocap to TAE, 7 Feb. 1883 and 19 June 1882; Rocap to Samuel Insull, 14 Nov. 1882; all DF (*TAED* D8334E, D8233ZCD, D8233ZEJ; *TAEM* 67:359, 61:1087, 1147); TAE to Arnold White, 27 July 1882, Lbk. 7:781 (*TAED* LB007781; *TAEM* 80:759); Edison Electric Light Co. estimate (p. 2), 3 Aug. 1882, DF (*TAED* D8227ZAT; *TAEM* 61:525).

Appendix 4

Cable Name Codes, 1881–1883

Edison and his transatlantic associates often used code names in their cable correspondence. To ensure the proper delivery of messages, codes for recipients were registered with the local office of a cable or telegraph company. Those used only in the body of messages were entered into books kept by each set of correspondents. The list below identifies most of the codes for personal or corporate names used during this period; not all appear in this volume but are presented as an aid to further research.[1] It does not include obvious variations, such as "Fifty seven" for "Fifty seven London." Some names appear more than once. Edward Johnson, for example, adopted different codes as his duties and office addresses changed. Different pairs of correspondents also used different names to refer to the same individual. Ciphers for equipment or specific transactions are not included.

Cipher	*Name*	*First used in Doc.*
Abaft	Edison	2141
Abatement	Egisto Fabbri	2155
Abdomen	Joshua Bailey	2155
Abduction	George Gouraud	2161
Abortion	George Barker	2150
Adequate[2]	Bergmann & Co.	
Adhere	U.S. Electric Light Co.	2141
Eknoside NY[3]	Edison	
Eknoside Paris[4]	Charles Batchelor	
Falsetto	Egisto Fabbri	2157
Fiftyseven London[5]	Edward Johnson	2197
Heraclite Paris	Joshua Bailey	2153

Knoside	Charles Batchelor	2141
Lowbatch	Grosvenor Lowrey and Charles Batchelor	2182
Noside[6]	Edison	
Noside London[7]	George Gouraud	2086
Phonos London[8]	Edward Johnson	2178
Puskabailey	Theodore Puskas and Joshua Bailey	2166
Sevenfour[9]	Edison Electric Light Co., Ltd. (London)	
Sevenfour NY[10]	Edison	

1. Charles Batchelor sent Samuel Insull additional lists of names related to the Paris Exposition and Edison's European electric light business. Batchelor to Insull, 15 Sept. and 31 Dec. 1881, both DF (*TAED* D8135ZBR, D8132ZCD; *TAEM* 58:1066, 581).

2. TAE to Batchelor, 10 Feb. 1882, Lbk. 11:248 (*TAED* LB011248; *TAEM* 81:326).

3. Batchelor and Lowrey to TAE, 31 Oct. 1881, LM 1:83C (*TAED* LM001083C; *TAEM* 83:913).

4. See, e.g., TAE to Batchelor, 2 Nov. 1881, LM 1:86A (*TAED* LM001086A; *TAEM* 83:915). This address was reserved for cables from Edison; see Batchelor to Insull, 1 May 1882, DF (*TAED* D8238ZAY; *TAEM* 62:391).

5. Johnson and Edison began using the cable addresses "Fiftyseven London" (for Johnson) and "Fiftyseven New York" (for Edison) about the same time. 57 Holborn Viaduct was the location of Johnson's new office.

6. Gouraud to TAE, 24 Sept. 1881, LM 1:46A (*TAED* LM001046A; *TAEM* 83:895).

7. Gouraud registered this cable cipher name ("Edison" spelled backwards) in 1880; see Doc. 2023.

8. Johnson used this cipher (or simply "Phonos") since late 1879; see Doc. 1865.

9. See Western Union to TAE, 11 May 1882, DF (*TAED* D8239ZBL; *TAEM* 62:887).

10. Edward Bouverie to TAE, 12 and 13 Oct. 1882, LM 1:256E, 257B (*TAED* LM001256E, LM001257B; *TAEM* 83:1000).

Appendix 5

Edison's Patents, April 1881–March 1883

Edison was exceptionally active in seeking patent protection for his inventions during the period of this volume. Having already secured the fundamental elements of his electric lighting system, he pursued patents for a variety of improvements, ancillary items, and manufacturing processes. In an unusual number of instances, Edison also permitted his associates to take patents on some minor inventions in their own names, or jointly with him.

Previous volumes of *The Papers of Thomas A. Edison* included a complete list of all patents originated by Edison in the relevant period. However, since this information is now available on the Edison Papers website, as described below, it has not been duplicated here. Instead, a chart (A) depicts Edison's successful and unsuccessful applications by month (according to execution or filing date, respectively).[1] The cumulative quarterly totals are shown in a table (B) by subject area and date. These are followed by a detailed list (C) of patents taken out by, or with, Edison's associates. Finally, because Edison and his financial backers regarded patent protection in Great Britain as second in importance only to protection in the United States, there is a list (D) of specifications filed there as well. Because British law permitted a single specification to aggregate multiple elements that would have required separate patents in the United States, Edison was able to file fewer patents there for essentially the same practical effects.

The full text and drawings of Edison's U.S. patents are on the Thomas A. Edison Papers website at http://edison.rutgers .edu/patents.htm. They can be searched by execution date, patent number, or subject; patents within particular subject

A. EDISON'S PATENT APPLICATIONS BY MONTH

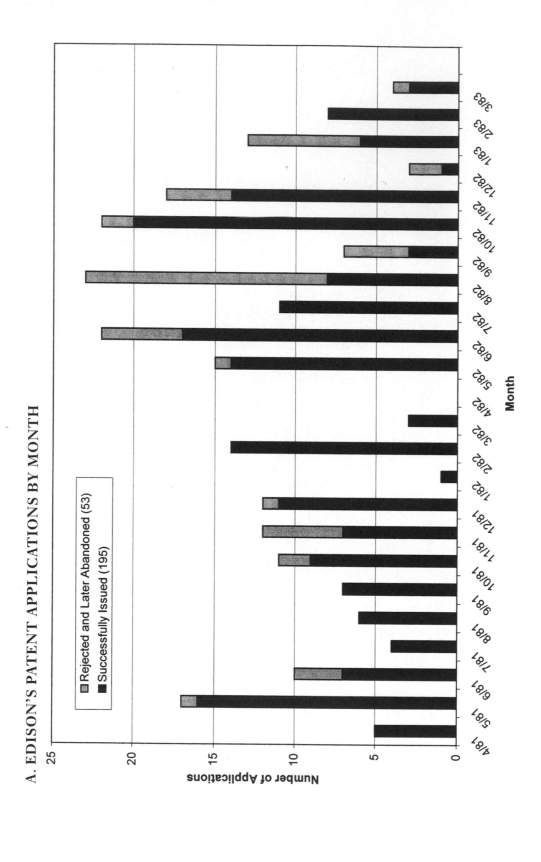

areas are further organized by execution date. The claims and drawings of Edison's unsuccessful applications (identified by application case number) are recorded in several casebooks (Patent Application Casebooks E-2536, E-2537, E-2538; Patent Application Drawings, (c. 1879–1886); all PS [*TAED* PT020, PT021, PT022, PT023; *TAEM* 45:698, 733, 787, 818]).

1. The large number of applications, many ultimately unsuccessful, attributed to August 1882 includes those filed in consequence of Zenas Wilber having failed to submit numerous cases to the Patent Office.

B. EDISON'S PATENTS AND UNSUCCESSFUL APPLICATIONS BY DATE AND SUBJECT

This table lists Edison's patent applications by quarter and subject matter. Patents that were successfully issued are tabulated by execution date of the application and tallied without parentheses. Unsuccessful applications are tabulated by date of filing and shown in parentheses. The table is arranged in descending order of the total number of applications within each subject classification. Applications for incandescent lamps are divided into manufacturing or design subcategories. Claims for methods or processes are generally considered to be for lamp manufacturing; those for lamp elements or combinations of elements are classified as lamp design.

	Apr.–June 1881	July–Sept. 1881	Oct.–Dec. 1881	Jan.–Mar. 1882	Apr.–June 1882	July–Sept. 1882	Oct.–Dec. 1882	Jan.–Mar. 1883	Total
Incandescent Lamp Manufacturing	6 (1)	2	6	1	3 (1)	7 (3)	9 (3)	6 (2)	40 (10)
Incandescent Lamp Design	4 (1)	0	5 (2)	3	1	1 (3)	12 (1)	2 (2)	28 (9)
Voltage Regulation and Engine Governing	2	2	4	11	8	4 (3)	11 (1)	1	43 (4)
Transmission and Distribution	3	3	2 (2)	0	8 (2)	1 (2)	2	6	25 (6)
Dynamo Design and Operation	3 (1)	5	3 (1)	0	3	4	0 (2)	2 (1)	20 (5)
Electric Traction	0	0	1	0	2 (1)	3 (4)	0	0	6 (5)
Battery	1	0	0 (1)	0	2 (2)	0 (2)	0	0 (2)	3 (7)
Consumption Meter	3	1	1	0	1	2	0	0	8
Arc Lamp	1	0	3 (1)	0	1	0	0	0	5 (1)
Fixtures and Interior Elements	2	0	1 (1)	0	0	0	0	0	3 (1)
Motor	2	0	0	0	0	0	0	0	2 (1)
Pyromagnetic Generator	0	0	0	0	1	0	0	0	1
Subtotal, Electric Light and Power	27 (3)	13 (0)	26 (8)	15 (0)	30 (6)	22 (17)	34 (7)	17 (8)	184 (47)
Telegraphy and Telephony	0	3	0	3	0	0	0	0	6
Mining and Ore Milling	1	1	1	0	0	0 (2)	0	0	3 (2)
Miscellaneous	0 (1)	0	0	0	1	0	1 (1)	0	2 (2)
Grand Total	28 (4)	17 (0)	27 (8)	18 (0)	31 (6)	22 (19)	35 (8)	17 (8)	195 (53)

846

C. U.S. PATENTS RELATING TO ELECTRICITY BY EDISON EMPLOYEES AND ASSOCIATES, EXECUTED APRIL 1881–MARCH 1883[1]

Patentee	Patent No.	Title	Executed	Filed	Issued	Assigned to
Sigmund Bergmann	257,146	Shade-Holder for Electric Lamps	20 Feb. 1882	25 Feb. 1882	2 May 1882	
Sigmund Bergmann	257,276	Switch for Electric-Light Circuits	20 Feb. 1882	25 Feb. 1882	2 May 1882	
Sigmund Bergmann	257,277	Socket for Electric Lamps	20 Feb. 1882	25 Feb. 1882	2 May 1882	Half to Edward Johnson.
Sigmund Bergmann	263,103	Electrical Chandelier	20 Feb. 1882	25 Feb. 1882	22 Aug. 1882	Half to Edward Johnson.
Sigmund Bergmann	262,271	Combined Gas and Electric Light Fixture	Not stated.	27 Apr. 1882	8 Aug. 1882	Half to Edward Johnson.
Sigmund Bergmann	262,272	Electrical Extension Chandelier	Not stated.	2 May 1882	8 Aug. 1882	Half to Edward Johnson.
Sigmund Bergmann	266,750	Telephone Switch	Not stated.	3 July 1882	31 Oct. 1882	Half to Edward Johnson.
Sigmund Bergmann	275,749	Connection for Electric Light Fixtures	21 Sept. 1882	9 Oct. 1882	10 Apr. 1883	Half to Edward Johnson.
Sigmund Bergmann	293,552	Socket for Electric Lamps	Not stated.	9 Oct. 1882	12 Feb. 1884	Bergmann & Co.
Sigmund Bergmann	293,553	Combined Gas and Electric Lamp Fixture	Not stated.	9 Oct. 1882	12 Feb. 1884	Bergmann & Co.
Sigmund Bergmann	298,658	Socket for Incandescent Electric Lamps	Not stated.	9 Oct. 1882	13 May 1884	
Sigmund Bergmann	275,748	Flexible Electric Conductor	Not stated.	6 Dec. 1882	10 Apr. 1883	Half to Edward Johnson.
Otto Moses	276,702	Voltaic Arc Light	8 Aug. 1882	13 Sept. 1882	1 May 1883	
Otto Moses	276,703	Voltaic Arc Lamp	5 Sept. 1882	13 Sept. 1882	1 May 1883	
Otto Moses	276,704	Voltaic Arc Lamp	11 Oct. 1882	26 Oct. 1882	1 May 1883	
Otto Moses	276,705	Voltaic Arc Lamp	11 Oct. 1882	26 Oct. 1882	1 May 1883	
Otto Moses	286,953	Voltaic Arc Lamp	5 Feb. 1883	17 Feb. 1883	16 Oct. 1883	

Name	Patent No.	Title				Notes
Otto Moses	292,840	Voltaic Arc Lamp	6 Mar. 1883	20 Apr. 1883	5 Feb. 1884	
Otto Moses	293,495	Voltaic Arc Lamp	6 Mar. 1883	20 Apr. 1883	12 Feb. 1884	
Edward Johnson	251,596	Socket or Holder for Electric Lamps	19 May 1881	27 May 1881	27 Dec. 1881	
Edward Johnson	256,701	Combined Gas and Electric Lamp Fixture	19 Sept. 1881	28 Nov. 1881	18 Apr. 1882	
Edward Johnson	264,298	Coupling Device for Electrical Conductors	19 Sept. 1881	28 Nov. 1881	12 Sept. 1882	Half to Sigmund Bergmann.
Edward Johnson	264,299	Connection for Electric Circuits	19 Sept. 1881	28 Nov. 1881	12 Sept. 1882	Half to Sigmund Bergmann.
Edward Johnson	314,582	Safety Catch	7 Oct. 1882	23 Nov. 1882	31 Mar. 1885	Edison Electric Light Co.
Edward Johnson	319,415	Electric Telephonic Transmitter	2 Feb. 1883	1 Mar. 1883	2 June 1885	
Samuel Mott	267,446	Electric Governor for Driving Engines for Dynamo Electric Machines	28 June 1881	1 Oct. 1881	14 Nov. 1882	
Samuel Mott	264,737	Incandescing Electric Lamp	19 Nov. 1881	29 Nov. 1881	19 Sept. 1882	Edison Electric Light Co.
Samuel Mott	283,270	Incandescing Electric Lamp	19 Nov. 1881	29 Nov. 1881	14 Aug. 1883	Edison Electric Light Co.
Samuel Mott	267,445	Means for Measuring Electric Currents	15 Apr. 1882	24 Apr. 1882	14 Nov. 1882	
Samuel Mott and William Stern[2]	250,094	Electric Cigar Lighter	23 Apr. 1881	25 Apr. 1881	29 Nov. 1881	
Samuel Mott and William Stern	251,126	Electric Cigar Lighter	23 Apr. 1881	25 Apr. 1881	20 Dec. 1881	
William Stern and Henry Byllesby	258,149	Lighting Cars by Electricity	11 Jan. 1882	16 Jan. 1882	16 May 1882	Half to Thomas Edison.[3]

Inventor	Patent No.	Title	Application Filed	Application Executed	Patent Issued	Assignment
Luther Stieringer	253,955	Lantern for Electric and Other Lights	12 Sept. 1881	15 Sept. 1881	21 Feb. 1882	
Luther Stieringer	266,550	Electric Light Chandelier	22 Dec. 1881	27 Dec. 1881	24 Oct. 1882	Equal partnership of himself, Charles Hanington, and Richard Dyer.
Luther Stieringer	259,235	Electrical Fixture	10 Mar. 1882	15 Mar. 1882	6 June 1882	Two-thirds to Charles Hanington and Richard Dyer.
Luther Stieringer	278,465	Support for Electric Light Conductors	10 Mar. 1882	15 Mar. 1882	29 May 1883	
Luther Stieringer	266,549	Electrical Fixture	31 May 1882	14 June 1882	24 Oct. 1882	
Luther Stieringer	272,169	Extension Electroliers	9 Aug. 1882	7 Sept. 1882	13 Feb. 1883	
William Holzer	264,698	Electric Lamp	27 Sept. 1881	28 Nov. 1881	19 Sept. 1882	Edison Electric Light Co.
William Holzer	289,838	Manufacture of Incandescing Electric Lamps	21 Mar. 1883	30 Mar. 1883	11 Dec. 1883	Edison Lamp Co.
William Holzer	305,191	Incandescent Electric Lamp	21 Mar. 1883	30 Mar. 1883	16 Sept. 1884	Edison Lamp Co.
William Holzer	289,837	Manufacture of Incandescing Lamps	28 Mar. 1883	30 Mar. 1883	11 Dec. 1883	Edison Lamp Co.
Edwin Greenfield and Charles Bradley	260,562	Switch and Indicator for Electric Lamps	Not stated.	22 Mar. 1882	4 July 1882	One-third to Sigmund Bergmann.
Edwin Greenfield	266,808	Safety or Cut-Off Switch	Not stated.	8 June 1882	31 Oct. 1882	Half to Sigmund Bergmann.
Edwin Greenfield	278,535	System of Electrical Distribution	Not stated.	8 June 1882	29 May 1883	Half to Sigmund Bergmann.
John Ott	265,858	Regulator for Dynamo Electric Machines	6 June 1882	14 Aug. 1882	10 Oct. 1882	Edison Electric Light Co.
John Ott	265,859	Regulator for Dynamo Electric Machines	6 June 1882	14 Aug. 1882	10 Oct. 1882	Edison Electric Light Co.
John Ott	271,654	Regulator for Dynamo Electric Machines	6 June 1882	14 Aug. 1882	6 Feb. 1883	Edison Electric Light Co.

Inventor(s)	Patent No.	Title				Assignment
Thomas Edison and Charles Clarke	287,525	Regulator for Systems of Electrical Distribution	4 Oct. 1882	20 Oct. 1882	30 Oct. 1883	Edison Electric Light Co.
Charles Clarke	284,382	Circuit and Apparatus for Electric Temperature and Pressure Indicator	9 Jan. 1883	19 Jan. 1883	4 Sept. 1883	Robert Hewitt, Jr.[4]
Richard Dyer and Henry Seely	259,115	Electric Toy	29 Oct. 1881	2 Nov. 1881	6 June 1882	One-third to Samuel Insull.
Richard Dyer and Henry Seely	287,758	Electric Flat Iron	Not stated.	15 Sept. 1882	30 Oct. 1883	One-third to Samuel Insull.
Henry Seely	259,054	Electric Flat Iron	6 Dec. 1881	6 Dec. 1881	6 June 1882	Two-thirds to Richard Dyer and Samuel Insull.
Calvin Goddard	287,532	Junction for Electrical Conductors	27 Nov. 1882	8 Dec. 1882	30 Oct. 1883	Edison Electric Light Co.
Calvin Goddard	287,533	Safety Catch for Electric Circuits	27 Nov. 1882	8 Dec. 1882	30 Oct. 1883	Edison Electric Light Co.
John Kruesi	273,859	Junction Box and Connection for Electrical Conductors	28 July 1882	5 Aug. 1882	13 Mar. 1883	
John Kruesi	275,776	Underground Electrical Conductor	23 Sept. 1882	9 Oct. 1882	10 Apr. 1883	
Alfred Haid	271,628	Secondary Battery	8 Aug. 1882	13 Sept. 1882	6 Feb. 1883	Edison Electric Light Co.
Thomas Edison and Patrick Kenny	479,184	Facsimile Telegraph	26 July 1881	6 Dec. 1881	19 July 1892	

1. This list is arranged by the number of patents held by an individual, then by execution date (or filing date, where absent), and patent number. For a list of U.S. patents relating to electricity from about this time see U.S. Patent Office 1882 and annual supplements. The U.S. Patent and Trademark Office maintains full-text images of all issued patents searchable by patent number at: http://patft.uspto.gov/netahtml/PTO/srchnum.htm.

2. William Stern (1860–1914), an associate from Edison's Menlo Park days, was with the Philadelphia printing firm Edward Stern & Co. in 1883. In 1884, he was connected with the Pittsburgh Electric Co. and in 1888 with the Pittsburgh firm of Stern & Silverman, both agents for Edison's electric light. In early 1883 Stern met with George Westinghouse and officials of the Pennsylvania Railroad to try to commercialize his patent on lighting railroad cars by electricity. He was apparently unsuccessful and in 1892 offered to sell his interest to Edison for $5,000. Edison replied that he would instead sell his interest to Stern. Stern to TAE, 13 and 16 Feb. 1883, 11 Aug. 1884, 26 July 1888, 13 Aug. 1890, 25 Mar. 1892; all DF (*TAED* D8303ZAQ, D8303ZAU, D8414ZAA, D8828ACN, D9050ACR, D9202ABD; *TAEM* 64:78, 84; 71:638; 122:911; 129:1077; 132:363); Insull to Stern, 6 Mar. 1883, Lbk. 15:427; Tate to Stern, 1 Sept. 1890, Lbk. 43:404 (*TAED* LB015427, LB043404; *TAEM* 82:212; 141:488); *Philadelphia Record*, 5 Sept. 1914, SC14.

3. TAE agreement with Stern and Henry Byllesby, 11 Jan. 1882, Kellow (*TAED* HK028AAA1; *TAEM* 144:1028).

4. A friend of Francis Upton, Hewitt was later associated with him in the Telemeter Co. in marketing process-control devices for the brewing industry. Hewitt to E. A. Pontifex, 29 May 1885; Hewitt to Upton, 14 July 1887; both Upton (*TAED* MU074, MU103; *TAEM* 95:724, 811).

D. EDISON'S BRITISH PATENTS

Edison regarded his patents in Great Britain as the most important in any country outside the United States. Those specifications distilled what he believed to be the most important elements of his many U.S. patents and served as templates for specifications in other countries and territories. This section identifies the patents filed for Edison in Britain from the beginning of 1881 through the close of this volume; it is derived from a list generously supplied to the editors by Brian Bowers, the former Senior Curator of the Science Museum in London. The dates and titles given below are taken whenever possible from official copies published by the Commissioners of Patents; data for patents not found by the editors is supplied from Dredge 1882–85. (Edison's British patents prior to 1881 are in Cat. 1321, Batchelor [*TAED* MBP; *TAEM* 92:3]; facsimiles of those found and listed below will be published at a later date on the Edison Papers website [http://edison.rutgers.edu]).

Specifications were filed at the Patent Office through the end of 1881 by Edward Brewer or Peter Jensen, patent agents in London originally hired by Edison's longtime patent attorney Lemuel Serrell and retained by his successor, Zenas Wilber. Richard Dyer, whom Edison hired to replace Wilber in early 1882, collaborated with Thomas Handford, a distinguished London patent attorney, to review and file Edison's specifications. Unlike U.S. patents, which bear unique sequential numbers, the British system numbered patents from 1 each year, making the year an essential part of any U.K. patent citation. As a "communication from abroad," each of Edison's patents was actually issued in the name of the agent who filed it.

Year Number 1881	Provisional Specification (Final)		Short Title
539	8 Feb. 1881	(8 Aug. 1881)	Electric Lamp
562	9 Feb. 1881	(9 Aug. 1881)	Manufacture of Carbon Burners, &c.
768	23 Feb. 1881	(23 Aug.1881)	Connecting the Carbon Ends to Conducting Wires in Electric Lamps
792	24 Feb. 1881	(24 Aug. 1881)	Electric Lamps, &c.
1,016	9 Mar. 1881	(9 Sept. 1881)	Veber-meters
1,240	21 Mar. 1881[a]		Armatures for Electric Machines[a]
1,783	25 Apr. 1881	(25 Oct. 1881)	Measuring Electric Currents
1,802	26 Apr. 1881	(26 Oct. 1881)	Electric Lights, &c.
1,918	3 May 1881	(3 Nov. 1881)	Manufacture of Carbon Conductors for Incandescent Electric Lamps
1,943	4 May 1881	(4 Nov. 1881)	Electric Lighting
2,482	7 June 1881	(7 Dec. 1881)	Electric Machines, &c.
2,492	8 June 1881	(8 Dec. 1881)	Electric Lamps
2,495	8 June 1881	(8 Dec.1881)	Electric Arc Lights
2,954	6 July 1881	(6 Jan. 1882)	Electric Machines
3,140	19 July 1881[b]		Facsimile Telegraphs[b]
3,231	23 July 1881[b]		Commutators for Dynamo or Magneto-Electric Machines, &c.[a]
3,483	11 Aug. 1881	(11 Feb. 1882)	Electrical Conductors, &c.
3,804	1 Sept. 1881	(1 Mar. 1882)	Commutators for Dynamo or Magneto Electric Machines and Electric Motors, &c.
3,932	10 Sept. 1881[b]		Dynamo or Magneto-Electric Machines[a]
4,034	19 Sept. 1881	(18 Mar. 1882)	Dynamo or Magneto Electric Machines and Electro Motors
4,174	27 Sept. 1881	(27 Mar. 1882)	Electric Lamps, &c.

4,552	18 Oct. 1881	(18 Apr. 1882)	Dynamo or Magneto Electric Machines
4,553	18 Oct. 1881	(18 Apr. 1882)	Charging and Using Secondary Batteries
4,571	19 Oct. 1881	(19 Apr. 1882)	Measurement of Electricity in Distribution Systems
4,576	19 Oct. 1881	(19 Apr. 1882)	Meters for Measuring Electric Currents

<u>1882</u>

1,023	3 Mar. 1882	(2 Sept. 1882)	Indicating and Regulating the Current of Electric Generators
1,139	9 Mar. 1882	(5 Sept. 1882)	Dynamo or Magneto Electric Machines
1,142	9 Mar. 1882	(8 Sept. 1882)	Regulating the Generative Capacity of Dynamo or Magneto Electric Machines
1,191	11 Mar. 1882	(11 Sept. 1882)	Regulating the Generative Capacity of Dynamo or Magneto Electric Machines
1,496	28 Mar. 1882	(28 Sept. 1882)	Dynamo or Magneto-electric Machines
1,862	18 Apr. 1882	(18 Oct. 1882)	Electrical Railways or Tramways
2,052	1 May 1882[b]		Electric Generators and Engines, &c.[a]
2,072	2 May 1882[b]		Electric Lights[a]
3,271	10 July 1882	(9 Jan. 1883)	Electrical Meters
3,355	14 July 1882	(13 Jan. 1883)	Apparatus for Supplying Electricity for Light, Power, &c.
3,752	5 Aug. 1882	(5 Feb. 1883)	Transmitting Electricity
3,755	5 Aug. 1882	(5 Feb. 1883)	Electrical Meters
3,756	5 Aug. 1882	(5 Feb. 1883)	Dynamo or Magneto Electric Machines
3,949	17 Aug. 1882	(17 Feb. 1883)	Apparatus for Supplying Electricity for Light, Power, &c.
3,955	18 Aug. 1882	(17 Feb. 1883)	Incandescing Electrical Lamps
3,961	18 Aug. 1882	(17 Feb. 1883)	Secondary Batteries
3,976	19 Aug. 1882	(17 Feb. 1883)	Electric Lights
3,991	19 Aug. 1882	(17 Feb. 1883)	Incandescing Conductors for Electric Lamps
3,995	21 Aug. 1882	(20 Feb. 1883)	Underground Conductors for Electrical Distribution
3,996	21 Aug. 1882	(20 Feb. 1883)	Dynamo and Magneto Electric Machines
4,446	19 Sept. 1882	(16 Mar. 1883)	Electrical Meters
4,674	2 Oct. 1882[b]		Steam Engines[c]

4,884	14 Oct. 1882	(7 Apr. 1883)	Electrical Distribution, &c.
6,183	27 Dec. 1882	(25 June 1883)	Electrical Generators and Motors
6,193	28 Dec. 1882	(27 June 1883)	Incandescing Electric Lamps
6,199	28 Dec. 1882	(28 June 1883)	Distribution of Electrical Energy for Light, Power, and other Purposes
6,206	29 Dec. 1882	(28 June 1883)	Incandescing Conductors for Electric Lamps, &c.

<u>1883</u>

1,019	24 Feb. 1883	(24 Aug. 1883)	Operation of Electrical Generators by Gas Engines
1,022	24 Feb. 1883	(24 Aug. 1883)	Construction of Electrical Railways

[a]Data from Dredge 1882–85. [b]Data from printed figures; patent text not found. [c]Title from Bowers typescript list.

Bibliography

Abbott, Lyman, ed. 1882. *How to Succeed.* New York: G. P. Putnam's Sons.

Acheson, Edward G. 1965. *A Pathfinder: Inventor, Scientist, Industrialist.* Port Huron, Mich.: Acheson Industries.

Ackerman, Kenneth D. 2003. *Dark Horse: The Surprise Election and Political Murder of President James A. Garfield.* New York: Carroll and Graf Publishers.

Aiken as a Winter Resort. 1883. Aiken, S.C.

Aikens, Andrew J., and Lewis A. Proctor, eds. 1897. *Men of Progress, Wisconsin.* Milwaukee: Evening Wisconsin Co.

Alglave, Emile, and J. Boulard. 1882. *La lumière électrique, son histoire, sa production et son emploi dans l'éclairage public ou privé.* Paris: Firmin-Didot.

Andreas, A. T. 1886. *History of Chicago. From the Earliest Period to the Present Time.* Vol. 3, *From the Fire of 1871 until 1885.* Chicago: A. T. Andreas Co.

Andrews, W. S. 1924. "The Belt Driven Edison Bipolar Dynamos." *General Electric Review* 27:163–66.

Ashmead, Henry Graham. 1884. *History of Delaware County, Pennsylvania.* Philadelphia: L. H. Everts & Co.

Atkinson, E., trans. and ed. 1893. *Elementary Treatise on Physics, Experimental and Applied.* From *Ganot's Éléments de Physique.* 14th ed. New York: William Wood and Co.

Baker, Edward C. 1976. *Sir William Preece, F.R.S. Victorian Engineer Extraordinary.* London: Hutchinson.

Baldwin, Neil. 2001. *Edison: Inventing the Century.* Chicago: University of Chicago Press.

Barker, George F. 1881. "Physics: Record of Recent Scientific Progress." In *Annual Report of the Board of Regents of the Smithsonian Institution,* 235–88. Washington, D.C.: Government Printing Office.

Batterberry, Michael, and Ariane Batterberry. 1999. *On the Town in New York: The Landmark History of Eating, Drinking, and Entertainments from the American Revolution to the Food Revolution.* New York: Routledge.

Bazerman, Charles. 1999. *The Languages of Edison's Light.* Cambridge, Mass.: MIT Press.

Beauchamp, K. G. 1997. *Exhibiting Electricity.* London: Institution of Electrical Engineers.

Beck, Bill. 1995. *PP&L: 75 Years of Powering the Future. An Illustrated History of Pennsylvania Power & Light Co.* Allentown, Pa.: Pennsylvania Power and Light Co.

Beckert, Sven. 1993. *The Monied Metropolis: New York City and the Consolidation of the American Bourgeoisie, 1850–1896.* New York: Cambridge University Press.

Beers, Diane L. 1998. "A History of Animal Advocacy in America: Social Change, Gender, and Cultural Values, 1865–1975." Ph.D. diss., Temple University.

Bell, Alexander Graham. 1880. "The Photophone." *Science* 1:130–34. In Cat. 1034:122, Scraps. (*TAED* SM034122a; *TAEM* 25:444).

———. 1881a. "Upon the Production of Sound by Radiant Energy." *Engineering* 31:481–82. In Cat. 1034:135, Scraps. (*TAED* SM034135a; *TAEM* 25:452).

———. 1881b. "Upon the Production of Sound by Radiant Energy." *Engineering* 31:514–15. In Cat. 1069:63, Scraps. (*TAED* SM069063a; *TAEM* 89:177).

———. 1881c. "Wheatstone's Microphone and Radiophonic Researches." *English Mechanic* 33:422–23. In Cat. 1052:34, Scraps. (*TAED* SM052034b; *TAEM* 26:482).

———. 1881d. "Radiophonic Researches." *Engineering* 32:11. In Cat. 1034:124, Scraps. (*TAED* SM034124a; *TAEM* 25:447).

Benedict, C. Harry. 1952. *Red Metal: The Calumet and Hecla Story.* Ann Arbor, Mich.: University of Michigan Press.

Bliss, George H. 1882. "The Electromotive: Interesting Description of Edison's Electric Railway." *Railway Age* 7:525. In Cat. 1008:66, Scraps. (*TAED* SM008066a; *TAEM* 23:360).

Bonsor, N. R. P. 1955. *North Atlantic Seaway: an illustrated history of the passenger services linking the old world with the new.* Prescot, Lancashire: T. Stephenson & Sons.

Böttinger, C. 1877. "Ein Absorptionsmittel für Kohlenoxyd." *Berichte der Deutschen Chemischen Gesellschaft* 10:1122–23.

Bourne, R. 1996. "The beginnings of electric street lighting in the City of London." *Engineering Science and Education Journal* 5:81–88.

Bowditch, John. 1989. "Driving the Dynamos: Origins of the High-Speed Electric Light Engine." *Mechanical Engineering* 111 (April): 80–89.

Bowers, Brian. 1982. *A History of Electric Light & Power.* Stevenage, U.K.: Peter Peregrinus Ltd., in association with the Science Museum.

———. 1990. "Edison and Hopkinson: Transatlantic relations in electrical engineering in the early 1880s." In *Électricité et Électrification dans le Monde: Actes du deuxième colloque international d'histoire d'électricité, organiseé par l'Association pour l'histoire de l'éctricité en France,* ed. Monique Trédé. [Paris]: Presses Universitaires de France.

———. 1991. "Edison and Early Electrical Engineering in Britain." *History of Technology* 13: 168–80.

Bright, Arthur A., Jr. 1972 [1949]. *The Electric-Lamp Industry: Technological Change and Economic Development from 1800 to 1947.* New York: Arno Press.

Bright, Charles. 1974 [1898]. *Submarine Telegraphs: Their History, Construction, and Working.* New York: Arno Press.

Brill, Debra. 2001. *History of the J. G. Brill Company.* Bloomington, Ind.: Indiana University Press.

Brody, David. 1960. *Steelworkers in America: The Nonunion Era.* New York: Harper & Row.

Brown, Ira V. 1953. *Lyman Abbott, Christian Evolutionist: A Study in Religious Liberalism.* Cambridge, Mass.: Harvard University Press.

Brown, John K. 1995. *The Baldwin Locomotive Works, 1831–1915: A Study in American Practice.* Baltimore: Johns Hopkins University Press.

Bruce, Robert V. 1973. *Bell: Alexander Graham Bell and the Conquest of Solitude.* Boston: Little, Brown & Co.

Buck, Solon Justus. 1963 [1913]. *The Granger Movement: A Study of Agricultural Organization and Its Political, Economic and Social Manifestations, 1870–1880.* Lincoln: University of Nebraska Press.

Bunsen, Robert. 1857. *Gasometrische Methoden.* Braunschweig, Germany: Friedrich Vieweg und Sohn.

Buss, Dietrich G. 1978 [1977]. *Henry Villard: A Study of Transatlantic Investments and Interests, 1870–1895.* New York: Arno Press.

Byatt, I. C. R. 1979. *The British Electrical Industry, 1875–1914: The Economic Returns to a New Technology.* Oxford: Clarendon Press.

Byrnes, Robert Francis. 1950. *Antisemitism in Modern France.* Vol. 1, *The Prologue to the Dreyfus Affair.* New Brunswick, N.J.: Rutgers University Press.

Calhoun, Charles C. 2004. *Longfellow: A Rediscovered Life.* Boston: Beacon Press.

Cameron, Rondo. 1961. *France and the Economic Development of Europe, 1800–1914.* Princeton, N.J.: Princeton University Press.

Carlson, W. Bernard. 1991. *Innovation as a Social Process: Elihu Thomson and the Rise of General Electric, 1870–1900.* Cambridge: Cambridge University Press.

Carosso, Vincent P. 1970. *Investment Banking in America: A History.* Cambridge, Mass.: Harvard University Press.

———. 1987. *The Morgans: Private International Bankers, 1854–1913.* Cambridge, Mass.: Harvard University Press.

Carr, William H. A. 1975. *From Three Cents a Week . . . the Story of The Prudential Insurance Company of America.* Englewood Cliffs, N.J.: Prentice-Hall.

Chirnside, R. C. 1979. "Sir Joseph Swan and the Invention of the Electric Lamp." *Electronics & Power* 45:96–100.

Clarke, Charles L. 1904. "Edison 'Jumbo' Steam-Dynamo." In Edisonia 1904, 27–55. Partially excerpted in Jehl 1937–41, 1041–46.

Cochrane, Rexmond C. 1978. *The National Academy of Sciences: The First Hundred Years, 1863–1963.* Washington, D.C.: National Academy of Sciences.

Columbia University. 1932. *Columbia University Alumni Register, 1754–1931*. New York: Columbia University.

Condit, Carl W. 1980. *The Port of New York*. Vol. 1, *A History of the Rail and Terminal System from the Beginnings to Pennsylvania Station*. Chicago: University of Chicago Press.

Condit, Kenneth H. 1952. *Cyrus Fogg Brackett (1833–1915) of Princeton: Pioneer in Electrical Engineering Education*. New York: Newcomen Society in North America.

Crookes, William. 1882. "Report on the Incandescent Lamps Exhibited at the International Exhibition of Electricity, Paris, 1881." *Journal of the Society of Telegraph Engineers* 10:229–43.

Crouch, Tom D. 1983. *The Eagle Aloft: Two Centuries of the Balloon in America*. Washington, D.C.: Smithsonian Institution Press.

Davenport, Neil. 1979. *The United Kingdom Patent System: A Brief History*. Portsmouth, U.K.: Kenneth Mason.

Davies, Edward J., II. 1985. *The Anthracite Aristocracy: Leadership and Social Change in the Hard Coal Regions of Northeastern Pennsylvania, 1800–1930*. DeKalb, Ill.: Northern Illinois University Press.

Davison, C. Wright. (Published annually.) *Minneapolis City Directory*. Minneapolis: C. Wright Davison.

De Borchgrave, Alexandra Villard, and John Cullen. 2001. *Villard: The Life and Times of an American Titan*. New York: Doubleday.

Derks, Scott, ed. 1994. *The Value of a Dollar: Prices and Incomes in the United States, 1860–1989*. Detroit and Washington, D.C.: Gale Research, Inc.

Dewar, James. 1927. *Collected Papers of Sir James Dewar*. Ed. Lady Dewar. Cambridge: Cambridge University Press.

Dewar, James, and P. G. Tait. 1874. "On a New Method of Obtaining Very Perfect Vacua." *Proceedings of the Royal Society of Edinburgh* 8:348–49.

———. 1875. "Farther Researches in Very Perfect Vacua." *Proceedings of the Royal Society of Edinburgh* 8:628.

Dod's Peerage, Baronetage, and Knightage of Great Britain and Ireland. (Printed annually.) London: Whittaker & Co.

Dredge, James. 1882–85. *Electric Illumination*. 2 vols. London: Engineering.

Du Moncel, Théodose. 1878. *Le Téléphone, le Microphone et le Phonographe*. Paris: Hachette.

———. 1881. "The International Exhibition of Electricity." *Scientific American Supplement* 12:4920–23. A typed transcript is "International Exhibition of Electricity, Incandescent Electric Lamps," n.d., DF (*TAED* D8135ZDA; *TAEM* 58:1136). The article originally appeared as "Les Lampes Électriques à Incandescence," *La Lumière Électrique* 5:1–16; in Cat. 1242, Batchelor (*TAED* MBSB31690X; *TAEM* 94:662). The English version was reprinted with additional illustrations in *Van Nostrand's Engineering Magazine* 25 (1881): 439–50, a copy of which is in Ser. 2, Box 27, Folder 5, WJH.

Dunbar, Willis F. 1980. *Michigan: A History of the Wolverine State*. Revised by George S. May. Grand Rapids, Mich.: Eerdmans Publishing.

Dunbaugh, Edwin L. 1992. *Night Boat to New England, 1815–1900*. New York: Greenwood Press.

Dyer, Frank, and T. C. Martin. 1910. *Edison: His Life and Inventions.* 2 vols. New York: Harper & Bros. Second edition, 1929, 3 vols.

Dyer, Richard N. n.d. "From Davy to Edison: An Historical Sketch of the Incandescent Electric Lamp." Typescript, NjWOE.

Eddy, John A. 1972. "Thomas Edison and Infra-Red Astronomy." *Journal for the History of Astronomy* 3:165–87.

Edison, Thomas A. 1882. "How to Succeed. As an Inventor." *Christian Union* 25:544.

Edison, Thomas A., and Charles T. Porter. 1882. "Description of the Edison Steam Dynamo." *Journal of the Franklin Institute* 64:1–12. Reprinted with the substitution of one illustration in *Electrician* 9 (1882): 199–202 and, with discussion from the ASME session but no figures, in *Transactions of the American Society of Mechanical Engineers* 3 (1882): 218–25; in Cat. 1327, item 2201, Batchelor (*TAED* MBSB52201; *TAEM* 95:196).

Edison Electric Light Co. 1881. *Instructions for the Installation of Isolated Plant.* New York: Edison Electric Light Co.

"Edisonia": A Brief History of the Edison Electric Lighting System. 1904. New York: Association of Edison Illuminating Companies.

Emden, Paul H. 1938. *Money Powers of Europe in the Nineteenth and Twentieth Centuries.* New York: D. Appleton-Century Co.

Fairburn, William Armstrong. 1945–55. *Merchant Sail.* 6 vols. Center Lovell, Maine: Fairburn Marine Educational Foundation.

Favre, Edouard. 1923. *Théodore Turrettini, 1845–1916.* Geneva: A. Kundig.

Feldenkirchen, Wilfried. 1994. *Werner von Siemens: Inventor and International Entrepreneur.* Columbus: Ohio State University Press.

———. 1995. *Siemens, 1918–1945.* Columbus: Ohio State University Press.

Finch, James Kip. 1954. *A History of the School of Engineering, Columbia University.* New York: Columbia University Press.

Fletcher, T. 1877. "Selenium." *Teleg. J. and Elec. Rev.* 5:295–97. In Cat. 1028:9, Scraps. (*TAED* SM028009a; *TAEM* 25:122).

Foreman, John, and Robbe Pierce Stimson. 1991. *The Vanderbilts and the Gilded Age: Architectural Aspirations, 1879–1901.* New York: St. Martin's Press.

Fournier d'Albe, E. E. 1923. *The Life of Sir William Crookes, O.M., F.R.S.* London: T. Fisher Unwin Ltd.

Fox, Edwin. 1879. "Edison's System of Fast Telegraphy." *Scribner's Monthly* 18:840–46.

Fox, Robert. 1996. "Thomas Edison's Parisian Campaign: Incandescent Lighting and the Hidden Face of Technology Transfer." *Annals of Science* 53:157–93.

Fox, Stephen. 2003. *Transatlantic: Samuel Cunard, Isambard Brunel, and the Great Atlantic Steamships.* New York: HarperCollins.

France. Ministère des Postes et des Télégraphes. 1881. *Exposition Internationale D'Electricité. Paris 1881. Catalogue General Officiel.* Paris: A. Lahure. A copy is in Ser. 2, Box 34, Folder 1, WJH.

Friedel, Robert, and Paul Israel. 1986. *Edison's Electric Light: Biography of an Invention.* New Brunswick, N.J.: Rutgers University Press.

Fry, Herbert. 1880. *London in 1880. Illustrated with Bird's-Eye Views of the Principal Streets.* London: David Bogue.

Gábor, Luca. 1993. *Telephonic News Dispenser.* Budapest: Hungarian Broadcasting Co.

Gijswijt-Hofstra, Marijke, and Roy Porter. 2001. *Cultures of Neurasthenia: From Beard to the First World War.* Amsterdam: Radopi.

Glaser, Gustave. 1881. "Magneto-Electric and Dynamo-Electric Machines in the International Electrical Exhibition at Paris." *Scientific American Supplement* 307:4894–95.

Gordon, F. W. Beauchamp. 1891. "The State and Electrical Distribution." In *A Plea for Liberty: An Argument Against Socialism and Socialistic Legislation,* ed. Thomas MacKay, 349–75. London: John Murray.

Gordon, Robert B. 1996. *American Iron, 1607–1900.* Baltimore: Johns Hopkins University Press.

Graham, Thomas. 1843. *Elements of Chemistry, Including the Applications of the Science in the Arts.* Philadelphia: Lea & Blanchard.

———. 1863. "On the Molecular Mobility of Gases." *Philosophical Transactions of the Royal Society of London* 153:385–405.

———. 1866. "On the Absorption and Dialytic Separation of Gases by Colloid Septa." *Philosophical Transactions of the Royal Society of London* 156:399–439.

Greig, James. 1970. *John Hopkinson: Electrical Engineer.* Science Museum Booklet. London: HMSO.

Griffiths, William H. 1959. *The Story of the American Bank Note Company.* New York: American Bank Note Co.

Grodinsky, Julius. 1957. *Jay Gould: His Business Career, 1867–1892.* Philadelphia: University of Pennsylvania Press.

Guagnini, Anna. 1987. "The Formation of the Italian Electrical Engineers: The Teaching Laboratories of the Politecnici of Turin and Milan." In *Un Siècle d'Électricité dans le Monde,* ed. Fabienne Cardot. Paris: Presses Universitaires de France.

Guillemin, Amédée. 1891. *Electricity and Magnetism.* Rev. and ed. by Silvanus P. Thompson. London: Macmillan and Co.

Halsey, Edmund D. 1882. *History of Morris County, New Jersey, with Illustrations, and Biographical Sketches of Prominent Citizens and Pioneers.* New York: W. W. Munsell and Co.

Hammer, Edwin W. 1904. "Edison Electric Railways of 1880 and 1882." In Edisonia 1904, 117–28.

Hammer, William J. 1889. "Edison and His Inventions." Typescript of lecture delivered at the Franklin Institute, 4 Feb. 1889. In Ser. 1, Box 16, Folder 6, WJH.

———. 1904. "The First Central Station for Incandescent Lighting." *Electrical World* 42:452–53.

Hammond, John Winthrop. 1941. *Men and Volts: The Story of General Electric.* New York: J. B. Lippincott Co.

Hannah, Leslie. 1979. *Electricity Before Nationalisation: A Study of the Development of the Electricity Supply Industry in Britain to 1948.* Baltimore: Johns Hopkins University Press.

Harding, Les. 2000. *Elephant Story: Jumbo and P. T. Barnum Under the Big Top.* Jefferson, N.C.: McFarland & Co.

Hartmann, Dagmar. 2001. "Henkenhaf und Ebert: Architekten der Stadthalle in Heidelberg und des Kurhauses in Scheveningen." Ph.D. diss., University of Heidelberg [Kunsthistorisches Institut]. www.ub.uni-heidelberg.de/archiv/3821.

Hawley, Gessner Goodrich. 1987. *Condensed Chemical Dictionary.* 11th ed. Revised by N. Irving Sax and Richard J. Lewis, Sr. New York: Van Nostrand Reinhold Co.

Haywood, William. 1883. "Reports to the Streets Committee . . . as to the Commission Undertaking to Supply Electricity in a District of the City." CLA/006/AD/07/059/10, UkLMA.

Heap, David Porter. 1884. *Report on the International Exhibition of Electricity Held at Paris August to November, 1881.* Washington, D.C.: Government Printing Office.

Heerding, A. 1986. *The History of N. V. Philips' Gloeilampenfabrieken.* Trans. Derek S. Jordan. Cambridge: Cambridge University Press.

Hicks, J. Maurice. 1979. *Roselle, New Jersey: Site of Thomas Alva Edison's First Village Plant.* Roselle, N.J.: Roselle Historical Society.

Higgs, William Henry Paget, and John Richard Brittle. 1878. "Some Recent Improvements in Dynamo-Electric Apparatus," with discussion. *Minutes of Proceedings of the Institution of Civil Engineers* 52 (pt. 2):36–98.

History of Cincinnati and Hamilton County. 1894. Cincinnati: S. B. Nelson & Co.

"A History of the Development of the Incandescent Lamp." [1910–1912?]. Typescript at NjWOE.

Hogan, John. 1986. *A Spirit Capable: The Story of Commonwealth Edison.* Chicago: Mobium Press.

Hopkinson, Bertram. 1901. "Memoir." In *Original Papers by the late John Hopkinson,* by John Hopkinson, 1:ix–lxii. Cambridge: Cambridge University Press.

Hopkinson, John. 1879. "On Electric Lighting. (First Paper)" with discussion. *Proceedings of the Institution of Mechanical Engineers* 30:238–65.

Hopkinson, John, and Edward Hopkinson. 1886. "Dynamo-Electric Machinery." *Philosophical Transactions of the Royal Society of London* 177:331–58.

Hounshell, David A. 1980. "Edison and the Pure Science Ideal in 19th-Century America." *Science* 207:612–17.

———. 1984. *From the American System to Mass Production, 1800–1932: The Development of Manufacturing Technology in the United States.* Baltimore: Johns Hopkins University Press.

Howell, John W. 1882. *Economy of Electric Lighting by Incandescence.* [New York?]: Privately printed. In Cat. 1243, item 1914, Batchelor (*TAED* MBSB41914; *TAEM* 95:93). An extract of the dynamo test, reprinted from *Van Nostrand's Engineering Magazine* 26 (1882): 51–59, is in PPC (*TAED* CA002B26; *TAEM* 96:150).

Howell, John W., and Henry Schroeder. 1927. *History of the Incandescent Lamp.* Schenectady, N.Y.: Maqua.

Hughes, Edan Milton. 1989. *Artists in California, 1780–1940.* San Francisco: Hughes Publishing Co.

Hughes, Thomas P. 1962. "British Electrical Industry Lag: 1882–1888." *Technology and Culture* 3:27–44.

———. 1983. *Networks of Power: Electrification in Western Society, 1880–1930.* Baltimore: Johns Hopkins University Press.

Hunter, Louis C. 1985. *A History of Industrial Power in the United States, 1780–1930.* Vol. 2, *Steam Power.* Charlottesville: University Press of Virginia.

Hunter, Louis C., and Lynwood Bryant. 1991. *A History of Industrial Power in the United States, 1780–1930.* Vol. 3, *The Transmission of Power.* Cambridge, Mass.: MIT Press.

Hyde, Francis E. 1975. *Cunard and the North Atlantic, 1840–1973.* London: Macmillan Press.

Incandescent Electric Lights, with Particular Reference to the Edison Lamps at the Paris Exhibition. 1882. New York: Van Nostrand.

Insull, Samuel. 1915. *Central-Station Electric Service: Its Commercial Development and Economic Significance as Set Forth in the Public Addresses (1897–1914) of Samuel Insull.* Edited and with an introduction by William Eugene Keily. Chicago: privately printed.

Israel, Paul. 1992. *From Machine Shop to Industrial Laboratory: Telegraphy and the Changing Context of American Invention, 1830–1920.* Baltimore: Johns Hopkins University Press.

———. 1998. *Edison: A Life of Invention.* New York: John Wiley & Sons.

Jarchow, M. E. 1943. "Farm Machinery of the 1860s in Minnesota." *Minnesota History* 24:287–306.

Jehl, Francis. 1882a. *The Edison Electric Light Meter.* [New York?]: Privately printed. In DF (*TAED* D8239ZBN; *TAEM* 62:889).

———. 1882b. "Electric Lighting at Brünn." *The Electrician* 10:38–40.

———. 1937–41. *Menlo Park Reminiscences.* 3 vols. Dearborn, Mich.: Edison Institute.

Jenks, W[illiam]. J. 1889. *Six Years' Practical Experience with the Edison Chemical Meter.* New York: Edison Electric Light Co. In PPC (*TAED* CA041G; *TAEM* 147:725).

Jones, Payson. 1940. *A Power History of the Consolidated Edison System, 1878–1900.* New York: Consolidated Edison Co. of New York.

Jonnes, Jill. 2003. *Empires of Light: Edison, Tesla, Westinghouse, and the Race to Electrify the World.* New York: Random House.

Josephson, Matthew. 1992 [1959]. *Edison: A Biography.* New York: John Wiley & Sons.

Kee, Fred. 1985. "The Electrical Manufacturing Industry." In *Electricity: The Magic Medium = L'électricité: cette prodigieuse énergie.* Thornhill, Ont.: IEEE, Canadian Region.

Kennedy, James G. 1978. *Herbert Spencer.* Boston: Twayne Publishers.

King, W. James. 1962. "The Development of Electrical Technology in the Nineteenth Century: 3. The Early Arc Light and Generator." In *United States National Museum Bulletin* 228:333–407.

King, William Harvey. 1905. *History of Homoeopathy and Its Institutions in America: Their Founders, Benefactors, Faculties, Officers, Hospitals, Alumni, etc.* New York: Lewis Publishing Co.

Kingsbury, J. E. 1915. *The Telephone and Telephone Exchanges: Their Invention and Development.* London: Longmans, Green, & Co.

Kirk, John Foster. 1891. *A Supplement to Allibone's Critical Dictionary of English Literature and British and American Authors.* Philadelphia: J. B. Lippincott Co.

Klein, Maury. 1986. *The Life and Legend of Jay Gould.* Baltimore: Johns Hopkins University Press.

Knight, Edward H. 1881. *Knight's American Mechanical Dictionary.* 3 vols. New York: Hurd & Houghton.

Koeppel, Gerard T. 2000. *Water for Gotham: A History.* Princeton: Princeton University Press.

Kohlstedt, Sally Gregory. 1980. "*Science*: The Struggle for Survival, 1880 to 1894." *Science* 209:33–42.

Lake Shore & Michigan Southern Railway System and Representative Employees. 1900. Buffalo and Chicago: Biographical Publishing Co.

Lewis, J. Volney. 1907. "The Newark (Triassic) Copper Ores of New Jersey." In *Annual Report of the State Geologist for the Year 1906*, Geological Survey of New Jersey, 131–64. Trenton, N.J.: MacCrellish & Quigley, State Printers.

[Lieb, John.] 1904. "The Historic Pearl Street New York Edison Station." In *Edisonia* 1904, 61–73. Reprinted in Jehl 1937–41, 1046–55.

Magie, W. F. 1915. "Cyrus Fogg Brackett." *Science* 41:523–525.

Marshall, David Trumbull. 1930. *Recollections of Boyhood Days in Old Metuchen.* Flushing, N.Y.: Case Publishing.

Martin, Thomas Commerford. 1922. *Forty Years of Edison Service, 1882–1922.* New York: New York Edison Co.

Mather, Loris Emerson. 1926. *The Right Honourable Sir William Mather.* London: Richard Cobden-Sanderson.

Matthews, Derek. 1986. "Laissez-faire and the London Gas Industry in the Nineteenth Century: Another Look." *Economic History Review* 39:244–63.

McCrea, Roswell C. 1910. *The Humane Movement: A Descriptive Survey.* New York: Columbia University Press.

McDonald, Donald. 1960. *A History of Platinum: From the Earliest Times to the Eighteen-Eighties.* London: Johnson, Matthey & Co.

McDonald, Forrest. 1962. *Insull.* Chicago: University of Chicago Press.

McMahon, A. Michal. 1984. *The Making of a Profession: A Century of Electrical Engineering in America.* New York: Institute of Electrical and Electronics Engineers.

McMurry, Sally. 1995. *Transforming Rural Life: Dairying Families and Agricultural Change, 1820–1885.* Baltimore: Johns Hopkins University Press.

Micale, Mark S. 1995. *Approaching Hysteria: Disease and Its Discontents.* Princeton: Princeton University Press.

Millard, A[ndre]. J. 1987. *A Technological Lag: Diffusion of Electrical Technology in England, 1879–1914.* New York: Garland Publishing.

Miller, Michael B. 1981. *The Bon Marché: Bourgeois Culture and the Department Store, 1869–1920.* Princeton: Princeton University Press.

Miller, William Allen. 1877. *Elements of Chemistry: Theoretical and Practical.* 3 vols. London: Longmans, Green, Reader, and Dyer.

Mosse, Werner Eugen. 1987. *Jews in the German Economy: The German-Jewish Economic Elite, 1820–1935.* Oxford: Clarendon Press.

Mott, Frank Luther. 1938–68. *A History of American Magazines.* 5 vols. Cambridge, Mass.: Harvard University Press.

Mull, Carleton M. 1969. "How Your Hobby Started Part V." *Gas Engine Magazine* (November). http://www.gasenginemagazine.com/complete-archive/362/.

Muller, H. G. 1991. "Industrial Food Preservation in the Nineteenth and Twentieth Centuries." In *'Waste Not, Want Not': Food Preservation from Early Times to the Present Day,* ed. C. Anne Wilson. Edinburgh: Edinburgh University Press.

Myllyntaus, Timo. 1991. "The Transfer of Electrical Technology to Finland, 1870–1930." *Technology and Culture* 32:293–317.

Norris, James D. 1978. *R. G. Dun & Co., 1841–1900: The Development of Credit-Reporting in the Nineteenth Century.* Westport, Conn.: Greenwood Press.

Ó Gráda, Cormac. 1994. "British Agriculture, 1860–1914." In *The Economic History of Britain Since 1700,* ed. Roderick Floud and Donald McCloskey. Vol. 2, *1860–1939.* 2d ed. Cambridge: Cambridge University Press.

Olson, Russell L. 1976. *The Electric Railways of Minnesota.* Hopkins, Minn.: Minnesota Transportation Museum.

———. 1990. *The Electric Railways of Minnesota Supplement.* St. Paul, Minn.: Minnesota Transportation Museum.

Pacinotti, Antonio. 1874. "Descrizione del gomitolo elettromagnetico, e di qualche esperimento per utilizzarlo nello construzione delle macchine magneto-electriche." *Il Nuovo Cimento* 12:140–48.

Partington, J. R. 1961–64. *A History of Chemistry.* 4 vols. London: Macmillan.

Partridge, Eric. 1984. *A Dictionary of Slang and Unconventional English.* 8th ed. New York: Macmillan Publishing Co.

Passer, Harold C. 1952. *Frank Julian Sprague: Father of Electric Traction, 1857–1934.* Cambridge, Mass.: Harvard University Press.

———. 1953. *The Electrical Manufacturers, 1875–1900: A Study in Competition, Entrepreneurship, Technical Change, and Economic Growth.* Cambridge, Mass.: Harvard University Press.

Pavese, Claudio. 1987. "La naissance et le développement de la Société Générale Italienne Edison d'Électricité." In *Un Siècle d'Électricité dans le Monde,* ed. Fabienne Cardot. Paris: Presses Universitaires de France.

Pender, Harold. 1922. *Direct-Current Machinery: A Text-book on the Theory and Performance of Generators and Motors.* New York: John Wiley & Sons.

Perry, P. J. 1974. *British Farming in the Great Depression, 1870–1914: An Historical Geography.* Newton Abbott, Devon, U.K.: David & Charles.

Petroski, Henry. 1990. *The Pencil: A History of Design and Circumstance.* New York: Alfred A. Knopf.

Phillips, Arthur Sherman. 1941. *The Phillips History of Fall River.* 3 vols. Fall River, Mass.: Dover Press.

Planté, Gaston. 1887. *The Storage of Electrical Energy and Researches in the Effects Created by Currents Combining Quantity with High Tension.* London: Whittaker & Co.

Platt, Harold L. 1991. *The Electric City: Energy and the Growth of the Chicago Area, 1880–1930.* Chicago: University of Chicago Press.

Pope, Franklin. 1869. *Modern Practice of the Electric Telegraph: A Handbook for Electricians and Operators.* New York: Russell Brothers.

Porter, Charles T. 1908. *Engineering Reminiscences: Contributed to "Power" and "American Machinist."* New York: John Wiley & Sons.

Preece, William. 1881. "Electric Lighting at the Paris Exhibition," with discussion. *Journal of the Society of Arts* 30:98–107; reprinted in *Van Nostrand's Engineering Magazine* 26 (1882): 151–63. In Cat. 1243, item 1836, Batchelor (*TAED* MBSB41836; *TAEM* 95:75).

Prescott, George B. 1884. *Dynamo-Electricity: Its Generation, Application, Transmission, Storage and Measurement.* New York: D. Appleton and Co.

Prindle, Edwin J. 1908. *Patents as a Factor in Manufacturing.* New York: Engineering Magazine.

Proctor, B. S. 1879. "On the Smoke of an Electric Lamp." *Chemical News* 39:283.

Prosser, Richard S. 1966. *Rails to the North Star: One Hundred Years of Railroad Evolution in Minnesota.* Minneapolis: Dillon University Press.

Pyne, Stephen J. 1982. *Fire in America: A Cultural History of Wildland and Rural Fire.* Princeton: Princeton University Press.

Raines, Rebecca Robbins. 1996. *Getting the Message Through: A Branch History of the U.S. Army Signal Corps.* Washington, D.C.: Center of Military History, United States Army.

Rand's New York City Business Directory. (Printed annually.) New York: Rand Directory Co.

Reade, A[lfred]. Arthur. 1883. *Study and Stimulants; or, the Use of Intoxicants and Narcotics in Relation to Intellectual Life, as illustrated by personal communications on the subject, from men of letters and of science.* Philadelphia: J. B. Lippincott and Co.

Reid, James D. 1879. *The Telegraph in America.* New York: Derby Bros.
———. 1886. *The Telegraph in America.* Rev. ed. New York: John Polhemus.

Révérend, Albert. 1974. *Titres et Confirmations de Titres: Monarchie de Juillet, 2e République, 2e Empire, 3e République.* Revised and edited by Jean Tulard. Paris: Librairie Honoré Champion.

Rolt, L. T. C. 1965. *A Short History of Machine Tools.* Cambridge, Mass.: MIT Press.

Russett, Cynthia Eagle. 1976. *Darwin in America: The Intellectual Response, 1865–1912.* San Francisco: W. H. Freeman and Company.

Sanders, Andrew. 1988. *The Companion to A Tale of Two Cities.* London: Unwin Hyman.

Satterlee, Herbert L. 1939. *J. Pierpont Morgan: An Intimate Portrait.* New York: Macmillan Co.

Scalfari, Eugenio. 1963. *Storia segreta dell'industria elettrica.* Bari: Laterza.

Schallenberg, Richard H. 1982. *Bottled Energy: Electrical Engineering and the Evolution of Chemical Energy Storage.* Philadelphia: American Philosophical Society.

Schellen, Heinrich. 1884. *Magneto-Electric and Dynamo-Electric Machines: Their Construction and Practical Application to Electric Lighting and the Transmission of Power*. Trans. Nathaniel S. Keith and Percy Neymann. New York: D. Van Nostrand.

Shaw, William H., comp. 1884. *History of Essex and Hudson Counties, New Jersey*. 2 vols. Philadelphia: Everts & Peck.

Shiman, Daniel R. 1993. "Explaining the Collapse of the British Electrical Supply Industry in the 1880s: Gas versus Electric Lighting Prices." *Business and Economic History* 22: 318–27.

Siemens, C. William. 1880. "On the Dynamo-Electric Current, and on Certain Means to Improve Its Steadiness." *Philosophical Transactions of the Royal Society of London* 171:1071–88 (with additional plates).

Siemens, Georg. 1957. *History of the House of Siemens*. Trans. A. F. Rodger. Freiburg, Germany: Karl Alber.

Siemens, Werner von. 1876. "On the Influence of Light Upon the Conductivity of Crystalline Selenium," *Teleg. J. and Elec. Rev.* 4:15–16. In Cat. 1028:8, Scraps. (*TAED* SM028008d; *TAEM* 25:122).

Sinclair, Bruce. 1974. *Philadelphia's Philosopher Mechanics: A History of the Franklin Institute, 1824–1865*. Baltimore: Johns Hopkins University Press.

———. 1980. *A Centennial History of the American Society of Mechanical Engineers, 1880–1980*. Toronto: University of Toronto Press.

———. 1989. "Technology on Its Toes: Late Victorian Ballets, Pageants, and Industrial Exhibitions." In *In Context: History and the History of Technology. Essays in Honor of Melvin Kranzberg*, ed. Stephen H. Cutcliffe and Robert C. Post. Bethlehem, Pa.: Lehigh University Press.

Singer, Charles, E. J. Holmyard, A. R. Hall, and Trevor I. Williams, eds. 1958. *A History of Technology*. Vol. 5. Oxford and London: Oxford University Press.

Smith, Grant H. 1943. *The History of the Comstock Lode, 1850–1920*. Reno: Nevada Bureau of Mines and University of Nevada.

Sobel, Robert. 1977. *Inside Wall Street: Continuity and Change in the Financial District*. New York: W. W. Norton & Co.

Southworth, P. J. M. 1986. *Some Early Robey Steam Engines*. Shirland, Derbyshire, England: Higham Press Ltd.

Spencer, Herbert. 1904. *An Autobiography*. New York: D. Appleton and Co.

Spewack, Bella. 1995. *Streets: A Memoir of the Lower East Side*. New York: The Feminist Press at the City University of New York.

Sprague, Frank Julian. 1882a. "The Demands of a System of Electrical Distribution." *Teleg. J. and Elec. Rev.* 11:177–79. In Cat. 1007, Scraps. (*TAED* SM007051a; *TAEM* 23:296).

———. 1882b. "The Edison System of Electric Distribution." *Teleg. J. and Elec. Rev.* 11:173–74. In Cat. 1007, Scraps. (*TAED* SM007046a; *TAEM* 23:294).

———. 1883. *Report on the Exhibits at the Crystal Palace Electrical Exhibition, 1882*. Office of Naval Intelligence, General Information Series. Washington, D.C.: Government Printing Office.

Strouse, Jean. 1999. *Morgan: American Financier*. New York: Random House.

Swan, J[oseph]. W. 1881. "On the 'Swan' Incandescent Lamp." *The Engineer* 52:229–30.

—————. 1882. "Electric Lighting by Incandescence." *Nature* 26:356–59.

Swan, Kenneth R. 1946. *Sir Joseph Swan and the Invention of the Incandescent Electric Lamp.* London: Longmans, Green and Co.

Swan, Mary Edmonds, and Kenneth R. Swan. 1929. *Sir Joseph Wilson Swan F.R.S.: A Memoir.* London: Ernest Benn, Ltd.

Swann, Leonard Alexander. 1965. *John Roach: Maritime Entrepreneur, The Years as a Naval Contractor, 1862–1886.* Annapolis, Md.: United States Naval Institute.

Szymanowitz, Raymond. 1971. *Edward Goodrich Acheson: Inventor, Scientist, Industrialist.* New York: Vantage Press.

Taltavall, John B. 1893. *Telegraphers of To-Day.* New York: John B. Taltavall.

Taylor, Jocelyn Pierson. 1978. *Grosvenor Porter Lowrey.* New York: Privately printed.

Teisch, Jessica B. 2001. "Great Western Power, 'White Coal' and Industrial Capitalism in the West." *Pacific Historical Review* 70:221–53.

Tenfelde, Klaus. 2005. "Krupp—the rise of a world-class German company." In *Pictures of Krupp: Photography and History in the Industrial Age,* ed. Klaus Tenfelde. London: Philip Wilson Publishers.

Thompson, Silvanus P. 1902. *Dynamo-Electric Machinery: A Manual for Students of Electrotechnics.* 8th American ed. New York: M. Strong.

Thurston, R[obert]. H. 1890. *A Handbook of Engine and Boiler Trials, and of the Indicator and Prony Brake.* New York: John Wiley & Sons.

Tritton, Paul. 1993. *The Godfather of Rolls-Royce: The Life and Times of Henry Edmunds, M.I.C.E., M.I.E.E., Science and Technology's Forgotten Pioneer.* London: Academy Books.

Trow's New York City Directory. (Printed annually.) New York: Trow City Directory Co. [Cited in previous Volumes under the name of H. Wilson, compiler.]

Tunbridge, Paul. 1992. *Lord Kelvin: his influence on electrical measurements and units.* London: Peter Peregrinus on behalf of the Institution of Electrical Engineers.

Turner, Gerard L'E. 1983. *Nineteenth-Century Scientific Instruments.* Berkeley: University of California Press.

United Kingdom. Parliament. House of Commons. 1882. *The Electric Lighting Act, 1882: Minutes of Evidence given before the select Committee of the House of Commons.* London: HMSO.

U.S. Bureau of the Census. 1965. *Population Schedules of the Ninth Census of the United States, 1870.* National Archives Microfilm Publication Microcopy M593. Washington, D.C.: National Archives.

—————. 1967. *Population Schedules of the Eighth Census of the United States, 1860.* National Archives Microfilm Publication Microcopy M653. Washington, D.C.: National Archives.

—————. 1970. *Population Schedules of the Tenth Census of the United States, 1880.* National Archives Microfilm Publication Microcopy T9. Washington, D.C.: National Archives.

U.S. Customs Service. 1962. *Passenger Lists of Vessels Arriving at New York, 1820–1897.* National Archives Microfilm Publication Microcopy M-237. Washington, D.C.: National Archives.

U.S. Patent Office. 1882. *Index of Patents Relating to Electricity Granted* by the United States Prior to July 1, 1881. With an Appendix Embracing Patents Granted from July 1, 1881, to June 30, 1882. Washington, D.C.: Government Printing Office.

Veith, Ilza. 1965. *Hysteria: History of a Disease.* Chicago: University of Chicago Press.

Villard, Henry. 1904. *Memoirs of Henry Villard, Journalist and Financier, 1835–1900.* 2 vols. Boston and New York: Houghton, Mifflin and Company.

Vinal, George Wood. 1955. *Storage Batteries: A General Treatise on the Physics and Chemistry of Secondary Batteries and their Engineering Applications.* 4th ed. New York: John Wiley & Sons.

Vreeken, Bert, and Ester Wouthuysen. 1987. *De Grand Hotels van Amsterdam: opkomst en bloei sinds 1860.* 's-Gravenhage: Sdu.

Walton, George. 1873. *The Mineral Springs of the United States and Canada.* New York: Appleton & Co.

Wanklyn, J. Alfred. 1866. "On the Action of Carbonic Oxide on Sodium-ethyl." *Journal of the Chemical Society of London* 4:13–14.

Western Union Telegraph Co. (Printed annually.) *Annual Reports.* New York: Western Union Telegraph Co.

Wheatley, Henry B. 1891. *London Past and Present: Its History, Associations, and Traditions.* London: John Murray.

Willard, X. A. 1877. *Willard's Practical Dairy Husbandry.* New York: American News Co.

Wilson, Robert. 1877. *Common Sense for Gas-Users. Being a Catechism of Gas-Lighting. For Householders, etc.* London: Crosby Lockwood & Co.

Wilson's Business Directory of New York City. (Printed annually.) New York: Trow City Directory Co.

Wise, George. 1985. *Willis R. Whitney, General Electric, and the Origins of U.S. Industrial Research.* New York: Columbia University Press.

Woodbury, David O. 1949. *A Measure for Greatness: A Short Biography of Edward Weston.* New York: McGraw-Hill Book Co., Inc.

Wunder, Richard P. 1991. *Hiram Powers: Vermont Sculptor, 1805–1873.* Vol. 1. Newark: University of Delaware Press.

Yates, W. Ross. 1987. *Joseph Wharton: Quaker Industrial Pioneer.* Bethlehem, Pa.: Lehigh University Press.

Youmans, Edward L., comp. 1973 [1883]. *Herbert Spencer on the Americans and The Americans on Herbert Spencer.* New York: Arno Press.

Credits

Reproduced with permission of Dun and Bradstreet and the Baker Library of Harvard University: Docs. 2097, 2394. Courtesy of the Milton S. Eisenhower Library (Henry Augustus Rowland Papers Ms. 6, Special Collections), Johns Hopkins University: Doc. 2391. From the collections of the Henry Ford Museum and Greenfield Village: Docs. 2094, 2098, 2126, 2238, 2243; illustrations on pp. 205 (acc. no. 29.1980.275), 677 (Trade Catalog Collection, Armington & Sims Engine Co., P.B. 1884), 412 (acc. no. 2003.0.23.1). From the collection of Ernie Hodgdon: Doc. 2223. Published with permission of the Pierpont Morgan Library: Doc. 2386. Published with permission of the Archives of the New York Stock Exchange, Inc.: Doc. 2331. Courtesy of the Postal and Telecommunications Museum Foundation, Budapest: Doc. 2103. Courtesy of the National Museum of American History, Smithsonian Institution: Docs. 2258, 2322; illustrations on pp. 247 (neg. 85-8768) and 406 (neg. 49.437). Published with permission of the Sterling Memorial Library, Yale University: Doc. 2367.

Reproduced from Alglave 1882, 170: frontispiece. Reproduced from Dredge 1882–85 (1:261; 2:331, 330): illustrations on p. 102, 115–16. Reproduced from Edisonia 1904, 162, 138: illustrations on pp. 664, 703. Reproduced from *Electrician* 8 (1882), 425, 202, 368 (color reversed): illustrations on pp. 211, 411, 472. Reproduced from Fry 1880, flyleaf opposite 176: illustration on p. 246. Reproduced from Guillemin 1891, 793: illustration on p. 410. Reproduced from *Harpers Weekly* 26 (1882), 433: illustration on p. 409. Reproduced from Howell and Schroeder 1927, 186: illustration on p. 271. Reproduced from the *Illustrated London News*, 80 (4 Mar. 1882), 204: illustration on p. 383. Reproduced from Jehl 1937–41, 1040, 807, 766: illustrations on pp. 424, 662, 665. Reproduced from *La Lumière Électrique* 6 (1882), 61–62 and 5 (1881), 419–20: illustrations on pp. 146, 453. Reproduced from *Scientific American* 47 (1882), 127: illustrations on pp. 425–27. Reproduced from Swan 1946, 12: illustration on p. 207. Reproduced from United States patents 263,143 and 264,646: illustrations on pp. 19–20 (*top*).

Courtesy of Edison National Historic Site (designations are to *TAED* notebook or volume:image number and *TAEM* reel:frame): illustrations on pp. 58 (N201:11; 40:465), 155 (QD008:16; 46:306), 156 (QD008:15; 46:305), 183 (N201:16; 40:470), 282 (CA002C:77; 96:261), 295 (D8132ZBV; 58:570), 347 *top* (MBN007:70; 363) *bottom* (NM017:43; 44:403), 348 (MBN007:70; 363), 576 *top* (N197:21; 40:353) *bottom* (N197:22; 40:354), 578 (N197:31; 40:363), 589 *top* (N304:43; 41:1101) *bottom* (N204:11; 40:585).

Index

Boldface page numbers signify primary references or identifications; italic numbers, illustrations. Page numbers refer to headnote or document text unless the reference appears only in a footnote.

Abbot, Lyman, **552**
Abbott, John Stevens Cabbot, 745
Abbott, William, 745
Abyssinia, 744
Académie des Sciences, 760 n. 3
Acheson, Edward Goodrich, **22**; agreement with, 22; and electric lighting in Milan, 401–2; and European lighting companies, 293; filament experiments, 6, 22, 38; and isolated lighting, 444 n. 17; and Ivry-sur-Seine factory, 402; and lamp factory, 22, 39, 43, 234–35, 402; lamp socket experiments, 39; lamp tests, 68 n. 2, 86; letter from Edison, 38; letter to TAE, 401–2; and Paris Exposition, 22, 45–46, 122 n. 1
Adams, Edward Dean, **274**
Adams, H. V., **11**
Adams, James, **431**
Adams, W. G., 659 n. 14
The Adventures of Gil Blas of Santillane, 745
Aerial balloon, 159 n. 3
Aetna Iron Co., 663, 813
Agreements with: Acheson, 22; Armington & Sims, 12; Babcock & Wilcox, 728 n. 5; Bergmann, 31–32, 581, 674; Dean, 448, 548–49, 798; Drexel, Morgan & Co., 492; Edison Electric, Ltd., 297 n. 2; Edison's Indian and Colonial Electric Light Co., 523 n. 1; Edison Spanish Colonial Light Co., 399; Fabbri, 706–7; Gou-

raud, 30, 395, 524 n. 1; Johnson, 31–32, 581, 674; Rocap, 448, 548–49, 798; Société Électrique Edison, 709; Villard, 91, 185, 292 n. 1
Aitkin, Sons & Co., **646**
Alaska, 221, 530, 656, 727, 775, 793–94, 798
Alden, William H., 655 n. 4
Alden & Sterne, **654**
Alexandre Pére & Fils, 295 n. 4
Alexandrovna, Marie, **404**
Allgemeine Elektrizitäts Gesellschaft, 821
Amalgamated Association of Iron and Steel Workers, 770 n. 10
American Bank Note Co., **646**
American District Telegraph Co., 484 n. 4
American Electric and Illuminating Co., 484 n. 4, 533
American Electric Light Co., 273 n. 4, 484 n. 4, 520 n. 10
American Journal of Science, 290
American Machinist, 290
American Queen, 291
American Society of Mechanical Engineers, 215 n. 2, 255, 416
American Swedes Iron Co., **766–67**
American Union Telegraph Co., 452 n. 3
Ampère, André-Marie, **807**
Ananias, 532
Andersen, Hans Christian, 745
Anderson, William, **554**
Andrews, Elisha W., 484 n. 4
Andrews, William Symes, **376**; and armatures, 526 n. 2; and dy-

namos, 376, 490 n. 1, 503, 508, 522 nn. 2 & 4, 569; and testing room, 663
Anglo-American Brush Light Co., 519 n. 3, 568 n. 9, 594–95
Anglo Pacific Electric Light, Telephone & Power Co., 500 n. 16
Anson, Thomas Francis, 399 n. 3
Ansonia Brass & Copper Co., 22 n. 3, 671, 701 n. 8
Appleton Paper and Pulp Co., 703 n. 3
Aquarium Exhibition, 656
Arago, François, **807**
Arizona, 388, 390
Armengaud, Charles, **96**, 129, 161, 486
Armengaud, Jacques-Eugéne, **96**
Armington, Pardon, 127 n. 7, 700
Armington & Sims, **12**, 373 n. 19, 759 n. 1; agreement with, 12; Insull at, 670; letter to TAE, 12; license to Machine Works, 659–60; production problems, 637, 659. *See also* Steam engines: Armington & Sims
Armstrong, William, **207**
Arrow, 586 n. 4
d'Arsonval, Arsène, 701 n. 7
Art Amateur, 291
Arthur, Chester, 761 n. 2
Art Interchange, 291
Aspinwall, John, **586**
Assyrian Monarch, 413
Australia: electric lighting in, 336, 455, 459–61, 493, 523–24, 546; electric pen in, 33; telephone in, 66 n. 6

Babcock & Wilcox, 111, 133, 425, 633, 646, 727

Bailey, Joshua, 77, 96, 132, 486, 828; in Berlin, 700; and Edison Telephone Co. of Europe, 119 n. 7; health, 197, 294, 435; in New York, 581, 630 n. 3, 673, 700, 707; and Paris Exposition, 81–82, 95–96, 98, 142; power of attorney, 251 n. 2; and telephone, 77, 186
—and electric lighting: in Britain, 744; dynamo orders, 563 n. 2; Edison Electric of Europe, 581, 673, 707; in Europe, 161, 314, 441–42, 743–44; European companies, 89, 163, 179, 189, 191–92, 197, 209–10, 219 n. 4, 221, 223–24, 251 nn. 1–2, 260, 435, 439–40; in France, 223; in Germany, 557 n. 9, 763; injunction against Maxim, 143 n. 3; isolated plants, 166 n. 1, 187 n. 4, 191–92, 334 n. 1; in Italy, 403 n. 5, 448, 455–56, 605 n. 5; manufacturing, 176, 186, 191–92, 314, 792; in Milan, 402 nn. 2 & 5, 728 n. 3, 780–82; Milan central station, 630 n. 3, 753; Paris central station, 186; Paris Opera, 303; patents, 763; Swan's claims as lamp inventor, 207 n. 3; in Switzerland, 709; three-wire system, 805 n. 3
—letters: from TAE, 592, 743–44, 780–82; to TAE, 81–82, 128–29, 187 n. 4, 439–42, 455–56
—telegrams: from TAE, 128, 186, 191–92, 228 n. 14; to TAE, 129, 186 n. 3, 209–10

Baldwin Locomotive Works, 213

Ballou, George, 804

Banker, James H., 217, 274, 678 n. 14

Bank of Italy, 401

Barker, George, 15; and dynamos, 272; and Electrical Congress, 15 n. 2, 69; and electrical engineering education, 719 n. 2; electric lighting plant, 6, 51, 69; and Faure battery, 507–8; and lamp tests, 51, 182, 357; and Maxim lamp, 272, 532; in Menlo Park, 51; in New York, 52 n. 12, 69, 272; and Paris Exposition, 6, 15, 69, 121, 166 n. 6, 221; relations with, 68–69, 182, 223, 273 n. 4, 581; retained as expert, 7 n. 4, 222–23; and Swan lamp, 416;

telegram to TAE, 221; and United States Electric Lighting Co., 182; visit to lamp factory, 583
—letters: from TAE, 15, 51; to TAE, 68–69, 272

Barnard, A. P., 719 n. 2

Barnes (A. S.) & Co., 645

Barnum, Marcus, 643 n. 1

Barnum, P. T., 413

Batchelor, Charles, 10, 59 n. 3, 262, 264, 397 n. 1, 418 n. 2, 436, 458, 463, 466, 554, 744; in Berlin, 763; in Britain, 437, 493; and Crystal Palace Exhibition, 301; and Paris Exposition, 6, 26 n. 3, 60–61, 62 n. 1, 77, 81, 81 n. 4, 88, 95–96, 120–21, 133–35, 142, 159–61, 165, 189, 197, 225, 240, 359, 500 n. 10; personal finances, 351, 370, 473–75, 497, 581, 660, 668, 672–75, 700, 707, 797, 809; returns to U.S., 790; and Tesla, 821; vacation, 3, 25, 62 n. 5
—and electric lighting: central stations, 700; dynamos, 158 n. 5, 178, 189–90, 197, 245 n. 3, 317–18, 333–34, 358, 489, 561, 570, 724, 808; Edison Electric, 351; in Europe, 401–2, 435, 557 n. 9, 570, 704–5, 763; European companies, 192 n. 3, 197, 209, 224, 250–51, 293–94, 353 n. 4, 440, 790; European patents, 486, 753, 763, 805; Isolated Co., 351; isolated plants, 162–65, 179, 187 n. 4, 189, 191 n. 2, 197, 254 n. 2, 333–34, 435, 724, 731 n. 3, 763; Ivry-sur-Seine factory, 176, 179–80, 189, 192, 197, 293–94, 314, 402, 485, 581, 590–91, 639–40, 660, 682, 700, 792–93, 821, 828; lamps, 86, 252 n. 1; manufacturing companies, 10, 348 n. 5, 351, 661, 663–64, 668, 674–75, 700, 707, 791 n. 1; Milan central station, 781; in Paris, 161–62, 239–40, 245 n. 1, 791 n. 1; proposed injunctions, 129; ship lighting, 658 n. 7; steam engines, 454, 700; Swan's claims as lamp inventor, 207 n. 3; three-wire system, 805; village system, 809
—letters: to Eaton, 142, 147 n. 7, 763–64; from Insull, 351–53, 473–76, 706–9, 808–9; to Insull, 668–75, 700; from Lowrey, 250–

51; to Puskas, 60–61, 70–71; from TAE, 168–82, 302–4, 454, 485–86, 563–64, 623–25, 639–40, 704–5, 746, 790, 792–93; to TAE, 134–35, 159–65, 189–90, 239–40, 250–51, 293–94, 333–34, 349, 435–36, 570, 590–91, 724, 738–40, 805
—telegrams: from TAE, 135 n. 1, 143, 167 n. 12, 207 n. 3, 805; to TAE, 147 n. 7, 189, 207 n. 3, 435, 792, 805

Batchelor, Emma, 122 n. 1

Batchelor, Rosa, 122 n. 1, 475, 709

Batchelor, Rosanna, 122 n. 1

Batteries: thermoelectric, 502; for voltage regulation, 550–51, 575;
—primary: Clark's, 69; Daniell's, 166 n. 6; Edison's, 741; tellurium, 807
—secondary (storage), 532–33; Edison's, 518, 579, 614, 636; Edison's opinion of, 192–93, 579, 730, 752, 757–59, 789; for electric lighting, 447–48, 757–59, 784–85; electrodes, 741–42, 757–58; experiments, 614, 681, 730; Faure's, 193 n. 1, 244, 280, 507–8, 518, 760 n. 3; patents, 192, 193 n. 3, 516 n. 4, 536 n. 19, 742; Planté's, 509 n. 8, 742; Sellon's, 518; Volckmar's, 518; for voltage regulation, 422, 502, 513–14, 536 n. 19

Beard, Dr., 397 n. 1

Becker, Christopher, 437

Becker & Sons, 437

Beebe, John, 771 n. 11

Beecher, Henry Ward, 552 n. 1

Bell, Alexander Graham, 119 n. 6; British patents, 326, 373 n. 17, 832; French patents, 97; photophone, 12 n. 1; and *Science*, 230 n. 2; telephone, 14, 66 n. 6

Bell Telephone Co., 112 n. 7

Belmont, Perry, 786

Bennett, James Gordon, 299 n. 3, 508 n. 3, 648 n. 13, 672, 816–17, 828–29

Benton, Charles Abner, 797

Berger, Georges, 81–82, 121, 132, 144–45, 163–64

Bergmann, Sigmund, 167 n. 10, 189, 591; agreements with, 31–32, 581, 674; Edison's opinion of, 830; and fixtures, 831; as partner in Bergmann & Co., 4, 31–32,

665; and patents, 430, 666; and purchase of Avenue B factory, 674; reputation, 666; and safety fuses, 604 n. 4; and theater lighting, 801; and voltage regulation, 376, 505 n. 9

Bergmann & Co., 178, 492, 659; and arc lights, 532; and carbonizing molds, 139; catalog, 257 n. 3, 800; and Edison Electric, 771–73; Edison's laboratory at, 675, 680, 708, 828, 831; finances, 674–75, 700, 707, 726; and fixtures, 581, 598, 652 n. 4, 665, 830–31; and isolated lighting, 731; and lamp sockets, 256, 262, 523 n. 7, 581, 665, 740, 830; locations of, 2, 665–66, 674, 707, 830; and meters, 490, 830; and Milan station, 727; orders for, 370, 734, 789; organization of, 4, 31–32, 665, 674; and stock printers, 830; and telephones, 416, 830; tools and machinery, 708; and voltage regulation, 275, 303, 376, 419 n. 5, 489, 505 n. 9, 782 n. 5; workforce, 666

Bergmann (S.) & Co., 32 n. 1. *See also* Bergmann & Co.

Berlin International Exhibition (1879), 409 n. 8

Bertholet, Pierre Eugène Marcellins, **612**, 616, 618–19

Berthoud Borel Cable Co., 257, 700

Betts, Atterbury and Betts, 655 n. 1

Betts, Frederic Henry, 207 n. 3, **319**, 566

Beveridge, Charles, 703 n. 3

Bidwell, Shelford, 268 n. 20, **340**–41, 371, 388, 467, 469, 495, 530 n. 2

Biedermann, Ernst, 266 n. 2, 558 n. 10, 711 nn. 15 & 16

Blaine, James G., **15**

Bliss, George, 199, **518**; at Machine Works, 636; and theater lighting, 681

—letters: from TAE, 722–23; to TAE, 559–60, 636–37, 702, 715–16

Bogart, Adam, **749**

Böhm, Ludwig, 41 n. 1, 483, 498

Bon Marché, 163–64

Bontoux, Eugène, 440

Borden, Spencer, **503**, 556 n. 2, 599 n. 6, 604; and Boston Fair, 599 n. 7; and electric lighting in

Massachusetts, 564–65; letter from TAE, 684–85; and Maxim voltage regulator, 684–85; and mill lighting, 503; and ship lighting, 678 n. 10; and storage batteries, 518

Boston Fair. *See* New England Manufacturer's and Mechanic's Institute Fair

Boston Herald, 759 n. 1, 789

Böttinger, C., **612**

Bouverie, Edward Pleydell, **391**, 399 n. 3, 492, 595, 656, 709

Brackett, Cyrus, 7 n. 4, 784–**85**

Bradley, James, **39**, 335; filament experiments, 139; and lamp factory, 478, 517, 553; lamp socket experiments, 39; and Paris Exposition, 46 n. 4

Brahms, Johannes, 138 n. 2

Bramwell, Frederick, 297 n. 2, 316, 345 n. 19, 382 n. 2, 462, 467, 469, 495

Brandon, David, **486**

Branner, John C., 90, **282**–83, 335

Breguet, Antoine, **60**–61

Brewer, Edward, **259**–60

Brewer & Jensen, **259**–60, 320, 392 n. 8, 472 n. 20, 486

Briggs, C. H. W., **135**

Bright, Charles Tilston, **121**

Brill, George, **476**

Brill, John G., **476**

Brill (J.G.) Co., **476**

Britain: Board of Trade, 501 n. 18, 594, 711–14; economic conditions, 30 n. 1, 596 n. 8; electrical manufacturing in, 215, 652, 696, 734; electric railway in, 708; gaslighting in, 581, 635; quadruplex in, 623, 623 n. 3; telephone in, 4, 640 n. 2. *See also* Crystal Palace Exhibition; Electric lighting; Electric lighting central stations: Holborn Viaduct [and] London; Electric lighting generators: direct-connected (C dynamo): for Holborn Viaduct [and] for London; Electric lighting isolated plants; London

British Association for the Advancement of Science, 66 n. 8, 208 n. 6, 228 n. 19, 259, 658 n. 1

British Electric Light Co., 437, 472 n. 17, 596 n. 5

British Gower-Bell Telephone Co., 14

British Museum, 308

British Post Office, 95 n. 2, 594 n. 9

Brittania, 225, 782

Brown, Frank T., **786**

Brown, James M., 750 n. 1

Brown & Sharpe Manufacturing Co., 592 n. 7

Brush, Charles, 161, **164**, 405, 535 n. 16, 759

Brush Electric Light Co., 416 n. 4; and British patents, 495; in Cleveland, 556 n. 4; and Gramme Electrical Co., 59 n. 4; license policy, 556 n. 2; lighting of Broadway, 511; organization of, 167 n. 7; and power transmission, 299 n. 2; and storage batteries, 532–33, 566

Brush-Swan Electric Co., 757

Builder & Woodworker, 290

Bulwer, Edward Lytton, 745

Bunsen, Robert, **615**–16

Bürgen, Emile, 652 n. 5

Burk, Addison, **213**

Burnett, John, **28**

Burrell, David, **719**

Burrell & Whitman, 719 n. 1

Buzzi, Felice, **401**

Byllesby, Henry, **429**, 567, 642 n. 3

California, 232, 273, 274 n. 1, 567 n. 3

Cambria, 481

Campbell, Charles, **503**

Canada, 456 n. 2; electrical manufacturing in, 532, 565, 666 n. 1; electric lighting in, 429 n. 3, 532, 565–66; Oscar Wilde in, 360 n. 21; patents, 47 n. 3, 532, 534 n. 8, 597

Canada, 102, 414 n. 16

Carlyle, Thomas, 745

Cases (patent): No. 290, 27 n. 2; No. 303, 26–27; No. 307, 337; No. 316, 50 n. 6; No. 322, 57 n. 7; No. 323, 46 n. 2; No. 324, 54; No. 335, 373; No. 339, 770 n. 3; No. 340, 770 n. 3; No. 370, 536 n. 18; No. 373, 536 n. 18; No. 379, 287 n. 2; No. 420, 516 n. 4; No. 428, 451 n. 4; No. 429, 450 n. 2; No. 432, 451 n. 6; No. 445, 543 n. 1; No. 450, 552 n. 3; No. 451, 552 nn. 3–4; No. 452, 516 n. 4; No. 453, 577 n. 7; No. 458, 516 n. 4; No. 466, 451 n. 7; No. 480, 516 n. 4; No. 493, 685–88; No. 504,

Cases (patent) (*continued*)
    698 n. 2; No. 509, 578 n. 18, 688–
    89; No. 511, 692; No. 530, 516
    n. 4; No. 531, 741
Casho, Joseph, 594 n. 3, **789**
Cassatt, Alexander, **136**
Cassatt, Mary, 137 n. 7
Caswell, Hazard & Co., 358
Cavanaugh, Mr., 566
Caveats: distribution system, 794;
    dynamos, 105 n. 5; electric light-
    ing, 447, 564 n. 2; incandescent
    lamps, 576 n. 2, 803 n. 1; meters,
    9, 148–55
Cecil-Gascoyne, Robert Arthur
    Talbot, **165**
Central District and Printing Tele-
    graph Co., 506 n. 1
Chamberlain, Joseph, **594**
Charles Speirs, **745**
Chatard, Alfred, **440**
Cheesman, W. H., 230 n. 3
Chemical News, 207 n. 2
Chemical Society of Newcastle,
    259
Chemistry, 808; and batteries, 741,
    742; electrochemistry, 479–80;
    and incandescent lamps, 612–19,
    687–88, 692, 806–7; lamp socket
    compound, 740; and meters, 9,
    154, 178, 201–3, 802, 806–7
Chicago Academy of Music, 637
Chicago Edison Co., 560 n. 1
Chicago Railway Exposition (1883),
    409 n. 8
Chimney Corner, 291
Chinnock, Charles, 790 n. 6, 820
Chipman, Naomi, **83**, 298
The Christian Union, 552 nn. 1–2
Church, J. B., **785**
Church, Melville E., 787 n. 2
City of Richmond, 147 n. 6
City of Rome, 198, 481
City of Worcester, 277
Clark, Josiah Latimer, 69
Clark, Muirhead & Co., **259**, 546
Clarke, Charles, 111, 554; and cen-
    tral stations, 111, 196, 447, 536
    n. 11, 566, 580, 593, 605, 629–30,
    726–27; as chief engineer, 111
    n. 2, 447, 520–21, 621–22; and
    dynamos, 89, 100–101, 112–14,
    177 n. 4, 304 n. 3, 447, 487, 489,
    503–4, 558 n. 14, 663, 722 n. 2;
    and electric lighting in Boston,
    752, 789; and electric railways,
    482; health of, 726; and incandes-

cent lamps, 556 n. 4; and isolated
    lighting, 731; at Machine Works,
    448; at Menlo Park, 447; and
    meters, 203 n. 9, 490, 503, 806;
    reminiscences, 13, 102 n. 2, 412;
    and underground conductors,
    212; vacation, 696; and voltage
    regulation, 275–76, 448, 503–4
—letters: from TAE, 376, 487, 503,
    621–22; to TAE, 489–90, 520–
    21, 561, 593, 629–30
Clarke, F. E., **510**
Claudius, Hermann, **99**, 111, 376,
    629, 643
Clerac, Hippolyte, **161**
Cochery, Louis Adophe, **332** n. 1
Colombo, Giuseppe, 581, 593 n. 3,
    630 nn. 2–3, **673**, 726–27, 781
    n. 2
Colt's Fire Arms Manufacturing
    Co., 100 n. 5
Columbia, 75 n. 36, 185 n. 2, 277,
    408 n. 5, 431
Columbia College, 718, 752, 795
Columbian Exposition (1893), 412,
    558 n. 11
Comitato per a Applicazioni
    dell'Elettricita Sistema Edison,
    401–2, 605 n. 5, 679 n. 14,
    781–82
Commercial Cable Co., 299 n. 3
Compagnie Continentale Edison,
    401–2, 593 n. 3, 780–82; direc-
    tors, 294; dynamo orders, 570
    n. 2; and German electric light
    companies, 554; investors, 81–82,
    128, 131, 163–64, 209–10, 223,
    251; organization of, 81–82, 179,
    209–10, 250–51, 314, 440; and
    Paris central station, 701 n. 5
Conant, Thomas, 694 n. 14, 740
    n. 5, 803 nn. 2 & 6
*Confessions of an English Opium-
    Eater*, 86 n. 10
Conley, M. R., 230 n. 3, 435 n. 4,
    767
Conservatoire National des Arts et
    Métiers, 164, 758
Consolidated Telephone Construc-
    tion and Maintenance Co., 14
    n. 3, 63, 119 n. 7, 546
Consolidated Virginia Co., 298 n. 1
Cooke, Conrad, **157**
Cooke Locomotive & Machine Co.,
    710 n. 8
Cooper, James Fenimore, 745
Cooper Union, 796 n. 2

Corning Glass Works, 42, 335, 590,
    659
Cowles, Alfred, **671**
Cox, Samuel, **786**
Crédit Lyonnais, **161**
Crerar, John, 560 n. 1
Critical and Historical Essays, 745
Crompton, Rookes Evelyn Bell, 652
    n. 5, **657**
Crompton & Co., 596 n. 1
Crookes, William, 166 n. 6, **222**,
    263–64, 543 n. 1
Crosby, George, 483
Crystal Palace Co., 382, 405, 461
Crystal Palace Exhibition (1882),
    402 n. 2, 656; arc lights at, 404;
    Brush's exhibit, 404–5; closes,
    493, 497, 546; Duke of Edin-
    burgh at, 404–5, 465; dynamo
    tests, 545; electric lighting at,
    404–5; Gladstone at, 438; John-
    son's lecture, 482 n. 1; jury, 461,
    493, 657 n. 14; lamp tests, 545;
    Lane-Fox's exhibit, 404–5;
    Maxim's exhibit, 404–5, 438;
    Siemens at, 405; Swan at, 404–5;
    Swan's exhibit, 404, 437–38;
    Weston's exhibit, 362
—Edison's exhibit, 198, 216, 241–
    44, 273 n. 3, 300–301, 312, 362–
    63, 379–*83*, 404–6, 415–16, 437–
    38, 458, 497; chandelier, 465, 656
    n. 9; dynamos, 241–43, 264–65,
    333, 381, *406;* electric pen, 381;
    electromotograph telephone,
    381; lighting plant, 458, 464–66;
    motors, 301; musical telephone,
    381; phonograph, 381
Cunningham, David, **493**, 498
Cutting, Robert, Jr., 824, 829

Daily Free Press, 360 n. 19
Darwin, Charles, 720 n. 1
David, Thomas B. A., **506**
Dean, Charles, **170**, 427 n. 4, 827–
    28; agreement with, 448, 548–49,
    798; and dynamos, 170–71, 174–
    75, 196, 311, 352, 427 n. 3, 489,
    504 n. 4, 505 n. 7, 520, 556; letter
    to TAE, 569; malfeasance of, 605,
    606 n. 10, 664; and steam yacht,
    586 n. 4; as superintendent of
    Machine Works, 3, 196, 311, 388,
    474, 509 n. 10, 663–64, 670, 754;
    telegram to TAE, 490 n. 1
Dechanel, Augustin Privat, **807**
Delany, Patrick, **257**

De[liland?], Count, 132
Delmonico's, 720
Demorest Magazine, 291
Deprez, Marcel, **121**, 701 n. 7
De Quincy, Thomas, **85**
Deutsche Edison Gesellschaft, 744
   n. 2, 764 n. 5, 821, 828
Devine, Arthur, 29 n. 1
Dewar, James, **365**
Dickens, Charles, 85 n. 3, 745
Dickerson, Edward, **319**, 520 n. 10,
   566
Dickerson, Edward, Jr., 520 n. 10,
   566
Direct conversion of coal, 396–97,
   455, 479–80, 502
Dixey, Henry E., 822, 827
Dixon (Joseph) Crucible Co., 641
   n. 4
Dolbear, Amos, **163**
Domestic Telegraph Co., 73 n. 14
Dominion Telegraph Co., 567 n. 4
Don Quixote, 745
D'Oyle Carte, Richard, **358**
Draper, Henry, **531**, 681, 717, 720
Dredge, James, 158 n. 1
Drexel, Anthony, 190 n. 2, **272**
Drexel, Harjes & Co., **189**
Drexel, Morgan & Co., **30**, 351;
   agreements with, 492; building,
   2, 645; as Edison's banker, 30–31,
   47, 80, 219; and Electric Tube
   Co., 665; loan to Edison, 581,
   660; and Oriental Telephone Co.
   stock, 796; partners, 258; telegram from Gouraud, 523; and
   telephone, 78 n. 7
—and electric lighting, 261; in
   Britain, 4, 59–60, 66, 215–16,
   371 n. 4, 391, 468, 478, 492, 796,
   821; in British colonies, 459–61,
   523–24; British patent rights, 200
   n. 4; British patents, 395 n. 4;
   Edison Electric, 708; Edison
   Electric of Europe, 682, 707; in
   Europe, 459–61; European
   patent rights, 200 n. 4; isolated
   plants, 255, 272; Pearl St. station,
   538–39, 645, 671; in Philadelphia, 536 n. 13
—letters: from TAE, 59–60, 215–
   16; to TAE, 523–24
Druggist Circular, 290–91
Du Boise-Reymond, Emil, **189**
Du Boise-Reymond, Paul, **189**
Dubos, C., **135**
Duchess of Edinburgh, **404**

Duke of Edinburgh, **404–5**, 465
Du Moncel, Théodose, 95–98,
   121–22, 131–32, 135, 145, 358,
   361
Dun (R. G.) & Co., **42**, 754, 771
Dyer, Frank, 611 n. 3
Dyer, George W., 98 n. 5, 250 n. 5,
   286, 312, 328 n. 25, 401 n. 2, 609
Dyer, Philip, **10**, 23, 139, 328 n. 25,
   473; and Lamp Co. finances, 10,
   353 n. 3, 553 n. 3; and lamp pric-
   ing, 80 n. 1; letter to Insull, 10
Dyer, Richard, **321**, 401 n. 2, 597;
   becomes Edison's patent attorney, 312, 369; and British patents,
   321, 325, 369, 371, 373–74, 450
   n. 1, 462; and Brush patents, 374;
   and electric railway patents, 486;
   and European patents, 486, 654;
   and lamp patents, 365; letter to
   TAE, 785–86; memoranda to,
   373–74, 377–78, 706; in Menlo
   Park, 446–47, 476, 513, 643;
   patent assignments to, 554–55;
   and patent for electroplating carbonized articles, 706; power of attorney, 611 n. 2; and Rowland's
   patent application, 762; and
   three-wire patents, 794 n. 7, 805;
   and U.S. Patent Office, 752–53,
   785–86; and voltage regulation
   patents, 418; work for Edison
   Electric, 609–10
Dyer and Seely, 558 n. 13
Dynamos. *See* Electric lighting
   generators

Eads, James Buchanan, **144**
East Newark, N.J. *See* Edison Lamp
   Co.: lamp factory, Harrison
Eaton, Sherburne, **98**, 161, 190 n. 2,
   225, 402 n. 2, 621, 785 n. 1, 796;
   memoranda, 447, 507, 531, 542,
   554, 605 nn. 1–2, 621; and ore
   milling, 230, 372 n. 13, 434 n. 1,
   752, 765–770; and patent law,
   538 n. 1; and press relations, 230
   n. 2; reports, 506–8; and steam
   yacht, 586 n. 2
—and electric lighting, 794 n. 4; in
   Boston, 752, 789; Brush Electric
   Light Co., 532–33; Bulletins,
   487–88, 498, 556 n. 4, 595; defect
   reports, 526–27; and dynamos,
   214, 555–56; Edison Electric,
   205–6, 447; Edison Illuminating
   Co. of New York, 206 n. 2; engi-

neering staff, 503, 621; experimental expenses, 184; in foreign
   countries, 199; foreign orders,
   783 n. 7; German lamp factory,
   792; infringement suits, 723 n. 2;
   isolated plants, 191 n. 2, 254 n. 2,
   273–74, 277, 352, 522 n. 3, 730–
   31; laboratory expenses, 710 n. 7;
   lamp business, 603 n. 6, 604, 676
   n. 3, 754, 755; manufacturing,
   569 n. 2, 583 n. 2, 753; New York
   Board of Fire Underwriters, 554;
   Pearl St. station, 111, 541 n. 1; as
   president of Edison Electric, 708,
   823; safety fuses, 523 n. 7; second
   district, 377 n. 3; steam engines,
   781; as vice-president of Edison
   Electric, 444
—letters: from Batchelor, 142, 147
   n. 7; from Insull, 217–18, 725–
   26; from Johnson, 771–73; from
   Morgan, 750; from TAE, 184,
   488, 507–8, 510–11, 518, 531–
   34, 542, 554–56, 604–5; to TAE,
   230, 273–74, 444–45, 478–79,
   564–65, 597–98, 609–10, 730–
   31, 747–48; from Upton, 771–73
Eckert, Thomas, **451**
Edison, Charles, 75 n. 34
Edison, Ellen J. Houlihan (Mrs.
   William Pitt), 749 n. 3
Edison, Marion ("Dot"), 84 n. 3,
   199, 680
Edison, Mary Stilwell, 33, 140
   n. 6, 225, 385 n. 6, 446, 482,
   579; in Florida, 313, 401, 417;
   gives party, 199, 348 n. 2; and
   Gramercy Park home, 655 n. 3,
   680; health, 313, 348, 655 n. 4;
   horses for, 449, 456; journal orders, 291; in Menlo Park, 199,
   298, 680; in Michigan, 91, 168;
   in New York, 313, 401 n. 2; in
   South Carolina, 313, 400, 401
   n. 2; telegram from Insull, 417;
   telegrams to Insull, 400, 417
Edison, Samuel, 91
Edison, Thomas Alva: article by,
   552; ASME paper, 215 n. 2, 417
   n. 2; book order, 745; in Boston,
   752, 789; cigars, 818, 828, 830;
   finances, 4, 13–14, 25, 30, 42, 60,
   90–91, 185 n. 4, 198, 219, 229,
   255, 261–62, 292 n. 1, 314, 351,
   367, 448, 460, 473–75, 492, 556
   n. 3, 581, 634, 651–52, 660, 662,
   665, 668–69, 672, 674–75, 707,

Edison, Thomas Alva (*continued*)
747 nn. 2–3, 749, 761, 770 n. 1,
809; Florida vacation, 313, 401,
403 n. 6, 418 n. 2, 432, 446, 454,
482, 823; genealogy, 85 n. 2;
health, 313, 401 n. 2, 446, 481–
84, 498, 696, 823; interviews,
509 n. 11, 552, 705 n. 2, 759 n. 1,
789; journal orders, 290, 448;
loan to Pitt Edison, 314; in Menlo
Park, 33, 197, 199, 252 n. 2, 292,
297–98, 313, 397 n. 1, 418 n. 2,
432, 446–48, 476, 506, 579, 662,
680; in Michigan, 91, 168; in
Montreal, 586 n. 4; in Newark,
3; patent policy, 430, 554–55;
portraits, 249; proposed trip to
Britain, 382, 405, 414, 483; prose
poem, 84–85; reputation, 21, 42,
65, 223–24, 334 n. 1, 391, 462,
495, 630, 771; residences, 2, 6, 83,
91, 199, 292 n. 3, 298, 313, 332 n.
2, 433, 446, 508, 579, 597, 654,
675, 680; Southern trip, 752, 785;
steam yacht trip, 449, 579, 586,
593 n. 2; in Tarrytown, 227 n. 12;
in Washington, D.C., 313, 401
n. 2; western trip, 272
—opinions on: alcohol, 453; chewing tobacco, 453; isolated lighting, 191; night work, 453; storage
batteries, 192–93, 579, 730, 752,
757–59, 789
Edison, Thomas Alva, Jr. ("Dash"),
84 n. 3, 680
Edison, William Leslie, 84 n. 3
Edison, William Pitt, 91, 314, 449,
456, 749
Edison Co. for Isolated Lighting,
370, 376, 399, 429 n. 3, 660;
agents, 789; agreement with
Edison Electric, 560 n. 4, 565;
agreement with Western Edison,
559; and American Electric &
Illuminating Co., 533; business
prospects, 352, 696, 797; catalog,
257 n. 3; Chicago office, 519 n. 8;
defect reports, 518, 520, 522,
526–27, 724; description of, 631–
33; directors, 274 n. 2, 730–31;
dynamos, 112; and Edison Electric, 198; finances, 598; and Herald building, 646; inside wiring,
554; instruction booklet, 267
n. 12; lamp orders, 776–77; and
Machine Works, 669; management of, 198, 274, 373 n. 19, 522

n. 3, 527, 598 n. 1, 673; New England agent, 505 n. 5, 510 n. 12;
number of plants, 731 n. 1; orders
for plants, 708, 789; policies,
730–31; prices, 297 n. 6; and
Roselle station, 642 n. 3; sales to,
585 n. 3; stock, 274 n. 2, 351,
353, 430, 475, 669, 672–73, 708;
wiring Edison's home, 597
Edison Effect, 289, 470 n. 5. *See
also* Electric lighting incandescent lamps: electrical carrying
Edison Electric Illuminating Co. of
New York, 232, 399, 407 n. 1, 598
n. 2, 631, 655 n. 1, 660, 728 n. 7;
accounts, 539–41; directors, 275
n. 3, 332 n. 1; finances, 708; investors, 157 n. 1; meter department, 490 n. 5; Morgan's isolated
plant, 750 n. 1; officers, 63 n. 3,
99 n. 7; and Pearl St. station, 90,
158 n. 2, 303, 368 n. 5, 426, 539–
41, 670, 708, 747–48, 780; policies, 206 n. 2; relations with Morgan, 796–97; stock, 353, 430,
475, 672
Edison Electric Illuminating Co of
Appleton, 703 nn. 1 & 3
Edison Electric Lamp Co. *See* Edison Lamp Co.
Edison Electric Light Co., 80, 137
n. 4, 226 n. 1, 232, 272, 394 n. 23,
437, 459–60, 467, 493, 556, 605
n. 4, 622 n. 1, 635, 649–51, 660,
823, 828; advance to Lamp Co.,
553 n. 3; agents, 487–88; bonds,
351; Bulletins, 254 n. 4, 360 n. 17,
413 n. 6, 426, 482, 487–88, 498,
518, 535 n. 11, 542, 556 n. 4, 595,
597, 605, 647 n. 5; Bureau of
Isolated Lighting, 373 n. 19;
Canadian lamp factory, 535 n. 8,
666 n. 1; canvassers, 377 n. 3;
and central stations, 593 n. 1,
606 n. 9, 737 n. 9; chief engineer,
111 n. 2, 498; defends patents,
268 n. 20; directors, 275 n. 3, 332
n. 1, 508, 598 n. 2, 609, 743; and
electric railways, 292 n. 1, 407;
engineering department, 503,
621, 728 n. 7; experimental expenses, 101, 184; and Gramme
Electrical Co., 509 n. 9; headquarters, 2, 23, 655 n. 4; instruction booklet, 267 n. 12; and isolated lighting, 113, 198, 274, 429
n. 3, 737 n. 9; licenses, 274 n. 2,

488, 554, 631, 634, 659, 809 n. 1;
and manufacturing, 4, 444–45,
659–60, 725–26, 753, 755, 771–
73, 822; New England Dept., 505
n. 5; officers, 28 n. 1, 63 n. 3, 69
n. 1, 99 n. 7, 205, 708; patent assignments, 432 n. 4; patent attorneys, 328 n. 17, 418, 447, 609–10,
654, 825–26; patent experts, 6,
785 n. 1; and patent infringement
suits, 722–23; patent policy, 806
n. 4; patents, 462–63, 609–10;
and Pearl St. station, 426; policies, 194, 205–6, 255; and press
relations, 230 n. 2; proposed
merger with Siemens, 743; and
Roselle station, 642 n. 3; San
Francisco agent, 542 n. 5; shareholders, 157 n. 1, 266 n. 7; stock,
138 n. 1, 185 n. 4, 273 n. 4, 348
n. 2, 353, 370, 430, 432 n. 5, 433
n. 6, 474–75, 671–72; and storage batteries, 518; and testing
room, 569, 663; and village system, 809; western agent, 519 n. 8
—agreements with: Isolated Co.,
565; Lamp Co., 11 nn. 5–6, 62,
234, 541 n. 5, 587, 662, 669, 707,
753–56, 822, 829; Stuart, 699
n. 2
Edison Electric Light Co., Ltd.,
821; agreements with, 297 n. 2;
defect reports, 774; directors,
462; disputes with, 581, 584–85,
649–51, 682, 695, 708–9, 788;
district companies, 546; dynamo
orders, 448, 492, 557 n. 8, 649–
50, 735, 839 n. 4; Edison's proposals for, 682, 711–14, 734–36,
738; and Electric Lighting Act of
1882, 501 n. 18, 594–95; engine
orders, 774; formation of, 311–
12, 369; infringement suits, 462;
and isolated lighting, 257 n. 3,
394–95, 738; lamp orders, 650,
776; licenses, 70 n. 3; manufacturing for, 581, 584–85; merger
with Swan, 547, 744; operations
of, 655–57; patent costs, 654;
patents, 735; shareholders, 311,
393 n. 21, 398–99; stock, 547
Edison Electric Light Co. of Europe, 218 n. 3, 655 n. 1, 753, 764;
agreement with Bailey, 581;
agreement with Italian company,
605 n. 5; agreement with Puskas
and Bailey, 673; and direct-

connected dynamo, 184; and European lighting companies, 190 n. 1, 192, 251 nn. 1–2, 435–36, 556 n. 3; finances, 554; and manufacturing, 238; officers, 99 n. 7; and Paris Exposition, 81 n. 4; stock, 119, 273 n. 4, 353, 430, 436, 475, 673–74, 682, 700, 706–7

Edison Electric Light Co. of Havana, 399 n. 1

Edison Electric of Cuba & Porto Rico, 399 n. 1

Edison Foreign Electric Light & Motive Power Co., 459

Edison General Electric, 662, 664, 666

Edison Lamp Co., 10, 159, 597, 659; credit report, 754, 771; Edison Electric, 662; and Edison Electric, 771–73; experimental expenses, 238; fiber search, 282–83, 335; finances, 10, 29 n. 3, 193–94, 238, 351, 473–75, 553, 587–88, 602, 627, 660, 668–69, 675, 700, 707, 726, 824, 827; foreign orders, 370; incorporation of, 587, 662; lamp orders, 755, 800; lamp prices, 10, 80, 90, 140, 187, 238, 256, 541 n. 5, 588 n. 4, 603 n. 6, 604, 650–51, 662, 668–69, 707, 755, 777, 793, 799–800, 822; letter from TAE, 517; letter to TAE, 86; manufacturing for Britain, 734; organization of, 661; partners, 10 n. 3, 32 n. 2, 348 n. 5; purchase of Harrison factory, 28–29; sales, 258; shareholders, 431; stock, 587

—agreements with: Edison Electric, 11, 62, 234, 541 n. 5, 587, 669, 707, 753–56, 822, 829; Holzer, 139; Lawson, 16

—lamp factory, 42; bamboo supply, 22, 234, 777; breakage department, 588; carbonizing department, 588 n. 7; carbonizing furnace, 601; carbonizing molds, 601; Edison at, 197, 252 n. 2, 263, 287 n. 1, 297, 303, 313, 370, 602 n. 2, 823; Edison's laboratory at, 829; experiments at, 5, 139–40, 197, 234–36, 313, 330–31, 335, 346, 390, 661, 757, 776; generator, 601–2; glassblowing department, 294 n. 2; glass supply, 335; Harrison, 4–5, 28–29, 75 n. 36,

236, 314, 346, 370–71, 447, 474–75, 478, 579, 581, 602 n. 2, 634, 660–62, 776, 822–23; manufacturing process, 16, 22–23, 197, 234–36; manufacturing superintendent, 313, 430, 478, 553, 600, 627, 661, 800; Menlo Park, 3–4, 8, 29 n. 2, 75 n. 36, 196–97, 234–36, 402, 475, 511 n. 2, 534 n. 8, 553–54, 590, 601, 661, 668, 822; night work, 25 n. 11, 39, 588 n. 6; output, 90, 140, 194, 236, 248, 314, 346, 352, 355, 475, 602 n. 2, 604, 634, 661, 668, 755, 799–800; photometer room, 22, 124, 252 n. 2, 370; platinum supply, 10, 176, 601; production costs, 39, 90, 141, 194, 234–36, 238, 262; production problems, 5, 22–23, 139, 187, 196–97, 234–36, 252, 263, 627, 638 n. 3; pump department, 587, 601, 627; shipments, 123, 142; testing department, 5, 234–35, 511, 739, 774; test reports, 23, 86, 139, 196; tools and machinery, 8, 517, 601, 662; vacuum pumps, 5, 8, 22–23, 39, 139–41, 187, 235–36, 335, 553, 590, 601–2; visitors to, 194, 583; workforce, 10, 29, 90, 124, 139–41, 194, 234–36, 313–14, 335, 346, 553, 661, 799–800; work stoppage, 314, 335, 492, 553, 800. *See also* Electric lighting incandescent lamps

Edison Machine Works, 3, 12, 75 n. 36, 88, 138 n. 1, 265, 463, 492, 659, 754, 761 n. 1, 823; absorbs Electric Tube Co., 665; and arc lights, 532; Bliss at, 636; British orders, 198, 311, 445 n. 3, 499 n. 5, 734; building, *664;* and central station dynamos, 100, 196, 232, 311, 352, 367, 410, 412, 426, 474, 670; Clarke at, 448; and disk dynamo, 99; dynamo deliveries, 605; dynamo inspections, 508; dynamo orders, 522 n. 2, 548–49, 571 n. 2, 707; dynamo prices, 299–300, 338–39, 584–85, 601, 838; dynamo tests, 333 n. 2; Edison at, 138 n. 1, 254, 333 n. 2, 564; and Edison Electric, 771–73; Edison's role, 35, 197; and electric railways, 304, 384 n. 1, 812, 828; expansion of, 581, 660, 813–14, 827–29; experiments at,

446, 448, 487, 503–4, 663; finances, 256, 448, 473–74, 634, 660, 663–64, 669–70, 674–75, 700, 725; foreign orders, 562; and Isolated Co., 351; and isolated dynamos, 112, 196, 248, 280; license from Armington & Sims, 702 n. 9; location of, *2;* management of, 3, 196, 661, 663–64, 754, 784, 791 n. 1; manufacturing for Britain, 734–35; meter experiments, 202; move to Schenectady, 664; night work, 60, 64; and ore separators, 771 n. 11; output, 248, 262, 448, 663, 669; Porter at, 417 n. 2; profit-sharing agreements, 448; Reményi at, 138 n. 1, 816; and second district, 377 n. 3; and steam engines, 262, 276, 670, 702 n. 9; strike at, 664; testing room, 376 n. 2, 437, 569, 663; tools and machinery, 99, 591; training courses, 522 n. 5, 663; underground conductor tests, 212; and voltage regulation, 197, 425; workforce, 474, 549 n. 3, 634, 663, 728 n. 7; work stoppage, 669

Edison Ore Milling Co., 569 n. 1; accounts, 433–34; directors, 230; patent rights, 250 n. 6; report on operations, 765–770; stock, 430, 475

Edison's Foreign Telephone Supply and Maintenance Co., 497

Edison's Indian & Colonial Electric Light Co., 448, 455, 523–24, 546, 675

Edison Spanish Colonial Light Co., 399

Edison Speaking Phonograph Co., 472 n. 15

Edison Telephone Co. of Europe, 77, 118

Edison Telephone Co. of Glasgow, 26 n. 2, 63, 76–77, 342

Edison Telephone Co. of London, 371 nn. 3 & 5; 394 n. 23; chief engineer, 472 n. 24; dispute with, 372 n. 6; Edison's reversionary share, 14, 63, 93, 117–18, 198; exhibitions, 158 n. 1; headquarters, 367 n. 1; shareholders, 316, 393 n. 21, 399 n. 3; stock, 460

Edmunds, Henry, 144–45, 165, 416 n. 4

Edson, Tracy, 274

Egyptian bonds, 656

Elbe, 789

Eldred, Horace, **494**

Electrical Congress (Paris), 15 n. 2, **64,** 122 n. 4, 132, 164, 344 n. 13

Electrical engineering education, 227 n. 12, 522 n. 5, 663, 718, 752, 795

Electric appliances: cigar lighter, 71, 178; fan, 70; iron, 555

Electrician, 36 n. 5, 135, 157, 308, 325, 472 n. 21, 746

Electric Light and Power Generator Co., 362, 596 n. 5

Electric lighting, 482 n. 1; caveats, 9, 148–55, 447, 564 n. 2, 576 n. 2, 794, 803 n. 1; compared to arc lighting, 510–11, 645, 777; compared to gaslighting, 34–36, 53–54, 64, 161–64, 203 n. 10, 233, 239, 245 n. 6, 279, 297, 305–7, 311, 322, 426 n. 2, 456 n. 3, 496, 508, 510–11, 533, 545, 559, 566, 581, 606 n. 9, 635, 642, 644, 671–72, 681–82, 701 n. 6, 711–14, 736; exhibitions, 23, 52 n. 6, 60 n. 2, 64, 136, 300; for railways, 244, 280; safety of, 371, 554; scientific opinions, 34, 95–98, 324; and storage batteries, 244, 258, 280, 447–48; tests, 164. *See also* Electric lighting central stations; Electric lighting generators; Electric lighting incandescent lamps; Electric lighting isolated plants; Exposition Internationale de l'Électricité; Patent applications

—arc, 213, 356, 423 n. 1; in Britain, 343 n. 1, 394, 439 n. 4; Brush's, 164–65, 213 n. 2, 416 n. 4, 532–33, 554; carbons, 575–76; combined with incandescent, 448; compared to incandescent lamps, 554, 636–37, 645, 777; Edison's, 178, *183,* 197, 349, 448; Fuller's, 637; in Germany, 189; high voltage, 759 n. 1; Jablochkoff's, 121, 163; Lampe Soleil, 210 n. 1; in Lawrence, 510–11; Mose's experiments, 448, 532, 636; patents, 534 n. 7, 537 n. 18; Pilsen, 121, 164; Siemens's, 164–65; Solaire, 395 nn. 1–2; and storage batteries, 757–59; Thomson-Houston's, 213 n. 2, 484 n. 4; Weston's, 363 n. 3

—distribution system: alternating current, 628; conductors, 21, 34, 341; high-voltage transmission, 423 n. 1, 446, 516 n. 1, 528, 537 n. 19, 561, 715–16, 722, 759 n. 1; inside wiring, 53–54; meters, 6, 9, 49, 90, 148–55, 178, 201–5, 305–6, 333, 388, 436–37, 490, 503, 521, 615, 625–26, 633, 802, 804, 806–7, 819–20; overhead wires, 673, 699, 752; power transmission, 298, 423 n. 1, 593, 633; safety fuse, 70, 142, 277, 306, 324, *485,* 489, 508, 521, 561, 591, 604; safety of, 628, 716, 722, 752; three-wire, 582, 699 n. 1, 722 n. 2, 753, 792 n. 1, 805; transformer, 528; underground conductors, 501 n. 18; village system, 446; voltage regulation, 324, 377–78, 489, 512–15, 528, 536 n. 19, 635, 684–85, 703 n. 3, 730, 791–92; water power, 702; wire insulation, 554. *See also* Electric lighting central stations

—in: Alsace (France), 678 n. 12; Australia, 336, 455, 459–61, 493, 523–24, 546; Austria, 441–42, 678 n. 12, 701 n. 6, 724; Belgium, 442, 678 n. 12, 801 n. 5; Brazil, 507; Britain, 4, 59–60, 64, 77, 135, 157, 163, 197, 215–16, 249–50, 258–61, 278–79, 369–71, 447, 581, 649–52, 655–57, 660, 682, 695–96, 708–9, 744, 753, 793–94, 800; British colonies, 448, 459–61, 801 n. 5; California, 232, 431, 565, 567 n. 3; Canada, 429 n. 3, 532, 565–66; Chile, 509 n. 5, 831; Cincinnati, 698; Cuba, 399 n. 1, 519 n. 7; Denmark, 442; Europe, 4, 88–89, 165, 440–42, 447, 570, 579–81, 660, 682, 705 n. 3, 743, 800; Finland, 349, 443 n. 13; France, 89, 570, 660, 678 n. 12, 682, 724, 744 n. 1; Germany, 165, 294, 440–41, 554, 678 n. 12, 682, 704–5, 724, 744 n. 1, 763, 792–93, 801 n. 5; Holland, 441–42, 678 n. 12; Hungary, 678 n. 12; Illinois, 559; India, 259, 455, 459–61, 474, 493, 523–24, 546; Iowa, 559; Italy, 261, 314, 401–2, 441–42, 448, 455–60, 604, 629–30, 678 n. 12, 753, 780–82; Lorraine (France), 678 n. 12; Massachusetts, 488 n. 1,

510–11, 564–65, 567, 599 n. 9, 681, 789, 829; Mexico, 531; New Jersey, 70 n. 3; New Zealand, 455; Norway, 200, 459–60; Ohio, 698; Oregon, 431; Pennsylvania, 532, 699 n. 1; Portugal, 459–60, 474; Puerto Rico, 399 n. 1; Romania, 442; Russia, 435, 441, 678 n. 12, 744 n. 1; South Africa, 460; South America, 127 n. 4, 156–57, 521, 604; South Carolina, 565; Spain, 314, 441; Strasbourg, 314, 333–34; Sweden, 200, 459–60, 474; Switzerland, 266 n. 2, 511, 557 n. 10, 709; Wisconsin, 559, 702

—patent rights: Australia, 474, 524 n. 2; Brazil, 508 n. 2, 542; Britain, 395 n. 4; Ceylon, 524 n. 2; Cuba, 400 n. 3; Europe, 200, 821; France, 182 n. 4, 821; Hong Kong, 91; India, 524 n. 2; Norway, 200; South Africa, 524 n. 2; Sweden, 200; Switzerland, 557 n. 10, 709

—patents, 447, 753; Australia, 336; Britain, 75 n. 36, 105 n. 5, 192, 193 n. 2, 200 n. 1, 259–60, 281 n. 9, 311–12, 315–22, 325–26, 336–38, 359 n. 4, 373–74, 380, 386–88, 398, 450, 476, 494–95, 529–30, 546–47, 564 n. 2; Brush's, 375 n. 6, 597; Crookes's, 271 n. 33, 543 n. 1; France, 75 n. 36; Gramme's, 4, 47 n. 3, 59 n. 4, 318; Hopkinson's, 753; Johnson's, 266 n. 4, 432 n. 1; Krizik's, 121 n. 8; Lane-Fox's, 182, 269 n. 21, 326, 438, 495; Maxim's, 89, 97, 128, 137 n. 4, 143 n.3, 182, 495, 543 n. 1, 685 n. 2; meters, 9 n. 1, 50 n. 3, 155 n. 5, 156 n. 11, 203 n. 3, 337 n. 2, 808 n. 9; Moses's, 534 n. 7; Muller's, 554; Norway, 200 n. 4; Piette's, 121 n. 8; Siemens's, 4, 47 n. 3, 191 n. 2, 198, 260, 316–17, 321, 387, 495; Swan's, 89, 181–82, 259–60, 320, 326, 495, 543 n. 1, 566, 599 n. 7; Sweden, 200 n. 4; Tasmania, 336; three-wire system, 794, 805; U.S., 4; voltage regulation, 418, 504 n. 2, 516 nn. 1–3 & 5, 528 n. 2, 685 n. 3, 791–92; Weston's, 543 n. 1. *See also* Electric lighting generators; Electric lighting incandescent lamps

Electric Lighting Act of 1882
(Britain), 383 n. 6, 496, 594–95,
711–14, 737 n. 2
Electric lighting central stations, 34,
305, 307, 681, 763, 806; Albany,
N.Y., 799 n. 4; Appleton, Wisc.,
702–3; Baltimore, Md., 606 n. 9;
canvasses, 314 n. 2; Cincinnati,
Ohio, 606 n. 9; Circleville, Ohio,
699 n. 2; compared to isolated
plants, 635; costs, 186 n. 2; design
characteristics, 713; distribution
system, 563–64; economic analysis, 536 n. 11; estimates for,
296–97, 593; in Europe, 743; Fall
River, Mass., 505 n. 5; Germany,
557 n. 9; Holborn Viaduct, 198,
243–47, 299, 306–7, 311–12,
363, 380, 382, 394, 410–13, 458,
463–66, 468–69, 472, 478, 490
n. 2, 493–94, 501 n. 18, 510,
545–46, 608, 624, 633, 714, 734,
774; inside wiring, 324; in Italy,
455; Lawrence, Mass., 565, 567;
London, 4, 59–60, 64, 77, 89,
92–93, 135, 296–97, 341, 545,
664, 734; in Massachusetts, 564;
Menlo Park, N.J., 35, 840 n. 7;
method of charging customers,
712–13; Middletown, Ohio, 699
n. 2; Milan, Italy, 402 n. 2, 412,
581, 629–30, 664, 673, 682, 726–
27, 753, 780–82; New York second district (Madison Square),
311, 376, 630 n. 1, 800; overhead
wires, 673; Paris, 186, 209, 664,
682, 700; Philadelphia, 532;
Piqua, Ohio, 699 n. 2; promotion
of, 531; Roselle, N.J., 569 n. 14,
580, 582 n.1, 630 n. 1, 642–43,
673, 704, 722 n. 2, 752; Santiago,
Chile, 412, 429 n. 3, 522 n. 5, 630
n. 1, 831; Shamokin, Penn., 630
n. 1; standard design of, 580;
Sunbury, Penn., 699 n. 1, 722
n. 2; Tiffin, Ohio, 699 n. 2; underground conductors, 508; village system, 373 n. 19, 569 n. 2,
569 n. 14, 580, 582, 582 n. 1,
600–601, 642–43, 673, 698–99,
704, 715, 722, 740 n. 5, 752, 793–
94, 796, 800, 804, 809; voltage
loss, 629–30; voltage regulation,
303, 349, 376, 418–20, 503–5;
voltages, 712–13
—Pearl St. (New York first district),
2, 34, 88–90, 191, 232–33, 307,

403 n. 6, 414, 423–27, 429, 446,
457, 482, 492, 559, 570, 604, 636,
703 n. 3; and arc lights, 534 n. 7;
begins operation, 426, 519 n. 9,
539 n. 1, 580, 644–45, 662; bill
collections, 789, 796, 819; boilers,
111, 158, 233, 248, 351, 425, 431,
632–33; building, 89–90, 111,
158, 196, 233, 248–49, 351, 424,
426, 429, 540, 632, 820, 828; canvasses, 34, 426 n. 1; coal supply,
111, 566; construction delays,
311; cost of, 425–26, 539–41,
633; customers, 423–24, 426,
538–39, 645, 671, 727, 747–48,
752, 798 n. 3; drawings, 424–26,
429; dynamos, 196, 303, 311, 351,
388, 410–12, 414, 424–25, 431,
447–48, 475, 489, 541, 580, 632,
670, 814–15; Edison at, 645, 815;
fire at, 412, 426; Herald article,
644–45; inside wiring, 34, 233,
427 n. 11, 482 n. 5, 554; lamp
supply from, 647 n. 5, 671, 708,
727, 833; map, 36 n. 5; meters,
111, 204–5, 448, 632–33, 819–
20; operation of, 631–32, 696,
705 n. 2, 707–8, 713, 726–27,
752, 788–89; power from, 636;
preparations for, 111; record-
keeping, 111; safety fuses, 448,
489; staff, 111, 521, 594 n. 3, 789,
820; steam engines, 489, 632–33,
696, 780, 788, 814–15; success of,
682; testing room, 632; tests of,
580; underground conductors,
34, 90, 111, 158, 196, 212, 248,
257, 311, 352, 367, 370, 425, 431,
447–48, 475, 482 n. 5, 518, 540,
580, 632–34, 664, 671, 696, 815,
817, 819, 825, 827–30; visitors to,
681, 718, 725, 823–24; voltage
loss, 624–25; voltage regulation,
419, 426, 448, 489, 632; water
supply, 566
Electric lighting generators, 483,
508; armatures, 260, 285, 288,
312, 321, 373, 415, 483, 513, 561,
569, 724, 762 n. 2; British
patents, 530; commutators, 276–
77, 285, 287, 322–23, 513, 526;
design of, 524–25; efficiency of,
355–56, 447, 604, 785 n. 1; for
electric railways, 502; gas engines
for, 681, 716–18; hydroelectric,
703 n. 3; manufacture by Mather
& Platt, 696; patent applications,

9 n. 1, 20 nn. 2–6; patents, 4, 19
n. 1, 20 nn. 2–6, 48–49, 50 n. 1,
57 n. 9, 58 nn. 10–11, 102 n. 9,
103 n. 11, 104 n. 2, 105 n. 5, 107
n. 3, 116 n. 11, 177 n. 4, 178, 288
nn. 5–6, 289 n. 10, 313, 373–74,
393 n. 10, 419 n. 4, 422 nn. 2–4;
pyromagnetic, 396–97, 455;
sparking, 322–23, 526; tests, 51,
356, 358, 701 n. 7; for village system, 722; voltage regulation, 168,
197, 275–76, 285–86, 288, 313,
355, 374, 418–22, 448, 502, 512–
15, 574–75, 577–78
—direct-connected (C dynamo),
102, 337–39, 474, 548, 593, 605,
663; armatures, 48–49, 88–89,
100–101, 103–10, 114, 181, 316,
352, 355, 410–11, 493, 498; commutators, 101, 168–75, 317, 352,
355, 389–90, 498; costs, 101, 184;
coupling, 317–18, 389, 463, 580,
681, 704, 707–8, 724, 814–15;
design changes, 101–5; drawings,
476, 489; efficiency of, 92–93,
101, 168, 173–75, 390; experimental, 4, 88, 100; experimental
costs, 357; heating of, 389; for
Holborn Viaduct, 198, 201, 212
n. 1, 214, 216, 242–43, 258, 276,
300, 304, 323–24, 342, 352–53,
367, 379–82, 391, 410–13, 463–
64, 493–94, 545, 775 n. 2; for
isolated plants, 258; "Jumbo"
as nickname, 413; for London,
59–60, 64, 135, 158, 175, 179,
737 n. 4; manufacture of, 88; for
Milan, 412, 727, 780; for Paris,
4, 13, 61, 65, 70–71, 88, 90, 92–
93, 100–107, 158, 166 n. 1, 168,
179, 225, 258, 317–18, 390, 410,
412, 663, 813, 827–29, 840 n. 8;
patents, 48–49, 102 n. 9; for Pearl
St., 196, 303, 311, 351, 388, 410–
12, 414, 424–25, 431, 447–48,
475, 489, 541, 580, 632, 670–71,
814–15; for Santiago, 412; sparking, 50 n. 1, 88, 101, 104, 114,
126–27, 134, 164, 168–74, 323,
390, 490 n. 1, 503–4, 544; specifications, 838; steam engines for, 8,
101, 104, 174–75, 214, 258, 264–
65, 333, 342, 352, 356, 367, 390,
410, 412, 416, 454, 670–71, 700,
724, 727, 814–15, 828; tests, 88,
101, 103–7, 133–34, 158, 168,
173, 181, 212 n. 1, 355, 426

Electric lighting generators
(*continued*)
—disk dynamo, 99, 260; armatures, 387–88; efficiency of, 51, 387; experimental designs, 4, 8, 17–20, 51, 55–56; for isolated plants, 178, 181, 303; for Paris Exposition, 387; patents, 55–56, 97, 102 n. 9, 103 n. 11
—isolated, 419 n. 5, 435; armatures, 113–14, 160, 178, 181, 303, 333–34, 340, 355, 457, 476, 487, 489, 521, 548; commutators, 113–14, 415, 489, 503–4, 607–8, 622–23; cost of, 415; designs, 89, 112–114, 556; drawings, 476; A dynamo, 840 n. 7; E dynamo, 112–14, 436 n. 2, 518, 531, 548, 592, 838; efficiency of, 181, 241–42, 264–65, 299, 415, 487, 809 n. 1; G dynamo, 114; half-light (B armature), 113, 135 n. 1, 159, 220 n. 1, 245 n. 3, 257 n. 3, 303, 388, 415, 457–58, 522 n. 1, 534 n. 2; H dynamo, 113–14, 570; heating of, 415; K dynamo, 112–13, *115*, 436 n. 2, 447, 476, 489, 503–4, 520, 545, 548, 554, 570, 584, 585 n. 3, 592, 601–2, 605, 607–8, 642 n. 3, 646, 648 n. 14, 663, 669, 672, 703 n. 3, 722 n. 2, 724, 746 n. 1, 750 n. 1, 775 n. 10, 781, 808–9, 838; L dynamo, 112–13, *116*, 393 n. 14, 419 n. 5, 476, 548, 585 n. 3, 592, 607–8, 651, 663, 669, 710 n. 4, 838; new styles, 800; for Paris Opera, 239–40; patents, 281 n. 9, 517 nn. 10–11; R dynamo, 839 n. 1; for ships, 508 n. 3; sparking, 490 n. 1, 607–8, 724; standard designs, 112–14; tests, 89, 126, 157, 164, 299, 355–56, 415, 457, 489; voltage regulation, 242, 277–78, *281–82*, 302–3, 323, 418–19, 514–15, 681, 684–85; Z dynamo, 70, 90, 112–14, *117*, 145 n. 2, 164, 215–16, 256 n. 1, 257 n. 3, 280 n. 6, 302–3, 308, 323, 333, 339–40, 349 n. 3, 381, 415, 419 n. 5, 433 n. 9, 436 n. 2, 476, 487, 520–21, 548, 570, 590, 592, 607–8, 622–23, 636 n. 7, 656 n. 11, 663, 669, 699 n. 5, 710 n. 4, 724 n. 2, 750 n. 1, 775 n. 3, 794 n. 4, 838
—other inventors': Bürgen, 649; Elias, 387; Farmer, 623 n. 3;

Field, 623 n. 2; Fuller, 771 n. 11; Gramme, 4, 19 n. 1, 47 n. 3, 58, 126, 163–65, 189, 260, 272, 300, 309 n. 11, 321, 386–87, 534 n. 2, 628; Hopkinson, 246 n. 9, 809 n. 1; Maxim, 534 n. 2, 684–85; Pacinotti, 47, 260, 308, 321, 387; Siemens, 4, 47 n. 3, 51, 260, 272, 299–300, 309 n. 11, 312, 316, 338, 340, 355, 379–80, 387, 584, 592, 608, 623 n. 1, 649, 704–5, 744, 777; Siemens & Halske, 762 n. 2. *See also* Patent applications
Electric lighting incandescent lamps: with absorption chamber, 346–*47;* A-lamps, 24 n. 4, 86, 113, 139, 203 n. 10, 212, 235, 257 n. 3, 304, 340–41, 393 n. 14, 401, 465, 522 n. 1, 545, 587, 600, 604, 627, 637, 673, 699 n. 5, 750 n. 1, 777, 794 n. 7, 840 n. 11; artificial fiber filaments, 447, 543, 571, 582, 681, 692; bamboo filaments, 22, 39, 42, *44*–45, 139, 234, 806; battery-powered, 159 n. 3; B-lamps, 24 n. 4, 86, 113, 139, 190, 236, 257 n. 3, 258, 280 n. 6, 303, 339–41, 388, 401, 457, 587, 627, 750 n. 1, 777, 794 n. 7; for Britain, 580; British patents, 530; carbonizing molds, 5, 38, 42, 139, 176, 187, 331 n. 1, 582, 639–40, 697; chemical treatment, 687–88, 692; coatings, 5, 23 n. 2, 284, 313; colored, 681; combined with arc light, 350 n. 7, 448; compared to arc lights, 554, 636–37; double chamber, 346, *348;* for dynamo tests, 504; efficiency of, 86 n. 1, 90, 166 n. 6, 175, 180, 234–35, 253–54, 263, 287 n. 1, 297, 303–5, 342, 356, 380, 390, 587, 600–601, 635, 776–77, 785 n. 1, 800; electrical carrying, 235, 289, 446–47, 549–50, 582 n. 4, 589–90, 681, 685–87, 698, 779, 802; electric light law, 287 n. 1; filament search, 282–83, 479 n. 2; fixtures, 4, 32 n. 4, 558 n. 11, 581, 598, 604, 649, 652 n. 4, 830–31; glass bulbs, 42–43, *45*–46, 54, 262, 338, 590, 640, 685–87, 700; high candlepower, 313, 356, 370, 380, 394, 448, 464–65, 510–11, 546, 572, 600, 777; high resistance, 197, 279, 287 nn. 1–3, 297, 304–5, 318–21, 325–26, 330,

337, 357, 370, 386–87, 391, 532, 542, 545, 585, 587, 627, 635–36, 650–51; hydrocarbon treatment, 5, 23 n. 2, 26–27, 38, 39, 52 n. 6, 338, 572, 582, 602, 687–88, 690, 721; inert gas atmosphere, 234, 338, 603, 681, 729; lamp life, 22, 35, 51, 61, 90, 93, 180, 182, 196–97, 235, 238, 252, 263, 303, 305, 322, 324, 337–39, 346–47, 356–57, 376, 390–91, 438, 532, 587, 602 n. 2, 604, 624, 662, 774, 776–77; lead-in wires, 11 n. 8, 16, 27 n. 3, 38–39, 42–43, 141, 165, 206, 252 n. 2, 260, 289, 338, 342, 391, 432 n. 4, 520 n. 10, 572–73, *577*, 639–40, 697–98, 748, 778, 802, 806–7; manufacturing process, 5–6, 42–46, 142, 286, 364–65, 603, 662, 680–81; multiple fiber filaments, 576, 688–89, 691, 721, 778, 806–7; other plant fibers, 90; paper filaments, 39, 318, 320, 438; patents, 25 n. 9, 26–27, 43 nn. 3–4, 46 n. 2, 49, 50 n. 6, 137 n. 4, 141 n. 8, 176, 234–35, 287 nn. 2–3, 288 n. 8, 315–22, 325–26, 347 n. 2, 365 nn. 1–2, 392 n. 5, 432 n. 4, 494, 685–92, 697–98, 721, 778 n. 4, 779 n. 1; platinum filaments, 46 n. 2, 137 n. 4, 159 n. 3, 279, 316, 321, 599 n. 6; pressed plumbago, 5, 9, 22, 26–27, 38; residual gases, 43, 386, 579, 603, 612–19, 681, 687, 698, 779; resistance lamps, 800–801; reusable bulb, 5, 285; for scientific community, 51, 80 n. 1, 95, 159, 200, 308 n. 1, 536 n. 21; small, 777; sockets, 32 n. 4, 38–39, 44 n. 5, 49, 142, 187, 197, 262–63, *271*, 306, 339, 349 n. 3, 380, 490, 505 n. 8, 521–22, 527, 581, 590–*91*, 601, 739–40, 774; spiral, 303, 319, 331 n. 4, 346, 357, 370; for street lighting, 313, 356, 370, 380, 394, 448, 464–65, 510–11, 546, 572, 636, 642 n. 3; tests, 67–68, 72 n. 8, 86, 139, 234–35, 252 n. 2, 490, 511, 545, 602 n. 1; vacuum seal, 5, 25 n. 11, 43, 46, 54–55, 67–68, 176, 235, 252, 318–19; for village plants, 580, 600–601; voltages, 24 n. 4, 81 n. 2, 213 n. 4, 235, 346, 489, 505 n. 8, 587, 600, 623–25, 774, 776–78; wood filaments, 330–31.

See also Edison Lamp Co.: lamp factory; Patent applications —other inventors': Crookes, 263–64; Freeman, 818, 828, 830; Goebel, 520 n. 10, 819, 830; King, 207; Lane-Fox, 145, 166n. 6, 191 n. 2, 259–60, 357, 438, 462, 532; Maxim, 35, 60 n. 2, 89, 95 n. 2, 121, 142 n. 2, 145, 160, 166 n. 6, 191 n. 2, 244, 272, 357–58, 380–81, 438, 532; Siemens, 531 n. 3; Swan, 25, 60 n. 2, 95 n. 2, 121, 142 n. 2, 144–45, 160, 165, 206–7, 244, 259–60, 320, 326, 357, 359, 381, 388, 416, 437–38, 459, 508, 532–33, 546–47, 585, 597, 627, 657

Electric lighting isolated plants, 358, 601; Aitkin, Sons & Co. (New York City), 646; *Alaska*, 656; American Bank Note Co. (New York City), 646; Baldwin Locomotive Works (Philadelphia), 213; Barker's home (Philadelphia), 6, 51; Bennett yacht, 508 n. 3; Berlin, 441, 724; Bijou Theater (Boston), 681, 829; Bon Marché (Paris), 163–64; Boston Fair, 599 n. 9; Britain, 113, 215–16, 244, 258–59, 299–300, 391, 394–95, 490 n. 2, 649–51, 738; British colonies, 191; British Parliament (London), 595, 656; Brünn (Austria), 441, 700, 746; Brush's, 594–95; Brussels (Belgium), 161, 164, 442; Bucharest (Romania), 441; Budapest (Austria), 441; Cafe Biffi (Milan, Italy), 402 n. 2, 441; California, 273; Canada, 427 n. 3, 565–66; Chicago, 637, 681; Chicago Academy of Music, 637; *City of Rome*, 198; *City of Worcester*, 277; *Columbia*, 185 n. 2, 277, 408 n. 5, 431; compared to central stations, 635; cost of operating, 592; Covent Garden (London), 388, 401; Croydon (London), 548 n. 5; Crystal Palace (London), 113, 241–43, 264–65, 322–23, 334, 379, *406*, 457–58; for docks, 431; Draper's home (New York City), 531; dynamos for, 584; Edison's home (New York City), 597; Edison's opinion of, 161, 191; Europe, 161, 180, 187 n. 4, 191 n. 2, 682; Ferrare (Italy), 441;

Finland, 349; Finlayson & Co. (Finland), 349 n. 3; Frankfurt (Germany), 440; geographical distribution of, 833, 835–37; Germany, 334 n. 1, 349, 557 n. 9, 753, 763; Hamburg (Germany), 441; Herald building (New York City), 646, 672, 817, 828–29; Hinds, Ketcham & Co. (New York City), 161, 253–54; Holborn Restaurant (London), 596, 656; Hotel Everett (New York City), 646; for hotels, 631, 646; Italy, 349, 455; Jaffray & Co. (New York City), 646; J.V. Farwell & Co. (Chicago), 637; King Philip Mill (Fall River, Mass.), 598; Krasnaplosky's Grand Cafe (Amsterdam), 442; Krupp Works (Germany), 349; Lancashire (Britain), 548 n. 5; Léon's home (Paris), 131; Lyons (France), 161; Max Ams Preserving Co. (New York City), 646; Maxim plants, 358; Metz (France), 334 n. 1; Milan (Italy), 441, 607; Mill Creek Distilling Co. (Cinncinnati), 699; for mills, 112, 188 n 1, 274 n. 2, 307, 349 n. 3, 430 n. 3, 503, 605 n. 2, 631, 758–59; Mills Building (New York City), 597; Morgan's house (New York City), 750; Mulhouse (France), 189, 334 n. 1; Munich (Germany), 724; Newburgh, N.Y., 112, 188 n. 1; New England, 509 n. 12; numbers of, 111–12, 833–37; orders for, 207, 219, 256, 708; Paris, 163–64, 724; Paris Exposition, 101, 114, 135, 145 n. 2; Paris Opera, 166 n. 3, 179–80, 190, 209, *211*, 239–40, 245 n. 1, 252 n. 1, 303; Pennsylvania Railroad, 136; *Pilgrim*, 672; prices of, 839 n. 4; *Queen of the Pacific*, 433 n. 9; for railway stations, 334 n. 1, 441, 595, 775 n. 10; Rheims (France), 163; Rome (Italy), 441; Rotterdam (Holland), 102, 442; Savoy Theatre (London), 357, 360 n. 29, 380, 393 n. 16; *Servia*, 584; for ships, 185 n. 2, 198, 277, 358, 433 n. 9, 508 n. 3, 584, 656, 672; Siemens's, 592, 594–95; 65 Fifth Ave. (New York City), 23, 93, 823, 827–28; South America, 126, 127 n. 4, 156–57, 168, 840

n. 7; St. Petersburg (Russia), 349 n. 3, 435, 441; steam engines for, 340, 584, 646; and storage batteries, 244, 258–59, 280, 758–59; Strasbourg (France), 164, 240 n. 2, 333–34, 441; Stuttgart (Germany), 441; Styria (Austria), 442; Swan's, 584, 594–95; Tammerfors (Finland), 349 n. 3, 435, 441; Teatro alla Scala (Milan, Italy), 350 n. 4, 401, 441; for theaters, 357, 380, 441, 637, 681, 700, 724, 801; Thurber Co. (New York City), 646; for train lighting, 248 n. 13, 441; types of, 833–34; underground conductors for, 672; for universities, 334 n. 1; Vanderbilt mansion (New York City), 6, 67, 567, 815–16, 827–29; Vienna (Austria), 161, 164, 724; voltage loss, 623–24; voltage regulation, 197, 608, 681; Waterloo station (London), 595, 775 n. 10; Winsted, Conn., 605 n. 2; wiring for, 112 n. 7. *See also* Electric lighting generators: isolated

Electric motors, 482 n. 1, 676 n. 4; armatures, 431; for electric railways, 476, 634, 733; for elevators, 634; five-horsepower, 8; Gramme generator as, 127 n. 3, 155 n. 4; in lamp factory, 236; for lathes, 634; for London, 391; for meters, 808 n. 9; for mining, 199; at Paris Exposition, 70–71; for railways, 407, 431; for sewing machines, 70, 634; for voltage regulation, 168, 418, 421–22, 513, 515, 574; Z dynamo as, 333, 339–40, 381

Electric pen: in Australia, 33; business, 519 n. 8, 33; at Paris Exposition, 64, 71, 381; royalties, 36 n. 3; at South Kensington Museum, 438

Electric Railway Co. of the United States, 409 n. 8, 733 n. 4

Electric railways, 75 n. 36, 272, 523 n. 6; article, 561 n. 6; for Britain, 708; dynamos for, 502, 697; for elevated lines, 407; elevators for, 366–67; experimental costs, 185, 292 n. 1, 506 n. 4; experimental line, 249, 370, 407, 431, 447, 476, 481, 506, 634, 732, 828; Field's, 733 n. 4; financing of, 292; for freight, 384 n. 1, 432, 447, 476, 481, 812–13, 828; locomotive,

Electric railways (*continued*)
304, 384, 407, *409*, 506; motors,
20 n. 6, 313, 384 n. 1, 385 n. 5,
476, 634, 733; for passengers, 384
n. 1, 431, 506; patent applications, 449–50; patent interference, 409 n. 8, 482, 609; patents,
449, 451 nn. 5 & 7, 486, 733 n. 4;
photographs of, 481; safety of,
732; Siemens's, 407, 432, 449–
50, 744; speed of, 481; Sprague's,
471 n. 13; in Switzerland, 557
n. 10; tests, 476; third rail, 732;
Van Depoele's, 733 n. 4
Electric Storage Battery Co., 519
n. 3
Electric Tube Co., 3, 377 n. 3, 634,
659, 828; absorbed by Machine
Works, 665; finances, 725; incorporation of, 665; location of, *2*;
manufacturing for Britain, 734;
and Milan station, 727; moves to
Brooklyn, 665; officers, 35, 111
n. 3; and other central stations,
664; and Pearl St. station, 158,
196, 664; shareholder, 266 n. 7;
workforce, 665
Electro Dynamo Co., 542 n. 3
Electromotograph: battery, 51, 71;
gong, 71; magnetic, 71; principle,
73 n. 17; relay, 71, 135; telephone
receiver, 51 n. 1, 71, 381, 416,
438, 459, 640 n. 2, 832
Elevators, 366–67, 407
Eliot, George, 745
Emerson, Ralph Waldo, 332
Emery, C. E., 817–18, 828, 830
Engineer, 206–7
Engineering, 157, 202
Estevez, Mr., 789, 797
Etheric force, 71
Evangeline, 86 n. 9
Evening Commercial, 566
Everett, Samuel H., 648 n. 14
Everett House, 446, 482 n. 4
Excelsior, 350 n. 4
Exhibitions. *See individual exhibi-
tions*
Exposition Internationale de l'Élec-
tricité (Paris), 359, 401, 535 n. 10;
arc lighting at, 120–22; catalog,
129; Colombo at, 678 n. 14; Com-
mission, 82, 145; Crompton's ex-
hibit, 658 n. 12; electric lighting
at, 61, 65, 120–22, 144–45, 159–
60, 210 n. 1, 240, 334 n. 1; electric lighting tests, 164, 189, 357;

and electric light investors, 81–
82, 128, 189; Elias dynamo at,
393 n. 9; galvanometers at, 701
n. 7; Lane-Fox's exhibit, 143,
145; Maxim's exhibit, 82, 89, 132,
142–43, 145, 160, 165, 189, 198;
Pond Indicator at, 77; press, 95–
96, 121–22; prize juries, 145, 189,
221, 344 n. 11; prizes, 220–22,
225–26, 249, 331; Siemens dynamo at, 299; Siemens's railway
at, 409 n. 8, *410;* Swan's exhibit,
82, 142–43, 145, 160, 165, 189,
198
—Edison's exhibit, 7, 60–61, 64–
65, 69–71, 77, 81–82, 95–96,
120–22, 131–33, 142, 144–*46*,
159–60, 216, 301, 361 n. 2, 458,
483, 493, 500 n. 10, 678 n. 14;
dynamos, 4, 13 n. 5, 61, 70–71,
82, 88, 101, 114, 135, 145 n. 2,
225, 840 n. 7; electric pen, 71,
135; electromotograph, 71, 135,
161; etheric force, 71, 135; fixtures, 70, 82; lamp exhibit, 5,
42–46, 70–71, 80, 142; meters,
70; motors, 70–71; odoroscope,
71; opinons of, 250; phonograph,
64, 88; safety fuse, 70, 142; telegraph instruments, 27, 64, 71, 88,
120–22, 135, 182; telephone, 64,
88, 135, 161

Fabbri, Egisto, **156,** 161, 221, 403
n. 5; agreement with, 706–7; and
Crystal Palace Exhibition, 216;
Edison's opinion of, 258, 261;
letters from TAE, 156–58; loan
to Edison, 314; and Oriental
Telephone Co. stock, 798 n. 2; at
Paris Exposition, 132, 409 n. 8
—and electric lighting, 157–58, 272,
358; in Australia, 336; in Britain,
135, 239 n. 4, 245, 261, 279, 296,
312–13, 321–22, 333, 342–43,
369, 393 n. 16, 457, 459–60, 466,
479, 547, 652 n. 7; in British
colonies, 461; Edison Electric of
Europe, 189 n. 1, 682, 700, 706–
7; in Europe, 191–92, 200, 260,
369; European companies, 89,
189; foreign isolated plants, 200,
459–60; Isolated Co., 274; Pearl
St. station, 538–39; in South
Africa, 460; in South America,
156–57; in Sweden, 460
Fabbri, Ernesto, **156,** 274 n. 2

Fabbri & Chauncey, 157 nn. 1–3,
509 n. 5
Falertoz, Mr., 129
Faraday, Michael, **807**
Farwell, John Villiers, **637**
Farwell (J.V.) & Co., **637**
Faure, Camille, **247 n. 10,** 518
Faure Electric Accumulator Co.,
244, 258, 263
Favier, Paul-André, **440**
Favre, Piere Antoine, **616**
Fiction, 291
Field, John W., **279**
Field, Marshall, **559**
Field, Stephen Dudley, 409 n. 8,
733 n. 4
Finlayson & Co., 349 n. 3
Fisk, Harvey, 360 n. 10
Fisk and Hatch, 358
Flagler, John H., **274**
Fleming, John Ambrose, **457;** and
Edison Electric, Ltd., 457, 467,
469, 656–67; and Holborn
Viaduct station, 466, 548 n. 4,
625 n. 2, 775 n. 2; insulating
compound, 432 n. 2
Fleming, Mr., 388
Florida. *See* Edison, Thomas Alva:
Florida vacation
Flower, Roswell, **786**
Fodor, Etienne de, 349 n. 2
Food preservation, 6, 30, 40–41,
75–76, 91, 92
Foote, Charles B., 360 n. 11
Forbes, George, **222,** 225
Force, Martin, **165;** and direct con-
version experiments, 397 n. 1;
and dynamos for telegraphy, 623
n. 2; and European lighting companies, 293; and Hamburg office,
349; and lamp experiments, 165,
748, 779, 802, 808; and meter experiments, 808; and Paris Exposition, 122 n. 1; and storage battery
experiments, 741 n. 2
Ford, Henry, 412
Fox, Edwin, 96 n. 1, 483, 498
France: electrical manufacturing in,
176, 402, 440; Minister of Fine
Arts, 210 n. 1; Minister of Posts
and Telegraphs, 331; U.S. Consul, 15. *See also* Electric lighting;
Electric lighting central stations:
Paris; Electric lighting generators: direct-connected (C dynamo): for Paris; Electric lighting
isolated plants; Exposition Inter-

nationale de l'Élecricité; Paris; Paris Opera; Société Électrique Edison: Ivry-sur-Seine factory

Frank Leslie's Illustrated Newspaper, 291

Franklin Institute, 73 n. 22, 213 n. 1

Frederick the Great, 745

Freeman, Frank, **145**, 225

Freeman, Walter K., 818, 828, 830

Frelinghuysen, Frederick, 332 n. 2

The French Revolution, 745

Fuller, James B., **637**

Fuller Electrical Co., 59 n. 4, 638 n. 5

Gallaway, Robert Macy, 407 n. 1

Galleria Vittorio Emanuele, 402 n. 2

Gallia, 191, 220, 326

Gambetta, Léon-Michel, 96, **144**, 144 n. 4, 161

Garbi, Alexander, **401**, 441

Garfield, James, 15 n. 4, 61 n. 3, 194 n. 3

Gaslighting, 631; at Bergmann's, 306; in Britain, 581, 635; compared to arc lighting, 554; compared to incandescent lighting, 34–36, 203 n. 10, 233, 239, 245 n. 6, 279, 297, 305–7, 311, 427 n. 1, 456 n. 3, 496, 508, 510–11, 533, 545, 559, 566, 581, 606 n. 9, 635, 642, 644, 671–72, 681–82, 711–14, 736; fixtures, 558 n. 12; in London, 36, 279, 305–7, 496; in Milan, 456 n. 3; in New York City, 828; in Paris, 161–64; for railways, 280; in Roselle, 642; in theaters, 681, 701 n. 6

Gas Light Journal, 279

Gasoline engines, 323, 337, 534 n. 1, 681, 716–18, 823, 827–28

Gebrueder Sulzbach, **763**

Gellert, 809

General Electric Co., 376 n. 2, 568 n. 6

Generators. *See* Electric lighting generators

Germany: electrical manufacturing in, 440, 557 n. 9, 792; Patent Office, 763. *See also* Electric lighting: in Germany; Electric lighting isolated plants

Gibbs, George, 740 nn. 4–5, 741 n. 2

Gilbert, William, 360 n. 20

Giovine, Pietro, **21**

Gladstone, William, 439 n. 5

Gladstone, William Henry, **438**

Glass, Louis, 231–33, 274 n. 1, 567 n. 3

Gmelin, Leopold, **617**

Goddard, Calvin, **62**, 351, 563 n. 2; and central stations, 698; goes to Europe, 566; and incandescent lamps, 187; and isolated plants, 187, 274; and Pearl St. station, 111, 538–39; as secretary of Edison Electric, 205–6; as secretary of Isolated Co., 710 n. 10; as vice president of Edison Illuminating Co. of New York, 206 nn. 2 & 4

—letters: from Insull, 538–39; from TAE, 642; to TAE, 62, 205–6

Goebel, Henry, **518**, 819, 828

Goerck St. *See* Edison Machine Works

Gold & Stock Telegraph Co., 15, 28 nn. 4–5, 351, 473, 628

Gold & Stock Telegraph Co. of California, 567 n. 3

Gordon, James, **532**, 656

Gould, Jay, **407**, 804 n. 1

Gouraud, George, **14**, 91, 157 n. 2, 369, 468–69; agreements with, 30, 394–95, 524 n. 1; British patent, 266 n. 4; and Crystal Palace Exhibition, 217 n. 5, 363, 415; as Edison's agent, 15, 25, 63–65, 117–19, 191, 262; and food preservation, 6, 30, 41, 75–76, 92; and ore milling, 76; in Paris, 30 n. 1; power of attorney, 30 n. 1; private secretaries, 15 n. 5, 36 n. 1; relations with, 261–62

—and electric lighting: in Australia, 474; in Britain, 64, 92–93, 249–50; in British colonies, 89, 459–61, 493, 546; dynamos, 391; Edison Electric of Europe, 119; in Europe, 459–61; foreign rights, 219; isolated plants, 89, 191, 207, 219, 255–56, 264, 394–95, 460; lamps for, 80; lawsuits, 157; manufacturing, 60 n. 2

—and telephone, 6, 117–18, 198, 368 n. 6, 459; British rights, 4; companies, 497; Edison's British stock, 13–14, 76–77, 90, 93–94; Edison's reversionary share, 63, 93, 117–18; foreign rights, 219; Oriental Telephone Co., 63–64, 483; stock, 460

—letters: from Insull, 63–65, 75–77, 92–94; to Insull, 248–49,

366–67; from TAE, 30, 94–95, 117–19, 192–93, 219–20, 255–56, 394–95; to TAE, 13–14, 94 n. 2

—telegrams: from TAE, 25, 119 nn. 2–3, 120 n. 9, 191, 394; to TAE, 119 nn. 2–3, 120 n. 9

Gower, Frederick Allen, **14**, 118

Gower-Bell Telephone Co. of Europe, 14 n. 3

Graham, Thomas, **613**, 618

Graham School for Young Ladies, 199

Gramme, Zénobe: British patents, 318, 321, 386–87; Canadian patent, 47 n. 3; European patents, 47 n. 3; U.S. patents, 4, 19 n. 1, 47 n. 3, 59 n. 4

Gramme Electrical Co., 59 n. 4, 508, 533, 538 n. 1, 542 n. 3, 723 n. 1

Grange (Patrons of Husbandry), 538 n. 1

Graves, W. A., 799 n. 4

Great Northwest (balloon), 159 n. 3

Grecian Monarch, 414 n. 16

Green, Norvin, **27**

Greenfield, Edwin, 111, 519 n. 9, 730

Green's (S.W.) Son, **645**

Grévy, François Paul, **131**

Griffin, Stockton, **431**

Grigg, F. R., **14**, 367 n. 2

Grower, George, 202, **490**, 503, 663

Guion, Stephen, **530**

Hagenbach, E., 166 n. 6

Haid, Alfred, 5, **22**, 99, 140 n. 1, 617

Haines, Henry, 37 n. 1

Hamilton, George, **622–23**

Hammer, William, **23**, 73 n. 22; and Crystal Palace Exhibition, 363, 466, 468; and dynamos, 113–14, 277, 324, 413, 457, 464, 544, 607–8, 840 n. 6; as engineer of British company, 497–98, 656; and Holborn Viaduct station, 306, 312, 458, 464, 466, 468, 493; and incandescent lamps, 44 nn. 5–6, 45, 236 n. 7; Johnson's opinion of, 467–68, 483; letter from TAE, 607–8; and Paris Exposition, 122 n. 1; and St. Louis World's Fair (1904), 46 n. 1; and Swan lamp, 459; and voltage regulation, 419

Hampson, Edward & Co., 157 n. 3

Hanford, Thomas, 312, 450 n. 1, **462–63**, 483

Hannington, Charles, **218**, 555
Harjes, John, 190 n. 2
Harper's Bazaar, 291
Harper's Weekly, 291
Harper's Young People, 291
Harrison, Mr., 797
Harrison, N.J. *See* Edison Lamp
  Co.: lamp factory, Harrison
Hatch, Daniel B., 360 n. 11
Hatch and Foote, 358
Hay, John, **61**
Hay, William Montagu, **523**
Hazard, Rowland, **507**
Hazen, William, **159**
Hearle, E., **565**
Henderson, John C., 408 n. 5
Hennis, Charles, **293**
Herresoff Manufacturing Co., 586
  n. 2
Herter, Christian, 67 n. 2, **332**, 815–
  16, 828–29
Herz, Cornelius, **58**, 97, 121–22
Hewitt, Abram, 230 n. 2, **786**
Hickman, George, 37 n. 3, **125**
Highland Park Hotel (Aiken, S.C.),
  400
Hinds, Ketcham & Co., 253–54
Hipple, James, 122 n. 1, **293**, 402
History of Frederick the Great, 745
Hitt, Robert, **81**
Hoffman, James, **564**–65
Hoffman House Hotel, 358
Holborn Viaduct. *See* Electric light-
  ing central stations: Holborn
  Viaduct
Holloway, James, **458**
Holmes Burglar Alarm Co., 112 n. 7
Holzer, Frank, **139**
Holzer, William, **139**, 294 n. 2;
  Lamp Co. shares, 431; and lamp
  experiments, 335, 346; as manu-
  facturing superintendent, 313,
  430, 478, 553, 600, 627, 661, 800;
  patents, 611 n. 6; vacuum experiments, 139, 235, 252 n. 2
Hood, John, 312, **318**, 363, 463–64,
  466, 545
Hopkinson, John, **243**; and British
  patents, 462, 495; and dynamos,
  338, 344 n. 7, 457, 809 n. 1; and
  Edison Electric, Ltd., 466, 469;
  and Holborn Viaduct station,
  466, 548 n. 4, 595, 625 n. 2; Johnson's opinion of, 466; report on
  Board of Trade rules, 711–14;
  and Siemens patents, 328 n. 27,
  380, 387–88; and telephone, 494;
  three-wire system, 753, 794, 805

Hornig, Julius, 111, 428 n. 7, 429
Houston, Edwin, 273 n. 4, 484 n. 4
Howard, Henry, 759 n. 1
Howell, John, **124**, 188 n. 2; dynamo
  tests, 358, 483; lamp experiments,
  331 n. 2; lamp tests, 23 n. 2, 38
  n. 1, 39 n. 1, 68 n. 2, 87 n. 3, 124,
  234, 252 n. 2, 304 n. 4; reminiscences, 331 n. 1
Howell, Wilson, 642 n. 3
Hubbard, Gardiner, 230 n. 2
Hughes, Charles, **41**, 85 n. 2, 99;
  and dynamos, 513; and electric
  railways, 91, 185 nn. 1 & 4, 304,
  370, 384, 408 n. 5, 447, 476, 481,
  506 n. 4, 697 n. 2; food preservation experiments, 6, 40–41, 91
Hughes, David, 95 n. 1, **121**, 227
  n. 10, 278
Hugo, Victor, 138, 745
Hunter, John, **618**–19
Husbands, José, **494**
Hussey, E. P, **531**

Illustrated News, 383 n. 4
Inflexible, 147 n. 6
Insulite, 430
Insull, Joseph, **797**
Insull, Samuel, **10**, 401 n. 2, 507,
  634, 658 n. 9, 752, 785 n. 1, 788
  n. 15; at Armington & Sims, 670;
  and Batchelor's finances, 351,
  473–75; and British telephone,
  14; and dynamos, 513, 724 n. 2; as
  Edison's business manager, 199,
  217–18, 230 n. 2, 275, 292, 312–
  13, 369, 430, 446, 482, 554, 562,
  660, 810 n. 2, 827–28; and Edison's reversionary share, 118; as
  Edison's secretary, 4, 6, 27, 35,
  37 n. 1, 52 n. 12, 78 n. 7, 91, 157,
  199, 216 n. 1, 260, 265, 280, 283,
  298, 304, 326, 334 n. 3, 339, 362
  n. 1, 363 n. 1, 392, 395, 403 nn. 5–
  6, 416, 433, 446, 453, 457, 459,
  470, 485, 491, 508, 518, 534, 556,
  560, 592, 592 n. 4, 593 n. 2, 605,
  642, 652 n. 1, 680, 697, 730, 744,
  793, 804; and electric lighting, 59,
  101, 246 n. 8, 274 n. 2; and European lighting companies, 210 n. 3;
  and isolated lighting, 731 n. 3; and
  Johnson's residences, 789, 797–
  98; and manufacturing shops, 10,
  35, 660–61, 663–64, 754, 784,
  798; in Menlo Park, 33, 292, 446–
  47, 476, 675; and Milan central
  station, 783 n. 5; and Paris Expo-

sition, 46 n. 4; and patents, 554–
  55, 599 n. 6; and Pitt Edison, 749;
  and Pond indicator, 79 n. 11; in
  Rahway, 371; in Roselle, 371; and
  Roselle station, 643; and steam
  engines, 333 n. 2, 775 n. 10; and
  village system, 809; on yacht trip,
  579
—letters: from Batchelor, 700; to
  Batchelor, 351–53, 473–76, 668–
  75, 706–9, 808–9; to Burrell,
  719; to Chipman, 298; to Draper,
  720; from Dyer, 10; to Eaton,
  217–18, 725–26; to Pitt Edison,
  456; to Glass, 231–33; to Goddard, 538–39; to Gouraud, 63–
  65, 75–77, 92–94, 248–49, 366–
  67; from Johnson, 239 n. 4; to
  Johnson, 369–71, 430–32, 481–
  82, 788–89, 796–98; to Kingsbury, 33–36; to Low, 718, 795;
  to McCrory, 732–33; to Moore,
  643; to Olrick, 695–96; to
  Twombly, 67; to Villard, 292
—telegrams: from Mary Edison,
  400, 417; to Mary Edison, 417;
  from TAE, 586; to TAE, 585–86
International Congress of Electricians. *See* Electrical Congress
  (Paris)
Iron Age, 518
Isolated lighting. *See* Edison Co. for
  Isolated Lighting; Electric lighting isolated plants
Italian Royal Opera, 393 n. 16
Italy. *See* Electric lighting: in Italy;
  Milan, Italy

Jaffray, Edward Samuel, 648 n. 14
Jaffray (E. S.) & Co., **646**
Janon, Camille de, **683**
Jehl, Francis, **21**; article by, 746;
  and Brünn lamp factory, 701 n. 6;
  and conductors, 21; and Crystal
  Palace Exhibition, 404–5; and
  dynamos, 19 n. 1, 52 n. 8, 101,
  104 n. 1, 108 n. 4, 127 n. 5, 168,
  389, 701 n. 7; and electric lighting
  in Milan, 402 n. 2; *Engineering* article, 202; and Holborn Viaduct
  station, 312, 493; letter from
  TAE, 21; letter to TAE, 436–39;
  meter pamphlet, 202; and meters,
  6, 9, 21, 155 n. 7, 156 n. 10, 202,
  305, 333, 388, 436–37; reminiscences, 22 n. 2, 102 n. 2, 539 n. 1,
  790 n. 6; and testing room, 663;
  and theater lighting, 701 n. 6

Jensen, Peter, **259**–60

Johnson, Edward, **31**, 34–35, 138, 189, 219, 418 n. 2, 584–85, 591, 635 n. 1, 675, 710 n. 8; agreements with, 31–32, 581, 674; British patents, 266 n. 4; and Crystal Palace Exhibition, 198, 208 n. 8, 216, 241–44, 264–65, 312, 322–23, 362–63, 379–82, 404–5, 415–16, 438, 458, 464–65, 482 n. 1, 493, 497; as Edison's agent, 197–98, 312, 362 n. 1; goes to Britain, 64, 191, 197, 248–49; and Oriental Telephone Co., 483, 796; in Paris, 466; patent assignments, 430; and Pennsylvania Railroad, 136; personal finances, 797, 809; power of attorney, 454–55; residences, 789, 797–98; returns from Britain, 483, 497, 530, 581, 753; and telephone, 93–94, 342, 494; telephone transmitter patent, 794

—and electric lighting: article, 358; Bergmann & Co., 4, 31–32, 581, 665, 674, 700; in Britain, 89, 223, 258–65, 311–12, 595–96, 774–75, 793, 796; in British colonies, 459; British patents, 197–98, 200, 311, 316, 318–19, 495, 529–30, 753; central stations, 713; dynamos, 116 n. 12, 158 n. 5, 201, 214, 241–43, 264–65, 317–18, 333, 352, 367, 391, 415, 457, 492, 526, 544, 554, 570, 808; Edison Electric, Ltd., 709; Electric Lighting Act, 496–97; electric railway, 828; in Europe, 459; Holborn Viaduct station, 92, 198, 243–45, 312, 380, 382, 458, 463–66, 469, 478, 493–94, 608, 624, 828; incandescent lamps, 62; isolated plants, 136, 207, 244, 256 n. 1, 264, 731, 750 n. 1; Lamp Co., 10 n. 3, 32 n. 2, 348 n. 5, 353 n. 3, 661; manufacturing, 660–61, 665; pamphlet, 337–38; Paris Opera, 240; Pearl St. station, 604, 655–57, 671, 815; safety fuse, 145; Swan's claims as lamp inventor, 206–7, 259–60; three-wire patents, 805; as vice president of Edison Electric, 708; village system, 794, 809; voltage regulation, 275–76

—letters: to Eaton, 771–73; from Insull, 430–32, 481–82, 788–89; to Insull, 239 n. 4; from Sprague, 655–57; from TAE, 136, 206–7, 257–65, 275–80, 296–97, 304–8, 317–26, 337–43, 355–59, 386–92, 414–16, 454–55, 483–84, 526, 793–94; to TAE, 200 n. 1, 241–45, 268 n. 20, 269 n. 24, 299–301, 329 n. 27, 379–82, 404–5, 457–70, 491–98, 529–30, 544–47; from Waterhouse, 268 n. 20

—telegrams: from TAE, 200–201, 238–39, 256 n. 1, 297, 309 n. 11, 333, 401, 499 n. 5; to TAE, 296, 499 n. 5

Johnson, John Henry, 316, 530 n. 1

Johnson, Margaret Kenney (Mrs. Edward), 371, **457**, 657

Johnson, Matthey & Co., 11 n. 8, 176, 519 n. 3, 601

Journal of Chemistry, 290

Journal of the Chemical Society of Newcastle, 259

Jumbo dynamo. *See* Electric lighting generators: direct-connected (C dynamo)

Kalakaua, David Laamea (King), 96, **160**–61

Karsten, Karl Johann Benhard, **612**

Kean, John, **786**

Kelly, James Edward, **249**

Kempe, Harry Robert, **135**

Kendall & Co., **507**

Kennard, Howard John, 250 n. 4

Kenney, Margaret, 470 n. 3

Kenny, Patrick, 73 n. 13, **451**, 629 n. 3

King, E. A., **207**

King, Samuel Archer, 159 n. 3

Kingsbury, John, **33**–36

Kline, William, **518**

Kolbe & Lindfors, 11 n. 8

Koster & Bial's, 138 n. 1

Krasnapolsky, Adolf Wilhelm, **442**

Krizik, Franz, 122 n. 8

Kruesi, John, **111**, 294, 660; and Electric Tube Co., 664–65; insulation experiments, 507, 518; and Machine Works, 664; and street safety fuse, 489; as superintendent of Electric Tube Co., 3, 111 n. 3, 196; and underground conductors, 22 n. 2, 90, 111, 196, 257, 352, 370, 426, 431, 475, 518, 538, 541 n. 1, 634, 672

Krupp, Alfred, 350 n. 6

Kundt, A., 166 n. 6

Laboratory (Menlo Park), 463; aluminum experiments, 480 n. 2; Barker at, 51; central station determinations at, 376; chalk cylinder production, 640 n. 2; chemical equipment, 493 n. 3; closes, 675, 680, 697, 708, 831; distribution system experiments, 794; dynamo experiments, 3–4, 19 n. 1, 51, 88, 100, 215 n. 2, 303, 487; Edison at, 7, 197, 313, 364–65, 369–70, 397 n. 1, 432, 446–48, 476, 506, 579; Edison Electric property at, 554; electric railway experiments, 313; handstamp, 508 n. 1; and lamp factory, 7, 8, 99, 517; meter experiments, 202; ore milling experiments, 767–68; and Paris Exposition, 7; policy on visitors, 255; Porter at, 4; reports, 7, 8; Sprague at, 471 n. 13; staff, 91, 99, 447, 467, 470 n. 6, 483, 484 n. 4, 500 n. 10, 523 n. 6, 643, 680, 697; stamp mill, 313, 370, 432; steam engines, 99, 327 n. 6, 556; tools and machinery, 99, 708; voltage regulation experiments, 313, 608

Laboratory (New York), 675, 827–29; annual cost, 680; begins operation, 680, 708; expenses, 710 n. 7; miscellaneous experiments, 680–81, 719, 802; phonoplex experiments, 831; staff, 740 n. 4

Ladd, George, 542 n. 5, 565

Lake Shore & Michigan Southern Railroad, 518

Lamp factories: Brünn, 701 n. 6, 753; Canada, 535 n. 8, 565; Germany, 792. *See also* Edison Lamp Co.: lamp factory; Société Électrique Edison: Ivry-sur-Seine factory

Lamps. *See* Electric lighting incandescent lamps

Landau, Jacob, **763**

Lane, Edward, **82**

Lane-Fox, St. George, **128**, 259; British patents, 182, 269 n. 21, 326, 438, 495, 595; and Crystal Palace Exhibition, 382; and isolated plants, 191 n. 2; and Paris Exposition, 143, 145, 221–22; proposed injunction, 128. *See also* Electric lighting incandescent lamps: other inventors'

Lane-Fox Electrical Co., 130 n. 3

Lapham, Elbridge, **786**

Laraway, Perrine & Co., 733 n. 3
Lavery, Mr., 594 n. 3
Lawrence, Frederick, **628**
Lawrence, Mass., 488 n. 1, 510–11, 565, 567
Lawson, John, **16**, 601; agreement with Upton, 16; and lamp experiments, 346, 602, 757; letter to TAE, 16; and Menlo Park copper mine, 125, 198–99
Lebey, George, **293**
Le Gendre, William, **63–64**
Legion of Honor (France), 331
Lehigh Valley Coal Co., 193
Lehigh Valley Railroad, 193 n. 2
Léon, Elie, **128**; and European lighting companies, 251, 293, 440; and isolated lighting plants, 131, 163; lighting syndicate, 81–82, 128, 131, 163–64, 209–10, 223, 251; at Paris Exposition, 160
Léon Syndicate. *See* Léon, Elie: lighting syndicate
Leslie's Ladies Magazine, 291
Leutz, Mr., 165
Lewis, J. Volney, 125 n. 1
Lieb, John, 196, 403 n. 6, 424, **727**
Liebig, Justus von, **615**
Lincoln, Robert Todd, 822–23
Lindsay, James Ludovic, **225**
Lippman, Gabriel, 626 n. 2
Liszt, Franz, 138
Literary and Philosophical Society of Newcastle, 207 n. 2
Littell, G. W., **135**
Locomotive Engine Safety Machine Co., 373 n. 19
Logan, James C., 274 n. 1, 404, 554
Logan, Thomas, 3–4, 7–8, 52 n. 8, 99, 697
London: City Temple, 312; Commissioner of Sewers, 547 n. 3; gaslighting in, 36, 279, 305–7, 496; Lord Mayor, 257; Post Office building, 312. *See also* Crystal Palace Exhibition; Electric lighting central stations: Holborn Viaduct [and] London; Electric lighting generators: direct-connected (C dynamo): for Holborn Viaduct [and] for London; Electric lighting isolated plants
London, Brighton and South Coast Railway, 248 n. 13
London and Globe Telephone and Maintenance Co., 500 n. 13
London Daily News, 300

London & South Western Railway, 595
London Stereoscopic & Photographic Co., 249
London *Times. See Times* (London)
Longfellow, Henry Wadsworth, 86 n. 9, 745
Long Island Railroad, 37
Lorraine, 557 n. 9
Louisiana Purchase Exposition (1904), 46 n. 1, 412, 558 n. 11
Louisville Exposition (1883), 409 n. 8, 429 n. 3
Low, Seth, 718, 795
Lowrey, Grosvenor, **220**; marriage of, 228 n. 16; in Paris, 243; and Paris Exposition, 220–26, 409 n. 8; telegrams to TAE, 223, 227 n. 13
—and electric lighting, 273 n. 4, 358; Australian patents, 336; in Britain, 239 n. 4, 313, 547, 793; central stations, 698; Edison Spanish Colonial Light Co., 399 n. 1; in Europe, 192 n. 3, 197, 219 n. 4, 221, 223–4, 250–51, 294, 436; European patent rights, 200; Isolated Co., 274 n. 2; lamp contract, 669, 707, 754, 755; manufacturing, 753; patent lawsuits, 89, 130 n. 3, 568 n. 12; ship lighting, 658 n. 7; Swan's claims as lamp inventor, 207 n. 3
—letters: to Batchelor, 250–51; to TAE, 220–26, 250–51
Lowrey, Kate Armour, 225, 228 n. 16
Lubbock, John, **391**, 399 n. 3, 469, 492, 547, 595
Ludwig, Carl F. W., **619**
La Lumière Électrique (Paris), 59 n. 3, 95–97, 121–22, 132, 208, 358, 361, 461
Lyndale Railway Co., 733 n. 1

Macaulay, Thomas Babington, 745
MacKay, John, **298**
MacKenzie, James, **77**
MacLean, T. C., **145**, 229 n. 21
Magnetite Mining Co., 37 n. 1
Magneto-Electric and Dynamo-Electric Machines, 313
Magnin, Claude, **440**
Mallory, Mr., 509 n. 7
Manchester & District Electric Light Co., 462
Manhattan Railway Co., 407 n. 1, 408 n. 3

Manufacturer & Builder, 290
Marble, Edgar M., **785**
March, Othniel, **725**
Marquess of Tweeddale, **523**
Marshall, Bryun, 124 n. 3
Marshall, David, 124 n. 3
Marshall, John, 87 n. 3, 124 n. 3
Marshall, William, **124**
Marshall Field and Co., 560 n. 2, 638 n. 6
Mascart, E., 166 n. 6
Mason, William, 385 n. 7
Mather, William, **630**
Mather & Platt, 557 n. 8, 635 n. 2, 696, 809 n. 1
Max Ams Preserving Co., **646**
Maxim, Hiram, **35**, 52 n. 2, 60 n. 2, 405, 554; British patents, 182, 495; European lighting business, 161, 163–4; French patents, 89; infringement suits against, 89, 128–29, 132, 143, 568 n. 11, 723; and isolated plants, 191 n. 2; and Paris Exposition, 82, 89, 121, 132, 142–43, 145, 160, 165, 189, 198, 221–22; patent interference with, 137 n. 4, 578 n. 18; U.S. lighting business, 238; U.S. patents, 137 n. 4, 543 n. 1. *See also* Electric lighting incandescent lamps: other inventors'
Maxim Co. (Britain), 594
Maxim-Weston Electric Light Co., 363 n. 3
Maxwell, James Clerk, 470 n. 5
May, E., **131**
McCarty, Mr., 542 n. 4
McCrory, William, 732–33
McGeorge, Mungo, **405**
McGowan, Frank, **479**, 567, 731
McLaughlin, Frank, **231**, 273, 346, 404–5, 567 n. 3
Meadowcroft, William, 274 n. 2, **433**, 556 n. 3, 731 n. 3, 788
Menlo Park, N.J., 483; Anderson in, 557 n. 6; Barker in, 51; Clarke in, 447; Congressional district, 786; copper mine, 125, 198–99, 828; description, 33–35; R. Dyer in, 446–47, 476, 513, 643; Edison in, 7, 33, 197, 199, 252 n. 2, 292, 297, 298, 313, 369–70, 397 n. 1, 418 n. 2, 432, 446–48, 476, 506, 579, 662, 675, 680; electric lighting in, 35, 136, 213; horses, 456 n. 2; Insull in, 33, 292, 446–47, 476, 675; lamp factory, 3–4; location of, *2;*

Mary Edison in, 199, 298, 680; P. Shaw in, 698; Sprague in, 471 n. 13; Stuart in, 698; Turrettini in, 711 n. 16; Upton in, 447, 662. *See also* Edison Lamp Co.: lamp factory, Menlo Park; Electric railways: experimental line; Laboratory (Menlo Park)

Metropolitan Telephone Co., 820

Michels, John, **229**

Microphone, 94–95, 122 n. 3, 161, 227 n. 10, 278, 362 n. 3, 398

Milan, Italy: electric lighting isolated plants in, 401–2, 441; gaslighting in, 456 n. 3. *See also* Electric lighting central stations: Milan

Mill Creek Distilling Co., 699

Miller, Warner, **786**

Miller, William, **612–13**

Miller's National Association, 803 n. 7

Mills, Darius Ogden, 332 n. 1, 598 n. 2

Mining: Calumet mine, 702; copper, 702; cost estimate, 433–34; electric power for, 298 n. 1; gold, 6, 37 n. 3, 76, 372 n. 13, 432, 765, 770; Hecla mine, 702; iron ore, 6, 37 n. 3, 76, 502, 765–770, 824–25; in Long Island, 434 n. 1; Menlo Park copper mine, 125, 198–99, 828; ore separators, 6, 37, 76, 230, 433–34, 765, 767–69; in Oroville, Calif., 274 n. 1; Osceola mine, 703 n. 2; patents, 7 n. 3, 76, 249, 770 n. 1; platinum, 6; in Quogue, 6, 37, 76, 768, 824–25; in Quonocontaug, 230, 434, 765–770; stamp mill, 313, 370, 432

Minneapolis, Lyndale & Minnetonka Railway, 733 n. 1

Minneapolis Tribune, 732

Mitchell, Vance & Co., 555

Montgomery, Thomas, 416 n. 4, 566

Moore, John Godfrey, 508 n. 2, **541**

Moore, Michael, 66 n. 5, **342**

Moore, Miller, **371**, 507; and dynamo orders, 558 n. 14; and Isolated Co., 198, 274 n. 2, 673; and lamp orders, 600; letter from Insull, 643; and Morgan's isolated plant, 750 n. 1; and Roselle station, 642 nn. 1 & 3, 643

Moore, William, **22**

Moore (J.G.) & Co., **539**

Mora, De Navarro & Co., 400 n. 2

Mora, Fausto, **399**

Mora, José, 399 n. 2

Morgan, John Pierpont, 391, 493, 539 n. 1, 580, 750, 797, 828

Morgan Iron Works, 667 n. 16

Morning Advertiser, 358

Morris, Joseph, 66 n. 6

Morris, Tasker, & Co., Ltd., **602**

Moses, Otto, **62**, 135, 431, 556; and aluminum experiments, 480 n. 2; and arc lights, 395 n. 2, 448, 532, 636; and Crystal Palace Exhibition, 362–63; and Edison's patents, 96–97; and European lighting companies, 224, 294; and incandescent lamps, 62; and Maxim injunction, 129; and Paris Exposition, 62 n. 1, 95–98, 120–22, 131–33, 224; and Paris Opera, 210 n. 1; patents, 534 n. 7 —letters: to TAE, 96–98, 120–22, 131–33, 144–45, 362–63

Mott, Charles, 7, 50 n. 1, 103 n. 11

Mott, Samuel Dimmock, **49**, 532; and patent applications, 49, 58 n. 11, 286, 289; patents, 554–55, 611 n. 6

Muirhead, Alexander, **259**

Muller, John, 554

Munroe's Fashion Bazaar, 291

Mutual Union Telegraph Co., 508 n. 2, 539 n. 4, 804 n. 1

Namouna, 508 n. 3

National Academy of Sciences, 52 n. 6, 272, 681, 718 n. 2, 725

Nationalbank für Deutschland, 763

National Board of Fire Underwriters, 358

Navarro, José de, **399**, 519 n. 7, 527

Newark, N.J.: Commissioner of Deeds, 29 n. 1; electrical manufacturing in, 674; location, 2; TAE in, 3. *See also* Edison Lamp Co.: lamp factory, Harrison

Newark Daily Advertiser, 291

Newark Sunday Call, 291

Newcastle Chemical Society, 206–7

New England Manufacturer's and Mechanic's Institute Fair, 597

New Orleans Exposition (1884), 429 n. 3

New York Board of Fire Underwriters, 277, 324, 358, 523 n. 7, 557 nn. 5–6, 597, 644

New York City: American Society of Mechanical Engineers meeting, 255; Bailey in, 581, 630 n. 3, 700; Barker in, 52 n. 12, 69, 272; Brooklyn Bridge, 424; City Hall, 424; Colombo in, 581, 630 n. 3, 728 n. 3; Commissioner of Public Works, 819; Croton aqueduct, 111; Department of Taxes and Assessments, 761; electrical manufacturing in, 3, 660; gaslighting in, 828; Gramercy Park, 579, 655 n. 4, 680; Kalakaua in, 162 n. 2; Madison Square, 311, 512 n. 4; Madison Square Theatre, 509 n. 7; National Academy of Sciences meeting, 681; Porter in, 416; Puskas in, 132; steam heating, 817–18, 828; Tammany Hall, 813–14, 827–29; Union Square, 512 n. 3; White in, 682, 714 n. 1, 736 n. 1. *See also* Electric lighting central stations: Pearl St. [and] second district (Madison Square); Electric lighting isolated plants

New York Daily Graphic, 75 n. 36, 629 n. 4

New York Daily News, 225

New York Edison Co., 728 n. 7

New York Elevated Railway, 408 n. 3

New York Evening Post, 185 n. 2

New York Herald, 75 n. 36, 274 n. 2, 644–46, 817

New York Journal, 759 n. 1

New York Steam Heating Co., 817

New York Stock Exchange, 628

New York Times, 645, 759 n. 1

New York Tribune, 230 n. 2, 566, 718 n. 1

New York University, 534 n. 2

New York Weekly, 291

Nichols, Edward Leamington, **22**; and lamp factory, 5, 22, 39, 43, 175–76, 234–35; lamp tests, 67–68, 86, 123–24

Northern Pacific Railroad, 185 n. 2, 292 n. 1, 812

Northwestern Telegraph Co., 560 n. 1

Nottage, George Swan, 250 n. 4

Nottbeck, Charles, 349 n. 3, **441**

Il Nuovo Cimento, 308

Nursery Monthly, 291

N.V. Nederlandsche Electriciteit-Maatschappij, 444 n. 17

Oakley, William, 643 n. 1

Odoroscope, 71

Oertling, Ludwig, **437**
Ohio Edison Electric Installation
Co., **698**
Olrick, Harry, **630**, 650, 695–96,
734, 738
Olrick (Lewis) & Co., 635 n. 1
Operator, 647 n. 11, 676 n. 4, 705 n. 2
Ophir Silver Mining Co., 298 n. 1
Oregon Railway & Navigation Co.,
185 n. 2, 281 n. 7, 431
Oregon & Transcontinental Co.,
292 n. 1
Ore milling. *See* Mining
Oriental Telephone Co., 118, 483;
investment in, 497; organization
of, 90–91, 198; patent rights, 63–
64; stock, 262, 460, 547, 796; tele-
phone manufacture for, 14 n. 3
Osborne, Robert S., **604**
Ott, John, **9**; and electric railway,
384; and isolated dynamos, 113;
and lamp experiments, 330–31,
346, 748, 779, 802, 808; letters to
TAE, 330–31, 384; and Menlo
Park laboratory, 710 n. 6; and meter experiments, 9, 625–26, 802,
804; voltage regulation experiments, 577 n. 9
Otto, Nikolaus, **716**
The Outlook, 552 n. 1

Pacific Bell Telephone Co., 567 n. 3
Pacinotti, Antonio, **47**, 308
Page, Samuel Flood, 217 n. 5
Paine, Sidney, **508**
Painter, Uriah, **461**, 532, 537–38,
539, 786
Pan American Exposition (1901),
558 n. 11
Paris, 454 n. 1; Armington in, 700;
Boulevard St. Antoine, 84; elec-
trical manufacturing in, 22, 165,
197; gaslighting in, 161–64;
Gouraud in, 30 n. 1; Johnson in,
466. *See also* Electric lighting
central stations: Paris; Electric
lighting generators: direct-
connected (C dynamo): for Paris;
Electric lighting isolated plants;
Exposition Internationale de
l'Électricité; Paris Opera; Société
Électrique Edison: Ivry-sur-
Seine factory
Paris Exposition. *See* Exposition
Internationale de l'Électricité
Paris Opera. *See* Electric lighting
isolated plants: Paris Opera

Parke Davis Co., 819, 828, 830
Parliament (Britain), 393 n. 21, 495–
96, 524 n. 3, 594–95, 714 n. 2
Parliament (Canada), 535 n. 8
Partridge (C.W. & E.) & Co., 638 n. 6
Patent applications, 53–56, 98 n. 5,
611 n. 6; aluminum, 480 n. 2; arc
lights, 197, 536 n. 18, 575–76,
638 n. 3; armatures, 288 n. 6; carbonizing molds, 141 n. 8, 640 n.
1, 641 n. 4; chemical stock quotation telegraph, 629 n. 3; direct
conversion of coal, 479–80; distribution system, 753; dynamos,
19 n. 1, 20 nn. 2–6, 48–49, 102
n. 9, 103 n. 11, 104 n. 2, 105 n. 5,
116 n. 11, 177 n. 4, 373–74, 525
nn. 1–2, 676 n. 4, 716–18, 753,
761–62; electric lighting, 4, 197,
279, 447, 572–76, 580; electric
railways, 449–50; for electroplating carbonized articles, 706; facsimile telegraph, 452 n. 2; high-
voltage transmission, 516 n. 1,
528 n. 2; incandescent lamps, 26–
27, 43 nn. 3–4, 46 n. 2, 287 nn. 1–
3, 290 n. 2, 347 n. 2, 364–65, 543,
571–73, 576, 582, 641 n. 7, 685–
92, 694 nn. 1–6, 721, 729 n. 2,
753, 778 n. 4, 779 n. 1; insulation,
582; meters, 9 n. 1, 156 n. 11,
202, 808 n. 9; mining, 770 n. 1; by
month, 844; primary batteries,
741; pyromagnetic generator, 397
n. 1; storage batteries, 193 n. 3,
516 n. 4, 742; telephone transmitter, 794 n. 3; three-wire system,
582; transformer, 528 n. 2; by
type, 846; vacuum pumps, 25
n. 9, 155 n. 5; voltage regulation,
288 n. 5, 289 n. 10, 374, 377–78,
418, 448, 502, 504 n. 2, 516 nn.
1–3 & 5, 517 nn. 10–11, 528 n. 2,
574–75, 685 n. 3, 753, 791–92
Patent interferences, 98 n. 5, 752–
53, 785–86; with Crookes, 543
n. 1; with Field, 409 n. 8, 482,
609; with Freeman, 818; with
J. Sprague, 9 n. 1, 155 n. 1, 156
n. 14; with Mather, 378 n. 1; with
Maxim, 137 n. 4, 543 n. 1, 578
n. 18; with Scribner, 378 n. 1;
with Short, 52 n. 7; with Siemens,
272 n. 1, 409 n. 8, 482, 609, 744
n. 1, 762 n. 2; with Swan, 543 n. 1
Patent Office Museum, 438
Patents: Brazilian law of, 542 n. 7;

Canadian law of, 534 n. 8; Edison's bound volumes, 75 n. 36;
Edison's policy, 430, 554–55;
French law of, 486, 537–38; German Patent Office, 269 n. 24; infringement cases, 89, 128–29,
143, 157, 371, 819; legal examination in France, 181; and monopoly, 325; royalties, 26 n. 2, 36 n. 3,
58, 66 n. 5, 76, 165, 189, 231 n. 5,
260, 312, 317, 351, 380, 445 n. 3,
496, 704–5, 743–44, 744 n. 2,
764 n. 8; Spanish law of, 441
Patents (Britain), 75 n. 36, 97,
200 n. 1, 483; 2,006 (1873), 328
n. 26; 4,226 (1878), 337; 5,306
(1878), 193 n. 2; 1,122 (1879), 269
n. 21; 2,402 (1879), 268 n. 20, 328
n. 22, 336; 4,576 (1879), 318–20,
325–26, 330 n. 39, 336; 5,127
(1879), 328 n. 20, 336; 18 (1880),
328 n. 21; 33 (1880), 336; 578
(1880), 336, 530 nn. 1–2; 602
(1880), 336, 359 n. 4, 383 n. 3;
1,385 (1880), 336, 530 n. 1; 3,494
(1880), 438; 3,765 (1880), 330
n. 39, 336; 3,880 (1880), 336, 530
n. 1, 564 n. 2; 3,894 (1880), 477
n. 10; 3,964 (1880), 105 n. 5, 336,
530 n. 1; 4,391 (1880), 337; 539
(1881), 337; 562 (1881), 337, 392
n. 5; 768 (1881), 337; 1,240
(1881), 337; 1,783 (1881), 337;
2,492 (1881), 338; 2,532 (1881),
266 n. 4; 2,612 (1881), 543 n. 1;
2,954 (1881), 374 n. 3; 3,799
(1881), 271 n. 33; 1,023 (1882),
281 n. 9; 1,142 (1882), 374 n. 4;
1,862 (1882), 450 n. 1; 2,052
(1882), 393 n. 10; 3,576 (1882),
794 n. 4; applications, 279, 449–
50, 717, 735, 805; disclaimers,
312, 316, 318–21, 325–26, 448,
495, 529–30, 546, 595; law of, 200
n. 1, 316, 318–21, 326, 327 n. 15,
382 n. 2, 832, 843; legal examination, 181, 198, 268 n. 20, 279–80,
311–12, 315–22, 325–26, 344
n. 11, 371, 382 n. 2, 386–88, 462,
495, 529–30; list of, 852–53;
phonograph, 530; telephone, 530
—other inventors': Bell, 326, 373
n. 17, 832; Crookes, 271 n. 33,
543 n. 1; Delany, 266 n. 4;
Gouraud, 266 n. 4; Gramme,
318; Hopkinson, 753, 794 n. 4,
805; Hunnings, 500 n. 15; John-

son, 266 n. 4; Lane-Fox, 182, 269 n. 21, 326, 438, 495; Maxim, 182, 495; Sellon, 519 n. 3; Siemens, 198, 316–17, 321, 387, 495; Swan, 182, 320, 326, 495; Volckmar, 519 n. 3

Patents (other): Australia, 336; Austria, 59 n. 4, 701 n. 6; Brazil, 508 n. 2, 542; British colonies, 91, 448; Canada, 47 n. 3, 532, 597; Europe, 47 n. 3, 59 n. 3, 486, 717, 753, 821; France, 75 n. 36, 97, 128, 143 n. 3, 176, 181, 350 n. 5, 486, 717, 805, 821; Germany, 191 n. 2, 682, 704–5, 763; Hong Kong, 91, 119 n. 6; South America, 4; Spanish colonies, 486

—other inventors': Bell, 78 n. 7, 90, 97; Gramme, 47 n. 3, 59 n. 4; Gray, 78 n. 7; Hopkinson, 805; Krizik, 121 n. 8; Maxim, 89, 97, 128, 143 n. 3; Otto, 718 n. 3; Piette, 121 n. 8; Pond, 79 n. 9; Rochas, 718 n. 2; Siemens, 47 n. 3, 191 n. 2, 682, 704–5; Swan, 89

Patents (U.S.): 214,636, 137 n. 4; 223,898, 46 n. 2; 224,511, 374; 227,229, 576 n. 4; 228,543, 685 n. 2; 240,678, 156 n. 11; 248,419, 43 n. 3; 248,428, 365 n. 3, 392 n. 5; 251,537, 58 n. 11, 374 n. 3; 251,544, 43 n. 3; 251,545, 203 n. 3; 251,549, 50 n. 6; 251,552, 337 n. 2; 255,311, 685 n. 2; 258,795, 554; 259,054, 558 n. 13; 259,115, 558 n. 13; 259,235, 558 n. 13; 261,077, 599 n. 6; 263,131, 7 n. 3; 263,132, 451 n. 5; 263,133, 102 n. 9, 103 n. 11; 263,138, 537 n. 18; 263,143, 19 n. 1, 20 n. 4; 263,144, 141 n. 8; 263,145, 26–27; 63,146, 116 n. 11; 263,147, 25 n. 9; 263,148, 20 n. 2; 263,149, 50 n. 1, 105 n. 5; 263,150, 19 n. 1, 20 n. 6, 57 n. 9, 58 n. 10; 264,298, 432 n. 1; 264,299, 432 n. 1; 264,646, 20 n. 2, 393 n. 10; 264,647, 50 n. 1, 104 n. 2, 107 n. 3; 264,653, 288 n. 8; 264,660, 419 n. 4; 264,662, 685 n. 3; 264,665, 423 nn. 2 & 7; 264,667, 423 n. 3, 517 nn. 10–11; 264,668, 685 n. 3; 264,671, 685 n. 3; 264,672, 422 n. 4, 517 nn. 10–11; 264,698, 432 n. 4; 264,986, 543 n. 1; 264,987, 543 n. 1; 264,988,

543 n. 1; 265,775, 537 n. 18; 265,776, 281 n. 9; 265,779, 289 n. 10, 374 n. 4, 419 n. 4; 265,780, 375 nn. 5–6; 265,783, 516 n. 3; 265,786, 516 n. 1; 266,447, 43 n. 4; 266,793, 792 n. 1; 268,205, 288 n. 6; 268,206, 693 n. 2; 269,805, 685 n. 2; 271,614, 676 n. 4; 271,615, 676 n. 4; 271,616, 676 n. 4; 273,486, 693 n. 2; 273,487, 577 n. 9; 273,491, 676 n. 4; 273,492, 516 n. 4; 273,493, 676 n. 4; 273,494, 451 n. 7; 274,290, 582 n. 1, 792 n. 1; 274,292, 516 n. 4; 274,293, 620 n. 2, 682 n. 4, 693 n. 4; 274,294, 721 n. 2; 274,295, 682 n. 4; 274,296, 571 n. 1; 275,612, 576 n. 6; 276,232, 718 n. 2; 278,413, 516 n. 2; 278,416, 365 nn. 1–2; 278,418, 516 n. 1, 528 n. 2; 278,419, 288 n. 5; 281,350, 577 n. 13; 281,353, 525 n. 1; 283,938, 791–92, 792 n. 1; 283,984, 792 n. 1; 287,515, 792 n. 1; 287,517, 792 n. 1; 287,518, 620 n. 2; 287,520, 694 n. 13; 287,522, 779 n. 1; 287,523, 525 n. 2; 287,525, 504 n. 2; 297,581, 620 n. 2; 304,082, 156 n. 11; 307,029, 641 n. 7; 314,115, 629 n. 3; 334,853, 583 n. 3, 640 n. 1; 353,783, 778 n. 4; 358,600, 287 n. 3; 365,465, 676 n. 4; 365,509, 543 n. 1; 370,123, 808 n. 9; 379,770, 46 n. 2, 287 n. 2; 379,771, 374 n. 4; 379,772, 516 n. 5; 379,777, 378 n. 1; 395,962, 620 n. 2; 400,317, 770 n. 3; 401,486, 537 n. 18; 401,646, 347 n. 2, 365 n. 2; 406,825, 9 n. 1, 155 n. 5; 411,016, 694 n. 10; 425,760, 50 n. 3; 425,763, 50 n. 1, 177 n. 4; 431,018, 104 n. 2; 435,687, 193 n. 3; 438,298, 729 n. 2; 439,390, 516 n. 1, 552 n. 3; 439,393, 641 n. 4; 446,666, 516 n. 1, 552 n. 3; 446,669, 779 n. 1; 460,122, 397 n. 1; 464,822, 516 n. 1, 552 n. 3; 466,460, 480 n. 1; 476,528, 749 n. 1; 492,150, 693 n. 5; 543,986, 682 n. 2, 694 n. 14; Design 13,940, 693 n. 6; by Edison associates, 847–50; law of, 534 n. 8, 537–38, 753, 832, 843

—other inventors': Bell, 819; Brush, 375 n. 6, 597; Crookes,

543 n. 1; Drawbaugh, 819; Faure, 509 n. 8; Field, 733 n. 4; Freeman, 818; Goebel, 819, 828; Gramme, 47 n. 3, 59 n. 4; Johnson, 794 n. 3; Maxim, 137 n. 4, 543 n. 1; Moses, 534 n. 7; Muller, 554; Pond, 79 n. 9; Rowland, 761–62; Seely, 558 n. 13; Siemens, 260, 743–44, 762 n. 2; Swan, 181–82, 259–60, 543 n. 1, 566, 599 n. 7; Weston, 543 n. 1

Paton, Mr., 671, 726

Patton, W. H., 298

Pearl St. *See* Electric lighting central stations: Pearl St.

Pemberton Mills, 511 n. 1

Pennsylvania Railroad, 136, 280, 812; ferries, 358; ferry terminals, 2; in New Jersey, 5

Perkins, George C., 274 n. 1

Perrie, Dr., 732

Perrine, Cyrus, 733 n. 3

Peters Manufacturing Co., 28

Philadelphia Electrical Exhibition (1884), 397 n. 1

Philadelphia Ledger, 75 n. 36

Philadelphia Photographer, 290

Philadelphia Public Ledger, 213 n. 1

Phillips, Charles, 745

Philosophical Transactions, 613

Phonograph, 519 n. 8; and British telephone patent, 358–59, 530; at Crystal Palace Exhibition, 381; Kelly portrait, 249; at Paris Exposition, 64, 71; recording, 21; at South Kensington Museum, 438

Photophone, 12 n. 1

Physical Treatise on Electricity and Magnetism, 535 n. 10

Pictorial War Record, 291

Piette, Ludwig, 121 n. 8

Planté, George, 742

Platinum: mining, 6; suppliers, 10, 176, 601. *See also* Electric lighting incandescent lamps: attachment to lead-in wires [and] platinum filament

Polhemus, John, 645

Pond, Charles Fremont, 628 n. 1

Pond, Chester, 77

Pond, Emma McHenry, 628

Pond Indicator Co., 79 n. 10

Popular Monthly, 291

Popular Science Monthly, 290

Porges, Charles, 83 n. 4, 130 n. 4, 131, 251 n. 2, 294, 440, 763–64

Porsh, L., 710 n. 8

Porter, Charles T., 8, 318, 352, 356, 827; ASME paper, 215 n. 2; Edison's opinion of, 333 n. 2; letters to TAE, 214, 416–17; at Menlo Park, 4; and Milan engines, 781; in New York, 416; production problems, 659. *See also* Steam engines: Porter-Allen

Porter, Lowrey, Soren, and Stone, 226 n. 1

Portfolio, 291

Port Huron, Mich., 91, 456 n. 1

Port Huron Railway Co., 749

Post, George, 598 n. 2

Postal Telegraph Co., 610 n. 1

Poughkeepsie Iron & Steel Co., 230 n. 3, 434, 765–66

Powers, Hiram, **47**

Powers, Longworth, **47**

Preece, William, **94**, 391, 467, 470 n. 5; and arc lights, 556 n. 4; and British patents, 316, 344 n. 11; and electric lighting, 222, 225, 239 n. 2, 278, 300, 324–25; and electric lighting article, 358; lamps for, 95; lectures, 243, 300, 324, 361; letter from TAE, 361; letter to TAE, 379; and microphone controversy, 94–95, 362 n. 5; at Paris Exposition, 121, 361; relations with, 95 n. 1, 312, 325, 379

Press relations, 6, 131 n. 5, 225, 230 n. 2, 746; in Britain, 222, 300, 325, 482 n. 1, 488; Crystal Palace Exhibition, 406 n. 1; Edison interviews, 509 n. 11, 552, 705 n. 2, 759 n. 1, 789; electric railway, 561 n. 6; in France, 222; Paris Exposition, 95–96, 121–22, 129, 132; Pearl St. station, 425, 580, 705 n. 2; scientific and technical journals, 89, 135, 157–58, 178, 181–82, 189, 202, 215 n. 2, 462, 488. *See also* individual newspapers and journals

Princeton University, 124, 785 n. 1

Progress, 290

Providence Journal, 759 n. 1

Pryor, Carolyn, 655 n. 1

Pryor, James Williamson, **654**

Puck, 291

Pullman, George, **559**

Puskas, Francis, **441**

Puskas, Theodore, **58**, 96, 121, 486, 828; at Crystal Palace, 437; in New York, 132, 217; and Paris

Exposition, 81–82, 95–96, 142; and telephone exchanges, 821–22
—and electric lighting: dynamo orders, 563 n. 2; Edison Electric of Europe, 673; European companies, 189, 192 n. 3, 209–10, 251 n. 1, 439–40; and Gramme dynamo, 58; isolated plants, 166 n. 1, 191–92, 314, 334 n. 1; in Italy, 448, 605 n. 5; and Maxim injunction, 143 n. 3; in Milan, 402 n. 2; organization of Paris companies, 179; Pacinotti's dynamo, 47 n. 3; Paris Opera, 303
—letters: from Batchelor, 60–61, 70–71; from TAE, 58, 592; to TAE, 81–82, 128–29, 439–42
—telegrams: from TAE, 129, 135 n. 1; to TAE, 129, 209–10

Quadruplex Case, 568 n. 12

Queen (James W.) & Co., 588

Queen of the Pacific, 433 n. 9

Quogue (Long Island), N.Y., 6, 37, 76

Quonocontaug, R.I., 230, 434

Rae, Thomas, 542 n. 5

Rae, Thomas Whiteside, **255**

Railway Age, 561 n. 6

Randolph, John, **99**, 125, 185 n. 4, 697 n. 2

Rathenau, Emil, 557 n. 9

Rau, Louis, **440**, 730

Reade, Alfred, **453**

Reed, James C., **761**

Reid, Whitelaw, 718 n. 2

Reiff, Josiah, **747**

Reményi, Eduard, 91, **138**, 816, 822, 827–29

Renshaw, Alfred George, **94**

Revue de la Mode, 291

Riley, William Willshire, 506 n. 2

Ritchie, Edward, 172

River Tyne Improvement Commission, 226 n. 6

Roach, John, **663**, 813

Roach (John) & Sons, 678 n. 10

Robey & Co., **362**

Robinson, Heber, **69**

Rocap, Charles, 100 n. 6, 477 n. 3, 505 n. 7, **548**, 754; agreement with, 448, 548–49, 798; as bookkeeper of Machine Works, 663; dismissal, 664; letter from TAE, 784; letter to TAE, 562

Rochas, Alphonse Beau de, 718 n. 3

Rogers, H. J., 703 n. 3

Roosevelt, Cornelius, 119 n. 7

Roosevelt, Hilborne, **119 n. 7**

Rose, Allen W., 468

Roselle, N.J., **5**, 371. *See also* Electric lighting central stations: Roselle

Rowland, Henry, 611 n. 6, 761–62

Royal Institution, 366 n. 6

Royal Society of Arts, 243, 259, 300, 324

Royal Society of Edinburgh, 366 n. 7

Royal Society of London, 344 n. 11, 406 n. 6

Russell, James A., **604**, 823, 829–30

Russell, Margaret, **433**

Russell, William Howard, 827–28

Rutgers College, 124 n. 3

San Francisco Chronicle, 542 n. 5

San Francisco Examiner, 542 n. 5

Satterlee, Herbert, 750 n. 1

Savoy Theatre, 244, 357, 380, 393 n. 16

Saxelby, F., **627**

Say, Léon, 96, **144**

Schellen, Heinrich, 313

Schleicher, Schumm & Co., **717**

Schönbein, Christian Frederich, **612**

Science, 229, 290

Scientific American, 51 n. 1, 290, 309 n. 11, 425, 561 n. 6, 597, 802

Scientific American Supplement, 96, 290

Scientific News, 290

Scott, Claud, **656**

Scribner's Monthly, 250 n. 3

Seely, George, 374 n. 4

Seely, Henry, **692**, 785, 806; and battery patents, 741–42; electric iron patent, 555; and lamp patents, 692, 721; and voltage regulation patents, 716–18

Self Instructor, 290

Sellon, John Scudamore, **518**

Serrell, Lemuel, **97**; and Australian patents, 336; and British patents, 268 n. 19, 320, 336–37; and European patents, 97, 486; and Gramme's patent, 47 n. 3; letter to TAE, 336–37; and patents for Spanish colonies, 486

Sérurier, Charles-François-Maurice, **440**

Servia, 584

Seubel, Philip, 122 n. 1, **165**, 240 n. 2, 293, 333, 349, 353 n. 1, 700

Sewell, William, **786**

Seyfert, Lucy, **747**

Seyfert, William, 747 n. 2

Seyfert v. Edison, 747

Shaw, Capt., 822, 828

Shaw, George Bernard, 832

Shaw, Phillips, **698**

Shepard, James, **401**–2

Shillito, Mr., 165

Shimer & Co., **766**

Short, Sidney, **51**

Shuster, Arthur, **616**

Siemens, Charles William, **228 n. 19**; article by, 358; at Crystal Palace Exhibition, 382, 405; dynamos, 328 n. 26, 379–80; and electric lighting in Britain, 469, 495–96; lecture by, 382

Siemens, Werner, **225**; Deutsche Edison Gesellschaft, 743, 777; dynamos, 4, 47 n. 3, 51, 328 n. 26, 379–80, 744; and electric lighting in Germany, 682; electric railway, 409 n. 8, 449–50, 482, 744; proposed merger in U.S., 743. *See also* Electric lighting generators: others', Siemens

Siemens Bros. & Co., 228 n. 19, 495, 531 n. 3, 596 n. 5

Siemens & Halske, 52 n. 9, 165, 228 n. 19, 401, 682, 705 n. 3, 744 n. 2, 762 n. 2, 763

Silliman, Benjamin, Jr., 682

Simmons, Zalmon Gilbert, 560 n. 1

Sims, Gardiner, **126**, 670

Singer Sewing Machine Co., 237 n. 19

65 Fifth Avenue, 75 n. 36, 138 n. 1, 331, 475, 567, 675; Bennett at, 816–17, 828–29; Dixey at, 822, 827; Edison at, 199; handstamp, 392 n. 1; hours of business, 369; isolated plant, 23, 93; Lincoln at, 822–23; location of, *2;* purchase of, 655 n. 4; Reményi at, 816, 822, 827, 829; Shaw at, 822; Stager at, 822–23; TAE's office, 3

Smith, Augustus Ledyard, **702**

Smith, Henry Daniel, 703 n. 3

Smithsonian Institution, 52 n. 5

Societa Generale Italiana di Elettricita, 728 n. 7

Société d'Appareillage Electrique, 511, 793 n. 2

Société du Téléphone Edison, 442 n. 3

Société Électrique Edison, 209, 250–51, 350 n. 5, 709; absorbed by Compagnie Continentale Edison, 701 n. 5; and direct-connected dynamos, 727; directors, 294; Ivry-sur-Seine factory, 485, 581, 590–91, 639–40, 660, 682, 700; letter from TAE, 654; management of, 440; and Milan central station, 783 n. 5; and orders from New York shops, 571 n. 2, 783 n. 7; patent costs, 654; and Switzerland, 709

Société Industrielle et Commerciale Edison: directors, 294; Ivry-sur-Seine factory, 176, 179–80, 189, 192, 197, 293–94, 314, 402; management of, 440; organization of, 179–80, 209–10, 250–51, 314

Society for the Prevention of Cruelty to Animals, 628 nn. 1–2

Society of Telegraph Engineers, 344 n. 11, 383 n. 6

Soldan, Gustav, **489**, 503, 663–64

Soren, George, 399 n. 1

South America: bamboo, 45; fiber search, 282–83; telephone in, 4. *See also* Electric lighting: in Chile [and] in South America; Electric lighting central stations: Santiago; Electric lighting isolated plants: in South America

South Kensington Museum, 244, 438

Southwark Foundry, 781

The Speeches of Charles Phillips, 745

Spencer, Herbert, **720**

Spottiswoode, William, **405**

Sprague, Frank, **461**–62, 655–57, 828

Sprague, John, 9 n. 1

Spring Valley Mining & Irrigation Co., 233 n. 1

St. Louis Exposition (1884), 429 n. 3

St. Louis Post-Dispatch, 482 n. 4, 509 n. 11

St. Louis World's Fair (1904). *See* Louisiana Purchase Exposition

St. Nicholas, 291

Stager, Anson, **542**, 559, 704 n. 4, 822–23

Stanley, William, **51**

Starr, J. W., 208 n. 5

Steam engines, 517 n. 1, 556; Armington & Sims, 4, 12, 101, 175, 253–54, 262, 280 n. 3, 301, 323, 333 n. 1, 356–57, 412, 454, 489, 558 n. 15, 584, 590, 633, 637, 646, 659–60, 670–71, 696, 700, 704, 724, 727, 774, 780–82, 788–89, 815, 827–28, 840 n. 10; Baxter, 99; Corliss, 674; donkey, 553; for electric railways, 732; experiments, 8; Hampson, 277; high-speed, 243, 280; for isolated plants, 340, 584; for London, 276; for Paris Exposition, 134; Porter-Allen, 4, 8, 13 n. 1, 36 n. 4, 214, 327 n. 6, 333 n. 1, 342, 352, 356–57, 367, 390, 410, 412, 416–17, 454, 466, 591 n. 3, 633, 659–60, 670–71, 696, 704, 727, 774, 780–81, 788, 814, 827; Robey, 362; tests, 12

Stephenson, James Cochran, **221**

Sterne, Morris E., 655 n. 4

Stevens Institute, 125 n. 4, 483, 728 n. 7, 796 n. 2

Stewart, Willis, **521**, 831

Stieringer, Luther, **555**

Stilwell, Alice, 140 n. 6

Stilwell, Eugenie, 680, **683**

Stockly, George, **532**–33

Stoddard, Mr., **342**

Stoddard's Musical, 291

Stuart, Archibald, 698–**99**, 704

Sullivan, Arthur, 360 n. 20

Swan, Joseph, **35**, 60 n. 2, 144, 416, 508, 554; British patents, 182, 319, 326, 495, 595; cellulose filaments, 694 n. 14; claims as lamp inventor, 206–7, 221, 259–60, 448; and Crystal Palace Exhibition, 382; and electric lighting in Britain, 279; *Engineer* article, 206–7; European lighting business, 161, 163–64; French patents, 89; infringement suits against, 128–29, 566, 723 n. 2; merger with in Britain, 744; and Paris Exposition, 82, 121, 142–43, 145, 160, 165, 189, 198, 221–22, 225; relations with, 221, 225, 238; reputation, 243–44; telegram to TAE, 221; U.S. patents, 181–82, 259–60, 543 n. 1. *See also* Electric lighting incandescent lamps: other inventors'

Swan's Electric Light Co., Ltd, 226 n. 7, 509 n. 7, 548 n. 8

Swanson, Alfrid, **99**, 513, 521, 697

Swan United Electric Light Co., 547, 584, 649, 657

Swinyard, Thomas, **565**

Tale of Two Cities, 85
Tales for Children, 745
Tasimeter, 11, 71, 438
Teatro alla Scala, 350 n. 4, 401, 441, 630 n. 2
Telegraphic Journal and Electrical Review, 135, 157, 325
Telegraphy: automatic, 64, 71, 182, 407 n. 3, 747 n. 4; chemical stock quotation, 628; district, 71, 135; duplex, 71, 451; dynamos for, 622–23; electromotograph relay, 71, 135; facsimile, 71, 451–53, 629 n. 3; patent applications, 452 n. 2; pressure relay, 71, 135; quadruplex, 27, 59 n. 3, 64, 71, 407 n. 3, 568 n. 12, 623 n. 3; recorder-repeater, 359; roman-letter automatic, 71, 629 n. 3; stock printers, 794; underground lines, 257; Universal private-line printer, 27, 71; Universal stock printer, 27, 64, 71, 438
Telephone, 519 n. 8; in Australia, 36 n. 3; Bell's, 14, 66 n. 6, 78 n. 7, 90, 819; Bell's photophone, 12 n. 1; in Britain, 4, 14, 25, 73 n. 18, 76–77, 90, 261, 311, 342, 398–99, 640 n. 2, 660, 832; British patents, 530; in Budapest, 822; in Canada, 351; carbon transmitter, 71, 94–95, 98, 278, 371, 398, 494, 832; Dolbear's, 163; Drawbaugh's, 819; electromotograph receiver, 51 n. 1, 71, 381, 416, 438, 459, 640 n. 2, 832; exchanges, 77; in France, 14 n. 2, 79 n. 8; Gower's, 90; Gray's, 78 n. 7, 79 n. 8; in Hong Kong, 66 n. 6, 91; Hunnings's, 500 n. 15; infringement suits, 398, 494; interferences, 358, 537 n. 17, 788 n. 12; musical, 71, 161, 381; at Paris Exposition, 64; patents, 371, 373 n. 17, 398, 494, 596 n. 9; and phonograph, 71, 358–59; repeater, 71; royalties, 76–77; in Russia, 443 n. 13; in Scotland, 371; switchboard, 71, 79 n. 9; transmitters, 794; transmitting with light, 11; using tasimeter, 11; and Western Union, 500 n. 15, 816, 828–29
—patent rights, 6; in Australia, 66 n. 6; British colonies, 90; in China, 66 n. 6; in Europe, 66 n. 6, 90; in

Hong Kong, 66 n. 6, 91; in Japan, 66 n. 6; in South America, 4
Teller, Henry, **785**–86
Tesla, Nikola, 821, 828
Thode & Knoop, **763**
Thomas A. Edison Construction Dept., 376 n. 2, 471 n. 13, 754, 809
Thompson, Frank, 828
Thompson, Sylvaus, **382**
Thomson, Elihu, 273 n. 4, 484 n. 4
Thomson, Frank, **358**, 812
Thomson, Hugo O., 819, 827–28
Thomson, William, **221;** and central stations, 297 n. 2, 299; and dynamos, 269 n. 24, 299, 355–56, 358, 380, 388, 415, 457; and Electrical Congress, 344 n. 13; and electric lighting in Britain, 221–22, 246 n. 8, 345 n. 19; galvanometer, 73 nn. 21–22, 419, 634; and lamps, 80 n. 1, 308 n. 1, 545–46; relations with, 95 n. 2
Thomson-Houston Electrical Co., 699 n. 4
Thurber, Francis Beattie, 648 n. 14
Thurber (H. K. & F. W.) Co., **646**
Times (London), 222, 483, 494, 498
Tracy, H. L., **747**
Transactions of the American Society of Mechanical Engineers, 215 n. 2
Transactions of the Newcastle Chemical Society, 207 n. 2
Tresca, Henri Edouard, **758**
Tufts, John, **642**
Turrettini, Théodore, 709
Twombly, Hamilton McKown, **67**, 816, 828–29

Union générale bank, 443 n. 9
United States Electric Lighting Co., 37 n. 7, 52 n. 2; and Barker, 69 n. 1, 182; and Bergmann & Co. shop, 674; and Gramme Electrical Co., 59 n. 4, 509 n. 7; injunction against, 132, 143; Paris Exposition awards, 222; and patents, 137 n. 4, 543 n. 1
United States Underground Cable Co., 266 n. 2
United Telephone Co.: and Edison's patents, 326; and electromotograph telephones, 416, 459; infringement suits, 371, 373 n. 17; stock, 13–14, 25, 65 n. 2, 342, 460, 497

University of Denver, 51
University of Pisa, 47
Upton, Francis, **16**, 370, 504, 660; and British lamp orders, 650–51; and conductors, 22 n. 2; and dynamos, 344 n. 7, 477 n. 7, 623 n. 1; and European companies, 436; and German lamp factory, 792; investment in Lamp Co., 10; and lamp contract, 755–56; and lamp experiments, 448, 807; as lamp factory superintendent, 3, 5, 16, 22–23, 28–29, 62 n. 5, 80, 123–24, 139–41, 159, 175–76, 187–88, 193–94, 234–36, 238, 252, 282–83, 313, 335, 339, 346–47, 430, 447, 478, 517 n. 1, 553, 579–80, 587–88, 590, 600–602, 627, 639, 661–62, 668, 739, 757, 776–77, 799–801; and lamp prices, 753; and lamp sockets, 527; loan to Lamp Co., 553, 662; in Menlo Park, 447, 662; at Princeton University, 785 n. 2; reports on lamp experiments, 22, 38–39, 139, 141, 579–80; reports on lamp tests, 87 n. 3, 123–24, 196–97
—letters: to Eaton, 771–73; to Johnson, 776–77; from TAE, 194, 252, 478, 527, 583, 757; to TAE, 22–23, 28–29, 80, 123–24, 139–41, 187–88, 193–94, 238, 282–83, 335, 346–47, 553, 587–88, 600–602, 627, 755–56, 799–801
U.S. Army Signal Corps, 159 n. 3
U.S. Commissioner of Patents, 137 n. 4, 785
U.S. Congress, 537–38, 786
U.S. Mint, 536 n. 12
U.S. Naval Academy, 471 n. 13
U.S. Navy, 461
U.S. Patent Office, 229 n. 21, 269 n. 24, 430, 580, 610 n. 1, 611 nn. 2 & 6, 752, 762 n. 2, 785–86, 794, 826
U.S. Patent Office Gazette, 556 n. 1, 826
U.S. Secretary of the Interior, 785–86
U.S. Torpedo Station (Newport, R.I.), 628

Vacuum pumps, 603; absorbent gases, 364–65; and food preservation, 6, 30, 40–41, 91; Goebel's improvements, 520 n. 10; hand, 729; at lamp factory, 5, 8, 22–23, 39, 141, 601–2; leakage of, 590; low vacuum experiments, 779; patent applications, 25 n. 9; patents, 25 n. 9; phosphoric chamber, 41, 365, 573; steam operated, 364–65

Vail, John, 513, 527, 599 n. 9

Van Cleve, Cornelius, **384**, 697 n. 2

Van Cleve, Hattie, 291, 385 n. 7

Vanderbilt, Maria (Mrs. William H.), 67 nn. 2–3, 816, 827

Vanderbilt, William H., 6, **67**, 332 n. 1, 567, 815–16, 827–29

Van Nostrand (D.) Co., 358

Van Nostrand's Magazine, 358, 361, 746

Verity, George, **656**

Verity, John B., **656**

B. Verity & Sons, **656**

Vick's Monthly, 291

Vienna Electrical Exhibition, 441, 458

Villard, Henry, **185**, 277; agreements with, 91, 185, 292 n. 1, 407 n. 3; and electric lighting in Boston, 752, 789; and electric railways, 91, 407 nn. 3 & 5, 812, 828; home isolated plant, 431; letter from Insull, 292; loan to TAE, 185 n. 4

Volckmar, Ernest, **518**

Volney W. Mason & Co., 553

Vroom, Garrett, 747 n. 1

Walker, George, **15**, 81; Paris Opera, 303

Wanklyn, James, **612**

Ward, Leslie, **348**

Washburne & Moen Manufacturing Co., **645**

Waterhouse, Theodore, 268 n. 20, 345 n. 19, **369**, 398–99, 493–95, 523, 547

Watson, Robert Spence, **221**

Waverly Magazine, 291

Webster, Richard, **494**–95, 832

Welles, Francis, 36 n. 3

Wells Fargo Express Co., 358

Welsh, Alexander, **139**, 188 n. 2, 346

Werra, 791

Western Edison Light Co. (Chicago), 488 n. 1, 543 n. 8, 556 n. 2, 559, 564–65, 636–37, 703 n. 3, 823

Western Electric Manufacturing Co., 33, 543 n. 8

Western Union Telegraph Co., 15 n. 3, 452 n. 3, 543 n. 8, 822; building, 424; dynamos for, 622–23; and European cables, 217; legal department, 226 n. 1; merger with Mutual Union, 804 n. 1; and Paris Exposition, 27; and Quadruplex Case, 568 n. 12; resistance lamps for, 304, 801; and telephones, 500 n. 15, 816, 828–29

Westinghouse Electric Co., 429 n. 3

Weston, Edward, 52 n. 2, 230 n. 2, **363**, 543 n. 1

Weston Dynamo Electric Machine Co., 363 n. 2, 674

West Orange, N.J., 332 n. 2

Wexel & De Gress, **531**

Wharton, Joseph, **639**

What to Wear, 291

White, Arnold, 158 n. 1, **369**, 465, 470 n. 5, 548 n. 5, 581, 635 n. 1, 793; defect reports, 774; as secretary of Edison Electric, Ltd., 656–67, 695, 709, 744, 788; visits New York, 682, 714 n. 1, 736 n. 1

—letters: from TAE, 584–85, 649–52, 774–75; to TAE, 594–96

Wilber, Zenas, 200 n. 1, **249**, 507; and arc light patents, 537 n. 18; becomes Edison's patent attorney, 98 n. 5; and British patents, 320; and European patents, 98 n. 5; and Holzer's patents, 611 n. 6; malfeasance of, 374 n. 4, 378 n. 1, 418, 450 n. 2, 580, 609–10, 825–27, 829; and Mott's patents, 554, 611 n. 6; and ore milling patents, 249; partnership with Dyer, 312, 609 n. 1; power of attorney, 611 n. 2; and Rowland's patent application, 611 n. 6, 761–62; and Swan's patents, 568 n. 11; work for Edison Electric, 447, 580, 597, 609–10

Wilde, Oscar, **358**

Winslow, Homer & Lanier, 707

Winslow, Lanier & Co., **518**, 518, 539 n. 1

Winterbotham, William, 500 n. 13

World's Industrial and Cotton Centennial Exposition (1884), 429 n. 3

Worthington, Henry, 699 n. 4

Wright, James Hood, **258**, 261, 272, 369, 538–39, 796

Wurtz, Charles-Adolphe, **440**

Young, Charles, 6, 785 n. 1

Young Ladies Journal, 291

Youthful Days of Frederick the Great, 745

Zetsche, Karl, **189**

Ziegfeld, Florenz, 638 n. 7